Evolutionary Biology: A Plant Perspective

Evolutionary Biology: A Plant Perspective

MITCHELL B. CRUZAN

OXFORD
UNIVERSITY PRESS

OXFORD
UNIVERSITY PRESS

Oxford University Press is a department of the University of Oxford. It furthers the University's objective of excellence in research, scholarship, and education by publishing worldwide. Oxford is a registered trade mark of Oxford University Press in the UK and certain other countries.

Published in the United States of America by Oxford University Press
198 Madison Avenue, New York, NY 10016, United States of America.

CIP data is on file at the Library of Congress
ISBN 978–0–19–088267–9

9 8 7 6 5 4 3 2 1

Printed by Sheridan Books, Inc., United States of America

Cover photo: *Scalesia pedunculata* forest (Galapagos daisy tree),
Los Gemelos, Isla Santa Cruz, Ecuador. Photo by the author.

For Carol

Thank you for your tremendous support

CONTENTS

PREFACE

This book was initiated as a result of my experience developing a course in plant evolutionary biology. The first day of class, I challenged the students to list unique characteristics of plants and to think about how these features affected the patterns and modes of plant evolution. This became the theme for the course and eventually for this text. As we worked through each topic, I continued to find examples of aspects of the biology and ecology of plants that had implications for evolutionary processes. I was motivated to begin writing reviews of topics that were not covered in the text we were using in class. As the term progressed, I continued writing to fill in gaps, and very soon, I had most of the outline for this book.

Writing this book has been one of the most rewarding experiences of my career. This project has given me an excuse to dive into many topics with which I previously had limited experience. For each topic, I have been pleased and amazed not only with the quantity of information, but also with the quality and creativity of the work. It was perhaps most enjoyable to explore areas of the literature with which I had the least familiarity. I have learned many new things along the way, and some of these have significantly changed my understanding of evolutionary processes in plants. Some of the ideas that were new to me include the pathways for the origins of eukaryotes and the first land plants, the processes affecting mutation accumulation in plant populations, and models for the evolution of sexual reproduction.

In writing this book, I have striven to provide discussions and clear explanations of difficult concepts, as well as new perspectives on topics that have already been thoroughly reviewed. It would not be worth writing such a volume if I did not have something new to say. I am sure that there will be other opinions, and in fact, I hope these ideas will promote more discussion, because otherwise we will learn nothing new. Science cannot progress if we just repeat what has already been said, and it is our duty to advance new ideas with the understanding that some will wane while others may flourish. If I can inspire just a few to engage in new avenues of investigation, whether they agree with me or not, then the book will have been a success.

This book is not solely focused on plants. Much of the discussion of evolutionary processes is applicable to a wide range of organisms. Moreover, the "plant perspective" only makes sense if we have something to compare it to—namely, animals. For each principle, I start out with a general discussion and in later sections, explore the ways that plant biology and ecology may modify our understanding of these principles with regard to plants. I was motivated to take this approach because too often I have noticed that concepts that have been developed from our understanding of animals have been blindly applied to plants without considering the implications of their biology. Virtually everything about plants is different—they are truly the most alien group of organisms we may ever have the privilege of encountering—and these differences should define the core of our understanding of their past, present, and future evolution.

It is probably not possible, and certainly not desirable, to put everything about plant evolution into a single book. I have attempted to be comprehensive without subjecting the reader to oppressive levels of detail. Of course, the topics selected necessarily represent my perspective, and I am sure there are many who will point out important topics that I overlooked. Similarly, I was limited on the number of excellent papers I could cite. My goal was to provide relevant examples from the literature, both historical and contemporary, so that anyone interested would have a window into the topic for further exploration. I purposefully tried to avoid over-citation of the work coming out of my own lab, but sometimes these papers provided the best examples. Out of the many thousands of papers I reviewed during the preparation of this work, I was only able to include a small fraction, and I am sure that many readers could provide a list of very important papers that could have been cited.

Completion of this book would not have been possible were it not for the contributions of many people, and most of all, the students with whom I have had the privilege of working. The best learning comes about through discussion, and I have benefited from discussions with all of my colleagues, and especially my mentors, fellow graduate students, postdocs, and every student who has ever spent time in my lab. Many others have had a direct impact on the content of this text. In particular, I deeply appreciate the contributions of Michelle Nicole Williamson, who generously provided many fine drawings and beautiful illustrations of paleo plants. Those who have provided edits, comments, and participated in discussions that have improved the presentation include Tina Arredondo, Tanya Cheeke, Carol Cruzan, Nic Diaz, Monica Grasty, Lizzi Hendrickson, Brendan Kohrn, Caitlin Lee, Caitlin Maraist, Gina Marchini, Trieste Musial, Jessica Persinger, Avery Pheil, Alisa Ramakrishnan, Jaime Schwoch, Pam Thompson, Michelle Williamson, and Rachael Workman, as well as many students from my courses on evolutionary biology and plant reproduction.

INTRODUCTION

Early in the history of life on Earth, the rise of bacteria capable of photosynthesis established a novel mode of energy capture. The new photosynthetic lineages embarked on an evolutionary trajectory that differed from the dominating consumers—one that was governed by the unique opportunities afforded by an autotrophic mode of resource acquisition. This simple difference had far-ranging implications for the future evolution of photosynthetic microorganisms and eventually, for land plants. Many characteristics of plants stem directly from the ability to obtain energy directly from the sun. The goal of this book is to improve our understanding of the origin and evolutionary consequences of the unique traits that characterize the vast and diverse group of organisms generally referred to as plants.

Developing an understanding of the consequences of evolutionary processes for proliferation and diversification of plants requires an appreciation of their unique biological features. The processes of mutation, selection, genetic drift, gene flow, and diversification remain the same for plants and animals, but specific characteristics of plants impact the way evolution is implemented. Unique traits of plants affect everything from the elevated potential for the accumulation of mutations during vegetative growth, to the distribution of genetic variation within and among populations, and ultimately to rates and patterns of diversification. My primary goal is to develop a deeper understanding of the origins of these features, and an appreciation of their implications for evolutionary processes.

Evolutionary biology is an exceedingly complicated and conceptually rich science. This field of study has tremendous depth, and I believe that all evolutionary biologists who are working in this field—no matter how experienced or knowledgeable—are constantly learning more of the nuances and developing a deeper appreciation of the complexity of plant evolutionary processes. This point has become even more relevant in recent decades as improved technologies allow us to explore genomes with increasingly greater resolution and as we strive to make appropriate evolutionary interpretations of the patterns we find.

A thorough appreciation of concepts in evolutionary biology requires familiarity with a range of fields. Students will find the discussions in this book most rewarding if they have some fluency in ecology, population genetics, physiology, cellular and molecular biology, and basic principles of evolutionary biology. This book is specifically designed for advanced undergraduates and graduate students and as a reference for researchers in plant evolutionary biology. My goal is to provide a balanced and thorough coverage of traditional and contemporary topics in evolutionary biology with reference to aspects of plant biology that are relevant to evolutionary processes. Evolutionary principles are discussed using classical and contemporary examples from the primary literature.

This book is organized by major topics in plant evolutionary biology, starting with the origin and diversity of plants, origins of genetic variation, principles of population and quantitative genetics, and processes of evolutionary diversification. To the degree that the information enhances our understanding of evolutionary processes, I have included discussion of more phytocentric topics, such as plant morphology and development and plant reproductive biology. Each chapter begins with a general introduction to increase accessibility to a broader audience. Basic principles are presented first, followed by expanded discussions that integrate more complicated aspects of plant biology. Several of the chapters include sections on the history and development of evolutionary thought to provide a contextual backdrop for discussion of contemporary issues in plant evolutionary biology. Many of the examples used to illustrate concepts are taken from applied biology (agricultural and conservation biology), and are informed by other fields such as plant developmental biology, anatomy, physiology, ecology, and molecular genetics and genomics.

I have provided readers with a list of questions for contemplation and discussion at the end of each chapter that should be particularly useful for students and instructors. The goal of these questions is to provide students with a tool for reviewing the information presented in each chapter and to provide topics for class discussion. Understanding of the topics addressed can be further developed by reading additional papers from the primary literature. Instructors using this text may consider incorporating regular discussions that utilize papers from the literature to expand coverage of each topic and provide examples of more recent advances in the field. The text is designed to provide an introduction to each topic, and it is up to the reader to determine which topics they wish to explore further.

PART I

ORIGINS OF DIVERSITY

1

ORIGIN OF PLANT LIFE

It is often said that all the conditions for the first production of a living organism are present, which could ever have been present. But if (and Oh! what a big if!) we could conceive in some warm little pond, with all sorts of ammonia and phosphoric salts, light, heat, electricity, etc., present, that a protein compound was chemically formed ready to undergo still more complex changes, at the present day such matter would be instantly devoured or absorbed, which would not have been the case before living creatures were formed.

Charles Darwin (1871) from a letter to J. D. Hooker

1.1 PRELUDE—ORIGIN OF LIFE

Some four billion years ago, a small planet that was just a few hundred million years old cooled sufficiently to allow water to condense from the atmosphere and form vast pools on its surface. By chance, the planet maintained an orbit around a star at a distance that provided a fairly narrow temperature range and sustained most of its water in a liquid state. During this violent period of early Earth's history, the tumultuous atmosphere was charged with electrical storms and the planet's surface was scorched by intense cosmic and ultraviolet radiation. Dry land was scarce because the Earth's surface was covered by a shallow sea that frequently vaporized due to volcanic activity and extraterrestrial impacts that continued for many hundreds of millions of years. Surges of energy, plus heat released from the molten mantle of rock that boiled beneath a thin crust, catalyzed chemical reactions to form organic molecules from the simple inorganic elements that were present in the water and the atmosphere. The emergence of self-sustaining prebiotic entities likely occurred many times over the course of trillions upon trillions of chance encounters among organic molecules over millions of years. Critical to this process was the eventual capture of singular combinations of organic components within stable microspheres consisting of lipids and proteins to form protocells (membrane-bound collections of organic molecules capable of simple metabolism). Organic molecules in the surrounding water contained valuable energy, stored in their chemical bonds, and formed a thin "primordial soup." This energy-rich solution provided sustenance for the new protocells, but the growth of these populations soon outpaced the regeneration of energy-rich molecules and competition ensued; this condition was necessary for evolution by natural selection, which, in turn, led to the refinement of metabolic systems and the eventual emergence of more complex organisms.

This process, while simple, had dramatic consequences; competition for limited resources favored the most efficient individuals and created an overwhelming force for change in the competing populations of protocells. As energy-containing organic compounds were depleted, the individuals that were able to glean more energy replicated at higher rates and soon replaced less efficient ones. This one simple principle—competition for limiting resources—was responsible for driving increases in the complexity of organic systems and the diversification of life on Earth. Competition selected for both efficiency and continuity, favoring the emergence of self-replicating entities that generated more individuals with the same capacity for efficient extraction of energy. Over many millions of years, innovations that improved energy extraction emerged and spread rapidly, so populations were subject to wave upon wave of improvements. Many forms enjoyed ephemeral success, but then fell behind and were lost as their molecular machinery proved too inflexible to meet new challenges from the organic mélange (Box 1.1).

Organic Matter—The lack of direct evidence for events that precipitated the origin of life has led to a proliferation of plausible scenarios. Much of the research in the field of *abiogenesis* (the study of how life could have developed from inorganic matter) is based on experiments designed to mimic different aspects of *Hadean Earth* conditions, which persisted for the first 700 MY (million years) or so of Earth's history. The most well-known of these were the *Miller-Urey experiments,* in which Stanley Miller (student of Harold Urey) exposed a mixture of water, hydrogen, methane, and ammonia to an electrical spark. By the end of a week, he found that 10–15% of the carbon had been converted to organic material, and the mixture contained eleven of the twenty amino acids that are the building blocks of proteins (Miller 1953). More recent analysis of the vials from these original experiments using improved methods revealed that more than twenty amino acids were produced (Parker et al. 2011).

The Miller-Urey experiments have been criticized because the gas mixture they used, which was based on ideas proposed by Alexander Oparin and J. B. S. Haldane (Knoll 2003), probably did not accurately reflect the composition of early Earth's atmosphere. Many permutations of these "primordial soup" experiments have been explored since then. Addition of gases that are commonly produced by volcanic activity to the Miller-Urey mixture, such as carbon dioxide, hydrogen sulfide, and sulfur dioxide, increases the diversity of molecules produced (Brooks et al. 2002). Incorporating inorganic substrates such as volcanic rock and silicate minerals in the solution promotes polymerization of organic subunits and the formation of microspheres (Fox et al. 1974) or chains of amino acids that are similar to short proteins (Cairns-Smith 1982). It is also possible that the Hadean atmosphere contained a much higher concentration of hydrogen than previously thought, which could have promoted the generation of a greater abundance and higher diversity of organic compounds.

In spite of these successes in the laboratory, other aspects of the *Primordial Soup Theory* (Box 1.1) have been questioned. For example, some nucleotides (subunits of DNA and RNA) require freezing conditions for efficient synthesis, while others may require boiling temperatures, making it difficult for the components necessary for the precursors of life to come together simultaneously. To overcome such limitations, it has been suggested that much of the organic material may have arrived extraterrestrially via comets, thereby jump-starting the process (Oro et al. 1992). This hypothesis is supported by observations that comets contain a variety of organic molecules, including amino and nucleic acids. On the other hand, it may be unlikely that extraterrestrial molecules could have arrived in a sufficient volume to be the sole source for the organic complexes that precipitated life. Life on Earth is characterized by *Chiral Biases;* life on Earth utilizes enantiomers of amino acids that are left-handed and enantiomers of sugars that are right-handed. It is interesting to note that the same chiral biases are present in mixtures of organic compounds isolated from extraterrestrial sources, perhaps due to the selective effects of polarized light (Clark 1999). It has been suggested that the enantiomers that ultimately became the foundation of modern metabolisms are inherently more stable. Whether due to improved stability, or a tendency to generate one enantiomer at higher frequencies during abiogenesis, once chiral bias was established, the more abundant enantiomer would have become dominant; selection would have favored the more abundant enantiomer because the self-replicating entities utilizing it would have been more successful.

At any one time in the early history of Earth there may have been a multitude of chemical systems that had the potential to generate life, but the record of the diversity of natural experimentation that might have existed has been lost. In spite of this, we can understand much about the early stages of the evolution of life on Earth using existing organisms to

Box 1.1 The Organic Mélange

Soon after the newly formed Earth cooled sufficiently to allow for the formation of liquid water, the range of environments available for natural experiments in organic chemistry began to proliferate. During early Earth's history, temperatures near boiling were common and may have promoted the formation of organic compounds, and evaporation in shallow pools would have led to the concentration of chemicals and the promotion of organic synthesis (Martin and Russell 2003). Models indicate that the sun was much dimmer—70% of its current energy output—so the super-hot planet may have cooled relatively quickly, to the point where there was potential for ice formation during the Hadean (Bada et al. 1994). While greenhouse effects due to the presence of high concentrations of carbon dioxide and ammonia in the atmosphere may have maintained warm conditions over much of the surface, the poles must have been much cooler. Meanwhile the planet was bombarded with comets and meteorites, exposed surfaces were irradiated with high levels of ultraviolet and cosmic radiation, and volcanic activity was widespread. The diversity of physical conditions and energy sources produced a multitude of opportunities for the production of organic compounds. The "organic mélange" refers to the diverse conditions that may have generated organic mixtures as well as the wide range of plausible scenarios that have been proposed for *abiogenesis*, the development of prebiotic autogenic systems.

Primordial Soup Theory—Proposed independently by Alexander Oparin and J. B. S. Haldane around 1924 and first tested by Stanley Miller in the early 1950s, who demonstrated that a wide array of organic molecules could be generated in oxygen-free vessels containing water, hydrogen, methane, and ammonia exposed to an electric spark (Miller 1953; Parker et al. 2011). Many additional experiments introduced various substrates, tested different gas mixtures, and used various energy sources to produce a wide range of organic compounds as they elaborated on these initial experiments. A variety of environments have been hypothesized to contribute to the generation of organic molecules, including hydrothermal vents, shallow pools of water affected by ultraviolet radiation (Powner et al. 2009), and tidal areas that had accumulated radioactive materials (the radioactive beach hypothesis; Dartnell 2008). Even with all these possible sources of organic material, the "soup" was probably quite thin, so many of the proposed scenarios include mechanisms that would have concentrated reactive components into small spaces. These include evaporation of water in shallow pools or the exclusion of impurities from ice crystals during freeze-thaw cycles, which caused the accumulation of organic molecules in the spaces between crystals.

Polymerization—Life as we know it depends heavily on the properties of chains of molecules referred to as polymers; proteins consist of chains of amino acids, and nucleic acids (DNA and RNA) are chains of nucleotides (adenine, guanine, thiamine, cytosine, and uracil). A drawback of the primordial soup theory is that the same energy that is capable of generating organic monomers has the tendency to cleave any polymers that are produced. One solution is the *iron-sulfur world*, where conditions similar to hydrothermal vents are hypothesized to have generated organic monomers and polymers through energy drawn from inorganic reactions involving hydrogen sulfide and various metal sulfides. Under the appropriate conditions, autocatalytic systems with the potential to self-replicate may have developed (Wächtershäuser 2007). The thermodynamic conditions for polymerization were made more favorable by the introduction of pyrite, which was generated as a byproduct of the redox reactions that may have been common in the iron-sulfur world. Other conditions that promote polymerization have been proposed, including concentration of mixtures of organic molecules through evaporation of water and the presence of lattices of electrical charges, such as those found on the surface of silica (clay) and other crystalline structures, which promote alignment and favor the bonding of monomers (Cairns-Smith 1982; Martin and Russell 2003). The formation of ice crystals has also been shown to concentrate impurities and promote the synthesis of organic molecules and polymers (Trinks et al. 2005).

Self-Organization—The tendency for elements of a complex to assemble without energy input would facilitate the establishment of autogenic systems that could easily be replicated. Self-assembly is central to virus replication within host cells, and self-organization of contained chemical reaction systems are the most conserved, and hence most ancient, features of living organisms (Martin and Russell 2003).

Autocatylitic Hypercycles—Catalytic cycles play a central role in the metabolisms of living organisms; for example, the citric acid, glyoxylate, and Calvin cycles are chemical reaction sequences that are critical for living cells. Eigen and Schuster (1978) hypothesized that abiotic systems consisting of multiple chemical pathways may have developed into autocatalytic cycles of reactions. Critical to the stability of such a system is the interdependence of chemical reactions in the sequence such that one reaction promotes others. When multiple cycles become interconnected, the system can be pushed further away from less stable equilibrium conditions to a hypercycle (multiple loops) that achieves a higher state of stability. At this point complexity may increase as an emergent property as more pathways are incorporated. Such systems that are self-organized and self-replicating may be subject to transmittable modifications, at which point prebiotic evolution ensues as these systems are subject to selection for increased efficiency due to competition for limiting substrates.

Protocells—When amphiphiles (molecules with both a hydrophilic and hydrophobic end) are present in water, any turbulence will result in the formation of bi-lipid bubbles containing aqueous mixtures of whatever compounds are present. The chance that catalyzing and self-generating complements of organic molecules might have been captured within these microspheres would have been increased as evaporation increased their concentration in shallow pools. While the vast majority of such bubbles would be expected to dissipate in a short time, the stability of the membrane could have been improved as proteins became integrated. The amphiphile and protein components in the bi-lipid membrane would have been ideal locations for establishing cyclic and autocatalytic reaction networks within a membrane-bound protocell (Martin and Russell 2003).

extrapolate backward in time. We can infer that at some point, the appropriate mixture of molecules was captured inside microspheres of amphiphiles (e.g., phospholipids) and proteinoids to form protocells with bilipid membranes, but exactly when and how this occurred, and the sequence of refinements that led to self-replicating entities, remains unknown. We can also infer that RNA played an important role in the development of early life (the *RNA World Hypothesis*—see the Protocells section) because of its 1) importance for critical cell processes, 2) close relationship to critical cofactors for metabolic processes (e.g., ATP, NADH, etc.), 3) catalytic potential, and 4) presence in a wide variety of viruses (retroviruses). These observations support the hypothesis that the first cells utilized RNA as the hereditary material. DNA is the hereditary material in contemporary cells, so presumably after basic cell processes had been established, RNA was replaced by DNA because of its greater stability and possibly as a consequence of selection pressure for immunity from RNA-based viruses (Forterre 2006). The improved stability of DNA apparently is due to infrequent switching from the normal pairing of bases between strands to *Hoogsteen Base Pairs*, which do not occur in RNA (Zhou et al. 2016). Cellular life retains the signatures of these early events, as DNA, RNA, and proteins are all critical players in the metabolic and hereditary processes of modern cells.

Metabolic Pathways—Envisioning the early stages leading up to the establishment of self-replicating organic systems challenges the limits of our imagination. Collections of organic compounds by themselves are lifeless, but salient features of living systems may emerge from interactions among organic and inorganic components to form simple metabolic pathways. A series of reactions may generate products that become substrates for other reactions, which in turn promote additional pathways. Simple networks of pathways develop to form sets of nested and interconnected cycles

(*Hypercycles;* Box 1.1) that are effectively autocatalytic (Eigen and Schuster 1978).

In some scenarios, molecules capable of producing hypercycles are first captured inside *Amphiphile Bubbles,* while in others they first develop in open systems where inorganic substrates act as important catalysts (e.g., Iron-Sulfur World; Wächtershäuser 2007), and the encapsulation in microspheres occurs later. In either scenario, we expect that hypercycles within lipid microspheres will accumulate molecules contributing to reaction pathways as organic compounds are absorbed from the environment. Microspheres containing hypercycle reaction mixtures would necessarily expand in size to accommodate the increase in contents, which would eventually make them unstable. These larger bubbles would be susceptible to cleaving as turbulence broke them into smaller spheres, each of which would have had a reasonable chance of containing a complement of compounds necessary for the replication of the hypercycle.

As communities of these organic bubbles developed, their growth would be limited by the availability of organic substrates; any chance event that produced even the slightest improvement in efficiency of metabolism or replication would tilt the balance so that improved bubbles would gain numeric dominance. This new collection of bubbles would eventually be replaced by the next improvement, and so forth. Complexity would increase along with efficiency, until something resembling a cell more than a bubble would emerge and embark on a trajectory of diversification and colonization of every habitable environment on the planet—a process that continues today.

We may never know whether anything like the scenario outlined above ever occurred, but we can be confident that natural processes of *Prebiotic Evolution* were critical to the emergence of life. Any limitation to growth or persistence imposes the potential for chance improvement and the enhanced success of the modified system. In an environment with limiting resources, selection for increased efficiency is a natural byproduct of competition. It does not matter whether the competitors are abiotic reaction systems, simple cells, or complicated organisms with the ability to exhibit sophisticated responses to their environment, the principle remains the same; natural selection is an emergent property of resource-limited environments that drives the inhabitants toward improvements in efficiency of resource acquisition and metabolism, and consequently, growth and reproduction. Once we comprehend the power of this simple principle, we come to the realization that the emergence of life and the diversification that ensued was an expected outcome of purely natural processes.

Protocells—A wide range of laboratory experiments have produced diverse groups of organic molecules, including common amino acids, which polymerize to form proteins, and nucleotides, which polymerize to form the nucleic acids, RNA and DNA. The observation that polymers of these molecules can be formed by physical mechanisms is a critical step for understanding the establishment of basic metabolic processes and the material of inheritance, but there is disagreement over the sequence of events that occurred during the early stages of the emergence of protocells. The proposed scenarios for the origin of the first life-like entities can be divided generally into the *RNA-World* and *Metabolism-First* (Protein-First) models. It has been suggested that RNA molecules were critical to the development of life because this single class of compounds can serve multiple functions; they have catalytic ability (ribozymes), the potential to store information, and the capacity for self-replication (Johnston et al. 2001). Under freezing conditions, nucleotide polymers can form and single strands of RNA can serve as templates for the generation of complementary strands (Trinks et al. 2005). On the other hand, RNA is relatively unstable, so alternative types of nucleic acids that are more stable have been suggested as precursors to RNA-based life (Nelson et al. 2000). Proponents of a protein-first model point out that nucleic acids would have had limited capacity to establish a primitive metabolism, and the centrality of proteins in basic metabolic processes suggest that they are more likely to have been the first to evolve (Smith and Morowitz 2004). The means to assimilate energy from the environment and to pass this ability on to progeny were necessary, and ultimately the salient features of both proteins and nucleic acids were necessary for the establishment of life.

As we consider the immense range of possible processes involved in the early steps that led to self-replicating organic forms, the difficulty of elucidating the actual sequence of events becomes clearer. Just as the first appearance of multicellular animals in the

Cambrian period reveals a spectacular range of natural experimentation by evolutionary processes, we can infer that prebiotic evolution led to an equally diverse range of organic forms. All of these natural experiments ultimately failed, save one—that one successful group of organisms, the ancestors of all life on Earth—referred to as the *Last Universal Common Ancestor* (*LUCA*). In one sense, the LUCA represents a biological singularity; the diversity of life preceding it may never be known because contemporary life only provides information about those ancestors that were successful.

Our knowledge of the history of life on Earth is limited by incomplete information; if we were able to rewind history to trace the history of life back through time, we would see an increasing accumulation of species and groups that are unknown because any record they may have left was lost to geological processes or remains undiscovered. As we travelled further back in time, we would note increasingly large gaps in the fossil records of successful lineages. We may gain insights by reconstructing minimalist cells based on features present in living organisms, but this approach is made difficult by the immense number of generations that have passed since the earliest stages—important intermediate stages may have been lost. For example, one may try to learn about the evolution of vehicles by deconstructing an automobile, but all that work would never lead to the conclusion that autos were predated by horse-drawn carriages, even though many of the components are homologous (e.g., wheels, axels, brakes, etc.). Nevertheless, the enormous number of processes uncovered by scientists that may have been involved in the origins of life on Earth is inspiring and secures our conclusion that the generation of life on this, and perhaps millions of other planets, was inevitable.

Rise of the Prokaryotes—The life forms that dominated Earth during the Archean were enormously complex compared to the first self-replicating protocells. These early prokaryotes (the *Domains of Archaea and Bacteria*) had resilient walls that defined the cell's structure and provided protection from the environment, and internal membranes that enclosed a chemical reaction vessel for extraction of energy from molecules that were absorbed from the environment. The success of these small wonders in the face of an onslaught of challenges depended on their ability to undergo modification (mutations), and then reliably transmit the beneficial changes to subsequent generations. This was accomplished by long polymers of nucleotides (adenine, guanine, cytosine, and thiamine or uracil) that were repeated in sequences of different combinations (RNA or DNA) for intergenerational transmission of information. This sequence of "letters" (nucleotides) formed "words" (codons) that inscribed the blueprints for the molecules (proteins) that sustained life and provided a means of proliferation. While there may have been many variants of genetic codes at one time or another, only one basic code survived and was carried forward by all the descendants of the first cells. This process was probably facilitated by rampant *Horizontal Gene Transfer* among lineages in the collection of organisms that predated the LUCA. The descendants of those earliest cells proliferated, diversified, and spread to occupy every imaginable habitat of the newly formed planet.

The diversification of these simple cells continued for several hundred million years with intense competition for scarce resources. Even though a vast amount of energy was absorbed from the nearby sun, nearly all was lost as heat that dissipated into space and never became available to the multitudes of bacterial lineages. A variety of innovations for the extraction of energy via inorganic chemical reactions arose (e.g., chemoautotrophy similar to that found in modern sulfur bacteria), but these sources of primary production made only limited contributions of energy to support bacterial communities. Eventually some lineages accumulated mutations making them capable of absorbing portions of the energy spectrum from the sun. The energy absorbed was transferred to electrons that were passed to neighboring molecules and perhaps promoted chemical reactions for more efficient metabolic pathways. Such simple systems of photosynthesis exist in some lineages of Archaea today. Ultimately this system was improved as molecules capable of absorbing light energy (pigments) were imbedded in membranes along with proteins (Goldsworthy 1987). The protein-pigment complexes efficiently converted the light energy into molecules of chemical energy (ATP and NADPH), which drive the conversion of carbon from its inorganic form (CO_2) to organic molecules (sugars). For the first time on the planet, there were organisms capable of generating a seemingly endless supply of energy for the support of life.

The advent of photosynthesis changed the course of evolution and had dramatic consequences for the living organisms that were present some 3.5 BYA (billion years ago). The conversion of light into chemical energy required a proton donor such as hydrogen, but these molecules were in short supply. The solution—hydrolysis (splitting of water)—while seemingly simple, was destined to become one of the most important chemical reactions in the history of life on Earth. The early Earth atmosphere was nearly devoid of oxygen, with only low levels generated by water molecules exposed to lightning, which induced hydrolysis. As photosynthetic prokaryotes proliferated, they released increasingly more oxygen into the aquatic environment and the atmosphere. The release of O_2 as photosynthetic cells became more abundant, increased the potential for oxidation and the efficiency for extraction of chemical energy; however, most of the existing organisms were not prepared for the onslaught of abundant oxygen. Exposure to free oxygen led to the oxidation of proteins and the demise of billions of living cells (*The Oxygen Catastrophe*) around 2.4 BYA.

Prokaryotes capable of photosynthesis probably first appeared as early as 3.5 BYA, but the free oxygen they produced was absorbed by metals and other reduced compounds (as evidenced by rusting of iron deposits from that period), so it took some time before free oxygen accumulated in substantial quantities in the atmosphere. Oxygen in the atmosphere interacted with ultraviolet light from the sun to initiate the oxygen-ozone cycle, which promoted the formation of the ozone layer in the stratosphere. Prior to the oxygen catastrophe ultraviolet radiation was intense, but aquatic life was protected by the filtering capacity of water. Development of the ozone layer was a critical step that made the emergence of terrestrial life possible.

Strong selection for oxygen tolerance led to numerous lineages that were not only adapted to the presence of oxygen but were also able to utilize it to improve the efficiency of the metabolic extraction of energy from organic molecules (aerobic respiration) to yield many times more energy than anaerobic respiration. Oxygen-free space for anaerobes became rarer, but these habitats were always regenerated when aerobic cells depleted the available oxygen in spaces where the diffusion of gases was restricted. The descendants of the original anaerobic bacteria that once dominated persist on our planet to this day, and still proliferate in water-saturated soils and benthic sediments where they are safe from oxidation.

We can get a glimpse of how life across the planet may have appeared without photosynthesis by observing organisms in deep-sea vent communities, which depend on extraction of energy from chemical reactions (*Chemoautotrophy*) as the foundation of their ecosystem. While there are some large animals in these communities (e.g., tube worms) that are mutualists with chemoautotrophic bacteria, these animals are very slow-growing and limited in their size and diversity. The food web is simple with only consumers that depend on the chemoautotrophic bacteria. Our general conclusion from observations of benthic vents and other chemoautotrophic ecosystems is that their evolutionary potential is stunted by a limited availability of energy for the support of complex biological communities.

It can easily be argued that this single event—the efficient capture of light energy by living cells—changed the course of life on Earth and made possible everything that followed. Prior to the advent of photosynthesis, the growth of living cells was severely curtailed by a lack of accessible energy, and the potential for development of more complex forms was limited. The appearance of organisms capable of an autotrophic existence led to the division of life into domains of primary producers and consumers and set the stage for these interdependent communities of organisms to embark on independent evolutionary trajectories that were dictated by their distinct modes of resource acquisition.

1.2 TROPHIC DOMAINS OF LIFE

The rise of bacteria capable of photosynthesis created a mode of energy capture that had not previously existed on Earth. The new photosynthetic lineages embarked on an evolutionary trajectory that differed from the dominant consumers—one that was governed by the unique opportunities afforded by an autotrophic mode of resource acquisition. The key feature of this new mode of life was the ability to harvest energy without significant expenditure. Mobility was not necessary, so adaptations for active movement (other than phototrophic movement in free-floating algae) were not favored. Once multicellular photosynthetic organisms became established in a site with good access to sunlight, they remained sedentary. This simple difference had far-ranging implications for the future evolution of photosynthetic microorganisms and eventually land plants. Most notably, the features that we view as the outstanding characteristics of plants and that are lacking in most animals, such as prolonged sedentary growth, prevalent hermaphroditism, and general dependence on biotic and abiotic vectors for reproduction and dispersal, all stem directly from their ability to obtain energy directly from the sun.

Developing an understanding of the consequences of evolutionary processes responsible for plant proliferation and diversification requires an appreciation of their biology. The basic processes of mutation, selection, genetic drift, and gene flow remain the same for consumers and producers, but specific features of plants (e.g., *Indeterminate Growth, Metameric Structure*, a multicellular haploid life stage, etc.) affect the way that evolution is implemented. For example, the tendency to remain sedentary through much of the life cycle has consequences for the fate of mutations, the distribution of genetic variation, and ultimately the rates and patterns of evolution within populations and diversification of lineages. These salient features of plants originated early in their evolution; some characteristics, such as their sedentary nature, are a consequence of selection for optimal allocation of resources. Other traits, such as retention of multicellular haploid stages, may be the product of historical contingencies. Our goal for the remainder of this chapter and the entirety of this book is to develop a deeper understanding of the origins of the unique characteristics of plants and their implications for patterns of diversification of producer lineages.

Origin of Eukaryotes—The earliest cells (dating from 3.8 BYA) gave rise to three phylogenetic domains of life that comprise biodiversity to this day. Two of those groups are prokaryotes, the Bacteria and Archaea. These organisms are unicellular, very small (0.5–5 microns), haploid, have a semi-rigid cell wall, and lack chromosomes. The prokaryote clades share many characteristics (e.g., similar genomes organized as a single circular strand of DNA, genetic code, and many metabolic processes), suggesting that they are derived from a single common ancestor. The third major group, the eukaryotes, is represented by a large number of unicellular microbes and macrophytic algae (traditionally Protista), plus all plants, fungi, and animals (Fig. 1.1; refer to color plate). Eukaryotic cells differ from prokaryotic cells in their size (being 10 to 100 times larger), their lack of a cell wall (except in fungi, plants, and many algae), and the presence of microtubules and filaments that define cell organization and shape. Eukaryotic cells also possess internal membrane-bound structures that are separate from the cell membrane (lacking in nearly all prokaryotes), organelles (mitochondria and chloroplasts), and a nucleus containing strings of DNA-protein complexes that comprise the chromosomes. The lack of a cell wall and the presence of internal compartmentalization allowed eukaryotic microbes to attain a relatively high level of complexity and to diversify into a broad range of habitats and modes of resource acquisition. It is from these early unicellular eukaryotes that much more complex multicellular organisms eventually arose.

The evolutionary significance of eukaryotes is reflected in the diversity of hypotheses proposed to explain their origin—*Eukaryogenesis* (Box 1.2). The traditional explanation for the origin of eukaryotes comes from the logic that simpler cell types must have originated earlier, so eukaryotes were assumed to have been derived from a prokaryotic progenitor. This general scenario is supported by analyses of DNA sequences that are common to all three domains, which place the LUCA between the two clades of prokaryotes, with eukaryotes arising later from within the Archaea clade (Gribaldo et al. 2010). The observation that transcription and translation of DNA in eukaryotes are more closely related to the same processes found in archaeans, while other cell processes have stronger associations with bacteria, led to the suggestion that nucleated lineages arose from the fusion of two or more cells, with a bacterium forming the cytoplasm and an archaean forming the nucleus (*Syntrophic Model*). On the other hand, the presence of many unique characteristics of eukaryotes makes their origin more difficult to decipher. Various scenarios have been suggested in which all three domains originated directly from the LUCA or the pre-karyote lineage (i.e., the precursor to eukaryotes) shares a later common ancestor with Archaea.

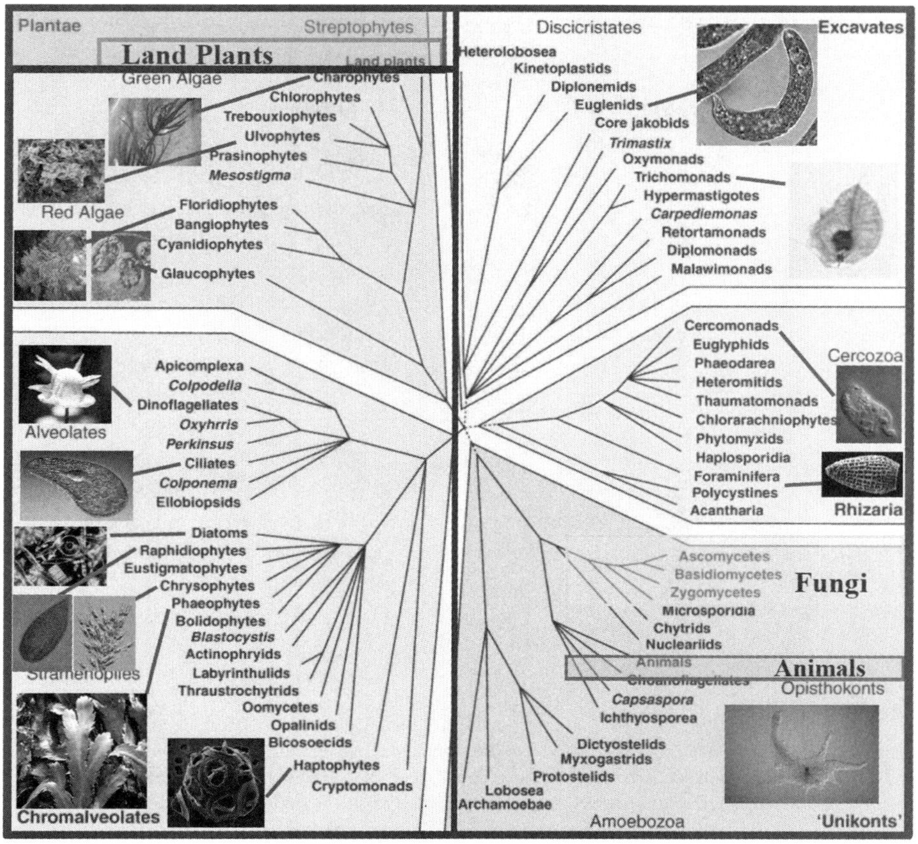

Fig. 1.1. Eukaryotic diversity. Adapted from Keeling et al (2005). Photos by Kristian Peters, Peter Southwood, Tim Evanson, Frank Fox, Gordon T. Taylor, Krengefors, Stemonitis, Neon, Deuterostome, Magnus Manske, and Tsukii Yuuji.

Attempts to reconstruct the exact sequence of divergences that produced the three domains are hampered by rampant horizontal recombination, and it is generally accepted that horizontal transfer among lineages with unstable genomic constitutions was prevalent during the time leading up to the LUCA (Vesteg and Krajcovic 2011). Some genetic reconstruction methods suggest that evolution during this early time period is best represented as a ring (referred to as the *Ring of Life* or *Fusion Circle*) rather than a branching tree structure, with different clades of Bacteria, Archaea, and Eukaryota originating from different parts of the ring (Rivera and Lake 2004). This model is consistent with the hypothesis that a pre-karyotic cell type (*Chronocytes*) originated much earlier and attained many of the unique eukaryotic characteristics, such as microtubules and internal membrane structures, prior to acquisition of a nuclear genome that displays a stronger similarity to Archaea.

The various models of eukaryote evolution reflect attempts to explain their different sets of unique features. While much emphasis has been placed on considering characters present in the diversity of life to be analogues of historical stages, the immense time span makes it likely that many intermediate forms are not represented among contemporary microbes. An alternative approach is to develop a probable sequence of events based on the most parsimonious (simplest) explanations. For example, it would be more parsimonious to assume that the pre-karyotic lineage was derived from a common ancestor with the prokaryotes that lacked a cell wall for three reasons: 1) Cell walls in Bacteria and Archaea are intrinsically

Box 1.2 Eukaryogenesis

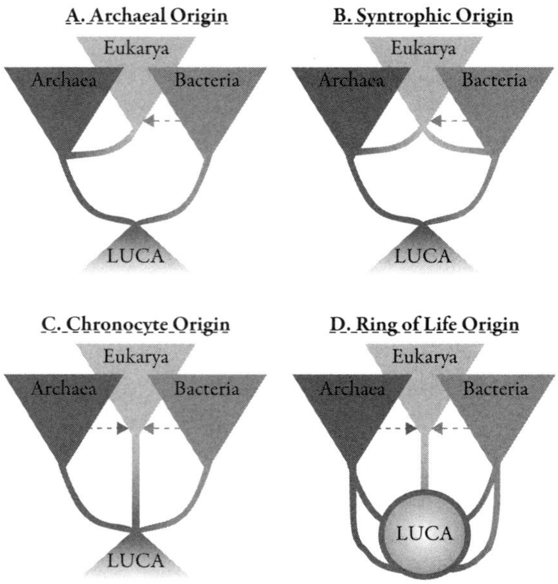

Fig. 1.2. Four hypotheses for the origin of Eukaryotes. A. For the Archaeal Origin model the primary ancestor is placed either at base of the Archaea clade or from subclades within the domain. Mitochondria and chloroplasts were acquired later from Bacteria (dotted arrow). B. In the syntrophic model, a mutualistic relationship among lineages results in the nucleus originating from Archaea and the cytoplasm from Bacteria. C. The Chronocyte model assumes that a pre-karyotic lineage originated directly from the LUCA and developed many Eukaryotic features prior to the acquisition of a nucleus (red dotted line) and mitochondria (blue dotted line) from Archaea and Bacteria, respectively. All of the models (A-C) assume that the LUCA was a single lineage whereas the Ring of Life model (D) depicts a collection of lineages that share genomic elements through horizontal gene transfer.

The origin of eukaryotes is complicated by their composite nature, as they share traits with Archaea, Bacteria, viruses, and possibly with cell lineages that are no longer present. Hypotheses for the origin of eukaryotes can generally be divided between A + B (Archaea + Bacteria) and A + B + C (+ an unidentified cell lineage; i.e., "Chronocytes") models. Hypotheses for the early evolution of the eukaryotes differ as to the timing of origin and acquisition of different features (Fig. 1.2).

It is generally acknowledged that mitochondria and chloroplasts have bacterial origins and were acquired via intracellular mutualisms. Nuclear genomic information has been used to place the root of the eukaryote clade more precisely, but conclusions reached depend on which set of proteins and methods of analysis are used. Several analyses of proteins that are common to all three domains place the pre-karyote either at the base of the Archaea clade or within one of the subclades in this domain (Gribaldo et al. 2010; Fig. 1.2.A, *Archaeal Origin*). The *Syntrophic Model* assumes the pre-karyote ancestor came from a bacterial lineage that had formed a mutualism with an archaean (Lopez-Garcia and Moreira 2006; Fig. 1.2.B). The bacterial contribution is supported by strong similarity between Bacteria and Eukarya for a number of metabolic processes occurring in the cytoplasm, while many features of the nucleus display stronger homology with archaeans.

The *Chronocyte Model* (Fig. 1.2.C) suggests that internal membrane structures and the cytoskeleton of eukaryotes developed in a third lineage that originated much closer to the LUCA from an ancestor that lacked cell walls. This proposal is supported by a lack of strong homology with either of the prokaryote groups for a set of eukaryote-specific proteins (Hartman and Fedorov 20021.). In this scenario, the nucleus and mitochondria were derived by intracellular mutualisms with Archaea and Bacteria after the three groups had undergone substantial divergence.

An independent pre-karyote lineage is also indicated in the *Ring of Life* model (Fig. 1.2.D). The ring is thought to be a consequence of rampant horizontal transmission of genomic elements among lineages that had RNA-based genomes. This amalgam of cell lineages became stabilized during a transition to DNA-based genomes leading up to a collection of related lineages that formed the LUCA. The depiction of the basal ancestor as a ring is supported by phylogenetic analyses using genome regions that are common to all three domains (Rivera 2007). Distance along the ring represents degree of relatedness, so some lineages of Archaea and Bacteria show closer relationships with Eukarya, while other lineages of the two prokaryotic domains have closer relationships to each other.

While several of the scenarios described here include mutualism with Archaea as the source of nuclei, Lester et al. (2006) point out that nuclear genes presumably derived via symbiosis arose at different times, indicating that the origin of nuclear genomic components via horizontal gene transfer is more likely.

The *Viral Eukaryogenesis Model* (not shown) is based on the observed similarities between viruses and eukaryotes, including a genome arranged as linear strands of DNA with specialized end sequences, associations between proteins and DNA (forming chromosomes), messenger RNA capping (modification to the nucleotide at the 5' end), and separation of the transcription and translation processes (Claverie 2006). Virus infection may have had influences on eukaryote genome structure and may have presented a motivating force for the conversion from RNA- to DNA-based genomes (Forterre 2006).

different, so they are likely to have been independently derived from a common ancestor that lacked a cell wall; 2) later origination of eukaryotes requires that we assume a secondary loss of the cell wall after divergence from archaeans; and, 3) the presence of a cell wall in Archaea presents a significant impediment for phagocytosis or other mechanisms that would allow the engulfment of intracellular bacterial mutualists (Hartman and Fedorov 2002; Martin and Russell 2003). The fact that eukaryotes have a significant number of proteins associated with intracellular structures that are unique to their group and that do not display homology with prokaryotic proteins supports the hypothesis that eukaryotes were derived from a pre-karyotic lineage (Hartman and Fedorov 2002). On the other hand, analyses of shared genomic regions (Vesteg and Krajcovic 2011) and the similarity of the translational processes supports a more recent common ancestor with archaeans. Fossils are not helpful in resolving these two alternative scenarios, since the rare fossils of bacteria dating from 3.5 BYA do not provide details of internal structures, and any cells without walls would have been extremely unlikely to leave a record.

While clear support for a single sequence of events is elusive, we can be confident that mutualisms and horizontal gene transfer were important for the development of the eukaryotic genomic architecture. Internal cellular structures could have developed along with divergence under either scenario, but it is probable that the lack of a cell wall and larger size would have been necessary for pre-karyotic cells to engulf bacteria that developed into obligate mutualists, which are represented in modern eukaryotes as mitochondria and chloroplasts (Vesteg and Krajcovic 2011). Much of the earliest history of life will likely remain obscure because of the confusion wrought by rampant horizontal recombination, the loss of ancestral and intermediate forms, and the lack of a fossil record.

Whatever the pathway, the new eukaryotic lineage that emerged provided a flexible template for adaptation and specialization. This time period, which extended from around 1,200 to 542 MYA (Millions of years ago), may aptly be referred to as a *Silent Explosion* of life (silent because these microbes did not leave a fossil record), the remnants of which are represented in diversity of eukaryotic microbes that have traditionally been classified as members of the kingdom Protista (Hedges 2002; Fig. 1.1; refer to color plate). This diversification necessarily predated the *Cambrian Explosion,* where a subset of the dazzling array of multicellular animals was sufficiently large enough, and had enough hard parts to leave fossil records. At the same time as this putative silent explosion of consumers, eukaryotic lineages that had developed mutualisms with cyanobacteria diversified in parallel into major groups of unicellular algae, the descendants of which are the dominant producers in oceans and freshwater systems to this day.

Life Cycles and Origins of Complexity—The large increase in chemical energy made accessible by photosynthetic microbes during the Proterozoic created conditions that promoted the emergence of a diversity of multicellular life forms. Advantages of multicellularity were a direct consequence of larger size and greater mobility, which improved predator avoidance on the one hand and predation effectiveness on the other. Multicellular organisms emerged in a single eukaryotic lineage of consumers that gave rise to the most recent common ancestor of all animals. Around the same time, multiple lineages of eukaryotic producers emerged, but only one eventually generated a lineage that diversified into all land plants.

One of the major differences between the new multicellular producers and the consumers was the inclusion of a multicellular haploid stage in their life cycles. Nearly all of the algal lineages developed only multicellular haploid stages (*Haplontic Life Cycles;* Fig. 1.3.A), or multicellular haploid and diploid stages (*Haplodiplontic Life Cycles;* Fig. 1.3.B), while the common ancestor to all animals had only a multicellular diploid stage (*Diplontic Life Cycle;* Fig. 1.3.C; Box 1.3). The emergence of diploid life stages and sexual reproduction occurred at least 1,200 MYA and set the stage for the silent explosion of eukaryotic life that followed (Butterfield 2000). Multicellular diploid stages would eventually dominate the life cycles of plants, but these lineages would not arise until long after plants had colonized land. It is interesting to note that all modern land plants retain a multicellular haploid life stage, but we will have to wait to explore the implications of their haplodiplontic life cycle in later chapters. Our question here is more basic: Why did multicellular diploid stages come to dominate the life cycles of all animals and the majority of land plants? There is a growing notion that diploidy and genome complexity are necessary precursors to the evolution of organismal complexity (Freeling and Thomas 2006). To understand why diploidy may have been necessary for the diversification of multicellular consumers, we need to examine the evolutionary consequences of different genomic architectures.

The question is whether the observed variation in genomic architecture across eukaryotic lineages is the result of adaptive responses, or whether the patterns we see today are the byproducts of historical contingencies. Several observations suggest that diploidy is a critical precondition for the development of complexity. First, there is tremendous variation in ploidy and life cycles across eukaryotes. For example, among the macrophytic algae, examples of all three of the life cycles described in Box 1.3 can be found, and the life cycles of many red algae (Rhodophyta) include two diploid stages (Mable and Otto 1998). There is even greater variation across microbial eukaryotes where a wide range of genomic compositions are found, including some radiolarians where genomes range from haploid (N) to around 1000N during their life cycle (Parfrey et al. 2008). Second, the haploid stages of algae, bryophytes, and ferns (*Gametophytes*) have relatively simple anatomical organization. The rise of dominance of the diploid life stage (*Sporophytes*) in land plants corresponds with an increase in the number of tissue types and specialized organs such as roots, stems, leaves, and complex reproductive structures. A variety of multicellular haploid forms exists among the eukaryotes, but none of these organisms possesses collections of tissues constituting complex organs.

This discussion suggests that the association between diploidy and complexity is not due to random accidents of history, but it does not identify a mechanism that would favor diploidy in more complex organisms. To understand this better, we need to examine the consequences of becoming larger and more complex. The dramatic increase in resources made possible by photosynthesis created opportunities for consumers at higher trophic levels, so selection favored larger and more mobile

Box 1.3 Cycles of Life

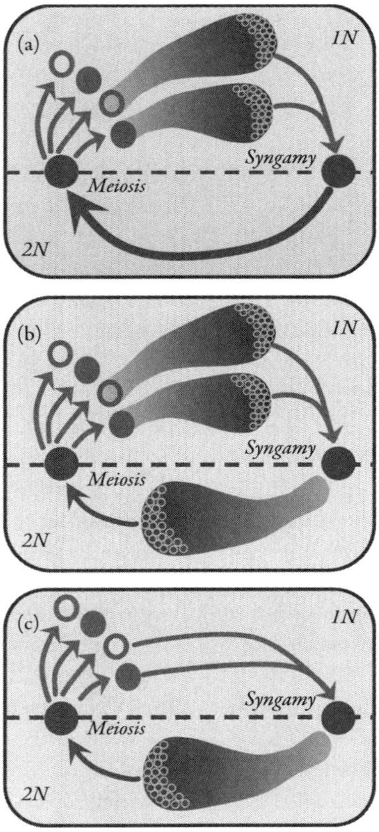

Fig. 1.3. Life cycle variation. A. For haplontic life cycles, mitosis occurs after meiosis but not after syngamy; B. Mitosis occurs after meiosis and syngamy in haplodiplontic life cycles; C. Mitosis occurs only after syngamy in diplontic life cycles.

Most eukaryotic organisms have two life stages separated by the fusion of gametes (*Syngamy*), which doubles the genome, and meiosis, which reduces the genome size by half. Except in rare cases, meiosis requires two cell divisions and is preceded by the doubling of the chromosomes. During the first division, homologous pairs form and may exchange genomic regions during crossing over (*Homologous Recombination*). The two sets of recombinant chromosomes in the diploid cells resulting from the first division are divided between two cells in a second division to produce a total of four haploid gametes.

Life cycles differ based on whether mitosis occurs in the haploid or diploid stage, or both. Mitosis may produce either large numbers of independent single cells or multicellular stages of the species. A species with a haplontic life cycle (Fig. 1.3.A) undergoes mitosis after meiosis to produce large numbers of clonal haploid unicellular organisms or multicellular individuals. Zygotes formed after the fusion of gametes undergo meiosis immediately to produce the next haploid generation. Haplontic life cycles occur in some fungi, some Protozoans, and in most green algae, including the Charophyte algae, which have a multicellular haploid stage and are considered to share a common ancestor with all land plants.

In species with haplodiplontic life cycles, mitosis occurs after both meiosis and syngamy (Fig. 1.3.B). Multicellular haploid and diploid stages occur in all land plants (everything from mosses and liverworts to flowering plants), some green algae, and some fungi. Species with diplontic life cycles have multicellular diploid stages that produce haploid gametes directly (Fig. 1.3.C). This type of life cycle occurs in nearly all animals, in some brown algae, and in some fungi, including yeast.

predators. Coincident with an increase in size was an increase in generation time and a decrease in reproductive potential compared to unicellular microbes. For unicellular organisms capable of rapid population growth, novel adapted lines can be generated by numeric dominance; during each generation, numerous individuals possessing unique mutations are subjected to selection. If the large majority of the daughter cells cannot survive challenging environmental conditions, it does not matter as long as there is at least one surviving individual with the appropriate mutations for the species to persist. The unique genetic complements of the successful lineages will be reliably transmitted to subsequent generations so they can continue to propagate and dominate until conditions change again.

For longer-lived organisms, a mode of adaptation requiring high rates of population growth is not an option. Higher complexity requires a longer generation time, and fewer offspring can be produced each generation. With a low reproductive potential, a haploid multicellular organism would have limited ability to generate adequate genetic variation in its offspring to meet the challenges of changing environments, because the haploid state sets a limit on the level of genetic variation maintained within a population.

In haploid organisms, every allele has an effect on the phenotype, so variation in expressed regions of the entire genome is subject to selection (Mable and Otto 1998). Any allele that is not favored in the current environment will be eliminated from a haploid population, so changes in conditions can only be matched by novel mutations or the acquisition of genetic material via gene flow or horizontal gene transfer. The rare instances of mutation producing adaptive alleles or capture of adaptive genetic elements from the environment work quite well for organisms with high reproductive output but would place severe limits on the evolutionary potential of longer-lived and more complex consumers because of their substantially lower rates of population growth.

For diploid organisms, high levels of variation among offspring can be produced by other mechanisms (Qiu et al. 2012). Dominance among alleles segregating at the same locus in diploids allows a much higher level of genetic variation to be sequestered in each genome as recessive alleles are effectively hidden from selection. The potential for variation in offspring is further enhanced by interactions among alleles across multiple loci (*Epistasis*). The combination of dominance and epistasis creates tremendous potential for phenotypic variation among offspring through the generation of novel combinations of alleles across chromosomes (*Recombination*). The rearrangement of chromosomes via recombination during sexual reproduction enhances the potential of organisms to glean the maximum advantage of diploid genome architectures, as a variety of novel genetic combinations can be produced. The ability to generate higher levels of genetic variation in the offspring of larger, more complex organisms may have been particularly important for keeping pace with their fast-evolving pathogens (the *Red Queen Hypothesis*; Chapter 9). Because larger and more complex organisms had longer generation times and lower reproductive potential, diploidy and recombination during sexual reproduction were probably necessary preconditions allowing them to diversify and adapt to novel conditions.

The explanation that longer generation times and reduced reproductive potential favored diploidy in complex consumers is not entirely satisfactory because multicellular haploid producers are also quite successful. For example, the haploid generations of a variety of macrophytic brown algae (kelp) are the dominant components of many coastal marine habitats, and mosses are often dominant in epiphyte communities of rain forests and terrestrial habitats in arctic regions. The common denominators across macrophytic haploids are their simplistic anatomical structure and the requirement of aquatic or moist environments for growth and reproduction. While one could argue that maritime and mesic habitats may present fewer environmental challenges, the observation that mosses have diversified into habitats characterized by extremes in temperature and moisture availability (Cox et al. 2010) challenges this notion. On the other hand, it is important to note that during times of limited water availability, the haploid moss thallus becomes dormant until water becomes available again, so these organisms effectively avoid stress from lack of water (*Poikilohydry*—see Chapter 2). This ability of mosses to become desiccated without losing viability has allowed them persist in many periodically dry habitats. On the other hand, mosses are limited to environments where moisture is abundant for a significant portion of the year. While they are dominant in some habitats such as tundra and epiphyte communities in wet forests, they are limited in size and organizational complexity, and their rate of diversification is stunted relative to groups with dominant diploid life stages (Stenoien 2008; Fiz-Palacios et al. 2011).

Consideration of genomic compositions across eukaryotes suggests that the association between diploidy and organismal complexity in the animal and plant kingdoms is probably not a coincidence. The mechanism that limits the morphological and ecological diversity of haploid macrophytes is most likely their genomic architecture, possibly because it constrains the level of genetic diversity that can be maintained within populations. While lower levels of genetic variation may impose constraints on the evolutionary potential of haploid populations, it is also likely that there is an intrinsic relationship between genomic and phenotypic complexity (Freeling and Thomas 2006). For example, genomic surveys across prokaryotes and eukaryotes indicate that the numbers of gene duplications, introns, and regulatory regions increase along with complexity (Lynch 2006). The observation that *Whole-Genome Duplications* (*Polyploidy*) are characteristic of major evolutionary transitions in ancestral chordates, and have occurred several times during the early stages of angiosperm diversification (Freeling and Thomas 2006), also supports this hypothesis. As discussed further in Chapter 3, genome duplication events have been critical for the development of genomic complexity and the origin of novel gene function. While diploid organisms have undergone manifold increases in genomic complexity, polyploidy events in haploids have been much rarer (e.g., in Bryophytes; Wyatt et al. 1988), suggesting that diploidy was a necessary precondition for the development of genomic and phenotypic complexity (Qiu et al. 2012). We are just beginning to understand how genomic architecture may set limits on organismal complexity, and new insights are likely to emerge as genomic information on a greater variety of organisms becomes available.

1.3 DIVERSIFICATION OF EUKARYOTES

Early in the Proterozoic, sexual reproduction arose in a lineage that gave rise to all descendent eukaryotes including the huge diversity of microbes, fungi, plants, and animals (Fig. 1.1; refer to color plate). This efficient mechanism for the generation of genomic variation through recombination within and among divergent lineages set the stage for the rampant episodes of diversification that would follow. In this section, we explore the origins of animal and algal diversity.

Origins of Animal Diversity—The transition to life cycles with dominant diploids in plant and animal lineages facilitated further increases in genome complexity and provided the necessary genetic material for dramatic phenotypic diversification. This process began perhaps 1.0 BYA during the Proterozoic for metazoans (Fig. 1.4), when the acquisition of a diplontic life cycle may have made possible a tremendous diversification of multicellular forms. Selection pressure from both competitors and predators was probably fierce as animal communities became more heavily populated. Green and red algae and photosynthetic prokaryotes (cyanobacteria and prochlorophytes) provided a foundation for a food web that became increasingly complex as ever-larger predators emerged to take advantage of the abundance of smaller consumers.

Evidence for the emergence of metazoan grazers can be seen from fossils showing traces of animal activity and from a decline in the abundance of stromatolites around 1.0 BYA (McNamara 1996). Stromatolites are fossilized accretion structures that are thought to have been formed by the activity of cyanobacteria (Fig. 1.5; refer to color plate; Lepot et al. 2008). In cross section, they display characteristic dark and light layers that represent the accumulation of material over time. Stromatolites probably developed in shallow water as granular material became trapped in the biofilm and was cemented by excretions from the microbes. These structures first appeared around 3.5 BYA and peaked in abundance by 1.25 BYA (Fig. 1.4). While this change in the frequency of stromatolites implies a change in the abundance of grazing animals, definitive fossil evidence of metazoans is not found until the Ediacaran period of the late Proterozoic (635–542 MYA; Knoll et al. 2006). Toward the end of the Proterozoic (542 MYA), the abundance of stromatolites had decreased considerably, probably as a consequence of grazing pressure by the increasingly abundant metazoans. Active stromatolites can still be found in hypersaline lakes where high salt concentrations exclude grazing animals.

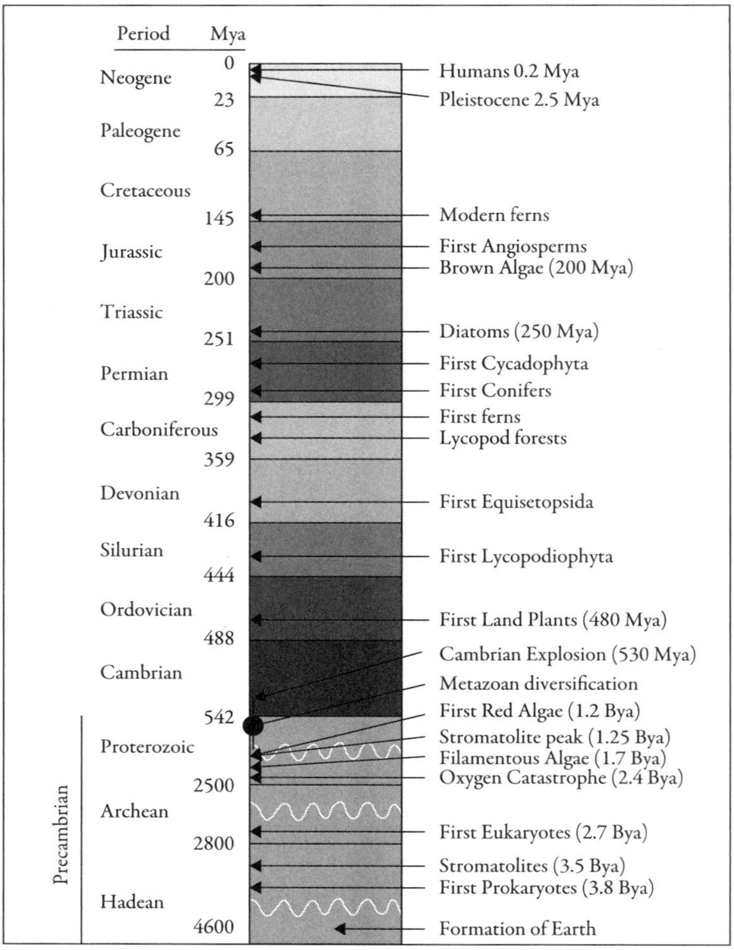

Fig. 1.4. Geologic time line (approximate).

Fig. 1.5. Fossil stromatolite (cross section—left) and modern stromatolites from Shark Bay in Western Australia (right). Photos by Didier Descouens and Paul Harrison.

During the early Cambrian around 530 MYA, the number of animal forms increased dramatically. This "sudden" appearance of a diverse fauna of many large-bodied species actually occurred over many millions of years, so the Cambrian explosion was actually a gradual increase that looks dramatic in the context of the impoverished fossil record that predated this period. Nonetheless, there was an exponential rate of diversification that was probably precipitated by a number of key innovations such as improvements in nervous systems, digestive tracts, musculature, and the appearance of hard body parts. While none of the species from that time period persist today, many of them displayed characteristics of modern animal phyla; there are also unusual forms that are only known from fossils.

Algal Diversification—During the late Proterozoic and early Cambrian, when consumers were undergoing spectacular diversification, changes in the number of producer lineages were modest by comparison. The earliest fossil evidence of eukaryotes dates from 2.7 BYA (Brocks et al. 1999; Hedges 2002), and the first definitive fossils of a multicellular filamentous algae are known from around 1.7 BYA (Bengtson et al. 2009). Phylogenetic reconstructions indicate that several independent lineages of photosynthetic eukaryotes were present during the Proterozoic (Fig. 1.6). All of the eukaryotic algae were likely derived from a single endosymbiosis event with cyanobacteria as early as 1.6 BYA (Hedges 2002; Yoon et al. 2004; Keeling 2010). The majority of photosynthetic lineages during the late Proterozoic were members of a single group referred to as the Archaeplastida, which today includes three clades of algae (red and green algae, plus the glaucophytes) and the land plants. A large increase in the number of archaeplastid species did not occur until much later during the *Devonian Explosion* (Chapter 2).

For much of the history of life on Earth the primary producers have been members of the Archaeplastida

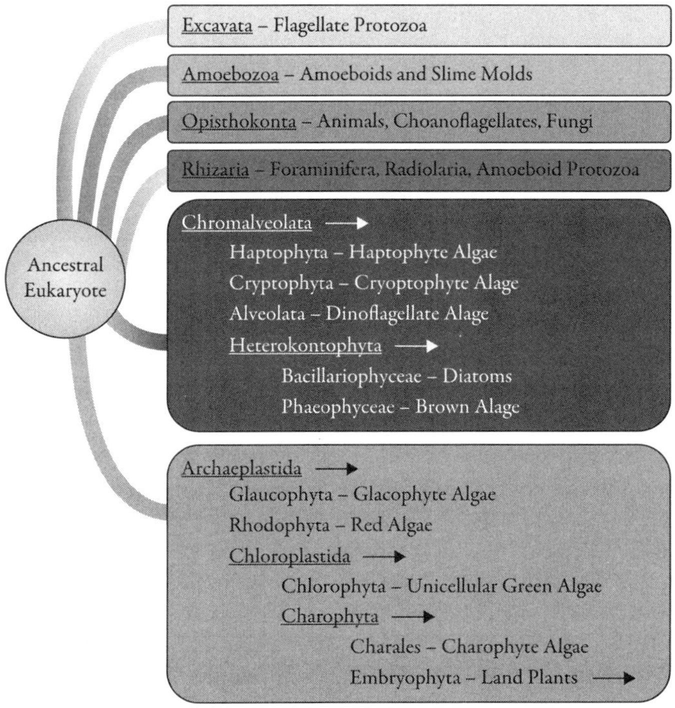

Fig. 1.6. Six monophyletic groups of Eukaryotes (in orange). Major animal and photosynthetic groups are in red. Subgroups follow arrows. Land plants are a subclade within the Charophyta. Branching order within clades shown remains unresolved.

including the glaucophytes, rhodophytes, and Chloroplastida (Fig. 1.6). The glaucophytes are less common today but were a major component of autotrophic communities during the Proterozoic. Glaucophytes are a small group of unicellular algae with plastids that differ from other algae and land plants by retaining the remnants of the peptidoglycan cell wall present in cyanobacteria. Photosynthesis in glaucophytes is accomplished using the pigment chlorophyll a (a trait shared with Cyanobacteria and green algae) and phycobilisomes (structures containing phycobiliprotein pigments), a trait they share with the Cyanobacteria and red algae. Note the basal origin of the glaucophytes among the algae in Fig. 1.1 (refer to color plate).

The red algae (Rhodophyta) were also important producers during the Proterozoic, and a large diversity of these algae continue to thrive. These producers differ from green algae (Chlorophyta) by retaining phycobiliproteins from the cyanobacterial ancestors of their chloroplasts as accessory pigments (Lee 2008), which gives them their characteristic reddish hue. Nearly all rhodophytes living in contemporary ecosystems are multicellular and the few known unicellular species lack flagella. Most rhodophytes have life cycles with multicellular haploid and diploid generations. One of the oldest known identifiable multicellular fossils (*Bangiomorpha*) displays strong morphological similarity to extant species of red algae and dates from around 1.2 BYA (Butterfield 2000). Rhodophytes, along with the green algae, glaucophytes, and cyanobacteria, were major primary producers starting in the Proterozoic and continue to be the most diverse aquatic macrophyte groups in modern marine and freshwater systems, with as many as 10,000 species. Red algae are common in near-shore and intertidal marine habitats and are the only macrophytes found in deeper water where light levels are extremely low.

The chlorophytes include several minor groups and two major clades; the Chlorophyta, which consist of only unicellular and colonial flagellated species, and the Streptophyta, which includes both unicellular and multicellular species of algae and all of the embryophytes (land plants). Because land plants share common ancestors with the charophytes, the green algae as a group are *Paraphyletic* (Fig. 1.6). All of the green algae and land plants (collectively referred to as the Viridiplantae—a *Monophyletic* group) utilize chlorophyll a and b as their primary photosynthetic pigments, lack the phycobiliprotein pigments present in cyanobacteria, glaucophytes, and rhodophytes, and store chemical energy as starch (Lee 2008). Excluding the embryophytes, the Viridiplantae include around 4,300 species and are major primary producers in aquatic ecosystems around the world.

The second major clade of eukaryotes that contains photosynthetic microbes and macrophytes is Heterokonta (brown algae and diatoms) within the Chromalveolata, which, along with the Archaeplastida (glaucophytes, rhodophytes, and chlorophytes), is one of the six major monophyletic clades of eukaryotes. The other four clades of eukaryotes are the Excavata (flagellate protozoa including *Euglena*, some species of which possess chloroplasts), Amoebozoa (amoeboids and slime molds), Opisthokonta (fungi, choanoflagellates, and metazoans), and the Rhizaria (foraminiferans, radiolarians, and amoeboid protozoa). The heterokonts obtained chloroplasts secondarily from symbiotic unicellular rhodophytes, so their plastids are surrounded by four instead of two membranes (Lee 2008). It appears that heterokonts obtained the ability to photosynthesize before they diversified, since they share similar chloroplasts with Haptophytes, a closely related group of phytoplankton. Heterokont chloroplasts usually contain chlorophyll a and c and the secondary pigment fucoxanthin, which gives these algae their characteristic golden-brown hue.

Heterokonts include the diatoms, a diverse group of unicellular phytoplankton characterized by having sculpted silica shells, and the brown algae (Phaeophyta; kelp), which includes up to 2,000 species of macrophyte algae and lacks unicellular members. Heterotrophic groups within the heterokonts (e.g., water molds) lost the ability to photosynthesize after they were derived from ancestors that possessed chloroplasts. The major group of the brown alga, Laminariales, includes species commonly referred to as kelp. These are some of the largest macrophytic algae known, and often form kelp forests in near-shore marine ecosystems in temperate and polar regions around the world. Unlike the other major groups of algae, the diatoms and phaeophytes diversified and became dominant components of aquatic ecosystems much later (only within the last 200 MY). Although much older fossils have been found that superficially resemble species of brown algae, all of these have been identified as other organisms (red algae or fungus) or are attributed to inorganic processes.

1.4 AQUATIC DIVERSITY

The diversity of algae that dominated marine and freshwater systems by the end of the Proterozoic provided the foundations for many diverse and productive ecosystems. Out of this tremendous diversity of multicellular algae, only one green algal lineage in the Charophyta (Streptophyta) managed to successfully colonize and persist in terrestrial habitats. We could imagine that macrophytic heterokonts may have been candidates for the invasion of dry land, but even if they had the propensity to venture into terrestrial environments, it may have been difficult because these habitats would have already been occupied by tremendously diverse descendants of the first land plants. Rhodophytes, on the other hand, diversified in parallel with the chlorophytes, so perhaps they could have colonized land along with the charophytes. Whether any red algal lineage ever colonized dry land is unknown, but it is also unlikely because this group probably lacked the particular characteristics that facilitated the success of charophytes outside of aquatic environments. In the next chapter, we will discuss the unique attributes of charophyte algae that were critical for their transition to dry land.

1.5 QUESTIONS FOR CONTEMPLATION AND DISCUSSION

1. What are the minimal conditions necessary for evolution by natural selection? Think about both the characteristics of organic entities and their environment.
2. Develop arguments to support the RNA-first versus protein-first scenarios for the origin of complex cells. Do you think one of these is more compelling than the other?
3. Compile a list of the unique biological features of plants that distinguish them from other eukaryotes. Describe each characteristic and argue whether it is more likely to be advantageous or neutral. What is the probable origin of each trait? Your ideas on this topic are likely to change as you proceed through the remainder of this book.
4. What were the consequences of photosynthesis for life on Earth? What were the necessary initial steps toward photoautotrophic life and why was the splitting of water to obtain hydrogen atoms so important? What were the consequences of increases in the abundance of free oxygen?
5. Develop arguments to support different scenarios for the origin of the eukaryotes. What is the evidence for the existence of a pre-karyotic lineage (chronocytes)?
6. What was the importance of the advent of sexual reproduction? How did the combination of diploidy and sex provide evolutionary advantages for larger and more complex organisms?
7. What are the disadvantages of a multicellular haploid life stage? Can you come up with reasons why the metazoans (animals) lost their haploid stage very early in their evolution while plants have retained it?
8. What is the link between genomic complexity and organismal complexity? Why do you think genome duplications were advantageous and perhaps necessary for the diversification of life? You may want to revisit this question after reading Chapter 3.
9. Why didn't plants diversify along with animals during the Cambrian explosion? What were the possible differences in selective pressures on these two groups of organisms? You may want to revisit this question after reading Chapter 2.

2

DIVERSIFICATION OF LAND PLANTS

The rapid development as far as we can judge of all the higher plants within recent geological times is an abominable mystery. Certainly it would be a great step if we could believe that the higher plants at first could live only at a high level. . . . Saporta believes that there was an astonishingly rapid development of the higher plants, as soon [as] flower-frequenting insects were developed and favoured intercrossing. I [should] like to see this whole problem solved. I have fancied that perhaps there was during long ages a small isolated continent in the S. hemisphere which served as the birthplace of the higher plants—but this is a wretchedly poor conjecture.

Charles Darwin (1879) from a letter to J. D. Hooker

2.1 THE DIVERSITY OF GREEN LIFE

During the Neoproterozoic, a diversity of producers and consumers were abundant in marine and freshwater environments, but other than bacteria and fungi, the dry land was nearly devoid of life. Closer examination may have revealed fragile rims of green along the shores of freshwater lakes and ponds—evidence of plants venturing into terrestrial environments. The establishment of charophyte algae (Streptophyta; Fig. 1.6 and Fig. 2.1; refer to color plate) that were capable of thriving out of water initiated the spread and diversification of plants across the landscape—a process that continued over the next several hundred million years. Adaptations enabled new plant lineages to radiate into drier habitats as their dependence on water for survival and reproduction incrementally decreased. As plants spread and diversified, they created the foundation for terrestrial ecosystems that ultimately supported diverse consumer and decomposer assemblages. These earliest *Embryophyte* (land plant) lineages have undergone tremendous diversification since they originated. In the present day we have around 400,000 species of flowering plants (*Angiosperms*); 900–1,000 species of conifers, cycads, and allies (*Gymnosperms*—used here to indicate the collective group of non-flowering seed plants); 11,000 species of ferns and horsetails (*Pteridophytes*); 1,200 species of clubmosses, spikemosses, and quillworts (*Lycopodiophytes*); and 12,000 species of mosses, liverworts, and hornworts (*Bryophytes*). In this chapter, we will discuss the origin and establishment of land plants and explore the diversity of embryophyte lineages that now dominate as the primary producers across all terrestrial habitats.

Toward the end of the Neoproterozoic, producers were so abundant in aquatic environments that the low availability of resources limited population growth. In deep-water habitats there was insufficient light, and in shallower water there was a severe lack of inorganic nutrients (primarily nitrogen and phosphorous). While nutrients were more abundant out of the water, they remained inaccessible for most aquatic life. Transition from aquatic to terrestrial environments required overcoming seemingly insurmountable obstacles: severe desiccation, large temperature fluctuations, intense solar radiation, and the effects of gravity, all of which rendered the terrestrial

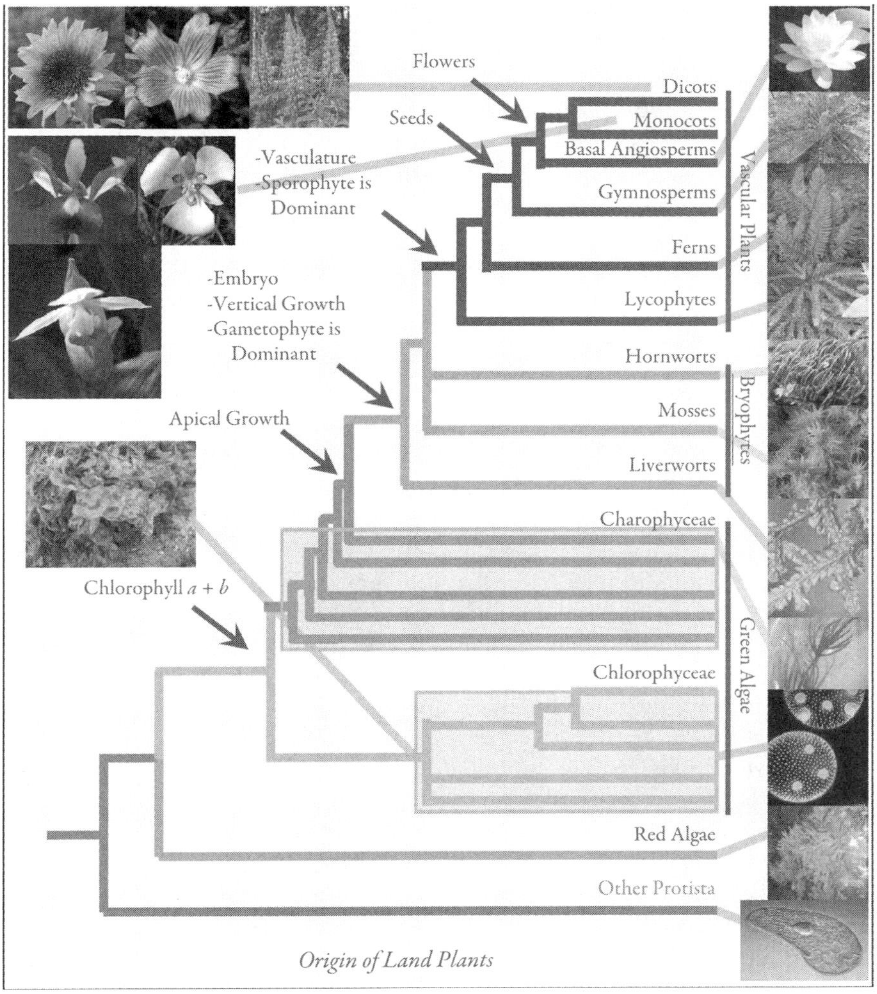

Fig. 2.1. A phylogeneny for the major groups of land plants and their algal ancestors. The land plants (Embryophytes) are a monophyletic group that includes the Bryophytes and vascular plants. Blue arrows indicate the origin of major transitional traits. Photos by Frank Fox, Peter Southwood, Kristian Peters, Hermann Schachner, KirinX, Edward Price, Jami Dwyer, T. R. Shankar Raman, M. Cruzan, Stan Shebs, Jon Sullivan, Oleg Yunakov, and Orchi.

environment deadly for most aquatic life forms. At the same time, there was strong selection to overcome these impediments as the ability to tolerate exposure to dry air afforded access to ample light and more abundant nutrients. The first algal lineages that ultimately persisted and thrived out of water sparked the diversification of numerous terrestrial groups (Fig. 2.1; refer to color plate). In the first sections of this chapter, we will discuss the origins of land plants and the challenges they faced as they made the transition to terrestrial habitats.

2.2 THE FIRST LAND PLANTS

The emergence of green life from the water was inevitable—the more abundant resources available on land were not likely to remain unexploited for long. The ancestors of land plants—the charophyte algae—were probably dependent on precipitation and runoff from dry land as the primary source of inorganic nutrients. With nutrient availability as a primary limitation to plant growth in the water, it was just a matter of time before the appropriate innovations appeared to allow colonization of terrestrial habitats. Survival on land required overcoming severe drying and exposure to sunlight; strong selection gradients existed at the water's edge where periodic exposure favored desiccation resistance. Under these circumstances any adaptations that improved tolerance to drying or the extraction of water and nutrients from the substrate would have spread, allowing early colonizers to incrementally invade drier habitats.

These first stages of transition to terrestrial habitats remain entirely unknown. There are no living plants that retain the morphological characteristics of the earliest land plants, and we do not have any fossils that can definitively be associated with transitional forms. While plants in the Bryophyta (mosses, liverworts, and hornworts) are often referred to as representatives of the earliest land plants, they actually are quite divergent and possess a number of complex traits that make them much more similar to other land plants than to streptophyte algae. For example, all bryophytes have a multicellular sporophyte stage (haplodiplontic life cycle; see Box 1.3) that is complex in contrast to the unicellular zygote of streptophytes (haplontic life cycle). Furthermore, each of the bryophyte lineages possesses suites of characteristics that are unique compared to each other and to other embryophytes, suggesting that they diverged from a much simpler common ancestor (Qiu et al. 2012). The characteristics shared among bryophyte groups include a multicellular sporophyte, parenchymous (i.e., undifferentiated cell) growth of the gametophyte, apical growth, and complex reproductive structures. The commonality of these traits among bryophyte lineages suggests that the common ancestor they share with other land plants (vascular plants, the *Tracheophyta*) probably possessed the same set of traits. On the other hand, Bryophytes differ from other vascular plants by having a dominant gametophyte stage, a dependent sporophyte stage, and the lack of true vascular tissue (*Tracheids*). Sporophytes and gametophytes of some mosses possess conducting cells (*Hydroids*) that serve as vascular tissue (the *Hydrome*; Ingrouille and Eddie 2006), but since most bryophyte lineages do not have this feature, it is more likely that hydroids are independently derived and not homologous to the vascular tissue of the tracheophytes. This scenario is supported by the observation that conducting cells in moss sporophytes and liverwort gametophytes lack the cell wall thickenings and lignification present in the tracheids of vascular plants (Ingrouille and Eddie 2006).

By considering the distribution of shared traits among extant lineages of embryophytes and streptophyte algae, we can make some logical guesses about the probable characteristics of the first land plants. We can infer that the first fully terrestrial lineages were small plants that had a dominant gametophyte stage, and a diploid stage that was either a unicellular zygote or a simple multicellular sporophyte that was dependent on the gametophyte (traits shared with bryophytes and streptophyte algae). The gametophyte stage was probably a *Thallus* (a flat plant body consisting of undifferentiated cells, lacking specialized tissues and organs) that had apical growth and unicellular *Rhizoids* (hair-like extensions of cells that serve the same function as roots)—traits shared with the gametophytes of hornworts, liverworts, lycopods, and ferns. These plants were restricted to areas of constant moisture, as they possessed little capacity to maintain their internal water status. Other traits present in different embryophyte lineages such as multicellular rhizoids, stomata, *Poikilohydry*, and *Vasculature*, are not shared across the entire clade, so they are unlikely to have been present in the earliest land plants and are probably independently derived in each lineage.

The picture that emerges for the first terrestrial plants is one of small, thalloid gametophytes with limited ability to maintain their internal water status and so remaining closely appressed to a moist substrate. These plants had a dominant gametophyte stage and probably produced motile sperm, so they required water for successful reproduction. The diploid stage would have been a unicellular zygote (similar to charophyte algae) or a reduced multicellular sporophyte (similar to liverworts). These first terrestrial plants may have been limited to locations with consistent moisture availability and some shade until adaptations appeared that allowed them to survive in more exposed sites.

As terrestrial lineages spread and became more abundant, competition would have ensued as habitat space with sufficient moisture became limiting to growth. Selection on terrestrial populations would have favored traits that contributed to their ability to colonize new habitat and to compete with other members of the plant community. For example,

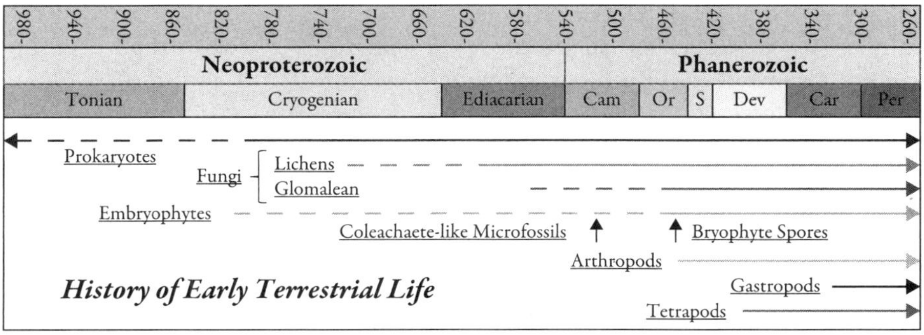

Fig. 2.2. Time scale for the appearance of terrestrial microbes, fungi, plants, and animals. Beginning of the solid portions of each line indicates the first fossil evidence. The earliest origins are based on estimates from molecular phylogenies (except prokaryotes) and are indicated by the dashed lines. The estimate for terrestrial prokaryotes is based on geochemical evidence. Scale is in millions of years ago (MYA).

improved resistance to desiccation would have allowed survival in drier habitats, better dispersal ability, and more drought-resistant spores would have promoted the colonization of new habitats.

The process of colonization of drier habitats may have been facilitated by the development of mutualistic associations with glomalean fungi, which had been present on dry land at least since the Ordovician (Redecker et al. 2000) and probably since the Neoproterozoic (Fig. 2.2). These fungi form symbiotic relationships with plants (*Mycorrhizae*) and may have been instrumental in the establishment and spread of plants across terrestrial habitats (Brundrett 2002; Bonfante and Genre 2008). Fossils of mycorrhizae date from the early Devonian, 460 MYA (Remy et al. 1994), but mycorrhizal fungi are likely to have been associated with the earliest land plants. The importance of plant-fungal mutualisms for the success of early land plants is supported by the observation that *DMI* genes, which are important for plant-microbe interactions, are present in all major embryophyte lineages (Wang et al. 2010).

2.3 ORIGIN OF THE LAND PLANTS

Dating the Origin of Embryophytes—Plants were not the first terrestrial inhabitants. Microbes and fungi were well established on dry land and had been there for some time before plants first ventured out of their aquatic environs (Fig. 2.2). There is geological evidence that microbial mats affected geochemical processes in terrestrial environments during the Paleoproterozoic, as early as 2,600 MYA (Watanabe et al. 2000), and the first microfossils of terrestrial bacteria date from around 800 MYA (Horodyski and Knauth 1994). The first eukaryote inhabitants of land were probably fungi that were associated with cyanobacteria to form lichen-like symbioses (Heckman et al. 2001), which may have been important for promoting the chemical degradation of rock (Schwartzman 1999). Inferences from molecular phylogenies are consistent with the hypothesis that lichens inhabited land as early as 720 MYA, but the first unambiguous fossil evidence of lichens comes from around 600 MYA (Yuan et al. 2005). These early terrestrial inhabitants, along with weathering and erosion by wind and water, contributed to the accumulation of fine-grained inorganic material and the release of nutrients (e.g., nitrogen and phosphorous), which would be important for the establishment and spread of early land plants.

The timing of the emergence of the first terrestrial plants remains equivocal, as estimates from molecular phylogenies for the origin of the embryophytes are consistently much earlier than the dates indicated by fossils (Fig. 2.2). The first fossil evidence of terrestrial plants comes from the early Paleozoic (~450 MYA), but molecular clock estimates date the split between embryophytes and streptophyte algae much earlier, between 568 and 815 MYA during the Neoproterozoic (Clarke et al. 2011). The wide range of dates for the origin of land plants from phylogenies is due to variation among gene regions sequenced and the use of different methods for date calibration; however, even the youngest estimates fall well before the age of the oldest embryophyte fossils.

There are several possible reasons for the discrepancy between dates of the oldest fossils and age estimates from phylogenies. One is the tendency for the fossil record to be biased toward younger dates. Fossilization is very rare and only occurs under specific chemical and physical conditions (Willis and McElwain 2002), so organisms that are abundant and have been present for some time are more likely to leave a record. Nearly all organic matter—especially from organisms that lack hard parts—will rapidly decay and leave no trace of the original form. In the rare cases where early land plants may have left a record, most fossils were lost due to geological processes occurring over tremendous spans of time. It is likely that streptophytes were present in terrestrial habitats for a considerable amount of time before geological processes captured evidence of their existence.

A second reason for the appearance of fossil land plants much later than their divergence from the algal streptophytes comes from difficulties with the identification of fossils that represent the earliest terrestrial plant forms. The early fossil record contains many unidentified objects referred to as *Acritarchs* (literally, "confused origin") that could represent key evidence of the earliest land plants or could be products of physical processes. For consistency, it is necessary to use unambiguous characters for the identification of fossils. For example, some of the first macrofossils that are clearly representative of embryophytes come from the early Devonian (e.g., *Cooksonia*; Fig. 2.3). In this case the presence of terminal sporangium was used to classify these fossils as representatives of early land plants, but the species represented by these fossils was far removed from the first land plants as they displayed more advanced traits such as a hydrome-like vascular tissue and complex reproductive structures (Ingrouille and Eddie 2006).

More recent discoveries of microfossil evidence have pushed back the appearance of land plants to the late Ordovician (around 450 MYA; Fig. 2.2); fossils that may once have been passed over as acritarchs were classified based on their close resemblance to the spores of modern bryophytes (Rubinstein et al. 2010). There may be other unidentified microfossils that could provide key insights into the origin of land plants. For example, a particular class of acritarchs from the Cambrian were not associated with Coleochaetophytes until these streptophyte algae were found to produce similar structures when grown under environmental conditions designed to mimic an emergent aquatic habitat (Graham et al. 2012). Identification of these microfossils provides important evidence that desiccation-resistant streptophytes may have been present on moist terrestrial substrates during the early Paleozoic around 515 MYA—a date that corresponds more closely with estimates for the origin of embryophytes from molecular phylogenies. Given the large diversity of unidentified fossils that occupy museums and labs around the world, it is likely that there are many more examples of acritarchs that have the potential to provide important clues about the nature of the earliest land plants.

Fig. 2.3. A depiction of the fossil plant *Cooksonia*. Plants were several decimeters tall with terminal sporangia, and vascular tissue in the stem and rhizomes. Original art by Michelle Nicole Williamson.

The Transition to Land—There was a long road of divergence—several hundred million years—from the time of the emergence of the first terrestrial streptophyte algae to fossils that provide evidence of the earliest land plants. We do not have fossils of the pivotal transitional stages from aquatic to terrestrial habitats, and there are no living descendants that possess the mixtures of traits we would expect to find in transitional forms. As mentioned above, the earliest definitive fossils are from the Ordovician and closely resemble the spore stages of modern bryophytes. Given the resemblance to modern bryophytes, it is likely that these plants had already acquired a complex life cycle and a high level of morphological complexity (e.g., modern bryophytes have a multicellular diploid phase, stomata, vascular tissue, and relatively complex reproductive structures; Graham et al. 2012). In contrast, the multicellular streptophyte algae bear little resemblance to bryophytes as they have unicellular diploid life stages and their gametophytes have simple anatomy and morphology.

From the difference in morphological complexity between streptophyte algae and bryophytes we can infer that the first terrestrial plants must have emerged much earlier than the oldest fossils of bryophyte-like spores. It is also possible that the lineage that ultimately produced terrestrial embryophytes remained in aquatic environments for some time before making the transition to land. During the Neoproterozoic (as early as 850 MYA) the green algae diverged into two major clades with lineages of streptophyte algae diversifying in fresh water habitats, while the chlorophyte algae were restricted to the oceans (Fig. 2.4; Becker and Marin 2009). The oceans were dominated by unicellular cyanobacteria and glaucophytes, unicellular and multicellular rhodophytes, and chlorophyte algae. While land plants share many traits with all of the green algae, phylogenetic evidence places the origin of embryophytes firmly among streptophyte lineages, which indicates that the transition to land occurred from freshwater instead of marine habitats (Fig. 2.4). During the late Neoproterozoic through the early Paleozoic, the streptophytes (including the charophytes) were the only algae occupying fresh water; it is likely that a combination of the ecological conditions present in lakes and ponds and a unique set of traits possessed by these algae facilitated the transition to dry land.

Living in freshwater would have provided an easier pathway to drier habitats because the change in salinity between aquatic and terrestrial environments was mild compared to that between the ocean, with its much higher salt content, and the land (Becker and Marin 2009). The combination of a strong salinity gradient along with dehydration, gravity, extremes in temperature, high irradiance, and ultraviolet radiation made the transition to life outside of water much more difficult for marine organisms. We can assume that the osmotic challenge for marine algae would have been substantial during the Neoproterozoic and early Proterozoic because the salinity levels in the oceans were at least as high as they are today (Knauth 2005). The ecological conditions afforded by freshwater habitats and the evidence from molecular phylogenies corroborate a scenario in which land plants originated from streptophyte ancestors in the lakes and ponds of continents rather than from green algae inhabiting the ocean.

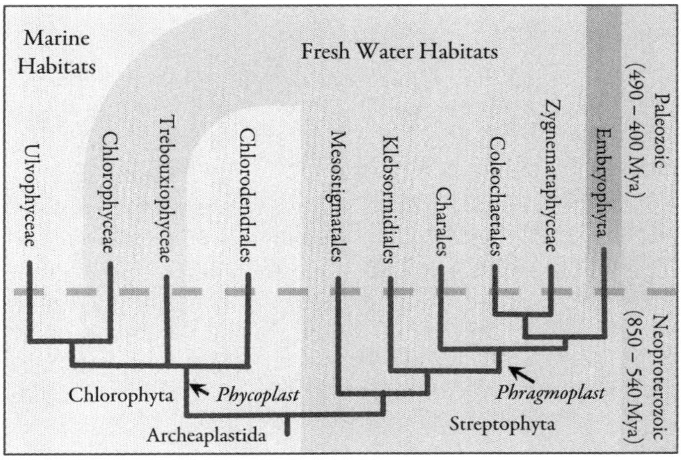

Fig. 2.4. Phylogenetic relationships within the Archeaplastida and historical habitat distributions of the major groups. Topology is based on Finet et al. (2012). Divergence dates are approximate.

Fossil evidence indicates that ponds and lakes were dominated by streptophyte lineages through the Cambrian, until their numbers declined as chlorophytes invaded and diversified in continental aquatic habitats starting around 400 to 490 MYA (Becker and Marin 2009). Divergence among the major clades of multicellular streptophytes (charophytes, coleochaetophytes, zygnemataphytes, and embryophytes) occurred much earlier during the Neoproterozoic according to molecular clock estimates (Fig. 2.4). With scant fossil evidence available, we can only infer that adaptations allowing colonization of drier habitats were acquired at some point during the Neoproterozoic, but we do not know whether the terrestrial transition began at origination of the embryophytes or at some time later during the early Paleozoic. It is possible that early embryophytes remained aquatic through the end of the Neoproterozoic, since the fossils of these earliest plants may be difficult to identify and descendants of the transitional forms did not persist. It would be difficult to know whether aquatic embryophytes had been around in the Paleozoic since freshwater streptophyte diversity declined by the end of the Permian (250 MYA) when they were replaced by chlorophytes. Descendants of the putative ancestral aquatic embryophyte lineages may have been lost due to competition with chlorophyte algae during the Permian, and other streptophyte algae continue to remain much less common than chlorophytes (Becker and Marin 2009).

Identifying the Closest Relatives of Embryophytes—The branching order among major streptophyte lineages remains ambiguous (Fig. 2.4 represents one hypothesis), so the closest ancestor to embryophytes could be any one of several algal streptophytes. The charophytes have been proposed as the most likely sister clade to embryophytes because they possess apical growth and other characters that superficially resemble land plants. On the other hand, members of the coleochaetophytes have been found to be desiccation tolerant and to share additional traits with embryophytes (Graham et al. 2012). By comparing characteristics of land plants with different groups of green algae we can begin to piece together the history and possible origin of the embryophyte clade.

All of the green algae share characteristics with land plants, including chlorophylls a and b as their primary photosynthetic pigments and storage of starch within plastids. This collection of characters alone is not diagnostic, but a number of other traits place the origin of land plants firmly within the streptophyte algae. The most prominent of these is the retention of the mitotic spindle during formation of the cell walls between daughter cells (Lee 2008; Fig. 2.5). This mode of cell wall construction (*Phragmoplast*) allows for the development of *Plasmodesmata* that connect the cytoplasms of adjoining cells and establishes the

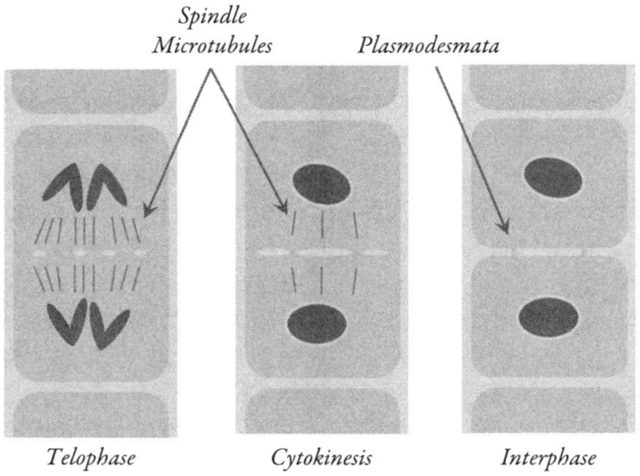

Fig. 2.5. During the development of the phragmoplast remnants of the spindle microtubules remain. The new cell wall develop as fragments along the equator of the mitotic spindle. Fragments fuse to form the new wall of each daughter cell, but adjoining openings remain to form the plasmodesmata that connect the cytoplasms of each cell.

Symplast (Chapter 6), which provides continuous communication of cytoplasm throughout the multicellular plant body. In the chlorophytes, the spindle fibers dissolve prior to telophase and plasmodesmata do not form. The chlorophytes also possess an extra external cell wall (the *Theca*), which restricts the expansion of daughter cells. The lack of a theca and the presence of plasmodesmata in streptophytes were critical pre-adaptations for the evolution of a multicellular plant.

A phragmoplast is shared by all the multicellular lineages of streptophytes (Fig. 2.4), but the Charophyta is the only group of macrophytes that also possess the capacity for apical growth. The morphological similarity of charophytes with land plants (e.g., *Chara*) has led to the assumption that they are sister clades; however, not all of the available evidence points to any particular streptophyte lineage as the closest relative to the embryophytes. For example, the chemical characteristics of the cell walls of several lineages of streptophyte algae (Zygnematales, Coleochaetales, and Charales) are similar to embryophytes, and lignin-like compounds, which are characteristic of embryophytes, have been found in species of Coleochaetales (Graham et al. 2012). There are also strong similarities between the transcriptomes of land plants and species in the Zygnematales and the Coleochaetales (Timme and Delwiche 2010), and a number of proteins formerly thought to be "plant-specific" have been found in several different lineages of streptophyte algae (Becker and Marin 2009).

This situation is further confused because a number of embryophyte genes thought to be important for adaptation to terrestrial environments are found in streptophyte algae, but not all of them are present in any single group (Leliaert et al. 2012). The distribution of land plant genes suggests that a single common ancestor possessed all of them, but different genes may have been lost in some algal lineages and not others. The stories told from molecular phylogenies based on genomic regions that are not associated with the terrestrial transition are just as conflated, with separate studies indicating the closest clades as the Charales (Qiu et al. 2006), the Zygnematales (Turmel et al. 2007), the Zygnematales combined with the Coleochaetales (Wodniok et al. 2011), while one broad phylogenomic survey of taxa placed the embryophytes nearest to the Coleochaetophytes (Finet et al. 2012).

Such phylogenetic confusion may indicate that the lineage with the closest affinity to the embryophytes has been long lost, and the phenotypic and genomic compositions of the remaining lineages have been modified as they diverged over hundreds of millions of years. Consequently, each remaining streptophyte shares some characteristics with others in the same clade and with the embryophytes. Given the scattered nature of shared characteristics across these two groups, it is likely that the divergence of the land plant lineage was basal within the clade, indicating they were derived from an earlier, and perhaps less complex, streptophyte alga (Fig. 2.4; Finet et al. 2012).

Because embryophytes share different sets of characters with several algal lineages, it is most likely that they diverged early in the history of the streptophyte clade. Under this scenario, ancestral embryophytes may have possessed many of their unique traits at the time they diverged and may have started occupying emergent habitats during the late Neoproterozoic. Terrestrial forms may have arisen later, so the green life surrounding ponds and extending into habitats with reliable water availability was probably present for some time before they left definitive fossil evidence of their presence.

The collective evidence from phenotypic and genomic studies supports an early divergence of the embryophytes from morphologically simple streptophyte algae in the Neoproterozoic. This scenario suggests that the first embryophytes may have been aquatic and transitioned to land somewhat later. It is also evident that the modern representatives of clades that have been viewed as the closest relatives to land plants (e.g., Zygnematales, Coleochaetales, and Charales) are substantially derived compared to common ancestors with embryophytes. For example, even though members of the Charales display superficial resemblance to some embryophytes—with node-like regions and whorls of leaf-like structures similar to land plants such as *Equisetum*—their development and anatomy is quite different from embryophytes (Ingrouille and Eddie 2006; Lee 2008).

It is also the case that we may not expect land plants to display strong similarities to any particular algal lineage because the likely timing of emergence of embryophytes does not correspond well with the history of modern streptophytes. For example, fossil evidence suggests that anatomically modern representatives of the Charales and Coleochaetales did not appear until the Silurian (excepting the Coleochaete-like acritarchs discussed earlier in this chapter)—well after the origin of embryophytes. These streptophytes remain aquatic even though they may have some potential for survival out of water for short periods (e.g., Graham et al. 2012).

Nonetheless, it is clear that the earliest land plants were derived from lineages that arose from within the streptophyte clade. It is also apparent that early streptophyte lineages possessed *Key Adaptations* (e.g., phragmoplast, plasmodesmata, apical growth, resistant spores, etc.) that were critical for the evolution of multicellular growth and success in terrestrial habitats. Because of the probable early divergence of streptophyte lineages, further pursuit of a single closest lineage to embryophytes may be difficult. There is much to learn about the process of adaptation to terrestrial conditions that occurred in the earliest land plants, and the best path forward in this endeavor may lie in genomic comparisons of these groups (Becker and Marin 2009; Leliaert et al. 2012).

2.4 DIVERSIFICATION OF LAND PLANTS

There may have been a number of algal lineages that made evolutionary excursions into terrestrial habitats, but only one diversified to form the major groups of land plants that remain today. The common ancestor of all embryophytes may have competed with other algal lineages that were taking advantage of the abundant resources afforded by terrestrial habitats, but if any others existed, they must have quickly died out because all of our early plant fossils display homologies with modern plant groups. As the original land plant lineage spread to colonize widespread regions across continents, it is likely that it underwent a major radiation that probably resulted in a large number of novel plant groups, but only four major clades from this initial radiation persist today. Three of these share a dominant gametophytic life stage and are grouped together as the bryophytes. The fourth group, the tracheophytes, which is characterized by a dominant sporophytic stage, diversified into the major plant groups that are present today including the ferns, lycophytes, gymnosperms, and the flowering plants. In this section, we examine some of the salient characteristics of these major plant groups, their patterns of diversification, and the phylogenetic relationships among the major groups of embryophytes.

The Bryophytes—The mosses, liverworts, and hornworts are often held up as representatives of the earliest land plants, but modern bryophytes possess many characteristics that render them much more complex than the first terrestrial embryophytes. While the bryophytes appear to express many of the characteristics we would have expected to find in the first land plants, they also have many more complex traits that were acquired as they diverged and diversified in terrestrial environments. Most notable is a gametophyte stage that has the ability to persist in habitats that are periodically dry. Many bryophytes, and mosses in particular, have the ability to survive episodes of minor to nearly complete drying, and then to rehydrate and resume metabolic activity when moisture becomes available. Some streptophyte algae, such as Coleochaetales, are desiccation resistant (Graham et al. 2012) but are not capable of surviving after drying. Many bryophytes, on the other hand, may survive extended periods of nearly complete desiccation (around 5–10% water content)—a trait they share with a small collection of tracheophytes, including some Lycopods, such as *Selaginella* (Proctor and Tuba 2002). All of the remaining tracheophytes are *Homoiohydric*, capable of internal regulation of water (Ingrouille and Eddie 2006), so it is likely that the lack of this ability, *Poikilohydry*, is derived independently in *Selaginella* and the bryophytes.

The ability to persist through dry periods as a mature thallus allows many moss species to resume growth almost immediately upon wetting. Other plants pass through inhospitable conditions as resistant seeds or spores that must germinate and complete their reproductive cycle while sufficient moisture is available. The advantage of poikilohydry is evident by the extreme habitats mosses occupy—for example, as epiphytes on trees or other impenetrable surfaces that are subject to many months of severe desiccation. These desiccation resistant lineages of mosses probably appeared much later when they diversified in the Cretaceous and expanded their ranges to occupy environments where other plants are incapable of living. Poikilohydry is a key adaptation that allows many of the most widespread bryophyte species to dominate habitats that are seasonally moist and subject to long periods of severe drying.

Another trait that differentiates the bryophytes from streptophyte algae is a multicellular sporophyte stage. In the mosses and liverworts, sporophytes are borne within archegonia and remain dependent (parasitic), while the gametophyte thallus is usually perennial and may live for many years (Shaw et al. 2011). Some hornworts are an exception since the gametophyte is typically ephemeral, and the sporophytes can persist after the gametophyte thallus withers. All three bryophyte groups have vertical growth of a stem that bears a *Sporocarp* (the sporophyte in mosses and hornworts) or an *Archegonia* that contains the sporophytes (the stalk is an extension of the gametophyte thallus in liverworts). Releasing large numbers of spores further from the ground promotes dispersal, as the spores are then more likely to be picked up by the wind.

Beyond these basic similarities, there are many more characteristics that differentiate the sporophytes of the three bryophyte groups and distinguish them from the sporophytes of the tracheophytes. Liverwort sporophytes remain mostly imbedded within the *Archegonium* so they are diminutive; they have simple morphology, and lack stomatal openings and vascular tissue. The moss sporophytes differ by being relatively complex with tissue differentiation in the stem (including conducting cells and stomatal-like openings) and by bearing a terminal sporocarp. When moss spores are mature, the terminal cap (*Operculum*) separates to expose peristome teeth that are extended to facilitate spore release in response to changes in humidity. Like mosses, sporophytes of the hornworts arise from the gametophyte thallus, but in this group, the sporocarp is an elongated photosynthetic stem that has stomatal-like openings and lacks vascular tissue. The sporocarp continues to elongate from the base and splits longitudinally exposing pseudo-elators and spores. The morphological distinctiveness among the sporophytes of the three bryophyte clades implies that they are quite derived relative to each other, and relative to the more complex sporophytes of the tracheophytes.

Phylogenetic Relationships among Bryophytes and Tracheophytes—The highly differentiated and complex morphology in bryophytes and tracheophytes suggests that they were derived from a single common ancestor that had relatively simple gametophyte and sporophyte stages. However, the relationships among bryophyte and tracheophyte lineages inferred from the distribution of phenotypic and genetic traits present a confusing picture. For example, several molecular phylogenies place the hornworts as the closest relative to the tracheophytes (Qiu et al. 2006; Chang and Graham 2011), but the observation that hornworts are the only bryophyte group that retains a *Pyrenoid* chloroplast structure (Shaw and Renzaglia 2004)—a characteristic they share with streptophyte algae—suggests that they occupy a more basal position. On the other hand, some evidence suggests that pyrenoids in hornworts are more recently derived and may not be homologous to algal pyrenoids (Villarreal and Renner 2012). Other traits such as the presence of stomatal-like openings and the ability to discriminate isomers of methionine (Ingrouille and Eddie 2006) would group hornworts and mosses together with tracheophytes, leaving the liverworts as the more basal lineage.

Different phylogenetic analyses alternatively place either the liverworts or hornworts as the earliest diverging embryophyte lineage and either hornworts, mosses, or clades consisting of these two groups combined, or clades including the liverworts and mosses together, as the closest relatives to the tracheophytes (Magallón and Hilu 2009). Several analyses suggest the bryophytes consist of a graded clade with liverworts most basal, followed by the mosses, and then the hornworts as the closest relative of the tracheophytes (Qiu et al. 2006; Chang and Graham 2011; Shaw et al. 2011). On the other hand, the more complex traits (i.e., leaf-like organs, stomata, and vascular tissue) in the bryophytes are probably independently derived as they display little homology to traits in the tracheophytes (Ingrouille and Eddie 2006; Renzaglia et al. 2007).

Phylogenies based on a substantial number of nuclear gene regions (77 genes) place the bryophytes and tracheophytes as two separate monophyletic clades (Finet et al. 2010; Finet et al. 2012; Fig. 2.6). While a number of the phenotypic traits, genes, and developmental programs in these groups were most likely inherited from a common ancestor of the four major embryophyte lineages (Shaw et al. 2011), it is also clear that each of the bryophyte and tracheophyte lineages have undergone extensive histories of specialization and divergence since their initial divergence.

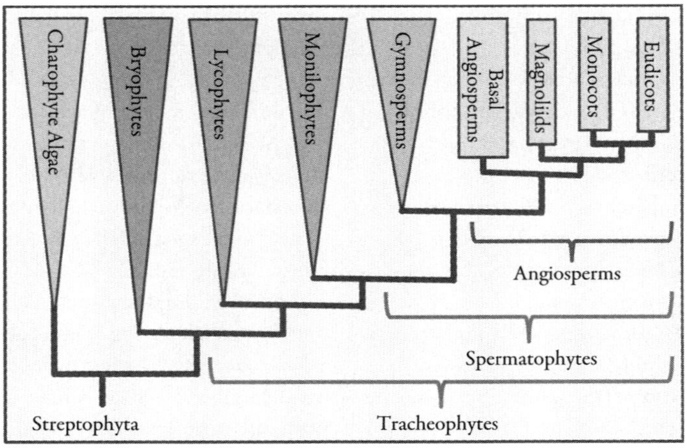

Fig. 2.6. Phylogenetic relationships among the major streptophyte lineages modified from Finet et al. (2012).

Over many hundreds of millions of years of diversification these plant groups may have lost some traits and acquired others, which makes the inference of ancestral traits and phylogenetic relationships among clades particularly difficult (Ligrone et al. 2012; Bowman 2013).

Origin of Tracheophytes—Out of the four original lineages of embryophytes that survived to diversify in terrestrial habitats one—the vascular plants (tracheophytes)—diversified into a much broader range of habitats than the other three. The substantially greater success of the tracheophytes may have been due to a number of unique characteristics. In addition to having vascular tissue that includes tracheids for water transport, the vascular plants share a number of other characters that differentiate them from their bryophyte relatives. For one, the sporophyte generation is much larger, produces multiple reproductive organs (so, tracheophytes can also be referred to as *Polysporangiophytes*), and is morphologically more complex—with multiple tissues and organ types. The bryophytes are characterized by *Ectohydry* (water transport along the external surface), poikilohydry (resistance to desiccation), and a limited capacity to regulate their internal water status. Vascular plant sporophytes, on the other hand, are nearly all *Endohydric* (some lycopods such as *Selaginella* being the exception), and most cannot survive even minor levels of dehydration of their tissues.

The gametophyte stage of vascular plants is also quite differentiated from the gametophytes of bryophytes. In the *Lycopodiophyta* (clubmosses, spikemosses, and quillworts) and *Pteridophyta* (horsetails and ferns) gametophytes are short-lived and diminutive thalloid plants (*Prothalli*) that are similar to very young liverworts. The gametophytes of these plants have unicellular rhizoids, flagellated sperm, and generally cannot survive unless they are in direct contact with a moist substrate. In the *Spermatophytes* (vascular plants with seeds), the female gametophytes (*Megagametophytes*—the egg sack within an ovule) are generated within the sporophyte tissue (i.e., within ovules). Male gametophytes (*Microgametophytes*—pollen) are capable of surviving harsh environmental conditions during dispersal but become parasitic on sporophytic tissue once they germinate and mature to complete fertilization of the megagametophyte. With the exception of pollen dispersal, most of the haploid phase of the life cycle in spermatophytes is completed within the protective tissues and organs of the sporophyte, so in general, the gametophytes of tracheophytes lack the higher degree of desiccation tolerance and poikilohydry that are characteristic of modern bryophytes.

The histories of both the bryophytes and tracheophytes are characterized by progression toward increased independence from water, but members of these groups have developed alternative mechanisms for persistence through dry periods (Proctor and

Tuba 2002). For the bryophytes, the strategy is one of avoidance as many mosses can become dormant and nearly dry when the environment is inhospitable. The success of tracheophytes, on the other hand, is marked by increases in morphological and physiological complexity that has enabled adaptations for the maintenance of internal water status when access to moisture is limited. Their small size and ability to adhere to solid substrates allows mosses to occupy habitats where poikilohydry provides a unique advantage, but these characteristics set limits on their distribution to specific types of environments where tracheophytes are unable to dominate. Mosses such as *Dawsonia* have vasculature (a hydrome), but their water conducting cell walls lack the lignin spiral thickenings that are present in the tracheids of vascular plants, so they are not effective when the moisture content of the substrate is low. Tracheophytes are capable of occupying a wider range of habitats, but under severely dry conditions, there are limits to their persistence as the inevitable drying of tissues will lead to death. Poikilohydry in some tracheophyte lineages such as *Selaginella* (resurrection plant) has a physiological basis that is distinct compared to bryophytes, so homoiohydry is assumed to be the ancestral condition in the vascular plants (Proctor and Tuba 2002). The independent derivation of poikilohydry in a few tracheophytes and the lack of desiccation resistance in the gametophytes of the basal tracheophyte lineages are consistent with divergence from an early common ancestor of bryophytes and tracheophytes that lacked desiccation resistance.

In spite of the distribution of a number of characteristics that would suggest an earlier divergence of the tracheophytes from a common ancestor with the bryophytes, the majority of phylogenetic treatments have concluded that a graded clade is more likely (Fig. 2.7), with liverworts basal and the hornworts as sister to the vascular plants (Chang and Graham 2011; Shaw et al. 2011) (but see Finet et al. 2010; Finet et al. 2012). This arrangement has much appeal as it is consistent with a progression of increasing sporophyte independence and dominance; from diminutive sporophytes of liverworts that are almost entirely embedded within gametophyte tissue to the larger sporophytes of hornworts, which are capable of photosynthesis and independent nutrient acquisition from the substrate (Qiu et al. 2006). The potential for

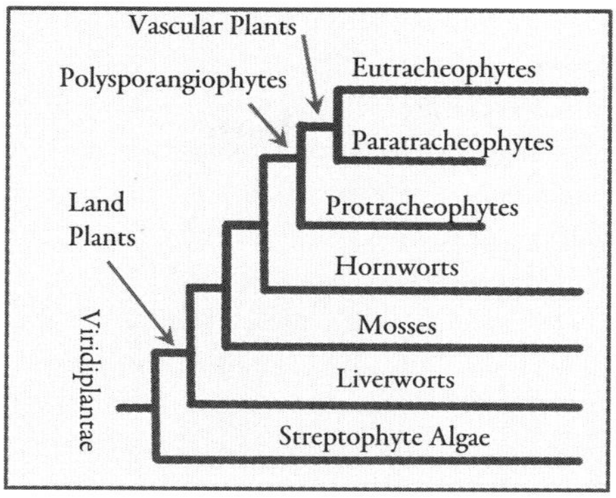

Fig. 2.7. Origin of vascular plants including the inferred positions of two extinct groups, the Protracheophytes and Paratracheophytes. Redrawn from Gerrienne and Gonez (2011).

independence of the sporophyte in hornworts, along with a number of other traits, supports their position as closest living group to vascular plants, but the order of divergence for the four major embryophyte groups remains an open question that will require comprehensive phylogenetic surveys and detailed analyses of shared traits to resolve.

The Transition to Sporophyte Dominance—The transition from a haploid-dominant, to a diploid-dominant life cycle appears to be a major step in the evolution of land plants, but the success of tracheophytes had much more to do with the increase in complexity of their tissue types, organs, and physiological responses, rather than just the relative size of the sporophyte. The change in life cycle and the increase in complexity are not necessarily independent, as diploidy and genome duplications facilitated adaptation to challenging environments.

As discussed in Chapter 1, diploidy is advantageous in multicellular organisms, which have long generation times and low population growth potential compared to unicellular microbes. Having two alleles at each locus allows for the accumulation of unexpressed genetic variation (due to dominance of one allele over the recessive allele) that contributes to broader ranges of variation in sexually produced offspring (Qiu et al. 2012). In diploid organisms, normally recessive mutations can be exposed to selection (i.e., made homozygous) as a consequence of segregation and recombination, and through interactions with alleles at other locations in the genome (epistasis).

Meiosis also promotes increases in genome complexity by elevating the frequency of partial- and whole-genome duplication events (Chapter 3). Duplication provides opportunities for selection to act on multiple copies of genes and is considered a major source of genetic novelty and specialization (Zhang 2003; Taylor and Raes 2004; Otto 2007). For haploid organisms, on the other hand, the accumulation of genetic variation within populations is greatly reduced; rates of evolution of haploid-dominant plants are slower (Stenoien 2008), and the potential for adaptation in response to environmental challenges is more limited. Some haploid organisms can attain a large size (e.g., some species of red, green and brown algae), but multicellular haploids are consistently simple in structure and are restricted to aquatic or wet terrestrial environments (i.e., in the bryophytes and the independent gametophytes of basal tracheophytes). Major innovations that allowed vascular plants to move into a wide range of terrestrial habitats and to attain a much larger size occurred concomitantly with dominance of the diploid sporophyte as well as a series of whole-genome duplication events (Jiao et al. 2011; Chapter 3).

Evolution of Gametophytic Life Stages—The transition to a sporophyte-dominant life cycle occurred by an increase in the size and morphological complexity of the diploid generation, while gametophytes may have remained small and anatomically simple. Observations of gametophytes in extant lineages of seedless tracheophytes—all of which are very small plants with a simple anatomical structure—suggest that the path to vascular plants may have never included species that had isomorphic generations (i.e., gametophytes and sporophytes of similar size and structural complexity). On the other hand, fossils of putatively transitional species found in the Rhynie chert from the early Devonian tell a very different story (Crane et al. 2004). The gametophytes from several species are relatively large and display similar levels of anatomical complexity as their presumed sporophyte generations (Gerrienne and Gonez 2011). For example, the fossil sporophyte species *Aglaophyton major* and gametophyte species *Lyonophyton rhyniensis* both have smooth vascular cells similar to hydroids in mosses (Fig. 2.8) and other anatomical features that place them together as representing life stages of a single protracheophyte species (Fig. 2.7). The

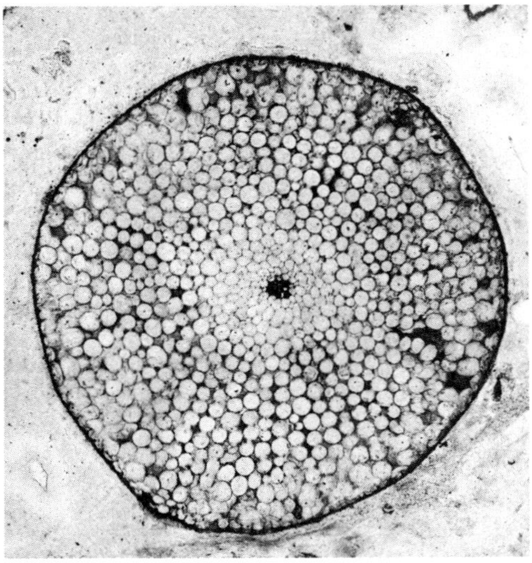

Fig. 2.8. *Rhynia* stem cross section showing hydroid-like vasculature. Photo by Plantsurfer.

sporophyte *Rhynia gwynne-vaughanii* is paired with the gametophyte *Remyophyton delicatum* based on their unique bi-layered vascular cells that place them as a single species in the *Paratracheophytes* (*Rhyniopsida*; Crane et al. 2004), which is an extinct sister clade to the tracheophytes (*Eutracheophytes*; Fig. 2.7). In addition to differences in vascular cells, these Rhyniophytes differ from eutracheophytes by having adventitious branching; the vascular strand of the branch is not connected to the strand in the central stem.

For both of these rhyniophyte species, the gametophyte generations are substantially smaller, but are nearly isomorphic with respect to their anatomical and morphological complexity. Assuming these fossil species are representative of the direct ancestors of modern vascular plants, the reduction in the complexity of the gametophyte generation must have occurred over a relatively short time frame before the appearance of the first fossil eutracheophyte, *Cooksonia*, which presumably had a diminutive gametophyte generation (Gerrienne and Gonez 2011). Alternatively, modern tracheophytes may have originated from an ancestor that retained a small, thalloid gametophyte; the fossil species from the Rhynie chert may represent lineages that developed in parallel with the eutracheophytes and ultimately were lost.

Whether the gametophyte generations of tracheophyte ancestors were nearly isomorphic with sporophytes and then diminished later, or were diminutive all along, the observed recession of haploid generations across the history of tracheophytes suggests there was a selective advantage for the reduction of the gametophytic life stage (Walbot and Evans 2003). To appreciate the probable reason for this trend, recall that the earliest eukaryotic producers retained a haplontic life cycle while consumer lineages (ancestors of metazoans—the animal kingdom) attained diplontic life cycles before they rapidly increased their morphological complexity leading up to the Cambrian (Chapter 1). The loss of a multicellular gametophytic stage may have been advantageous for metazoans during the Proterozoic when they were under pressure from intense competition and predation, but the embryophytes may not have experienced intense levels of competition until they came to dominate terrestrial environments in the Devonian (see section 2.5). The evolutionary advantages of diploidy for multicellular organisms may not be fully realized unless there is no post-meiotic expression of genes in the haploid life stage; for multicellular organisms we may generally expect that a diplontic life cycle would have long-term advantages compared to haplontic and haplodiplontic cycles.

The retention of a multicellular haploid life stage in plants may represent an evolutionary constraint; the haplodiplontic life cycle of plants may have ultimately retarded the development of more complex sporophytes that were capable of meeting the stringent challenges presented by terrestrial environments. J. B. S. Haldane recognized this limitation and the stark contrast to metazoans when in his 1932 monograph he quipped, "A higher plant species is at the mercy of its pollen" (Haldane 1932). Haldane's insight was the realization that selection on microgametophytes was potentially quite strong, so the evolutionary response of any gene expressed in both generations would be overwhelmed by selection in the haploid stage. Consequently, we can predict that the weight of constraints imposed on the evolution of the sporophyte is a direct function of the level of gene co-expression in gametophytes; we expect expression in the haploid generation will impede the evolution of the sporophyte. Any mutation that led to a reduction in the complexity of the gametophyte generation may have been favored if it reduced the liability of co-expression between the life stages. The ultimate product of this long history of evolution toward gametophyte reduction is a microgametophyte (pollen) and megagametophyte that consist of only a few cells and undergo development entirely within the protective confines of the sporophytic tissue (the ovule). We will discuss the consequences of gametophytic gene expression and possible reasons for the retention of a multicellular haploid stage later in this chapter, and in Chapter 5 (selection) and Chapter 9 (reproduction).

The First Tracheophytes—The earliest fossil evidence of tracheophytes dates from the late Silurian, with the appearance of polysporangiophytes such as *Cooksonia* (Figs. 2.3, 2.9), the Zosterophylls, and *Baragwanathia* (Fig. 2.10), the latter of which is the oldest identifiable fossil lycopod (Lycopodiophyta). These earliest vascular plants were characterized by stems with a central core of vascular tissue (protostele) that consisted of tracheids surrounded by phloem. *Cooksonia* was a genus of very small plants with leafless stems that were up to 3 mm in diameter, branched dichotomously, and bore terminal sporangia (Fig. 2.3).

The class Zosterophyllopsida included a large diversity of plants with dichotomously branching stems that were smooth or covered with small *Enations* (leaf-like projections that lacked vasculature). The zosterophylls are distinguished by their stem tips, which grow by unrolling (circinate vernation), similar to the unrolling of fiddleheads into fern fronds. Plants in the genus *Baragwanathia* looked very much like modern lycopods with thick stems (up to 1 cm in diameter and 1 m in length) that were densely covered with spirally arranged *Microphylls* (leaves with a single vascular strand) and bore sporangia in the axils of their leaves toward the end of the stem. These plants grew as stems that trailed along the ground with branches that extended upward and bore sporangia. The stems of these early tracheophytes grew along one axis (*Monopolar*—growing from a single point) and lacked true root systems (plants with true roots display *Bipolar* growth). Instead, adventitious rhizoids formed wherever the stems made contact with the substrate. Several similar lycopod species are known from the early Devonian including *Halleophyton*, *Drepanophycus*, and *Asteroxylon*. The gametophyte stages of these fossil species are unknown but are assumed to have been small prothalli and independent of the sporophyte—much like the gametophytes of modern lycopods and ferns.

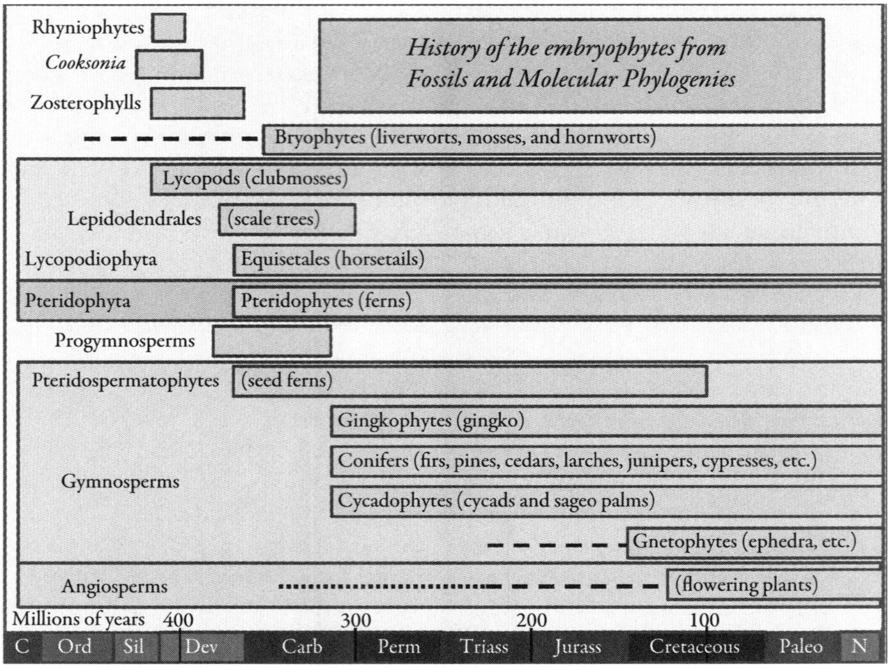

Fig. 2.9. The fossil history of land plants covering five hundred million years from the mid-Cambrian to the Neogene. Boxes indicate the age range of fossils associated with each group. The dashed line for Bryophytes indicates the first occurrence of microfossil spores. Inferences for origins from molecular phylogenies are indicated by dashed lined for gnetophytes and angiosperms. The dotted line indicates the age of the split between seed fern ancestors of the angiosperms and extant gymnosperms.

Fig. 2.10. Artist depictions of the Zosterophyll *Sawdonia ornate* (left) and the lycopod-like *Baragwanathia* (right). Original art by Michelle Nicole Williamson.

The early emergence of lycopodiophytes and other early vascular plants is consistent with the hypothesis that tracheophytes diversified in parallel with, and then eventually displaced, the rhyniophytes. Under this scenario, the gametophyte stage of early tracheophytes remained small and continued to diminish while the sporophyte increased in size and complexity. Species of *Cooksonia* and *Baragwanathia* and representatives of the zosterophylls spread to attain worldwide distributions by the early Devonian and were the dominant terrestrial primary producers along with bryophytes between 433 to 393 MYA. During this time, plant life was restricted to moist environments and must have appeared as a carpet of green that grew only a few decimeters in height. Other terrestrial inhabitants included arthropods such as myriapods (ancestors of centipedes and millipedes) and fungi, including members of the Glomeromycota, which are mutualists with the large majority of modern embryophytes.

2.5 THE DEVONIAN EXPLOSION

During the Devonian, the number of species competing for space and light began to increase. For the first time in their history, the producers were faced with intense competition for a limiting resource (light) that could be obtained through phenotypic changes (vertical growth) that conferred advantages over other species. What resulted was an increase in phenotypic diversity, and in particular, a number of morphologies with enhanced ability for vertical growth that allowed individuals to overtop their competitors.

By 395 MYA, the increase in the diversity of plants began to accelerate as a rich variety of new forms appeared through the late Devonian and into the Carboniferous. This explosion of diversity was comparable to the increase in the number of new types of animals seen in the Cambrian, both in terms of the variety and the range of unusual forms. Much of this flora was unfamiliar and bizarre compared to modern plants. It included such things as giant horsetails that resembled enormous bamboo shoots, huge clubmosses that appeared as herbaceous trees with tufts of leaves at the top, conifer-like trees with canopies of fern-like fronds instead of needles, and fern-like plants that bore seeds on their fronds instead of spores (Fig. 2.11; refer to color plate). Amid this jungle of unusual plants, there were also much more familiar species, such as the ancestors of modern ferns, cycads, and gingkoes, as well as some of the earliest conifer-like trees. Many major components of modern floras, such as the ferns and most gymnosperms, had originated by the end of the Carboniferous (by about 300 MYA). Others, such as the flowering plants and gymnosperms in the Gnetophyta (*Gnetum, Welwitschia,* and *Ephedra*), probably originated in the Triassic (around 220 MYA) and did not become common until the Cretaceous (see section 2.7).

Competition for light must have been intense in the early Devonian as low-growing plants crowded into habitats that provided reliable moisture. By the middle Devonian, the earliest polysporangiophytes, such as *Cooksonia,* and the rhyniophytes had disappeared while increasingly large lycopods and giant horsetails (Equisetales) emerged to dominate the extensive swamp forests. One of the largest and most common types of lycopod were the Lepidodendrales (Isoetales; *Lepidodendron* and *Sigillaria*), which are commonly known as scale trees (Fig. 2.12). These were fast-growing arborescent plants with photosynthetic trunks that sometimes exceeded 1 m in diameter, covered with "leaf cushions" where microphylls were attached, and which continued to expand as the trunk increased in diameter. The trunks of these "trees" contained a woody layer toward the surface of the cortex, which surrounded a soft central pith region and a cylinder of vascular tissue (*Siphonostele*). The trunk ended in dichotomously branched stems that were covered with microphylls and bore a terminal sporangium. It is thought that many scale trees may have been *Monocarpic* (died after reproduction; Chapter 6). These plants grew in very high densities and along with other early tracheophytes, contributed to the accumulation of tremendous amounts of organic material that eventually became the extensive coal and oil deposits associated with the Carboniferous.

The first "true" trees, that is plants with solid wood trunks (the *Progymnosperms Wattieza* and *Archaeopteris*; Fig. 2.13) appeared in the mid to late Devonian. These plants were similar to conifers, but they differed by having canopies of large fern-like fronds and by producing spores instead of seeds. These medium- to large-sized trees were wide at

Fig. 2.11. A Devonian scene. An artist depiction of an early Devonian forest. Original artwork by Eduard Riou.

the base, with girths up to 1.5 m in diameter. They had well-developed root systems and *Monopodial* growth of the stem (a central stem with branches like conifers). Unlike scale trees, which had a cylinder of wood filled with soft tissue, the progymnosperms accrued diameter from a vascular cambium so their entire trunk was woody and similar to that of modern conifers. The progymnosperms were some of the first plants to have true leaves (*Megaphylls*), which had a branching vascular system instead of the single vascular strand present in microphylls. The fronds bore sporangia and may have been deciduous. During the late Devonian and the early Carboniferous, these were dominant trees in the wet soils of the flood plains, but they died out by the mid-Carboniferous.

Among the large diversity of plants that first appeared in the late Devonian and the Carboniferous, many were classified into a loosely defined group known as the "seed ferns" (Pteridospermatophyta; Fig. 2.14). The label "seed fern" was first applied to fossil species in the order Lyginopteridales after it was discovered that its fronds bore ovules (a megasporangium surrounded by leaf-like structures that formed a divided integument). After this discovery, more plants were identified that had fern-like leaves and appeared to produce ovules instead of sporangia. A number of ovule-bearing fossil species have been referred to as pteridospermatophytes, including plants with entire leaves as well as those with divided fronds. This group of perhaps nine orders is held together by few common features and is more accurately included with gymnosperms—plants with ovules exposed on a stalk or a leaf instead of enclosed in a *Carpel* (megasporophylls fused to form a container) as in the flowering plants. Seed ferns first appeared in the Devonian, and were abundant through the Carboniferous and Permian, but disappeared by the end of Cretaceous (Fig. 2.9). In this text, we will use the term "seed fern" to designate spermatophytes that have the characteristics described above.

While many of the plants that first appeared in the Devonian and the Carboniferous are not present today, representatives of some groups that originated during this time persist in modern floras (Fig. 2.9). For example, many types of ferns and their relatives (Pteridophyta) were present during the Carboniferous, and some were dominant components of the vegetation. A good example of the latter is the giant horsetail (Equisetales), such as *Calamites*, which grew to 30 m and was abundant in the "coal swamp" forests. Modern

Fig. 2.12. Artist depiction of a scale tree. Original art by Michelle Nicole Williamson.

species of the horsetails are small herbaceous plants. While modern representatives of some groups are quite different from their ancestors, others are seemingly unchanged. A number of the ferns (Polypodiopsida) from the Carboniferous were thought to grow in the canopies of trees as epiphytes and are strikingly similar to modern species. Some Gymnosperms, including cycads (Cycadophyta) and gingko (Gingkophyta), first appeared in the Carboniferous, and modern specimens appear to be very much like these early ancestors.

2.6 ORIGIN OF SEED PLANTS

Toward the end of the Carboniferous, precipitation declined and the widespread swamp forests began to dry up. As more arid climates became prevalent, selection for resistance to desiccation and for morphological changes that allowed for reproduction independent of water increased. This problem was eventually solved with the independent evolution of pollen and seeds in the separate lineages of gymnosperm and angiosperm ancestors toward the end of the Carboniferous, but the stage was set for the emergence of this novel mode of reproduction well before, during the Devonian explosion.

Fig. 2.13. Artist depiction of the progymnosperm *Archaeopteris*. Original art by Michelle Nicole Williamson.

Fig. 2.14. Artist depiction of the seed fern *Medullosa*. Original art by Michelle Nicole Williamson.

The Importance of Heterospory—Plants of the late Silurian and early Devonian produced spores that developed into independent gametophytes much like modern lycopodiophytes and ferns. Some species were *Homosporous* and produced one type of spore—similar to modern lycopods, so their gametophytes were bisexual, and a single haploid individual would produce both *Archegonia* containing egg cells and *Antheridia* containing flagellated sperm. Other species were *Anisosporic* or *Heterosporous*, producing two different sizes of spores that grew into separate male and female gametophytes (as in modern quillworts, horsetails, and some ferns). Heterosporous species produce two different sizes of spores in separate sporangia. Two different sizes of spores can also be produced within one sporangium—anisospory, more common in the Devonian is rare in extant species (Bateman and Dimichele 1994).

Homospory is advantageous because it allows self-fertilization, which promotes colonization of habitats over long distances. Homosporous species produce spores that germinate into gametophytes that can produce both sperm and eggs, so a single spore is sufficient to colonize a new site. A major disadvantage of homospory is inbreeding; bisexual gametophytes may frequently engage in intra-gametophyte matings, which would produce completely homozygous sporophytes (because both sperm and egg are produced by the same haploid individual). Heterospory, on the other hand, is more likely to produce novel genetic variants through recombination among individuals (outbreeding), which could facilitate adaptation to novel environments (Chapter 9). While heterospory improves the chances for outbreeding, both micro- and megagametophytes have to disperse and produce reproductive structures for successful production of the next sporophyte generation (Fig. 2.15.A; refer to color plate). While outbreeding probably accelerated adaptation to novel conditions, ecological pressures may have been more important for the origin of heterospory and ultimately for the origin of pollen and ovules.

In heterosporous species, both micro- and megaspores could respond independently to selection mediated by ecological conditions to improve reproductive success through male and female function. Increased competition for light and other resources in the overcrowded swamps of the Devonian produced strong selection for larger megaspores. After fertilization, the female gametophyte supports the early growth of the sporophyte until the young plant can develop adequate root and shoot systems for the acquisition of resources. Competition produces a strong selective advantage for young sporophytes that receive an extra boost of resources from the supporting megagametophyte (indirectly from the sporophyte parent); larger megagametophytes are better able to ensure the success of their sporophytic offspring. This was probably the case for plants occupying the coal swamp forests of the Carboniferous as canopy species, such as scale trees and giant horsetails, were strongly heterosporous (Ingrouille and Eddie 2006).

Producing different spore types (*Microspores* and *Megaspores*) in separate *Sporangia* (heterospory) was a necessary pre-condition for the evolution of spermatophytes; sporangia producing the larger spores became the precursor to ovules, and sporangia producing the smaller spores eventually evolved into anthers. Selection for improved female function in response to competition results in a reduction in the number of spores produced per sporangium and a correlated increase in the size of megaspores (Bateman and Dimichele 1994). The endpoint of this process is single large spore within each sporangium; one of the four haploid products of meiosis from a single spore mother cell develops into the spore while

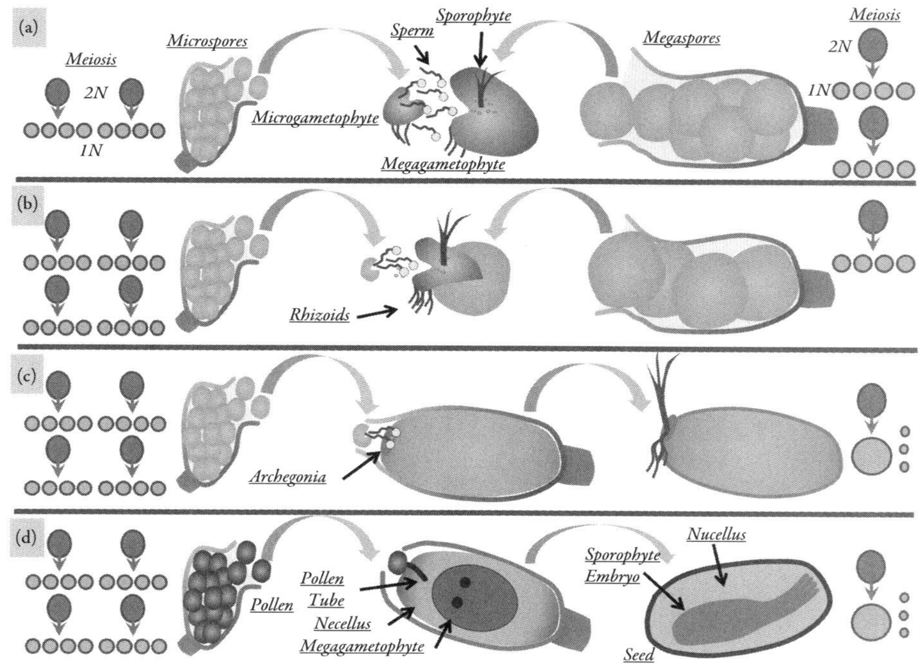

Fig. 2.15. Stages of transition from heterospory (A) the production of pollen and ovules (D). Endospory (B) develops by the retention of gametophytes within the spore walls. (C) The mature endosporic megagametopyte is retained so fertilization occurs while still attached to the sporophyte. Microspore size decreases (A-D) and the number increases as megaspore size increases until only there is only one (C, D)

the other three degenerate (Niklas 1997; Fig. 2.15.C; refer to color plate). For microspores the selection pressure is just the opposite; large numbers of small spores increase dispersal ability and improve the chances that some will encounter megagametophytes bearing archegonia.

Selection for improved sexual function and avoidance of harsh conditions could ultimately result in *Endospory* (Fig. 2.15.B; refer to color plate): the gametophyte develops partially or entirely within the spore wall (as in the Isoetales and Equisetales). Megagametophytes of endosporic species only extend rhizoids and the archegonium from the confines of the spore wall and often do not produce chlorophyll. Endosporic microgametophytes complete their maturation and produce antheridia entirely within the spore wall and then release flagellated sperm after dispersal. In many respects, the degree of reduction in the gametophyte stages in endosporic species is similar to the production of pollen and ovules by spermatophytes; however, endosporic species retain a megagametophyte stage that must grow independently from the sporophyte, and they still rely on the presence of water for successful reproduction.

Pollen and ovules differ from the spores of heterosporic and endosporic species because the gametophytes and the new sporophyte remain protected from harsh conditions throughout critical stages of growth and reproduction (Fig. 2.15.D; refer to color plate). Pollen grain development begins while still in the anther where multicellular microgametophytes are encapsulated within a protective coat (pollen grain wall consisting of sporopollenin). After arrival on a receptive surface the grain releases flagellated sperm (in cycads and gingkoes; Fig. 2.15.C; refer to color plate) or germinates to produce a *Pollen Tube* (in gnetophytes, conifers, and angiosperms; Fig. 2.15.D; refer to color plate) that delivers sperm directly to the megagametophyte to effect fertilization. The megagametophyte completes its life entirely within the confines of an ovule, which is a sporangium contained within either one (gymnosperms) or two (angiosperms) *Integuments* (see the Origin of Ovules section). The sperm or pollen tubes enter the ovule through an opening in the integuments that forms a pore (*Micropyle*) where the haploid tissue of the megagametophyte (*Nucellus*) is exposed.

Origin of Ovules—In modern seed plants, the new sporophyte remains within the sporangium following fertilization and is provisioned by the parental sporophyte during its early growth to the embryo stage. Embryos are protected from harsh conditions by a seed coat that forms from the integuments of the ovule. This combination of innovations—separate mechanisms for the protection of gametophytes and the sporophytic embryo from harsh conditions and seed provisioning—allowed early spermatophytes to become independent from the presence of water for reproduction and dramatically increased their range of habitats.

The origin and diversification of *Euphyllophytes* (predecessors to spermatophytes and pteridophytes) occurred in parallel with lycopodiophytes, as both groups are monophyletic and share a common ancestor that dates from as far back as the Cambrian (Soltis et al. 2002). The earliest fossils of euphyllophytes with megasporangia that resemble ovules are from seed ferns of the mid to late Devonian (Willis and McElwain 2002; Gerrienne et al. 2004; Linkies et al. 2010). Across fossil species, there is a veritable continuum of sporophyll morphologies that can be interpreted as early stages of ovule development; from multiple sporangia in the axils of sterile filaments (*Telomes*) along a terminal branch, to a single sporangium surrounded by multiple telomes (Fig. 2.16). This arrangement of telomes into a cupule with the filaments surrounding the top of the megasporangium may have slowed air movement for the more effective capture of wind-dispersed microspores (Niklas 1997). The widely accepted model for the development of the ovule assumes the telomes fused to form an integument, which was modified in the mature seed to form a seed coat. Megasporangia surrounded by filamentous cupules occur as early as the mid-Devonian, such as *Runcaria* (Gerrienne et al. 2004), but definitive examples of fossil ovules and seeds do not appear until the late Devonian, such as in *Elkinsia* (Willis and McElwain 2002). It is not clear whether all of the early seed ferns with ovule-like megasporangia retained their embryos to become seeds or dispersed the fertilized megasporangium so the megagametophyte could grow and support the sporophyte during establishment and early growth.

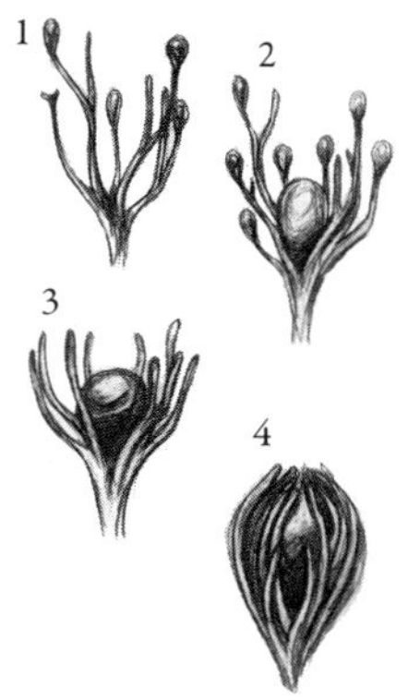

Fig. 2.16. Hypothesized stages for the origin of the integument: (1) multiple sporangia; (2) One or a few megasporangia in leaf axils with microsporangia on stem tips; (3) a single megasporangium surrounded by sterile fronds that form a cupule; and, (4) fusion of fronds to form an integument. Original art by Michelle Nicole Williamson.

Origin of Seeds and Pollen—There is a natural progression of stages from a megaspore (heterospory; Fig. 2.15.A; refer to color plate), to independent end osporic gametophytes that complete development within the spore wall (Fig. 2.15.B; refer to color plate). This stage is followed by the retention of the megagametophyte for fertilization by flagellated sperm while it is still attached to the sporophyte (similar to the process in ginkgoes and cycads; Fig. 2.15.C; refer to color plate). Finally, the production of pollen and ovules that afford protection of the gametophytes and the sporophyte during its early stages of development (Fig. 2.15.D; refer to color plate). It is not clear which of these stages generated a lineage with a continuous integument that developed into a protective covering (seed coat) for dispersal. We can infer that the evolution of a seed coat was advantageous for protecting the sporophytic embryo, but the same logic could be applied to megagametophytes of species that lived in periodically dry habitats, so protective integuments could have been a feature of early species that dispersed their megasporangia. Independence from water was improved with the advent of seeds and pollen grains, which protected gametophytes during development and dispersal (Fig. 2.15.D; refer to color plate).

While plants with seeds and pre-pollen (sporopollenin-coated microgametophytes with some characteristics of spores) were present by the late Devonian, it was not until the late Carboniferous that we see the origin and diversification of seed plants with living representatives (i.e., gingkophytes, cycadophytes, and conifers; Fig. 2.7). The sudden emergence of several gymnosperm groups around 310 MYA corresponded with the appearance of modern pollen morphologies (Willis and McElwain 2002) and was predated by a whole-genome duplication event around 319 MYA (Jiao et al. 2011).

Drying conditions toward the end of the Carboniferous precipitated one of the few extinction events that had a major effect on the compositions of floras around the world. Before we proceed with our review of the emergence of modern plant groups we will digress for a discussion of episodes of extinction and diversification that have affected floras and faunas over the history of life on Earth.

2.7 MAJOR EXTINCTIONS OF TERRESTRIAL PLANTS AND ANIMALS

Paleontologists have used large changes in the composition of the fossil fauna as markers to denote extinction events and transitions between geological periods. From the Cambrian through the present there have been a number of extinction events of varying magnitude, but five extinctions stand out as substantially more significant periods during which greater than half of the animal species disappeared (McElwain and Punyasena 2007; Fig. 2.17). The first of these (Ordovician–Silurian extinction event) had a large impact on marine animals, but the effects on terrestrial life is unknown because of a lack of fossils. Other major extinctions occurred toward the end of the Devonian (Late Devonian extinction event), at the end of the Permian (Permian–Triassic extinction event), at the end of the Triassic (Triassic–Jurassic extinction event), and at the end of the Cretaceous (Cretaceous–Paleogene extinction event). While these major extinction events had severe consequences for animal diversity, their

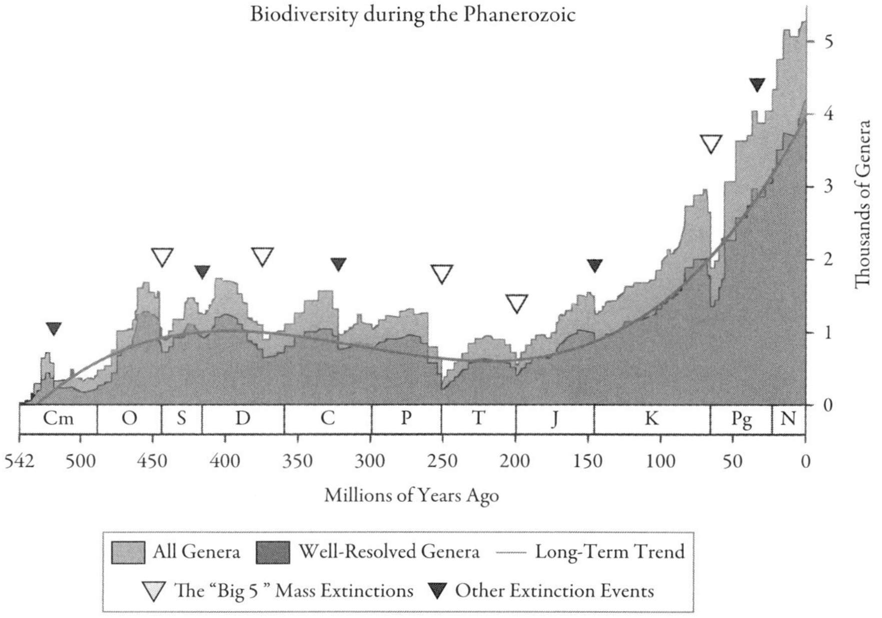

Fig. 2.17. Changes in diversity over time. Vertical axis scale is thousands of genera. Minor (dark) and major (light) extinction events are indicated by the triangles.

impact on plants was not nearly as large (McElwain and Punyasena 2007); the ability of plants to persist through dormant vegetative states and resistant spores and seeds may have allowed them to survive periods of severe conditions. Events having major effects on floras have more often been gradual climatic shifts and changes in faunas that have occurred between the major animal extinctions. Just as we see with major extinctions of animals, the loss of dominant plant taxa was generally followed by the diversification of new groups that replace older assemblages.

The increasing abundance and diversification of plants on land in the Devonian and Carboniferous provided resources for the establishment of herbivore communities and, eventually, for more complex ecosystems that included several trophic levels. By the end of the Devonian, the biomass of terrestrial plant life was substantial and provided habitat for a high diversity of arthropods, including the first insects. The abundance of grazers and detritivores presented opportunities for higher trophic level consumers. The first tetrapods were amphibians that diversified to become the dominant top predators during the Carboniferous. Like many of the plants living at this time, these animals required wet environments to complete their reproductive cycles as their eggs could not survive outside of water.

The first of the major extinction events that occurred after animals became abundant on land may have been caused by the spread and diversification of land plants. During the late Devonian, the increase in terrestrial plants presented a significant carbon sink, which may have produced a period of global cooling. This decrease in temperatures is thought to have precipitated the loss of more than 50% of terrestrial animal species during the Late Devonian extinction event. Over the same time period, the impact on terrestrial plants was much less as there was only a minor change in diversity and little turnover of species.

A much slower climatic shift that had a greater impact on the composition of vegetation—known as the *Carboniferous Rainforest Collapse*—occurred toward the end of the Carboniferous. This time period is marked by the disappearance of the extensive swamp forests as the climate became cooler and drier. Large lycopod species, such as the scale trees and the giant horsetails that had dominated the coal swamps during the Carboniferous, were replaced by tree ferns, along with conifers and other gymnosperms. The loss of many seedless tracheophytes corresponded with the disappearance of the swamp forests, while many species of ferns, seed ferns, and gymnosperms that were better adapted to dry conditions dominated forests through the Permian.

During the Permian the fauna underwent a major transition as the amphibians, which had dominated through most of the Carboniferous, where mostly replaced by early reptiles—particularly synapsids (ancestors of mammals) and later in the Permian, some very large species of anapsids. The third major group of tetrapods, the diapsids, originated during this same period but remained less common until the Triassic. The synapsid pelycosaurs that were common in the early Permian were later replaced by the therapsids, another synapsid lineage, which persisted until the early Cretaceous. The first mammals emerged from one of these therapsid lineages (cynodonts) during the late Triassic, and now mammals are the only surviving group of synapsids. The anapsids were originally thought to be ancestors of modern turtles, but molecular phylogenies indicate that turtles are more closely related to descendants of diapsids, which includes crocodilians, lizards, snakes, and birds.

Dry conditions presented strong selection for independence from water during the Permian. For animals, the amniotic egg was a key adaptation that allowed reptiles to complete their reproductive cycle outside of aquatic habitats. Similarly, early gymnosperms possessed desiccation-resistant gametophyte stages in the form of pollen and megagametophytes that developed within the protective tissues of the ovule. Unlike the bryophytes, lycopods, and ferns, the reproduction of these early spermatophytes was not constrained to times when water was present. As conditions became dryer, early gymnosperms, including the Cordaitales (an early conifer-like tree), cycads, ginkgoes, and a variety of seed ferns, became more abundant. Seeds and pollen provided crucial adaptations as they allowed these plants to diversify into a broad range of increasingly drier habitats.

The end of the Permian period, and of the Paleozoic era, was marked by the most severe faunal extinction event on record (Permian–Triassic extinction event). During this time, more than 90% of marine and 70% of terrestrial animal species were lost. The large synapsid and anapsid reptile fauna of the Permian disappeared, and it took some time before those animals were gradually replaced by the radiation of diapsid reptiles (early archosaurs), which were subsequently supplanted by dinosaurs (a subgroup within the Archosauria), after another extinction event at the end of the Triassic. While the Permian–Triassic extinction precipitated a prolonged period of low faunal diversity and major shifts in the dominant groups of animals, the impact on plant diversity was minimal.

The climate during the Triassic was generally hot and dry, but conditions became wetter starting in the Jurassic and remained warm and humid through the Cretaceous. Floras during the Mesozoic era (Triassic, Jurassic, and Cretaceous periods) continued to be dominated by a diversity of ferns and gymnosperms, including seed ferns (especially the Glossopteridales in the southern hemisphere), which may have given rise to flowering plants during the Triassic. The origin of the angiosperms is denoted by a second genome duplication event that occurred around 192 MYA (Jiao et al. 2011), but the first fossils of flowering plants do not appear until much later in the Cretaceous, 130–120 MYA (Willis and McElwain 2002). The angiosperms probably originated much earlier, but did not begin to spread until the Cretaceous, and became dominant components of the vegetation after the Cretaceous–Paleogene (K–Pg) extinction event. By the end of the Cretaceous, many gymnosperm groups that had been common in the Mesozoic including the last of the seed ferns had disappeared. Like previous extinction events, the K–Pg extinction had only minor effects on plant diversity, but changes in the climate and the fauna that followed created conditions that favored the diversification and spread of flowering plants along with some groups of mosses, ferns, and conifers.

2.8 THE RISE OF FLOWERING PLANTS

Conditions that favored the origin of the earliest flowering plants during the Triassic may have existed near the limits of tolerable habitat conditions for most gymnosperms and ferns. There are a number of outstanding features of angiosperms that represent major innovations compared to their seed fern ancestors, but only a handful of these may have been instrumental in establishing the first angiosperm lineages. It is important to remember that the fossil record presents us with a stuttered and incomplete view of the history of life; the sudden appearance of angiosperms in the Cretaceous with all the appropriate features already in place represents the culmination of a long journey of the slow accumulation of these traits. The apparent rapid emergence and diversification of flowering plants after their appearance in the mid-Cretaceous was so dramatic that Charles Darwin referred to it as "the abominable mystery" (Friedman 2009).

The most obvious unique feature of angiosperms is the flower. The flower consists of several layers (whorls) of leaves stacked closely together (nodes are compressed and internodes are lacking) that are modified into *Sporophylls* that produce microgametophytes (pollen) or megagametophytes (ovules), or remain sterile and are modified for the attraction of animal pollinators (petals) and the protection of the developing bud (sepals) (Fig. 2.18; refer to color plate). The ovules are contained within a carpel composed of fused sporophylls (see the Closing of the Carpel section), which is the defining characteristic that differentiates these plants from gymnosperms (*angio* = container or vessel).

In addition to traits associated with sexual reproduction, there are characteristics of the vascular anatomy (the presence of vessel elements) and vegetative morphology (leaves with reticulate vasculature) that are thought to have been important for the success of the flowering plants. By the mid-Cretaceous, all of these characteristics were already assembled into a single lineage, so our challenge is to understand the selection regimes that may have favored each set of traits found in early flowering plants. Without fossils or descendants of early transitional forms, it may not be possible to reconstruct the exact sequence of events, but we can examine the selective pressures that may have favored sets of traits that differentiate angiosperms from their gymnosperm relatives.

Closing of the Carpel—Changing conditions during the Triassic produced biotic and abiotic challenges for plants that may have been pivotal for the origin of angiosperm traits. The hot and dry climate that was prevalent during this period would have presented constraints on the distribution of plants. At the same time the plant consumers—herbivorous beetles in particular (Powell 2009)—were becoming more diverse and abundant. Pressure from herbivores may have been particularly intense in drier regions

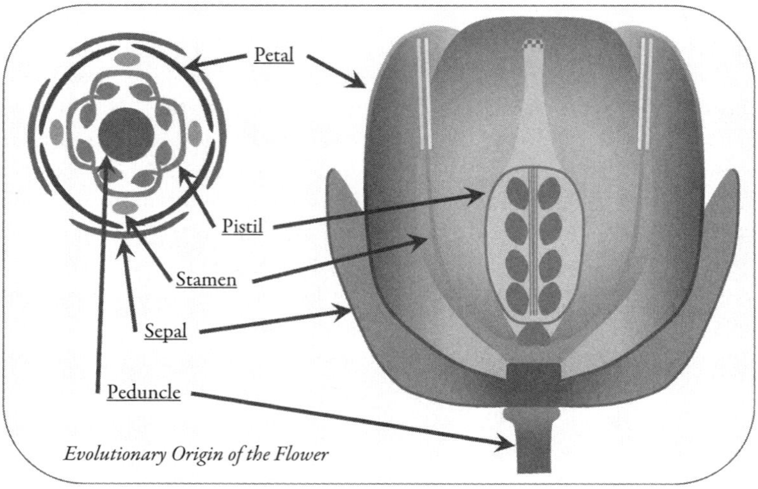

Fig. 2.18. The flower as a series of stacked nodes with each whorl of leaves modified into sepals, petals, anthers (microsporophylls), and the pistil (with megasporophylls fused to form a carpel).

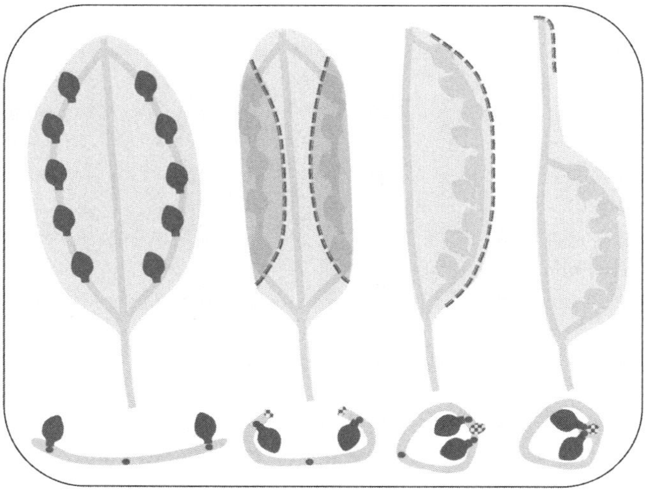

Fig. 2.19. Stages in the conversion of a megasporophyll to a closed carpel. The dashed region represents the location of stigma tissue in second through fourth stages, and transmitting tissue within the carpel in the final stage.

where vegetation was sparse. Removal of leaf material by herbivores may reduce the growth of an individual plant, but the destruction of ovules diminishes reproductive capacity and population growth as fewer individuals are recruited into the next generation. We expect that gymnosperm species with ovules exposed on leaves (i.e. seed ferns) would have been negatively affected to a much greater degree than species with their ovules concealed within cones (e.g., conifers, cycads, etc.), or within the ovaries of the first angiosperms (Stewart 1983). Selective pressure by herbivores led to species that produced their ovules within tight clusters of leaves at the ends of branches (e.g., plants in the Bennettitales), or by curling the edges of the sporophyll inward (e.g., seed ferns in the Glossopteridales; Fig. 2.19). The importance of herbivory for the evolution of ovules enclosed within carpels (angiosperms) or woody cones (gymnosperms) is supported by the observation that seed ferns did not persist past the end of the Cretaceous. On the other hand, there are several genera of extant plants with carpels that are not completely closed (e.g., *Drymis, Degeneria,* and *Exospermum*; Niklas 1997).

Sequestering ovules within carpels presented additional challenges as they became less accessible for fertilization (Zavada and Taylor 1986). As carpels became closed it was necessary for the site of pollen receipt to be moved away from the opening of the ovule (micropyle), as occurs in the gymnosperms, to a specialized tissue on the sporophyll (the *Stigma* of angiosperms). Stigmas are glandular regions of epidermal tissue that lack a cuticle and have sticky papillae or produce exudates that promote the germination and growth of pollen grains. We can infer that stigmatic regions were initially produced along the edge of the sporophyll with traces running along the inside to provide transmitting tissue for the passage of pollen tubes to the ovules (Fig. 2.19. Movement of receptive sites away from the ovules would require pollen to grow distances that were several hundred times the diameter of the grain. Microgametophytes had to obtain adaptations for the elongation of the pollen tube over these distances and for the uptake of nutritive substances from the exudates produced by sporophytic cells along the transmitting track. Some of these traits may have already existed—pollen tubes of some gymnosperms digest the nucellus tissue during their growth to the egg nucleus. We expect that the sequence of reciprocal adaptations in the pollen and sporophyte to accommodate displacement of the stigma further away from the ovules would have occurred incrementally and may have started before the carpel was completely closed. Ovules may have originally been placed near the fused edge of the sporophyll so the initial distance would have been minimal, and this distance increased as selection favored the movement of the stigma toward the tip of the sporophyll to improve the chances for pollen capture (Fig. 2.19).

Isolating megagametophytes within carpels and away from the location of pollen receipt had implications well beyond the initial advantage of protecting ovules and seeds from herbivores (Whitehouse 1950; Zavada and Taylor 1986). The advent of a carpel required that the microgametophyte pass through sporophytic tissue prior to delivering sperm to the ovule. Moreover, there was a potential for the delivery of a relatively large number of pollen grains to stigmas—particularly once insect pollination became more effective with the diversification of bees in the late Cretaceous. The potential for intervention by the sporophyte to discriminate among pollen was realized early in angiosperm evolution with the origination of self-incompatibility systems that inhibited the growth of microgametophytes produced by the same individual and thereby reduced the level of inbreeding (Igic and Kohn 2001; Chapter 9).

Ecological Range—Many of the features that differentiate angiosperms from many gymnosperms appear to be responses to selection for tolerance to chronically or seasonally dry conditions. It is widely accepted that xylem containing vessel elements was a major innovation that allowed angiosperms to colonize a wider range of habitats (Sperry 2003). Vessels lack the terminal cell walls of tracheids (Fig. 2.20) and are capable of linking together to form continuous tubes that are meters long, so vessels have lower resistance and can generate much stronger tension for the extraction of water from dry soil. The larger diameter vessels require greater reinforcement via bands of lignin, so wood with high vessel content tends to be more expensive to construct. This difference is also apparent in the strength and other characteristics of the lumber products that can be obtained (i.e., angiosperms produce "hardwood" and gymnosperms produce "softwood"). Convergent evolution of vessels occurred in several groups during the Triassic, including the gnetophyte genera such as *Ephedra* and *Welwitschia*, which inhabit extremely arid regions. On the other hand, the most widespread and abundant gymnosperms, the conifers, do not produce vessels and are capable of dominating the vegetation in a number of habitats, including some regions that are periodically dry. While conifers dominate contemporary plant communities in cooler regions, the large majority of plants in temperate to tropical biomes, including the most arid regions of the world, are angiosperms. Xylem architecture certainly contributed to the success of angiosperms, but the ecological breadth and dominance of flowering plants is a consequence of a composite set of characteristics that differentiate them from the gymnosperms.

The other major innovations of the angiosperms that appear to have contributed to their ecological

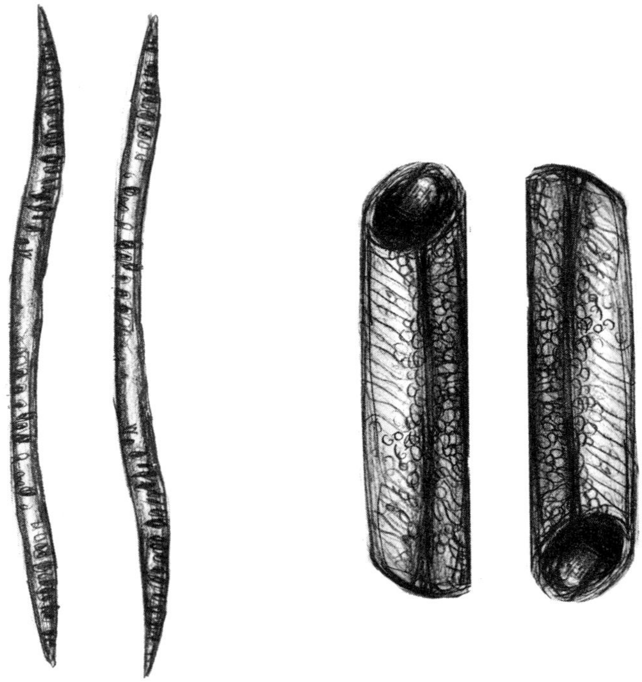

Fig. 2.20. Components of xylem including tracheids (left) and vessels (right). Original art by Michelle Nicole Williamson.

amplitude is the production of a broad range of leaf and root morphologies compared to gymnosperms. Leaves with branching and reticulate venation first appear in seed ferns such as the Glossopteridales from the Permian, but diversification of leaf types did not accelerate until the Cretaceous (Willis and McElwain 2002). The diversity of leaf morphologies in flowering plants is tremendous and includes differences in size, shape, and indumentum characteristics (e.g., leaf hairs). Foliar adaptations reflect strategies for tolerating or avoiding stressful conditions including leaves that are sclerophyllous (stiff leaves with hardened cell walls), deciduous (abscising in response to the onset of drought or freezing weather), succulent, and with a wide range of epidermal features (Ingrouille and Eddie 2006). At the other extreme, the leaves of plants from wet tropical regions are often very large and possess physiological adaptations that allow them to maintain plant growth under the low light conditions of the forest understory.

The life histories of flowering plants are also much more diverse than those of gymnosperms, which are mostly long-lived evergreen woody species. Angiosperms range from woody evergreen or deciduous trees to herbaceous plants that endure inhospitable conditions by producing seeds or by surviving as underground storage structures such as roots, tubers, and rhizomes (geophytes). These morphologies are accompanied by anatomical and physiological adaptations that have allowed flowering plants to diversify into nearly every habitat that offers minimal levels of liquid water, light, and inorganic nutrients. Their success has been facilitated by associations with soil fungal mutualists (mycorrhizae), which also occur in many lycophytes, ferns, and gymnosperms (Brundrett 2002; Bonfante and Genre 2008; Winther and Friedman 2008).

Origin of Angiosperms—The origin of the angiosperms is obscure because of the lack of fossil evidence of the early stages when the distinguishing characteristics of this group were being assembled into a single common ancestor. Molecular phylogenies indicate that the flowering plants first diversified in the late Triassic, but the lack of fossil evidence and extant species with transitional features challenges our abilities to make robust inferences about the characteristics of the earliest angiosperms. The traditional view was that members of Gnetophyta such as *Gnetum* were transitional between gymnosperms and angiosperms because they share *Double Fertilization* (Box 9.1; Carmichael and Friedman 1996). Molecular phylogenies provide two contrasting views; one is that both angiosperms and gymnosperms are monophyletic within the spermatophytes, and that the Gnetophyta arose in parallel with the flowering plants (Soltis et al. 2002; Qiu et al. 2006; Fig. 2.21). Other phylogenies indicate that the gnetophytes are basal within the seed plants so the flowering plants are imbedded within a larger clade of plants referred to as gymnosperms (Rai et al. 2008). The exact branching order of seed plant groups remains unsettled, but more recent genomic analyses indicate that the Gnetophyta is more closely related to the Coniferophyta so both gymnosperms and angiosperms are monophyletic (Lee et al. 2011; Wu et al. 2013; Fig. 2.21).

The extended lineage that ultimately led to angiosperm diversification in the Triassic split with gymnosperms around 350 MYA in the Carboniferous (Soltis et al. 2002), so the ancestor of flowering plants was one of the many different types of seed ferns (Pteridospermatophyta) that were common during this period (Fig. 2.21). Several fossil species that share similarities with flowering plants have been proposed as candidates for the likely ancestor. For example, plants in the Bennettales had flower-like reproductive structures that included both male and female sporophylls, two integuments around their ovules, and similar wood anatomy as angiosperms (Willis and McElwain 2002), but these plants may have closer affinities with the cycads (Crepet 2000). Candidate angiosperm ancestors include seed ferns in the genera *Furculla* (leaves with net venation similar to angiosperms), *Gigantopteris* (net venation, vessel elements, and secondary compounds), *Caytonia* (carpel-like cupules that surround the ovules), and *Crinopolles* (angiosperm-like pollen, among others). While these examples are tantalizing, they lack sufficient combinations of traits to make them the single, most-likely flowering plant ancestor.

Several hypotheses have been proposed to provide an ecological context for the origin of angiosperms. The most basal lineages within the angiosperm clade are the monotypic genus *Amborella* and the family Nymphaeaceae (water lilies). The observation that *Amborella* is a shrub and water lilies are herbaceous has led to the suggestion that early angiosperms were fast-growing shrubs or herbaceous plants (Willis and

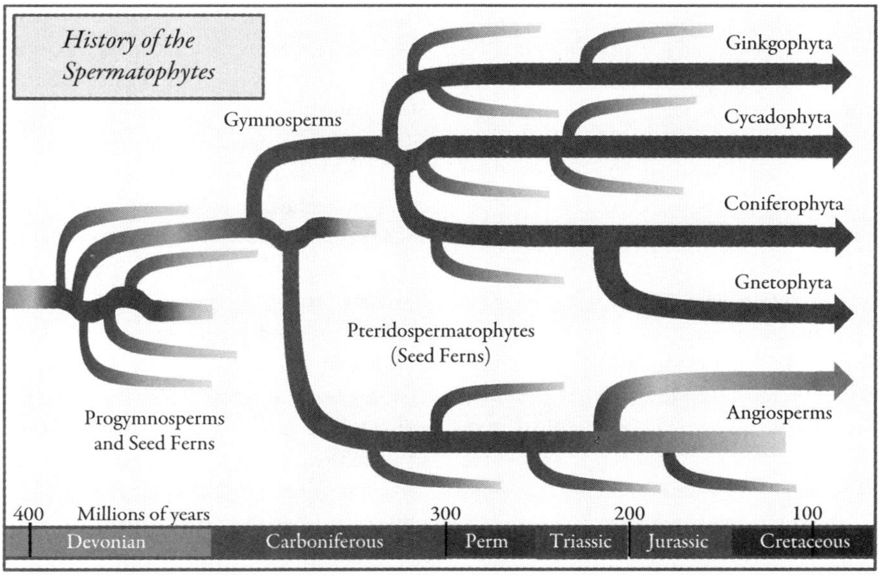

Fig. 2.21. Origin of the major spermatophyte groups. The time scale is expanded for the Devonian and Carboniferous.

McElwain 2002). This scenario is feasible since there was a diversity of seed fern species present during the Triassic that were shrubby or herbaceous (Greb and DiMichele 2006). It is also notable that all of the angiosperm species found in the early Cretaceous were either small woody or herbaceous plants with relatively small leaves and seeds. The collective evidence supports a scenario in which angiosperms arose as short-lived, early successional plants that occupied ephemeral habitats or lived in regions with limited growing seasons.

Finding definitive fossil evidence of early transitional stages between seed ferns and flowering plants may be difficult if angiosperms originated in geographically isolated or arid regions. Biomes that were mostly dry or had limited periods of adequate precipitation for plant growth were common in the Permian through the Cretaceous, but fossil evidence of the flora that may have occupied these habitats is sparse or non-existent (Willis and McElwain 2002). The lack of fossils may reflect the low frequency of fossilization in dry regions and the possibility that the occupants were largely short-lived herbaceous species, which would reduce their chances of leaving a record. The observation that several early angiosperms possessed a number of characteristics that appear to be adaptations to dry conditions, such as vessel elements and small leathery leaves, is consistent with the idea that they may have originated in arid habitats. While drought-associated traits are not exclusive to angiosperms, it is possible that these characteristics, combined with a weedy growth form, allowed early flowering plants to proliferate in arid regions that were inhospitable for long-lived species.

Biotic Interactions—Several hypotheses for the origin and diversification of angiosperms focus on biotic interactions. One proposal posits that changes in the composition of the types of herbivorous dinosaurs from the late Jurassic through the end of the Cretaceous created conditions that promoted angiosperm diversification (Bakker 1978; Barrett and Rayfield 2006). High-browsing dinosaurs of the Jurassic were replaced by low-browsing species in the early Cretaceous, which may have reduced seedling survival for woody gymnosperms and thereby reduced the density of canopies. The diversification of low-browsing dinosaurs may have been enabled by the increased abundance of herbaceous dicots and grasses (Prasad et al. 2005). The foraging activities of dinosaurs may have led to increased levels of disturbance, which could have facilitated the establishment and spread of short-lived angiosperm species. The activity of large herbivores along with abiotic disturbances may have been important sources of habitat creation for the diversification of flowering plants as they moved out of the arid regions into wetter biomes where they began to displace ferns and gymnosperms.

The second major category of biotic interaction that was important for the diversification of angiosperms was their association with animal pollinators, insects in particular. Fossil evidence, including gut contents of beetles and flies, suggests that plants may have relied on insects for pollination as far back as the Triassic. Early pollinators may have been beetles, flies, and wasps, while bees did not appear until the mid-Cretaceous (Danforth et al. 2006; Glover 2007; Thien et al. 2009). The earliest fossils of angiosperms from the early Cretaceous indicate there were two major types of flowers (Willis and McElwain 2002; Friis et al. 2006); larger, magnolia-like flowers with multiple parts, which may have been beetle pollinated (Fig. 2.22; refer to color plate), and reduced flowers similar to extant species in the Chloranthaceae or Piperaceae (Fig. 2.23; refer to color plate), which may have been pollinated by flies (Thien et al. 2009).

Early pollinators were relatively ineffective, and it was not until bees began to diversify that more efficient pollinators became common. Bees have specific adaptations for nectar and pollen harvesting; their bodies are covered with branching hairs that render them more effective for picking up pollen compared to the relatively hairless bodies of other insects. Bees also display high levels of constancy (visiting multiple flowers of the same species in one foraging bout), so they tend to deposit larger numbers of intraspecific pollen grains on stigmas than do other types of insects (see also Chapters 9 and 10). Even though much of the early history of angiosperms occurred in the absence of bees, the early animal pollinators such as beetles, flies, and wasps were probably an improvement over abiotic pollen vectors such as wind and water.

Bees and angiosperms diversified together starting in the Cretaceous. Many families of plants with more specialized floral morphologies, such as the

Fig. 2.22. Examples of magnolia-like flowers similar to those appearing in the early Cretaceous that were probably pollinated primarily by beetles. Left to right: Fragrant water lily, *Nymphea odorata* (photo by the author), Austrobaileyales, *Schisandra rubriflora* (photo by Scott Zona), and *Magnolia wieseneri* (photographer unknown).

Fig. 2.23. Examples of reduced flowers similar to those appearing in the early Cretaceous that were probably pollinated primarily by flies. Left to right: Black pepper, *Piper nigrum* (photo by H. Zell) and *Sarcandra glabra* (Chloranthaceae; photo by Kenpei).

legumes and orchids, began to spread in the mid to late Cretaceous, but this diversification did not necessarily occur in lockstep with their pollinators. For example, orchids and legumes, as well as a number of other plant families, have distributions across Africa, Europe, and the Americas, suggesting that they began spreading prior to the substantial breakup of Laurasia and Gondwana into the modern configuration of continents (Fig. 2.24). While several families of bees such as the Apidae and Colletidae have worldwide distributions, some groups that have strong associations with particular types of plants have much more limited ranges. For example, the orchid bees (Euglossini) and the plants they pollinate are restricted to the American tropics even though orchids are distributed across several continents.

Associations with animals for seed dispersal may have originated in the late Cretaceous and early Paleogene when mammals and birds were becoming more common (Tiffney 2004; Eriksson 2008). Animal dispersal of seeds and fruits may have been a critical force for the early diversification of the angiosperms during the Cretaceous (Eriksson et al. 2000; Bolmgren and Eriksson 2010). This time frame corresponds to the early diversification of mammals including primates (Tavare et al. 2002; Martin et al. 2007) and birds (Chiappe 2007). Several groups of birds and primates are frugivore specialists, so the

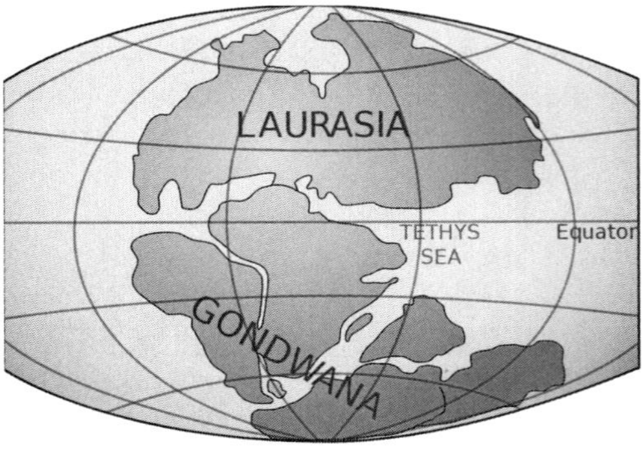

Fig. 2.24. Laurasia and Gondwana prior to the separation of contemporary continents during the Triassic (~200 MYA). Image by Lenny at Wikidata.

evolution and diversification of these groups were strongly influenced by the availability of new resources presented by early angiosperms with fleshy fruits. Animal-mediated dispersal also occurs in some gymnosperms for seeds (fleshy fruits occur in Gingkoes and yews) and pollen (beetle pollination has been noted in some cycads); while associations with animals was certainly influential, this may not have been the single most important factor that enabled the diversification of the flowering plants.

It is clear that the associations between flowering plants and animals for pollination and seed dispersal engendered each other's diversification during the Cretaceous, but this coevolution was largely diffuse. With some notable exceptions (e.g., yucca moths and yuccas), most of the specialized pollination associations between plants and animals have developed relatively recently, within the last few million years. For instance, pollinator associations can define individual species, and many genera include species with highly divergent *Pollination Syndromes* (Chapter 10): for example, *Aquilegia* (columbines), *Penstemon* and *Antirhinum* (snapdragons), *Solanum* (nightshades, tobacco, and tomatoes), and *Banksia* in Australia (Fig. 10.6; refer to color plate). In contrast, several large families that may have diversified early on have more generalized pollination syndromes; plants in the Asteraceae (sunflowers) and Papaveraceae (poppies) have open, bowl-shaped flowers, and plants in the Poaceae (grasses) and Amaranthaceae (amaranths and chenopods) are wind pollinated. Animal associations with seed dispersal (e.g., fleshy fruits) also tend to be consistent across species within larger groups; for example, within genera such as *Rubus* (blackberries), pomeaceous genera in the Rosaceae (apples, pears, hawthorns, etc.), and berries in many species of *Solanum* (tomatoes and nightshades), to name a few. Apparently, non-specific associations between plants and animals were influential for the early diversification of the flowering plants, and more specialized associations did not become important until much more recently.

2.9 LIMITS AND SOURCES OF PLANT DIVERSITY

Over the history of the embryophytes the sources of selection that have motivated diversification (i.e., in response to competition) have shifted away from adaptations to biotic and abiotic conditions to an increasing importance of interactions with animals. The earliest terrestrial lineages occurred only in wetter habitats, while expansive stretches of drier regions remained devoid of plant life. The early history of land plants is marked by responses to intense competition in these over-crowded wet environments. Improvements in vertical growth, canopy structure, and leaf morphology allowed better access to light. Until the end of the Devonian, much of the land may have remained uninhabited by plants, as their radiation into drier habitats was exceedingly slow. The journey from the first microfossil evidence of terrestrial plants in the Ordovician until the emergence of plants with seeds and pollen-like spores in the late Devonian required more than one hundred million years.

During the early history of land plants, there is little evidence for the effects of animals on their evolution. Fossils suggest that arthropods may have fed on spores and pierced stems of land plants during the Devonian (Labandeira 1998), but herbivory did not become common until much later. Evidence for the rising importance of herbivores as a source of selection shaping plant evolution appears in the late Permian with fossils of seed ferns in the Gigantopteridales associated with secondary metabolites that are similar to anti-herbivore compounds found in extant angiosperms. As described earlier in this chapter, the evolution of the carpel may have been triggered by the diversification of herbivorous beetles beginning in the Triassic and extending through the Cretaceous (Powell 2009). Inferences about the morphology of early flowers and their possible responses to selection mediated by herbivores remains speculative; from their origin in the Triassic through the mid-Cretaceous the angiosperms apparently had a limited geographic distribution or were restricted to arid environments where they remained invisible to the fossil record.

The rate of diversification of flowering plants since the early Cretaceous has been much greater than that of other plants; angiosperm lineages have accumulated at a rate several orders of magnitude faster other embryophyte groups (Fiz-Palacios et al. 2011). During the late Cretaceous and early Paleogene, the diversity of mosses, ferns, and conifers also increased, but at much more subdued levels compared to the rampant spread of the flowering plants. The dramatic increase in the number of angiosperm families and genera coincided with a series of genome duplication events (Otto 2007; Soltis et al. 2009), including a tripling event in the early Cretaceous that marks the origin of Eudicots (Jiao et al. 2012), which comprise 75% of flowering plant species. Similarly, the diversification of ferns is associated with a series of whole-genome duplication events (Otto and Whitton 2000). In contrast, polyploidy is rare in gymnosperms, and experimental evidence suggests that whole-genome duplication events in conifers more often have negative consequences (Ahuja 2005). The tolerance of the angiosperm genome to duplication, along with characteristics of their life history, may have been critical for the diversification of vegetative and floral morphologies and their rise to become the dominant primary producers across the majority of terrestrial habitats.

A unique combination of characteristics may have colluded to spawn the diversification and dominance of flowering plants. As discussed earlier, a weedy life history along with hydrological and morphological adaptations may have enabled early angiosperms to spread into habitats that were ephemeral or that had short growing seasons. The acquisition of these adaptations and the rapid diversification of new flowering plant forms were almost certainly enabled by increases in genomic complexity. Ferns had the same potential for genome duplication, so we might have expected they would have been just as successful; however, ferns retained independent gametophyte generations that required water for successful reproduction, so the range of habitats they could occupy was more limited. Moreover, the relatively large proportion of the genome expressed in the independent gametophyte of ferns may have thwarted more rapid evolution. In this context, we may view the large number of duplication events and relatively high chromosome counts in modern ferns (Otto and Whitton 2000) as a mechanism to overcome extensive post-meiotic expression of haploid genomes; duplication results in gametophytes that are effectively diploid, which increases the potential for the accumulation of genetic variation. Similarly, diversification of mosses in the genus *Sphagnum* (peat mosses) appears to be associated with multiple bouts of genome duplication (Ricca and Shaw 2010). The angiosperms, on the other hand, had reduced gametophytes with limited post-meiotic expression. Furthermore, angiosperm gametophytes are contained within the sporophytic tissue where they are buffered from environmentally mediated selection (Walbot and Evans 2003). A diminished gametophyte stage combined with a genome that was tolerant of duplication may have been key features that promoted the rapid diversification of angiosperms.

The other factor that may have been instrumental for the emerging dominance of flowering plants was the closing of the carpel, or more specifically, the potential it created for intervention by the sporophyte on microgametophytes prior to fertilization (Whitehouse 1950; Mulcahy 1979; Chapter 9). The sporophytes of angiosperms are unique relative to other plants because of their ability to prevent self-fertilization (*Self-Incompatibility*), which apparently arose very early in angiosperms (i.e., *Amborella* is self-incompatible; Williams 2009), and has not been observed in gymnosperms (Zavada and Taylor 1986). Gingkoes and cycads avoid self-fertilization by being *Dioecious* (separation of sexes between individuals), but many gymnosperms are *Monoecious* (male and female cones produced by the same individual) and lack self-incompatibility, so the potential for inbreeding is high. In addition to dioecy, there are other mechanisms that can reduce or prevent self-fertilization, such as differences in the timing of pollen dispersal and receptivity of female flowers on the same individual (*Herkogamy*; Chapter 9). Self-incompatibility may have advantages over other mechanisms that enforce outcrossing, because it is more effective than herkogamy, and it allows for the production of pollen and ovules in the same flowers.

Curtailing self-fertilization in hermaphroditic individuals was a major advantage of closed carpels, but it has also been argued that this new arrangement created the potential for increases in fitness through competition among microgametophytes (*Gametophytic Selection* by *Pollen Competition*; chapters 5 and 9) and through sporophyte-mediated selection (Mulcahy 1979; Zavada and Taylor 1986). Pollen competition occurs when the number of pollen grains arriving on a stigma exceeds the number of ovules, in which case we would expect only the fastest growing gametophytes to be successful. For example, pollen competition may be an effective way to eliminate deleterious mutations because all expressed alleles in the haploid microgametophytes will be subject to selection (Walbot and Evans 2003); any mutations that cause slower growth will be rapidly eliminated if they are expressed in pollen tubes. Another suggestion is that the sporophyte may be capable of assessing pollen quality, so only a selected subset of microgametophytes would be responsible for siring seeds (*Selective Fertilization*). While these ideas are intriguing, the potential for gametophytic selection (pollen competition and selective fertilization) to increase fitness may be marginal, and its evolutionary importance is not widely accepted. We will revisit this topic in more detail in Chapters 5 and 9.

The idea that selection on the haploid generation may be beneficial seems to contradict the thesis developed earlier in this chapter that the haplodiplontic life cycle slowed the diversification of land plants. We can view these hypotheses as complementary by making a distinction between selection mediated by the environment and selection on aspects of metabolic efficiency that may be largely independent of environmental conditions. In the first case, we infer that independent gametophytes have restricted the response of seedless tracheophytes to environmental challenges because gene expression in the haploid stage limited levels of genetic variation for ecologically relevant traits. On the other hand, we expect that genes relevant to environmental adaptations are not expressed during the haploid life stage of angiosperms, so only genetic variation that affects growth rates and other aspects of metabolic performance would be subjected to selection on microgametophytes. This scenario raises the possibility that while a haplodiplontic life cycle stalled evolution of early land plants, more limited post-meiotic gene expression in the gametophytes of angiosperms may have been beneficial.

While we have considered the retention of a multicellular haploid life stage as an evolutionary constraint, it is important to note that it is possible to transition from a haplodiplontic to a diplontic life cycle. For example, the evolution of diplontic life cycles from haplodiplontic ancestors has been observed in other eukaryotes, such as brown algae (Lee 2008). If retention of a gametophytic life stage was not constrained, then there is the possibility that gametophytes were never completely reduced to just sperm and eggs because of the advantages provided by post-meiotic gene expression. While these ideas are speculative, evidence for the beneficial effects of gametophytic selection continues to accumulate, and its potential effects on plant evolution continue to be active topics of discussion (Mulcahy et al. 1996; Walbot and Evans 2003; Seoighe et al. 2005).

2.10 THE DIVERSITY OF PLANT LIFE

Descendants of the first embryophytes have managed to invade virtually every nook and cranny that offers even the smallest measures of adequate light, nutrients, and liquid water. The explosion of evolutionary experimentation of plant forms in the Devonian ultimately produced an expansive array of morphologies that are now represented as the most abundant primary producers in terrestrial environments. While the flowering plants are clearly the most diverse producers, they are by no means dominant across all habitats. In fact, many conifers prevail in the cooler biomes and in some seasonally dry temperate regions where they are important sources of forestry products. The bryophytes—mosses in particular—are dominant components of the vegetation in far north and south tundra regions and in the epiphyte communities in the canopies of wet temperate and tropical forests. While angiosperms attained the greatest diversity and occupy the widest range of habitats, it is appropriate to view all modern plant groups as finely tuned products of a long history of evolution.

In many regions of the far northern and southern latitudes there are floras in wetter regions that seem like they are left over from the Cretaceous (e.g., the temperate rainforests of western North America). These forests are dominated by conifers with diverse understory assemblages of ferns, horsetails, and carpets of bryophytes that extend into canopies to form dense epiphyte communities. Angiosperms are limited to the understory in coniferous forests but increase in abundance and dominance toward the warmer latitudes. In milder temperate regions with wet growing seasons, deciduous trees dominate the canopy and other flowering plants form the large majority of the shrub and herbaceous understory communities. While flowering plants are prevalent in these habitats, a diversity of ferns, bryophytes, and lycophytes can also be found in the understory and epiphyte communities. Moving into more arid temperate regions, we find grassland and shrub communities that consist almost entirely of angiosperms. Similarly, with some notable exceptions (e.g., *Ephedra*), the driest regions of the world—the deserts—are dominated by flowering plants, and *Welwitschia* occupies extremely dry regions in southern Africa. As we move closer to the equator, flowering plants continue to be dominant in savanna regions and the warm, humid tropical forests. Wetter climates create more opportunities for the bryophytes and ferns, which are the foundation of the diverse epiphyte communities in wet tropical forests. In many of the more tropical regions around the world we can also find gymnosperms represented by many species of cycad and trees and vines in the genus *Gnetum*.

Terrestrial habitats have sometimes offered conditions that challenged the evolutionary potential of plants, but descendants of embryophytes have mustered the necessary adaptations to flourish under all but the most inhospitable conditions. The hands-down champions of extreme conditions would have to be the mosses, because of their ability to grow on impervious substrates where they have to endure extended periods of desiccation. Mosses are also capable of thriving at extremely high temperatures near geothermal features, and sphagnum mosses are foundational species responsible for the formation of bogs, in situations where concentrations of inorganic nutrients are extremely low. Among the gymnosperms, the conifers are notable for their ability to tolerate extended periods of freezing temperatures and then resume photosynthesis as soon as conditions become warmer. The members of the Gnetophyta mentioned earlier are outstanding for their ability to grow in some of the driest regions in the world, though most desert species are angiosperms. The flowering plants are the sole occupants of habitats characterized by edaphic extremes, with species and local ecotypes capable of surviving in soils with high concentrations of salt, calcium, and heavy metals. A diverse group of angiosperms—including some of the most basal lineages—occur as life forms that are submerged, floating, or emergent aquatic species in freshwater lakes and ponds, and the only embryophytes that have successfully invaded marine habitats are flowering plants (e.g., species of *Zostera, Thalassia*, and *Posidonia*). Perhaps the only places plants cannot grow is where the ground is covered by ice and snow year-round, on rock surfaces that are not exposed to enough atmospheric moisture to support mosses, or in the deeper and darker reaches of the ocean.

The evolutionary processes that enabled adaptation to this impressive array of habitats are diverse and complex. In the coming chapters we will describe the diversity of living plants, explore the foundations and development of modern evolutionary biology, and engage in detailed discussions of each evolutionary process in turn, their relationships to each other, and their collusion for the generation of the spectacular diversity of plant life around us.

2.11 QUESTIONS FOR CONTEMPLATION AND DISCUSSION

1. What traits are required for the invasion of emergent and terrestrial habitats? Can you develop a composite phenotype that has the basic characteristics of streptophyte algae plus a minimal set of adaptations that allow survival in open air?
2. Outline the characteristics that differentiate the earliest known terrestrial embryophytes from their algal ancestors.
3. What are the different scenarios for the timing of the origin of land plants? Develop a list of observations that support early (Proterozoic) versus later (Cambrian or later) emergence of terrestrial plants. Do you think that early embryophytes were aquatic or terrestrial?
4. What is the evidence that embryophytes diverged from streptophyte algae in freshwater systems? What characteristics of streptophytes provided pre-adaptations for the establishment and

diversification of multicellular terrestrial plants? Can you identify one particular streptophyte group that is more likely to be the closest relatives of embryophytes?
5. Describe the diversification of early land plants. List the traits shared between tracheophytes and each bryophyte lineage. How does this list differ if you consider traits that might be independently derived or not? Is there a single bryophyte lineage that is the most likely sister group to tracheophytes?
6. Why was the transition to a sporophyte-dominant life cycle important for the diversification and success of embryophytes?
7. Why didn't photosynthetic algae undergo rapid diversification in parallel with the animals during the Cambrian? What were the conditions that precipitated the explosion of plant diversity in the Devonian?
8. How did heterospory develop and how was it important for the evolution of pollen and ovules?
9. Describe the origin of ovules with one or more integuments from naked megasporangia. Why was this development critical for the origin of seeds?
10. How is it that plants survived through many of the major extinction events with only minor effects on their diversity? What conditions can cause plant extinctions? Based on this history, how well do you think plant diversity will be maintained through contemporary environmental challenges?
11. When did the seed fern ancestors of flowering plants initially diverge from other lineages, and when did true angiosperms eventually emerge? What were the likely conditions that promoted the evolution of flowering plants? Why did they remain invisible to the fossil record until the Cretaceous?
12. What were the selective pressures that led to the closing of the carpel? What were the implications of this change in the morphology of flowers? You might want to revisit this question after reading sections of Chapters fiv5e and 9.
13. How did biotic interactions promote the diversification of the flowering plants?

PART II

MECHANISMS OF DIVERSIFICATION

3

ORIGINS OF VARIATION

The process of mutation is the only known source of the raw materials of genetic variability, and hence of evolution. It is subject to experimental study, and considerable progress has been accomplished in this study in recent years. An apparent paradox has been disclosed. Although the living matter becomes adapted to its environment through formation of superior genetic patterns from mutational components, the process of mutation itself is not adaptive. On the contrary, the mutants which arise are, with rare exceptions, deleterious to their carriers, at least in the environments which the species normally encounters. Some of them are deleterious apparently in all environments. Therefore, the mutation process alone, not corrected and guided by natural selection, would result in degeneration and extinction rather than in improved adaptedness.

Theodosius Dobzhansky (1957)

3.1 ORIGINS OF VARIATION

The generation of genetic variation through mutation is fundamental to the diversification of life. This diversity is recorded in the genome of every species, which reflects individual histories and relationships to others. Each lineage's genetic profile represents an accumulation of modifications that have been filtered by processes of random and directed change, which we refer to as *Genetic Drift* and *Selection*, respectively. These processes require the continuous creation of genetic variation through changes in the sequence of nucleotides (*Mutations*), which arise regularly through DNA damage from environmental stressors and errors occurring during DNA replication. As suggested in the quote at the beginning of this chapter, the preponderance of mutations are detrimental, but the small fraction that have proven advantageous over the last 3.8 billion years are responsible for the incredible diversity of all life that has been generated since the first protocells arose in the early Proterozoic.

The process of mutation is random but occurs regularly with each cell generation. With each cell division, the entire genome is copied, but the replication of DNA strands is not perfect so sequence mistakes can be passed on. DNA damage caused by chemical toxins, reactive oxygen species, and ultraviolet radiation is usually repaired if one strand remains intact, but mistakes that are not repaired correctly will be transmitted to the next generation of cells. Mutations can occur during the mitotic cell divisions of *Somatic* growth and during meiotic divisions that result in haploid gametes. While changes in DNA arising during somatic growth typically involve only a single base (nucleotide) change or short insertions and deletions (*Indels*), the magnitude of mistakes can be amplified during meiosis as crossing over can result in the duplication or deletion of longer stretches of sequence. The sum of these processes results in a complement of mutations within each generation that may be filtered by selection or genetic drift—mutations are the foundation of the process of evolution as they create opportunity for adaptation and set an upper limit to rates of change across generations.

The accumulation of genetic variation and transmission of mutations from one generation to the next is fundamentally different in plants than in metazoans (Klekowski 2003). After an animal zygote is formed, a specific set of cells is set aside as the *Germline.* Germ cells maintain a high level of telomerase activity, which maintains *Telomeres* and suppresses mutations. The germline undergoes a limited number of mitotic cell divisions that result in gametogonia (spermatogonia and oogonia), which produce gametocytes that divide by meiosis to produce sperm and eggs. Plants do not have germlines in the same sense as animals; cells in their *Meristem* regions undergo numerous divisions in long-lived plants. In animals, *Somatic Mutations* (acquired mutations) generally have little consequence because they are not passed on to the next generation. Plants, on the other hand, have indeterminate growth, and meiosis occurs in flowers toward the ends of stems and branches, so there are a large number of mitotic cell divisions occurring within generations that contribute to the accumulation of heritable mutations (Fagerstrom et al. 1998; Klekowski 2003; Schultz and Scofield 2009).

Mutation is not the only source of heritable variation; environmental conditions experienced during growth can lead to changes in DNA methylation and associations with histones, which can affect gene expression and result in transient or heritable *Epigenetic* variation (Kalisz and Purugganan 2004; Schmitz and Ecker 2012). Both changes in the DNA sequence and epigenetic modifications acquired during somatic growth of plants can be passed on to their gametes, and the next sporophytic generation. The fate of this heritable variation will depend on the phenotypic effects of mutations and epigenetic modifications in the sporophytic and gametophytic life stages, and ultimately the dynamics of selection and genetic drift in populations. In this chapter, we explore the origins and evolutionary consequences of genetic variation.

3.2 TAKING A STROLL ALONG THE GENOME

If we were able to obtain a view of the genome at the molecular level, any particular strand of DNA we might encounter would seem to extend endlessly in both directions. The large majority of most plant genomes consist of long stretches of *Noncoding DNA* (so called "junk" DNA), so we would likely land in one of these regions and would have a long walk before we might reach a *Reading Frame* (consisting of exons and introns; one of several types of *Functional Sequence Regions,* also generically referred to as *Genes*). Nested within the noncoding regions are transcribed, but untranslated, sections of functional DNA that code for transfer RNA, ribosomal RNA, and regulatory RNAs, which include miRNA (microRNA), siRNA (small interfering RNA), and rasiRNA (repeat associated small interfering RNA) (Jones-Rhoades et al. 2006; Ruvkun 2008). In most species, most of the vast stretches of noncoding/non-functional DNA consist of dormant retrotransposons and DNA transposons that include long interspersed and short interspersed nuclear elements (LINES and SINES). We also expect to find many regions of repeated sequences (motifs) of two to six nucleotides, which may occur in groups of two to ten or more repeat motifs (*Simple Sequence Repeats,* also referred to as *SSRs* or *Microsatellites*). Other types of noncoding DNA that we might encounter include *Pseudogenes,* which were formerly active genes that are either not transcribed or produce non-functional proteins. As we approach an active gene region we might encounter *Cis*-Regulatory Elements, which affect the expression of a nearby upstream or downstream reading frame. Scattered throughout the genome there are also *Trans*-Regulatory Elements, which regulate the expression of more distant genes.

Once we find a reading frame, we note that embedded within it there are unexpressed regions (introns or intragenic regions) that are transcribed and spliced out of the mRNA sequence prior to translation. Other regions (exons) are transcribed and spliced together to produce the translated sequence after the introns are removed. Because only the exons are used to produce the final mRNA molecule that defines the protein, only mutations occurring in these regions have the potential to affect growth, form, and function of the organism (i.e., the *Phenotype*). Mutations occurring within reading frames and regulatory regions have the potential to be expressed in the phenotype, so these *Expressed Mutations* may be affected by both selection and genetic drift (see Chapter 4 and Chapter 5). In contrast, mutations occurring within introns and other non-functional regions of DNA will not be expressed; these *Silent Mutations* (neutral mutations) will only be affected by genetic drift and not by selection.

The potential for mutations within reading frames to be exposed to selection depends on whether they cause a change in the structure or function of a protein, but we do not always have enough information to know whether any base change may be subject to selection. To circumvent this problem, we can categorize base changes based on whether they cause an amino acid substitution. The genetic code is redundant; more than one codon may code for the same amino acid. *Synonymous Changes* within codons indicate the same amino acid so these base substitutions will not have an effect on protein structure (i.e., these are silent mutations). Many *Non-Synonymous Changes* will also remain silent to selection because of the similarity in the structures of amino acids; changes among amino acids of the same class are less likely to affect the structure or function of the protein. We do not always know the structure of the protein, so base substitutions within reading frames are generally categorized as synonymous or non-synonymous.

As we make our way along the genome and pass various transcribed regions, we may notice an overlay of modifications to the DNA; regions of higher or lower methylation and histone association. These epigenetic modifications can be transient within a generation (metastable) or may be inherited across generations (Schmitz and Ecker 2012). Methylation occurs as cytosine bases adjacent to guanine (CpG sites) are modified by the addition of methyl groups. As the density of methylated sites increases, the propensity for transcription decreases. Epigenetic modifications occur within a single generation at a rate that is four times greater than genetic mutations. In general, epigenetic modifications are promoted by environmental conditions; the ability to undergo directed adaptations while growing may have been a critical advantage for plants, because their sedentary growth form limits the potential for behavioral responses to environmental challenges. We will explore this topic further later in this chapter, and when we discuss phenotypic variation in Chapter 6.

Effects of Codon Position—As mentioned previously, the consequences of mutations depend on whether the base change causes an amino acid replacement. Recall that the genetic code consists of sets of three letters (codons) that code for different amino acids. The potential effect of a DNA base change varies among codon positions (Fig. 3.1). Mutations at the first base position will always cause a change in major class of amino acids, and so are more frequently subject to selection. Differences at the second position indicate different amino acids within a class. The impact of these amino acid changes on protein structure and function is not as great, consequently, base substitutions at the second codon position are subject to selection less often than changes at the first position. Most of the redundancy in the genetic code occurs at the third position; these nucleotide changes are more often synonymous so they will have a greater chance of being silent to selection. The potential effect of non-synonymous mutations at all three codon positions will depend largely on the impact on protein structure and function; mutations in regions coding for active sites or critical structural components of proteins will be much more likely to be subject to selection.

Mutations and Substitutions—The potential for a single base-change to have an effect on the phenotype will determine its chances of persisting in the population. The per-nucleotide mutation rate (μ_T) varies across the genome and is determined primarily

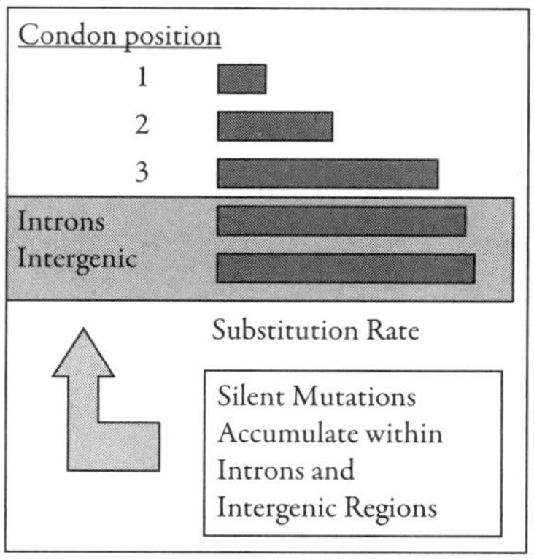

Fig. 3.1. Relative frequencies of base substitutions expected for the three codon positions, introns, and intergenic (non-functional) DNA.

by various sources of DNA damage, and the propensity for mistakes to be made during DNA replication and recombination. The *Substitution Rate* (K), on the other hand, represents the chance that mutations will persist and increase in frequency; we consider a substitution to have occurred when a new mutation has become fixed (i.e., all individuals in the population are homozygous for the mutation; $p_m = 1.0$, where p_m is the frequency of the mutation). The substitution rate reflects the chance that any single mutation will go to fixation—difference in substitution rates among genomic locations can be the consequence of variation in mutation and recombination rates (see Effects of Recombination later in this chapter) and the effects of selection.

Nearly all non-synonymous mutations are deleterious (i.e. have negative effects on fitness), so we expect that selection will rapidly eliminate nearly all that have a measurable effect on the phenotype. Consequently, we generally observe lower substitution rates for expressed mutations, and substitution rates for any particular location or region in the genome generally reflects the chance that nucleotide changes have an effect on the phenotype. This line of argument leads to the prediction that, for silent mutations, substitution rates should approach the base mutation rate (see section 3.3). The substitution rate should decline as the chance of phenotypic effects of mutations increase (i.e., substitution rates are lowest for the first base position of codons, higher for the second and third positions, and higher still for introns and intergenic regions; Fig. 3.1).

The differences in substitution rates that we observe among codon base positions and unexpressed regions are a consequence of the effects of selection on new mutations. Selection on expressed mutations is often referred to as *Purifying Selection* because the "wild type" allele will generally be favored. Consequently, expressed regions in general, and those portions of the gene that are most critical to protein function in particular, will be more conserved across generations and lineages than introns and intergenic regions. More rarely, expressed mutations have positive effects on fitness and will be favored by selection. Later in this chapter, we will discuss methods for estimating the frequency of advantageous mutations, and in Chapter 5 we discuss genomic signatures of positive selection in more detail.

3.3 ESTIMATING MUTATION RATES

Mutations are difficult to study directly because they are infrequent, and a large proportion of new mutations are lost within a few generations due to the effects of selection and genetic drift. While we cannot normally observe mutations when they first occur, it is much easier to quantify sequence differences (substitutions) between lineages. Consequently, much of the information that we have on mutation rates comes from sequence comparisons among species and lineages within species. Making suitable comparisons for the estimation of substitution rates, and ultimately mutation rates, requires identifying regions that are *Orthologous* (gene copies in divergent lineages that are derived from the same copy in the most recent common ancestor), a task that is made more difficult by small-scale and whole-genome duplication events (Gaut et al. 2011). Nonetheless, a number of useful comparisons have been made for both closely and more distantly related species of the same ploidy level. Our task is to ascertain how to use data on substitution rates to estimate the base mutation rate.

As described earlier, the fate of single-base mutations within reading frames depends largely on whether they cause changes in the amino acid sequence. Base substitutions that do not cause changes in the amino acid sequence are effectively neutral and are generally referred to as synonymous substitutions (K_S). The use of "synonymous" usually refers to changes within a reading frame, while base changes in introns and non-functional regions would be considered silent mutations. We expect the synonymous substitution rate will be higher than the non-synonymous rate (K_A) because the large majority of non-synonymous changes will be eliminated by purifying selection. Consequently, except in uncommon cases of adaptation to new gene function across a number of sites, K_A should be much less than K_S.

Underlying both of these substitution rates is the total number of mutations occurring within a particular genomic region per generation (μ_T). Some fraction of these mutations (f_0, which varies from 0 to 1.0) will occur at base positions that are synonymous, and the remainder will occur at the non-synonymous sites. For the moment, we will assume that the mutation rate is the same across all sites, so we can predict that the fraction of mutations occurring each generation that are effectively neutral to selection is the product of both these parameters ($f_0\mu_T$—the proportion of neutral mutations per genome each generation).

To understand how synonymous substitutions and mutations are related we need to explore the fate of new mutations within populations. Population processes—including the effects of selection and genetic drift—will affect the accumulation of mutations within populations over time. The mutation rate per haploid genome region (μ_T) is an average and includes both transitions (swapping a purine molecule for another purine, A ↔ G, or a pyrimidine molecule for another pyrimidine, C ↔ T) and transversions (swapping a purine molecule for a pyrimidine, or a pyrimidine molecule for a purine).

Each offspring established in a population will carry novel mutations, so the total number of mutations introduced is directly proportional to population size. Since the number of mutations entering a population depends on the mutation rate and the number of individuals present (N), we can express the total number of mutations entering a population each generation as a product of the mutation rate and the number of genome copies (i.e., $\mu_T 2N$ for a diploid population). This provides us with the simple prediction that larger populations will contain more new mutations each generation than small populations (Fig. 3.2).

The fate of synonymous mutations is also a function of population size. Each new mutation entering a population occurs in one genome, so it will exist as a single copy and will be carried by a single individual in the heterozygous state (e.g., AA', where allele A' is a new mutation). The probability that a new mutation will become fixed by genetic drift directly corresponds to its frequency; any neutral mutation will have a probability of $1/2N$ to reach a frequency of 1.0 (because there is one copy of the mutation out of $2N$ genomes in the population), and once fixed, it will contribute to the synonymous substitution rate.

By combining these two observations—$2N$ as the number of mutations, and $1/2N$ as the chance each one will go to fixation—we can predict the total substitution rate (K) for any genomic region that contains synonymous and non-synonymous sites as

$$K = (f_0\mu_T 2N)(1/2N),$$

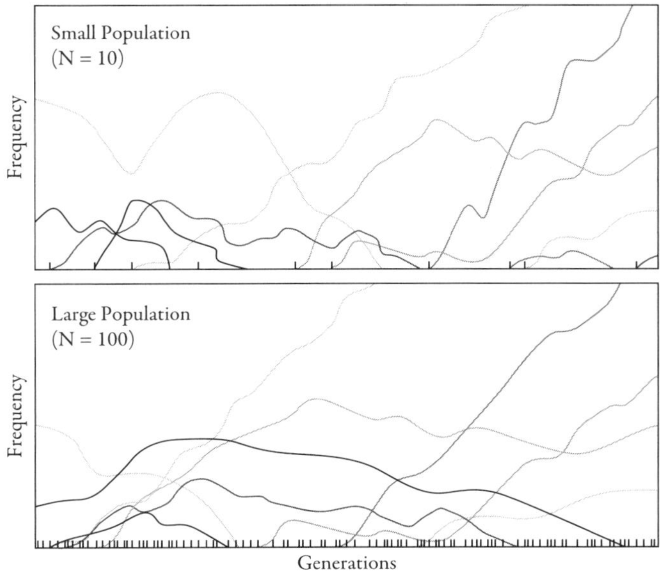

Fig. 3.2. Fates of mutations in small and large populations. Each tick mark on the baseline is a single mutation. Ten times more mutations enter the population when N = 100, but only a small frequency persist more than a few generations. The same numbers of mutations become fixed in both populations.

and we can simplify this equation by eliminating $2N$, in which case

$$K = f_0\mu_T$$

In other words, the substitution rate (K) is equal to the frequency of neutral mutations ($f_0\mu_T$). This relationship is quite powerful as it allows us to predict K for different genomic regions without having to consider the effects of population size, because the same number of mutations will become substitutions over time regardless of the number of individuals (Fig. 3.2). As the proportion of the region that is non-synonymous and silent increases, f_0 declines and consequently K is lower (see Box 3.1 for an example and further discussion). For non-functional sequence regions, f_0 is equal to one (all mutations are synonymous or silent), and we expect the substitution rate for the entire sequence to be the same as the mutation rate ($K = \mu_T$; and in general, $K_S = \mu_T$).

The utility of this relationship is that K_S provides us with an estimate of base mutation rate, so we can use it to standardize the non-synonymous substitution rate (K_A) for comparing rates of protein evolution. This is necessary because, as we will learn in the next section, mutation and substitution rates can vary substantially among sites, genomic regions, and lineages.

3.4 SUBSTITUTION RATES

There are substantial differences in evolutionary rates among the nuclear and organelle genomes of plants and among genes within these genomes (Wolfe et al. 1987). On the average, plant mitochondrial genes evolve the slowest—roughly six times slower than plant nuclear genes and three times slower than chloroplast genes (Gaut et al. 2011). The substitution rate in nuclear genes is similar between plants and animals, but the rate of mtDNA evolution is much higher in animals than the nuclear rate in plants and animals. The slower rate of evolution for plant mitochondrial (mtDNA) and chloroplast (cpDNA) genomes is the opposite of what we would expect; the generation of reactive oxygen species within these organelles should increase rates of DNA damage and consequently the rates of mutation. While the lower rate of evolution for plant organelle genomes is widely recognized, there are no obvious explanations for these mutation rate differences.

The one overriding factor that may have the largest influence on the evolution mutation rates among genes and genomes is the average fitness effects of base substitutions; we expect that genomic regions where sequence changes are more likely to cause severe fitness losses will have the lower overall mutation rates. On the other hand, mutations are also important for the generation of novel genetic variation and adaptation to challenging environments, so we might expect the evolution of mutation rate to reflect the sum effects of the potential for positive and negative consequences of base changes. We also recognize that the large majority of mutations—including synonymous and silent changes—are neutral to selection, but as we will see later, their chances for fixation may be governed by linkage with sites that are subject to selection. In the remainder of this section we will discuss variation in substitution rates starting with differences among sites, then among genes, and finally among divergent lineages.

Site Effects—Substitution rates vary substantially among nucleotide sites for both functional and non-functional genomic regions. First, as described earlier, purifying selection reduces substitution rates at the first and second codon positions to a greater extent than the third position (Fig. 3.1). Second, there is a bias in substitution rates that depends on nucleotide base composition. In particular, cytosine and guanine bases undergo transitions (C → T and G → A) at rates that are ~1.6 times higher than transitions in the opposite direction (Gaut et al. 2011). To some extent the higher substitution rates are associated with levels of methylation, but more frequent transitions for G and C sites may also be due to greater sensitivity to UV damage (Ossowski et al. 2010). The third general pattern is the tendency for substitution rates to be higher near sites that have been subject to insertions or deletions (indels). The site-specific differences in substitution rates associated with G and C bases and indels seem to be due to biases in the base mutation rate since these patterns persist across both functional and non-functional genomic regions (Gaut et al. 2011).

Variation among Genomic Regions—When we expand our view to focus on stretches of DNA we find large amounts of variation in divergence rates (substitutions) among genomic regions. We can

Box 3.1 Mutations and Substitutions

The mutation rate per haploid genome (μ) can be estimated from the synonymous substitution rate as described in the text (i.e., we expect that $K_S = \mu$). This relationship holds because the effects of population size on the number of new mutations produced each generation is cancelled out by the chance that any one mutation will increase in frequency until it becomes fixed. For example, say we had a population size of $N = 100$ and a known mutation rate of $\mu = 0.5$ per genome/generation. If we assume that we are working with an annual species with constant population size then we expect that, on the average, each individual will be carrying one novel mutation and a total of $\mu 2N = 100$ new mutations will be present in the population each generation. If these mutations are completely neutral to selection, then each one of them has a probability of $1/2N = 0.005$, or 0.5% chance of becoming fixed. This is not to say that any new mutation will go to fixation, but if we average the effects of population size across all genomes present then we find that the effects of population size cancels out. Consequently, the number of synonymous substitutions that we observe when we compare orthologous genes from divergent lineages can provide a reliable estimate of the base mutation rate.

The power of the synonymous substitution rate (K_S) is derived from its potential to standardize the non-synonymous rate (K_A). As explained in the text, mutation rates vary across the genome, so robust comparisons of the effects of selection on rates of evolution requires that we scale K_A with K_S for the same genomic region.

To calculate these substitution frequencies, we could compare a gene sequence in a derived lineage to a known ancestral sequence (Fig. 3.3). In this example, we count up the number of possible synonymous base changes according to how many will not change the amino acid sequence. In the first codon, three of three possible base changes at the third position will be synonymous ($3/3 = 1$). In the second codon zero of three third base changes are synonymous ($0/3 = 0$), and in the third codon one of three changes in the third base are synonymous ($1/3$), etc. Summing across these sites, we note that there are two possible base changes that are synonymous for this sequence ($3/3 + 0/3 + 1/3 + 1/3 + 1/3 = 6/3 = 2$). Similarly, there are 13 possible non-synonymous base changes ($[11 \times 3/3] + 2/3 + 2/3 + 2/3 = 39/3 = 13$). Now we can compare actual substitutions to the number of possible changes. In our derived sequence we have one synonymous substitution (GTC → GTA) out of two that are possible, so $K_S = 1/2 = 0.5$. We also see one non-synonymous substitution (AAA → AAC; lysine changes to asparagine) out of 13 that are possible, so $K_A = 1/13 = 0.077$. Based on these estimates we can calculate the ratio of non-synonymous to synonymous base changes; $K_A/K_S = 0.077/0.50 = 0.154$.

	Amino Acids				
Ancestral	Val	Met	Cys	Lys	Glu
Sequence	GTC	ATG	TGT	AAA	GAA
Degeneracy	**004**	**000**	**002**	**002**	**002**
Synonymous	003	000	001	001	001
Non-Synon	330	333	332	332	332
Substitutions	333	333	333	333	333
Derived	↓			↓	
Sequence	**GTA**	**ATG**	**TGT**	**AAC**	**GAA**
	Val	Met	Cys	Asn	Glu

Fig. 3.3. Calculation of the number of possible synonymous and non-synonymous substitutions, and the actual substitutions observed in a derived sequence.

In practice, we often do not have information about the ancestral sequence, but we can compare orthologous regions from two or more lineages to obtain a sum of substitutions. For an accurate assessment of K_S and K_A we should first infer the ancestral sequence and then count up the substitutions in each lineage. Ancestral conditions can generally be inferred from the phylogeny of multiple species. More often than not, we are primarily interested in the ratio of K_A to K_S so the extra step of comparison to a putative ancestral sequence is not necessary; we need to just estimate total frequencies of both substitution types across all the sequences and then calculate the ratio. The actual substitution rates are not necessarily as straightforward as described above and can be affected by other factors such as the G + C content and transition bias. A number of inference methods have been developed for the estimation of K_A/K_S and these have been implemented in a number of software applications (Yang and Bielawski 2000).

estimate divergence rates between two or more lineages by counting the number of single-base sequence differences for orthologous regions from a pair of species or lineages within a species. Of course, more distantly related taxa tend to have more sequence differences (substitutions and indels), but we also find substantial variation in the frequency of substitutions among regions within the genome. As described earlier, when we are comparing orthologous reading frames, these base substitutions can be divided up into synonymous (K_S) and non-synonymous (K_A) changes. Levels of synonymous divergence among nuclear genes in a single genome can vary as much as ~14 fold, but generally fall within a narrower range of ~2.6 fold (Gaut et al. 2011). As you might expect, the non-synonymous rate of evolution is generally lower than K_S, but the breadth of variation is much greater, as its range is three to five times higher than that of the synonymous rate. The higher variance in K_A is not surprising since these base changes can be affected by selection, while K_S should only be affected by genetic drift (but see the discussion below on the effects of linkage). Many non-synonymous changes will be rapidly eliminated by purifying selection, and a few may become fixed if they are neutral, or in the rarer case that the mutation proves to be advantageous. The high variance in K_S across gene regions may reflect differences in the base mutation rate; as discussed earlier, mutation rate differences among genes can be explained by higher G and C content. On the other hand, a large proportion of the variance in evolutionary rates across chromosomes may be more difficult to explain—we will explore this topic further in the next section.

Variation among Genes—Since the base mutation rate places an upper limit on the rate of protein evolution it is not surprising that K_A and K_S are positively correlated across the genome (Gaut et al. 2011). If variation in μ is the primary factor that governs differences in substitution rates among genes and the mutation rate varies across genomic regions, then the mutation rate needs to be taken into account when interpreting differences in the rate of non-synonymous substitution (protein evolution). We cannot make direct estimates of the mutation rate, but based on the argument outlined above, K_S should provide a valid estimate of μ_T. By using K_S as a scaling factor, we can make more robust comparisons of the evolutionary rates of amino acid substitutions. Consequently, we generally discuss rates of protein evolution in terms of the ratio of K_A to K_S ($K_A/K_S = \omega$ and is often referred to as *dN/dS*; Yang and Bielawski 2000).

As you might expect, K_A is typically much lower than K_S so ω averages around 0.2, which indicates that around 80% fewer non-synonymous mutations become fixed within lineages compared to synonymous base changes. If a particular gene region were subject to recurrent fixation of non-synonymous changes because of positive selection, then K_A could exceed K_S resulting in a $\omega > 1.0$. At first glance, the idea that non-synonymous substitution rates could exceed synonymous rates seems counterintuitive, since K_S should reflect the base mutations and limits the rate of evolution; however, recall that under drift, each new mutation has only a $1/2N$ chance of becoming fixed in a population (the neutral expectation). Consequently, the large majority of neutral mutations are lost from the population within a few generations. Typically, non-synonymous mutations are deleterious or neutral so the proportion that contribute to the substitution rate is much less than $1/2N$, but if a mutation were advantageous, then the chance of fixation under selection would be much higher than the neutral expectation. If a number of advantageous mutations are concentrated in a particular gene region—say a new gene copy that is under selection to fulfill a novel function—then K_A could exceed K_S, but such events seem to be rare or difficult to detect.

Using K_S to standardize rates of non-synonymous substitution for comparison across genomic regions requires that synonymous changes should reliably reflect the mutation rate—in other words, that they are affected only by genetic drift and not selection. However, there are a number of circumstances where the levels of polymorphism for silent substitutions can be affected by selection, so the assumption of neutrality may not hold true.

One example of selection on synonymous base changes is codon usage bias: some codons for the same amino acid have greater translational accuracy and are weakly favored by selection (Gaut et al. 2011). However, the effect of codon bias is very subtle and is expected to influence substitution rates only over long periods of evolutionary time—across phylogenetic time scales.

In contrast to codon bias, selection on linked non-synonymous changes has much more potential to affect levels of synonymous divergence at shorter time scales (Cutter and Payseur 2013). For example, rapid increase in the frequency of advantageous mutations in response to selection (*Hard* and *Soft Selective Sweeps*; Chapter 5) results in *Genetic Hitchhiking* and reduces levels of polymorphism in adjacent regions (Fig. 3.4; refer to color plate).

Selective sweeps are uncommon, but gene regions are more frequently subjected to selection due to repeated deleterious mutations (*Background Selection*; Fig. 3.4; refer to color plate). Removal of deleterious mutations by selection is expected to reduce levels of polymorphism at linked neutral sites (Charlesworth et al. 1993). This reduction of regional polymorphism due to background selection is also a consequence of genetic hitchhiking, but in this case, neutral mutations are effectively lost due to the fixation of a smaller proportion of the total variation present. Background

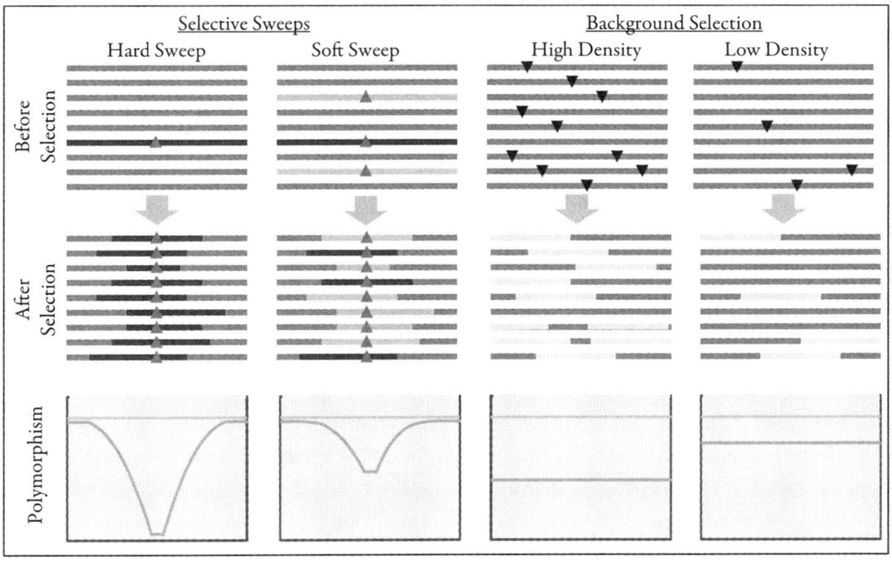

Fig. 3.4. Effects of selection on nucleotide variation at synonymous and silent sites across a genomic region. Selective sweeps due to advantageous mutations (orange triangles) reduce variation in adjacent regions. Hard sweeps (new mutations) reduce polymorphism more than soft sweeps (from standing genetic variation). The effect of background selection on levels of polymorphism at neutral sites depends on the density of deleterious mutations. Recombined regions are the light green sections of the sequence. Adapted from Cutter and Payseur (2013).

selection has the effect of eliminating closely linked neutral variation as recombination facilitates the removal of genomic regions containing deleterious mutations. As the density of deleterious mutations increases, the level of polymorphism across the entire genomic region declines (Fig. 3.4; refer to color plate).

The combination of direct selection (due to codon bias) and *Associative Selection* (including selective sweeps and background selection) reduces the level of neutral polymorphism and ultimately the rate of synonymous substitution. The intensity of selection may vary among genes because of differences in the fitness consequences of amino acid substitutions, so we expect variation in the intensity of background selection among gene regions. Consequently, the rates of synonymous and non-synonymous substitution will tend to be correlated. The first implication of this revelation is that, unless we are examining pseudogenes or intergenic regions that are far removed from functional genomic regions, K_S will chronically underestimate μ. Because of the effects of associative selection on synonymous substitution rates, obtaining estimates of K_S that accurately reflect base mutation rates is intrinsically difficult (e.g., DeRose-Wilson and Gaut 2007).

The second implication of the effects of associative selection on levels of polymorphism is the possibility that ω (K_A/K_S) will be overestimated—lower K_S could produce inflated values for rates of protein evolution compared to estimates we might obtain if non-synonymous changes were standardized by the actual base mutation rate. At first glance, this seems problematic for comparison of evolutionary rates of proteins across genomic regions; however, it is important to note that associative selection reduces levels of polymorphism across the entire region, so there will generally be fewer mutations available for the forces of selection and genetic drift. If the overall effect of associative selection equally reduces the number of synonymous and non-synonymous mutations entering the population then ω may provide adequate estimates of the rate of protein evolution for comparisons among genes.

Effects of Recombination—The influence of associative selection on synonymous substitution rates depends heavily on the rate of recombination. Genome regions with low rates of recombination will more often be subject to *Selective Interference* due to linkage among advantageous and deleterious mutations, and neutral mutations will be more likely to be swept to fixation or extinction via hitchhiking in response to selective sweeps or background selection (Fig. 3.5; refer to color plate). As the recombination rate declines, larger genomic regions are subject to associative selection, so the synonymous substitution rate declines. In general, we can interpret the association between higher rates of recombination and higher estimates of synonymous substitution as a consequence of the alleviation of the effects of associative selection.

An alternative hypothesis for the positive correlation between K_S and the recombination rate posits that crossing over has a direct effect on the mutation rate due to the effects of chromosome breakage (Pal et al. 2006). If this were the case then increased recombination would be directly responsible for higher rates of mutation and consequently increased levels of K_A and K_S, but closer examination of patterns of mutation has led to the conclusion that, except for some specific circumstances, evidence for increased mutation rates due to the direct effects of crossing over is lacking (Webster and Hurst 2012). While there may be some small effect of crossing over on mutation rates, it is clear in most cases that the observed increase in K_S with higher rates of recombination is more likely a consequence of the disruption of associative selection.

A reasonable conclusion is that higher rates of synonymous substitution in regions with more recombination is due to the reduction in the influences of selective interference and hitchhiking, which allows the effects of selection to be focused on smaller genomic regions. As neutral mutations are freed from the effects of associative selection, their fates are more likely to be governed by genetic drift, so rates of synonymous substitutions will increase. With the base mutation rate held constant, we expect that higher rates of recombination will increase the efficacy of selection and consequently rates of protein evolution will be higher.

3.5 PROTEIN EVOLUTION

The complement of proteins expressed during the life cycles of plants ranges from essential components of metabolic pathways to enzymes with more specific

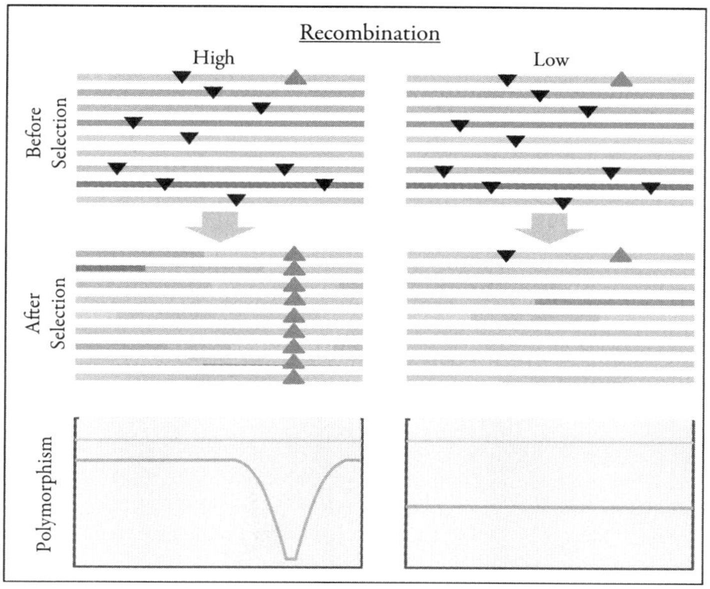

Fig. 3.5. The effects of recombination on levels of polymorphism. High recombination breaks up associations and allows a beneficial mutation (blue triangle) to go to fixation (selective sweep). With low recombination larger genomic regions are lost due to background selection and the beneficial mutation remains at low frequency due to selective interference.

functions that are expressed in a limited range of tissues or under specific conditions. The more essential a protein is to cell function, the more likely it is to have a core function in metabolic pathways and to interact with larger numbers of other proteins (Fig. 3.6). These essential proteins generally have higher levels of expression in a broader range of tissue types and display slower rates of evolution (Wright et al. 2004; Gaut et al. 2011). Proteins with higher levels of expression are generally coded in genomic regions of low recombination and consequently have lower synonymous and non-synonymous substitution rates. The fact that many of these essential proteins tend to be conserved across highly divergent lineages suggests that their observed lower rates of substitution is a consequence of selection; perhaps because they generally have earlier phylogenetic origins, the chance that novel mutations could improve their structure or function is greatly reduced. If this were true then we would expect that nearly all non-synonymous changes in essential proteins would be deleterious, and selection would favor reduced mutation rates for essential genes.

At present, our understanding of the primary drivers for the rate of protein evolution is imperfect, and the strong correlations among the characteristics associated with variation in amino acid substitutions renders the separation of causal factors difficult (Fig. 3.6). While the essential function, number of interactions, and expression level often top lists of influential correlates of protein evolution, a number of studies have identified expression level as the most important determinant of low substitution rates (Pal et al. 2006). This observation has led to speculation that transcriptional and translational fidelity are more critical for proteins that are highly expressed. In particular, misfolding can lead to protein aggregation and toxicity—a problem that is amplified for proteins that are in high abundance within cells. On the other hand, it is notable that the machinery involved in transcription and translation is distinct from DNA replication, and error rates in translational processes are considerably higher (Lynch 2010). Given the separation of these cellular processes, it is unlikely that transcription and translation fidelity are directly responsible for the observed lower substitution rates for genes with high expression levels. The effects of translational fidelity, functional roles, and expression levels on rates of protein evolution remain speculative, and many more studies will be necessary before general patterns emerge—much work remains to be done to

Rate of Protein Evolution	Faster	Slower
Expression Level	Low	High
Number of Interactions	Low	High
Essential Function	Low	High
Recombination Rate	High	Low

Fig. 3.6. Proteins with higher expression levels and more interactions with other proteins tend to have essential functions, to be coded by genes in regions of low recombination, and to evolve at slower rates.

improve our understanding of patterns of processes of protein evolution.

Substitution Rates and Protein Evolution—Much of the observed variation in rates of evolution among proteins is probably due to differences in the recombination rate, and its consequences for associative selection, rather than actual differences in the base mutation rate among genomic regions. It may be the case that mutation rates are similar across genomic regions, and the variation that we see in K_S is more likely due to regional differences in recombination. The main factors that could have direct effects on the base mutation rate are G and C content, levels of methylation, and the frequency of indels. We expect that rates of synonymous substitution will generally be affected by associative selection due to selective sweeps and genetic hitchhiking, which compromises the usefulness of K_S for estimating regional mutation rates. While K_S may generally be an imperfect indicator of mutation rates, it still can be used as an adequate control for rates of protein evolution if we accept that it reflects the amount of mutational variation that is made available for the influence of selection. Consequently, at the present time, ω (or dN/dS) is the comparator of choice for discussions of protein evolution within and among species (Pal et al. 2006; Gaut et al. 2011).

Examination of the general patterns of variation in rates of protein evolution leads us to the conclusion that it would be advantageous to organisms if regional genomic mutation rates were amendable to the influence of selection. On the other hand, the magnitude of the potential effects of selection on the frequency of random base changes among genomes and genomic regions remains to be determined. As we will discuss in the next section, the frequency of random base changes does appear to differ among lineages, but other characteristics such as population size and life history traits may place limits in the evolution of mutation rates.

3.6 VARIATION AMONG LINEAGES

It is clear from the discussion in the previous section that the mutation rate places an upper limit on substitutions and ultimately on the rate of protein evolution. While gene flow may introduce variation that is already present in other populations, mutation is the only source of novel genetic variation that contributes to the evolution of improved protein function and has the potential to promote the origin of enzymes that might allow the exploitation of novel resources. On the other hand, random genetic change in DNA sequences is a double-edged sword; most mutations are deleterious and are expected to contribute to the *Genetic Load* of a population. While many of these mutations are recessive, inbreeding within populations increases homozygosity, which exposes recessive deleterious mutations and reduces fitness (see Chapter 4 for more discussion of the effects of inbreeding). Populations that have higher mutation rates will accumulate advantageous mutations, but they will also be segregating for larger numbers of deleterious mutations. Surveys across organisms indicate that the mutation rate can vary widely (Fig. 3.7), so we might expect that it could be subject to evolutionary processes. For example, individuals from different populations and species may differ with

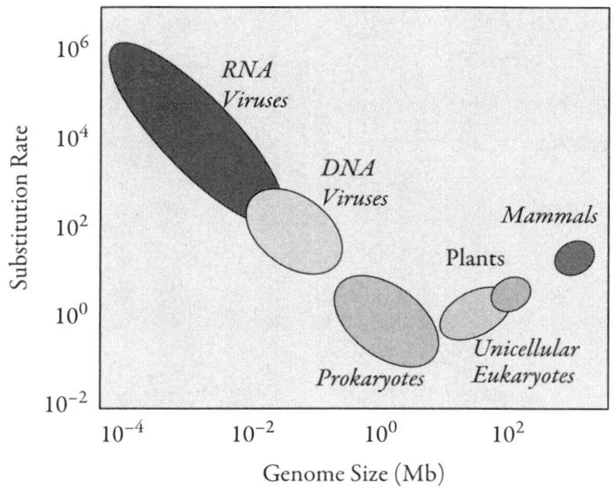

Fig. 3.7. The relationship between genome size and substitution rate (per 10^9 bp/generation) across the diversity of life. Invertebrate animals are included with one estimate for plants (*Arabidopsis*). Adapted from Lynch (2010).

respect to their susceptibility to DNA damage and the efficiency of DNA repair. Our goal here is to understand the biological and evolutionary processes that govern the accumulation of genetic changes within populations.

Mutation rates vary across organisms by several orders of magnitude from extremely high levels in RNA viruses to much lower rates in Archaea, Eubacteria, and eukaryotes (Lynch 2010; Fig. 3.7). If we just focus on viruses and bacteria, we see the mutation rate decreasing with increasing genome size. This trend is generally explained by selection on viruses to rapidly respond to host defenses, which is enabled by extremely high reproduction rates that are coupled with a high rate of sequence replication error. In other words, viruses do not suffer from increased genetic load as a consequence of high mutation rates because of the large number of offspring that are produced each generation—only a small fraction need to survive to sustain the population.

On the other hand, if the analysis of mutation rate is restricted to cellular organisms—from prokaryotes and unicellular eukaryotes to mammals—then the opposite trend is observed; base mutation rate increases with genome size (Fig. 3.7). One major difference between prokaryotes and eukaryotes is that genomes of the latter group contain much more repetitive DNA, which is subject to more errors during replication so we expect the mutation rates in eukaryotes will generally be higher.

The other major difference across the cellular species in Fig. 3.7 is ploidy; prokaryotes and some unicellular eukaryotes are haploid, while the multicellular species are diploids. Since mutation rate estimates are based on detectable changes in functional regions of the genome, new mutations will contribute to substitution rates only if they persist within populations. We expect that a large portion of the differences between haploid and diploid organisms can be explained by the phenotypic effects of new mutations. Haploid unicellular organisms typically reproduce by cell division and accumulate mutations during mitosis. Any cell lineage possessing a deleterious mutation would have lower rates of replication and would be unlikely to persist, so those mutations would be unlikely to be observed. As deleterious mutations are lost, the overall level of polymorphism among individuals is reduced across the entire genome because there is no recombination during cell division. In other words, for asexual organisms, associative selection affects the entire genome and genetic variation is lost from populations as selection intensity increases. Excepting for occasional conjugation among asexual cell lineages, we expect prokaryotes to have lower substitution rates.

Diploid species, on the other hand, may sequester larger numbers of recessive mutations in their genomes due to the effects of dominance and recombination. Recessive variation is sequestered in diploid lineages and recombination allows deleterious mutations to be selectively removed while maintaining higher levels of polymorphism across the genome. While these differences in genome structure explain a portion of the variation in detectable mutations, the difference across diploid multicellular eukaryotes is nearly two orders of magnitude, so there may be other factors at play that affect the evolution of mutation rates.

Evolution of Mutation Rates—If we view mutation rate as a trait that may respond to selection then we can gain some insights into the pressures and limitations on the evolution of the frequency for DNA replication errors. One approach is to assume that sequence errors are a side effect of selection for rapid DNA replication (Baer et al. 2007). The assumption here is that high fidelity is not possible when DNA replication is rapid (*Fidelity/Replication-Rate Trade-Off Hypothesis*). If this were the case then we would expect lineages with larger genomes to have higher mutation rates because selection for rapid cell replication would outweigh selection for high fidelity. The high mutation rates that are associated with exceptionally large mitochondrial genomes in some species of the plant genus *Silene* (catchfly; Sloan et al. 2012) would seem to be consistent with this hypothesis; however, in this case it is also possible that the elevated mutation rates occurred first and contributed to the accumulation of noncoding DNA.

The fidelity/replication-rate trade-off hypothesis seems to explain a good proportion of variation in mutation rates. On the other hand, Lynch and others have argued that replication fidelity is generally not an adequate explanation for higher mutation rates with larger genome sizes because rapid replication in eukaryotes is enabled by the division of the genome into multiple chromosomes (Lynch 2010). Compared to prokaryotes, large-genome species could undergo rapid cell division without a loss of fidelity because DNA replication occurs in parallel across all of the chromosomes. At present, evidence is inconclusive, and the fidelity/replication-rate trade-off hypothesis remains an active topic of discussion.

The second hypothesis that has been promoted to explain variation in mutation rates across lineages assumes persistent directional selection for lower mutation rates, but its effectiveness is offset by the counterbalancing influence of genetic drift (Lynch 2010). In this case we assume that genome size scales with the *Effective Population Size* (N_e—see Chapter 4), with microorganisms having extremely large populations and multicellular organisms having much smaller effective population sizes. Lynch argues that there will always be selection to reduce the DNA replication error rates through mutations that improve fidelity of the polymerase enzyme or increase the efficiency of repair mechanisms. However, as N_e declines, the effects of selection are compromised, and the effects of drift dominate the fates of mutations that represent incremental improvements in repair efficiency. This *Drift-Barrier Hypothesis* adequately predicts mutation rates across prokaryotes and eukaryotes, among mammals that vary in genome size (Lynch 2010), and for some comparisons among plants (e.g., among species of checker mallow; Andreasen and Baldwin 2001). On the other hand, it is often difficult to obtain reliable estimates of the effective population size, so a limited amount of data is available to assess the veracity of the drift-barrier hypothesis for the evolution of mutation rates.

Another general pattern that emerges from comparisons within groups of multicellular organisms is the tendency for long-lived species to have much lower rates of protein evolution. The *Generation-Time Hypothesis* was first proposed based on observations across species of animals, but the same general trend appears to hold for plants (Gaut et al. 2011). For example, within the monocots both bromeliads and palms have much slower substitution rates than the grasses. Similar patterns—with higher substitution rates for annual species—were found in a broad survey across the eudicots (Smith and Donoghue 2008). In a broad genomic survey across a large number of genes, rates of evolution were much faster for two annual species (*Arabidopsis thaliana* and *Medicago trunculata*) compared to grapes (*Vitis vinifera*) and poplars (*Populus trichocarpa*) (Yue et al. 2010). This last example is particularly interesting because both of the annual species used in this survey habitually self-fertilize, so their effective population sizes should be very small—consequently, this result is also consistent with the drift-barrier hypothesis.

The pattern that is evident from these surveys is that protein evolution rates are consistently low for long-lived woody plants but range widely among herbaceous species (Smith and Donoghue 2008). It is notable that mating systems also vary to a much greater extent across herbaceous plants, while nearly all woody species are outcrossers (Barrett and Eckert

1990; Chapter 9), so it is possible that levels of inbreeding may have consequences for the evolution of protein evolution.

The generation-time hypothesis appears to have the greatest explanatory power for variation in substitution rates among species across both plants and animals; however, the process of growth and reproduction is intrinsically different for plants. In fact, as we will see later in this chapter, an evaluation of the process of mutation accumulation during the plant life cycle would predict the opposite pattern: long-lived species have much greater potential for mutation accumulation during somatic growth so substitution rates should be much higher. To understand the basis of this puzzle we need to take a closer look at the processes of plant growth and reproduction, which is a topic we will take up later in this chapter (see section 3.10).

3.7 EPIGENETIC VARIATION

Changes in nucleotide sequences have traditionally been identified as the sole source of heritable variation, but this view has changed in the last few decades with the realization that epigenetic modifications of DNA control a substantial portion of heritable phenotypic variation in plants (Kalisz and Purugganan 2004). Epigenetic variation is broadly defined as induced phenotypic variation that is not associated with changes in nucleotide sequences. The best studied form of epigenetic modification is the addition of methyl groups to cytosine-guanine pairs that are separated by a single phosphate molecule (CpG sites). In plants, the range of methylation sites is much broader and includes CpNpN sites (where N can be A, T, G or C; Angers et al. 2010). Epigenetic modifications can also be due to associations with histone proteins and can be induced or maintained by the action of small RNA molecules that promote methylation activity (Law and Jacobsen 2010). Transcription is repressed in regions with higher densities of methylation and histone association, so epigenetic modifications alter phenotypes through gene expression rather than changes to protein structure.

Phenotypic changes controlled by methylation and histones are sometimes referred to as epialleles. In animals the epigenome is reprogrammed during reproduction so *Transgenerational Epialleles* tend to be uncommon, while in plants the large majority of phenotypic variation under epigenetic control can be transferred from parent to offspring (Becker and Weigel 2012). Along with somatic mutations, epigenetic modifications in plants may contribute to the heritable variation acquired during vegetative growth, but in some cases epigenetic marks are effectively fixed, and are responsible for major phenotypic differences in floral morphology (e.g., the peloric epimutant of common toadflax; Fig. 11.2; refer to color plate).

Epigenetic modifications differ from genetic mutations in two important ways. First, phenotypic changes can be induced during the lifespan of an individual in response to environmental cues—stress in particular (Angers et al. 2010; Grativol et al. 2012). Epigenetic responses to stress from pathogens, herbivores, or abiotic conditions can be directed to produce adaptive phenotypes that are passed on the next generation. These transgenerational epigenetic adaptations are particularly important for plants that have limited seed dispersal since their offspring are likely to experience the same environment as the maternal plant.

The second way that epigenetic modifications differ from genetic changes is that epigenetic responses are transient; they may be reversible within an individual's lifetime or within a few generations (Law and Jacobsen 2010). If an environmental challenge continues then epigenetic modifications will be maintained, but if stress is reduced in subsequent generations than methylation patterns will return to their original condition (Holeski et al. 2012). Thus, epigenetic modifications provide a mechanism to improve the fitness of offspring experiencing the same environment as the parent and allow the flexibility for future generations to respond as conditions change. As epigenetic changes can be directed and are reversible, they represent *Phenotypic Plasticity* responses that may facilitate tolerance to biotic and abiotic challenges across generations.

A small number of studies have evaluated the potential for epigenetic modifications to control adaptive phenotypic responses under field conditions. For example, much of the success of the highly clonal invasive plant *Fallopia japonica* (Japanese knotweed) appears to be due to epigenetically induced changes in phenotypes (Richards et al. 2012). In a separate case, differences in herbivory levels recorded over twenty years for a species of violet (*Viola cazorlensis*) varied substantially among individuals and populations and were strongly associated with a small number of methylation sites (Herrera and Bazaga 2011).

The significant associations found in both of these studies suggest a causative relationship between patterns of DNA methylation and adaptive phenotypes, but this hypothesis was not directly tested in either case. A more robust test of the adaptive value of environmentally induced responses was made by manipulating the light environment of maternal plants of the understory herbaceous plant, *Campanulastrum americanum* (American bellflower; Galloway and Etterson 2007). Offspring from these plants exhibited transgenerational plasticity responses—presumably based on epigenetic modifications—that caused changes in the life history of offspring (annual versus biennial) and elevated their fitness when grown in the same light environment as the maternal parent. The results from these studies shed light on an unexpected revelation; epigenetic modifications can provide flexible frameworks that promote adaptive transgenerational plasticity responses and may control substantial portions of phenotypic variation in plants.

Exposure of plant populations to different levels of environmental heterogeneity may have consequences for the adaptive value of the transmission of epigenetic marks between generations. We expect that epigenetic transmission would not be favored in environments where the predictability of environmental conditions across generations is low. The transmission of methylation patterns between sporophyte and gametophyte (pollen) generations has been found to be high (around 75%) for plants growing in controlled growth conditions. If transfer of methylation patterns to offspring can respond to selection then we expect to see variation among populations for transmissibility. In natural populations of stinking hellebore (*Helleborus foetidus*) there were substantial differences in transmissibility among populations found in different regions in Spain (Herrera et al. 2013). This suggests that there is variation in the transmissibility of epigenetic modifications that could be subject to selection. To this point, we have only small vignettes on the evolutionary consequences of the epigenome, and we are just beginning to understand its implications for phenotypic variation, its adaptive capacity, and the potential for the evolution of epigenetic transmissibility.

3.8 GENOME DUPLICATIONS AND REARRANGEMENTS

Diversification of life on earth has proceeded through the acquisition of unprecedented structures and functions that enabled novel modes of resource utilization as organisms exploited new niches. Such transitions required creation of completely new protein functions. Up to now, we have discussed the generation of variation that modifies the expression patterns or functions of existing proteins, but the potential for these refinements to generate novel functions may be limited. The emergence of novelty in evolution is facilitated by a simple principle; catalysts are usually capable of promoting multiple chemical reactions. Catalytic proteins (enzymes) coded in plant genomes are no exception; enzymes often have catalytic ability for more than one reaction—a property that renders cell metabolisms intrinsically redundant. Duplication of reading frames that code for enzymes capable of catalyzing multiple reactions facilitates specialization for the refinement of alternative biochemical pathways or the generation of novel functions (Conant and Wolfe 2008). These duplication events have the potential to generate novel phenotypes by the co-option of gene copies for the evolution of new products. Thus, when we consider the *Multicatalytic Capability* of enzymes combined with *Gene Duplication*, a simple paradigm emerges for understanding the generation of novel protein functions and ultimately for phenotypic traits that enable major evolutionary transitions.

In this section we will discuss major genomic aberrations including *Small-Scale Duplication (SSD)*, *Whole-Genome Duplication (WGD; Polyploidy)*, and chromosomal rearrangements (*Translocations, Inversions, Fusions*, and *Fissions*). Duplication events may generate the potential for the evolution of novel phenotypes through gene duplication, while rearrangements contribute to diversification through the creation of novel chromosomal lineages. These events differ from the base substitutions and small indels that were discussed above because they involve genomic regions that are large enough to contain one or more reading frames. At the same time, all of these mutational events share the same population dynamics when they first appear—there is a single copy of the new variant in one individual. Just as with point mutations, events that involve larger portions of the genome will generally be lost within a few generations or may increase in frequency through drift or the action of selection. On the other hand, alterations that involve larger portions of the genome may create lineages with low fertility or may result in reproductive isolation in a single generation. To understand the implications of major genomic aberrations we will first discuss their consequences for meiotic irregularities and their potential for persistence in populations.

Origins of Cytogenetics—Studies of chromosome morphology and behavior—*Cytogenetics*—was an active field research over much of the last century. The field of cytogenetics finds its beginning in the mid-nineteenth century when chromosomes were first observed by the Swiss botanist Karl (Carl) Wilhelm von Nägeli; later, in 1882 their behavior during mitosis in salamanders was described by Walther Flemming. In the early twentieth century, the correspondence between chromosomes and genes became widely accepted, and the field of cytogenetics expanded as methods for the characterization of the *Karyotype* was established. Beginning with the work of Gregor Mendel in the late nineteenth century, studies of the inheritance of traits—the field of modern genetics—developed in parallel, but not independently from, cytogenetics.

While it was generally recognized that chromosomes contained genes, there are no generalities concerning the correspondence between the morphological features of chromosomes and phenotypes. For example, heterochromatin banding, translocations, inversions, and changes in DNA content are not consistently associated with detectable phenotypic differences (Stebbins 1971; Levin 2002; Young et al. 2012). On the other hand, there are some well-documented examples of chromosomal aberrations that are associated with pigment variation in corn (*Zea*) and tobacco (*Nicotiana*), and morphological characters in *Oenothera* (Grant 1978). While linkage and recombination were well understood from studies of inheritance, associations between chromosome behavior and patterns of inheritance were often difficult to make. For example, the observation of a correspondence between crossing over (formation of *Chiasma*) and recombination of traits was first made in corn by Harriet Creighton and Barbara McClintock in 1931 (Creighton and McClintock 1931).

When new translocations or duplications occur they occasionally result in large-scale differences in the morphology of individual offspring. For example, observations of novel variants within a population of *Oenothera lamarkiana* in the late nineteenth century inspired the Dutch botanist Hugo de Vries to refer to these novelties as mutants (the origin of the term mutation). Investigation by de Vries on phenotypic variation in *Oenothera* became the foundation of his mutation theory, which was promoted as a challenge to Darwinian gradualism in the waxing years of the twentieth century. This work bolstered claims by *Saltationists* such as William Bateson, who considered large mutational events to be the primary material that natural selection worked on to elicit the emergence of new species.

After the rediscovery of Gregor Mendel's work by de Vries and Carl Correns, the saltationists felt they were vindicated because they now had a model for the genetic inheritance of large-scale phenotypic differences in populations (*Polymorphisms*). After this point, they referred to themselves as Mendelians and criticized the work of gradualists (*Biometricians*) such as Karl Pearson and Raphael Weldon, claiming that such minor variations had no genetic basis (also see Chapter 7). The controversy between the Mendelians and the biometricians was effectively resolved when Fisher demonstrated mathematically that both small and large variations could be controlled by loci that segregated according to Mendelian laws of inheritance (Fisher 1918; Fig. 7.2). Unfortunately, most biologists did not appreciate the implications of Fisher's mathematical arguments, and misconceptions about the implications of Mendelian genetics for Darwin's theory of evolution by natural selection persisted well into the 1930s (Chapter 4).

Chromosomal Rearrangements—Throughout the twentieth century, physical aspects of karyotypes, including chromosome lengths, chromatin banding patterns, position of centromeres, deletions, insertions, translocations, inversions, and the presence of supernumerary chromosomes (also referred to as accessory or B chromosomes; Stebbins 1971) provided convenient genetic markers for population, taxonomic, and phylogenetic studies. Species possessing chromosomes that are more susceptible to breakage and rearrangement became the foundation of cytogenetic studies, including plants such as *Paeonia* (peony), *Trillium* (trilium), *Clarkia* (godetia), *Gaillardia* (blanket flowers), and several other genera that became the focus of analyses of intra- and interspecific variation in karyotypes (Grant 1978; Levin 2002). Studies of chromosomal polymorphisms in the mid to late twentieth century contributed significantly to our understanding of many principles of genetics.

Out of the broad range of rearrangements observed in plants, permanent translocation heterozygotes are the most intensely studied (Stebbins 1971; Harte 1994; Levin 2002). These cases of reciprocal translocations between non-homologous chromosomes result in the formation of chromosome rings during meiosis. Recombination may be limited to the ends of these chromosomes where homologous regions overlap or may be suppressed across the entire genome in cases of multiple-translocation heterozygotes, which is the case for several species of *Oenothera* (evening primrose) in the subgenus *Euoenothera*.

Restricted recombination enforced by multiple translocations may cause the co-segregation of whole blocks of chromosomes or entire haploid genomes (*Renner Complexes*; Harte 1994; Rauwolf et al. 2008). Species of *Oenothera* in the *Euoenothera* group appear to have been formed by pairing Renner complexes in different combinations; each species is closely related to two others by the entire haploid genomes that they share, and new hybrids can be formed by making novel combinations of haploid genomes. While this system of haploid genome co-segregation is intriguing, it appears to be unique to a few genera in the evening primrose family (Onagraceae) including *Oenothera* and *Clarkia*.

Cytogenetic studies continued to be abundant until the late twentieth century when karyotypes were largely replaced by genetic markers based on protein and DNA sequence variation (Chapter 4). Karyotypes and other cytogenetic methods—such as fluorescent in situ hybridization (FISH) for the localization of specific sequences on chromosomes—continue to be valuable tools that complement plant breeding programs and genomic investigations (e.g., Oleszczuk and Lukaszewski 2014). While the number of papers focusing on cytogenetics has declined in the last few decades, there are a number of unresolved questions remaining from the rich cytogenetic literature that developed over the last century—many of which may be amendable to investigation with more recently developed genomic techniques.

Gene Duplication—As long as environments remain constant, the function of enzymes will generally be maintained through purifying selection as non-synonymous mutations are eliminated from populations. We expect that the large majority of amino acid changes will be either neutral or detrimental primarily because these enzymes have been under selection for extensive periods of time, so any mutations that could improve function are likely to have been fixed in ancestors of living organisms. On the other hand, as changes in biotic and abiotic environments present new challenges, altered or novel enzyme function may be favored.

The foundation of adaptation originates from proteins that perform new roles in cell function, development, and structure, so ultimately the potential for organisms to adapt to new conditions may be limited by the arsenal of genes available to be modified by selection. Gene duplication provides raw material for selection to refine enzyme specialization and to generate novel functions. Intrinsic to this process is the natural tendency for individual proteins to catalyze multiple reactions—not only does this provide redundancy for essential cellular processes, it also creates opportunities for specialization and co-option into novel roles that serve to promote diversification and adaptation.

Gene duplication may occur by large-scale WGD (polyploidy), or by small-scale duplication (SSD) events that arise during DNA replication and recombination. Small genomic regions that include reading frames may also be duplicated through transposon activity. Genes can be captured within a transposon and copied during transposition events (*Transduplication*), or retrotransposons containing genes may be copied by reverse transcription to generate cDNAs that are capable of reintegration into the genome (*Retropositioning*). Comparison of *A. thaliana* (Brassicaceae—mustard family) to representative genomes from the Vitaceae (*Vitis vinifera*—wine grapes) and the Caricaceae (*Carica papaya*—papaya) indicated that up to 75% of genes were subject to transposition soon after the origin of the mustard clade (Freeling et al. 2008).

The potential for SSD events via transposition is enhanced because gene copies (*Paralogs*) generated by transposition often lose their regulatory context so they may have altered expression. Different mechanisms of gene duplication are not mutually exclusive as increased rates of transposition are often observed following WGD (Flagel and Wendel 2009). The large number of pseudogenes and gene fragments associated with transposable elements suggests that transposition is a major source of gene duplication in plant genomes.

Models of Gene Diversification—Models for the generation of gene diversity through duplication fall into broad categories of *Neofunctionalization, Subfunctionalization,* and *Dosage Selection* (Conant and Wolfe 2008; Fig. 3.8). The neofunctionalization category includes *Mutation During Non-Functionality (MDN)*, where duplication occurs prior to co-option to a novel role. In this case, the two copies may diverge for some time before a mutation in one copy enables a new function. Once partial functionality for a new role is introduced, further refinement of the duplicate gene occurs by mutation and selection (M/S in Fig. 3.8). In the MDN model, the accumulation of mutations may occur during a period when both copies are transcribed, or one copy may become non-functional and then become reactivated (resurrected) at a later date. In either case, the neofunctionalization model assumes that the original protein did not have functionality in the role adopted by the new copy of the paralogous pair. After duplication, selection may be relaxed so mutations that degrade the original function for one copy will not be detrimental as long as the other paralog remains intact. In the MDN model, the novel function of one paralog arises by random chance through the accumulation of non-synonymous mutations after duplication.

Subfunctionalization models assume that the original protein had at least partial functionality in both roles prior to gene duplication. Two models of specialization after duplication by subfunctionalization have been proposed. In the first, *Duplication, Degeneration, and Complementation,* (*DDC*), the process of refinement for different roles of each paralog may occur by "degeneration"—accumulation of mutations or loss of sequence regions that are relevant to the alternative function. In this model, the role of selection is minimal so much of the degeneration occurs by mutation and genetic drift (M/D in Fig. 3.8). Note that in the DDC model, the dual functionality of the original protein was high and little or no refinement

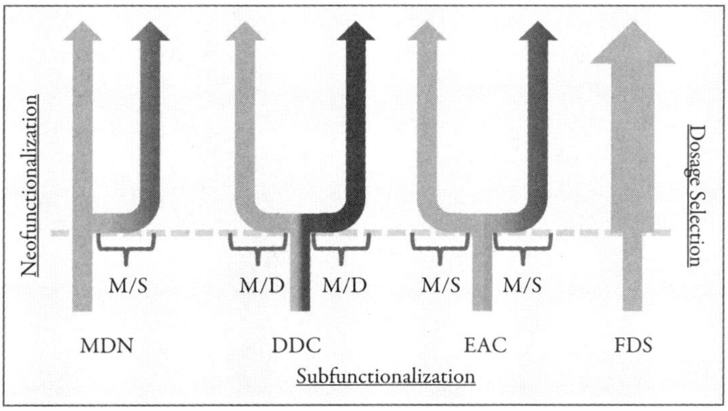

Fig. 3.8. Models for evolutionary processes favoring gene duplication: MDN—Mutation during non-functionality leading to neofunctionalization. After gene duplication selection favors mutations that improve a novel function. DDC—Duplication, degeneration, and complementation. Two copies of the gene lose the ability to serve alternative roles through genetic drift. EAC—Escape from adaptive conflict. Specialization is only possible after duplication because mutations improving one function compromise the other function. FDS—Fixation due to dosage selection. Gene duplication is favored through selection for increased dosage. In these models mutations are favored by selection (M/S) or are subject to genetic drift (M/D) leading to gene divergence.

by selection occurs after gene duplication. Cases of changes in regulation that result in tissue- or organ-specific expression of multiple copies of the same gene may generally fit the DDC model (see examples in the Evolution of Gene Function section).

In the second subfunctionalization model, *Escape from Adaptive Conflict,* (*EAC*), selection plays an active role to improve the functionality of both gene copies in different roles after duplication (Fig. 3.8). This model assumes that specialization for each function requires trade-offs; improvement in one function leads to degradation of performance in the alternative role. Thus, there is an "adaptive conflict," which prevents a single protein from having high functionality in both roles. In this case, duplication is necessary to achieve high functionality, and directional selection on each member of the paralogous pair after duplication allows specialization by mutation and selection (M/S) as the function of each protein is improved.

The fourth model of adaptation through gene duplication (Dosage Selection) assumes no change in the function of paralogs after duplication. The *Fixation due to Dosage Selection* (*FDS*) model of gene duplication assumes limitations on transcription rates and selection for up-regulation of a particular gene (Fig. 3.8). Duplication provides two sites for transcription and consequently a mechanism to increase the availability of the protein. We expect purifying selection will maintain the original function of both gene copies, but each sequence may diverge through the accumulation of synonymous mutations in exons and silent mutations within intron regions. This model of gene duplication differs from the other three because it does not increase gene function diversity.

Evolution of Gene Function—The three models of gene duplication that lead to new protein function or expression patterns through neo- and sub-functionalization (MDN, DDC, and EAC) provide a robust theoretical framework for understanding gene diversification and the evolution of novel function. Since both the MDN (neofunctionalization) and EAC models require an active role for selection, we expect these cases will be distinguishable from DDC subfunctionalization by evidence of larger frequencies of non-synonymous substitutions (Conant and Wolfe 2008; Flagel and Wendel 2009). The MDN and EAC models may be further distinguished from each other by evidence that selection acted on one or both paralogs; in the neofunctionalization model, we assume that one gene copy will retain its original function so higher rates of non-synonymous substitution would only be evident in one paralog. For the EAC model, on the other hand, we assume that functionality of the original gene in both roles was imperfect, so there should be evidence of selection acting on both gene copies after duplication.

While the expectations for neofunctionalization and EAC subfunctionalization seem clear, distinguishing between cases of gene duplication under these models may be difficult without information on the function of the ancestral gene. When we make a sequence comparison for a paralogous pair of genes we are able to estimate divergence between them, but we do not know whether evolution has occurred in one or both copies unless we have information on the function of the gene from which they were derived.

In one study of stress response genes in *Arabidopsis*, phylogenetic methods were used to infer the ancestral gene functions for paralogous pairs (Zou et al. 2009). In this case, a large proportion of genes displayed partitioning of function where one copy retained, and the other lost, the ancestral stress response, so these appear to be cases of neofunctionalization. Loss of ancestral function for both copies (presumably coupled with acquisition of novel function) was found for only a small fraction of gene pairs, so we can conclude from these data that subfunctionalization via EAC in these stress response genes is less common. For another example, see Chapter 6 on the developmental control of flowers.

In spite of the differences in the expectations of models for the roles of selection and genetic drift acting on members of a paralogous pair, it may be difficult to identify cases with clear evidence that supports one model to the exclusion of all others. Subfunctionalization under the DDC model may be found for a large proportion of paralogous genes that display tissue-specific expression in *A. thaliana* (Duarte et al. 2006) and *Gossypium* (cotton; Flagel et al. 2008). In these cases, subfunctionalization of paralogous pairs appears to have occurred primarily through complementary expression in different organs more often than non-synonymous changes within reading frames. Similarly, for paralogous pairs of stress response genes in *A. thaliana* both copies retained the ancestral function (Zou et al. 2009)—a pattern that is consistent with subfunctionalization by DDC. On the other hand, we expect that duplicate copies of genes will be difficult to maintain without some influence of selection; tissue-specific expression provides opportunities for each paralog to respond independently to selection for optimal function, so refinement of each gene copy to serve organ-specific roles is probably common. Consistent with this expectation, Zou et al. (2009) found many cases that superficially resembled DDC in stress response genes that had evidence of higher rates of non-synonymous substitution following subfunctionalization, suggesting that each paralog was refined as selection improved function in their new roles.

Models of gene duplication provide us with clear sets of expectations and tools for understanding evolutionary processes of novel gene function and diversification. On the other hand, the examples of gene duplication that are available often do not fall cleanly into one category or the other. For example, it would be difficult to unequivocally identify cases of subfunctionalization by EAC without evidence that the dual function of the ancestral protein was compromised by adaptive conflict. Partitioning of ancestral gene function in derived gene copies is often asymmetrical to varying degrees (Zou et al. 2009), so it may be better to view neofunctionalization and EAC as endpoints in a continuum of completely symmetrical to asymmetrical evolution of paralogous pairs. It is also likely that examples of subfunctionalization occurring purely by DDC are rare, because duplication is often followed by refinement of gene function. Implicit in our view of gene duplication is the understanding that new gene copies are unlikely to be retained in populations unless they confer some selective advantage; new paralogs most often are lost through selection or genetic drift, or they degenerate through mutation accumulation to become pseudogenes. Different types of genes may be more likely to persist as paralogous pairs after either SSD or WGD, and as we will discuss in the next section, a large part of their fate appears to be due to selection for dosage balance.

Dosage Selection—The fate of new gene copies often depends on the immediate consequences of changes in relative transcript abundance. If metabolic networks are highly constrained, duplication could produce unfavorable enzyme abundances—a concept known as the *Dosage Balance Hypothesis*. In general, gene duplication is more common for genes that code for enzymes with low levels of pathway connectivity and less severe consequences for knockout (artificially induced loss of function or expression; Conant and Wolfe 2008). Duplication of genes coding for proteins that are more essential or have high levels of connectivity may result in dosage conflicts and consequently reduced fitness for individuals carrying them. These generalities and the consequences of copy number variation depend largely on the mode of duplication; paralogs within constrained networks will be more likely to persist after WGD events because dosage balances are maintained. Paralogs of genes that are less constrained by dosage conflicts will be more likely to be generated by SSD events. A number of plant groups, including the modern ferns and many angiosperms, have recent histories of multiple WGD events, and gene duplication through polyploidy is much more prevalent for plants compared to other eukaryotes such as the metazoans (Flagel and Wendel 2009).

There is evidence that WGD provides an important pathway for replication of entire networks that avoids the negative effects of dosage imbalances. Individual genes tend to fall into functional groups that are more likely to be duplicated by either WGD or SSD events, but not vice versa. Paralogs produced by WGD tend to retain higher proportions of their gene interactions, but often display changes in expression patterns following duplication (Conant and Wolfe 2008)—such cases are consistent with DDC subfunctionalization. This pattern of replication of entire gene networks through polyploidy is presumably due to constraints imposed by dosage selection; duplication of individual genes within highly connected networks may produce unfavorable changes in relative protein abundance. While connectivity provides some predictive power for persistence after polyploidy, gene network duplication is not a universal phenomenon, as there is an equal number of essential or highly connected genes whose paralogs are not retained after WGD. In general, doubling an entire complement of interconnected genes may provide a mechanism to increase genome complexity that otherwise would not be possible by SSD.

While dosage selection can constrain duplication of individual genes within highly connected networks, there are also many cases where selection for increased dosage may promote retention of paralogs. Evidence of SSD events in plants is quite common, with upward of 20% of the genomes of *A. thaliana* and rice consisting of tandemly arrayed gene clusters (Flagel et al. 2008). These regions are characterized by multiple SSD events within arrays, or entire clusters may be duplicated in different genomic regions. Tandem pairs within arrays often retain one regulatory region so their expression is synchronized. Adjacent paralogs within these gene clusters also diverge much more slowly than expected, apparently due to homogenization processes such as unequal crossing over and *Gene Conversion*. Given the coordinated expression and low rates of divergence for tandem paralogs, these events appear to be examples of response selection for increased dosage and are consistent with the FDS model of gene duplication.

Dosage selection may also be the primary driving force behind cases of somatic WGD including *Polytene Chromosomes* (Carvalheira 2000) and *Endopolyploidy* (Barow 2006). Polytene chromosomes form through multiple repeated rounds of DNA replication within cells to form giant chromosomes and are found in specific tissues; for example, in the developing megagametophyte tissue during ovule development and in the tapetum of anthers during pollen development. Endopolyploidy occurs in a wide range of plant tissues and organs; it has been observed in trichomes, endosperm (up to 96n), the root nodules of legumes (up to 128n), and leaves, fruits, and stems of a wide range of flowering plants, but appears to be absent in the gymnosperms. While some cases of endopolyploidy may not serve specific functions (e.g., in some pith and cortex cells of the stem), in other cases its repeated occurrence in specific tissues across plant species suggests some physiological advantage (Barow 2006).

Dosage selection often determines the immediate fate of paralogs, and a good proportion of gene-copy mutations are probably lost from populations as selection acts to remove individuals with unfavorable dosage imbalances. On the other hand, gene duplication by SSD and WGD is very common in many plant groups. New paralogs that are not lost by genetic drift or removed by dosage selection may subsequently be subject to alterations in expression patterns or function through subfunctionalization and neofunctionalization. From examination of various cases of evolution through gene duplication, we have learned that the alternative models of gene evolution are not necessarily mutually exclusive, but they do provide useful constructs that improve our understanding of processes that are important for gene diversification. In the next section, we will discuss the processes and outcomes of whole genome duplication in more detail.

3.9 WHOLE GENOME DUPLICATION

Polyploidy (resulting from WGD) is more prevalent in plants than any other major group of eukaryotes. Study of the taxonomic distribution of polyploidy—and chromosome numbers in general—has been a major focus of cytogenetics since its inception more than a century ago. Much of the cytological work in the twentieth century was closely tied to the field of systematics, as WGD was associated primarily with the generation of new hybrid lineages though *Allopolyploidy*. In contrast, it was thought that WGD within lineages (*Autopolyploidy*) was rare (Stebbins 1950; Grant 1971); however, this perception may have been due to the general lack of phenotypic distinction among ploidy levels within species, which makes cases of autopolyploidy difficult to detect (Soltis et al. 2010). More recent appraisal of the relative frequencies of WGD suggest that autopolyploid formation by triploid bridges (Fig. 3.7) may be more common than allopolyploidy (Ramsey and Schemske 1998). The point may be somewhat muted by the fact that auto- and allopolyploidy are more accurately considered as the endpoints of a continuum that depends on the level of divergence between combined genomes. While it is generally recognized that the categorical classification of WGD as auto- or allopolyploidy is artificial, these terms remain in wide use because they are convenient for referring to the expected patterns of meiotic behavior of chromosomes in newly formed polyploids. In this section, we will discuss the prevalence and distribution of polyploidy, and the conditions that may favor the origin and spread of auto- and allopolyploid lineages.

There is no doubt that polyploidy has had a major role in the diversification of plants (Fig. 3.8). Variation in chromosome numbers within clades indicates that between 47% and 70% of all living plants have had recent histories of WGD (Ramsey and Schemske 1998). Moreover, the history of flowering plants is characterized by multiple ancient WGD events (Soltis et al. 2009), a phenomenon referred to as *Paleopolyploidy* (Box 3.2). These ancestral instances of polyploidy are detected by analyzing substitution rates for paralogous gene pairs. The occurrence of clusters of duplicated genes of similar age indicate that ancestral polyploidy characterizes nearly all of flowering plants (possibly excepting *Amborella*), and multiple WGD events occur in some clades (Fig. 3.8; Soltis et al. 2009). Evidence for multiple ancient polyploidy events also comes from the fossil record; based on changes in guard cell size, which is a strong indicator of DNA

Box 3.2 Paleopolyploidy

Paleopolyploidy generally refers to ancient WGD events that are not apparent from cytogenetic evidence. These genome duplication events typically occurred at least two million years in the past and resulted in a large number of paralogous gene pairs. Plants with polyploidy in their ancestry undergo normal meiosis with bivalent chromosome pairing and distribution of separate chromatids

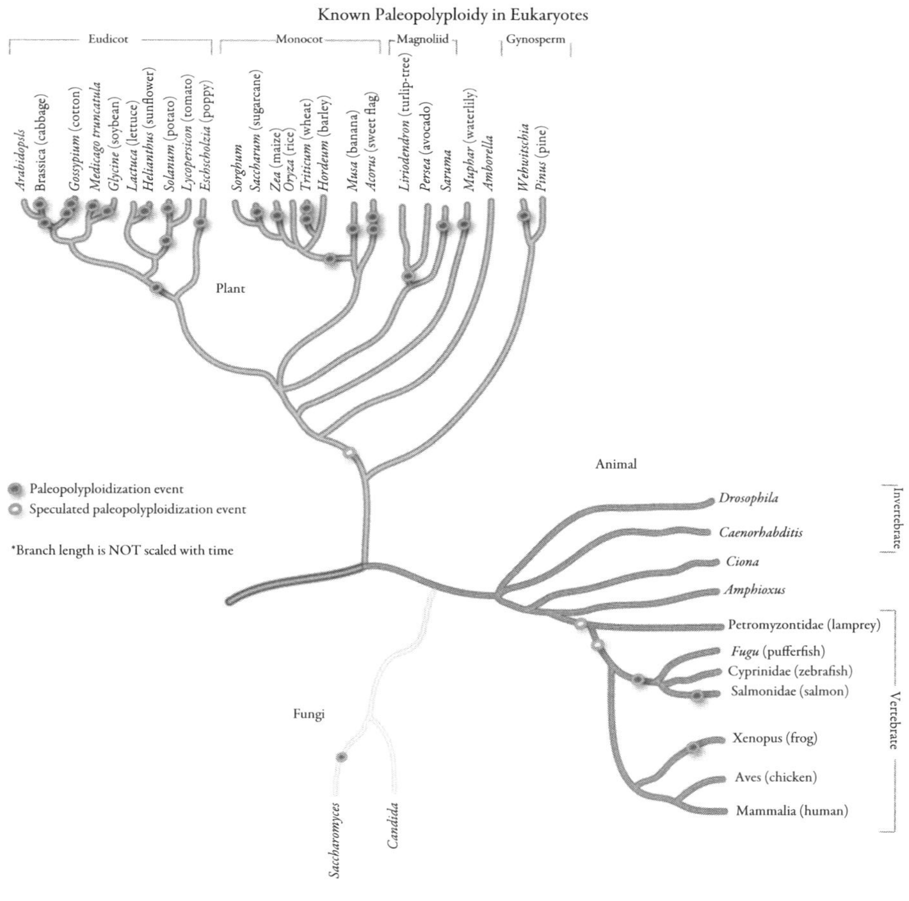

Fig. 3.9. The distribution of ancestral polyploidy events across the phylogenies of multicellular eukaryotes. Image created by 5dPZ at English Wikipedia.

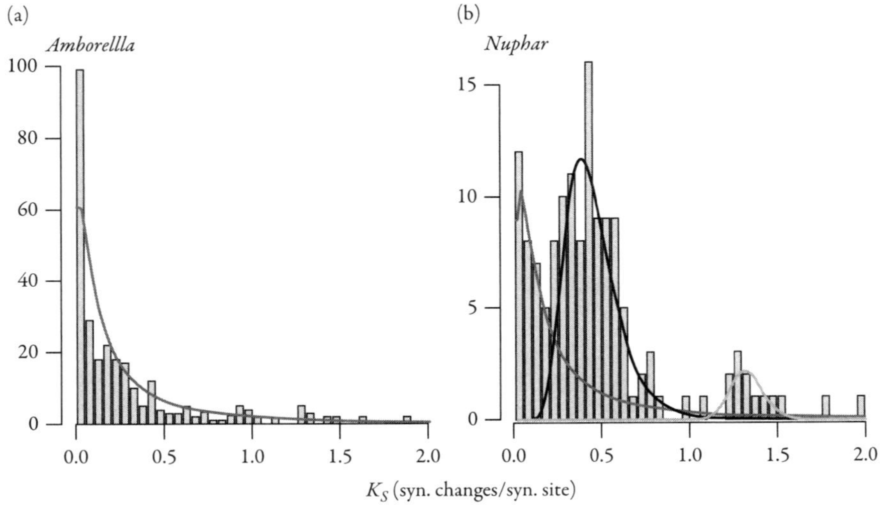

Fig. 3.10. Numbers of synonymous substitutions for pairs of paralogous genes in *Amborella* (A) and *Nuphar* (B). Used with permission (Soltis et al. 2009).

into gametes. Paleopolyploidy is apparent in the ancestry of a wide range of eukaryotes including vertebrates, but the flowering plants display the highest density of ancestral WGD (Fig. 3.9).

Ancestral genome duplication is detected by comparing pairs of paralogous genes across a single genome for numbers of synonymous substitutions (K_S; Fig. 3.10). In this analysis the relative age of any particular duplicate gene pair is indicated by the number of synonymous base differences between them. This analysis assumes that after duplication each gene copy accumulates mutations independently. We expect most pairs of paralogs to be due to recent SSD events and to display low levels of divergence (green lines in Fig. 3.10; the highest frequencies of gene pairs are for smaller values of K_S). WGD events will be evident as clusters of gene pairs that are of similar age, which creates the peaks seen for K_S in the aquatic plant *Nuphar* (blue and yellow lines in Fig. 3.10). We also expect that duplicates from WGD will occur as large, non-overlapping blocks, while paralogs from SSD events will be scattered across the genome. Following WGD, much of the duplicated genome is lost through deletions during the process of diploidization so the gene pairs we observe are duplication events favored by selection after polyploidy.

Analyses of several species of flowering plants reveal paleopolyploidy events shared among lineages and clades (Fig. 3.9). All seed plants (gymnosperms and angiosperms) share a WGD event from around 310 MYA, and the flowering plants are all characterized by a more recent event around 192 MYA (Jiao et al. 2011). Note that these early duplication events are not apparent in the analysis presented by Soltis et al (2009; Fig. 3.10), which is based on a large number of *Expressed Sequence Tags (ESTs)*. Within the flowering plants a triplication event predates the diversification of the core eudicots (Jiao et al. 2012), and a number of other WGD events have occurred across many lineages since then (Fig. 3.9).

content, a large proportion of extant species are marked by historical polyploidy (Masterson 1994). These repeated rounds of WGD have contributed to wide variation in flowering plant genome size; nearly 2,400 fold from around 63 Mb (megabases, millions of base pairs) for carnivorous plants in the genus *Genlisea* from Madagascar to more than 148,000 Mb in *Paris japonica*, a close relative of trilliums that is native to subalpine regions of Japan (Bennett and Leitch 2011).

While it is evident that polyploidy events have been frequent throughout the history of many plant lineages, it is also clear that genome size reduction has been common (Knight et al. 2005; Soltis et al. 2010). Following polyploidy, genomes typically undergo a series of deletion and gene silencing events that reduce functional redundancy and stabilize the polyploid genome in a process referred to as *Diploidization.* This process also serves to reduce the tendency for *Homeologous* chromosomes to form multivalent clusters (rather than the normal bivalent pairs of homologous chromosomes) during meiosis. Multivalence often results in incomplete or aneuploid complements of chromosomes in gametes. Consequently, most extant species have normal pairing of chromosomes during meiosis and behave entirely as diploids in spite of the fact that they have polyploidy ancestors.

From the discussion in the previous sections, we understand that WGD has been important for the duplication of metabolic pathways and the evolution of complexity, but there may also be more immediate ecological and genetic advantages to polyploidy that account for it prevalence. To develop a better understanding the factors that may promote or inhibit the success of polyploids, we will first take a closer look at the pathways leading to their formation.

Origins of Polyploidy—Polyploidy can be generated directly by the doubling of chromosome complements during somatic growth, or polyploid individuals may be produced through the union of unreduced gametes (Fig. 3.11). Somatic WGD has been documented in a wide range of species (Ramsey and Schemske 1998). As discussed in previous sections of this chapter, endopolyploidy occurs outside of meristems and only affects specific tissues or organs that generally do not contribute to gamete production. On the other hand, there are documented examples of plants consisting of cell lineages of different ploidy levels (i.e., chimeras—discussed later in this chapter). These instances of polyploidy may be

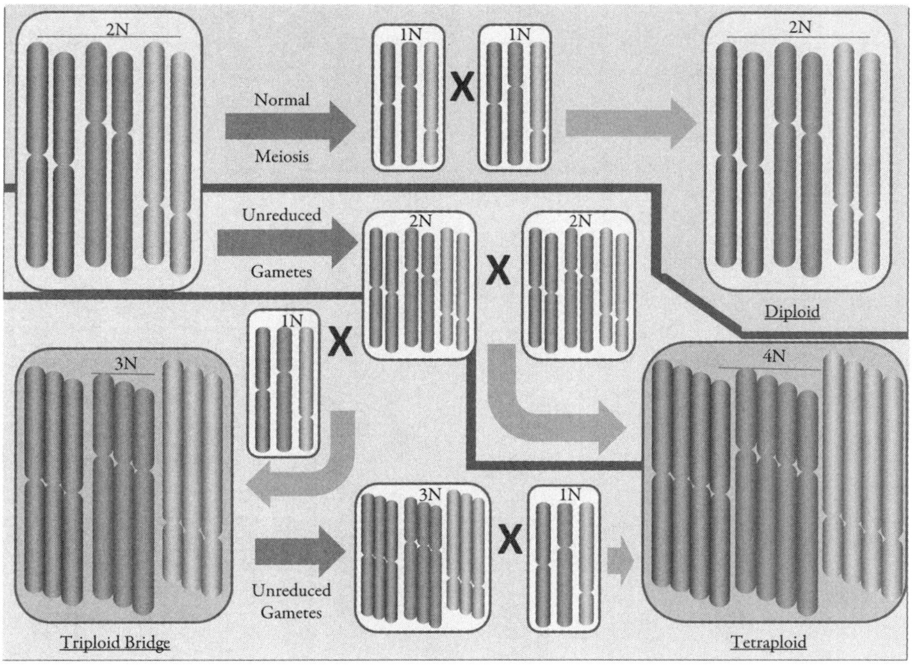

Fig. 3.11. Pathways to polyploidy (tetraploidy). Normal meiosis is contrasted with the production of unreduced gametes that may combine with each other to produce tetraploids, and may combine with haploid gametes to produce triploid plants. Triploids then serve as bridges to the formation of tetraploids.

due to mitotic aberrations that result in the transfer of the entire complement of duplicated chromosomes to one of the two daughter cells. In this way tetraploid (4N) cell lineages may be formed in a plant that is otherwise diploid. Once a polyploid chimera is established, the tetraploid cell lineages may generate floral primordia and produce gametes that carry twice the chromosome complement compared to other individuals in the population (e.g., 2N gametes are produced by tetraploid individuals). While somatic polyploidy clearly has the potential to generate autopolyploid offspring, we have limited understanding of the contribution of this mechanism to the generation of historical and contemporary instances of WGD.

It is generally assumed that the large majority of polyploid lineages are generated through meiotic aberrations that result in unreduced gametes (Ramanna and Jacobsen 2003; Soltis et al. 2010). Unreduced gametes (e.g., 2N micro- and megagametophytes in a diploid population) can combine with normal (1N) gametes to generate triploid offspring, which often occur at low frequencies in populations (Ramsey and Schemske 1998). Triploid plants may suffer low fertility due to the formation of multivalent chromosome clusters during meiosis, but some fraction of the gametes they produce will be functionally haploid or diploid, so these plants may function as triploid bridges to the formation of tetraploids (Fig. 3.11). Low frequencies of triploids may serve as pathways for the recurrent formation of tetraploids and provide avenues of gene flow from the diploid population to tetraploids, which may otherwise be reproductively isolated.

Unreduced gametes may also combine to produce tetraploids directly without a triploid bridge (Fig. 3.11). This process is facilitated by selfing of first-generation hybrids from crosses between divergent lineages, which are expected to have univalent chromosomes aligning during meiosis (Fig. 3.12). Lack of bivalent pairing can result in *First Division Restitution* (*FDR*) through *Single Division Meiosis* (*SDM*), which produces unreduced gametes after the first division of meiosis (Oleszczuk and Lukaszewski 2014). Alternatively, *Second Division Restitution* (*SDR*) may occur through a variety of mechanisms that result in unreduced gametes (Ramanna and Jacobsen 2003). Note that FDR is more likely to occur when homeologous chromosomes do not pair during meiosis so the gametes produced retain high levels of heterozygosity (Fig. 3.12). In contrast, in SDR homeologous chromosomes do form bivalents but the pairs of chromatids do not separate in the second division of meiosis; consequently, the gametes will be highly homozygous except for chromosome regions that have recombined (Ramanna and Jacobsen 2003).

While FDR and SDR mechanisms may consistently generate unreduced gametes, it is also common for hybrids from divergent crosses to produce a range of gametes with varying chromosome complements,

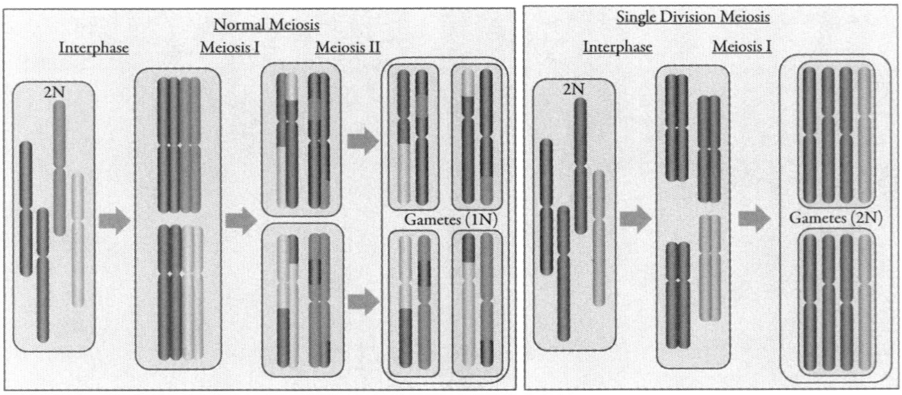

Fig. 3.12. Segregation and recombination in normal meiosis contrasted with the production of unreduced gametes through Single Division Meiosis (SDM). In SDM univalent chromosomes line up prior to cytokinesis and First Division Restitution (FDR) of normal meiosis occurs when gametes are produced after meiosis I.

which reduces their fertility. The observation that the doubling of the entire chromosome complement in hybrids between divergent lineages provides a direct pathway to restitution without loss of fertility led to the prediction that allopolyploids would generally be the consequence of wide crosses (Darlington 1937). More recent analyses do not uphold this prediction as tetraploids appear to arise after hybridization between pairs of lineages that have diverged to varying degrees (Soltis et al. 2010). Ranging levels of divergence between pairs of hybridizing taxa results in partial to full pairing of homeologous chromosomes during meiosis and facilitates different pathways for the generation of unreduced gametes. Hybrids from wide crosses may be more susceptible to the production of unreduced gametes through SDM, while partial pairing of homeologs in hybrids from closer crosses may generate unreduced gametes through asymmetrical allocation of chromosomes. The formation of multivalents during meiosis in polyploids leads to reduced fertility as larger proportions of gametes possess incomplete or partially duplicated sets of chromosomes (Soltis et al. 2010).

Predicted loss of fertility in tetraploids that combine closely related genomes prompted early geneticists to assume that nearly all tetraploids originated from hybrids between more distantly related lineages and that autopolyploidy should be rare (Stebbins 1950; Grant 1971). More recent evidence suggests autopolyploidy may be relatively frequent, and the meiotic behavior and levels of fertility of auto- and allopolyploids may be more similar than originally thought (Ramsey and Schemske 2002). Since tetraploidy is common in flowering plants there must be other circumstances that offset the lower fertility that may be expressed in newly formed polyploids.

Maintenance of Polyploids in Populations—Polyploidy is generally associated with hybridization as genome doubling can restitute normal meiosis in hybrids that have only partial homolog or homeolog pairing (Ramanna and Jacobsen 2003). Like any other mutation, newly formed polyploids face loss from populations through genetic drift, but their chances of persistence may be diminished if they also suffer lower reproductive success (because they are unable to mate with diploids). On the other hand, the generation and persistence of tetraploids may be favored in populations that have a high tendency for the production of unreduced gametes. Diploid gamete production varies substantially among plant genera and families and appears to be under genetic control (Ramsey and Schemske 1998). Populations with higher frequencies of unreduced gametes will be subject to recurrent formation of tetraploids, which may increase their chance of persistence and spread.

The maintenance of polyploids in populations may be further facilitated by the alleviation of inbreeding depression and the acquisition of self-fertility (Ramsey and Schemske 2002; Rausch and Morgan 2005). Offspring from self-fertilization of newly formed tetraploids should suffer less inbreeding depression as the effects of deleterious recessive mutations will remain masked because of the presence of normal alleles on the other homologous or homeologous chromosomes. Moreover, the spread of tetraploids is often favored because their diploid pollen is not inhibited by self-incompatibility mechanisms acting during pollen-tube growth. The combination of lower inbreeding depression and higher rates of selfing may be adequate to counterbalance fertility losses in newly formed tetraploids and would explain their prevalence in many flowering plant genera and families. This conclusion is further supported by the observation that polyploids generally display higher rates of selfing than closely related diploids (Barringer 2007).

Diploidization—Newly formed polyploids are subject to relatively rapid genome reorganization as selection favors stabilized chromosome pairing during meiosis. Early-generation tetraploids often have high frequencies of aneuploidy, higher levels of modifications to DNA methylation patterns (Liu and Wendel 2003), and high levels of phenotypic variation among offspring (Ramsey and Schemske 2002). Selection on gametophyte and sporophyte generations favors deletions that improve meiotic stability. As discussed in previous sections of this chapter, this process is non-random with respect to genes as paralogs are favored in some cases but not in others. This process of diploidization ultimately results in normal bivalent pairing during meiosis as the previous history of polyploidy is effectively erased, at least from a cytogenetic perspective. There may also be selection against large genomes that result from WGD, as plants with larger genome sizes tend to have slower growth and photosynthesis rates, are less common in extreme environments, and tend to have lower rates of diversification (Knight et al. 2005). After polyploidy we expect specific sets of paralogs to be retained as selection favors bivalent pairing of chromosomes during meiosis and overall genome reduction over longer time frames.

3.10 MUTATION ACCUMULATION IN PLANTS

The generation of genetic variation is substantially more complex in plants compared to other organisms. The basic processes of genome sequence change and epigenetic modification remain the same, but in plants, these mutations must be transferred through numerous cell generations of somatic growth, and then pass through the gametophytic stage before entering the population of sporophytes. Because of the potential for somatic mutations to contribute to genetic variation, we expect that the potential for mutation accumulation is substantially greater for plants compared to animals; however, out of the many mutations occurring each generation, only a small, non-random subset is passed on to offspring (*Somatic Mutation Accumulation Hypothesis*). The implications for selection on mutations occurring during vegetative growth is only just beginning to be fully appreciated.

The potential for mutation accumulation is much greater for plants compared to animals because plants produce gametes after extensive somatic growth. In animals, a germline is established soon after formation of the zygote. Germline cells divide a limited number of times prior to the production of gametes and have much lower mutation rates than somatic cell lineages (Sabour and Scholer 2012). The separation of germlines in animals provides for more reliable transfer of genetic information across generations without the contaminating effects of somatic mutation and epigenetic modifications—a phenomenon referred to as the *Weismann Barrier* (Buss 1983). Consequently, the primary source of mutations entering animal populations is DNA replication errors occurring during meiosis. For plants, on the other hand, meiotic and mitotic mutations, and epigenetic modifications all contribute to genetic variation within populations, because flowers are borne at the ends of stems that were produced by many cell divisions. Based on this observation it has been suggested that mutation accumulation during vegetative growth of long-lived species could be substantial (Whitham and Slobodchikoff 1981; McKnight et al. 2002; Klekowski 2003; Schultz and Scofield 2009). To understand the processes affecting mutation accumulation in plant populations we need to examine aspects of vegetative growth in some detail.

Meristem Structure and Stem Growth—Plant stems and roots elongate by mitotic division of a small population of cells located in apical meristems (Box 3.3). The meristem is divided into separate layers and zones where cells divide and differentiate into the separate tissues of the stem or root. At the center of apical meristems are clusters of cells that divide more slowly than other meristematic regions; in the stem this region is referred to as the *Central Zone* (Fig. 3.13; refer to color plate) and in the root it is called the *Quiescent Center* (Fig. 3.14). Cells in these slowly dividing meristem zones are often referred to as *Stem Cells* because they remain undifferentiated and *Totipotent* over the life of the plant; they are capable of producing cells that can differentiate into any cell type. Plant stem cells may have enhanced DNA repair mechanisms, so mutation rates may remain lower than in other regions of somatic growth (Jones et al. 2012; Heyman et al. 2013). Outside of the stem cell zones are regions of active cell division where foundation cells are produced. Foundation cells divide to produce cells that differentiate into the separate tissues that constitute the stem, leaves, flowers, and axillary buds.

As plant stems elongate, mutations may accumulate through both DNA damage and replication errors during mitosis. After many cell divisions the separate stems on the same plant, and even cell lineages within stems, may become genetically differentiated (Klekowski 2003; Schultz and Scofield 2009). Consequently, plants will consist of a mosaic of genetically distinct cell lineages that are characterized by unique complements of mutations. Most of these mutations are not expressed so they are not easily detected, but occasionally mutations will cause changes in the appearance of some portions of an individual plant. These *Chimeric* plants are morphologically heterogeneous for flower color, chlorophyll pigmentation, or other characteristics. The patterns of coloration produced in chimeric plants depend on the location of the original mutation. For example,

Box 3.3 Plant Apical Meristems

Plant apical meristems are responsible for stem and root elongation. These are very small regions that occur at the tips of stems and roots and are characterized by centers of active cell division. Apical meristems are organized into groups of cells (cell layers and zones) that serve different functions. The stem apical meristem (SAM) is covered by one (monocots) or two (dicots) layers of cells (L1 and L2) that are each one cell thick and form the *Tunica*. These cells divide along a two-dimensional plane across the surface of the stem tip and contribute to the formation of the epidermis (Figs. 3.13 color plate and 3.14). The L3 or *Corpus* layer consists of cells that divide in three dimensions to differentiate into tissues that constitute the tissues of the stem. As the stem elongates, the three meristem layers contribute to the formation of leaf, flower, and axillary bud primordia. For example, the epidermis of the stem and leaves is typically derived from the L1 layer and the *Mesophyll* cells of the leaf from the L2 layer, while the L3 layer produces cells that differentiate into the vascular tissue of both the leaf and stem and into the cortical and pith tissues of the stem.

The three cell layers that comprise the tunica and corpus are organized into zones of cells that are defined by their roles in meristem structure and cell differentiation (Fig. 3.13; refer to color plate). Cells in the central zone remain undifferentiated throughout the life of the plant and are characterized by their small size and large vacuoles. They are not stationary or immortal through the life of the plant, but the turnover rate of stem cells is relatively slow compared to the surrounding regions. Cells in the *Organizing Center* are thought to be responsible for cell signaling that controls rates of cell division and prevents cell differentiation for cells in other zones. Cell division is relatively rapid in the *Peripheral* and *Rib* zones. Cells produced in the peripheral zone become founder cells for the formation of the separate tissues that form leaf, flower, and axillary bud primordia, while cells in the Rib zone differentiate into the separate tissues of the stem. The central and peripheral zones include cells from the L1, L2, and L3 layers while the organizing center and rib zones are all L3.

Mutations occurring in stem meristems sometimes cause changes in in the color or other morphological characteristics of the leaves or flowers. Most often, these are chlorophyll deficiency

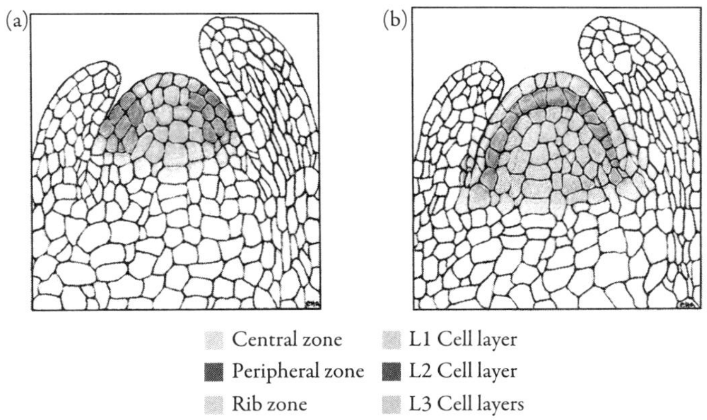

Fig. 3.13. Two depictions of the shoot apical meristem. A) Radial domains. Lateral organs originate from cells recruited from peripheral zone. Cells from the rib zone produce the bulk of the stem tissues. Stem cells reside in the central zone, which replenish the peripheral and rib zones. B) The layers of cells across the apical meristem. The epidermal (L1) and subepidermal (L2) layers undergo anticlinal (sideways) cell division, and are referred to as the tunica. Interior cells contribute to growth in all directions. Used by permission (Sharma and Fletcher 2002).

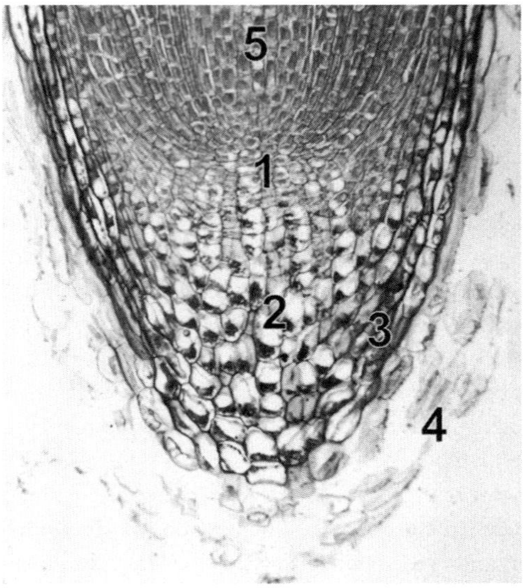

Fig. 3.14. Cell zones in the root apical meristem: 1) Quiescent center; 2) live root cap cells; 3) root cap; 4) sloughed off root cap cells; and, 5) procambium. Photo by SuperManu.

mutations that can produce different shades of green or white portions of leaves. Common cultivars with variegated leaves are the result of periclinal chlorophyll mutations, which occur in a single layer of the meristem (L1, L2, or L3). If the mutation occurs in a stem cell then it may affect all three meristem layers on one side of the stem to produce a sectoral chimera.

The root apical meristem (RAM) is protected by a *Root Cap* (Fig. 3.14). The root cap cells are constantly replaced as they are sloughed off due to friction as the root forces it way among soil particles. The quiescent center represents a reservoir of stem cells that serves a similar purpose as the central zone of the SAM. Cells in the quiescent center remain totipotent, divide slowly, and provide replacement cells for the meristem regions of more active cell division. Cells in the quiescent center apparently have enhanced DNA repair mechanisms and are thought to have lower mutation rates compared to other somatic cells (Jones et al. 2012; Heyman et al. 2013).

if the mutation occurs in a single layer of the meristem (*Periclinal Chimera*; Box 3.3), it will produce variegated patterns on all of the leaves produced after the mutation occurs (Fig. 3.15; refer to color plate). Mutations that occur in the central zone of the meristem may affect all layers of the meristem and consequently the entire section of the stem. In these *Sectoral Chimeras*, whole leaves or sections of leaves are uniformly distinct from other leaves on the same plant (e.g., Fig. 3.16; refer to color plate).

Mutations occurring during mitosis will only affect descendent cells, so in the examples of chimeras given here (Figs. 3.15 and 3.16; refer to color plates), the lower portion of a plant would have normal coloration and only the branches containing the mutation would be chimeric. Depending on the position of the mutation (in central or peripheral zones of the meristem; Box 3.3) it may or may not be passed on to sexually produced offspring. For example, variegation that is due to periclinal chimeras can only be propagated asexually; seeds produced from these plants will result in seedlings that are either uniformly green or white (the latter being non-viable since these plants are incapable of photosynthesis). These natural mutations and experimentally induced mutations have been instrumental for the development of models of meristem structure and cell differentiation processes (Poethig 1987).

Fig. 3.15. A periclinal chimera results in variegated leaves in this *Ficus benjamina*. Photo by Kurgus.

Fig. 3.16. A sectoral chlorophyll mutation results in the leaf coloration pattern of this chimeric buckeye (*Aesculus hippocastanum*). Photo by Elektryczne jabłko.

Mutation Accumulation During Somatic Growth—If the somatic mutation accumulation hypothesis were correct then we would expect mutations to accumulate in branches and roots over the life of the plant. As stems elongate, they produce flowers toward their tips so accumulated mutations could be passed on to the next generation. To understand the potential for somatic mutations to be passed on to sexually produced offspring, and consequently to contribute to the genetic load of the population, we need to consider the dynamics of cell division and growth within the apical meristem.

The central zone of the stem apical meristem acts as a reservoir for the replacement of cells in the peripheral zones and is the ultimate source of cells that constitute all of the above-ground vegetative and reproductive organs. Further away from the central zone, cells differentiate and maintain the same position in the stem while elongation continues, so any mutations occurring in more distal regions will not contribute to the somatic mutational load of the meristem. The population of central zone cells at the tip of each stem could be carrying a load of mutations that affects large portions of the plant and would be inherited by sexually produced offspring.

By counting the number of cells present in the internodes of mature stems, we can estimate the number of cell divisions required for stem growth and presumably the number of divisions that could contribute to heritable somatic mutational load (Scofield 2006). On the other hand, division rates are low in the stem cell regions of the meristem, and many more cell divisions are occurring in peripheral zones. For example, based on the frequency of cells undergoing mitosis, the cell division rates were nearly three times greater in peripheral zones compared to the central zone in wild-type *Arabidopsis* meristems (Laufs et al. 1998). Below the meristem, cell division continues as internodes elongate (Brown and Sommer 1992). Since stem growth proceeds by a sequence of divisions—starting from the central zone to peripheral zones and then continuing during tissue differentiation—the number of cell divisions contributing to somatic mutational load will be far less than estimates of the number of cells present in mature stems. If we assume the number of cells in the mature stem is the product of the cell division rates in each region, it is possible that only a few central zone cell divisions are required to produce each centimeter of stem length. Considering this much slower rate of division along with the possibility that plant stem cells have much lower mutation rates (e.g., Heyman et al. 2013), we might conclude that these groups of cells effectively prevent excessive mutation accumulation in long-lived plants (Jones et al. 2012); an idea that we can refer to as the *Plant Germline Hypothesis*.

Cell Lineage Selection—Since a large proportion of the genome is expressed during stem growth, we would expect that any mutations affecting growth rate would be subject to selection. Cells within the meristem may vary in rates of cell division, so the opportunity exists for cell lineages with faster growth to displace the slower ones (Poethig 1987; Otto and Orive 1995; Klekowski 2003). If these differences in division rates are at least partly determined by mutation accumulation, then we would expect that *Cell Lineage Selection* would contribute to the purging of somatic mutational load, and possibly the acquisition of advantageous mutations during stem growth.

The potential for cell lineage selection to affect the genetic composition of the apical meristem depends on the phenotypic expression of mutations, and the number of cell lineages present in the meristem (population size). The majority of mutations are deleterious and generally have only a small effect (around 2%) on fitness (Lynch et al. 1999). Since the population of cells within the central zone is quite small—less than thirty cells in *Arabidopsis* (Laufs et al. 1998)—we expect the effects of weak selection could easily be overcome by genetic drift. Consequently, a large proportion of these weakly deleterious mutations could be retained and passed on to sexually produced offspring. The large majority of deleterious mutations are recessive so they may have only small effects when they first appear; however, after self-fertilization they are more likely to be homozygous and to have more severe effects for inbred offspring—an idea will explore further in the next section.

Estimating Levels of Somatic Mutational Load—If it were true that plant stem cells provide a reservoir for the preservation of genetic integrity during vegetative growth (the plant germline hypothesis), then the consequences of somatic mutation accumulation would be minimal and difficult to detect. Alternatively, the somatic mutation accumulation hypothesis predicts that individual branches will develop substantial mutational loads, and separate stems on the same plant will become differentiated. While we expect that cell lineage selection would eliminate mutations that reduced growth rates, there could be substantial accumulation of mutations that are not exposed to selection when they first appear. For example, a large proportion of deleterious mutations will be recessive, but after inbreeding they may be made homozygous and reduce the fitness of offspring. Additionally, there is a substantial array of genes with more specialized functions that contribute to fitness. Somatic mutations will continue to accumulate in genes regardless of whether they are expressed during stem growth, but unexpressed mutations would not be subject to cell lineage selection. To determine whether somatic mutations will contribute to population genetic variation we need to develop a more thorough understanding of the amount of mutational load that accumulates during the vegetative growth of individual plants.

From our observations of chimeric plants, it is clear that somatic mutations can occur. Models suggest that somatic mutations will continue to accumulate during stem growth, so the genetic load of older stems could be substantial (Schultz and Scofield 2009). If mutation accumulation were a function of stem length, we would expect that large-statured plants such as trees should have much greater mutational load than herbaceous species. For example, in one survey the frequency of chlorophyll mutants was ten times greater in trees compared to herbaceous annuals (Klekowski 1998; Fig. 3.17), and somatic mutations have been shown to contribute to genetic variation in a long-lived shrub (Gross et al. 2012). These data confirm the expectation that mutations should be more frequent in long-lived species, but this study did not directly address the question of whether somatic mutation contributes significantly to the genetic load of populations.

One way to make a direct test of the levels of somatic mutational load would be to make comparisons of seedling fitness from self-pollinations within the same flower (*Autogamy;* w_a) with seedlings from self-crosses between flowers on different stems of the same plant (*Geitonogamy;* w_g) to estimate levels of *Autogamy Depression* ($\delta_{AD} = 1 - w_a/w_g$; Schultz and Scofield 2009; Fig. 3.18). Both of these crosses are self-pollinations so seedlings will be highly homozygous, but offspring from autogamous crosses will be homozygous for a proportion (~25%) of the somatic mutations being carried by each stem, while seedlings from geitonogamous crosses will be heterozygous for somatic mutational load. If mitotic mutation rates are substantial, the fitness of autogamous offspring will be lower and the somatic mutation accumulation hypothesis will be supported. The difference in fitness for offspring from autogamous and geitonogamous crosses can provide direct estimates of genetic load accumulation through somatic mutation during stem growth.

A number of studies comparing the fitness of offspring from autogamous and geitonogamous crosses have been conducted, and a survey of estimates of δ_{AD} for herbaceous and woody species reveals that somatic mutation accumulation is common (Bobiwash et al. 2013). In this survey, five out of ten herbaceous plants

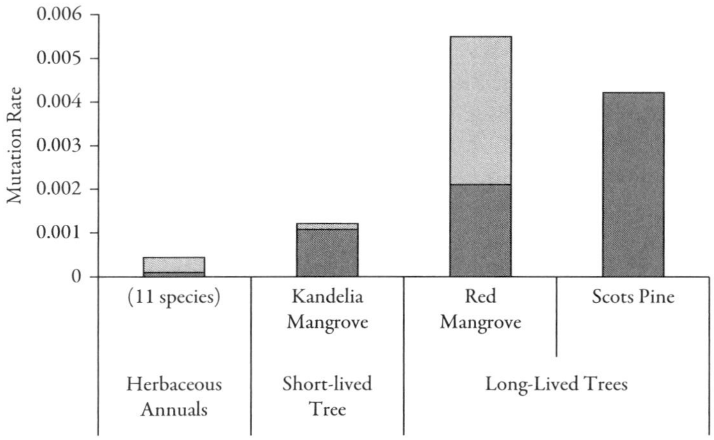

Fig. 3.17. Rates of chlorophyll mutants per generation for plants with a range of life histories. Top of the dark blue bar represents the minimum (or only) estimate, and the top of the light blue bar indicates the maximum. Modified from Klekowski (1998).

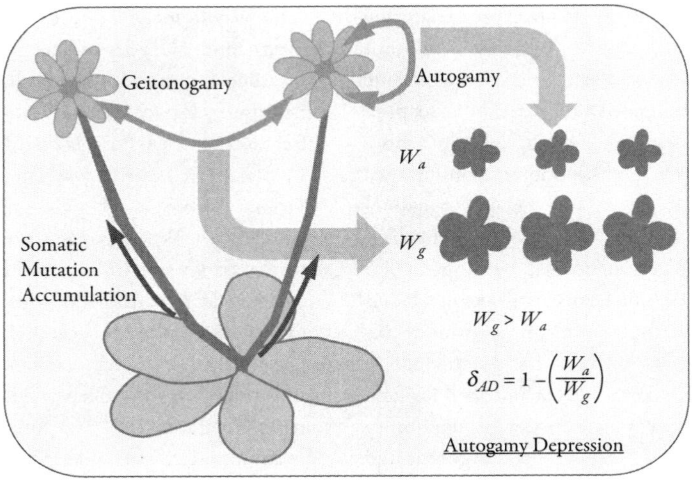

Fig. 3.18. The fitness of seedlings from autogamous and geitonogamous pollinations are compared to estimate levels of autogamy depression. We expect that somatic mutation accumulation will lower the fitness of offspring from autogamous crosses ($w_g > w_a$).

demonstrated measurable levels of autogamy depression, with values of δ_{AD} ranging from 0.12 to 0.69. For woody species (shrubs and trees combined), levels of autogamy depression greater than zero were found for eight of twelve species studied, and estimates ranged from 0.49 to 1.00 (Bobiwash et al. 2013). Note that when δ_{AD} is equal to 1.00 the autogamous pollinations failed to produce seed—presumably because mutation accumulation caused high rates of ovule abortion—while geitonogamous pollinations resulted in at least some seed set (e.g., Gomez and Zamora 1996). The wide range of estimates of autogamy depression probably reflects variation in the methods used, but there is a general pattern of higher levels of δ_{AD} in woody species, which is consistent with the hypothesis of higher mutation accumulation in long-lived plants, and seemingly opposite the prediction of the generation-time hypothesis described earlier in this chapter. Whether offspring from these plants are produced through selfing or outcrossing among genets, these experiments demonstrate that mutations accumulated during vegetative growth can be passed on to the next generation and may contribute to the genetic load of populations.

Another approach to examining the effects of somatic mutation accumulation uses pollen viability as a test for the expression of deleterious alleles in the haploid state. Since a large proportion of the genome is expressed in the gametophyte stage (Seoighe et al. 2005; Arunkumar et al. 2013), we would expect that accumulation of recessive mutations would affect the integrity of pollen. A study of male fertility in trembling aspen (*Populus tremuloides*), a long-lived clonal species, found a decline in pollen viability with clone age (Ally et al. 2010; Fig. 3.19). This is expected under the mutation accumulation hypothesis, as older stems should have higher mutational load. Clones of this species can become quite large, sometimes covering forty or more hectares, and some extremely large clones may be 80,000 years old. In the study by Ally et al. (2010), clones with age estimates up to 10,000 years had pollen viability estimates approaching zero, so we might expect older clones to be nearly sterile. In this species, somatic mutation accumulation has had a dramatic effect on clone fertility, but it is not known whether somatic mutation has detrimental effects on other extremely long-lived plants (e.g., giant bamboo, bristlecone pines, creosote bush, etc.).

Somatic Evolution—Many plant species have extensive vegetative growth by runners or rhizomes on the soil surface, or underground by structures such as tubers, bulbs, corms, or occasionally by suckering—adventitious stems arising from roots (Chapter 9). Asexual reproduction can also occur through seeds—a wide range of plants reproduce by *Apomixis* including dandelions (*Taraxacum*), flea banes (*Erigeron*), and hawkweeds (*Hieracium*) in the

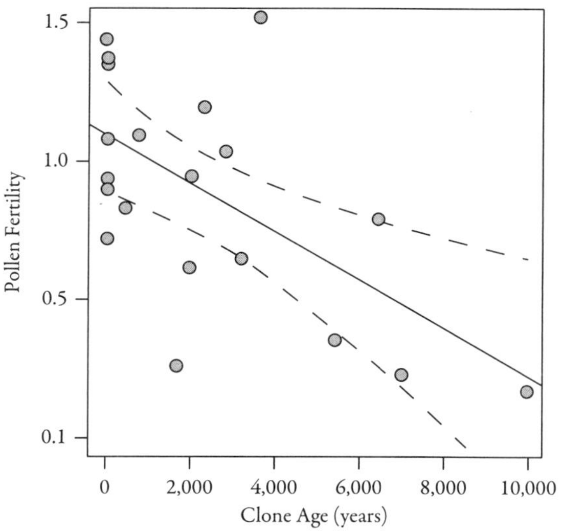

Fig. 3.19. Decline in pollen fertility with clone age in the quaking aspen. Modified from Ally et al. (2010).

aster family (Asteraceae); the hawthorns (*Cretaegus*), blackberries (*Rubus*), and shadbush (*Amelanchier*) in the rose family (Roseaceae); and the meadow grasses (*Poa*) in the grass family (Poaceae), to name a few. Whether asexual offspring are produced through seeds or by vegetative propagation, all products of mitosis are considered *Ramets* (vegetative parts) of the same *Genet* (all of the vegetative parts that are derived from the same seed). Vegetative propagules (ramets) are all members of a single genet whether they remain attached to the original plant or not; new genets can only originate through sexual reproduction. Plant species that are considered to be apomictic or highly clonal typically consist of many distinct genets and may have infrequent sexual reproduction to produce new genets (e.g., in dandelions; Majesky et al. 2012).

Asexual reproduction is quite common in plants (Chapter 9). For example, many highly successful weeds such as dandelions and blackberries are apomictic, so single genets can achieve extensive geographic ranges. Similarly, genets of some aquatic invasive plants that spread by vegetative propagules may be distributed across multiple continents (e.g., water hyacinth; Zhang et al. 2010). Many other plants spread through a combination of sexual and asexual reproduction. For example, wild strawberry (*Fragaria*) often establishes by sexually produced seeds and then spreads by runners. This strategy is employed by some very successful invasive plants that colonize open sites by sexually produced seeds and then spreading vegetatively (e.g., invasive species such as privet, honeysuckle, and English ivy). Asexual propagation is widely used in horticulture and agriculture; varieties of many ornamental plants such as irises, tulips, roses, variegated plants, and important crops such as varieties of apples, citrus, bananas, and wine grapes can only be propagated vegetatively to retain their unique and desirable characteristics. It is a little astonishing to realize that grape varieties such as Chardonnay or Pinot Noir are represented by single genets that are planted around the world and continue to produce wines with the same general characteristics as when the vines of these unique varietals first arose from seed.

With such extensive amounts of asexual reproduction in plants, the potential exists for divergence among ramets through the accumulation of mutations. While the accumulation of neutral variation can be substantial (e.g., Majesky et al. 2012), the modular structure of plants also lends itself to somatic evolution in response to selection; as ramets encounter new environments, cell lineage selection will favor any mitotic mutations that enhance growth. If selection on cell lineages were effective, we might expect plants with extensive vegetative propagation and apomictic formation of seeds to undergo adaptive somatic evolution as their ramets are exposed to new environmental challenges.

While the idea of somatic mutation leading to the adaptive divergence of ramets is attractive, the number of cases of clonal adaptation is limited. For example, Breese et al. (1965) succeeded in selecting for increased tillering ability (production of new grass stems) within genets of perennial ryegrass (*Lolium perenne*). Similarly, artificial selection effectively improved branching in the red seaweed, *Asparagopsis armata* (Monro and Poore 2009). While these examples are impressive, we do not know whether the differences in the phenotypes of ramets in response to selection are due to DNA sequence changes or to epigenetic modifications. For example, the invasive Japanese knotweed (*Fallopia japonica*) is represented by a small number of genets and may owe its success in a range of environments to differences in DNA methylation patterns (Richards et al. 2012). Whether the somatic evolution observed in these cases is due to the generation of random sequence changes during mitosis, or to more directed changes in differential methylation of genome regions, the variation observed is heritable and has the potential to enable clonal adaptation in response to selection.

It is clear from the discussion above that individual ramets can undergo substantial genetic and epigenetic differentiation. This observation has led some to disparage the concept of clonal uniformity (Loxdale and Lushai 2003); such concepts as somatic cell lineages made widely available for medical research cannot hold much value for experimentation if mutation rates are as high as they appear to be. On the other hand, some fields of plant breeding have embraced the idea of clonal diversification for the development of new varieties. For example, the Pinot Noir wine grape is ancient compared to modern grape varieties so it has diversified by somatic mutation into more than 1,000 clonal lineages—all of which are promoted for their unique characters. Somatic mutations in Pinot Noir clones have also generated several new varieties such as Pinot Gris, Pinot Blanc, and Pinot Meunier, the latter of which is a periclinal chimera.

In summary, highly clonal plants present us with an enigma. On the one hand, we rely on them for their uniformity, without which valuable commercial products such as Cavendish bananas and McIntosh apples would not be possible. On the other hand, we celebrate them for their potential to diversify into new varieties. The difference between these two views may be largely a product of the preferences of horticulturalists and farmers; those interested in producing a consistent crop would cull the odd types, whereas others may select out novel ramets for further testing and possible cultivation. For wild plants, the potential for clonal adaptation is clear, but generally advantageous mutations may be overwhelmed by the somatic accumulation of deleterious variation. Whether the variation garnered during the somatic growth of individuals contributes to longer-term rates of substitution at the population level is another question entirely—one that we will address in the next section.

Resolving the Generation-Time Conundrum—As pointed out in a previous section in this chapter, long-lived plants display slower substitution rates than annual and perennial herbaceous species over phylogenetic timescales (the generation-time hypothesis for differences in mutation accumulation among lineages). At first glance, this observation appears to be in direct conflict with the observed higher levels of somatic mutation accumulated in long-lived plant species—if substitution rates are primarily limited by the number of mutations entering the population each generation, then large-statured plants should have higher rates of evolution. The resolution of this puzzle lies in the realization that substitution rates over phylogenetic time scales may ultimately be limited by the combined effects of selection and recombination.

To understand how the frequency of recombination limits substitution rates, it may be best to view plants as populations of mitotically dividing cells that periodically reproduce sexually (e.g., diatoms and yeast have diploid vegetative stages, occasional sexual reproduction, and similar substitution rates to land plants; Sorhannus and Fox 1999; Hall and Joseph 2010). By analogy, individual plants consist of populations of meristem cells that periodically produce floral meristems and gametes. It may be the case that individual stems of herbs and trees undergo similar numbers of mitotic divisions in a single growing season. If somatic mutation rates are similar for herbaceous and woody plants we would expect that all plants should have similar annual rates of protein evolution, so the reason for slower substitution rates in woody plants remains unexplained.

From the perspective of a meristem cell population, the primary difference between herbaceous and woody plants is the frequency of sexual reproduction or more to the point, the frequency of recombination. Recall that selection interacts with the recombination rate to affect levels of polymorphism for different regions of the genome (Fig. 3.5; refer to color plate). The same effect—reduction of variation under selection—would be expected across cell generations. Consequently, the rate of evolution should be positively correlated with the frequency of sexual reproduction (Lynch and Gabriel 1983). With higher rates of recombination, associations among deleterious and advantageous mutations will be disrupted—alleviating selective interference and allowing for selection to act on individual mutations (Hartfield and Otto 2011); recombination allows more effective purging of genetic load and fixation of beneficial variation. The accumulation of somatic mutations on an annual basis may be similar in woody and herbaceous plants, but substitution rates over phylogenetic timescales are limited by the frequency of recombination, which increases the opportunities for selection to more effectively determine the fates of individual mutations. The combination of a longer time to initiate sexual reproduction and the consequential lower frequency of recombination may slow rates of evolution in long-lived plants.

Phenotypic Effects of Mutation Accumulation—Mutations can be quantified through sequence comparison among individuals and lineages as described earlier in this chapter, but this information may not be adequate to understand the consequences of nucleotide base changes for phenotypic trait expression and fitness. An alternative approach is to analyze variation in traits to assess the effects of mutations that have phenotypic effects: the expressed mutation rate—U (Eyre-Walker and Keightley 2007). We can describe the relationship between U and the mutation rate per nucleotide site per generation (μ) by defining a proportion of base positions that are subject to selection (C) and the total number of nucleotides in the haploid genome (G; Haag-Liautard et al. 2007). For a diploid species, the number of expressed mutations appearing each generation is a product of the per-nucleotide mutation rate and the expressed portion of the genome ($U = 2GC\mu$). Note that U should be closely related to K_A—they are both aimed at quantifying variation that is potentially subject to selection—but not all nonsynonymous substitutions will have effects on protein structure or function, so they will not affect the phenotype. The expressed mutation rate (U) will also capture the effects of base changes across all types of functional regions, while K_A is typically limited to variation within reading frames.

Estimates of the expressed mutation rate can be obtained by quantifying the increase in variance of phenotypic traits among individuals across generations. This is best achieved by initiating a set of homozygous lines from a single maternal lineage that has been inbred by selfing for several generations. *Mutation accumulation (MA)* is tracked over generations by measuring a phenotypic trait that is closely associated with fitness (e.g., number of fruits per plant). Lines are propagated each generation using single seeds, and then after some number of generations (usually at least ten), a large number of offspring from each line are grown and quantified for the fitness-related trait (Fig. 3.20; refer to color plate). In the first generation, all of the lines will be nearly identical, but they will begin to diverge for average fitness (as estimated by the chosen traits) as they accumulate unique mutations over a number of generations. Trait expression will change as mutations accumulate so the variation among lines will increase over time.

Mutations may appear in gametes produced by meiosis and in vegetative cells produced by mitosis during somatic growth. The process of mutation accumulation and the distribution of fitness effects of base changes entering the population each generation may differ depending on whether the germline is separated from somatic growth; in other words, we expect the mutation accumulation process to be fundamentally different in plants. It is generally thought that the large majority of mutations will be deleterious, but in a species with extensive somatic growth, we expect cell-lineage selection to eliminate a portion of the deleterious mutations (Fig. 3.20; left panel; refer to color plate). Selection among mitotically dividing cell lineages also has the potential to capture beneficial mutations that appear during somatic growth, so ramets may undergo adaptive evolution. In contrast, species with separate germlines lack intra-generation selection, so we expect to see a more substantial decline in the average fitness of MA lines (Fig. 3.20; right panel; refer to color plate). The numbers of beneficial and deleterious mutations occurring in animals and plants is probably similar, but extensive somatic growth provides an opportunity for selection to affect the distribution of mutations prior to gamete

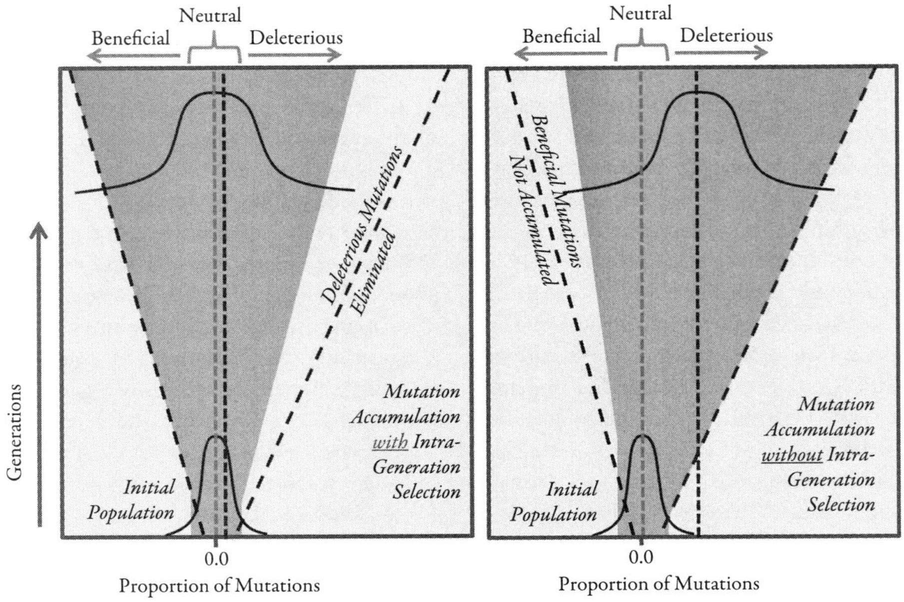

Fig. 3.20. Predicted effects of intra-generation selection on fitness distributions for sets of mutation accumulation lines. In both scenarios the initial populations display low phenotypic variance (distributions are narrow). As mutations accumulate, the variance among lines increases (upper normal curves). Space between the angled blue and red dashed lines indicate the potential increase in variance among lineages due to mutation accumulation. The shaded area indicates that actual increase in variance due to mutation accumulation; blue shading indicates the proportion of beneficiation mutations and red shading is the proportion of detrimental mutations. Left panel—a portion of deleterious mutations is eliminated and advantageous mutations retained in a species with intra-generation selection, but not in a species with a germline (i.e. intra-generation selection is minimal or absent; right panel). Vertical orange-dashed line indicates the mean fitness before mutation accumulation and the vertical red-dashed line is the mean after mutation accumulation.

production. If the effects of intra-generation selection were substantial, the process of mutation accumulation may be fundamentally different in organisms with separate germlines (i.e., metazoans) compared to plants and other diploid organisms that produce germ cells after extensive somatic growth (Otto and Orive 1995; Rutter et al. 2010).

The large majority of MA experiments have been conducted using species of metazoans such as insects and other invertebrates (primarily *Drosophila melanogaster* and *Caenorhabditis elegans*). As expected, these studies have found declines in average fitness across generations (Halligan and Keightley 2009). Notable exceptions to this trend include results from MA in diploid yeast, *Saccharomyces cerevisiae* (Joseph and Hall 2004; Dickinson 2008) and the plant, *A. thaliana* (Shaw et al. 2002). Yeast reproduces primarily by mitosis with infrequent sexual reproduction so these fungi are similar to plants in that mutation accumulation could be affected by cell lineage selection.

When high rates of beneficial mutation accumulation were first observed in yeast and *Arabidopsis* the results were controversial (Bataillon 2003). It was generally assumed that most proteins had a long history of selection so mutations with potential to improve structure or function must already have been acquired. Consequently, the relatively high rates of beneficial mutation in species lacking germlines were treated with skepticism; it was suggested that either the traits chosen may not have been closely tied to fitness, or that these unexpected results must be an artifact of the experimental conditions (Halligan and Keightley 2009). These criticisms were addressed in subsequent MA experiments that confirmed high frequencies of beneficial mutations under field conditions and quantified the effects of specific base changes in *A. thaliana* (Rutter et al. 2010; Rutter et al.

2012). Similarly, high rates of beneficial mutation accumulation were found for a wider range of traits when the original MA experiment with yeast was continued for 1,000 generations (Hall and Joseph 2010). Additional evidence for high frequencies of beneficial mutations was found for selection response experiments with corn (*Zea mays*; Durand et al. 2010). Results from MA experiments with yeast and two plant species are consistent with the hypothesis that intra-generation selection affects patterns of protein evolution by reducing the load of deleterious mutations and perhaps by favoring the retention of beneficial mutations that arise during somatic growth.

While the intra-generation selection scenario portrayed in Fig. 3.20 (refer to color plate) is the best explanation available for the higher rates of beneficial mutation accumulation in plants and yeast, this idea remains speculative; we do not have direct estimates of somatic mutation rates or evidence for the influence of cell lineage selection on the distribution of mutations. Furthermore, not all MA studies with plants find large proportions of beneficial mutations—declines in fitness have been observed for two annual species of fiddlenecks (*Amsinckia*) over ten generations (Schoen 2005).

The outcomes from MA studies in *Arabidopsis* and corn may seem somewhat contradictory to the results from the autogamy depression experiments described earlier, which suggested that deleterious mutation accumulation during somatic growth could be substantial. On the other hand, autogamy depression experiments focus on loss of fitness due to inbreeding and the effects of beneficial somatic mutations could be masked by the expression of genetic load. In spite of inconsistencies in results among MA studies in plants, the arguments for the potential effects of cell lineage selection are compelling, so the processes and consequences of somatic mutation accumulation in plants continues to be an active area of research.

3.11 MUTATION ACCUMULATION IN PLANTS—SUMMARY

Mutation accumulation in plants differs from organisms with separate germlines in a number of ways. First, the potential for accumulation of genetic mutations and epigenetic modifications is much greater in plants because of extensive somatic growth intervening between zygotes and the formation of gametes. The large number of mitotic divisions occurring during vegetative growth generates a large mutational load that may depress the fitness of inbred offspring. *Inbreeding Depression* can be particularly severe after autogamous pollinations because of mutations that accumulated during the growth of each stem. The potential for somatic mutation accumulation is enhanced by the lack of recombination during mitotic cell division and the potential for masking the effects of deleterious alleles by dominant normal alleles.

Second, somatic mutational load may be alleviated, and beneficial mutations retained, by cell lineage selection (Fagerstrom et al. 1998; Otto and Hastings 1998; Monro and Poore 2009). The effects of selection in populations of cell lineages contributing to vegetative growth may be responsible for the observation of relatively high frequencies of beneficial mutations in plants and diploid unicellular organisms. Competition among mitotically dividing lineages of cells constitutes an intra-generation level of selection that could reduce genetic load and accelerate adaptive responses to challenging environmental conditions.

Third, the exposure of mutations to selection during pollen growth has the potential to provide an additional episode of intra-generation selection (Mable and Otto 1998; Arunkumar et al. 2013) —a topic we will explore further in Chapter 5. A substantial proportion of the plant genome is expressed during the haploid life stage (Seoighe et al. 2005; Arunkumar et al. 2013), while animals have little to no gene expression in sperm (da Costa-Nunes and Grossniklaus 2004). *Gametophytic Selection* has much more potential to purge deleterious variation because mutations are expressed in the haploid state, so they are not masked by dominant alleles. Expressed lethal mutations will probably be eliminated during pollen development (e.g., those affecting pollen viability as described above for aspens), pollen germination, and pollen-tube growth, and much of the deleterious mutational load may be filtered out by competition among pollen prior to fertilization.

The effects of extensive somatic growth and intra-generation selection have the potential to alter the process of mutation accumulation in plants by removing deleterious mutation and favoring the retention of beneficial changes. Lack of a separate germline and inclusion of post-meiotic transcription in the gametophyte may increase the potential for the generation and filtering of mutations. The separation of germlines in metazoans provides for reliable transmission of genetic information that is unadulterated by the vagaries of genetic and epigenetic alterations that accrue during somatic growth (although there may be some exceptions to this generality; Solana 2013). The reason for the contrast in germline transmission between plants and animals remains unresolved but may be due to much greater sensitivities to gene dosage imbalances during animal development. Plants, on the other hand, may benefit from the combination of somatic mutation accumulation and intra-generation selection as a consequence of their extensive vegetative growth and sedentary life form. The evolutionary consequences of somatic mutation accumulation and intra-generation selection in plants have been the topic of much speculation for some time, but only recently have we been able to apply high-throughput sequencing technologies to examine these processes in unprecedented detail.

Whether genetic and epigenetic modifications are generated during meiosis or somatic growth, this variation is the foundation of evolution in populations and sets an upper limit on rates of adaptation and diversification. The fate of novel mutations will depend on a wide range of ecological and demographic processes affecting population growth and the dispersal of seeds and pollen. In the upcoming chapters we will explore how these processes function independently and in concert to manifest evolution within and among populations and ultimately for the diversification of lineages.

3.12 QUESTIONS FOR CONTEMPLATION AND DISCUSSION

1. Think about synonymous (K_S) versus non-synonymous (K_A) base changes. What type of sequence data do you need to estimate these parameters? Why do you expect K_S to generally be greater than K_A, and when could the reverse be true?
2. Why might a proportion of non-synonymous base changes not be subject to selection? Think about all aspects from protein structure/function to phenotypic trait expression.
3. Describe in your own words why the substitution rate for synonymous and neutral sites is expected to be the same regardless of population size.
4. Describe the reasons why K_S might be a poor estimate of the base mutation rate (μ). Consider the effects of linkage, selection, and recombination rates.
5. Explain in your own words how K_A could be higher than K_S under positive selection.
6. How do selective sweeps and background selection affect K_S? How do these processes differ?
7. What are the general patterns of substitution rates, recombination, essential function, and network connectivity for proteins?
8. Think about variation in substitution rates among different groups of organisms. How are they consistent or inconsistent with the fidelity/replication rate hypothesis, the drift-barrier hypothesis, and the generation-time hypothesis?
9. How do epigenetic modifications differ from genetic mutations? How is the inheritance and importance of epigenetic marks different between plants and animals?
10. How have genome duplications contributed to the diversification of life? What is the evidence supporting the importance of duplications?
11. What type of evidence would you need to discriminate among the subfunctionalization and neofunctionalization hypotheses for the evolution of novel gene functions?
12. How does the process of mutation accumulation differ between plants and animals? What is the potential for cell lineage selection, gametophytic selection, and selective ovule abortion for filtering mutations entering the population each generation?
13. Consider the plant germline and somatic mutation accumulation hypotheses. What is the evidence for both? How can we explain the relatively moderate levels of mutation accumulation in long-lived plant species?
14. How can we reconcile somatic mutation accumulation during vegetative growth with the generation-time hypothesis for plants?

4

DISTRIBUTION OF GENETIC VARIATION

THE FOUNDATIONS OF population genetics were laid chiefly by mathematical deduction from basic premises contained in the works of Mendel and Morgan and their followers. Haldane, Wright, and Fisher are the pioneers of population genetics whose main research equipment was paper and ink rather than microscopes, experimental fields, Drosophila *bottles, or mouse cages. Theirs is theoretical biology at its best, and it has provided a guiding light for rigorous quantitative experimentation and observation.*

Theodosius Dobzhansky (1955) "Cold Spring Harbor Symposium" in Quantitative Biology

4.1 THE GENETICS OF POPULATIONS

A dynamic confluence of interacting processes governs the distribution of genetic variation within and among populations. Almost as quickly as novel genetic variation is generated it is lost from populations due to the small chance that new mutations (alleles) will persist more than a few generations. This progression of allele turnover is largely random; by chance some alleles will succeed while others are lost—a process referred to as *Genetic Drift*. Over time, there is constant renewal of genetic variation as new mutations enter the population while others disappear. Some small fraction of this newly generated variation may succeed—solely by chance—to rise in frequency until virtually every individual in the population is homozygous for the new variant (i.e., the allele is *Fixed* in the population).

While the large majority of mutations are invisible to selection, a small fraction will have phenotypic effects. Nearly all expressed mutations are detrimental, so, as discussed in Chapter 3, they are likely to be eliminated by purifying selection. More rarely, a new mutation may confer some selective advantage. For these beneficial mutations, the chances of maintenance and eventual fixation may increase with the positive effects of selection, but even beneficial mutations can be easily lost when they are at low frequency due to the effects of drift (Chapter 5). Along with selection, genetic drift and gene flow are the primary processes determining the fate of novel mutations. In this chapter, we focus on gene flow and drift, and we will reserve our discussion of selection for the next chapter.

Population genetic processes do not act in isolation, so the fate of a new mutation may simultaneously be affected by gene flow and genetic drift as well as by selection if the mutation affects the phenotype. As mutations increase in frequency, they have a greater chance of being carried by pollen or seed that may be moving among populations. When seeds disperse across the landscape, they may arrive in existing populations, or in suitable habitat to establish new populations. If seeds survive to reproduction, or pollen succeeds in parenting offspring, then these dispersal events will affect the genetic composition of the recipient population, in which case we consider *Gene Flow (Migration)* to have occurred.

New mutations are likely to occur at low frequencies after incorporating into recipient populations, so the influences of genetic drift and selection on their persistence begins anew. Even though most of these migration events may fail, continued gene flow can repeatedly reintroduce the novel variant, so the chances for the ultimate establishment and fixation of new mutations—especially beneficial mutations—increases over time. This sequence of processes—gene flow followed by genetic drift and selection—occurs repeatedly as novel variants spread among populations. Eventually a small fraction of the new mutations that first occurred in single individuals may spread until they dominate a geographic region and eventually the entire range of a species.

This scenario could apply to any organism, but several unique aspects of plants render the processes of genetic drift, selection, and migration intrinsically more complex. First, the sedentary nature of plants often limits opportunities for mating to a local group of individuals that may be a small subset of the entire population, which will enhance the effects of genetic drift on a local scale. Second, plants must rely on biotic or abiotic vectors for pollination and the dispersal of seeds. Consequently, the behavior of dispersal agents will have a tremendous influence on population genetic structure and the distribution of genetic variation across geographic scales. Third, gene flow in plants can occur during the sporophytic (seed) or gametophytic (pollen) life stages, and depending on the vector, dispersal distances can be extremely limited or quite extensive. Differences in dispersal potential between pollen and seeds will have implications for the distribution of genetic variation within and among populations and for the relative contributions of pollen and seed dispersal to gene flow. Fourth, most plants are hermaphrodites, and many have the potential to self-fertilize. In species with high frequencies of selfing, pollen-mediated gene flow is negligible, and loss of genetic variation occurs through both genetic drift and inbreeding. Even for plants that routinely outcross, dispersal distances may be constrained. Limited dispersal of seeds and pollen and the potential for selfing can facilitate local inbreeding, and widespread populations can become substructured into patches that differ in their genetic composition on a small spatial scale.

While the basic features of evolution are effectively the same for plants as for most other organisms, the biological and ecological characteristics of plants alter the implementation of evolutionary processes and the relative influences of genetic drift, selection, and gene flow within and among populations. The sedentary life form, dependence on vectors for dispersal and gene flow, and propensity for self-fertilization ultimately have consequences for the distribution of genetic variation and for the diversification of lineages. Evaluating the effects of genetic drift, inbreeding, migration, selection, and mutation on the distribution of genetic variation, and ultimately for lineage diversification, can be done with the most clarity at the population level—a field of study known as *Population Genetics*.

In this chapter, we explore methods for the quantification of genetic variation and discuss analyses of the distribution of genetic variation. We will establish conceptual frameworks for the study of population genetics—based on the *Hardy–Weinberg Principle*—through mathematical and graphical analyses. This chapter focuses on neutral genetic processes (genetic drift, gene flow, and inbreeding); we will take up the topic of selection in Chapter 5. Each evolutionary process is discussed in turn; we will describe basic principles first, and then broaden our discussion to integrate the effects of plant biology on these evolutionary processes. This chapter culminates with a broader discussion of the interplay among population genetic processes and the consequences of gene flow, genetic drift, and inbreeding for the distribution of genetic variation within and among plant populations.

THE ORIGINS OF POPULATION GENETICS

We can quantify variation among individuals in a number of ways. The most obvious level of variation—the phenotype—will be discussed in Chapters 6 and 7. The phenotype is perhaps the easiest to quantify but is more distantly removed from mechanisms of genetic change at the genomic level and consequently is much more complex. The most fundamental level of variation can be found in the sequence of nucleotides and its organization into the chromosomes that comprise the genome. As discussed in Chapter 3, variation in the genome may be represented by single base substitutions (*Single Nucleotide Substitutions—SNPs*), gain or loss of one or more consecutive bases (*Insertion/Deletion Mutations—Indels*), or chromosomal rearrangements (*Translocations, Aneuploidy,* and *Polyploidy*). Most alterations of the genome occur in non-functional regions; they have no effect on the phenotype and consequently are hidden from the influence of selection (*Silent Mutations* in intergenic regions, and *Synonymous Mutations* within expressed regions). Genetic variation that is invisible to selection (*Neutral Variation*) is important as its accumulation provides us with a gauge of divergence among lineages, populations, and individuals within populations. Neutral genetic variation also establishes a baseline for the comparison of levels of divergence for genomic regions that might be under the influence of selection. As we will see below, the utility of neutral genetic variation at the molecular level was not always widely appreciated.

In this chapter, our goal is to explore the processes of neutral evolution. We first explore the history and development of the diversity of *Genetic Markers* that are available, because these are the tools of population genetics—the study of processes affecting the distribution of genetic variation among populations of the same species. A genetic marker is any heritable polymorphism that displays discrete differences in phenotypic traits, proteins, DNA sequences, or epigenetic modifications. The variety of genetic markers available has changed dramatically over just a few short decades from the 1960s to the early 2000s. Some of these methods are no longer commonly used, but they have contributed to a substantial literature that provides a foundation for contemporary investigations in plant evolutionary biology. In this section, we describe the history of genetic marker development and the advantages and limitations of different markers and discuss applications of genetic marker data for estimating the levels and distribution of genetic variation within and among populations.

Ecological Genetics and the Selectionist Program—In recent decades a considerable amount of research in population genetics has been conducted using DNA-based markers, but this has not always been the case. Starting in the mid-twentieth century, morphological markers were used to assess patterns of selection and the distribution of variation in populations. Early studies in *Ecological Genetics*—a field established by E. B. Ford (Ford 1971)—used both *Polymorphic* and *Quantitative* morphological traits. The large majority of this work was conducted with polymorphic animal species such as the grove snail (*Cepea nemoralis*), and various species of butterflies and moths. During this time much of the work on morphological variation in plants focused on problems in systematics and hybridization (e.g., Clausen et al. 1948; Anderson 1949; Stebbins 1950; Baker and Stebbins 1965; Grant 1981), but there was also some interest in plant population ecology and ecological genetics, including investigations into the effects of selection and gene flow on adaptation to novel environments (e.g., Bradshaw and Snaydon 1959; McNeilly and Antonovics 1968; Hamrick and Allard 1972; Hamrick and Holden 1979).

Access to methods for assaying population genetic variation changed dramatically in the 1960s with the advent of gel electrophoresis of enzymes (Appendix). Thealleles observed at each allozyme locus are based on the differential migration rates of protein variants through hydrolyzed starch or cellulose acetate gels. Specific proteins (enzymes) are selectively visualized by providing the appropriate substrate for the enzyme reaction, along with a stain that is activated in the presence of chemical reactions (e.g., reductase activity). The multiple bands that appear on allozyme gels are due to variation in amino acid sequences and represent alternative protein variants that differ in size or surface charge. These protein variants can be interpreted as alleles segregating at single loci (*Allozymes*) or at separate loci coding for proteins that display the same catalytic ability (*Isozymes*, representing paralogous genes). Alleles at each locus segregate in offspring as expected from Mendelian principles, and consequently these markers are useful for a broad range of applications in population genetics.

To appreciate the diversity of applications and limitations of allozymes and DNA-based genetic markers it is useful to review the early days of their development and the controversies that emerged as evolutionary biologists struggled to understand the evolutionary processes governing variation at the molecular genetic level.

The Classic and Balanced Schools of Selection—The introduction of allozyme electrophoresis in 1966 brought into clear focus a controversy that had been brewing since mathematically inclined biologists first entered the field of evolutionary biology in the early part of the century. At stake was the prevalent view among evolutionary biologists that selection was the dominant force shaping all aspects of organisms and that the best way to understand this process was through empirical investigation of phenotypic variation within and among populations. This perception was derived from the experiences of biologists working with organisms in their natural environments where phenotypes generally appeared to be the consequence of their biotic and abiotic interactions—virtually all characteristics of plants and animals were interpreted as adaptations and were assumed to be the products of selection (a view referred to as *Panselectionism* or *Adaptionism*). Empiricists of the mid-twentieth century saw little value in the mathematical illustrations of evolutionary processes that R. A. Fisher, Sewell Wright, and J. B. S. Haldane had been developing since the early 1930s. Ernst Mayr referred to the mathematical approach as "beanbag genetics," and doubted that the consideration of genes as independent units could accurately reflect the evolution of complex phenotypic traits (Mayr 1963; Haldane 1964). Similarly, G. L. Stebbins made a clear distinction between biologists who were working with organisms in their natural environments and those whose experience was limited to the laboratory and theoretical investigations (Stebbins 1974). This broiling debate came to a head as the first data on molecular genetic variation became available in the 1960s.

While evolutionary biologists of the mid-twentieth century unanimously agreed that natural selection was the prevailing force affecting all characteristics of organisms, there was disagreement over the fundamental basis of phenotypic variation and the origins of adaptations. The *Classical School* assumed that the large majority of mutations were deleterious and were removed quickly from populations by purifying selection. In this *Neo-Darwinian* view, advantageous mutations were rare and adaptation occurred through the action of positive selection that favored beneficial variants in populations. This model of adaptive evolution was supported by the observation that mutations found in nature and generated in the lab were usually deleterious, and beneficial mutations were extremely rare.

The *Balance School* viewed variation among individuals in natural populations as the product of *Heterosis* (*Overdominance*). This interpretation is supported by the observation that phenotypic variation in populations is common, so researchers assumed that substantial amounts of genetic variation must be segregating at loci controlling these traits. Advocates of the balanced school were also influenced by results of breeding programs, where "hybrid vigor" (heterosis) is often observed after crossing among inbred lines. The prevalence of heterosis is well established in the breeding literature, with roots in crossing experiments conducted by Darwin in which he contrasted the vigor of offspring from self- and cross-fertilization for a large number of plant species (Darwin 1876). The idea that heterozygosity was favored by selection was compelling, and many leading evolutionary biologists such as Theodosius Dobzhansky favored this view (Dobzhansky 1970).

When the first surveys of molecular genetic variation became available in the mid-1960s, the high levels of variation detected was unexpected and led to much discussion and disagreement among evolutionary biologists of the day. Molecular genetic variation was first visualized by applying a method developed by molecular biologists for the characterization of enzyme variants (Hunter and Markert 1957). This method was adapted by Lewontin and Hubby (1966) to generate estimates of allozyme variation in natural populations of *Drosophila pseudoobscura*. Most biologists assumed that variation at the molecular level would correspond directly to the type of phenotypic variation typically found among individuals; mutations that generate new alleles would be rare, and any polymorphisms detected would represent deleterious mutations that had not yet been eliminated or beneficial mutations that were in the process of going to fixation.

The level of allozyme variation found in the first surveys was much higher than anyone expected; of the eighteen loci assayed in five populations, 30% were polymorphic with an average of 11% heterozygotes. At first glance, these data appeared to support the position of the balance school since it was assumed that this variation must be maintained by selection for heterozygosity; any explanation other than selection was unthinkable to most evolutionary biologists at that time. The alternative view—that the observed variation in allozymes was neutral to selection—was promoted by Motoo Kimura (1968) and later by King and Jukes in a paper with the provocative title "Non-Darwinian Evolution" (King and Jukes 1969). These propositions elicited strong reactions from many evolutionary biologists, and it was some time before the neutrality of variation at the molecular level gained wide acceptance.

Rise of the Neutral Theory—The development of the *Neutral Theory of Molecular Evolution* as proposed by Kimura in 1968 had its beginnings from his time as a graduate student with James F. Crow at the University of Wisconsin, where he had the opportunity to interact with Sewell Wright. Kimura extended the work of Wright using advanced mathematical approaches that accounted for the effects of genetic drift, migration, and mutation. Later, in collaboration with his graduate mentor, he developed the *Infinite Alleles Model* for the maintenance of neutral genetic variation in populations (Kimura and Crow 1964). This model was particularly useful for understanding the dynamics of genetic variation at the molecular level and became widely used as the foundation for more elaborate models in molecular evolution. After returning to Japan, Kimura continued his work and along with his students, defended and expanded the neutral theory. Eventually he published a book expounding on his life-long thesis, *The Neutral Theory of Molecular Evolution* (Kimura 1983), and he has come to be considered the founder of the modern field of molecular evolution.

Kimura's early work was almost clairvoyant in its anticipation of the molecular data, which would soon become available. By the time the first data for levels of amino acid sequence divergence among species was published in 1965 (Zuckerkandl and Pauling 1965), Kimura had already established the foundation of his theory of neutral evolution. He quickly realized that selection could not explain the high rates of divergence seen among species, because selection for change in allele frequencies comes with a cost; each substitution requires substantial amounts of death of individuals in the population to allow the alleles that are favored by selection to increase in frequency. J. B. S. Haldane was the first to introduce this idea as "the cost of natural selection" (the *Cost of Selection*; Haldane 1957), which later became known as "Haldane's dilemma" (Valen 1963). Haldane pointed out that if selection were acting on a large number of alleles across the genome, it could lead to high rates of mortality and declines in population size (for more discussion on the cost of selection see Chapter 5).

Kimura proposed that the cost of selection would be too high—and would likely lead to the extinction of populations—if we were to assume that it accounted for the number of substitutions occurring at the molecular level. Based on comparisons among species of *Drosophila*, Zuckerlandl and Pauling (1965) established that protein divergence was occurring at a relatively constant rate. From these data, Kimura estimated that about one nucleotide substitution every two years had occurred since the divergence from a common ancestor. Under the classical model of selection, the demographic consequences of a high rate of allele frequency change would be required for deleterious alleles to be removed from the population, or for advantageous mutations to increase from their initial frequency of $1/2N$ to fixation (Chapter 3). The cost of allele substitution due to selection would be so severe that a species would not be expected to survive. Even though Kimura's arguments were well supported by mathematical analyses and clearly articulated, the vast majority of evolutionary biologists remained unconvinced.

The view that natural selection was the prevailing force shaping all aspects of evolution was deeply engrained and opinions were slow to change. During the 1960s and 1970s many evolutionary biologists working with animals, including Ernst Mayr, Theodosius Dobzhansky, G. G. Simpson, and E. B. Ford, as well as leading plant biologists such as Verne Grant and G. Ledyard Stebbins, championed this position. Some of these biologists seemed completely unwilling to admit any role for neutral evolution. In the 1971 edition of his tome *Ecological Genetics*, Ford inserted an extensive diatribe that was highly critical of neutral evolution. Stebbins was slightly more gracious in admitting that neutral variation at the molecular level could occur, but he was of the opinion that it only amounted to "evolutionary noise" and was of no consequence for adaptive processes at the phenotypic level (Stebbins 1974). Selectionism was the prevailing view when the first molecular genetic markers for the study of populations were introduced in the mid-1960s.

4.2 THE MOLECULAR QUANTIFICATION OF NATURE

The Allozyme Era—Soon after the introduction of starch gel electrophoresis of protein variation in 1966, an increasing number of researchers began using these markers for population and ecological genetic investigations (Fig. 4.1). Evolutionary biologists were hungry for methods to track the consequences

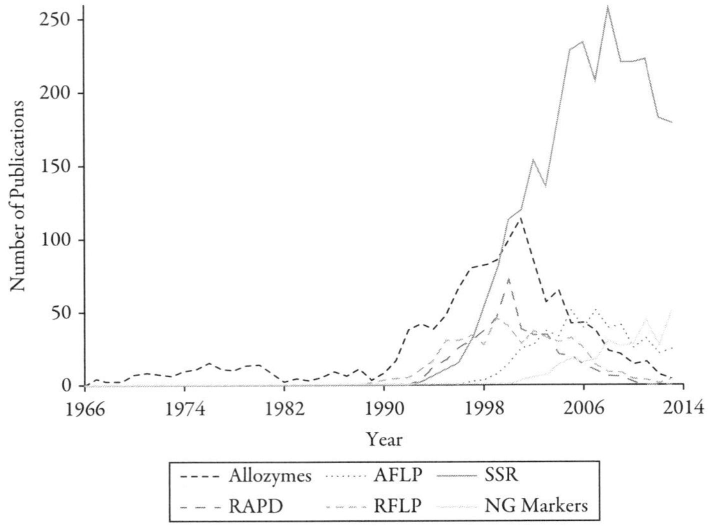

Fig. 4.1. Numbers of genetic marker publications by year for Allozymes, RAPD, AFLP, RFLP, microsatellites (SSRs), and next generation marker techniques (GBS, RAD, SNP, and sequence capture; NG Markers). Data are from the Web of Science citation database polled from the journals *Genetics, Heredity, The American Naturalist, Evolution, Molecular Ecology,* and the *American Journal of Botany.*

of evolutionary processes, and allozymes seemed to be the perfect solution. Starch gel electrophoresis offered an inexpensive technique to quantify molecular genetic variation in hundreds of individuals for up to a few dozen loci. Since virtually all evolutionary biologists of the day firmly believed that all features of organisms were fashioned by selection, they quickly adopted allozymes as the newest tool for investigations in ecological genetics.

A good proportion of the allozyme studies on plants published in the 1970s were intent on demonstrating that natural selection was acting directly on the allelic variants visualized on gels (e.g., Allard et al. 1972; Clegg and Allard 1972; Hamrick and Allard 1972), and many of these papers presented their results as a challenge to the neutral theory. Most of these studies focused on a species of grass (*Avena barbata*) that is completely selfing (*Autogamous*). Selfing lineages become homozygous within a few generations, so alleles under selection are inherited together with allozyme marker alleles. In other words, alleles across the genome remain associated across generations (*Gametic Disequilibrium*, also referred to as *Linkage Disequilibrium*; Box 10.1), so it is not possible to determine which genes are the targets of selection. In a subsequent paper, this association among alleles at different loci was invoked to explain putative examples of selection on allozyme alleles in highly selfing species (Hamrick and Holden 1979). In this later paper, the authors retracted their earlier statements about selection on allozyme alleles and concluded that non-random associations between allozyme genotypes and environments were the consequence of Associative Selection on allelic variants that were co-segregating with the allozyme markers.

Some allozyme studies reported correlations between heterozygosity and overall vigor—an observation that appeared to support balancing selection—and this encouraged a few labs to become focused on demonstrating the positive effects of heterosis (i.e., heterozygote advantage; Mitton 1989). What at first appeared to be overdominance at allozyme loci was later shown to be due the effects of inbreeding because allozyme markers were indicative of genome-wide heterozygosity (*Associative Overdominance*; Ohta 1971; Fig. 4.2). This idea was first proposed by East (1936) and has been referred to as both associative overdominance and *Pseudo-Overdominance*. The latter of these terms specifically refers to the cumulative effects of loci that are closely linked to a genetic marker locus (Delph and

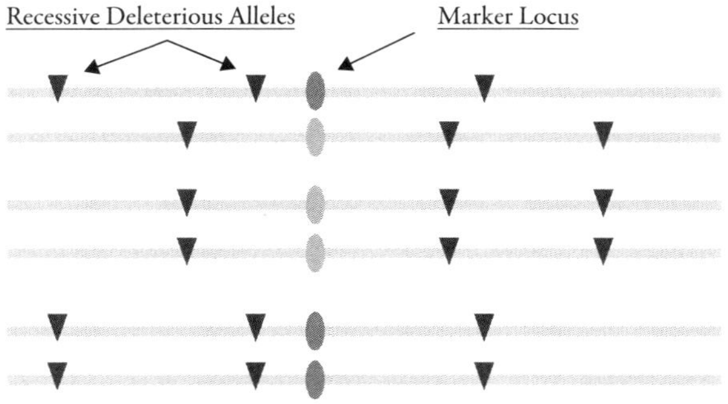

Fig. 4.2. Associative overdominance at a genetic marker locus. Each marker allele (dark and light ovals) becomes associated with different sets of linked deleterious recessive alleles (triangles) that cause reductions in fitness that appear to be due to homozygosity at the marker locus.

Kelly 2014), while associative overdominance is a more general term that encompasses the effects of genome-wide associations in heterozygosity that develop in inbreeding populations (Ohta and Cockerham 1974).

Arguments for heterosis at allozyme loci were criticized because the allelic variants that were visualized on starch gels were mostly due to differences in surface charges and were unlikely to have effects on enzyme function. On the other hand, it was pointed out that individuals that were more homozygous at allozyme loci were likely to be more inbred and consequently more homozygous across their entire genomes. Therefore, the associative overdominance hypothesis predicts that lower rates of growth and survival is due to inbreeding, and associations between heterozygosity and fitness should be absent in very large outbreeding populations. This prediction was confirmed in a critical test of the hypothesis of heterosis at allozyme loci conducted by Houle (1989) who found no effects of heterozygosity on fitness in large and outbreeding populations of *Drosophila*. While there have been a few cases where selection may have been acting on allozyme variants (e.g., Hilbish and Koehn 1987), it has become widely accepted that alleles visualized on starch gels are neutral (or nearly neutral) to the effects of selection.

The tendency for early explorers of molecular genetic variation to jump to the conclusion that the allelic differences they observed were under the influence of selection reflected the widely held opinion of the day. When allozymes were first introduced, the nature of their allelic variation was not well understood; most of these researchers later acknowledged that enzyme migration rates in gels were controlled by non-functional amino acid variation. By the early 1980s, attitudes were changing and nearly all of the early advocates of selectionism had come to recognize the value of allozymes as neutral genetic markers.

The range of applications of allozymes as neutral genetic markers expanded in the 1990s as the uses of genetic marker techniques increased at an exponential pace (Fig. 4.1). The growth in interest in genetic markers among plant biologists was spurred by the publication of two edited volumes that presented applications for a wide range of questions concerning plant systematics, population genetics, mating, demography, and ecology (Soltis and Soltis 1989; Brown et al. 1990). One contribution in particular (Hamrick and Godt 1990) established a foundation for our understanding of the effects of a range of biological and ecological characteristics of plants on the distribution of genetic variation within and among plant populations. Allozymes have continued to be valuable genetic markers for a wide range of investigations through the early twenty-first century, but since then their use declined to the point where publications using these genetic markers became rare after 2010 (Fig. 4.1).

The PCR Revolution—The landscape of genetic markers began to change dramatically in the early 1990s as methods utilizing thermostable DNA polymerase in *PCR* (Polymerase Chain Reaction) for assays of DNA sequence variation became more common. During this period allozymes remained a valuable tool for population genetic studies due to their low cost and ease of use; however, there was growing frustration, because allozymes offered limited levels of variation that were often inadequate for robust estimates of population genetic parameters.

One of the first hypervariable genetic marker approaches used restriction enzyme digestion (Restriction Fragment Length Polymorphism—*RFLP*) of whole-genome samples (e.g., Meagher 1986). Methods for whole-genome RFLP analysis were laborious, so few population studies successfully applied this method. RFLP analysis of genomic DNA was later supplanted by digestion of PCR amplification products (*PCR-RFLP, CAPS*; Appendix). This method was particularly valuable for assaying sequence variation in chloroplast genomes (Cruzan 1998), and remained popular until its use declined after 2004, as DNA sequencing methods became increasingly accessible and economical.

While PCR-RFLPs were valuable for the smaller genomes of chloroplasts and mitochondria, this method was not easily applied to nuclear genetic variation because most genes were present as multiple copies. Consequently, PCR primers designed for one gene region often amplified multiple paralogs as well as pseudogenes, resulting in banding patterns that were difficult to interpret.

Two methods were introduced in the late 1980s and early 1990s that circumvented these difficulties with PCR-RFLP approaches. One of these methods, *RAPDs*, provided a relatively easy and inexpensive procedure for sampling sequence variation randomly across the genome (Welsh and McClelland 1990; Williams et al. 1990). This genetic assay was an attractive alternative to allozymes because single PCR reactions could generate large amounts of information on variation across many loci with minimal investment in the development and refinement of protocols (Appendix). Because of the low initial investment and ease of use, RAPDs became the genetic marker of choice for a large number of labs in the mid-1990s.

A second method that became popular in the 1990s and remains in wide use today was based on the discovery of large numbers of DNA sequence regions consisting of multiple repeats of short sequences (bi-, tri, tetra-, or penta-nucleotide motifs) that are referred to as *Microsatellites* (microsatellite DNA regions—also known as *SSRs, STRs*, or *VNTRs*; Avise 2004; Appendix). These short-repeat regions are subject to relatively high mutation rates (approximately 10^{-3}, compared to 10^{-5} or lower for base substitutions in non-repeat regions) due to slippage during DNA replication. Consequently, microsatellite loci often segregate for large numbers of alleles. High allelic variation renders these markers particularly suitable for identity profiling (*DNA Fingerprinting*) and for estimating relatedness among individuals (Appendix).

Microsatellite variation (SSRs) was first discovered as a valuable tool for quantifying genetic variation in humans and a handful of other species (Jeffreys et al. 1985; Burke 1989), but the first usage in plants did not occur until some years later after it was confirmed that plant genomes contained microsatellite regions (Morgante and Olivieri 1993). This method was slow to gain wide usage because of the relatively large investment required for marker development. In general, unique PCR primers were required for each SSR locus in each species, and developing these markers required a significant investment of time and resources for the screening of DNA clones, sequencing, and the designing and testing of primers. By chance some of the primer pairs originally developed for humans also worked for a few other species including some birds, so some of the earliest papers using microsatellites in population biology took advantage of this *Cross-Amplification* ability (Burke 1989; Reeve et al. 1992).

In the late 1990s, it became apparent that genetic markers based on the RAPD technique were often unreliable and sometimes not reproducible in different labs. This is largely due to the small size of the oligonucleotide primers used for this method, which are typically only ten to twelve bases in length. Shortening the length of primers increases the number of bands appearing on gels, but also results in unstable and nonspecific primer binding. Consequently, this method is susceptible to small variations in PCR reaction conditions, so many labs began abandoning RAPDs in favor of alternative methods in the early twenty-first century. The demise of RAPDs and a similar method that used primers that were anchored in a microsatellite region (*ISSR*; Appendix) was formalized when journals such as *Molecular Ecology* established policies stating that manuscripts based solely on these markers "rarely pass the scrutiny of peer review."

The relatively large investment required for the development of microsatellite markers during the late 1990s and early 2000s continued to present an impediment that motivated researchers to seek alternative methods. One of the most successful methods combined RFLP analysis with selective PCR amplification of the cut fragments (*AFLP*; Vos et al. 1995; Appendix). By ligating oligonucleotides that were twenty or more base pairs in length to the ends of fragments and using complementary sequences as primers, AFLPs circumvented the priming difficulties inherent to RAPDs (Mueller and Wolfenbarger 1999). Variation in these markers is due to DNA base pair changes in specific restriction sites rather than priming affinities (as in RAPDs and ISSRs), so AFLPs are more effective at sampling variation in defined portions of the genome. The utility of AFLPs (and microsatellite markers) was further enhanced by assaying samples with the automated capillary electrophoresis instruments that became available in the late 1990s. The higher reliability, low start-up costs, and high levels of polymorphism of markers sampled from across the genome contributed to the continued popularity of AFLPs (Meudt and Clarke 2007).

In spite of the improved reliability of AFLPs, these markers share several limitations with other random fragment techniques such as RAPDs and ISSRs (Appendix). All three of these marker types are biallelic—variation is based on the presence or absence of a PCR product. These random fragments have dominant expression (i.e., a band is present on the gel whether the genotype is homozygous or heterozygous) so the range of applications is more limited than for codominant markers such as SNPs (see the next section), microsatellites, and allozymes. AFLP assays do not provide allelic data so the bands (or peaks) are often referred to as phenotypes. Moreover, since these assays do not sample specific locations in the genome, peaks that appear to be the same size in the genetic fingerprints of two individuals may not represent the same location. Because of these problems, AFLPs and other random fragment methods have been criticized because of the low confidence for the homology of similarly sized fragments. In contrast, for microsatellites and other codominant marker methods there is an a priori expectation of banding patters (i.e., one or two peaks for homozygous and heterozygous diploid genotypes, respectively), so banding artifacts appearing in these assays can be more confidently discarded. The ease of interpretation of fragment patterns and high level of variation obtained from microsatellites contributed to their rapid increase in popularity in the 1990s, which continued as more efficient protocols for primer development became available in the 2000s (Fig. 4.1).

The NGS (Next Generation Sequencing) Revolution—DNA sequencing as a tool for investigations in evolutionary biology started to become common in the late 1980s. Early methods were laborious and expensive, but sequencing became more accessible with the introduction of automated capillary electrophoresis instruments in the 1990s. During this time, DNA sequencing developed as an important method for comparison of orthologous regions in phylogenetics, phylogeography, and population genetics. This technology was also important for the development of microsatellite markers; one of the largest obstacles for the generation of new markers is screening (sequencing) large numbers of clones for the identification of repeat regions and the design of primers. Access to large amounts of sequence data became a reality after 2004 when *Next Generation Sequencing* (*NGS*) instruments became widely available. These technologies have continued to diversify and have enabled the collection of large amounts of population genetic data for the resolution of questions in population and evolutionary biology.

Many NGS methods for the collection of population genetic data are based on locating *Single Nucleotide Polymorphisms* (*SNPs*; Appendix). These are nucleotide base positions that are homologous and differ in nucleotide identity (A, G, T, or C) among individuals, populations, or lineages. Unlike random fragment methods, SNPs are codominant so they provide allelic information (Morin et al. 2004). SNPs may represent silent variation (within non-functional DNA) or within reading frames, they provide us with information on synonymous and non-synonymous differences (see Chapter 3 for discussion).

Data on SNPs and indels can be gathered using a variety of techniques designed to create reduced libraries of DNA fragments that are then barcoded with unique oligonucleotide sequences to identify samples and pooled for mass sequencing. Methods amendable to these procedures include the isolation of fragments associated with restriction cut sites such as *Restriction Site Associated DNA Markers* (*RAD* or *RAD-Seq*; Baird et al. 2008; Davey et al. 2011) and *Genotyping by Sequencing* (*GBS*; Elshire et al. 2011). Unlike previous restriction site marker methods (RFLP and AFLP) these sequencing methods allow comparison of genomic regions that are known to be homologous. SNPs can also be scored for specific regions by fragmenting DNA using complementery RNA baits to capture known sets of DNA segments (*Targetted Capture*; Cronn et al. 2012; Jones and Good 2015). This method is particulary amendable to *Exome* (spliced exon) sequencing using gene sequences obtained from mRNA (*Expressed Sequence Tags; ESTs*), and for sequencing of the chloroplast genome (Stull et al. 2013).

GBS and RAD techniques provide samples from across the genome while methods based on EST sequences will target expressed regions. These NGS-based methods enable quantification of variation at thousands of variable sites for many hundreds of individuals at a relatively low cost. Since many of these methods require little investment in time and financial resouces for development, they may replace microsatellites as the markers of choice for many population and ecological genetic studies on non-model species—a trend which may already be underway (Fig. 4.1).

In the next section we will discuss the history and origin of the field of genetics and the integration of Medelian principles with Darwin's theory of evolution. We will then return to the application of genetic markers to answer questions in population genetics as we explore each evolutionary process in turn.

4.3 MEASURING GENETIC VARIATION

Origins of Modern Genetics—To appreciate the types of information that different measures of genetic diversity can provide, it is useful to review some fundamental aspects of the distribution of genetic variation in nature—the Hardy–Weinberg (HW) Principle. After the rediscovery of Mendel's work in 1900, there was considerable confusion over the meaning of his findings. Some researchers proposed that Mendelian laws did not apply to inheritance of most phenotypic characters, and there was disagreement over the long-term fate of dominant and recessive traits (Provine 1971). A few researchers interpreted Mendel's results to imply that dominant alleles would always be favored, while others thought that heterozygotes would be superior. A geneticist, Reginald C. Punnett (inventor of the Punnett's square), brought the problem to the attention of Godfrey H. Hardy, a fellow cricket enthusiast and mathematician. Hardy described the solution to the problem in a letter to *Science* in 1908 (Hardy 1908). Using what he considered "simple" mathematics, Hardy concluded that, in the absence of any opposing force, allele frequencies would be expected to remain the same from one generation to the next. It was later pointed out that Wilhelm Weinberg had arrived at the same conclusion in a separate paper (Weinberg 1908), after which the prediction of constant allele frequencies across generations was referred to as the Hardy-Weinberg principle (or HW equilibrium).

Hardy's insight was to predict diploid genotype frequencies from allele frequencies given some simplifying assumptions. His argument followed from Mendel's observation that the two alleles at each locus were passed on to offspring at equal frequencies (i.e. half of the gametes produced would be expected to carry each allele). If most individuals in a population are carrying one of two alleles then the majority of the gametes produced will have the most common allele. In a randomly mating population, the chance (probability) that any two alleles would be combined corresponds to their frequencies; gametes carrying alleles that are more common are more likely to combine to form genotypes.

We can predict the probability that two events will occur together (events A and B occur at the same time) by multiplying the frequencies of each event; in this case the chance of combining alleles carried by two different gametes. Hence, the probability of allele A_1 from two gametes combining to form a homozygous diploid in a randomly mating population is the product of the allele frequencies ($p * p = p^2$; where the two allele frequencies sum to $1.0 = p + q$; Fig. 4.3). For example, if allele frequencies for two alleles segregating at one locus are both 0.5, then the probability of forming a homozygous A_1A_1 genotype will be 0.25 (0.5*0.5), or if p were 0.9 then the homozygous genotype frequency would be 0.18. Similarly, heterozygous genotypes may be formed by combining gametes carrying two different alleles, but this may happen in two different ways (i.e., A_1 and A_2 or A_2 and A_1). With two different ways to get the same thing (or), the combined probabilities are added ($p*q+q*p=2pq$; 2*0.5*0.5 = 0.5; and 2*0.9*0.1 = 0.18). For a locus segregating for two alleles we can predict all three genotype frequencies in a population from the allele frequencies (p^2, $2pq$, and q^2 for A_1A_1, A_1A_2, and A_2A_2, respectively; note that genotype frequencies sum to 1.0; Fig. 4.3).

Hardy argued that without some type of perturbation, the allele frequencies in the next generation could be predicted from genotype frequencies in the current generation. If we start by predicting allele frequency in the next generation (p_1; $t = 1$) based on genotype frequencies in the current generation ($t = 0$) we can state that

$$p_1 = p_0^2 + ½(2p_0q_0)$$

Note that we use the expected genotype frequencies in the current generation (p_0^2 and $2p_0q_0$) to predict the allele frequency in the next generation. Homozygotes carry two A_1 alleles while heterozygotes carry only one so only half the heterozygote frequency is included. By substituting $1 - p_0$ for q_0 (i.e. because $p_0 + q_0 = 1$) we get a version of this equation that can be simplified more easily;

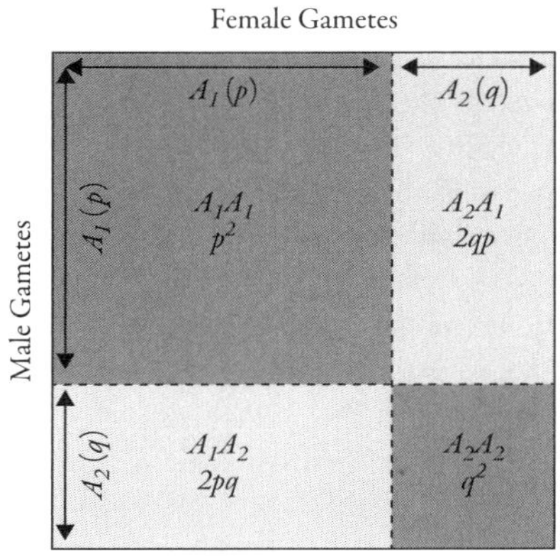

Fig. 4.3. Change in allele frequencies in male and female gametes and predicted changes in genotype frequencies under the Hardy-Weinberg principle.

$$p_1 = p_0^2 + ½(2p_0(1 - p_0)),$$

which becomes

$$p_1 = p_0^2 + ½(2p_0 - 2p_0^2).$$

Note that the "2" cancels out so now we have

$$p_1 = p_0^2 + (p_0 - p_0^2),$$

and finally

$$p_1 = p_0$$

This relationship will remain true—p will remain constant across generations—as long as any force or other perturbation (such as selection, genetic drift, gene flow, or mutation) does not influence genotype frequencies.

At the time that Hardy's paper was published equivalence conditions allowing constant allele frequencies were not thoroughly understood, but we recognize them today as the primary assumptions for HW equilibrium. These include: 1) the organism is diploid; 2) reproduction is sexual; 3) mating is random; 4) the population size is very large (to avoid random fluctuations in allele frequencies—genetic drift); 5) allele frequencies are equal between the sexes; 6) there is no selection; 7) there is no gene flow from other populations; and 8) there is no mutation. In most population genetic studies, this last condition is often ignored because the effects of mutation are expected to be extremely small. We also note that, for simplicity, population genetic models often assume that generations are discrete and not overlapping, but this condition is not necessary for HW equilibrium to be maintained. Given these conditions, we expect allele frequencies to remain constant over many generations, and with no mutation, indefinitely.

In the next section, we will discuss how the HW principle can be applied for the estimation of genetic diversity within populations using different types of genetic markers. In later sections, we will explore the consequences of violations of the HW assumptions for the levels and distribution of genetic variation within and among populations.

Measures of Genetic Diversity—With the increasing availability of genetic marker technologies after the

1970s, the application of principles from evolution theory for the description of levels and distribution of genetic variation became more common. Analytical methods for the quantification of genetic diversity and relatedness among individuals and populations developed in step with characteristics of available markers. Depending on the type of genetic marker data available, different measures of genetic diversity can be applied. At the most basic level—say for a polymorphic phenotypic trait—we can count the numbers of different variants present. For flower color polymorphism in snap dragons (*Antirrhinum majus*), individual plants have white, red, or intermediate (pink) flower types, which correspond to homozygous and heterozygous genotypes for a locus that is segregating for two alleles. From these data it is relatively straight forward to estimate allele frequencies from observed genotype counts:

$$p = (2G_{11} + G_{12}) / 2N.$$

Where G_{11} and G_{12} are the numbers of homozygous (white) and heterozygous (pink) genotypes, and N is the population size (note that we use $2N$ as the denominator because each individual carries two alleles). In this case we are estimating the allele frequencies (p and $q = 1 - p$) directly from the observed genotype frequencies, so the assumption of HW equilibrium is not necessary. We can use these allele frequency estimates to generate genotype abundances predicted under HW (Np^2, $2Npq$, and Nq^2 for $\hat{G}_{11}$, $\hat{G}_{12}$, and $\hat{G}_{22}$ respectively). The observed genotype numbers can then be compared to the expected values using standard tests such as the chi-squared test for heterogeneity to identify significant departures from HW expectations. We will explore the causes of various departures from expected genotype frequencies later in this chapter.

Most genetic markers studies differ from this example, because data are available for multiple loci and often for multiple alleles per locus. When comparing levels of genetic diversity among populations within and among species we seek standardized measures that are not affected by population genetic processes. Comparisons are generally valid only for the same types of data because markers differ for mutation rates, which translates into different levels of variation (i.e., variation in SSRs > AFLPs > allozymes). Markers also differ with respect to allelic dominance, numbers of alleles per locus; random fragment methods such as RAPDs and AFLPs are bi-allelic and have dominant expression, while allozymes and SSRs are codominant and multiallelic, and SNPs are codominant and typically treated as bi-allelic. The number of loci typically available for use also differ, with up to a dozen or so for allozymes, between eight and thirty for SSRs, from fifty to a few hundred for random fragment techniques, and Next Generation Sequencing (NGS) techniques typically yield variation at thousands of SNP loci.

Differences in the characteristics of markers also affect the types of parameters that are generally used to measure variation. For example, the *Proportion of Polymorphic Loci* (P) ranges from 0.0 to 1.0; a locus is considered polymorphic if the most common allele frequency is less than an arbitrary threshold such as 0.95 or 0.99. This was a routinely used measure of variation for allozyme data because similar sets of enzymes could be compared among species. The proportion of polymorphic loci is less valuable for DNA-based markers because standard sets of loci cannot be compared (markers are unique to each species), the number of loci assayed may differ among different studies and species, and loci with little or no polymorphism (i.e., monomorphic loci) are generally not assayed because of cost considerations (e.g., for SSR studies).

For multiallelic allozyme and DNA-based markers a simple measure of diversity is the *Average Number of Alleles* detected across loci ($A = (1/k)\Sigma n_j$; where n_j is the number of alleles detected for locus j, for j = 1 to k loci). This parameter provides an estimate of allelic richness, and it reflects differences in diversity because of variation in allele frequency. For example, if two alleles at a locus occur at equal frequency then the diversity is higher than if one occurred at a high frequency (e.g., $p = 0.9$).

The lack of sensitivity to allele frequency differences can be remedied by calculating the *Effective Allele Number* (A_e), which captures differences in diversity due to allele richness and evenness. We can calculate A_e as the average of the reciprocal of heterozygosity across loci:

$$A_e = \sum_j \left(1/\left(1-h_j\right)\right)/k = \sum_j \left(1/\sum p_i^2\right)/k.$$

Where h_j is the heterozygosity at each locus ($h_j = 1 - \Sigma p_i^2$; summed across alleles at each locus), and k is the number of loci. The effective allele number represents a weighted average that is larger when more alleles are present at equal frequency.

Another commonly used parameter that captures both allelic richness and evenness is the *Expected Heterozygosity* (H_e). For a single locus with two alleles the expected heterozygosity is equal to expected frequency under HW conditions ($H_e = 2pq$). With multiple alleles the expected heterozygosity is most conveniently calculated as the difference between unity and the sum of the expected frequencies of homozygous genotypes ($h_j = 1 - \Sigma p_i^2$). Averaging across loci we obtain

$$\bar{H}_e = \sum_j \left(1 - \sum p_i^2\right) / k.$$

The average expected heterozygosity is also referred to as *Nei's Genetic Diversity* (H_s; Hedrick 2009). This parameter represents the sum of the product of allele frequencies without the influence of population genetic processes so it reflects the level of genetic diversity expected under HW conditions. Like the effective allele number, H_e is maximized when allele frequencies are equal (Fig. 4.4).

As a measure of genetic diversity, H_e is particularly useful because it can be used for both codominant and dominant genetic markers. For the latter, the frequency of the dominant allele at each locus (p) is calculated from the frequency of the recessive genotype ($p = 1 - \sqrt{(G_{22} / N)}$; where G_{22} is the number of individuals lacking a PCR product—no band on the gel—for the locus). Again, we are assuming that the allele frequencies are in HW equilibrium. While this is unlikely to be true under most circumstances, H_e is valuable for diversity comparisons among populations and species because it provides a reliable estimate of the total genetic variation present.

Estimates of genetic diversity such as A_e and H_e for genetic markers represent the level of neutral variation available in a population at any particular time, which will reflect the historical effects of population size and migration on levels of genetic diversity. Estimates of diversity that assume HW conditions avoid biases that are introduced due to contemporary population processes. For example, as we will see later in this chapter, the observed heterozygosity (H_o) is often lower than the expected heterozygosity (H_e) due to the effects of inbreeding. Since the intensity of inbreeding may vary among populations, we would expect H_o to provide a less reliable comparator for genetic diversity among populations and species than the expected heterozygosity.

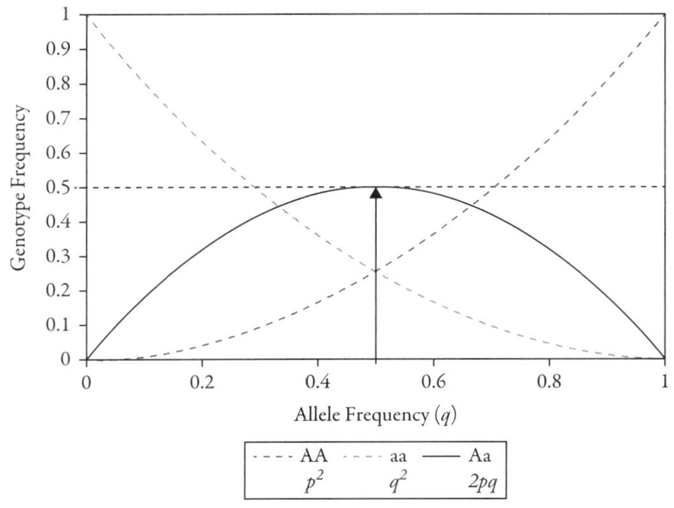

Fig. 4.4. Predicted changes in genotype frequencies with allele frequency (q) under the Hardy-Weinberg principle.

For sequence data (SNPs) or restriction enzyme polymorphisms, estimates of genetic diversity can be made at the nucleotide level (π) using the formula,

$$\pi = \frac{n}{n-1}\sum_{i=1}^{n}\sum_{j=1}^{i-1} p_i p_j \pi_{ij}.$$

Where p_i and p_j are the frequencies of the ith and j^{th} alleles (sequence variants), π_{ij} is the number of base differences between the ith and j^{th} alleles, and n is the number of haploid sequences sampled (Hedrick 2009). The summations are designed to compare all possible pairs of alleles for $i \neq j$. The correction for small sample sizes $(n/(n-1))$ approaches 1.0 for large samples. This estimate of nucleotide diversity is similar to expected heterozygosity and can be used for comparisons of diversity within and among populations.

Estimates of genetic diversity such as A_e and H_e are sensitive to differences in the type of genetic markers used and the complement of loci chosen, which renders the comparison of levels of genetic diversity among different studies or species more difficult. For example, random fragment markers generally provide lower estimates of heterozygosity than SSR loci (Nybom 2004) simply because there are only two alleles per locus for random fragment loci. Even for studies using the same type of technology, the number of loci included and the choice of different sets of markers will affect diversity estimates. In addition, different researchers may choose to either include or exclude less variable loci (monomorphic markers) in their estimates of diversity. In spite of these difficulties, the expected heterozygosity remains as the most commonly used and most reliable estimate of genetic diversity for allozyme-, PCR-, and NGS-based genetic marker studies for comparison among populations and species.

These genetic diversity parameters, and especially estimates of heterozygosity, provide measures for estimating the effects of evolutionary processes on the distribution of genetic diversity within and among populations—the field of study known as population genetics. In the next section, we will discuss the origin and development of the field of population genetics and then explore the effects of population size, inbreeding, and gene flow on levels of genetic diversity.

4.4 POPULATION GENETICS

Most genomic regions of organisms are subject to forces that violate the assumptions of HW. As discussed in Chapters 3 and 5, transcribed regions will be subject to purifying selection and occasionally will be affected by selective sweeps. Population-level processes such as genetic drift and gene flow (migration) from other populations will affect the entire genome, and typically, multiple processes simultaneously affect the distribution of genetic variation within and among populations. In this section, we will first discuss the development of the field of study focused on genetic processes at the population level—population genetics. We will then explore each process that represents a departure from HW conditions.

The Origins of Population Genetics—After the rediscovery of Mendel's work in the dawn of the twentieth century, many biologists interpreted the revelations from his studies on the inheritance of phenotypic traits in plants as the end of Darwinism. This perception finds its roots in skepticism expressed by Darwin's contemporaries such as Thomas Huxley and Francis Galton concerning the importance of *Gradualism* versus larger evolutionary jumps (*Saltations*). The views of saltationists gained momentum toward the end of the 1800s from investigations by Hugo de Vries on populations of Lamarck's evening primrose, *Oenothera lamarckiana* (Fig. 4.5; refer to color plate). From his observations of spontaneously generated varieties of *O. lamarckiana*, de Vries developed his *Mutation Theory*, which was promoted as an alternative to Darwinian evolution. Mutation theory and the saltationists' views were catapulted into prominence when Mendel's experiments came into the spotlight in 1900 (Provine 2001). In spite of the efforts of the biometricians (Raphael Weldon, Karl Pearson, and Francis Galton; also see Chapter 7), Darwin's version of evolution by gradualism was under siege and doubted by many biologists of the day who followed William Bateson and his Mendelian view of evolution.

The disagreement between Mendelians and gradualists focused on the primary mode of evolution: whether speciation occurred by saltations according to the Mendelian laws of inheritance, or whether the slow accumulation of small variations could be the foundation of evolutionary transitions.

Fig. 4.5. Lamarck's evening primrose, *Oenothera lamarckiana*. Image by Hugo de Vries.

This controversy was quieted to some degree after Ronald A. Fisher published a paper demonstrating mathematically that continuous traits could be controlled by Mendelian loci (Fisher 1918; Fig. 7.2), but due to the broiling controversy and the uncertainty over the role of gradualism, advances in evolutionary biology remained largely stalled through the first few decades of the twentieth century (Provine 2001).

During the 1920s three geneticists were working to clarify the importance of gradualism and to integrate Darwin's ideas on evolution by natural selection with Mendelian laws of inheritance (Provine 2001). The first of these was R. A. Fisher, who continued his work to develop a general theory for the effects of natural selection; this was first published in a series of papers during the 1920s and later in its entirety in his 1930 monograph *The Genetical Theory of Natural Selection* (Fisher 1930). The second was Sewall Wright, a geneticist working with coat color inheritance in Guinea pigs. Wright's experience with breeding experiments had a strong influence on the development of his ideas on the population genetics of small populations.

Unlike Fisher, who was a mathematician, Wright was an empiricist who developed many of his ideas on evolution and genetics from the inheritance of phenotypic traits. Wright described a view of selection that focused on interactive complexes of genes controlling traits that conferred adaptation. Fisher, on the other hand, assumed that adaptation occurred via selection on single loci. The other important differences in the approaches of Wright and Fisher were their assumptions concerning the roles of population size and genetic drift. Fisher modeled selection in large populations where the influence of genetic drift was minimal, while Wright thought the consideration of many small populations subject to inbreeding and genetic drift was more realistic. In Wright's view, beneficial mutations could become fixed more easily in small populations so evolution was limited by gene flow among populations more than by mutation. Wright published the compendium of his modeling efforts in population genetics and his views on evolution in a paper that appeared in 1931 (Wright 1931).

The third geneticist working to reconcile Mendelian genetics with Darwinian natural selection during the late 1910s and 1920s was J. B. S. Haldane (Provine 2001). Haldane focused on problems centering on the estimation of linkage and recombination among loci. He generally agreed with many points emphasized by both Fisher and Wright and wrote a series of papers that examined different aspects of selection in populations. Haldane summarized much of his work in a book titled *The Causes of Evolution*, which was published in 1932 (Haldane 1932).

Fisher, Wright, and Haldane all shared the same goal; to dispel the perception that there was a contradiction between Mendelism and Darwinism. All three authors also agreed on the primary importance of natural selection in evolution but differed with respect to the importance of genetic drift. Their work successfully integrated Mendelian genetics with Darwinian evolution and established a mathematical foundation for population genetics, which became known as the *Modern Synthesis of Evolutionary Biology*.

Many biologists of the early twentieth century did not fully appreciate the implications or importance of the mathematical treatments presented by Fisher,

Wright, and Haldane. It was not until some years later when evolutionary biologists such as Theodosius Dobzhansky (Dobzhansky 1937), E. B. Ford (Ford 1971), Ernst Mayr (Mayr 1942), G. G. Simpson (Simpson 1944), and G. Ledyard Stebbins (Stebbins 1950) provided interpretations and illustrations of the mathematical principles of population genetics that the modern synthesis came to fruition and was established as the foundation of modern evolutionary biology. In the upcoming sections, we explore several of the major issues addressed by the mathematical architects of the modern synthesis, including the effects of small population size on genetic drift and inbreeding, gene flow among populations, and later in Chapter 5, the effectiveness of natural selection.

4.5 EFFECTIVE POPULATION SIZE

The sedentary nature and two-dimensional spatial distribution of terrestrial plants restricts pollen and seed dispersal to local areas, which limits the number of individuals able to interact as an effective population. Wright defined the effective population size as the number of individuals that could readily mate with each other—a concept he referred to as an *Idealized Population* (Wright 1931). Limited dispersal imposes restrictions on mating opportunities and has consequences for population genetic processes (Loveless and Hamrick 1984). Because of spatial limitations, the effective population size (N_e) will always be smaller than the census number of individuals occupying any particular geographic region (*Census Population Size*). With smaller effective population size, the effect of genetic drift becomes prevalent so that genetic variation is more likely to be affected by random changes in allele frequencies across generations (Fig. 4.6). Smaller populations are characterized by larger fluctuations in allele frequency each generation and a greater chance of *Fixation* (allele frequency becomes 0 or 1) over shorter time frames.

Genetic drift is a consequence of unequal contributions of each individual to offspring in the next generation, but these differences are random with respect to genotype. For any particular population, individuals will vary in their reproductive success due to the vagaries of random events; larger fluctuations in allele frequencies for smaller populations are a direct consequence of random events acting on a smaller group of individuals. By analogy, we could imagine flipping a coin ten times in a row, in which case it would not be that unusual to get "heads" seven out of ten times (70%; a deviation of 0.2 from the expected frequency of 0.5). In contrast, if we were to flip a coin 100 times, we would expect the frequency of heads would generally be much closer to 50%, and seeing 70 heads would be extremely rare. Translating this process to ecological processes, we would expect that random events during reproduction, seed germination, and survival would have much larger effects on allele frequencies when the sample size (N_e) is smaller. This principle is of universal importance as it affects the levels and distribution of genetic variation across populations and ultimately the rates of lineage divergence over longer time scales.

While local dispersal is probably the primary factor limiting N_e for most plants, the effective population size is often further reduced by other circumstances contributing to unequal reproductive success. For example, variation in plant size results in corresponding differences in seed and pollen production. Larger individuals will produce more flowers, which reduces the diversity of the gamete pool and the level of genetic variation among offspring in the population. We can estimate the effects of unequal reproduction on N_e by taking into account the census population size (N), the average number of offspring produced per individual (n), and the variance in offspring produced per parent (V_n):

$$N_e = \frac{Nn}{n + \frac{V_n}{n}}.$$

(equation is modified from Waples 2006 and Hedrick 2009 to accommodate hermaphroditic species). For a stable hermaphroditic population ($n = 1$), the effective population size is the same as the census ($N_e = N$) when $V_n = 0$ (Fig. 4.7). With the exception of controlled breeding conditions, we expect that V_n will be greater than zero, and increasing the variance in offspring numbers (V_n) dramatically reduces the effective population size. For example, for a population size of $N = 10$, if five individuals produce two progeny each then $V_n = 1.1$ and N_e is slightly less than five ($N_e = 4.8$). In the case where only one of ten individuals produces all of the offspring, $V_n = 10$ and the effective population size is slightly less than one

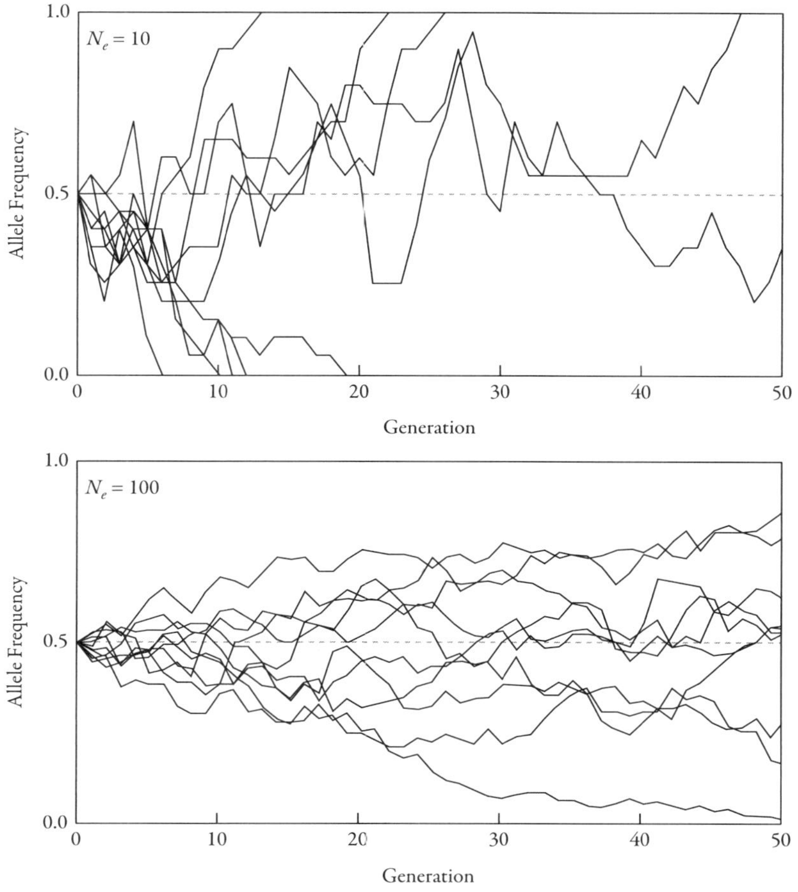

Fig. 4.6. Simulations of the effect of population size on changes in allele frequencies across generations. Each line represents a single population with a single locus that is segregating for two alleles. Scenarios for ten populations with N_e = 10 (top panel) and 100 (bottom panel) are shown. Simulations created using PopG (http://evolution.gs.washington.edu/popgen/popg.html).

($N_e = 0.9$). Note that the effect of the variance in offspring numbers is moderated by the average number of offspring per individual (n), so differences in reproductive success among individuals has a reduced effect on N_e in a growing population ($n > 1$), and an exaggerated effect on N_e when population numbers are declining ($n < 1$).

Other factors can create unequal contributions to the progeny pool and reduce the effective population size. In dioecious species, an unequal sex ratio means that fewer individuals of one gender are parenting offspring so the effective population size will be lower. We can estimate the effective population size in a dioecious population from the census numbers of males (N_m) and females (N_f) as

$$N_e = \frac{4N_m N_f}{N_m + N_f}.$$

As an example, if there were nine males and one female in a population then N_e would be $(4*9*1)/(9+1) = 3.6$. The proportional decrease in N_e due to unequal sex ratio is not dependent on total population size (e.g., for 90 males and 10 females $N_e = 36$). When N_m is equal to N_f then N_e becomes $N_m + N_f = N$, the census population size.

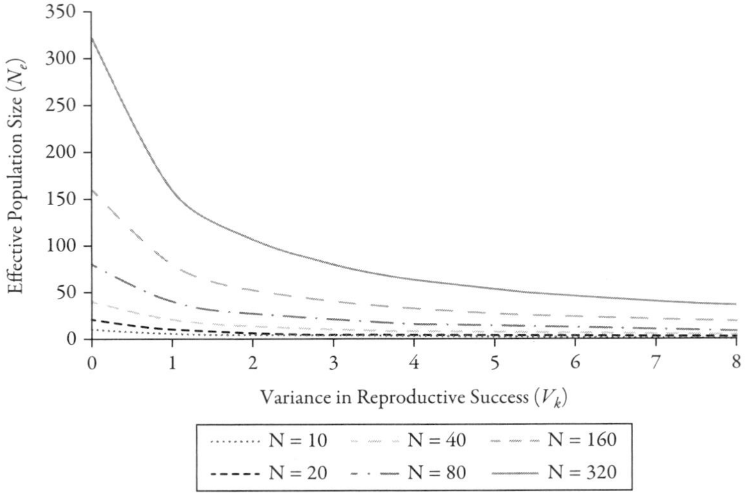

Fig. 4.7. The effects of the variance in reproductive success on the effective population size. Average number of offspring $n = 1$.

A number of methods have been developed for the estimation of effective population size using genetic marker data from multiple samples across generations and from a single sample in one generation (Luikart et al. 2010; Do et al. 2014). While we could consider population size across long time frames, most of the available methods focus on contemporary estimates of N_e over the past few generations. Luikart et al. (2010) also make a distinction between the inbreeding effective population size (N_{eI}) and the variance effective size (N_{eV}). The former parameter focuses on loss of observed heterozygosity (see the section on inbreeding later in this chapter), while the latter is concerned with loss of allelic variation. Estimates of N_e based on multiple samples generally assume discrete generations (i.e., annual life history) and no gene flow or selection. Single-sample methods use non-random associations among alleles at different loci (gametic disequilibrium; Box 10.1), which are generated due to genetic drift. The larger number of genetic marker loci obtained from NGS analyses have allowed for the development of methods that account for the effects of selection and gene flow for more robust estimates of N_e from single samples (Luikart et al. 2010; Do et al. 2014).

The effective population size is also affected by the action of other evolutionary processes. Strong selection produces unequal reproductive success and consequently smaller number of individuals contributing to the next generation. Selection acts on specific genomic regions, but if fewer individuals are successfully reproducing, then the effective population size will be reduced across the entire genome. As a consequence, genetic drift and selection do not act independently, and the potential effects of drift need to be accounted for when designing experiments that apply selection across multiple generations (Hill and Caballero 1992)—a topic we will explore further in Chapter 5.

4.6 GENETIC DRIFT

Principles of Genetic Drift—In very large populations the loss of individuals through accidents of nature will hardly be noticed, but random events are more pronounced when numbers are few. Solely by chance, allele frequencies may fluctuate as the fate of individuals carrying them is played out; to the extent that survival and reproductive success is independent of genotypes, allele frequencies will change randomly—a phenomenon referred to as genetic drift. In this section, we describe five principles that govern changes in allele frequencies due to genetic drift.

The first principle of genetic drift is that, for collections of populations, the mean allele frequency is predicted to remain the same as the initial (expected) allele frequency across generations. The random fluctuations in allele frequencies among populations are seemingly intractable (Fig. 4.6), but there are some predictions we can make for the distribution and levels of genetic variation. As a first estimate, we could generate a distribution of probable allele frequencies from one generation to the next, but our confidence in these predictions would rapidly decline as we attempt to project further into the future. The initial allele frequency (p_0) provides a theoretical prediction of the expected allele frequency ($\hat{p}$) in all subsequent generations ($\hat{p} = p_0$), but this expectation is never met in real populations. On the other hand, note that a roughly equal number of populations diverge above and below the initial allele frequency (the blue dashed lines in Fig. 4.6). If we were to estimate the average allele frequency across all of the populations, we would find that it remains close to the initial allele frequency (i.e., p_0). From these considerations, we can conclude that for a collection of populations the mean allele frequency is expected to remain the same—close to the expected allele frequency—across generations.

The second genetic drift principle is that random changes in allele frequencies across generations will lead to divergence in allele frequencies among populations, and levels of divergence will increase over time. Note that in Fig. 4.6 the allele frequencies for individual populations diverge independently; they tend to drift away from each other, and over generations, allele frequencies will continue to diverge among populations.

The third genetic drift principle is that the variance in allele frequency among populations will increase over time until all loci become homozygous. We can characterize the level of divergence due to genetic drift as an increase in the variance in allele frequencies among populations (V_q; Fig. 4.8). As more generations pass, the range of divergence away from the initial allele frequency increases in a predictable pattern; the variance in allele frequency increases more rapidly for populations with small effective size, while frequencies in larger populations tend to remain closer to their initial values. If we assume that initial frequencies are the same for all populations, we can predict the variance among populations as a function of generations (t), the initial frequencies of both alleles (p_0 and q_0), and the effective population size (N_e);

$$V_{q \cdot t} = p_0 q_0 \left[1 - \left(1 - \tfrac{1}{2}N_e\right)^t\right]$$

Note that, given enough time, V_q will approach a maximum value ($V_{q \cdot max} = p_0 q_0$). We can see with the relationship for $N_e = 10$ in Fig. 4.8 where the line is

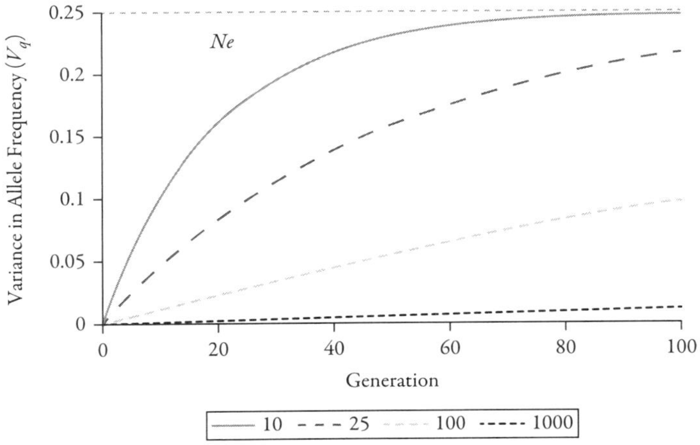

Fig. 4.8. Expected increases in the variance in allele frequency across generations for populations of different size subject to genetic drift. Initial allele frequencies are $p = q = 0.5$ for all four effective population sizes.

making an asymptotic approach to a value of 0.25 (blue dashed line) as more populations become fixed for one allele or the other. If we extended the time frame long enough all of the lines in Fig. 4.8 would eventually approach 0.25 until nearly all populations were homozygous. If allele frequencies were not equal, then $V_{q \cdot max}$ will be lower (e.g., $V_{q \cdot max} = 0.09$ for $p_0 = 0.9$).

The fourth genetic drift principle is that random changes in allele frequencies across generations lead to loss of genetic variation within populations. The asymptotic relationship for V_q over time reflects the fact that higher proportions of populations are becoming homozygous as t increases. Under genetic drift with no mutations and no gene flow, the ultimate fate of all loci is homozygosity regardless of population size; however, when the population is large, fixation will require an exceedingly long time. If we are tracking a limited number of populations, we expect all of them will eventually become fixed. The relationship described by the equation above is asymptotic because it assumes an infinite number of populations are evolving; some infinitesimally small fraction of populations continues to segregate for both alleles as V_q approaches, but never becomes equal to, $V_{q \cdot max}$. Any individual population will eventually become completely homozygous regardless of the effective population size.

The fifth genetic drift principle is that the probability of fixing an allele is equivalent to its initial frequency. As genetic drift proceeds, roughly equal numbers of populations diverge above and below the initial (expected) allele frequency (Fig. 4.6). Note that if the initial allele frequencies were not equal this would naturally lead to more populations fixing for the allele with the lower frequency simply because the starting point is closer to zero or one. From this observation, we can make the prediction that the probability of fixing an allele in a population is directly proportional to its initial frequency. The other way to express the same idea is that, for any collection of populations, the proportion eventually fixing for each allele is equal to the initial frequencies. So, if the initial allele frequency for all populations was $p_0 = 0.3$, we would expect 30% of them to eventually become fixed for that allele.

Drift in Small Populations—In the absence of other population genetic processes, genetic drift will ultimately lead to the loss of genetic variation. Consequently, we can expect that small, isolated populations will become genetically depauperate over time. We can conclude from the principles outlined above that the effects of small population size are much more severe for rare alleles—primarily because their initial frequency is low. We would expect that loss of rare alleles in small populations would proceed more rapidly than declines in heterozygosity, but ultimately all loci will become homozygous. This expectation is borne out by results from surveys of rare and widespread plants, which typically find lower levels of genetic diversity in species with restricted distributions (Gitzendanner and Soltis 2000; Furches et al. 2013). Species that are only represented by one or a few small populations present a challenge for conservation geneticists because levels of genetic variation are expected to erode over time (Barrett and Kohn 1991; Ellstrand and Elam 1993). Later in this chapter, we will explore the roles of gene flow and mutation for the replenishment of genetic variation in populations subject to genetic drift.

Variation in Population Size—So far we have assumed that populations remain at a constant size over time, but real populations often fluctuate in numbers for a variety of reasons. For example, the effects of weather conditions, herbivory, and seed predation may vary across years causing variation in the numbers of seedlings recruited into the population each growing season. *The first principle of the effects of population size on genetic drift is that populations with larger fluctuations in size will have a smaller effective size and consequently will be more subject to the effects of genetic drift.* Larger fluctuations in population size are more likely in annual or short-lived perennial species, and these changes in the number of individuals reproducing across years have consequences for the average effective population size. In particular, years with low census numbers have a disproportionate effect on N_e because these smaller populations cause greater losses in genetic diversity (Hedrick 2009).

When population size fluctuates among years, the effective population size approaches the harmonic mean ($N_e \approx t / [\Sigma 1 / N_i]$; where t is the number of years and N_i is the population size each year), which is smaller than the arithmetic mean ($\Sigma N_i / t$) when the variance in population size is greater than zero. As an example, we can consider a population with large fluctuations across years (e.g., $N_i = 10$, 100, 50, 75, and 15 for $i = 1$ to 5). In this case, the arithmetic mean is equal to 50, and the harmonic mean is equal to 26.1. When changes in population size among years are less dramatic (e.g., N_i = 40, 60, 40, 60, and 50), the two means are more similar (50 and 48.7 for the arithmetic and harmonic mean, respectively). Years with smaller population numbers have a larger influence on the average effective population size across generations, and consequently, the effects of genetic drift will be greater when fluctuations in population size are large.

Severe reductions in population size that persist for multiple generations can have long-lasting effects on levels of genetic diversity. Catastrophic events may lead to loss of habitat and sudden reductions in population numbers (*Population Bottlenecks*). Similar reductions in population size occur during colonization and establishment of new populations (*Founder Events*). In either case we assume that the genetic composition of the reduced population represents a random collection of individuals from the source, so allele frequencies in the original population play a large role in determining levels of genetic variation immediately after a bottleneck or founder event. We can predict the probability of loss of an allele due to a bottleneck as $(1-p)^{2N}$; where p is the allele frequency in the original population, and N is the size of the reduced population (Hedrick 2009; Fig. 4.9).

The second principle for the effects of population size on genetic drift is that rare alleles are more likely to be lost during population bottlenecks. The first conclusion we can glean from consideration of the probability of allele loss is that rare alleles are extremely unlikely to be retained in bottlenecked populations; even with a population size of N = 50 an allele that had a frequency of p = 0.01 in the source population has a ~36% chance of being lost. Common alleles, on the other hand, are retained at high frequencies even when the bottlenecked population is only ten individuals or less (Fig. 4.9). For example, if all loci were segregating for two alleles at equal frequency, a single survivor would have a 50% chance of carrying both alleles (i.e., $2pq = 0.50$), and a bottlenecked population of only three individuals would be adequate to retain >98% of variation from the original population ($[1-p]^{2N} = 0.56 = 0.016$ for the probability of allele loss).

The third principle of the effects of population size on genetic drift is that after a bottleneck, heterozygosity will

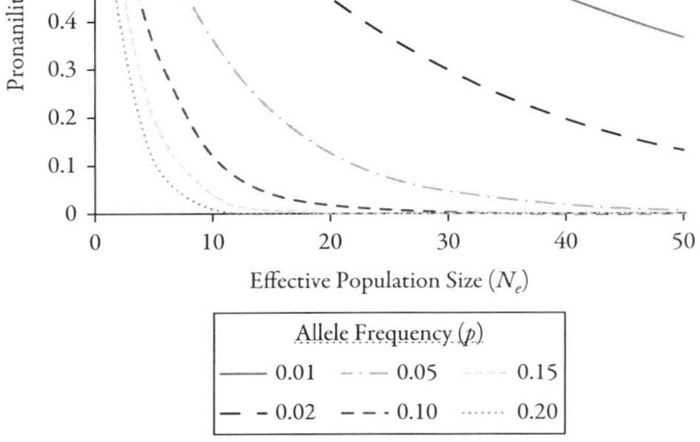

Fig. 4.9. The effects of initial allele frequency (p) on the probability of loss after a population bottleneck (N_e).

decline more quickly if population growth rate is slow. If the population recovers quickly then we predict that nearly all the genetic diversity captured by a bottleneck event will be retained. In fact, when population growth is rapid the effective population size can approach the census size (N) even when the variance in reproductive success is high.

The maintenance of genetic diversity in recently bottlenecked populations will be governed by how quickly they can recover and return to their original size. If the population growth rate is slow after a bottleneck the effects of genetic drift will prevail and we can expect continued loss of heterozygosity (H_e) due to small population size; $H_e = 1-(1-1/2N_e)^t$, where t is the number of generations (Fig. 4.10). Severe declines in heterozygosity can occur in just a few generations when N_e is small (i.e., when $N_e < 10$), but loss of diversity will be minimized if population growth is rapid or if the bottlenecked population is not too small.

Another way to view this process is to consider the effects of variation in progeny numbers. From the equation described above, we predict that when $n > V_n$, the number of surviving progeny is large so the chances of losing alleles becomes diminished. For example, in a population of $N = 10$, if $V_n = 1$ and $n = 1$ the effective population size $N_e = 5$, but increases to $N_e = 8$ when $n = 2$. As the population grows, N_e rapidly increases so most of the genetic diversity is retained, and the effects of genetic drift are diminished. Even with relatively high levels of variance in reproductive success among individuals, high survival rates of progeny can increase N_e and the chances of retaining alleles.

We can draw several general conclusions from consideration of the effects of population bottlenecks and founder events. These include: 1) the genetic variation from the original population can be retained even when the bottleneck is severe; 2) rare alleles are much more likely to be lost during a bottleneck (probability of retention scales with allele frequency); and, 3) population growth rate after the bottleneck will have an overriding effect on the long-term maintenance of genetic diversity (heterozygosity). In general, population bottlenecks of a short duration will lead to the loss of rare alleles but will have minimal effects on overall heterozygosity. Longer bottlenecks, on the other hand, will lead to the loss of rare and common alleles and heterozygosity will continue to decline as long as the population remains small.

Population bottlenecks have become relatively common over the history of life on Earth as species are routinely pushed to the brink of extinction due to natural events, and more recently, human activities.

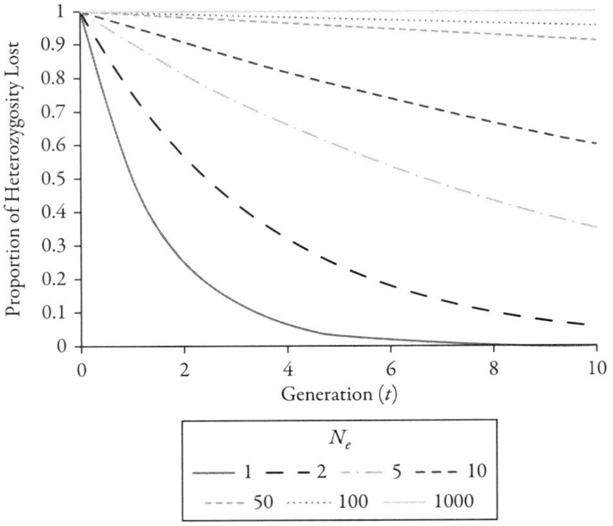

Fig. 4.10. Loss of heterozygosity (H_e) over generations (t) as a function of the effective population size (N_e).

Some of the well-known animal examples include the giant pandas, whooping cranes, California condors, and Florida panthers. Instances of bottlenecked plant populations are much less common. One example is the Wollemi pine (*Wollemia nobilis*) from Australia (Fig. 4.11; refer to color plate), which is not a pine at all but as a member of the family Araucariaceae, is more closely related to other southern hemisphere gymnosperms such as the monkey puzzle tree (*Araucaria araucana*). The Wollemi pine is a "living fossil"; it was known from fossils for some time before living individuals were discovered in a steep narrow canyon in Wollemi National Park, Australia. Surveys of genetic diversity in *Wollemia* using allozymes, AFLPs, and SSRs revealed extremely low levels of genetic diversity (Peakall et al. 2003), so we can conclude that this bottlenecked population has remained small for an extended period of time.

In a separate example, one rare species of triggerplant (*Stylidium coroniforme*; Stylidaceae) from Australia was known to have been reduced to only three individuals in recent history (Coates 1992). In spite of this severe bottleneck, recovered populations of this species retain high levels of genetic diversity, suggesting that population growth has been rapid. Other spectacular cases of recently bottlenecked plants include several species of European feather grass (genus *Stipa*; Durka et al. 2013) and the Belgian Gagea lily (*Gagea spathacea*). This latter species appears to be represented by one or a few clones across much of its range (Pfeiffer et al. 2012). In most of these cases, the recovered populations appear to have recovered quite well and often retain greater levels of genetic diversity than expected given the severity of the bottleneck, possibly because of the buffering effects of seed banks and other forms of dormancy (see discussion on seed banks later in this chapter).

Population bottlenecks and small population size result in the loss of genetic variation due to drift, but based on the examples listed above, it appears that plants are less susceptible to severe bottlenecks than animals. While species-wide homozygosity is rare in plants, individual populations can be subject to loss of genetic variation. Without gene flow, the genetic variation in these populations will continue to decrease over generations as relatedness increases, which results in inbreeding. In the next section, we explore the consequences of mating among related individuals for the levels and distribution of genetic variation within populations.

4.7 INBREEDING

For any population of limited size (i.e., not infinitely large) some proportion of matings will occur

Fig. 4.11. The Wollemi pine and young male cone (inset). Photos by Fritz Geller-Grimm and Velella (inset).

among individuals that are more closely related than the average of pairs of individuals drawn at random. Mating among individuals that share recent common ancestors elevates the level of inbreeding in the population. We generally consider matings within twenty generations to be *Consanguineous,* because the chance of sharing alleles inherited from older ancestors is exceedingly small (see the next section). In plants, inbreeding may occur by *Biparental* (consanguineous) mating between related individuals or by self-fertilization (*Selfing*) in species that are hermaphroditic. For biparental matings, shared ancestry increases the chance that both members of a mating pair will carry an allele derived from a common ancestor. We identify alleles with common ancestry by referring to them as *Identical by Descent (IBD)*. Alleles IBD are distinguished from those that appear to be the same but are not derived from a common ancestor (i.e. they are *Identical by State, IBS*). Note that alleles IBS may be derived from independent mutations, or if alleles are derived from the same mutation, they may become IBS after many generations of outbreeding, which reduces relatedness until they approach the average among random pairs of individuals in the population. Quantifying the proportion of alleles that are IBD indicates the chance that unique mutations inherited from a common ancestor will be homozygous in the offspring of a mating pair of individuals; in other words, the probability of homozygosity for alleles IBD provides an estimate of the level of inbreeding. Higher levels of inbreeding will result in lower fitness offspring due to increased homozygosity of deleterious recessive mutations (*Inbreeding Depression*; see the Evaluating Inbreeding from Pedigrees section).

Evaluating Inbreeding from Pedigrees—We can estimate levels of inbreeding from known pedigrees by calculating the probability that alleles IBD will be homozygous in offspring (Fig. 4.12). *The first inbreeding principle is that offspring from pairs of individuals that share recent common ancestors will have a higher chance of being homozygous for alleles IBD, and consequently will be more inbred than the average for the population.* In the scenario described in Fig. 4.12), two full siblings (i.e., individuals X and Y share both parents, A and B) mate to produce an offspring (Z). From this pedigree, we can estimate the probability that Z will be homozygous for alleles IBD (i.e., the proportion of loci across the

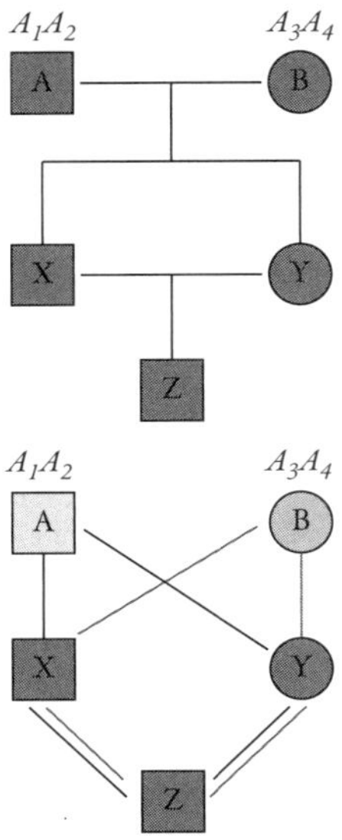

Fig. 4.12. Pedigree for a full-sib mating (above), and the same pedigree modified to show gamete pathways (below). Squares indicate pollen parents and circles are seed parents.

genome that are homozygous for alleles IBD). The upper panel displays a traditional pedigree, which is translated into the same pedigree showing gamete pathways in the lower panel (i.e., following red lines indicating inheritance of alleles from A and blue lines from B). If we assume equal segregation then there is a 50% (½) chance that × will inherit allele A_1 from parent A. It is the same chance for the transfer from X to Z, so the probability that Z will inherit A_1 from parent A is ½ × ½ = 0.25. For Z to be homozygous for A_1 this allele would have to be inherited through both X and Y, so the chance of homozygosity for A_1 is (½ × ½) × (½ × ½) = 0.0625. Note that Z could be homozygous for A_1 or A_2 from parent A and A_3 or A_4 from parent B, so the chance of homozygosity for alleles IBD is $f = 4(\frac{1}{2} \times \frac{1}{2} \times \frac{1}{2} \times \frac{1}{2}) = 4(\frac{1}{2})^4 = 0.25$ (where *f* is the inbreeding coefficient). Even though we only focused on a single locus in this example, the inbreeding coefficient is a general parameter that provides an indication of homozygosity for alleles IBD across the genome.

An alternative method for estimating *f* that provides a short-cut for more complicated pedigrees is the chain counting technique (Hedrick 2009). To apply this method, we first examine the pedigree to identify "loops" defined by a common ancestor (*CA*) that contribute to inbreeding (ancestors connected to the descendent by two gamete pathways). For the example in Fig. 4.12, both A and B represent *CAs*. Next, we count the number of ancestors in each *CA* loop (N_i; excluding the descendent—in this case Z); for Z that would include X-A-Y for one loop and Y-B-X for the other ($N_i = 3$ for both the A and B loops). We then apply the formula

$$f = \sum_{i}^{m} \left(\frac{1}{2}\right)^{N_i} \left(1 + f_{CA(i)}\right),$$

where m is the number of loops and $f_{CA(i)}$ is the inbreeding coefficient for each *CA*. Note that if the *CAs* were inbred ($f_{CA(i)} > 0$) then the level of inbreeding in the descendent would be elevated further, so f will increase across generations as mating occurs among individuals sharing alleles that are IBD.

The second inbreeding principle is that individuals that share more recent common ancestors will be more closely related, and mating between related individuals will result in offspring that are more inbred (i.e., a higher proportion of loci will be homozygous for alleles IBD). For the pedigree in Fig. 4.12 we assume f is zero for the common ancestors, so $f = (\frac{1}{2})^3 + (\frac{1}{2})^3 * (1+0) = \frac{1}{8} + \frac{1}{8} = 0.25\ (\frac{1}{4})$. We can use this formula to calculate f for a wide variety of pedigrees. For example, for offspring of first cousins the inbreeding coefficient would be $f = (\frac{1}{2})^4 + (\frac{1}{2})^4 * (1+0) = 0.0625\ (\frac{1}{16})$. For mating between half-sibs (e.g. if X was parented by A, plus a different seed parent—C—instead of B) then Z would only have one common ancestor from which alleles IBD could be inherited (B) and $f = (\frac{1}{2})^3 * (1+0) = 0.125\ (\frac{1}{8})$. The highest rate of inbreeding occurs when hermaphroditic species self-fertilize; offspring produced by self-fertilization of an outbred parent ($f_{CA} = 0$) will have a 50% chance of being homozygous for alleles IBD (i.e., there is only one common ancestor so $f = \Sigma(\frac{1}{2})^1 * (1+0) = 0.50$. Note that the rate of increase in f directly corresponds to how recently the mating pair shares a common ancestor—the last example being the most severe case of inbreeding because a selfing individual is its own common ancestor.

Inbreeding in Populations—Under most circumstances we do not have access to pedigrees to estimate levels of inbreeding, but we can estimate the inbreeding coefficient for a population (f) from neutral genetic markers. *The third inbreeding principle is that the inbreeding coefficient can represent the average level of inbreeding and can be applied as a population parameter.* To estimate the population-wide level of inbreeding we need to take into account both the frequency of each allele and the probability that it will be homozygous for copies that are IBD (i.e., f, the inbreeding coefficient). The chance that any individual will inherit an allele is equivalent to p, its frequency in the population. We are specifically interested in the proportion of homozygous genotypes where both copies are IBD, but with genetic marker data, we have to assume that alleles that are IBS are also IBD. The probability that an individual will inherit a second allele that is IBD is the same as the inbreeding coefficient (f), so the frequency of homozygosity for alleles IBD is the product of the allele frequency and the inbreeding coefficient (i.e., pf). Consequently, higher levels of inbreeding result in a reduction in expected heterozygosity ($fH_e = 2fpq$; Fig. 4.13) and reflect the average effects of inbreeding for a population. The reduction in heterozygotes will be greatest for alleles that are more common ($p = 0.5$; Fig. 4.13) because they have a higher chance of being homozygous for two alleles that are IBD.

The fourth inbreeding principle is that inbreeding will result in a decline in the frequency of heterozygous genotypes and corresponding increases in the frequencies of homozygous genotypes. To evaluate the overall level of homozygosity in an inbred population we need to take into account genotypes that are homozygous for alleles IBD. We can assume that the chance of drawing two alleles that are not IBD is equivalent to p^2 less the proportion of homozygous genotypes that are IBD ($1 - f$), so the fraction of homozygous alleles IBS would be equivalent to $p^2(1-f)$. Now we can estimate the frequency of genotypes in the population that are homozygous for allele A_1 (G_{11}). From the argument above, we know the IBD homozygote frequency is pf, so the total frequency of homozygous genotypes should be $G_{11} = pf + p^2(1-f)$, and with some rearrangement we get

$$G_{11} = p^2 + fpq.$$

From this line of reasoning, we conclude that the frequency of each homozygous genotype in an inbred

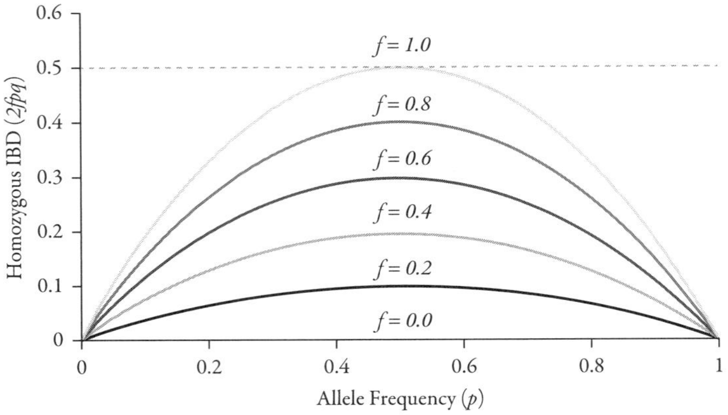

Fig. 4.13. The effects of inbreeding (f) and the allele frequency (p) on the frequency of genotypes that are homozygous for alleles identical by descent (IBD).

population will exceed the HW expectation by fpq. The frequency of homozygous genotypes—both G_{11} and G_{22}—will exceed the HW expectation in an inbred population so the total homozygosity will be $2fpq$ (Fig. 4.13). As homozygous genotypes increase in frequency, there must be a corresponding decline in heterozygotes, so we predict that the effects of inbreeding on genotype frequencies will be

$$G_{11} = p^2 + fpq$$

$$G_{12} = 2pq - 2fpq$$

$$G_{22} = q^2 + fpq.$$

From these relationships, we can predict that the frequency of heterozygotes will decline with higher levels of inbreeding as f increases over generations. Homozygous genotypes will approach a combined frequency of 1.0 and heterozygotes will decline to zero as f approaches unity (e.g., when $f = 1.0$ and $p = q = 0.5$; $G_{12} = (2*0.5*0.5)-(2*1.0*0.5*0.5) = 0$). Regardless of the allele frequency, the total homozygosity in the population will increase until no heterozygotes are present when $f = 1.0$.

The Effects of Small Population Size—High levels of inbreeding are characteristic of small populations because relatedness among individuals increases over time. The inevitable increase in inbreeding is clear if we consider a small population with discrete generations (Fig. 4.14; refer to color plate). To evaluate inbreeding in a population of limited size we assume that the initial colonists are unrelated in the first generation ($f=0$). Outcrossing among the unrelated colonists would not increase the level of inbreeding, but this would soon change if the population remains small. In the second generation every individual will share one parent with two other individuals (i.e., they are half-sibs), so random mating will produce many inbred individuals in the third generation (in Fig. 4.14; refer to color plate, $f = 0.125$ for an offspring in the third generation from mating of half sibs). As subsequent generations are produced, the average relatedness among individuals—and consequently the level of inbreeding—will increase. Note that we can apply the above equation for inbreeding in pedigrees using the chain counting technique to predict that a fifth generation offspring from a second generation of full-sib inbreeding would have $f = 0.141$ because its *CA* was inbred at a level of $f = 0.125$ in the third generation (Fig. 4.14; refer to color plate).

The scenario in Fig. 4.14; refer to color plate) illustrates inbreeding in single lineages, but the rate of increase in f over generations corresponds to the chance that related individuals mate, which depends on population size. If the population were excessively large and *Panmictic* (characterized by random mating) then related individuals would be much less likely to encounter each other, and the population will remain highly outbred. In large populations, the level of inbreeding can be elevated if mating is

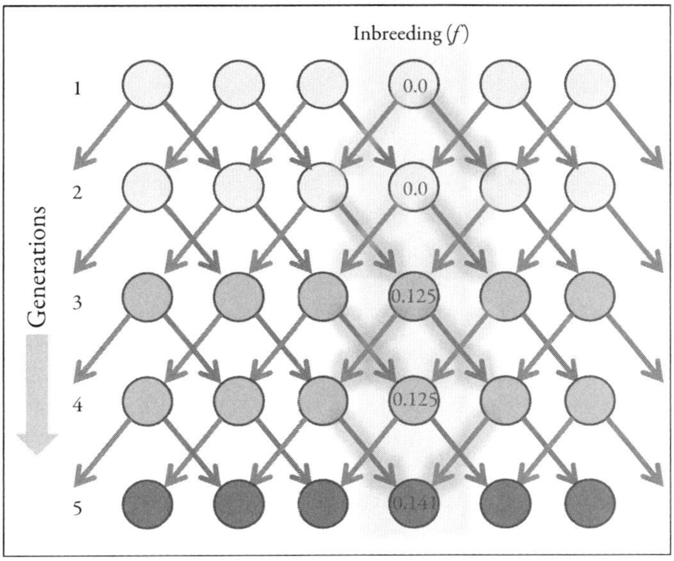

Fig. 4.14. Inbreeding after a population bottleneck. Arrows indicate gamete contributions through ovules (green) and pollen (orange) to each generation. Highlighted arrows illustrate the pattern of mating and effects on the inbreeding coefficient (*f*).

non-random—if mating is *Assortative* by genetic relatedness. More commonly, plant populations may be subject to some degree of *Disassortative Mating,* because related individuals will be more likely to share the same *Self-Incompatibility* alleles that enforce outbreeding (Chapter 9). Even for outbreeding species, small population size will increase the level of inbreeding, and the average level of inbreeding will increase even if matings are random because the average level of relatedness will increase over time.

The fifth inbreeding principle is that the level of inbreeding is higher in smaller populations. We can predict the effect of population size on the rate of increase in inbreeding by starting with a population with individuals sharing no alleles IBD so f = 0. In each generation, the probability of drawing any one allele (with replacement) is equivalent to its frequency $(1/2N)$. Under random mating the chance of drawing the same allele twice is the product of its frequency $(1/2N * 1/2N = 1/4N^2$, which is equivalent to p^2 when every allele is unique). Since any one of $2N$ alleles could be made IBD the expected level of inbreeding with random mating (including selfing) each generation is $2N/4N^2 = 1/2N$. This relationship demonstrates that the frequency of inbred progeny produced each generation is an inverse function of population size; if $N = 10$ there would be a 1 in 20 chance (5%) of producing inbred offspring, and inbreeding would be reduced to 1 in 200 (0.5%) for a population size of $N = 100$.

The sixth inbreeding principle is that the effects of inbreeding will accumulate so all populations become more inbred over time. From the discussion above we conclude that inbred progeny are generated at a rate of $1/2N$ each generation; however, inbreeding is incremental across generations so to estimate inbreeding levels in populations we need to account for the contribution of f in the previous generation $(f_{(t-1)})$ to the level of inbreeding in the current generation (f_t). Since we already know the probability of generating offspring homozygous for alleles IBD is $(1/2N)$, it follows that the frequency of progeny generated by outbreeding in the current generation would be $(1-(1/2N))$. Taking into account the level of inbreeding in the previous generation, we obtain

$$f_t = (1/2N) + (1-(1/2N))f_{(t-1)}.$$

From this equation, it becomes clear that without the influence of other evolutionary processes, f will increase over generations (Fig. 4.15). In very small populations, the increase in the level of inbreeding is rapid, but even for larger populations the frequency of genotypes homozygous for alleles IBD will increase over time. While these predictions seem reasonable, it is important to recognize that this is an analytical evaluation of the effects of inbreeding in populations; given enough time, f will asymptotically approach a value of 1.0 for any population size. These incremental increases may not seem reasonable in small populations where the small number of offspring produced each generation would result in large changes in the level of inbreeding for individual loci. On the other hand, if we consider the genome as a whole we can view these relationships as the probability of homozygosity for alleles IBD across a large number of loci.

If inbreeding only affects genotype frequencies without eliciting a change in allele frequencies then we would expect that a population would become composed of only homozygotes as f approaches 1.0; however, we recognize that outcrossing among individuals homozygous for different alleles would regenerate heterozygous genotypes, so the prediction depicted in Fig. 4.15 may hold true only under specific circumstances. For example, the inbreeding coefficient could approach 1.0 under complete selfing, but even low levels of outcrossing would regenerate heterozygotes so f would approach equilibrium at a lower level that would depend on the level of inbreeding, and particularly on the frequencies of selfing and outcrossing.

The other way a population may become completely homozygous is if inbreeding and genetic drift act together. Inbreeding on its own will reduce heterozygosity, but without the influence of other evolutionary processes, heterozygous genotypes will persist in the population even if they are rare. With only inbreeding, the frequency of homozygotes will increase, but genetic diversity will not be lost because each allele will be maintained as homozygous genotypes. It is important to recognize that this is a theoretical expectation; whenever population size is small, we expect that genetic drift and inbreeding will act in concert to reduce genetic diversity. Moreover, inbreeding enhances the effects of genetic drift because it results in lower effective population size. Under inbreeding we predict that $N_e = N/(1+f)$, so N_e will approach half the census population size as f increases to 1.0; with complete selfing $(f=1.0)$, half as many parents are generating offspring compared to complete outcrossing (Charlesworth 2003). In small populations, levels of inbreeding will increase as relatedness increases over generations, which will decrease the effective population size and accelerate the loss of genetic diversity. Genetic marker surveys of

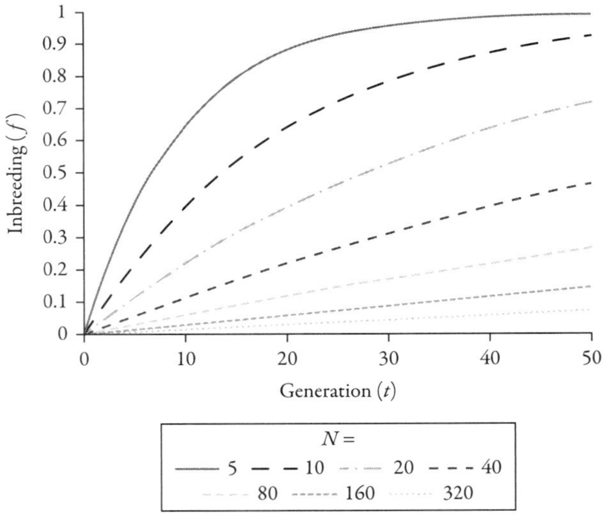

Fig. 4.15. The effect of limited population size on increase in the level of inbreeding (f) across generations (t).

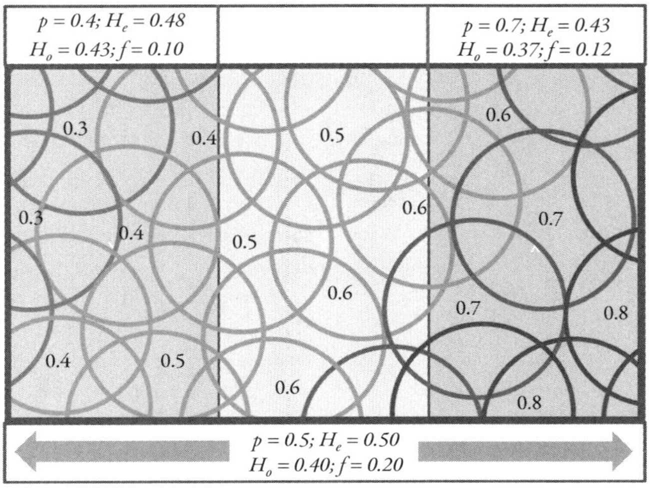

Fig. 4.16. The effects of local breeding on the distribution of allele frequencies across a population. Circles represent a subset of all possible overlapping neighborhoods. Upper panels provide estimates of *f* for the two subgroups and the lower panel shows *f* for the entire population.

plant populations are consistent with this expectation as more inbred populations typically display lower levels of genetic diversity (Hamrick and Godt 1990; Charlesworth 2003).

Inbreeding in Plant Populations—While inbreeding will occur in all organisms, this process is often enhanced in plant populations because their sedentary lifestyle leads to limited mating opportunities. Seed dispersal is often restricted, so siblings from the same maternal parent will germinate and flower in close proximity to each other. Consequently, biparental inbreeding is common in plants and can result in deficiencies of heterozygotes on local scales (Wright 1943; Crawford 1984). Many species of plants also self-fertilize to some degree, so the combination of selfing and biparental inbreeding will contribute to elevated levels of homozygosity. Over time, extensive plant populations will become a patchwork of overlapping, inbred neighborhoods (Fig. 4.16), and small populations will become uniformly homozygous. The collective effects of inbreeding and genetic drift will have consequences for the fitness of individuals and the levels and distribution of genetic diversity within and among populations.

We can use inferences from genetic marker assays to study the effects of inbreeding. Estimation of levels of inbreeding in populations is made difficult because *f* in any one generation reflects the cumulative effect of non-random mating over generations. We could use genetic markers to determine the paternity of seeds collected from individual plants, but without pedigree information, this would only estimate the level of inbreeding in the current generation. Inspection of the predicted effects of inbreeding on the frequency of heterozygous genotypes (i.e. $G_{12} = 2pq - 2fpq = (1-f)2pq$) presents an alternative approach to estimate *f* from codominant genetic markers. To convert this relationship to a form that will be useful for estimating levels of inbreeding we first recognize *2pq* as the level of heterozygosity expected under HW conditions (H_e). Second, if we assume that the decrease in heterozygosity in a population is solely due to inbreeding we can equate G_{12} to the observed heterozygosity (H_o). By substituting the expected heterozygosity (H_e) for *2pq* and the observed heterozygosity (H_o) for G_{12}, we convert the formula for the effect of inbreeding on heterozygosity to $H_o = (1-f)H_e$, and by rearrangement we obtain

$$f = (H_e - H_o)/H_e.$$

In other words, the proportional difference between the observed and expected heterozygosities can provide an estimate of inbreeding. For a population of hermaphrodites, this estimate of *f* reflects contributions from both biparental inbreeding and selfing. It is important to note that this estimate of f is for a single point in time so it does not indicate whether the population is close to equilibrium or whether the level of inbreeding is continuing to increase.

This method of inferring the level of inbreeding from population genetic data has several limitations. For small populations inbreeding may be underestimated if the genetic marker loci being used are nearly homozygous; H_e and H_o may both be very low and very similar, resulting in less reliable estimates of f. In larger populations, heterozygote deficiency may be exaggerated because genetic marker estimates are sensitive to the spatial distribution of samples. As noted earlier, limited dispersal results in genetic differentiation at local spatial scales, so a large population may be subdivided into genetic neighborhoods that are characterized by different allele frequencies (Fig. 4.16). Adjacent neighborhoods will tend to have similar allele frequencies because a large number of overlapping mating groups connects them (only a subset of possible neighborhoods is shown in Fig. 4.16). Note that the estimate of inbreeding across this population depends on the area sampled; when subgroups containing neighborhoods with similar allele frequencies are sampled, f ranges from 0.10 to 0.12, but the difference between the observed and expected heterozygote frequencies is much greater when we sample across the entire population, so the inbreeding coefficient is much larger ($f = 0.20$). The overestimate of f occurs because the average of H_o across subgroups will always be lower than H_e calculated from the average allele frequency. This phenomenon is referred to as the *Wahlund Effect* (Hedrick 2009) in honor of the Swedish geneticist Sten Wahlund, who first described this principle in 1928.

The effect of sampling on the apparent heterozygote deficiency depicted in Fig. 4.16 is an extreme case, but the Wahlund effect will hold true even when allele frequencies across subpopulations are more similar. For example, if $f = 0.10$ in two subpopulations with different allele frequencies ($p = 0.4$ and 0.1, respectively), then after combining the first ($H_e = 0.42$, $H_o = 0.378$) with the second ($H_e = 0.18$, $H_o = 0.162$) we get an average $H_o = (0.378 + 0.162)/2 = 0.27$, an expected heterozygosity of $H_e = 0.32$ (based on $p = 0.2$), and an inbreeding coefficient of $f = 0.16$. The inbreeding coefficient will always be inflated when samples from different subpopulations are combined—even when genotype frequencies of each subpopulation match HW expectations. Furthermore, the degree of bias for f will be minor when allele frequencies are similar and will increase along with the variance in allele frequencies across the sampled area. Similarly, differences in allele frequencies across life stages will produce a Wahlund effect in the same way as sampling across spatially separated panmictic groups. Depending on dispersal distances relative to the scale of population sampling, we can generally expect that estimates of f from genetic marker data will be biased upward to varying degrees.

Inbreeding Depression—The disparity between the observed and expected heterozygote frequencies can decline as selection removes more inbred individuals from populations. Recall from Chapter 3 that deleterious mutations accumulate over generations, so all populations will be segregating for some amount of *Genetic Load*. We expect deleterious mutations contributing to genetic load to have low levels of expression when heterozygous, especially those with large effects. As inbreeding ensues, larger proportions of deleterious alleles will be homozygous and the average fitness of individuals in the population will decline. Offspring from consanguineous matings and selfing will be homozygous for higher proportions of deleterious alleles and consequently will exhibit slower growth, lower reproductive success, and reduced rates of survival (inbreeding depression). While we do not expect neutral genetic markers to be directly affected by selection, under inbreeding the chance of homozygosity will be higher across the entire genome, so genetic marker genotypes will become associated with genotype frequencies for loci under selection—even if they occur in different linkage groups. As selection acts to eliminate individuals homozygous for deleterious alleles, there will be a corresponding decrease in homozygosity across the genome, so individuals heterozygous at neutral genetic marker loci will appear to have higher fitness (associative overdominance; Fig. 4.2).

The effect of selection on deleterious recessive alleles within inbreeding populations varies across different life stages and among plants with different mating systems (Husband and Schemske 1996). For outcrossing plants, the effects of inbreeding depression are generally more severe for early life stages; after self-pollination, seed set is much lower due to higher rates of ovule abortion, and the seeds produced tend to have lower rates of germination. Selfing taxa generally display much lower levels of inbreeding depression and the negative effects of inbreeding are more likely to be expressed later in the life cycle—perhaps because the effects of selection are weaker late in life (because individuals have already successfully reproduced; Williams 1957; Hamilton 1966; Chapter 6).

The expression of genetic load in inbreeding populations will reduce the frequency of homozygous genotypes and elevate the frequency of heterozygotes. If later life stages are sampled then H_o will be more similar to H_e because more homozygous individuals have been eliminated, which leads to an underestimate of the degree of inbreeding in the population. For example, in some forest tree species such as ponderosa pine (*Pinus ponderosa*) there is strong evidence of inbreeding (i.e., high frequencies of homozygous genotypes) at the seedling stage, but sampling adult canopy trees can reveal an excess of heterozygotes compared to the HW expectation (Mitton 1993). While such cases were originally interpreted as evidence for heterosis at individual allozyme loci, it is generally recognized that this apparent overdominance is due to association with loci across the genome that are homozygous for deleterious recessive alleles (associative overdominance; Fig. 4.2). Selection on homozygous individuals can also lead to the elimination of deleterious recessive alleles from populations—a topic we will explore further in Chapter 5.

4.8 GENE FLOW

The movement of seeds and pollen by biotic and abiotic vectors creates the potential for modification of the genetic composition of recipient populations. We make a distinction between *Dispersal* (movement of spores and propagules) and gene flow (*Migration*), which is the integration of genetic material carried by immigrants into the recipient population. This difference reflects the reality that the large majority of pollen grains and seeds leaving plants do not successfully reach receptive stigmas or suitable germination sites. Much of the pollen produced by bee-pollinated plants is collected as food for rearing larvae; when bees are not foraging for pollen they simply groom it off their bodies as they travel between flowers (Thomson 1986; Ne'eman et al. 2010; Holmquist et al. 2012). Ultimately, only a small fraction of pollen produced by animal-pollinated plants is deposited on receptive stigmas. The situation for wind-dispersed pollen is much worse since receptive stigmas represent infinitesimally small targets, and dispersal is much less directed compared to animal pollination. Similarly to pollen, most dispersed seeds are deposited in unsuitable locations for germination, or germinated seedlings do not survive to adulthood (Harper 1977). Attempts to directly measure dispersal by tracking pollen and seeds generally underestimate dispersal distances, while more reliable estimates of migration can be obtained using methods that are based on genetic markers (Loiselle et al. 1995; Ouborg et al. 1999; Gerber et al. 2014).

In this section, we will examine simplified models to develop our understanding of the effects of gene flow on the distribution of genetic variation within and among populations.

Models of Migration—As seeds and pollen are dispersed their genetic contributions impact local genetic composition and provide a mechanism to integrate genetic variation across populations. From a population-genetic perspective, the primary effect of gene flow is to homogenize allele frequencies. We can see this effect for a recipient population in the *Continent–Island Model* (Fig. 4.17). In this model, we assume that immigrant individuals from the continent replace some fraction of the island population, as defined by the migration rate (m, which ranges from 0 to 1), each generation (t). Gene flow is unidirectional and the continent population is very large, so the allele frequency of migrants (q_m) remains constant (i.e., the allele frequency on the continent does not change). Each generation, the allele frequency on the island (q_t) is modified by migration from the continent at a rate that is proportional to m and the continent allele frequency (mq_m), while the remaining proportion of the island population ($1-m$) maintains the same allele frequency:

$$q_t = mq_m + (1-m)q_{(t-1)},$$

where $q_{(t-1)}$ is the allele frequency on the island in the previous generation.

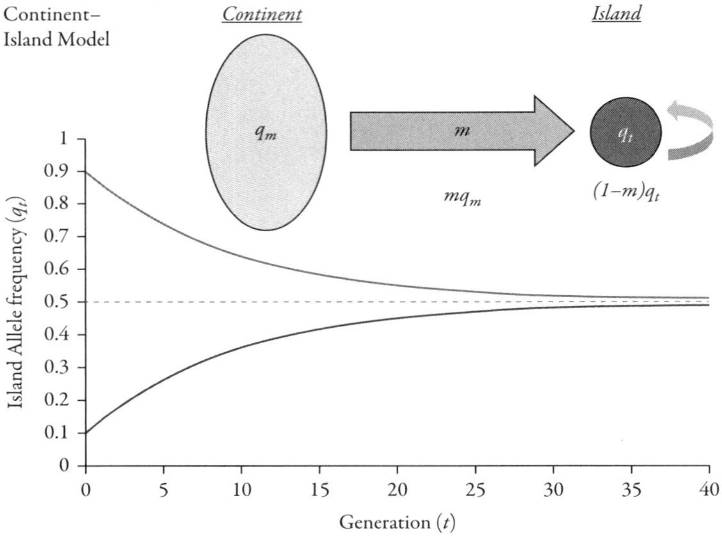

Fig. 4.17. The Continent-Island Model of gene flow ($m = 0.10$).

Under this model, the allele frequency on the island will asymptotically approach the frequency on the continent regardless of the initial allele frequency (Fig. 4.17). The rate of change in the allele frequency on the island is a function of the migration rate and the absolute value of the difference in frequencies ($\Delta q = |q_m - q_t| = \sqrt{(q_m - q_t)^2}$, where Δq is always positive). For any particular migration rate, the rate of change in the allele frequency on the island will be greater when the difference in allele frequencies is larger and will decline as the frequency on the island approaches the continental allele frequency.

The continent–island model can be viewed as a process of dilution. Imagine that the continent is a source of pure water, and the island is a bottle containing 70 proof alcohol. If we repeatedly remove a small amount of liquid (m) from the bottle and replace it with water, we can predict that the concentration of alcohol will decline. Over time, the concentration will approach zero, but there will always be some small fraction of alcohol left in the bottle no longer how many times we replace some of the contents with water (i.e., the approach to pure water is asymptotic).

The continent–island model is unrealistic in some respects (e.g., allele frequencies in source populations are likely to change over time), but it clearly illustrates the equalizing effect of gene flow. We can apply this same principle to groups of populations that are connected by dispersal (the *Island Model*; Fig. 4.18). If we assume that all populations are of equal and constant size, then we can expect that allele frequencies will approach the average allele frequency across islands (Hedrick 2009).

Note that the island model does not incorporate any spatial structure—populations are assumed to be equidistant to each other regardless of their number. The *Stepping Stone Model*, on the other hand, assumes spatial structure in one or two dimensions (Fig. 4.18). Including the more realistic assumption of local gene flow produces a pattern of *Isolation by Distance* (*IBDist*; Wright 1943), where allele frequencies in nearby islands become more similar as local exchange occurs among neighboring islands. When gene flow is restricted to adjacent neighbors, undulating patterns of alelle frequencies may develop across an extensive archipelago, but in the absence of other evolutionary forces, and given enough time, we expect the frequencies on all islands will eventually approach the collective mean allele frequency. *The primary principle we can derive from these models of migration is that gene flow will homogenize allele frequencies within and among populations.*

The movement of seeds and pollen within and among populations plays a crucial role for the spread of advantageous mutations and the maintenance of genetic diversity. We will continue our discussion of gene flow in the next section as we explore the counterbalance between migration and genetic drift. In later sections of this chapter and in Chapter 8 we will discuss the processes and consequences of dispersal for levels of gene flow among populations.

4.9 GENETIC DIFFERENTIATION

Multiple evolutionary processes can simultaneously affect the distribution of genetic variation within and among populations. For neutral genetic variation, genetic drift and gene flow working in opposition to each other govern levels of genetic differentiation among populations (Fig. 4.19; for a version that includes selection see Fig. 5.17; refer to color plate). In this metaphorical model of a balanced lever (i.e., a seesaw or teeter-totter), any change in the angle of the lever represents allele frequency change. The influence of gene flow depends on the level of dispersal (high rates of dispersal have more leverage) and the influence of genetic drift depends on population size (small population size has more leverage). This model emphasizes the idea that gene flow and drift can counteract each other; change in allele frequencies can be dominated by low rates of gene flow in large populations, but it will take higher rates of dispersal to overwhelm the effects of genetic drift in small populations. In this chapter, we are focusing on neutral genetic variation and we will take up the effects of selection in the next chapter.

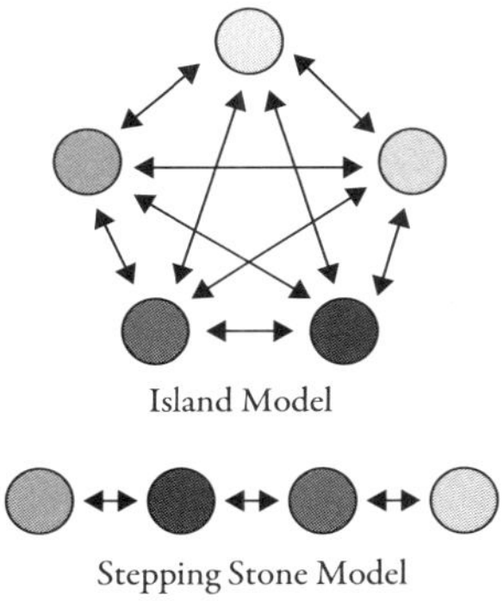

Fig. 4.18. The island and stepping stone models of gene flow.

From on our discussions so far, we expect that genetic drift will lead to greater divergence among populations while gene flow will have a homogenizing effect (Fig 4.19). While a number of circumstances can affect genetic differentiation, the one factor that has an overriding effect on the distribution of genetic variation is dispersal ability, because it affects both rates of gene flow and the effective population size. For example, pollen dispersal will affect both gene flow and access to potential mates; with more restricted pollen movement, the number of accessible mates is lower so there will be a correlated reduction in N_e and gene flow (m). Seed dispersal primarily affects gene flow, but limited movement of seeds increases the potential for inbreeding and consequently reduces the effective population size, as siblings are more likely to remain

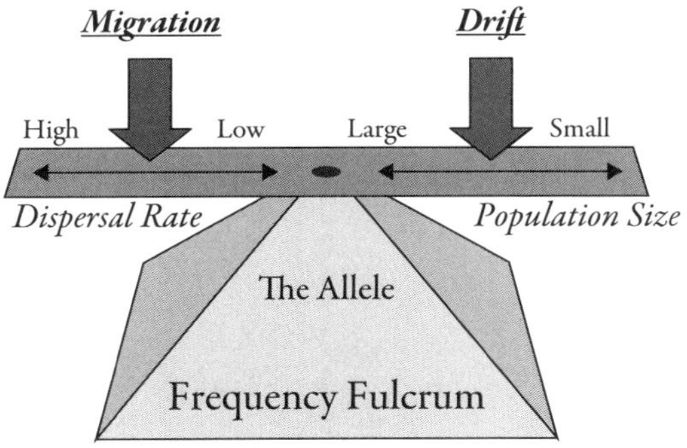

Fig. 4.19. A metaphorical model for allele frequency change in populations under the influences of gene flow (migration) and genetic drift. Higher rates of dispersal increase the leverage of migration and smaller population sizes increase the leverage of genetic drift. Both of these processes acting together will affect allele frequency change, but one can dominate if its leverage is substantially greater. For a version of this model that includes selection see Fig. 6.17.

in close proximity to each other. Levels of pollen and seed dispersal have consequences for gene flow, genetic drift, and inbreeding, and ultimately govern the levels and distribution of genetic variation within and among populations.

For simplicity, models of genetic structure assume multiple discrete and stable populations that are subject to genetic drift and connected by gene flow. Note that this framework is different from ecological models of *Metapopulations* where demographic stochasticity and dispersal govern frequencies of colonization and extinction (Hastings and Harrison 1994). Population genetics focuses on the consequences of population subdivision for the distribution of genetic diversity without extinction and colonization; in this text, we will use the term metapopulation in a broader sense to indicate a group of populations.

The population genetic approach (without considering the effects of colonization) works well for many species where stable populations are distributed across patches of suitable habitat. For many plants—especially those with more specific habitat requirements—populations are separated by a matrix of inhospitable conditions at some spatial scale, and depending on the rate of gene flow, genetic divergence will occur among isolated populations. Later in this chapter, we will discuss the consequences of limited dispersal for the development of genetically structured neighborhoods across larger populations, including cases where species are continuously distributed across thousands of hectares. In this section, we will examine the consequences of gene flow and genetic drift for the distribution of genetic variation among discrete populations.

Population Structure—Recall that in our previous discussion of levels of inbreeding estimated from genetic marker data, we discovered that the apparent heterozygote deficiency (f) increased as we sampled across groups of individuals that differed for allele frequencies—a phenomenon referred to as the Wahlund effect. We also noted that the difference between the observed and expected heterozygosity increased as the difference in allele frequencies among subgroups increased. Wright (1950; 1965) recognized this characteristic of heterozygote deficiencies as a convenient method to quantify levels of genetic differentiation among populations. When multiple populations (i.e., stable populations that are connected by gene flow) are included in the same sample the heterozygote deficiency increases with differences in allele frequencies among populations. The apparent heterozygote deficiency for samples of individuals from multiple populations provides a robust method to describe levels of genetic differentiation.

Wright quantified levels of variation within and among populations by using a set of hierarchical measures referred to as *F-statistics*. For one or more loci that are segregating for two alleles, Wright's measure of the differentiation among subpopulations (F_{ST}) is calculated as the difference between the mean expected heterozygosity (i.e., $H_e = 2pq$) across loci for the pooled sample of all populations ($\bar{H}_T$), and the expected heterozygosity for each population individually, and then averaged across all populations ($\bar{H}_S$; Fig. 4.20);

$$F_{ST} = \frac{\bar{H}_T - \bar{H}_S}{\bar{H}_T}.$$

In the example in Fig. 4.20, the two populations have an average expected heterozygosity of $\bar{H}_S = 0.40$ (the average of 0.48 and 0.32 for the red and blue populations, respectively). The population sizes are equal, so the pooled allele frequency is $p = 0.70$ (the average of $p = 0.60$ and 0.80), and the total heterozygosity is $\bar{H}_T = 2pq = 2 * 0.7 * 0.3 = 0.42$. From these values we can calculate $F_{ST} = (0.42 - 0.40) / 0.42 = 0.05$ for this small difference in allele frequencies between the two populations. Note that this estimate of F_{ST} is low because allele frequencies are similar, but this measure of differentiation increases by the square of the difference in allele frequencies among populations. For example, if the allele frequencies were $p = 0.5$ and 0.9, then $H_S = 0.34$ and $F_{ST} = (0.42 - 0.34) / 0.42 = 0.19$.

F_{ST} can be used to measure differentiation in cases where there are only two alleles per locus and values obtained should always be greater than zero. For any stable set of populations, F_{ST} will reflect the collective influences of genetic drift and gene flow; populations that are regularly undergoing high rates of genetic exchange will have similar allele frequencies across loci and consequently low estimates of genetic differentiation. Populations that are more isolated will deviate in allele frequencies due to genetic drift in the absence of gene flow and will have larger estimates of F_{ST}.

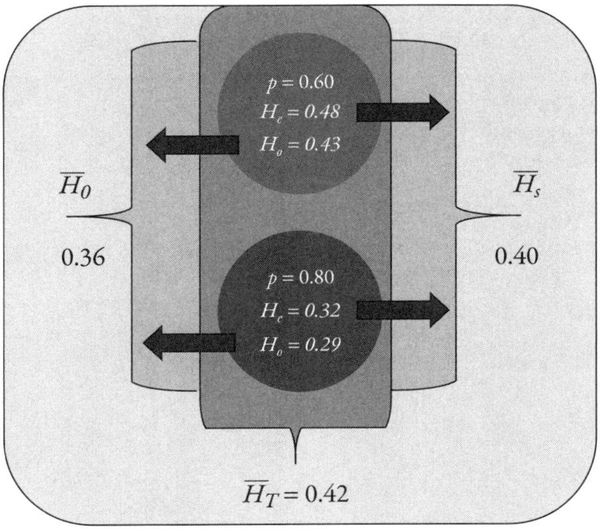

Fig. 4.20. Measures of heterozygosity within and among populations. For this example there are two populations (subpopulations) and one locus.

The average heterozygote deficiency due to inbreeding within populations (F_{IS}) can be calculated as the difference between the average expected ($\bar{H}_S$) and observed ($\bar{H}_O$) heterozygosities across populations and loci:

$$F_{IS} = \frac{\bar{H}_S - \bar{H}_O}{\bar{H}_S}.$$

This parameter is analogous to the inbreeding coefficient (f) that we discussed earlier in this chapter and represents the average effect of inbreeding on heterozygote deficiencies across populations. Positive values of F_{IS} indicate that populations are inbred, while negative values are due to an excess of heterozygotes, which could be due to higher rates of mortality for homozygous genotypes or to recent gene flow. For the example in Fig. 4.20, the average observed heterozygosity is $\bar{H}_O = 0.36$ (the average of $\bar{H}_O = 0.43$ and 0.29), and $F_{IS} = (0.40 - 0.36)/0.40 = 0.10$, which suggests a substantial level of inbreeding. Note that for calculations of F_{IS} we assume each population is panmictic—that there is no substructure within populations. This will often not be the case for large populations of plants that have limited dispersal, so interpretation of this parameter should recognize the potential for biased estimates due to Wahlund effects within populations.

The total genetic variation for a collection of subpopulations is given by F_{IT}, which is defined as the difference between the pooled expected ($\bar{H}_T$) and average observed homozygosity across loci and populations ($\bar{H}_O$):

$$F_{IT} = \frac{\bar{H}_T - \bar{H}_O}{\bar{H}_T}.$$

While F_{IT} is called the "total" variation, its relationship with F_{ST} and F_{IS} is more complicated than adding these two values together because each parameter represents the proportion of variation relative to variation at the next higher level (Hedrick 2009). The interpretation of F_{IT} is complex and can include effects of inbreeding and population differentiation, so population genetic studies generally focus on F_{ST} and F_{IS}.

The relationships among Wright's F statistics are hierarchical and additional levels of variation can be evaluated. Variation among regions ($\bar{H}_R$; groups of populations) could be estimated as F_{SR} ($(\bar{H}_R - \bar{H}_S)/\bar{H}_R$) as variation among populations within regions, and F_{RT} ($(\bar{H}_T - \bar{H}_R)/\bar{H}_T$) as variation among regions (in addition to F_{IS} and F_{ST}). For example, F_{SR} could be used to quantify variation among watersheds and F_{RT} would represent variation among populations within watersheds. This ability to nest *F* statistics in hierarchical subgroups provides a flexible and convenient set of parameters for the description of levels and distribution of genetic variation within and among populations and geographic regions.

Evaluating Population Genetic Structure—Wright's measure of genetic differentiation among populations assumes two alleles per locus, which is too restrictive for many applications, because codominant genetic markers such as allozymes and SSRs tend to have multiple alleles. A statistic that allows for multiple alleles (G_{ST}) was developed by Nei (1973). For this measure of genetic differentiation we use a more general formula for the expected heterozygosity (H_e) averaged across populations ($\bar{H}_S$; where $H_e = 1 - \sum p_i^2$, for $i = 1$ to n alleles at each locus). G_{ST} is then calculated as:

$$G_{ST} = (\bar{H}_T - \bar{H}_S) / \bar{H}_T.$$

Nei's method for the calculation of G_{ST} is similar to the calculation of F_{ST} since average heterozygosities are used, and G_{ST} behaves the same as F_{ST} in cases where all loci are bi-allelic (Hedrick 2009). Difficulties arise when there are more than two alleles; with multiple alleles the maximum value of G_{ST} will be less than 1.0 because the expected heterozygosity within populations will always be greater than zero (i.e., $\bar{H}_S > 0$ when $n > 2$ alleles). For example, with more than two alleles and only two populations sampled there has to be some level of expected heterozygosity because at least one of the populations has to have more than one allele.

In general, the maximum value that G_{ST} can achieve declines with the number of alleles present at each locus, which may constitute a problem for highly variable markers such as microsatellites. In fact, it is not uncommon for SSR studies to yield very low estimates of G_{ST}, because the maximum possible value may be only 0.20 or less—a range of values for F_{ST} that would be predicted to lead to high levels of genetic differentiation (Fig. 4.21; see later discussion on alternative measures of differentiation). In these cases, interpretation of population differentiation is made difficult because G_{ST} appears to be low relative to estimates from allozyme data even though the populations are highly differentiated (Balloux and Lugon-Moulin 2002). The influence of allele number on G_{ST} renders comparisons among studies difficult when genetic markers with different mutation rates are used (e.g., allozymes versus SSRs).

A second problem may arise for genetic differentiation estimates with G_{ST} and F_{ST}, because calculations are made based on heterozygosities without reference to allelic identity. For example, in the extreme case where two populations possess several alleles per locus and do not share any alleles in common, the estimate of G_{ST}—at its maximum—will be well below 1.0. This is expected because G_{ST} depends solely on the difference in expected heterozygosities within populations. This problem is amplified for marker loci with high mutation rates because estimates of G_{ST} will not reflect genetic differentiation due to *Private Alleles* (i.e., alleles unique to a single population) and alleles that characterize groups of populations.

The Coalescence F_{ST}—Both F_{ST} and G_{ST} quantify genetic variance among populations, so they are analogous to the F-ratio used in the *Analysis of Variance (ANOVA)* for differences among means. These measures of genetic variation are focused on differences in heterozygosity (a measure of genetic variance) without regard to the relationships among alleles. Heterozygosity-based estimates of differentiation work well for samples of populations that are connected by gene flow and have similar allelic compositions, but G_{ST} does not perform well in cases where gene flow is much lower than the mutation rate of markers, which results in the accumulation of alleles that are unique to populations. As mentioned above, lower levels of allele sharing among populations may inflate H_s, which would result in estimates of G_{ST} that underestimate the levels of differentiation.

As an alternative approach for cases of higher levels of divergence among populations, we can derive a more robust estimate of F_{ST} that is based on the ancestry of individual alleles (i.e., for alleles that are identical by descent, IBD). With information on the history of alleles, we would predict that the time back to the most recent common ancestor would be shorter for a pair of alleles drawn from the same population ($\bar{t}_0$) compared to a pair drawn randomly from the entire collection of populations ($\bar{t}$). With adequate information we can infer these parameters—the history of individual alleles—using *Coalescence* methods (Whitlock 2011; Box 5.1). In this case, we

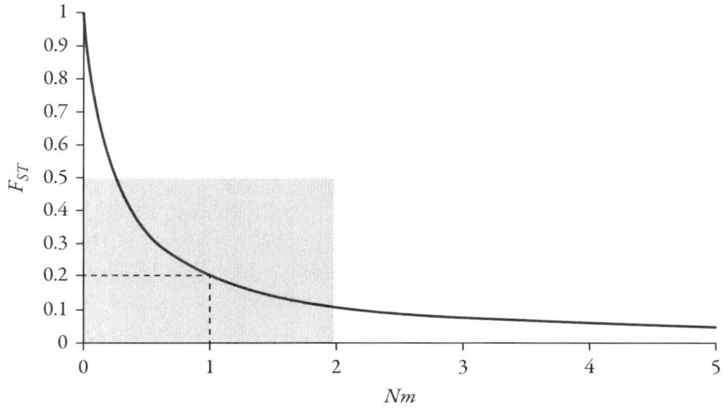

Fig. 4.21. The relationship between *Nm* and F_{ST}. The dotted line indicates a widely accepted threshold value of *Nm* < 1 that allows population differentiation. The gray box indicates a region where the estimation of *Nm* from F_{ST} is more reliable.

define population structure in terms of a coalescence estimate of F_{ST}:

$$F_{ST-coal} = \frac{f_0 - f_1}{1 - f_1} = \frac{\bar{t} - \bar{t}_0}{\bar{t}}.$$

Where f_0 is the probability of IBD for two different alleles from the same population, and f_1 is the probability of IBD for alleles from two different populations. $F_{ST-coal}$ differs from the heterozygosity methods described earlier by considering the co-ancestry of individual alleles. The coalescence method utilizes estimates of the age of alleles to evaluate levels of differentiation within and among populations, and variation in $F_{ST-coal}$ should reflect differences in levels of gene flow in cases where $m < \mu$ (i.e., where migration is less than the mutation rate) for stable populations that are in equilibrium with respect to gene flow and genetic drift (Whitlock 2011).

The information provided by a coalescence analysis of allelic variation goes well beyond a generalized genetic structure analysis as the ancestry of alleles reflects individual population history and levels of connectivity among divergent populations. When alleles first originate by mutation their frequency will be very low, and consequently they have little chance of being dispersed to neighboring populations. This aspect of $F_{ST-coal}$ is similar to the private alleles method of estimating gene flow (i.e., the frequencies of alleles unique to populations is negatively associated with the rate of gene flow; Slatkin 1985). Older alleles will tend to occur at higher frequencies, so the chance of dispersal is elevated, and they are more likely to be found in multiple populations. Using a coalescence approach for the estimation of F_{ST} requires that individual alleles possess adequate information for the inference of ancestry. This approach can work well for tightly linked sequence regions that contain multiple SNPs across individuals or for genomes that have little or no recombination such as the chloroplast and mitochondrial.

The $F_{ST-coal}$ can also be estimated for SSR loci if a *Stepwise Mutation Model* (*SMM*—mutations increase allele size by a single motif) for the relationship among alleles can be assumed (Whitlock 2011). A similar measure of differentiation (R_{ST}; Slatkin 1995) also takes advantage of the predictable nature of SSR mutations by comparing differences in allele sizes within and among populations. Slatkin's R_{ST} method considers allele identity by assuming SMM as a framework to infer historical relationships among alleles. Estimates of R_{ST} are susceptible to biases due to independent origins of alleles IBS and these estimates often suffer from high variances (Balloux and Lugon-Moulin 2002). The coalescence F_{ST} may be a preferable method for SSRs and sequence data as it can provide a more robust measure of population differentiation that reflects the effects of gene flow and population size—particularly in cases where levels of divergence are relatively high (i.e., $\mu > m$).

Alternative Measures of Differentiation—In cases where the history of individual alleles cannot be reliably inferred or for cases where m is much greater than μ, the heterozygosity method for estimating G_{ST} can be applied; however, as noted earlier, high mutation rates tend to inflate H_s leading to low estimates that do not reflect actual levels of differentiation. To circumvent this problem Hedrick (2005b) standardized this measure of differentiation (G'_{ST}) by dividing G_{ST} by its maximum possible value. This measure of differentiation ranges from 0 to 1.0 regardless of the number of alleles segregating at each locus. A similar measure (D) developed by Jost (2008) estimates the fraction of allelic variation that differentiates populations. Both G'_{ST} and D have the advantage of providing estimates of genetic differentiation that range up to 1.0 even when H_S is large.

Standardized measures of differentiation have been criticized because they do not directly reflect the population genetic processes of greatest interest (Whitlock 2011). In particular, both F_{ST} and G_{ST} were developed within analytical frameworks that clearly define their relationships to rates of gene flow and effective population size. Under equilibrium conditions, the product of population size and migration rate (Nm) determines F_{ST}:

$$F_{ST} = \frac{1}{4N_e m + 1}.$$

Note that $N_e m$ is generally written as Nm for simplicity. From this formulation, we can clearly see that F_{ST} (and G_{ST}) will directly reflect the effects of gene flow and genetic drift (Fig. 4.21). By contrast, both G'_{ST} and D are sensitive to migration and mutation rates and have no direct links to the effects of population size; consequently, these measures of differentiation do not accurately reflect the effects of genetic drift (Whitlock 2011).

Standardized measures of genetic differentiation may also be biased across a wide range of conditions. We expect genetic differentiation to decline as levels of migration increase, but G'_{ST} and D overestimate differentiation compared to $F_{ST\text{-coal}}$—especially when there are more than two populations and migration rates are high (Whitlock 2011). These standardized measures are sensitive to mutation rates and overestimate differentiation when $\mu > m$; under these conditions isolated populations accumulate parallel mutations that are identical by state (i.e., are not IBD) so estimates of differentiation will be artificially low, which will give an impression of a high migration rate. High mutation rates generally render the use of markers such as SSRs difficult, so some authors have cautioned their use for the estimation of population genetic parameters—especially in cases where migration rates are expected to be very low (Whitlock 2011). In general, the application of G'_{ST} and D can be problematic because of their sensitivity to mutation rates; variation in these parameters may reflect characteristics of genetic marker loci (i.e., differences in mutation rates) more than the effects of gene flow and genetic drift (e.g., Nm).

Interpreting Genetic Differentiation—For a stable collection of populations that are connected by dispersal, there is a clear equilibrium relationship between effective gene flow (Nm) and F_{ST} (Fig. 4.21). We refer to Nm as a measure of the effectiveness of migration because it takes into account the level of gene flow relative to population size. For example, a population size of $N = 10$ could have a migration rate of $m = 0.1$, which gives an $Nm = 1.0$. The same value for Nm for a population size of $N = 100$ requires a migration of $m = 0.01$. This tradeoff between N and m reflects the fact that a higher rate of gene flow is required to overcome the effects of genetic drift when population size is small (Fig. 4.19). A level of effective gene flow of $Nm = 1.0$ is often identified as a threshold level below which populations are expected to diverge independently (dashed line in Fig. 4.21). As Nm increases above 1.0, separate demes are more likely to behave like a single large population, and levels of differentiation are reduced until F_{ST} approaches zero. This characteristic of F_{ST} makes it valuable for the information it can provide about the potential effects of gene flow and drift on the distribution of genetic variation.

The relationship between F_{ST} and Nm will only hold true if the populations sampled have been stable long enough for the tradeoff between gene flow and drift to approach equilibrium conditions (Pannell and Charlesworth 1999; Whitlock and McCauley 1999). When populations are first established by colonization, we expect that population bottlenecks will govern allele frequencies, so levels of genetic differentiation will initially be high. After some number of generations—depending on population size—the effects of gene flow will begin to equalize allele frequencies, but this approach to equilibrium may be quite slow. Moreover, additional colonization and extinction events are likely to occur, so in general, we expect that groups of populations may be far from equilibrium and F_{ST} will underestimate *Nm*. For this reason, F_{ST} should not be used to estimate levels of gene flow among populations unless (near) equilibrium conditions can be confirmed, for example, by demonstrating isolation by distance (Hutchison and Templeton 1999; Vekemans and Hardy 2004).

Even for sets of populations that are close to equilibrium, criticisms of the use of F_{ST} for the estimation of *Nm* have been raised based on the statistical properties of these parameters; namely, that the confidence limits are generally too large to provide reliable estimates of gene flow (Whitlock and McCauley 1999). Other authors have found strong support for the use of F_{ST} for the estimation of *Nm* under a range of conditions (Slatkin and Barton 1989). This discrepancy in conclusions about the utility of F_{ST} for estimating gene flow comes from differences in the assumptions made by different authors and the non-linear relationship between F_{ST} and *Nm* (Neigel 2002; Fig. 4.21). In particular, when *Nm* is greater than about 2.0, very small differences in F_{ST} translate into large differences in gene flow. Similarly, when *Nm* is much less than 1.0, a wide range of estimates for F_{ST} translate into very small differences in gene flow. Consequently, estimates of gene flow will be most reliable for intermediate values when F_{ST} lies between approximately 0.1 and 0.5 (the blue region in Fig. 4.21). Note however, that the accuracy of these estimates still relies on conditions of near equilibrium for the effects of gene flow and genetic drift on levels of population divergence. Furthermore, highly polymorphic markers will restrict the maximum value for G_{ST}, so estimates will be biased downward as described above. Given the restrictive conditions under which F_{ST} (or G_{ST}) provides reliable estimates of *Nm*, alternative methods that estimate gene flow more directly using genetic marker data (e.g., coalescence methods, assignment methods, and private allele analysis) are generally considered to be more reliable (Slatkin and Barton 1989; Neigel 2002; Paetkau et al. 2004).

Application of Genetic Differentiation Estimates—In spite of the difficulties associated with the estimation and interpretation of F_{ST} and its analogues, differentiation among populations remains a primary focus of population genetic studies. Combined information on the levels of genetic diversity within populations and differentiation among populations provides insights into salient evolutionary processes (Hamrick and Godt 1996; Balloux and Lugon-Moulin 2002). The development of a number of freely available programs (Excoffier and Heckel 2006) has facilitated estimation of Θ (Θ_p) as an unbiased estimator of F_{ST} that is suitable for a wide range of applications. The distribution of genetic variation provides information on levels of connectivity and the effects of population size on the origin and spread of novel genetic variation. Estimation of population genetic differentiation remains as an important endeavor for a broad range of investigations in evolutionary ecology, population genetics, and conservation biology.

The choice of method for estimation of differentiation will depend largely on the type of genetic marker employed and levels of divergence among populations. In cases of low divergence, which would be characterized by high levels of allele sharing and few private alleles, it is preferable to rely on heterozygosity-based estimates of differentiation (F_{ST} and G_{ST}). Under these conditions, we would expect heterozygosity comparisons to perform better than coalescence-based methods because the collection of alleles sampled from within and among populations—and estimates of $\bar{t}$ and $\bar{t}_0$—would be very similar. Consequently, the comparisons of coancestry would yield $F_{ST\text{-}coal}$ estimates that are less sensitive to more subtle differences in allelic composition. Heterozygosity-based estimates should provide accurate estimates of differentiation when the differences in allele frequencies among populations are more similar. As levels of divergence among populations increase (i.e., mutation rate is high relative to migration) there will be greater accumulation of unique alleles within populations and $F_{ST\text{-}coal}$ may become the method of choice.

The large majority of genetic structure studies in recent years have used SSRs, even though the high mutation rates of these markers sometimes render interpretations problematic. Many SSR-based studies report multiple measures of differentiation including G_{ST} and R_{ST}, which provides insights into the partitioning of genetic variation, as the strengths and weaknesses of different parameters can be appreciated. Under intermediate levels of divergence, R_{ST} estimates from microsatellites have been shown to perform better than heterozygosity-based estimates

(Balloux and Lugon-Moulin 2002). On the other hand, the higher mutation rate of SSRs may become a liability when sampling populations that display greater divergence; over longer periods, we expect stepwise mutations will lead to higher frequencies of convergence and homoplasy.

Under high levels of divergence, that is, between subspecies, (Whitlock 2011) it would be advisable to choose a genetic marker with a lower mutation rate. In general, there is a tradeoff between the negative consequences of high mutation rates and levels of marker diversity; low-mutation rate markers may not provide adequate variation to accurately reflect levels of differentiation among populations while high-mutation rate markers lose their utility when migration rates are very low. If the alleles sampled provide adequate information (i.e., they are based on sequence data), coalescence-based methods for estimating differentiation are preferable when populations are highly divergent, while heterozygosity-based estimates may be more applicable with low to intermediate levels of divergence and for genetic markers that are based on DNA fragment size.

Regardless of the method chosen for estimating differentiation, it is advisable to report multiple parameters. Comparison of different estimates may provide more insights into the processes affecting the values obtained. For example, estimates based on G_{ST} may be low compared to R_{ST} in cases of high allelic diversity. Similarly, comparison of G_{ST} standardized measures such as D or G'_{ST} will provide information about the proportions of variation distributed among populations. Understanding the advantages and limitations of these parameters will allow a more thorough characterization of the influences of evolutionary processes on distributions of genetic variation.

Sampling Strategies—When designing population genetic surveys, the level of divergence among populations may require consideration of different sampling strategies—particularly when using coalescence-based methods. In cases where migration is low relative to the mutation rate, private allele frequencies may be too low to be detected unless a large number of individuals are sampled from within populations. For example, a sample size of ten would be expected to detect private alleles only if they were present at frequencies of at least 10%. Low within-population sample size will limit the accuracy of estimates of population differentiation when limited numbers of loci are assayed—for example when using SSRs, sequence data from a limited number of nuclear regions or chloroplast DNA (cpDNA) sequences. With fewer loci, more intensive sampling is required to detect low-frequency alleles to estimate within-population coalescence times. This problem can be alleviated by using NGS-based methods such as RAD or GBS (Appendix) that provide information on very large numbers of loci. Sampling larger numbers of loci across the genome effectively compensates for limited numbers of individuals within populations (Willing et al. 2012) but does bias sampling toward more common alleles. If adequate numbers of SNPs are present at each locus, then $F_{ST\text{-}coal}$ may be the method of choice if $\mu > m$.

Regardless of the method used to estimate levels of differentiation, interpretation of these genetic parameters remains problematic as demographic instability will generally cause departures from equilibrium levels of genetic diversity (Pannell and Charlesworth 2000). Because population history will have unquantified effects on differentiation estimates, most authors refrain from calculating migration rates from F_{ST} or G_{ST} and present population genetic parameters as "rough" assessments of population connectivity (Neigel 2002). These measures of population genetic differentiation represent the best available descriptors of evolutionary processes affecting the distribution of genetic variation. Later in this chapter, we will discuss the application of population genetic parameters for describing patterns of genetic differentiation in plant populations. In the next section, we will examine the consequences of limited dispersal and local inbreeding for the generation of within-population spatial genetic structure.

4.10 GENETIC STRUCTURE WITHIN POPULATIONS

For discrete populations it is easy to see how isolation will lead to genetic differentiation, but recall that dispersal is generally restricted, so variation in allele frequencies among local groups of populations commonly develops across a continuous range (Fig. 4.16). Even for plants that occur in isolated patches, individual populations are often large enough for the development of genetic substructure. Limited dispersal across a continuous population results in substructure defined by overlapping genetic neighborhoods, which will have consequences for estimates of levels of inbreeding and population genetic structure (Hedrick 2009). In this section, we will discuss methods to estimate levels of genetic structuring within populations.

Neighborhood Size—The potential for limited dispersal to promote the development of structure within populations resulting in local genetic neighborhoods was first recognized by Wright (1946). Estimates of neighborhood area (A_N) and neighborhood size (N_{Ns}) in a continuous population can be made from the variances in dispersal distances for seeds (σ_s^2) and pollen (σ_p^2). If we assume that dispersal follows a normal distribution in a population of uniformly distributed individuals, we can estimate the effective neighborhood size as $N_{Ns} = A_N d$; where d is the density of reproductive individuals and A_N is the neighborhood area. Neighborhood area is then calculated as

$$A_N = 4\Pi\sigma^2,$$

where σ^2 is the total dispersal variance and Π (pi) is the constant for the ratio of a circle's circumference to its diameter (Wright 1946). Both pollen and seed dispersal contribute to the total dispersal variance, but since pollen is haploid its contribution is diminished by half (i.e., $\sigma^2 = \frac{1}{2}\sigma_p^2 + \sigma_s^2$; Crawford 1984). Using this estimate of σ^2, we expect >50% of parents of local offspring to have parents within the same neighborhood area.

The genetic neighborhood was originally defined by Wright (1946) as an area of panmictic mating where the parents of individuals at the center are ~86% likely to be contained within a circle of diameter 2σ (two times the standard deviation of the dispersal distance). This estimate assumes that dispersal via male and female function are equivalent, but this is unlikely to be true for plants (Crawford 1984). Pollen and seed vectors differ for most plant species so the shapes of the dispersal functions will differ. Except in cases where seeds have mechanisms for long-distance dispersal by wind, the majority of gene flow may occur via pollen—especially in cases of dispersal by animals and wind. Moreover, gene flow through female function occurs only through seed dispersal whereas male function combines dispersal by pollen and seeds ($\sigma_♂ = \sigma_p + \sigma_s$), because half of the seed genome is obtained through fertilization by pollen. For both of these reasons, gene flow will almost always be greater via male compared to female function, and a much smaller proportion of male parents will be likely to be included in a circle defined by 2σ (because many pollen donors are likely to be outside of this circumference). The one exception is plants that completely self (autogamy), in which case gene flow through male and female function will be the same, so the assumptions of Wright's neighborhood model are met. In species where there is at least some degree of outcrossing, we predict that at least 50% of individuals will be derived from parents outside the neighborhood.

Estimates of the variance in seed and pollen dispersal have been used to estimate neighborhood sizes ranging from 4 to 95 individuals for herbaceous perennials when pollinator flight distance between plants was used to estimate pollen dispersal distances (Kerster and Levin 1968; Schaal 1980). Actual pollen dispersal may be much greater than the estimate from nearest-plant flight distances due to pollen carryover as bees visit a series of flowers (Thomson and Thomson 1989), in which case N_{Ns} may include several thousand plants. In another study, N_{Ns} based on the paternity of seeds was relatively small (between 1.6 and 5.5) and only about 1.5 times greater than the size estimated from pollinator flight distances (Karron et al. 1995), which suggests that the effects of pollen carryover may be more limited in some cases.

We can see from the discussion above that estimates of neighborhood size depend heavily on the accuracy of pollen and seed dispersal measures. Even in the case where pollen carryover is taken into account, direct measures of dispersal chronically underestimate actual dispersal distances; for example, Thomson and Thomson (1989) found some degree of pollen dispersal at the limits of the distances they measured (>40 m), which suggests that actual dispersal distances in the glacier lily (*Erythronium grandiflorum*) can be much greater. Seed dispersal is also difficult to quantify accurately, and long-distance dispersal events are routinely missed (Ouborg et al. 1999) but can have a disproportionate effect on range expansion and rates of gene flow (Cain et al. 2000).

Furthermore, direct measures of dispersal do not indicate whether pollen and seeds succeed and affect the genetic composition of recipient subpopulations (e.g., Gerber et al. 2014). Even as a small fraction of dispersed pollen and seeds are able to succeed, it is likely that direct measures of dispersal will chronically underestimate actual dispersal distances and the size of genetic neighborhoods. Most of the gene flow exceeding the neighborhood will be to adjacent areas where allele frequencies are likely to be very similar; however, recognizing the high level of genetic exchange that occurs across populations helps emphasize the fact that genetic neighborhoods are fluid and overlapping rather than discrete entities.

Neighborhood size (N_{Ns}) should be associated with effective population size (N_e), but there are a variety of reasons why these parameters are not equivalent. We expect pollen and seed dispersal to be key factors that determine both N_e and N_{Ns} because they both reflect the size of the local mating pool, which is the primary determinant of stochasticity in allele frequencies across generations. On the other hand, there are factors such as the mating system that have large effects on N_e but do not influence estimates of N_{Ns}. As we saw above, selfing and biparental inbreeding reduce N_e by as much as half for a population that is highly selfing.

While limited seed movement is known to increase inbreeding, dispersal does not accurately reflect the consequences of inbreeding for N_e, because it does not take into account the mating system. Consequently, we expect N_{Ns} to overestimate N_e in populations that have any degree of selfing and limited seed dispersal. Furthermore, neighborhood size will chronically underestimate the effective population size because it assumes equal reproductive success, which is virtually never the case, because fitness typically follows a log-normal distribution where only a minority of individuals produces the majority of offspring (Herrera and Jovani 2010; Chapter 5). While we could refine estimates of N_{Ns} to more accurately predict N_e, this approach has largely been abandoned as the use of estimates of the effective population size from genetic markers has become prevalent (Luikart et al. 2010; Do et al. 2014).

Spatial Genetic Structure—Evidence of local inbreeding and genetic differentiation across continuous populations can be inferred from analyses of fine-scale *Spatial Genetic Structure* (*SGS*; Vekemans and Hardy 2004). These studies use measures of relatedness across varying geographical distances to estimate levels of spatial autocorrelation for genetic similarity (Fig. 4.22). Autocorrelation analysis estimates the strength of the association (correlation) between pairs of individuals separated by different distances. We expect that neighborhood structure will lead to higher average genetic similarity among individuals growing in close proximity compared to those that are further away, due to the development of IBD across a population. Spatial genetic autocorrelation provides information on genetic structure, which allows inferences of levels of gene flow and inbreeding in populations.

Outcomes of analyses of spatial structure using spatial autocorrelation are consistent with our expectations; highly selfing species tend to have smaller neighborhood areas, whereas outcrossing species growing at low densities and large species such as trees display positive correlations for genetic similarity across larger areas (Vekemans and Hardy 2004; Fig. 4.22). These correlograms reflect effective gene flow on a local scale and should provide information on the effective population size.

Under some rather restrictive assumptions of equilibrium conditions for genetic drift and gene flow across continuous populations (i.e., a pattern of IBDist), and with equidistant spacing of individuals, spatial correlograms can be used to estimate genetic neighborhood size (N_{Ns}; Vekemans and Hardy 2004). Under these conditions, the regression slope (b_F) of kinship with distance between samples (i.e., slope of the line in Fig. 4.22 where $r > 0$) provides an estimate of a statistic that is equivalent to the inverse of N_{Ns} ($1/4d\Pi\sigma^2$). There are a number of caveats to this approach since the shape of the correlogram may depend on the spatial distribution of samples as well as departures from the simplifying assumptions mentioned earlier. On the other hand, the mating system as well as dispersal distances will affect estimates of N_{Ns} that are based on SGS analysis, and these estimates should be more closely related to N_e compared to dispersal-based estimates. These fine-scale genetic analyses generally provide estimates of neighborhood structure that qualitatively reflect our expectations based on demographic and mating system characteristics.

Our understanding of the consequences of various estimates of neighborhood size and effective

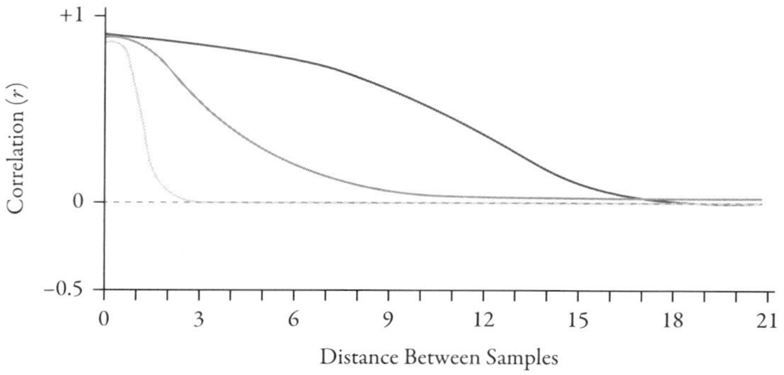

Fig. 4.22. Spatial autocorrelation (correlogram) for kinship at different distances across a continuous population. Hypothetical examples are for selfing (blue), biparental inbreeding (red), and a low density outcrossing population with limited dispersal (green line). More positive *r* values (Moran's I) indicate higher levels of genetic similarity.

population size for evolution at different spatial scales remains incomplete. In particular, estimates of effective population sizes may differ depending on whether we are considering population (short term) or lineage-wide (extended time scale) processes. In a large continuous population, we may consider the total population size to be more relevant to the potential for adaptive responses to challenges over longer time scales; however, the effectiveness of gene flow across long distances in continuously distributed plants remains poorly understood. For example, one may expect gene flow to be extensive in continuous populations, but genetic marker studies indicate that the large majority of matings often occur among nearest neighbors (Linhart and Grant 1996; Young et al. 1996).

Over generations, successive rounds of mating spread the effectiveness of gene flow across ever-larger areas, so the effective population size increases when viewed across longer time scales. This lineage-level effective population size (*Coalescence Effective Population Size*) can be estimated as the expected heterozygosity divided by the mutation rate (Lynch and Conery 2003). Whether a species is distributed continuously over large areas, or is fragmented into smaller populations, the question remains as to which spatial/temporal scale—and consequently what population size—is most important for the fate of novel mutations and the balance between the effects of gene flow and genetic drift. In the next section, we will focus on specific aspects of the biology of plants and their dispersal vectors affecting the distribution of genetic variation.

4.11 PLANT POPULATION GENETICS

The distribution of genetic variation within and among plant populations is governed by their unique biology. The sedentary growth of plants elevates the importance of dispersal vectors, and their behaviors in response to plant densities and habitat characteristics, as the primary determinants of levels and patterns of gene flow. Most plants produce both seeds and pollen, and the proportion of offspring produced through selfing and outcrossing can have strong influences on effective population sizes and levels of inbreeding. Plants persist in their habitats through perennial growth and dormancy of seeds, which creates a genetic memory that stabilizes population genetic processes. In this section, we will summarize patterns of genetic variation that have been revealed through genetic marker studies and explore the effects of biotic and abiotic interactions for the patterns of gene flow. We evaluate the consequences of dormancy for the dynamics of population genetic processes and discuss the effects of asexual reproduction and mating systems on the distribution of genetic variation in plant populations.

Patterns of Genetic Differentiation—Since the late 1960s, genetic markers studies have been used to quantify the distribution of genetic variation for thousands of different plant species, and Hamrick and Godt in 1990 and 1996 published comprehensive reviews of the large number of allozyme studies conducted during the last decades of the twentieth century (Hamrick and Godt 1990, 1996). The lack of more recent reviews of genetic structure may reflect the difficulties of comparing parameters derived from different genetic marker techniques (Balloux and Lugon-Moulin 2002). The patterns reported by these early reviews largely reflect our expectations for the effects of life-history traits and breeding systems on the distribution of genetic variation.

Hamrick and Godt compiled more than 2,200 allozyme studies to evaluate general patterns of association between levels and distribution of genetic variation, and individual plant life-history characteristics (Hamrick and Godt 1996; Fig. 4.23). Of the traits examined, breeding system had the strongest effect on levels of diversity and genetic differentiation among populations. Levels of genetic diversity within populations (H_e) increased, and differentiation among populations (G_{ST}) decreased, with an increase in the frequency of outcrossing. This pattern is expected because higher rates of selfing increase levels of inbreeding and consequently should enhance the effects of genetic drift because of the lower effective population size. The influence of breeding system on the distribution of genetic variation was maintained across growth forms, geographic distributions, seed dispersal vectors, and major taxonomic groupings (dicots, monocots, and gymnosperms).

Variation in population genetic parameters reflected the overriding effect of breeding system across the other characteristics examined. For example, endemic species tended to have lowered H_e and higher G_{ST}, but this difference was most pronounced for selfing species. The effect of breeding system was much greater for within-population diversity and did not consistently affect levels of differentiation among populations. For example, among growth form, levels of differentiation ranged from "Mid" ($G_{ST} = 0.191$) for outcrossing annuals to "High" ($G_{ST} = 0.553$) for selfing annuals, but the effect of breeding system on G_{ST} was less apparent in perennials (G_{ST} ranged from 0.218 to 0.442 in short-lived perennials; no studies on selfing in long-lived perennials were available; Fig. 4.23).

While the effects of breeding system on the distribution of genetic variation were clear, effects of other life-history traits were not as apparent. For example, there were no clear effects of dispersal vectors on the distribution of genetic variation (Fig. 4.23). Offhand, we might expect that species with seeds lacking specialized dispersal mechanisms (i.e., gravity/attached) would have lower dispersal distances, and consequently these species would have higher levels of differentiation among populations compared to those

Trait	State	H_e	G_{ST}
Breeding	*Selfing*	Low	High
	Mixed	Mid	Mid-High
	Outcrossing	High	Mid
Growth	*Annual*	Mid	Mid-High
	SL Perennial	Mid	Mid-High
	LL Perennial	Mid	Mid
Distribution	*Endemic*	Low-Mid	Mid-High
	Regional	Mid	Mid-High
	Widespread	Mid	Mid-High
Dispersal	*Attached/Gravity*	Mid	Mid-High
	Ingested	Low-Mid	Mid-High
	Wind	Mid	Mid-High

Fig. 4.23. General patterns of plant population genetic diversity (H_e) and among-population differentiation (F_{ST}) values associated with breeding system, growth form, distribution, and seed dispersal characteristics. Modified from (Hamrick and Godt 1990; Hamrick and Godt 1996).

with fleshy fruits (ingested; Fig. 4.23), or structures that enhance dispersal by wind. All of the dispersal syndromes reported by Hamrick and Godt (1996) displayed wide variation in H_e and G_{ST} that depended more on the growth form and breeding system than on the dispersal vector. The lack of effects of the seed dispersal syndrome on levels and distribution of genetic variation is surprising but suggests that pollen dispersal may overwhelm the effects of seed movement for the determination of genetic structure (Ennos 1994; McCauley 1994). As we will see in the next section, the effects of dispersal by seeds and pollen on gene flow and genetic structure can be complex as vector behavior may depend on the spatial distribution of populations and characteristics of the landscape.

Gene Flow across Landscapes—In the island model of gene flow described previously. we assumed populations were equidistant from each other, so migration rates across the metapopulation were the same. The stepping stone model added spatial structure, but we still assumed that gene flow between neighboring populations was constant—that populations were separated by the same distance, and the intervening landscape was homogenous. Movement of seeds and pollen is generally much more complex than was assumed in the simplistic models of gene flow that we have considered up to now, since these processes depend heavily on the behaviors of biotic or abiotic vectors. In particular, animal vectors may display preferences for particular types of habitats, and abiotic vectors such as wind may have characteristic directionality and effectiveness that depend on regional weather patterns, local topography, and vegetation architecture (e.g., the presence of hedgerows and forest fragments in grasslands and prairies). Furthermore, the spatial distribution of suitable habitat for a plant may interact with vector behavior to have substantial consequences for rates and patterns of gene flow within and among populations. In this section, we will explore the consequences of the spatial distribution of populations and landscape features on migration rates.

The spatial distribution of individuals and populations can have substantial effects on rates of gene flow (Young et al. 1996). For high-density populations the majority of mating typically occurs on a local scale, so nearest neighbors will be the pollen parents for the majority of seeds produced. Gene flow distances increase when plant density is lower, as animals or wind moves pollen further. For animal-pollinated plants, energy costs dictate that flight distance between plants should be minimized (Essenberg 2012); the distribution of plants affects pollinator behavior and consequently gene flow distances. Similarly, densities of wind-dispersed pollen are much greater closer to sources (Friedman and Barrett 2009). So, both biotic and abiotic vectors promote local inbreeding and the development of within-population genetic structure. When plant distributions are sparse, animal pollinators may be motivated to move pollen further and wind can carry pollen over long distances between plants. When populations are fragmented into isolated patches, isolated individuals may routinely receive pollen from distant sources, which can lead to genetic homogenization over large areas (Young et al. 1996).

Long-distance gene flow may also be promoted through the introduction of genetic variation into inbred populations, which results in higher-fitness offspring due to heterosis and consequently promotes gene flow (Ingvarsson and Whitlock 2000). Since plant populations are often inbred, matings within populations generally result in some degree of inbreeding depression and lower-fitness offspring. Pollen or seeds arriving from greater distances are likely to be carrying alleles that are not locally common. Hence, their progeny will be heterozygous across a larger number of loci and will be more likely to display higher fitness, which would facilitate the incorporation of migrant genomes and enhance rates of gene flow.

As distances among populations increase, the effectiveness of dispersal will decline until population genetic processes are dominated by genetic drift. Increases in population isolation through natural or anthropogenic causes will eventually reach a *Fragmentation Threshold*, beyond which the level of connectivity is too low for the maintenance of genetic diversity within populations (Young et al. 1996). We could infer levels of connectivity by tracking the movement of seeds and pollen, but it is difficult to detect dispersal over long distances using direct measures (Nathan et al. 2008). As an alternative, genetic markers allow us to assess the effects of dispersal on levels of genetic diversity and connectivity among populations (Cruzan 2001; Fig. 4.24). For these fragmentation analyses, the densities of individuals

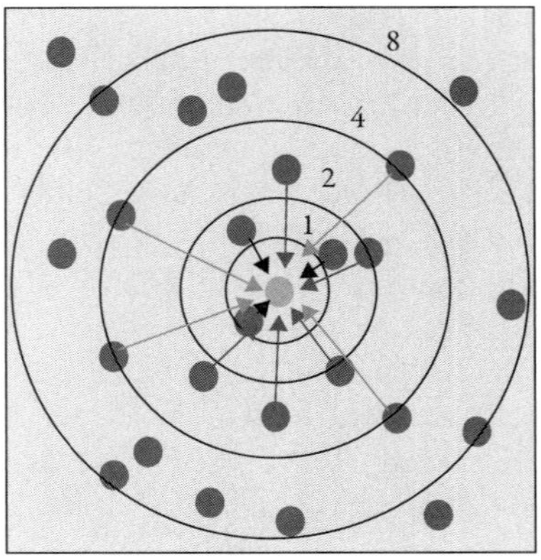

Fig. 4.24. Spatial analysis of the influence of gene flow on levels of genetic diversity. The number of populations or individuals (dots) within each concentric circle are tallied for a target population (central dot). The radius of each circle is given in km. Arrows indicate the influence of gene flow on genetic diversity in the target population. This design is repeated for multiple target populations.

are used to predict levels of genetic diversity within target populations (Fig. 4.25). Comparison of fragmentation thresholds among species provides information on the effectiveness of dispersal agents. In this example, effective gene flow occurs over longer distances in the moth-pollinated large-flowered skullcap (*Scutellaria montana*; 4 to 8 km) than in the bee-pollinated perennial Eggert's sunflower (*Helianthus eggertii*; 1.0 km; Fig. 4.25). The difference in fragmentation thresholds between these two species probably reflects differences in the average flight distances of their pollinators.

Fragmentation analysis provides information on the consequences of spatial distribution for the maintenance of genetic diversity; however, this approach assumes that the intervening matrix separating populations is homogeneous, so the rate of gene flow depends only on geographic distance. We expect that variation in habitat types and geographic features will affect the behavior of dispersal vectors and consequently the rates of gene flow. By applying methods of *Landscape Genetics*, the effects of biological and physical characteristics of environments on levels of gene flow can be quantified to estimate *Isolation by Resistance* (McRae 2006; Jaquiery et al. 2011). Resistance (the reciprocal of conductance) for each landscape character is estimated using multiple regression models that predict levels of genetic relatedness among populations (Fig. 4.26). This approach uses data from *Geographical Information Systems* (*GIS*) and genetic markers to predict the resistance of habitat types and landscape features to gene flow (McRae et al. 2008). For the example in Fig. 4.26, resistance is lowest (conductance is highest) across a plant's habitat (brown region). Lower rates of gene flow to disjunct populations are due to higher resistances across the purple and blue habitats in addition to the effect of geographic distance. In these models each landscape variable can be represented with resistance response functions that are linear or quadratic (i.e., with intermediate optima; Fig. 4.26).

The majority of landscape genetic analyses have been conducted on animals (Storfer et al. 2010), while fewer have been attempted using plant species (e.g., Lee and Mitchell-Olds 2011; Dyer et al. 2012). For animals, the distribution of suitable habitat and behavioral responses to landscape features are of primary concern. Plants present a more complicated challenge, because the effects of landscape characters on dispersal vectors also need to be accounted for. Suitable habitat promotes higher densities of plants and may promote the movement of dispersal agents. On the other hand, gene flow via pollen in dense populations may be limited due to local mating, while seed dispersal may be less sensitive to plant density.

We generally expect differential rates of gene flow via seeds and pollen; for most plants, different types of vectors disperse seeds and pollen, and landscape characteristics may impose different levels of resistance. For example, rivers and roads may present barriers to the movement of animals but may have little effect on wind dispersal. While we generally assume that the majority of gene flow is due to pollen movement, this is not always the case; for example, the large majority of gene flow in a tree with wind-dispersed fruits was due to seed rather than pollen movement (Bacles et al. 2006).

The separate effects of pollen and seed dispersal on gene flow may be determined by conducting separate analyses on the genetic structure of nuclear and chloroplast (cpDNA) markers. This is possible because nuclear markers are inherited through pollen and seeds, and cpDNA markers are maternally inherited (i.e., only through seeds) in most plants (Birky 1976; McCauley 1997; Sork and Smouse 2006; White et al.

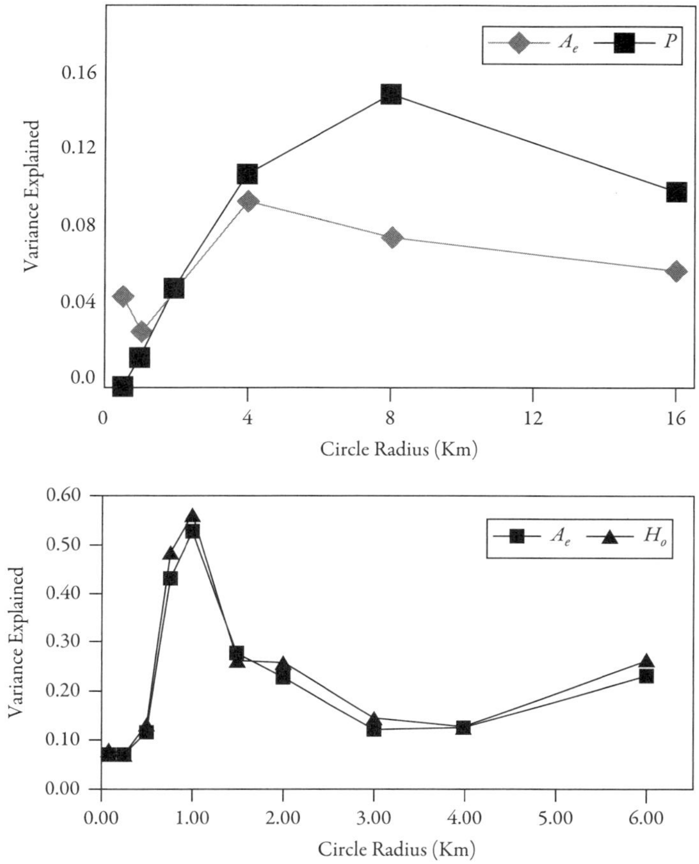

Fig. 4.25. The variance explained by circle size is as illustrated in Fig. 4.24 is used to determine the distance that best predicts genetic diversity in target populations. These analyses are for *Scutellaria montana* (upper panel; modified from Cruzan 2001) and *Helianthus eggertii* (unpub. data).

2007). Application of landscape genetic models to understand pollen and seed movement may provide critical information on the future success of plants under pressure from altered climates (Sork et al. 2013).

The sedentary nature of plants and the diversity of associated dispersal vectors present unique challenges to population geneticists trying to understand the impacts of the environment on the distribution of genetic variation. We expect plants to be more sensitive to subtle differences in *Edaphic* (soil) traits, moisture availability, and microclimates compared to animals, which results in fragmented distributions across heterogeneous landscapes. Animals and wind may be less sensitive to soil characteristics and minor differences in temperature, but dispersal vectors can have unique responses to landscape features that result in unexpected patterns of genetic relatedness. For example, genetic divergence may be elevated for populations in close geographic proximity that are separated by features that impose high-resistance barriers or due to environmental conditions that affect flowering time (e.g., Hirao and Kudo 2004). Sedentary growth also has consequences for changes in genetic composition over time; individual plants of long-lived species may persist for an extensive period, which creates a more intransigent genetic structure and some resilience for the maintenance of genetic variation. Additionally, many plant populations sustain high levels of genetic variation over extended periods through seed dormancy—a topic that we will explore further in the next section.

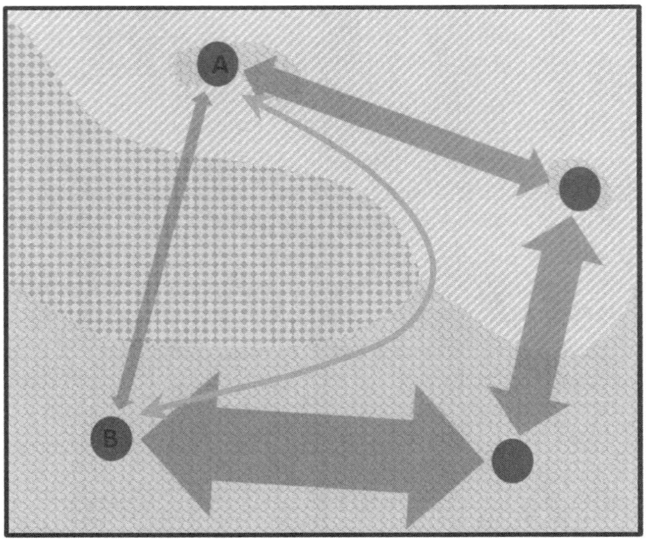

Fig. 4.26. Landscape genetics. Conductance across different habitat types is indicated by arrow widths (wide = low resistance). The brown habitat is suitable for the plant, while the blue and purple habitats impose different levels of resistance. Dark green circles are sampled populations. In this example the lowest resistance path between populations A and B is indicated by the dashed line arrows.

The Genetic Consequences of Dormancy—The genetic composition of the population of seeds produced each growing season reflects the mating success of individuals in the previous generation. Population genetic models generally assume that evolutionary processes will immediately affect the next generation, but this is not the case for many plants, as seed germination may be delayed from one to many generations. With the exception of some tropical species, seed germination is typically postponed for some period that is determined by characteristics of seeds and the environmental conditions they are exposed to (Harper 1977; Baskin and Baskin 2014). In temperate regions, germination may occur in response to changes in day length (*Photoperiodism*): some species germinate in the fall in response to shortening day lengths (*Short-Day Plants*), while others germinate in the spring (*Long-Day Plants*). Germination in response to photoperiodism requires adequate exposure to light and moisture; without appropriate conditions, seeds may remain ungerminated and persist in the soil for many years. In this case, latency is induced by external environmental conditions, so these seeds are effectively in hibernation. In other species, seeds may possess innate physical or physiological mechanisms that maintain them in a state of *Dormancy* until conditions are met that break these barriers. For example, seeds of many temperate species require exposure to cold and moist conditions for some period before their physiological dormancy is broken (*Vernalization*). Many plants, such as some legumes (Fabaceae), have hard seed coats that must be ruptured before the seed can germinate—an example of physical dormancy. Because of seed hibernation and dormancy, many plant populations consist of adult individuals as well as *Seed Banks* that may be composed of seeds produced over a number of growing seasons.

The timing of germination can be critical for seedling survival and is generally under strong selection to match local climatic conditions; consequently, germination requirements may vary substantially across a species' geographic range (Baskin and Baskin 2014). For seeds of temperate species germinating in the fall, late seedling emergence may lead to higher over-winter mortality, while seeds germinating in the spring run the risk of mortality due to late frosts. Most plants produce seeds that vary in their germination requirements to "spread the risk" so at least some offspring survive; a strategy referred to as *Bet Hedging* (Imbert 2002). In some species, such as the herbaceous composite, *Heterotheca latifolia*, seeds are heteromorphic for size and dormancy; larger seeds are more often dispersed close to maternal parents and have strong dormancy, while smaller seeds that disperse further lack dormancy (e.g., Lu et al. 2013). This strategy effectively enhances local seed banks while promoting the colonization of new sites.

In species with mechanisms of hibernation and dormancy, seeds can accumulate in soils and may reach densities of many hundreds per square meter. Seed banks are particularly important for the persistence of populations in unpredictable and harsh environments where they serve as a demographic buffer. There are also population genetic implications of seed banks, as their composition reflects contributions from many growing seasons. Each season, some fraction of seeds representing a number of past growing seasons may germinate and potentially contribute to the genetic composition of the adult population. In this way, the seed pool in an annual species has some similarities to an age-structured perennial population (Templeton and Levin 1979)—an important difference being that the genetic composition of the adult population will reflect historical environments and episodes of selection. Seeds, on the other hand, are effectively buffered from selection, so a broader range of genotypes may persist in the seed bank. As allele frequencies in populations of reproductive individuals may fluctuate over time, the seed bank serves as a "genetic memory" and a reservoir of genetic variation.

The expectation that seed banks should possess more genetic variation than above-ground populations may occur only under specific circumstances. For neutral variation, the genetic composition of adult and seed bank populations should be similar if germinating seeds are a random subset of the seed population and adult populations are stable. Models of seed banks assume that germinating seeds will be drawn proportionally from cohorts produced across a number of growing seasons, so the genetic composition of adult populations should directly reflect the variation available in the seed bank (Templeton and Levin 1979; Vitalis et al. 2004). This prediction is upheld in some studies (e.g., Tonsor et al. 1993; Cabin et al. 1998), but in other cases higher levels of genetic diversity were found in seed banks compared to adult populations (e.g., McCue and Holtsford 1998; Morris et al. 2002). It is interesting to note that both of the studies with higher diversity in seed banks were of rare and endemic plants, and seed banks may often serve to mitigate loss of genetic diversity in species that have recently undergone habitat fragmentation (Honnay et al. 2008). Loss of genetic diversity in above-ground populations of plants with limited distributions may be due to selection (Templeton and Levin 1979), genetic drift (Vitalis et al. 2004), or recent population bottlenecks. Under these circumstances, seed banks may serve as important sources of genetic variation that may facilitate evolutionary responses of populations to novel biotic and abiotic challenges.

When seeds from several seasons contribute to a new cohort of seedlings, estimates of inbreeding may be inflated by temporal variation in allele frequencies. Changes in allele frequencies across generations due to gene flow, selection, or drift can produce a *Temporal Wahlund Effect* in seedling populations (Templeton and Levin 1979; Vitalis et al. 2004). In this case, an apparent deficit of heterozygotes is due to differences in allele frequencies across seed cohorts of different age and will result in overestimates of inbreeding.

As seed banks consist of populations of seeds of different age, they effectively contain an historic record of population genetic processes. The activity of soil invertebrates and the effects of frost heaving (i.e., cycles of freezing and thawing) promote the movement of seeds deeper into the soil over time, so the seed bank may become stratified according to the age of seed cohorts. Sampling of stratified seed banks can provide a record of changes in population genetic processes. For example, Morris et al. (2002) sampled the stratified seed bank of Pyne's ground plum (*Astragalus bibullatus*), an endemic plant restricted to cedar glades of Central Tennessee, USA (Fig. 4.27). In this study, levels of genetic differentiation among older (deeper) seed populations at different sample locations were substantially greater than for the youngest layer, suggesting that populations had recently become more isolated. This pattern is consistent with historical records for the region, as fire suppression and urbanization has led to loss of ground plum habitat in recent decades (Morris et al. 2002). Younger seed populations also displayed higher levels of inbreeding, which is consistent with increased isolation (Fig. 4.27). In addition to increased inbreeding, the larger difference between H_e and H_o could be due to a temporal Wahlund effect, if the youngest seed populations consisted of

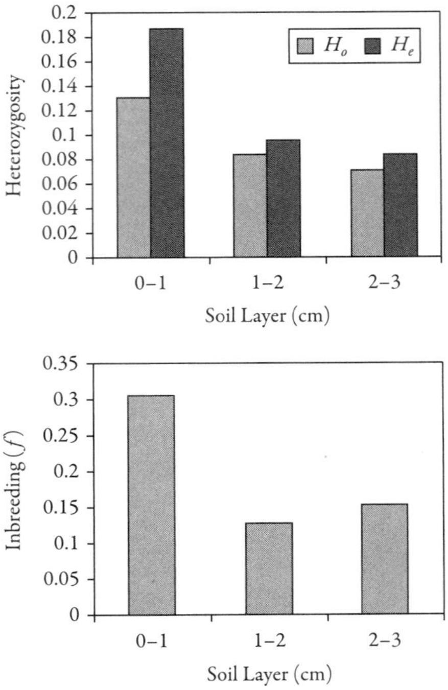

Fig. 4.27. Effects of habitat fragmentation on heterozygosity and inbreeding in the stratified soil seed bank of *Astragulus bibullatus*. Seeds are from three successively deeper 1 cm thick soil layers. Observed and expected heterozygosities (upper panel) and inbreeding ($f = (H_e - H_o)/H_e$) are based on 10 allozyme loci. Modified from Morris et al. (2002).

contributions from a broader range of years when allele frequencies were changing. This scenario is consistent with the higher overall heterozygosity in the youngest seed populations, which suggests an increase in long-distance gene flow as populations became more isolated. Both inbreeding due to increased fragmentation and a temporal Wahlund effect may be responsible for the large difference between expected and observed heterozygosities in the youngest seed populations of this species.

Seed banks provide demographic insurance for persistence of annual and short-lived perennial populations in harsh and unpredictable environments, and they form a genetic reservoir that has consequences for evolutionary processes. While the primary selective advantage of seed dormancy may be derived from demographic stability, another positive consequence of seed banks is their potential to replenish genetic variation in species that are in decline due to habitat loss or climate change. For population genetic processes, the presence of latent seeds will impose a degree of genetic resilience as changes in allele frequencies will be slower compared to species with no seed dormancy. Genetic resilience through seed banks and other forms of dormancy may have contributed to the persistence of plant diversity during major extinction events that resulted in the loss of many animal species (Chapter 2). These mechanisms of dormancy may also account for the high levels of genetic variation in plant species that recently experienced severe population bottlenecks, as was discussed earlier in this chapter.

The genetic memory of seed banks may buffer plant populations from abrupt changes in allele frequencies. Because of this lag effect, annual plants with persistent seed banks may display similar rates of evolution as longer-lived species. However, we should also note that many of the predictions for the genetic consequences of seed banks rely on simplifying assumptions that remain to be tested. For example, it may not be the case that seedling cohorts are a direct reflection of the genetic composition of seed banks. For perennial plants, asexual reproduction may facilitate the development of a genetic reservoir and the long-term persistence of populations—a topic we will discuss further in the next section.

Sexual and Asexual Reproduction—Flowering plants are unique in their diversity of reproductive modes found across species, genera, and families. Plants vary in their expression of sexual function, the proportion of offspring produced through self-pollination (selfing) and outcrossing, and the degree of asexual reproduction through vegetative spread and production of clonal propagules (Holsinger 2000; Chapter 9). Outcrossing may be enforced by the production of pollen- and ovule-bearing flowers on separate individuals (*Dioecy*), or for individuals having both sexual functions, through *Self-Incompatibility* mechanisms. Autogamy (automatic self-pollination) is found in some annual species, several of which are successful weeds (e.g., cheat grass, *Bromus tectorum*, and mouse-eared cress, *Arabidopsis thaliana*). Other species, including some violets (*Viola*), impatiens (*Impatiens*), some legumes (e.g., peanuts), and many grasses have a mixed strategy: individual plants produce some flowers that are open-pollinated (*Chasmogamous*), plus others that remain closed (*Cleistogamous*) and are also autogamous.

Many plants that reproduce sexually also have various capacities for asexual reproduction. These range from vegetative spread by roots, rhizomes (underground stems), and runners (stems that extend horizontally along the soil surface) to asexually produced propagules, including seeds that are produced by *Apomixis* (parthenogenesis), and small plantlets or bulbils that are clonal replicates of the parent plant. This variety of reproductive strategies promotes the persistence and spread of plant populations and can have strong effects on the distribution of genetic variation.

Asexual reproduction and selfing are similar to the degree that both result in the transfer of two copies of the genome from parent to offspring, but they differ because selfing allows for recombination during meiosis and for the potential effects of *Gametophytic Selection,* which we will discuss more thoroughly in the Chapter 9. Self-fertilization is generally considered advantageous in species that occupy early successional habitats and need to frequently colonize disturbed sites. High colonization potential can be also achieved through apomixis, for example in plants such as dandelion (*Taraxacum*) and fleabane (*Erigeron*). Both the ability to self and production of asexual propagules provide *Reproductive Assurance* that allows a new population to develop from a single colonist. We would expect this strategy to result in populations that consist of a single asexual or selfing lineage, but populations of clonal and selfing species often consist of a number of lineages (Holsinger 2000; Eckert 2001).

Lineage diversity (selfing or clonal) within populations probably has more to do with dispersal ability; species that reproduce by vegetative spread are more likely to have populations consisting of one or a few clones, while high rates of dispersal can elevate within-population genetic diversity. The genetic structure of selfing and clonal populations differ because asexual reproduction results in the maintenance of fixed heterozygotes while selfing lineages are often completely homozygous (Hamrick and Godt 1997). In both cases the number of distinct genotypes present will be less than the number of individuals, so both selfing and asexual reproduction will reduce the effective population size.

Both autogamous and asexual species challenge our view of populations; these collections of lineages happen to occur within close proximity, but interbreeding among neighboring lineages may be rare or nonexistent. Without recombination, these plants do not form populations in the same sense as outcrossing species. Consequently, it may not be useful to report standard population genetic parameters of genetic differentiation and inbreeding. For example, a collection of clonal lineages may give the impression of a large deficit of homozygotes due to inbreeding, even though each lineage has a high proportion of heterozygous loci.

To the extent that recombination among lineages is lacking, selfing and asexual lineages are independent entities and should not be approached with the same population genetic methods as outcrossing species. At the same time, it is important to recognize that there may be low frequencies of recombination among lineages, which has the potential to generate high levels of genetic variation and the origin of novel lineages. Analyses of clonal and selfing plant populations should account for lineage persistence and differences in the frequencies of recombination over longer time frames to develop more accurate evaluations of the influences of gene flow, drift, and selection on the distribution of genetic variation.

4.12 POPULATION GENETIC PROCESSES

The unique biological characteristics of plants present us with some difficult and exciting challenges. In many ways, plants are ideal for population genetic analyses as their sedentary growth facilitates sampling. The sedentary growth of plants also facilitates fine-scale analyses of the effects of gene flow, mating patterns, and limited population size on population structure. At the same time, development of a conceptual framework of population genetics for plants requires a thorough understanding of their biology; their mating systems, dispersal vectors, longevity, the presence of seed banks, and varied growth forms can all have substantial effects on the levels and distribution of genetic variation within and among populations. We have seen throughout this chapter that this is clearly the case for neutral genetic variation. In the next chapter, we will expand our discussion to include selection, which, along with gene flow and genetic drift, represents the third pillar of processes affecting the genetics of plant populations.

4.13 QUESTIONS FOR CONTEMPLATION AND DISCUSSION

1. As you go through this chapter, keep a list of how biological characteristics of plants that differ with respect to animals affect the population genetic processes described in each section.
2. How did the prevalent views of evolution (schools) differ with respect to the role of selection for the maintenance of genetic variation? Why do you think it was so difficult for evolutionary biologists from this period to accept the tenants of the neutral theory? Has this controversy has been resolved?
3. Are neutral genetic markers useful? Some ecologists have argued that we should focus on phenotypic traits that are likely to confer adaptation to changing environments. Frame an argument for the utility of genetic markers for conservation biology.
4. In this chapter, we introduced equations as a mechanism to illustrate basic evolutionary principles. Did you find this useful to improve your understanding of the processes being discussed? Why do you think it is important that the modern synthesis of evolutionary biology had a mathematical foundation?
5. What are the advantages of using expected heterozygosity compared to other measures of genetic diversity for diploid organisms?
6. Why is effective population size almost always lower than census size? What are the major contributors reducing effective population size for plants? Why is effective population size important for predicting the influence of genetic drift?
7. What predictions can we make about the effects of genetic drift?
8. What are the consequences of inbreeding? Does inbreeding contribute to the loss of genetic variation? If so, then how?
9. What is the Wahlund effect and why is it particularly important for plant populations (spatially and temporally)? How is the Wahlund effect related to F_{ST} and G_{ST}?
10. What does the inbreeding coefficient (*f*) represent? How is it related to F_{IS}?
11. What is associative overdominance? How could you differentiate associative overdominance from true overdominance at a marker locus?
12. Describe how F-statistics can be used to describe population genetic structure. What kind of useful information can be gleaned from the application of these statistics? Under what circumstances would it be more appropriate to use heterozygosity F_{ST} versus the coalescence F_{ST}?
13. What is neighborhood size? How is it related to effective population size?
14. What is isolation by resistance? How does it differ from isolation by distance? Why is the estimation of gene flow by dispersal more complicated for plants compared to animals?
15. What are the population genetic consequences of seed banks?

APPENDIX

GENETIC MARKERS

Allozymes (Isozymes)—Enzymes are separated in an electric field as they migrate through hydrolyzed potato starch gels (Fig. 4.28). Migration rates of individual proteins are based on their size and surface charge. Tissue is crushed in a buffer and soaked onto small filter-paper wicks that are inserted along a line cut through the gel. After the separation of proteins for several hours, gels are sliced into thin layers, each of which is stained for a different enzyme. Enzymes are stained by adding a specific substrate, cofactors (e.g., NADP, ATP, etc.), and stains that react in the presence of enzyme activity (Soltis and Soltis 1989).

Genotypes are scored based on observed banding patterns in each lane, which corresponds to one filter-paper wick (representing one individual; Fig. 4.28). Allozyme alleles are traditionally scored based on the distance they travel after a set time—either in millimeters or just qualitatively as fast (F), medium (M), slow (S), etc. Heterozygotes appear as one of two banding phenotypes (Fig. 4.28); monomeric enzymes appear as two bands while dimeric enzymes have three—the middle band being the heterodimer form of the protein. Variation is based on differences in amino acid sequences, so these markers display lower levels of polymorphism than DNA-based methods.

Allozymes remain as one of the most economical genetic assay procedures in cases where just a few polymorphic loci are needed. The cost of chemicals and

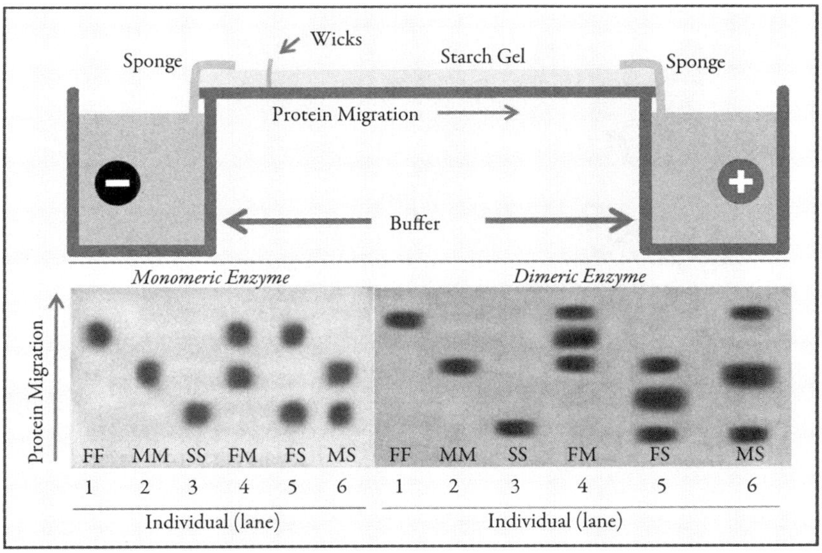

Fig. 4.28. Allozyme starch gel electrophoresis. Plexiglas boxes hold buffer with immersed electrodes (upper panel) and the gel is suspended between them. Filter paper wicks soaked with tissue extract are inserted into the gel. An electric field is applied using sheets of sponge material soaked in buffer as conductors and proteins migrate from the wicks towards the anode (+). Enzymes are visualized by selective staining. Three homozygous (FF, MM, and SS) and heterozygous (FM, FS, and MS) genotypes are shown for a monomeric and a dimeric enzyme. Gel images by Steven M Carr, Terra Nova Genomics Inc.

time investment per sample are minimal compared to other methods. Forty or more individuals can be assayed on each gel, and individual gels can be sliced into layers and stained for several different enzymes (Soltis and Soltis 1989).

Random Fragment Methods—These markers are based on different combinations of random priming and restriction enzyme digestion. Restriction Fragment Length Polymorphisms (RFLPs) were originally scored from whole-genome digestion using radioisotope labeling and southern blot techniques, but this approach has largely been abandoned. Current methods use polymerase chain reaction (PCR) of specific genome regions followed by restriction digest (PCR-RFLP), or whole-genome digest followed by selective PCR to achieve library reduction (i.e. a smaller subset of fragments is selected for scoring; AFLP; Fig. 4.29). All of these methods result in dominant banding (or electropherogram peak) phenotypes, so it is not possible to estimate the frequency of heterozygotes present in populations (observed heterozygosity; H_o). Allele frequencies and expected heterozygosity (H_e) can be inferred using the frequency of the null-banding phenotype (lack of a band) as an estimate of the homozygous recessive genotype frequency, and assuming the genotype distribution is in HW equilibrium (see text for more discussion).

Random fragment methods were improved in the late 1990s using fluorescent labels and scoring bands on automated capillary gel electrophoresis instruments. Incorporating size standards with each sample allowed researchers to adapt automated sequencers for fragment analysis, which provided a means to size fragments to within a few nucleotide bases. Because these assays are based on fragment size, problems can arise when similarly sized bands are amplified from multiple locations within a genome or from non-homologous regions in divergent genomes. Several random fragment methods are described below.

PCR-RFLP (also called Cleaved Amplified Polymorphic Sequences—CAPS; Konieczny and Ausubel 1993) has generally been applied to nuclear genomes and smaller genomes such as cpDNA and mtDNA (chloroplast and mitochondrial DNA, respectively). For the more slowly evolving chloroplast genome, the most successful assays have used

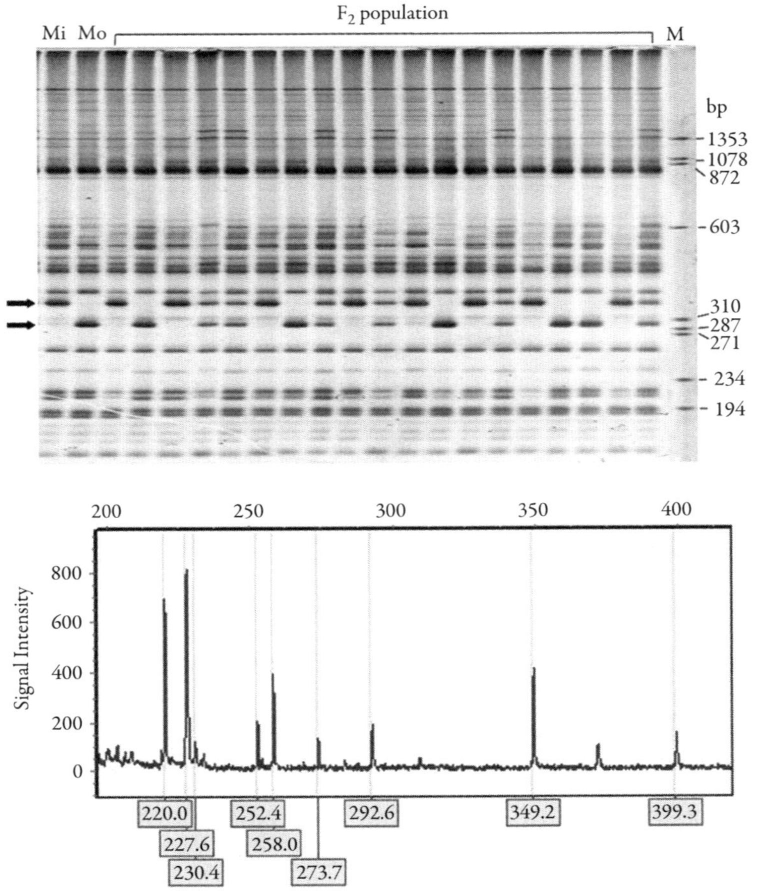

Fig. 4.29. AFLP fragment fingerprints for parents and a segregating F2 population (upper panel) and the banding pattern revealed by an electropherogram of a single lane (one individual) from capillary electrophoresis. Labels indicate the size of each peak. Gel image is from Xia et al. (2007).

some hypervariable non-functional regions that are interspersed among tRNA coding regions (Taberlet et al. 1991; Demesure et al. 1995). These regions generally contain a number of SNPs and indels that can be used for phylogenetic and phylogeographic analyses (e.g., Vijverberg and Bachmann 1999; Maskas and Cruzan 2000). PCR-RFLPs have largely been replaced by DNA sequencing, which allows a more thorough assessment of levels of genetic variation across each genomic region. Sequencing has become increasing more economical than restriction enzyme assays since the early 2000s.

RAPDs (Randomly Amplified Polymorphic DNAs) use random oligonucleotide sequences as primers for PCR amplification of fragments (Welsh and McClelland 1990; Williams et al. 1990). This method relies on random primer sequences that are 10–12 bases in length, which usually generate large numbers of amplification products in the range of 50–2,000 bp. This method relies on random priming of opposite DNA strands that happen to fall within ~2,000 bp. Genomic locations of markers are unknown and multiple products can appear as similar-sized bands on gels, so this method is not ideal for investigations where confidence in the homology of markers is necessary (e.g., phylogenetic similarity). This method has largely been abandoned due to poor consistency and reproducibility of band phenotypes.

ISSRs (Inter-Simple Sequence Repeat, ISSR-PCR; Godwin et al. 1997) takes advantage of the high density of microsatellite regions in the genome. Anonymous primers correspond to specific SSR regions so they amplify DNA, separating a pair of microsatellite sequences. Bands are scored by their size so this method has the same homology problems as RAPDs and AFLPs. The primers for ISSR analysis are longer so banding patterns tend to be more reliable than RAPDs. Sequencing of ISSR amplification products provides a method for the development of microsatellite markers (Godwin et al. 1997).

AFLPs (Amplified Fragment Length Polymorphisms; Vos et al. 1995) are based on polymorphisms in restriction enzyme cut sites. Total genomic DNA is first digested with two enzymes and oligonucleotide sequences that are 20 or more bp in length are ligated to each end of the fragments. Subsets of the fragment library are amplified by adding bases to the 5' end of primers that complement the ligated oligonucleotides. Fluorescent labeling allows fragments to be accurately sized using capillary electrophoresis. AFLPs are considered more reliable than random-priming methods (RAPD and ISSR) because polymorphisms are based on restriction enzyme recognition of short sequences (4–6 bp in length). AFLPs have been successfully used for studies in population genetics, genome mapping, and shallow phylogenetics (Meudt and Clarke 2007). Levels of homoplasy for AFLP bands have been estimated to be around 5% for pairs of species within the same genus, but simulations suggest up 30% of bands are not scored because non-homologous bands amplified from a single genome are too similar in size to be distinguished.

Microsatellites—While random fragment methods sample bi-allelic variation from a large number of sites across the genome, microsatellites (Simple Sequence Repeats—SSRs; Variable Number of Tandem Repeats—VNTRs; Simple Tandem Repeats—STRs) provide information on larger numbers of alleles from a smaller number of locations. These markers are codominant so frequencies of heterozygous genotypes can be estimated directly. Development of microsatellites is more labor intensive because specific primers have to be designed for each locus. Whole genomic DNA is first digested with one or more restriction enzymes or fragmented using sonication, and the resulting library is enriched for fragments containing repeat regions using bead- or glass-slide-capture techniques (Glenn and Schable 2005). Prior to 2005, fragments were combined with vector DNA used to transform *Escherichia coli* cells, which are then grown and screened for successful transformation. Cloned DNA was then sequenced and primers designed for fragments with adequate regions flanking microsatellites. This methodology has been modified in recent years by mass-screening of enriched libraries without the cloning step, which is enabled by NGS technologies (Zalapa et al. 2012). In either case, amplification conditions need to be optimized for each primer pair developed and individuals screened to test for polymorphism. Assaying larger samples can be made more efficient by amplifying multiple loci in single PCR reactions (*Multiplexing*), and combining products labeled with different fluorochromes for screening on capillary or next-generation sequencing instruments. Because substantial effort is required for the development of these markers, primer sequences are often reported in short papers (*Primer Notes*) that provide information on optimal amplification conditions and levels of polymorphism. Several journals specializing in genetic resources provide venues for reporting microsatellite primer sequences (e.g., *Molecular Ecology Resources* and *Applications in Plant Sciences*), which are also deposited in open-access online genome resource libraries such as the *National Center for Biotechnology Information's (NCBI) GenBank*.

Expressed Sequence Tag (EST) Approaches—An alternative method of obtaining a genome-wide sample of variation for the development of expressed sequence tag microsatellites (EST-SSRs) and SNP markers uses messenger RNA (*mRNA; Transcriptome*). Total RNA is isolated and converted to DNA (cDNA) using reverse transcription. These libraries are typically reduced to select for mRNA (poly-A RNA). Libraries can be normalized prior to sequencing, or normalization can be conducted at the bioinformatics stage to reduce the abundance of the most common transcripts (e.g., *RuBisCO* in plants). Full-length transcripts can be assembled from overlapping regions of short sequences (reads) to form *Contigs* using methods such as *RNAseq* (De Wit et al. 2012). This sequence information can be used to develop genetic markers that are distributed across the genome. With short read lengths, alleles (SNPs) at individual marker loci can be scored, but there are limits to the characterization of haplotypes (allelic associations across a chromosome) unless single reads overlap multiple polymorphic sites, a problem which is referred to as *Phasing*. The association with functional DNA raises the possibility that EST markers may not be neutral to selection.

Targeted Intergenic Sequencing Approaches—These methods are based on known sequence regions (targetted capture; Grover et al. 2012) or restriction-site associated sequencing approaches such as *GBS* and *RAD-seq*, which are analogous to sequencing of AFLP fragments (Elshire et al. 2011; Rowe et al. 2011). Using NGS with these methods, a large number of samples can be multiplexed for simultaneous assay of population samples. With adequate sequencing depth (numbers of overlapping reads) to overcome base-call errors (at least 30–50x; Straub et al. 2012), these methods have the potential to quantify SNP and indel variation for among-individual genotyping (Cronn et al. 2012). Sequence capture requires some investment in oligonucleotide "baits" that are used to capture fragments from digested or fragmented genomic DNA libraries, but once bait libraries are designed hundreds of samples can be processed at a relatively low cost. Capture methods work well for targeted nuclear regions and whole chloroplast genomes for a relatively large number of samples (Stull et al. 2013).

The cost of nuclear genome sequencing remains high, but RAD and GBS will often yield thousands of useful SNP markers at a very low cost per individual. Individual SNPs can be selectively sampled from each restriction cut site (to avoid tightly linked markers), or with information on phasing, multiple SNPs can be combined as haplotypes (alleles), which would allow the application of coalescence-based methods (e.g., $F_{ST\text{-}coal}$; see text for more details). With current technologies, hundreds of individuals can be sequenced at the same time, so large amounts of data can be collected at a relatively low cost. These NGS methods are best applied in cases where samples of variation from large numbers of locations across the genome are desirable.

5

SELECTION

A PLANT WHICH annually produces a thousand seeds, of which on average only one comes to maturity, may be more truly said to struggle with the plants of the same and other kinds which already clothe the ground.

Charles Darwin (1859) On the Universal Struggle for Existence; On the Origin of Species, page 63.

5.1 NATURAL SELECTION

As suggested by Charles Darwin in this chapter's epigraph, individual plants often produce many hundreds or even thousands of seeds each year, but very few will manage to germinate, become established, and eventually reproduce. Darwin was using this example to emphasize that his view on the "struggle for existence" was not so much about physical combat as it was about the struggle for survival under miserable odds and adverse conditions. Most seeds land in sites that offer no hope for germination or survival, and many others may be subject to predation by granivorous animals including rodents, birds, and insects. The large majority of this culling of the annual seed crop is random with respect to genotype as it depends on the chance of landing in a suitable location or of escaping predation. As the small fraction of seeds that land safely in favorable sites establish themselves and grow, they are likely to encounter a myriad of biotic and abiotic challenges. High rates of mortality may be due to physical constraints such as limited soil water or exposure to high temperatures or to biotic stressors such as competition with other plants and attacks by herbivores and pathogens.

As this drama plays out, the stand of young plants becomes thinned, as less fit or less fortunate individuals are lost. Many young plants may hang on to successfully reproduce, but only to a lower degree, while a select few will dominate and eventually parent a disproportionate share of future offspring. Often the success of individuals is by chance events; seeds may happen to arrive in sites that are particularly favorable (Harper 1977). On the other hand, when the sorting of fates is nonrandom—that growth and reproduction depend on the characteristics of individuals—we can identify cases where *Natural Selection* has some influence over success. If a trait conferring superior performance can be transferred to the next generation, the offspring produced will also be more likely to experience better than average growth and reproduction, and we can say that *Evolution by Natural Selection* has occurred. Through this process, superior characteristics spread as they are inherited across generations, and as the new trait becomes prevalent populations undergo evolution by natural selection.

The idea of evolution by selection is compelling because it is a natural outcome of challenges organisms face; under difficult conditions, any small modification that confers an advantage will be favored and will increase in frequency over generations. This concept captured the imagination of Darwin and nearly all who have become familiar with his work. Natural selection operating over long time spans offers a powerful paradigm for understanding the origins of the tremendous diversity of life forms inhabiting our planet. Since Darwin, the prevailing view has been that the forms of plants and all other organisms found in nature are the outcomes of natural selection. As we discussed in the last chapter, our understanding of evolution has become more sophisticated as we recognized the conciliatory effects of random changes along with natural selection as complementary processes that contribute to diversification of plants among habitats, regions, and continents.

Intrinsic to our understanding of selection is the concept of *Fitness*; the potential for an individual to survive and to produce offspring that have high rates of survival and reproduction in the next generation. The concepts of selection and fitness complement each other as low fitness individuals are more likely to be removed from populations by selection, and the degree to which mortality and reproductive success are associated with phenotypic characteristics of organisms reflects the strength of selection. A related concept—*Adaptation*—refers to fitness in the current environment; well-adapted individuals have high fitness in the environment they occupy because the characteristics they possess have been favored by selection in the past. Selection, on the other hand, is generally discussed with reference to single characteristics (*Traits*) that may affect fitness. Traits represent the expression of a feature (e.g. flower size), and traits may be considered to be adaptive if they contribute to higher fitness. Each trait may be subject to selection, and fitness is determined by the entire complement of traits that defines each individual.

Fitness is difficult to quantify because it is a composite measure that includes the probability of survival, asexual reproduction and vegetative spread, sexual reproductive success through male and female function (in hermaphrodites), and the fitness of sexual and asexual offspring in subsequent generations. Fitness is a statistical parameter that reflects genetic contributions to future generations—it is associated with a specific genotype (the entire complement of genes possessed by an individual) or, in population genetic models, with alleles segregating at a single locus.

Estimates of fitness based on one or a few parameters measured in a single growing season are not always reliable indicators of long-term success. For example, if an individual plant has high reproductive success because it happened to become established in a particularly favorable spot, then it may appear to have high fitness, but the offspring it produces will be unlikely to have the same level of success. On the other hand, if fitness were genetically determined—perhaps due to an advantageous mutation—then the offspring produced will be more likely to express the same trait and consequently, to have higher fitness.

Plants that are better adapted to current environmental conditions—that is, those that have higher fitness—will tend to grow faster, obtain a larger size, and to have greater sexual reproductive success (i.e., flower production) and asexual propagation over their lifetime. As size hierarchies are established in plant populations the large majority of individuals will tend to be smaller than average and a small number will attain the largest sizes (highest fitness). This sorting of individuals by fitness generates a log-normal distribution (Fig. 5.1). Characteristics of plants that contribute to fitness, such as survival, growth rate, size, reproductive success, and physiological traits that contribute to higher rates of carbon fixation, all tend to be positively correlated (Molina-Montenegro et al. 2013), and such fitness-related traits will also tend to be log-normally distributed.

Strong size hierarchies and log-normal distributions for size are common in plant populations. Long-term demographic studies demonstrate that the largest individuals in populations tend to have the highest flower production across seasons, which suggests that plant size is a good indicator of fitness (e.g., Herrera and Jovani 2010). Flower production (display size) is generally associated with both greater seed production (female reproductive success) and increased pollen

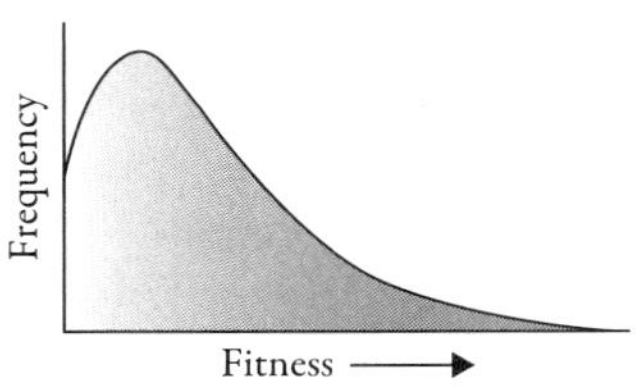

Fig. 5.1. A log-normal distribution for fitness in a plant population. The horizontal axis represents the expression of a fitness-related trait. Individuals with higher values for the trait have the highest fitness.

export, which tends to result in greater siring success (male reproductive success; Mitchell et al. 2009). Larger individuals have greater sexual reproductive success through flower production and may have higher levels of asexual reproduction through vegetative growth and clonal spread.

Plant growth and reproductive success are strongly affected by the prevailing biotic and abiotic environments, and consequently, fitness is heavily dependent on the ecological context. The sedentary growth of plants engenders strong associations with their local environments, so most plants are sensitive to differences in climatic (precipitation and temperature) and edaphic (soil) conditions that affect growth and reproduction. Recognition that environmental conditions define the selection regime and have a strong influence on the evolution of organisms in their natural environments led to the initial establishment of the field of *Ecological Genetics* by Clausen et al. (1947), which was later formalized by E. B. Ford (Ford 1971). In the ecological genetics view, the phenotype serves as the interface between the genome and its environment—the outcomes of phenotype–environment interactions across individuals are translated into changes in allele frequencies. Ecological genetic studies focus primarily on natural selection as the product of the prevailing environmental conditions.

In this chapter, we explore the processes and consequences of natural selection in plant populations. A related process, sexual selection, is often included in discussions of evolution in animal populations, but it is not clear whether this concept can be fruitfully applied to plants. We will take up aspects of selection with respect to reproductive success, including sexual selection, in Chapter 9.

In the following sections, we first discuss the origin and development of Darwin's views on evolution by natural selection. We then explore the consequences of selection by analyzing single-locus models that illustrate its interplay with other population genetic processes. In later sections, we expand our discussion to include different approaches for the measurement of the effects of selection on molecular genetic variation. In this chapter, we focus on the processes and consequences of selection on genes and genomes—we will save our discussion of the influence of selection on phenotypic traits for the next two chapters.

5.2 DEVELOPMENT OF THE THEORY

The ideas that Darwin expressed on the origins of biological diversity in 1859 represented a stark departure from the widely accepted views of the time, which assumed the Earth was very young, and that species had remained unchanged since their creation by a supreme being. Darwin's ideas were revolutionary, but the proposition of evolution by natural selection was not completely novel. Early philosophers such as Empedocles and Lucretius promoted the idea that nature randomly produces variety and only those forms that successfully survive and reproduce persist in nature, but Aristotle later rejected this idea. The ideas of early philosophers were speculative and not directly derived from observations of nature.

It was not until the late eighteenth century that more information on natural variation in nature became widely available and the discussion of the foundational ideas for evolution became more common. Georges-Louis Leclerc, Comte de Buffon (Count de Buffon) was the first to point out the patterns of similarities among plants and animals corresponded to their relatedness in his multi-volume work *Histoire Naturelle*, which was published between 1749 and 1789. While Buffon's work illustrated the problem of evolution and brought it to the attention of the scientific world, he did not propose any mechanism. Buffon's writings had a large impact on Jean-Baptiste Lamarck who, in lectures and writings made between 1800 and 1822, proposed the first comprehensive theory of organismal variation that provided a mechanism for change from one generation to the next. Lamarck's idea that characteristics acquired during an organism's life could be passed on to the next generation was widely ridiculed, but his theory contained an idea that enticed scholars and natural historians of the day—species can transform (change) over time. While Lamarck used the French *Transformisme*, use of "transmutation"—a term borrowed from alchemy in reference to the transmutation of base metals to gold—was more commonly used to describe the instability of species. This new view—that species were capable of changing over time—was provocative and was a popular topic of writings and debate among scholars during the early 1800s.

One of the participants in early discussions of transmutation was a physician and natural philosopher by the name of Erasmus Darwin—Charles's grandfather—who at one point formulated his ideas on evolution in a poem. Erasmus had a strong interest in plants and was the first to depict them as collections of independent individuals (stem tips); an idea that has become important for understanding plant structure and evolution. Around 1800, Erasmus Darwin produced a book entitled *Phytologia,* which was a compendium of botanical knowledge of the day and became a standard text for studies in plant biology. Erasmus's ideas on transmutation contributed to a growing murmur in the scholarly community, which was becoming increasingly dissatisfied with theological views for the origin of biodiversity.

As a medical student and later a student of theology, Charles Darwin was not swayed by the ideas proposed by Lamarck or by the writings of his grandfather and other proponents of transmutation. He demonstrated a passion for natural history by reading widely on the topic. He was particularly fond of books by John Stevens Henslow and William Paley on "natural theology"—a popular idea of the day that purported that evidence of God's work could be found in nature and the design of organisms. Charles was not a good student, as he demonstrated a greater passion for riding and hunting than for his studies, but he had a keen interest in the collection and classification of things from the natural world, including fossils, plants, and beetles. When Darwin did pay attention to his studies, he enjoyed attending lectures by John Stevens Henslow. Darwin clearly impressed his professor as Henslow arranged for an offer of a post as companion and Gentleman Naturalist for Captain FitzRoy on the HMS Beagle for a voyage to map the coastline of South America. In December of 1831 at the age of 22, Charles Darwin embarked on a voyage that lasted nearly five years with the primary purpose of studying geology and making collections of the plants and animals he encountered.

The growth of the ideas that ultimately led to Darwin's theory of evolution stemmed directly from observations he made during his voyage on the Beagle. His struggle to fit the biological variation he observed into accepted religious views is reflected in his speculations about separate creations to explain the unique faunas of different geographic regions such as the Galapagos Islands and the continent of Australia. The challenge of extinct species and the idea that faunas may change over time came into sharp focus for Darwin as he made notes on fossils in South America. Such fossil beds do not contain any specimens that are recognizable as representatives of living species—an observation that was difficult to reconcile with special creation. Darwin was familiar with this paradox, as it had been contemplated in writings by geologists such as Georges Cuvier and Charles Lyle, who speculated about the challenge it presented to the accepted history of life based on scripture. As Darwin organized his notes on the return voyage, the small differences he saw among species on islands stuck in his mind, and he wrote, "such facts would undermine the stability of species."

It was not until Darwin had returned to England and showed his collections to expert zoologists that he realized the full magnitude of the challenge. Birds he had collected from the Galapagos Islands had striking morphological differences, which Darwin had assumed were different types of finches, wrens and grosbeaks, all turned out to be closely related species of finches. How could such closely related species be so strikingly different in their appearance? It was within a few months of this revelation, with all the information he had amassed still fresh in his mind, that he sketched in his notebook the now famous drawing of a branching tree, which showed species replacing each other and diversifying into more species along with the words "I think" (Fig. 5.2). The seed of an idea had germinated, but it would be many years before it would come to fruition.

When Darwin returned to England in the fall of 1836, he was greeted as something of a celebrity. The numerous specimens and notebooks he brought back were extremely valuable, and he spent much of his time in these early years preparing his journal for publication. In spite of recurring illnesses, Darwin eagerly pursued his work on transmutation of species by gathering information from farmers and animal breeders and discussing his ideas with other scholars at every opportunity. In 1939, Charles Darwin received one of the highest honors an English scholar of his day could achieve when he was elected as a Fellow of the Royal Society.

Even though Darwin had a nearly complete sketch of his ideas on evolution by natural selection within a few years of returning from his voyage, it would be more than twenty years before he would publish a complete description of his theory. During this time, Darwin was not idle; he was engaged in writing and gathering information and developing his ideas on

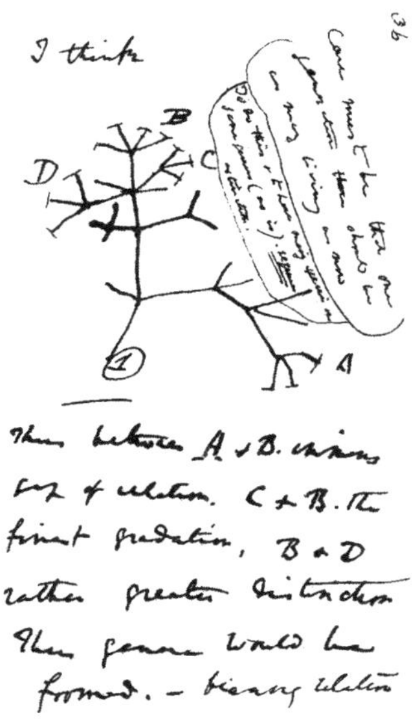

Fig. 5.2. From Darwin's notebook on transmutation of species, page 36 (mid-July 1837).

evolution. He had lengthy discussions with his close friend Charles Lyle, a prominent geologist of the day, concerning evidence that the Earth was much older than a few thousand years. He also read Thomas Malthus's *An Essay on the Principle of Population,* which highlighted the fact that the numbers of plants and animals remain constant even though the potential for population growth is much greater. This principle—that many more offspring are born than survive to adulthood—Darwin later referred to as "the universal struggle for existence," and it became a critical part of the theory he was assembling.

While his immersion in scholarly activities was paramount in the development of his work, the idea that Darwin considered the most "beautiful" part of his theory was born out of his discussions with farmers who strove to perfect breeds of plants and animals through selective breeding. From these experiences, he learned first-hand the power of selection to modify organisms over time. Darwin was fascinated with the prospect that selection in nature would favor characters that would be passed on to the next generation just as selection by breeders did in the stockyards of farmers. In the years leading up to the publication of his theory, Darwin conducted many of his own experiments on breeding. The results of these experiments, along with the breeding successes he observed in the domestication of plants and animals and the observations he made during his voyage, established the foundation of evidence he needed to support his arguments for the theory of evolution by natural selection.

During the years leading up to the publication of his ideas on evolution several events occurred that prompted Darwin to rush to present his ideas for public scrutiny. First, was the anonymous publication of a book entitled *Vestiges of the Natural History of Creation* by the Scottish journalist Robert Chambers. Chambers's primary motivation was to challenge natural theology, then the prevailing view among many scholars and the public and endorsed by the clergy. In stark contrast to the widely accepted assumption that special creation was responsible for the origin of all life, *Vestiges* emphasized a theme of progressive transmutation of species. This somewhat radical book quickly became popular with the public but was dismissed by the scientific community for its amateurish and inaccurate presentations of biology and geology.

While Darwin did not view Chambers's book as a threat to his own ideas, he felt that perhaps it laid some ground for acceptance of his arguments as it brought the ideas of transmutation into the public psyche. Darwin received more of an impetus when he received a letter describing natural selection from Alfred Russell Wallace, an accomplished naturalist who had been exploring and collecting in Southeast Asia for some years. Wallace's essay was not nearly as detailed or as articulate as Darwin's, but it caused Darwin a bit of a panic because its ideas were very similar to his own. The work to complete his book, which Darwin referred to as an "abstract" of a much larger, planned book, was difficult , as his son was then ill with scarlet fever, from which he subsequently died. When the book was finally published, it was unexpectedly popular and all 1,250 copies sold out immediately.

The text that Darwin produced in 1859, *On the Origin of Species by Means of Natural Selection, or the Preservation of Favoured Races in the Struggle for Life* (shortened to *The Origin of Species* for the sixth edition published in 1872), is littered with numerous examples of plants used to illustrate his ideas (Darwin 1859). In fact, in the third chapter Darwin used more references to plants than to animals to illustrate the "universal struggle for existence"

(Kutschera and Briggs 2009). This is not surprising, as Darwin had always had a keen interest in botany and made numerous botanical collections and drawings of plants during his voyage. After returning to England, many of his discussions with farmers focused on the breeding of vegetable crops, and he conducted a large number of garden experiments with plants that involved crosses and evaluations of the effects of inbreeding and responses to light and mechanical stimuli. During his life Darwin produced six different books that focused on various aspects of plant biology including *The Effects of Cross and Self Fertilization in the Vegetable Kingdom* (Darwin 1876) and *The Different Forms of Flowers on Plants of the same Species* (Darwin 1877).

The inclusion of examples from a range of organisms provided broad-based support for evolution by natural selection and emphasized that Darwin intended his theory to be general and apply to all forms of life. He used contrasts between battles among animals for resources and the survival of plants growing in harsh conditions to emphasize that the action of natural selection was due to interactions between organisms with their biotic and abiotic environments and not to physical conflict. While there are generalities for evolution by natural selection that apply to all organisms, the ecology and biology of plants have specific implications for the processes and consequences of selection. In particular, the sedentary growth of plants renders them sensitive to selection mediated by climatic conditions and the local biotic environment. Furthermore, their dependence on vectors for mating and dispersal has resulted in selective forces that have molded the morphology of their reproductive structures. In the next section, we explore fundamental characteristics of selection acting on populations of plants.

5.3 SELECTION—CHARACTERISTICS AND DEFINITIONS

The Process of Natural Selection—Natural selection is the primary force promoting adaptation to novel conditions and generating the seemingly endless forms and expressions of life around us. The process of biological diversification has been almost imperceptibly slow; the majority of changes promoted by natural selection have been incremental, while larger jumps are less common and perhaps exceedingly rare (Fisher 1930). This is a logical conclusion when we consider that more extreme aberrations of the phenotype are likely to overshoot the target and result in underperforming and dysfunctional individuals. It is generally more reliable to advance toward the "sweet spot" of optimal fitness by small steps than by leaps. This gradualist view was held by Darwin, and with the notable exception of the saltation mode proposed by the Mendelians of the early twentieth century (Chapter 3), and the derivative idea of *Punctuated Equilibrium* proposed by Gould and Eldredge (1977), gradualism is the primary mode of evolution assumed by nearly all biologists. Even *Key Innovations* that have precipitated adaptive radiations by enabling novel modes of resource acquisition were most likely initiated as minor modifications of form and function. For example, initial stages of the capture of light energy by proteins (similar to those present in some Archaea; Chapter 1) were sorely inefficient and required extensive refinement to develop the more elaborate systems of photosynthesis present today. Similarly, the protective films exuded onto the surfaces of the earliest amphibious streptophyte algae to prevent drying were likely to have originated by minor modification of existing molecular pathways. The subsequent refinement of these coatings to allow persistence in terrestrial habitats was probably a gradual process requiring extended periods of time. We can see the effects of this incremental progress of evolution in the history of land plants; from the time of their origin in the late Proterozoic eon, land plants required hundreds of millions of years to extend their ranges into the most arid regions of the planet during the Permian and Triassic periods.

While the large majority of plant evolution has occurred by incremental steps, there are some notable examples of more dramatic changes that have simple genetic determination. One of the best documented cases is the origin of domestic corn (*Zea mays mays*) from hybridization among subspecies of teosinte (*Zea mays* ssp.) in Central America (Doebley 2004; Fig. 5.3; refer to color plate). This origin stands out because the progenitors of corn are much more spindly in stature and possess loose inflorescences with few seeds compared to the robust stalks and large ears of domesticated corn. The identification of parental taxa for the origin of domestic corn remains unresolved,

Fig. 5.3. Teosinte (left), domestic corn (right), and their F_1 hybrid (center). Photo by John Doebley.

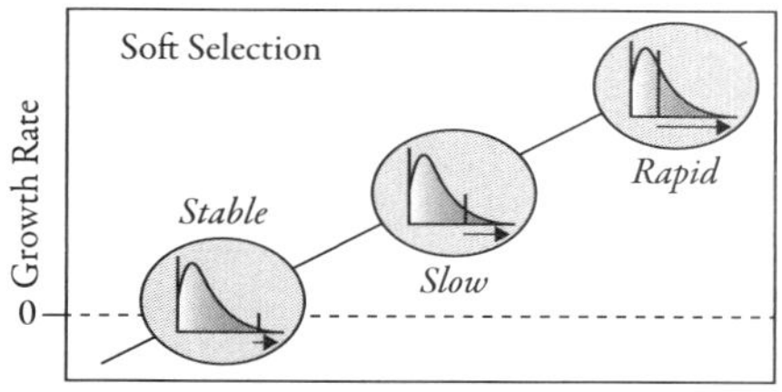

Fig. 5.4. The effect of population growth on the strength of soft selection. Growth rate varies from zero (Stable) to rapid population expansion (Rapid). Selection intensity is depicted as the proportion of distributions (in this case log normal distributions) that represents successful individuals—selection intensity increases and the cutoff threshold moves further out to include a smaller fraction of the distribution's tail.

but it is clear that the change in the inflorescence architecture is primarily due to two genes (Fig. 5.25). Clearly, other refinements were made after initial domestication through selective breeding, but that does not negate the impressive morphological difference between teosinte and corn and the fact that it is under simple genetic control.

Evidence of evolution by natural selection due to gradual modifications is often difficult to discern, because the paths of evolution promoted by selection are wandering and not linear. Studies of selection have led us to the realization that most environments are marked by stochastic fluctuations; the fitness optimum is a moving target determined by current conditions, so traits may be pushed along undulating courses that display little directionality. Exceptions to this generality can be found for species that are distributed on environmental gradients—where the range has not yet extended into more extreme conditions—and in cases where the climate is undergoing a directional change. Under more stable conditions, and when environmental change is slow, directional change may be difficult to detect as fluctuating selection obscures long-term trends.

Hard and Soft Selection—Selection operates in all populations whenever resources or physical conditions limit population growth. The large majority of populations are relatively stable with equivalent rates of mortality and establishment of new individuals each generation. Under stable conditions, only a small fraction of offspring becomes established and successfully reproduces. To the degree that success is non-random with respect to phenotypes, *Soft Selection* may continue to operate in such stable populations even when environment conditions are relatively benign and consistent from one year to the next. Differences in the intensity of soft selection among populations are due to the rate of replacement of adult individuals (Fig. 5.4). Soft selection does not depend on physical limitations imposed by environmental conditions—it operates solely on the basis of demographic conditions. The action of soft selection is density independent; the density of individuals may vary based on availability of resources, but the intensity of soft selection is a function of the rate of replacement of individuals. In other words, in rapidly growing populations the establishment rate is high as the intensity of soft selection is weak. In a stable population, the intensity of soft selection is high as individuals are replaced (Fig. 5.4). Hence, soft selection is primarily due to competition among individuals striving to fill the few open spaces that might be available.

Hard Selection differs from soft selection because it culls phenotypes based on limitations imposed by the environment. One way to think about hard selection is as a condition that sets a limit on success; only individuals with expression of a trait that falls above

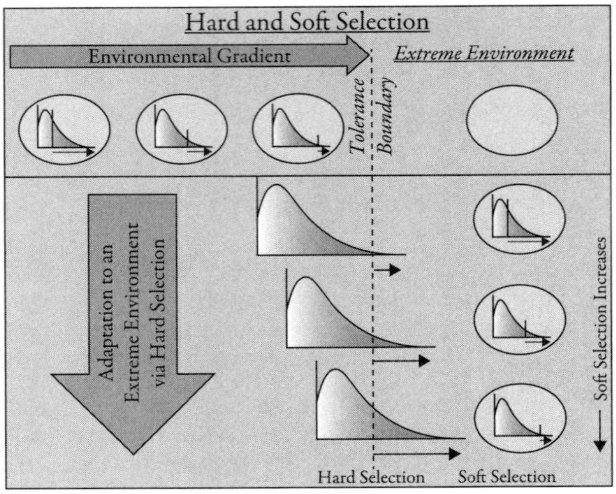

Fig. 5.5. The distinction between hard and soft selection. Populations are blue ovals and purple curves refer to the range of trait values among individuals within populations. Red arrows under the curves represent the proportion of successful individuals. Upper panel: Hard selection varies across an environmental gradient. Stronger hard selection occurs when the environment is harsher (right side of the gradient). A tolerance boundary prevents colonization of an extreme environment. Lower panel: Adaptation to the extreme environment via hard selection (over time). Only individuals with the highest trait values are able to survive in the extreme environment. As more individuals become adapted to the extreme environment soft selection within populations becomes stronger.

a threshold value will survive and reproduce. Unlike soft selection, the threshold imposed by hard selection is not sensitive to demographic conditions, so it is consistent across populations that occupy the same environment. As the environment becomes harsher, hard selection becomes more intense so a smaller fraction of the population will survive (Fig. 5.5 upper panel).

Adaptation to novel conditions may occur through both hard and soft selection. At the point in time represented in the upper panel of Fig. 5.5, survival is not possible past a tolerance boundary (vertical dotted line). Adaptation to the more extreme environments past this boundary occurs via hard selection (Fig. 5.5 lower panel). This process may be facilitated by soft selection, as individuals best adapted to less severe conditions are favored, so their offspring may be the first to colonize the extreme environment. As the colonization of the new habitat ensues, population growth is initially high so soft selection in the extreme environment is less intense. Over time, a larger proportion of individuals are adapted to the extreme conditions so population growth rates slow and soft selection becomes stronger (moving downward in the right side of the lower panel).

As illustrated in Fig. 5.5, hard selection and soft selection typically act simultaneously in populations, and when both forms of selection are acting on the same traits, the intensity of soft selection will depend on the strength of hard selection. Hard selection may establish a minimum for a trait, so only those individuals with trait expression above the survival threshold will be subject to the effects of soft selection. In environments where hard selection is weak, a larger number of individuals will survive and soft selection on the same trait will be more intense. As the intensity of hard selection increases, the number of survivors declines, and soft selection is proportionally more mild. Consequently, hard selection and soft selection are interdependent processes; the intensity of soft selection is contingent both on the severity of hard selection and demographic characteristics of the population (i.e., whether the population is growing or stable).

While the definition of hard and soft selection is useful to illustrate concepts associated with adaptive evolution, in practice it is difficult to distinguish the separate effects of these different modes of selection. The large majority of studies of fitness-related traits estimate selection as a composite variable without attempting to distinguish between the effects of biotic interactions and physical limitations imposed by the environment.

Performance and Viability Selection—In the earlier discussion, we used thresholds of success (Figs. 5.4 and 5.5), and we implicitly assumed that all individuals with trait expression above a particular value had equal fitness. This depiction of selection is convenient for illustrative purposes and for developing models, but it does not accurately reflect the effects of selection in populations. We expect that selection would continue to operate on the fraction of individuals that survive to reproduce; those with the highest trait expression should achieve the largest size and higher reproductive success, a pattern we will refer to as *Performance Selection.*

Treating fitness as a continuous variable to measure performance selection more accurately reflects the variation observed and the types of data that are typically collected to estimate selection intensity. Selection is often discussed in terms of survival (i.e., Darwin's "struggle for survival"), but in practice it is difficult to obtain information on the association between traits and survival. Few studies have attempted to include pre-reproductive mortality (*Viability Selection*), but this "invisible fraction" can have substantial effects on estimates of selection and fitness (Mojica and Kelly 2010). In spite of the general recognition that estimates of fitness should include the effects of viability selection, the large majority of studies that estimate selection focus on traits of adult individuals.

Performance and viability selection are somewhat analogous to soft and hard selection, respectively, but there is not a direct correspondence among these processes. Hard selection may be similar to viability selection if environmental conditions increase mortality rates for a subset of genotypes, but hard selection may also reduce the fecundity of individuals to the point where they make little or no contributions to the next generation (performance selection). Similarly, soft selection may be similar to performance selection by its expression as variation in reproductive success or as intraspecific competition that leads to higher rates of mortality (viability selection). Ultimately, selection will act to affect the probability of survival and the reproductive success of each genotype, but whether the action of selection is through environmental tolerances or biotic interactions will be difficult to discern based on patterns of survival and reproductive success that we observe in populations.

Forms of Selection—Variation among individuals within populations provides the raw materials for the process of evolution by natural selection. Change in the distribution of a trait subjected to viability selection effectively reduces the range of variation. Selection may act to favor individuals expressing traits at both tails of the distribution (*Disruptive Selection*), toward one extreme (*Directional Selection*), or toward the center (*Stabilizing Selection*; Fig 5.6). These forms of selection are typically illustrated as viability selection using frequency distributions—according to the type of selection acting on a trait, only individuals under some portion of a curve are likely to survive (Fig. 5.6 lower panels).

When fitness-related traits are measured in adult individuals, a selection gradient can be estimated that may correspond to one of these forms of selection (Fig. 5.6 upper panels). In this case, the region of each curve where data points are highest along the vertical axis indicates the highest fitness. To estimate selection gradients an individual trait is measured (e.g., leaf area; horizontal axis) and its relationship to a trait that is considered to be strongly associated to fitness is estimated (e.g., flower production; vertical axis). The estimation and interpretation of selection gradients will be discussed in Chapter 6.

Cost of Selection—The alteration of allele frequencies by selection requires turnover of mature individuals in the population; allele frequencies change as low-fitness individuals are replaced by those possessing traits favored by selection. Haldane recognized that the need for replacement of individuals in a population (*Cost of Selection*) sets an upper limit on the rate of change in the frequency of a selected allele (Haldane 1957). He concluded that simultaneous selection on multiple alleles could impose a demographic burden that may result in population decline; strong selection on multiple traits could elevate the danger of population extinction. This paradox, which is sometimes referred to as *Haldane's Dilemma,* is largely the consequence of some of the simplifying assumptions made for the convenience of calculation (Grant and Flake 1974). First, Haldane assumed a constant population size, but population growth may occur along with the spread of an advantageous allele so "replacement" of individuals is

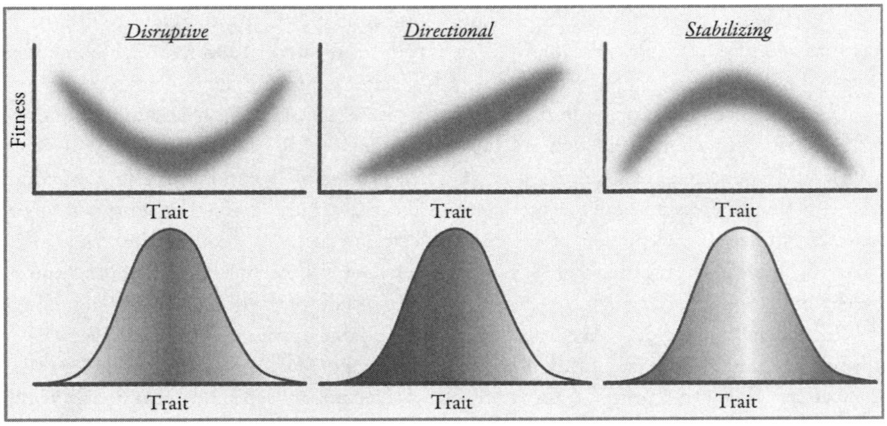

Fig. 5.6. Different modes of selection. Selection gradients (upper panels) are shown as clouds of points, with darker regions indicating higher densities of data. Lighter areas of frequency distributions (lower panels) indicate higher fitness.

not necessary to increase allele frequency. Second, Haldane did not account for recombination, so individuals may be carrying multiple alleles that are subject to positive selection. Haldane also noted that if selection acts on large numbers of offspring, then the demographic cost is greatly reduced. Since most plant species produce large numbers of seeds and seedlings, we expect there is little or no demographic cost for the culling of detrimental alleles. While we expect the rate of allele substitution to be lower in long-lived species, there does not appear to be any substantial demographic cost of selection that would limit the rate of adaptive evolution over longer time scales.

Artificial Selection—To a large degree, Darwin's inspiration for the proposition of natural selection came from his observations of the consequences of *Selective Breeding* (artificial selection) for plants and animals under domestication. Selective breeding is effectively equivalent to viability selection because only a small subset of the population reproduces. Unlike viability selection in natural populations, the remaining individuals not selected for breeding may continue to survive, but by failing to reproduce, they are not contributing to the genetic composition of the offspring generation. From an evolutionary perspective there is no distinction between the failure to reproduce and failing to survive; in either case genetic material is not passed on to the next generation.

Selective breeding differs from natural selection because it imposes only viability selection on the population, and all breeding individuals make equal contributions to the next generation. With natural selection, a larger number of individuals representing a much wider range of fitness successfully reproduce and contribute to the next generation. Consequently, selection will be much stronger during selective breeding, but with only a few individuals reproducing these populations are also more susceptible to the effects of inbreeding and genetic drift. Over successive generations of selection for desirable traits, overall fitness of the selection lines may diminish due to increases in homozygosity and the expression of *Genetic Load* (see section 5.6). Selective breeding programs circumvent the negative effects of small effective population size by using hybrids generated from crosses between parallel selection lines as the final product, which is why vegetable seeds in retail markets are often labeled as F_1 Hybrids.

Units of Selection—Selection generally operates on individuals—the total collection of cells, tissues, structures, and physiological processes that constitute the phenotype. The primary unit of selection is the phenotype and the targets of selection are individual traits, which are controlled by the genotype, and may be influenced to varying degrees by the environment. As local biotic and abiotic conditions affect trait expression, the consequences of selection for variation at the genome level will be reduced. Moreover, genetic control of traits is often complex—involving many genes that act independently and through interactions with other genes—so selection pressure at the phenotypic level may be distributed across a number of genomic locations. Hence, the translation of selection on traits to genetic change in populations is imperfect to a degree that depends on environmental influence and the genetic architecture of the trait—a topic we will explore further in Chapter 6.

In animals, the primary unit of selection is the diploid phenotype, but plants present us with a more complex problem because selection can operate during both the haploid (gametophyte) and diploid (sporophyte) life stages and at multiple levels of organization. Unlike animals, which have little gene expression in their haploid sperm (da Costa-Nunes and Grossniklaus 2004), the majority of the plant genome is transcribed in pollen (the microgametophyte)—the male multicellular haploid life stage of flowering plants (Seoighe et al. 2005; Arunkumar et al. 2013). Selection on pollen (gametophytic selection) has the potential to substantially affect the genetic composition of plant populations because alleles are expressed in the haploid state and are not masked by dominance (Mable and Otto 1998).

Plants also differ from animals because they lack a germline. As discussed in Chapter 3, selection may affect the genetic composition of growing stems as cell lineage selection acts on the population of meristem cells responsible for stem elongation (Otto and Hastings 1998; Monro and Poore 2009). Individual stems (ramets) arising from a single genet will become differentiated from each other as somatic mutations accumulate and are filtered by selection among meristem cell lineages (Poethig 1987; Klekowski 2003). Since flowers are produced at stem tips, the cumulative effects of mutation and selection during stem growth will be passed on to the gametophyte stage and may ultimately be incorporated into the next sporophyte generation.

Selection is possible at levels of organization other than the individual phenotype, as any group of individuals that display variation and are capable of replication are subject to sorting based on fitness (Williams 1966; Lewontin 1970). Selection can occur at the level of the gene, so entities such as retrotransponsons may replicate within genomes and spread in populations if they do not adversely affect the fitness of their hosts. At the cellular level, neoplasms (tumors) in animals may evolve by selection as individual cell lineages compete for resources (Merlo et al. 2006), but tumors can only survive until the host perishes. In contrast, successfully replicating genes and genes conferring advantages to cell lineages in plants have the potential for long-term persistence and spread within and among diploid populations. Hence, the process of selection in plants may be viewed as hierarchical—plant cell lines possessing advantageous mutations may be favored due to their faster growth rate (cell lineage selection), which would then increase the fitness of the entire individual (*Direct Selection; Individual Selection*). At all levels of organization, the fate of individuals is governed by selection on the traits they possess and the collective effects of their traits on fitness.

At higher levels of organization, selection among phylogenetic lineages due to differences in the effects of *Group Selection* (*Indirect Selection*) is possible; those lineages that are more prolific will diversify at higher rates and will have lower chances of extinction (Jablonski 2008). In animals, group selection is generally discussed in the context of behavior and social interactions that benefit the group—sometimes at a cost to individual fitness. This sacrifice on the part of individuals is often ascribed to fitness benefits accrued through *Kin Selection* (*Inclusive Fitness*); behaviors that reduce individual fitness are favored if close relatives are likely to benefit. In plants, positive interactions among individuals that increase the fitness of neighbors may be favored through indirect selection (Stevens et al. 1995), and such "social" interactions in plants have been considered to be analogous to group selection in animals. Evidence for indirect selection has been found in some plants such as the Great Lakes sea rocket (*Cakile edentula*; Donohue 2004). In this plant both indirect and direct selection (*Multilevel Selection*) on plant fitness was detected, and indirect selection was more prevalent in high-density populations. For indirect selection to be favored, it would need to produce a higher total fitness for the population, but the potential for indirect selection to affect the long-term success of plant populations and lineages remains poorly understood.

In this chapter, we focus on the processes and consequences of individual selection on diploid individuals (sporophytes). We explore the fate of novel genomic variation that affects fitness and segregates as alleles at specific genomic regions (loci). Regardless of whether novel variation (mutation) is generated during meiosis or somatic growth, the ultimate fate of new alleles will be determined at the population level. In Chapter 4, we focused on the population dynamics of neutral genetic variation. Here, we continue our discussion of population genetics as we explore the process and outcomes of selection and its interaction with other population genetic processes.

5.4 POPULATION GENETICS OF SELECTION

Selection acts on variation among individuals within populations, so it is most fruitful to begin our exploration of the effects of selection from a population genetics perspective. To develop a better understanding of the process and outcomes of selection we will start by focusing on single loci that are segregating for two or more alleles. We introduced this type of single-locus model in the last chapter, so we will now expand our discussion of population genetics to incorporate the effects of selection. In this section our goal is to first examine the dynamics of allele change within populations in response to selection and then to explore the interactions of selection with other population genetic processes.

Fisher on Selection—Our understanding of the effects of selection on allele frequencies in populations is largely due to the efforts of Sir Ronald A. Fisher, which are summarized in his book *The Genetical Theory of Natural Selection* in 1930 (Fisher 1930). As discussed in Chapter 4, Fisher viewed natural selection as the primary force of evolution, and in contrast to Wright, assumed that population sizes were very large, so the effects of genetic drift were minimal.

Early in his career, Fisher was primarily focused on statistics and applied mathematics (Provine 2001). He became interested in genetics and evolution after reading papers on the mathematics of evolution by Karl Pearson. During the tumultuous debate between the Mendelians and biometricians in the early twentieth century (Chapters 3 and 7), Fisher maintained an independent view; he accepted the Mendelian laws of inheritance but was convinced that the primary mode of evolution was by Darwinian gradualism. Fisher's perspective differed from Pearson and the other biometricians, who assumed that the types of large phenotypic differences (polymorphisms) studied by Mendel were not responsible for adaptive evolution. Fisher's views diverged from the Mendelians, because he did not accept their assertion that Mendel's work was a challenge to Darwinian evolution and gradualism.

In the years leading up to the First World War, Fisher worked with Pearson on the derivation of statistical characteristics of the correlation coefficient, which Pearson was primarily interested in as a measure of non-Mendelian heritability. Their relationship became strained when in 1918 Fisher published a paper demonstrating the Mendelian basis of continuously variable traits (Fisher 1918; Fig. 7.2). Pearson rejected Fisher's ideas on inheritance and wrote that he was "not a believer in cumulative Mendelian factors as being the solution of the hereditary puzzle" (Provine 2001, page 143).

Throughout the 1920s, Fisher largely ignored the bickering between the biometricians and Mendelians as he worked to develop a mathematical basis for evolution by natural selection. In his 1930 book, Fisher states that one of his motivations for writing it was to demonstrate that there was no conflict between the Mendelian mode of inheritance and Darwin's theory of evolution by natural selection, and his *Genetical Theory of Natural Selection* was the first substantial synthesis of Mendelian inheritance and Darwinian evolution. The foundation he established was further developed by contributions from Sewell Wright, J. B. S. Haldane, and other contributors to the modern synthesis of evolutionary biology.

Selection and Dominance—Fisher's interest in natural selection centered on the effects of dominance relationships among alleles and their responses to selection, and we will follow Fisher's lead as we explore the population genetics of selection. Like Fisher, we assume a very large population (i.e., no genetic drift) of diploid individuals with random mating and no gene flow. For simplicity, we also use an annual life cycle where all individuals perish at the end of each season (Fig. 5.7). The only opportunity for changes in allele frequencies occurs as differences in survival of juvenile plants, as viability selection occurs in the population prior to reproduction. Since all mating is random, each adult individual makes equal contributions to the next generation, so there is no

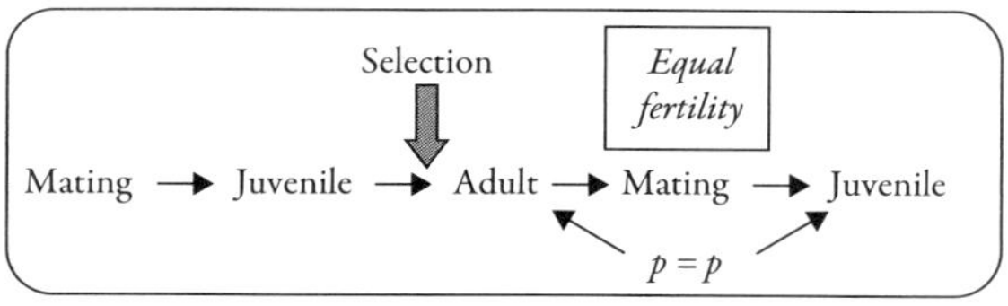

Fig. 5.7. Life history diagram for selection models. Viability selection occurs prior to reproduction each generation. The allele frequency (p) is influenced by selection as juvenile plants grow to become adults.

Genotype	A_1A_1	A_1A_2	A_2A_2
Fitness	W_{11}	W_{12}	W_{22}
Lethal Recessive	*1.0*	*1.0*	*0*
Deleterious	*1.0*	*1–hs*	*1–s*
Beneficial	*1.0*	*1+hs*	*1+s*
Overdominance	$1-s_1$	*1.0*	$1-s_2$
Underdominance	$1+s_1$	*1.0*	$1+s_2$

Fig. 5.8. Fitness profiles for models of selection. Fitness of the heterozygous genotype (w_{12}) is a function of the strength of selection (s) and dominance of the allele under selection (h).

	A_2A_2	A_1A_2	A_1A_1	h_2
$W=$	*0*	*1.0*	*1.0*	*0.0*
$W=$	*0*	*0.5*	*1.0*	*0.5*
$W=$	*0*	*0*	*1.0*	*1.0*

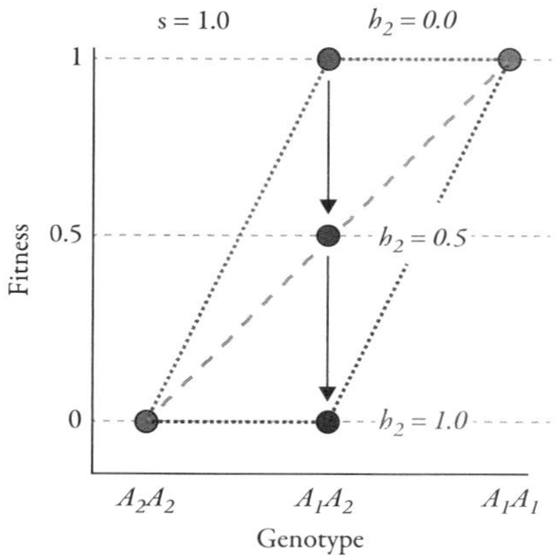

Fig. 5.9. The effects of dominance on the fitness of the heterozygous genotype for a recessive lethal allele.

change in allele frequencies from adults to seeds. All seeds have equal chances of germination, and so the cycle of selection on juveniles begins again each generation.

When two alleles segregating at one locus differ in their effects on fitness, the rate of change is a function of the exposure of alleles to selection (Hedrick 2009). To evaluate the outcomes of selection on an allele that is deleterious or beneficial (A_2), we need first to define the relative fitness (w) of each genotype (Figs. 5.8 and 5.9). In this model of we assume that the fitness of the A_1A_1 (w_{11}) genotype is always *1.0*. The fitness of the A_2A_2 genotype is diminished or elevated to the degree that selection acts on an individual allele; $w_{22} = 1 - s$ for the deleterious case, and $w_{22} = 1 + s$ when A_2 is beneficial (Fig. 5.8). The fitness of the heterozygous genotype (w_{12}) is affected by both the strength of selection (s) and the dominance of the A_2 allele (h_2). When A_2 is deleterious and completely recessive ($h_2 = 0$) the fitness of A_1A_2 will be the same as the A_1A_1 genotype. In general, the fitness of the heterozygous genotype is $w_{12} = 1 - hs$ (Figs. 5.8 and 5.9). Dominance varies continually between 0 and 1.0, and the "dominant" and "recessive" conditions are special cases where h is equal to 1.0 and 0.0, respectively. If h is greater than zero and less than one then the dominance of one allele is incomplete (*Incomplete Dominance*). When $h = 0.5$ the two alleles make equivalent contributions to the phenotype of heterozygotes and the effects of each allele are said to be *Additive* (Fig. 5.9).

The strength of selection and the degree of dominance will have a substantial influence on the dynamics of allele frequency change in populations. We can predict allele frequencies in the next generation (p_t; $t = 1$) based on genotype frequencies in the current generation ($t = 0$) from the fitness of each genotype. As illustrated in Fig. 5.8, the fitness of each genotype (w_{ii}) can be determined from the strength of selection (s) and the level of dominance for the allele under selection (h). The proportion of each genotype present in the population after selection can be calculated from the product of the genotype frequency and fitness; the fitness of the A_1A_1, A_1A_2, and A_2A_2 genotype will be p^2w_{11}, $2pqw_{12}$, and q^2w_{22}, respectively. We can predict the frequency of a deleterious A_2 allele in the next generation (q_1) as:

$$q_1 = \frac{1}{2}\left(\frac{2p_0q_0w_{12}}{\overline{w}}\right) + \frac{q_0^2w_{22}}{\overline{w}}.$$

Where $\overline{w}$ is the mean fitness. Note that the heterozygous genotype possesses only one A_2 allele so the consequences of change in its frequency is one-half compared to the effects of change in the frequency of the homozygous genotype. In this case, the A_2 allele is deleterious so selection leads to higher rates of mortality for the A_1A_2 and A_2A_2 genotypes. Genotype frequencies are standardized by the mean fitness to account for changes in genotype frequencies after selection, and the mean fitness is calculated as:

$$\overline{w} = p_0^2w_{11} + 2p_0q_0w_{12} + q_0^2w_{22}.$$

The first selection principle is that allele frequency change across generations will be more rapid when selection is stronger. By applying the equation for change in the frequency of an allele under selection, we can see the effects of increasing the strength of selection over generations (Fig. 5.10). In these scenarios we are assuming additive effects, so each allele has an equal contribution in the heterozygote fitness ($h = 0.5$). The deleterious allele is exposed to the effects of selection in the heterozygote so its removal from a population is more rapid when selection is stronger (Fig. 5.10).

The second selection principle is that the effectiveness of selection is reduced when the allele frequency is close to zero or one. The rate of change in allele frequency between generations ($\Delta q = q_1 - q_0$) is greatest when the allele frequency is close to $q = 0.5$ (Fig. 5.10; for $h = 0.5$, see Fig. 5.11 upper panel). As the allele frequency declines the frequencies of the A_1A_2 and A_2A_2 genotypes become rarer (Fig. 4.3), so the potential for selection to change the frequency of the A_2 allele is diminished. Consequently, Δq declines as q approaches zero (extinction) and the A_1 allele approaches fixation. In this model, we are assuming an infinitely large population, so q will continue to asymptotically approach extinction but will never become equivalent to zero no matter how many generations we project into the future. In finite populations, the A_2 allele can be lost due to selection, and q will continue to equal zero until the allele is reintroduced by mutation or gene flow.

The third selection principle is that the effectiveness of selection is diminished for recessive alleles when they occur at low frequency. We can use the model just described

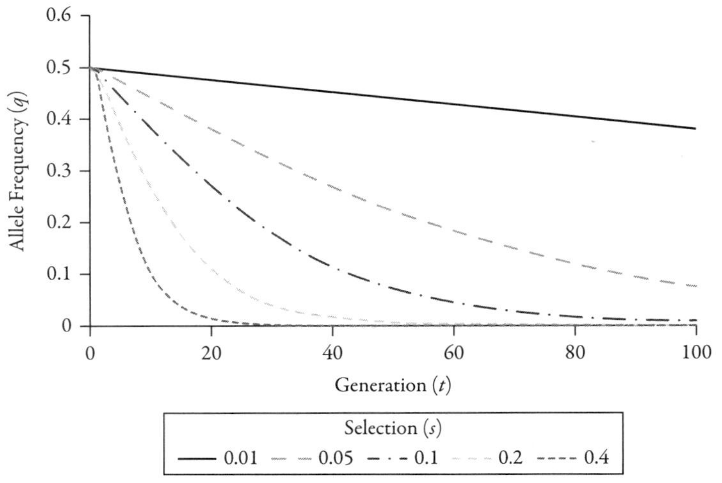

Fig. 5.10. Consequences of selection on a deleterious allele with additive effects ($h = 0.5$).

to explore the effects of dominance on changes in allele frequencies over generations (Fig. 5.11). The upper panel of Fig. 5.11 illustrates the effect of allele frequency (q) on the rate of change between generations (Δq) for a deleterious allele. In the additive case ($h = 0.5$), change across allele frequencies is symmetrical and Δq is highest at around $q = 0.5$. Lower expression of the deleterious allele in the heterozygote ($h < 0.5$) leads to higher rates of change when its frequency is high, but lower rates of change as q approaches zero (Fig. 5.11 short-dashed light gray lines). We can see this effect as Δq is more negative when q is close to 1.0 and is very close to zero for values approaching $q = 0$ (Fig. 5.11 upper panel). This occurs because if the deleterious allele is recessive, then the A_1 allele is dominant; high expression of the A_1 allele in heterozygotes increases the effectiveness of selection when q is high. As q approaches zero the frequency of the A_1A_2 genotype becomes vanishingly small—it becomes less visible to selection—so the deleterious allele has a good chance of being maintained in the population at a low frequency.

The fourth selection principle is that Δq is very small when dominant alleles occur at high frequencies. The change in allele frequency between generations for the dominant case ($h = 1.0$) is minimal when $q > 0.9$ (Fig. 5.11 long-dashed lines). In this region, the frequency of the A_1 (normal) allele is low, and

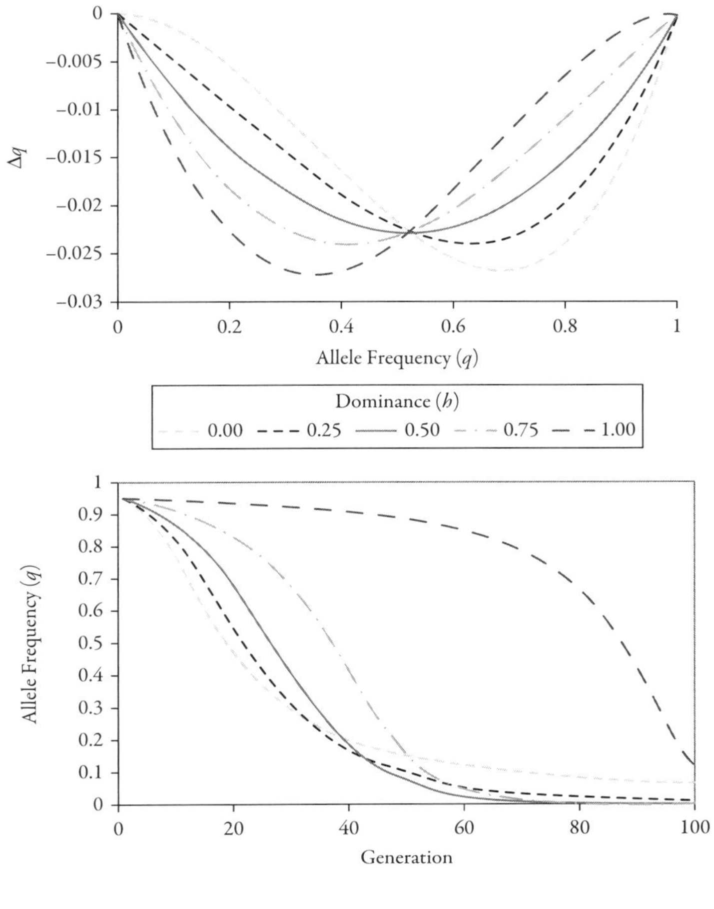

Fig. 5.11. The effects of dominance (h) on rates of frequency change for a deleterious allele. Upper panel shows Δq in response to the allele frequency (q) for five different levels of dominance. The lower panel shows allele frequencies across generations for the same levels of dominance.

because it is completely recessive, it is not exposed to the effects of selection. As the deleterious (A_2) allele declines in frequency, its rate of removal from the population increases, because the heterozygote frequency is higher. The rate of change in allele frequency declines again as q approaches zero; Δq is more negative when $q < 0.5$ and remains low when q is very close to zero (Fig. 5.11 long-dashed line, upper panel). Deleterious dominant alleles will be removed slowly when they are at very high frequency but will rapidly decline and have the greater potential to be purged from the population because they are expressed in heterozygotes even when they occur at very low frequencies.

Dominance affects the potential for selection to increase the frequency of beneficial alleles and to purge deleterious ones. From consideration of the simple model of selection described above, we predict that selection will be more effective on dominant alleles because of their high levels of expression in heterozygous genotypes. The change in the frequency of recessive alleles will be rapid when they are at high frequency but will become exceedingly slow at low frequencies. This principle of selection has implications for recessive beneficial mutations, which are likely to be invisible to selection as they only occur in heterozygotes when at low frequency. Similarly, it will be difficult to purge deleterious recessive alleles from populations because they will be sheltered from selection at low frequencies (Chapter 4). The level of dominance together with its interaction with allele frequencies are the most influential factors governing the dynamics of allele frequency change in response to selection. We will continue our exploration of dominance for the effectiveness of selection when we discuss genetic load later in this chapter.

Overdominance—Up to this point, we have discussed cases of simple dominance where one of the two homozygous genotypes displays the highest fitness. If selection acts against both homozygotes then the heterozygote will have the highest fitness and will display overdominance (heterosis). The idea of the superiority of progeny from outcrossing and hybridization has permeated biologists' view of nature at least since Darwin. In *The Effects of Cross and Self Fertilization in the Vegetable Kingdom* (1876), Darwin noted that offspring from crosses between plants had improved vigor compared to offspring resulting from self-fertilization. Darwin also observed that offspring from crosses between varieties, which he called "mongrels," often exhibited improved growth (Darwin 1859). In the early twentieth century, some of the first explorations of the mathematics of Mendelian inheritance by Udny Yule and William E. Castle included the misguided notion that allele frequencies would naturally gravitate toward equality (Provine 2001), which implicitly assumes the superiority of the heterozygote.

In the early twentieth century, results of breeding experiments with inbred strains of corn provided motivation for the proposal that heterozygote superiority (heterosis) was a widespread and general phenomenon (East 1908; Shull 1908). In those early days of genetics, the idea of heterozygote advantage was derived from the observation that hybrids from crosses between inbred lines often displayed superior vigor. Since the main difference between offspring from crosses within and between inbred lines was the level of heterozygosity, the apparent conclusion—and the view that was adopted by many—was that the heterozygote condition at individual loci (heterosis) was directly responsible for the observed increased vigor of offspring. East and Shull did not agree with this view and referred to hybrid vigor as "heterozygosis" (East 1936; Shull 1948). This was not a widely held view, and most biologists favored the idea that hybrid vigor was due to the effects of heterozygosity at individual loci. The idea of heterosis was compelling, and observations of heterozygote superiority became the foundation of the *Balance School* that dominated views of evolution by natural selection for most of the twentieth century (Chapter 4).

Interest in heterosis was reinvigorated in the late twentieth century when the results of some allozyme studies showed evidence of associations between heterozygosity and fitness (e.g., Mitton 1989). As discussed in Chapter 4, almost all of these putative cases of heterosis were due to the association of variation at neutral allozyme marker loci with levels of inbreeding and genome-wide heterozygosity (associative overdominance; Fig. 4.2).

While studies detecting the effects of associative overdominance are common, clear cases for the direct effects of heterozygote advantage based on individual loci (true overdominance) appear to be rare. Over the last century only a handful of cases of overdominance have been found in animals (e.g., sickle-cell anemia and warfarin resistance in Norway rats; Hedrick

2009), and definitive evidence of increased fitness due to heterozygosity at a single locus appears to be lacking in plants (but see Wang et al. 2013 for a possible exception).

In the early twenty-first century, new tools for genomic analyses became available and facilitated critical tests of the associative overdominance and heterosis hypotheses. For example, a *QTL* (quantitative trait locus) analysis (see Chapter 7) in rice found evidence of overdominance at several neutral loci, but as the authors suggest, this could easily be due to pseudo-overdominance as a large number of genes were in the vicinity of each marker locus (Li et al. 2008). A separate study analyzed levels of gene expression in corn and found a small proportion of loci that appeared to display overdominance for gene expression (Chen 2010); however, none of these loci were associated with heterosis for growth and yield (i.e., fitness). Even though conclusive evidence for the effects of overdominance on fitness in plants has been elusive, it remains an active topic of discussion and speculation (Charlesworth and Willis 2009; Birchler et al. 2010; Delph and Kelly 2014).

One reason for the rarity of clear examples of heterosis at individual loci is that this condition is inherently unstable because fixed heterozygosity after gene duplication is expected to confer higher fitness than overdominance at a single locus (Spofford 1969). Under this hypothesis, any locus displaying heterozygote advantage would be subject to duplication followed by the fixation of alternative alleles at each duplicated locus. These fixed heterozygotes would display higher fitness than single-locus heterozygotes, because half the offspring of the latter genotype will have low fitness, whereas all the offspring of the duplicated fixed heterozygote will have high fitness.

Models of Overdominance—Even though clear cases of overdominance in plants may be uncommon, the evaluation of how the dynamics of allele frequency change for loci exhibiting heterozygote advantage is a fruitful endeavor, as it contributes to our understanding consequences of selection operating in populations. Moreover, we expect that as individuals experience selection in different environments due to spatial and temporal heterogeneity, *Balancing Selection* will be more common and will have the same effects as overdominance (Delph and Kelly 2014). Several putative examples of balancing selection have been identified in *Arabidopsis* and other plants, although at least some of the patterns of molecular diversity that were interpreted as outcomes of balancing selection could be artifacts of demographic changes or sampling design (Wright and Gaut 2005).

To develop a population genetic model to evaluate the effects of overdominant selection on allele frequency change in populations, we assume that both homozygotes are subject to selection of strength s_1 and s_2 for the A_1A_1 and A_2A_2 genotype, respectively (Figs. 5.8 and 5.12). Since the change in allele frequencies is due solely to selection on homozygotes, we no longer need to be concerned with the frequency of heterozygote genotypes (Hedrick 2009). The consequences of overdominance selection are best viewed with Δq as a function of q (Fig. 5.12):

$$\Delta q = \frac{pq(s_1 p - s_2 q)}{1 - s_1 p^2 - s_2 q^2}.$$

The fifth selection principle is that overdominance will favor the maintenance of multiple alleles at intermediate frequencies. When Δq is positive, selection is favoring the A_2 allele, so q is increasing up to the point where the line crosses zero, which indicates that an equilibrium has been reached. Conversely, when Δq is negative it indicates that selection results in a decline in the frequency of the A_2 allele. The result is a *Stable Equilibrium* for allele frequencies ($\hat{q}$) that can be predicted from the selection coefficients:

$$\hat{q} = \frac{s_1}{s_1 + s_2}.$$

At the point of equilibrium, the forces of selection on the two homozygous genotypes balance each other so there is no net change in allele frequency. From this relationship we can predict that symmetrical selection ($s_1 = s_2 = 0.2$) results in a stable equilibrium at $\hat{q} = 0.5$. When selection is asymmetrical the equilibrium is shifted in the direction of the weaker selection coefficient, so when $s_1 = 0.2$ and $s_2 = 0.05$ the equilibrium is at $\hat{q} = 0.8$ (Fig. 5.12).

Allele frequencies will be in equilibrium at any point where the line for Δq becomes equal to zero so two additional equilibria exist at $q = 0$ and $q = 1$. Under the assumptions of this model (no mutation or gene flow) these equilibria are stable, but in general, they are unstable; as soon as the alternative allele is

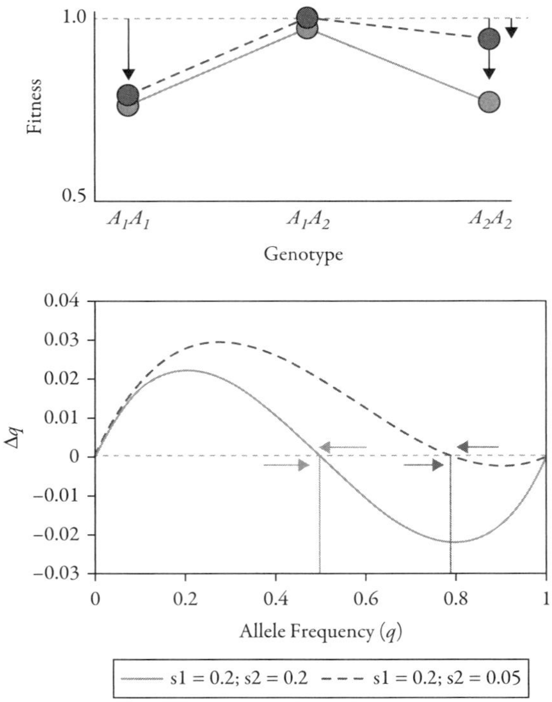

Fig. 5.12. The effects of overdominance on change in allele frequency. Equal selection on each homozygous genotype results in a stable equilibrium at $q = 0.5$ (solid lines). Asymmetrical selection moves the equilibrium to favor the allele subject to weaker selection (dashed lines).

introduced by mutation or gene flow, selection will push allele frequencies toward the stable equilibrium at an intermediate value of $\hat{q}$. The maintenance of an intermediate equilibrium by selection is resilient to other evolutionary processes; any perturbation by genetic drift or gene flow will be counteracted by selection, so the system will always gravitate toward equilibrium. In finite populations, genetic drift will generate random departures from equilibrium, and selection will act to push it back, so depending on the population size, we expect the allele frequency to fluctuate to varying degrees around the equilibrium value across generations.

Underdominance—Just as selection may favor the heterozygous genotype, it can also favor both homozygotes so their fitness is higher than the heterozygote (*Underdominance; Heterozygote Disadvantage*; Fig. 5.13). This could occur if the two alleles segregating at a locus conferred higher fitness in two different habitats, while the heterozygotes had low fitness in both habitats. Underdominance differs from balancing selection in that the average fitness of heterozygotes is lower across habitats, whereas with overdominance the average fitness of the heterozygote would be higher than the homozygous genotypes. Cases of underdominance are assumed to be rare because selection would oppose the spread of an allele causing lower fitness of the heterozygote (Wright 1940). On the other hand, it has been suggested that fixation of alleles causing underdominance may be promoted by the effects of genetic drift in very small populations (Wright 1941; Hedrick 1981).

Underdominance has also been hypothesized to contribute to reproductive isolation between divergent lineages and species, in which case the F_1 hybrid is heterozygous and displays lower fitness (Chapter 10). In cases where crosses between lineages have revealed genes with underdominant effects, there are generally additional loci involved that are capable of restoring fitness (e.g., Chen et al. 2008; Smith et al. 2011). This observation is

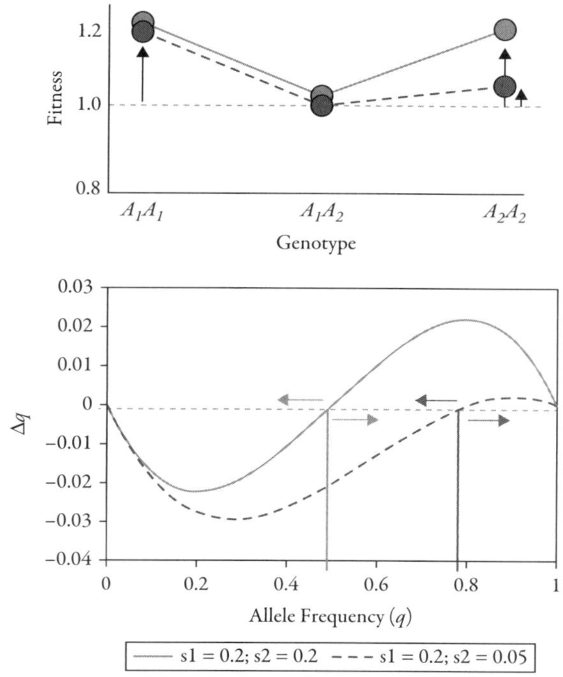

Fig. 5.13. The effects of underdominance on change in allele frequency. Equal selection on each homozygous genotype results in an unstable equilibrium at $q = 0.5$ (solid lines). Asymmetrical selection moves the equilibrium to favor the allele subject to weaker selection (dashed lines).

consistent with our expectation that the initial fixation of underdominant alleles occurred in genetic backgrounds where the new allele did not have negative effects on fitness (Coyne and Orr 2004). In this case, underdominance is due to gene interaction (epistasis) and is only apparent in a hybrid genetic background where a large number of loci are heterozygous. We will discuss the importance of epistasis and the evolution of reproductive isolation in more detail when we discuss reproductive isolation in Chapter 10.

We can model underdominance by assuming that the heterozygous fitness is equal to 1.0 and the homozygous genotypes are favored, so selection increases the fitness of A_1A_1 and A_2A_2 to be $1+s_1$ and $1+s_2$, respectively (Figs. 5.8 and 5.13). To evaluate the dynamics of allele frequency change with underdominance selection, we can examine the response of Δq (Fig. 5.13; lower panel). For heterozygote disadvantage the change in allele frequency can be predicted from an equation that is very similar to the one for overdominance:

$$\Delta q = \frac{pq(s_2q - s_1p)}{1 - s_1p^2 - s_2q^2}.$$

The sixth selection principle is that underdominant selection will result in the fixation of one of the two alleles depending on the initial allele frequency. As with heterozygote advantage, symmetrical selection results in equilibrium at an allele frequency of $q = 0.5$ (Fig. 5.13 solid lines), and the equilibrium is moved in the direction of weaker selection in the case of asymmetrical selection (red lines). In contrast to overdominance, the values of Δq above the equilibrium are positive, so selection will push the A_2 allele to fixation if $q > \hat{q}$. Alternatively, if q is less than the point of equilibrium, then selection results in a decline in allele frequency until the A_2 allele is lost from the population. Therefore, in the case of underdominant selection we predict that an Unstable Equilibrium will exist for an intermediate value of q where selection favoring the two homozygous genotypes is in balance. The equilibrium can be determined from the selection coefficients using the same equation described for $\hat{q}$ for overdominance.

In general, with heterozygote disadvantage we expect selection will drive one of the two alleles to fixation depending on their initial frequencies. In the unique case that $q = \hat{q}$ the effects of the two selection coefficients are perfectly balanced; in the absence of mutation, gene flow, and drift both alleles will be maintained in the population and the allele frequency will remain unchanged. If q deviates from $\check{q}$ to even the smalles,.t degree then selection becomes unbalanced and one allele will go to fixation. Furthermore, once either allele is fixed the introduction of the second allele by mutation or gene flow will be counteracted by selection, so only one allele will be maintained in the population. In contrast to overdominance where the intermediate equilibrium is stable, with underdominant selection the intermediate equilibrium is unstable and the equilibria at $q = 0.0$ and $q = 1.0$ are stable.

Selection on Beneficial Alleles—Up to this point we have been concentrating on lethal and deleterious alleles because, as we emphasized in Chapter 3, we expect the majority of mutations to be detrimental. However, the immense diversity of plant life on Earth is due to the accumulation of beneficial mutations that first became fixed within populations and then spread to entire lineages. In the absence of gene flow and genetic drift, we can predict that selection will inevitably lead to the fixation of a beneficial allele at a rate that depends largely on the level of expression in the heterozygote (Fig. 5.14).

Recall that the initial allele frequency for a beneficial mutation will be $1/2N$. The scenarios in Fig. 5.14 represent a new mutation in a population of $N = 10$, so the initial frequency is $q_0 = 0.05$. With any degree of expression in the heterozygote $(h > 0)$, fixation is relatively rapid; however, when the beneficial allele is completely recessive it is maintained in the population for an extended period of time before it begins to increase in frequency. In finite populations, the fate of recessive mutations is more likely to be determined by the vagaries of genetic drift because they need to reach a relatively high frequency to be exposed to selection in the homozygous condition. For example, the frequency of a recessive allele would need to exceed 0.30 before we can expect at least one out of ten individuals to be homozygous (i.e., if $q^2 = 0.1$, then $q = 0.316$). Since the chance of fixation is equivalent to the initial frequency (see Chapter 3), we expect the large majority of recessive beneficial mutations never reach a high enough frequency to be expressed in homozygotes and most will be lost due to genetic drift.

Mutation-Selection Balance—Mutations occur each generation and accumulate in populations over time. While the large majority of new mutations are lost within a few generations, some will persist at intermediate frequencies for some time so that many expressed regions across the genome will be segregating for multiple alleles. As we discussed in Chapter 3, the large majority of expressed mutations are expected to be deleterious to varying degrees. In finite populations, mutations will be subject to selection once they reach high enough frequencies to be expressed in heterozygous and homozygous genotypes, but the effects of genetic drift will also complicate their fate.

We can evaluate the effects of selection on mutations with more clarity if we limit our analysis to populations that are infinitely large and not influenced by gene flow (Hedrick 2009; Fig. 5.15). In this model, we assume that mutations are completely recessive and their frequencies are only affected by the mutation rate (μ) and the strength of selection (s). At equilibrium, new alleles are generated by mutation as quickly as they are lost from the population due to the effects of selection. Note that in this model, the population is infinitely large so there will always be a low frequency of genotypes homozygous for deleterious mutations even when the allele frequency is very close to zero.

For any particular mutation rate, we expect the equilibrium $(\hat{q})$ to depend solely on the strength of selection (Fig. 5.15). From these relationships, we can see that mutations that are more weakly deleterious are maintained at a higher frequency than those that are subjected to stronger selection. Because the population is infinitely large, we expect a very low frequency of mutations to be maintained even when selection is very strong. From consideration of just the mutation rate and strength of selection, we can conclude that populations will always be segregating for expressed genetic variation due to mutations that are deleterious to varying degrees, and that genomic regions subject to higher mutation rates will be segregating for more alleles.

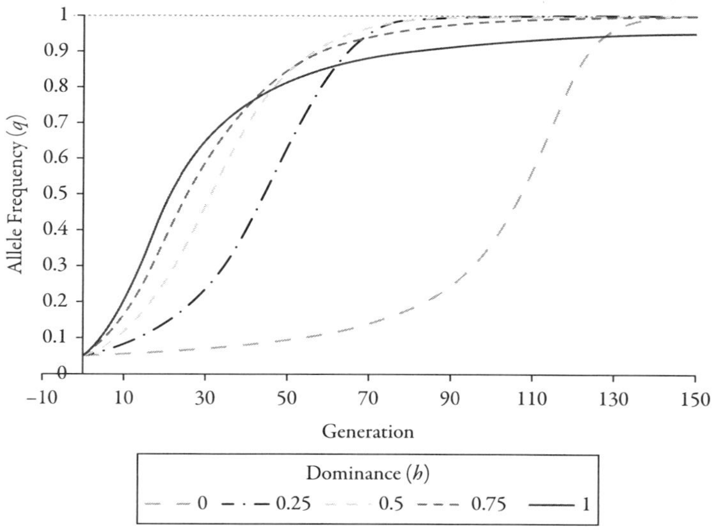

Fig. 5.14. Effects of dominance on selection for a beneficial allele. Selection (s) is 0.2 in all cases.

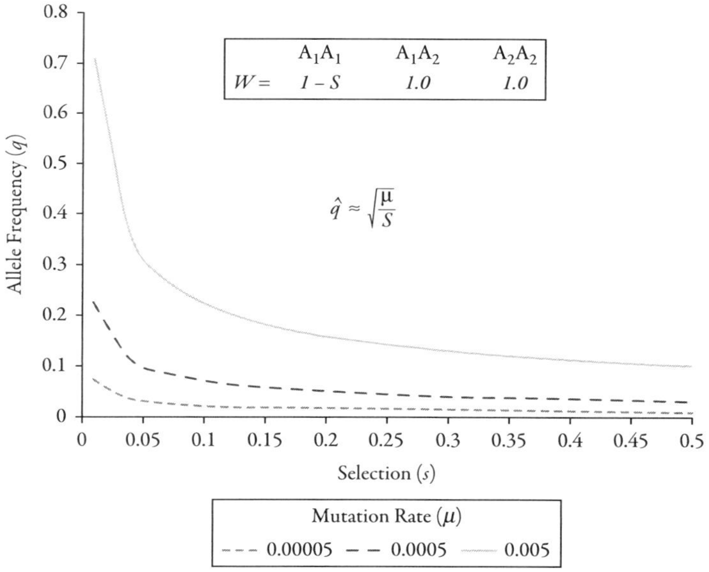

Fig. 5.15. The effect of selection on the equilibrium frequency of mutations in an infinitely large population with no migration.

We can extrapolate from the limited scenario described above to make general inferences about the expected distribution of mutations in finite populations that are subject to a wider range of conditions. While populations are generally not large enough to be segregating for many deleterious mutations at single loci, the relationship in Fig. 5.15 predicts the same distribution of mutations would occur for a large number of loci distributed across the genome. This mutation-selection balance model tells us that across genomes we should expect to find few mutations that are strongly detrimental and a larger number that have only mildly deleterious effects.

In real populations, the two factors that have the largest influence on the equilibrium frequencies of deleterious mutations are dominance (h) and the level of inbreeding (f; Hedrick 2009). The general effects of dominance will be to increase the level of expression in heterozygotes and consequently, to reduce the equilibrium allele frequency compared to that shown in Fig. 5.15. This depression of the equilibrium frequency becomes much more substantial when dominance exceeds the minimal value of $h > 0.10$, which is much less than the expression of additive alleles (i.e., $h = 0.5$). We can conclude that even low levels of expression of deleterious alleles in heterozygotes depresses the equilibrium curve to the point where the asymptote for $\hat{q}$ approaches very close to zero when selection is strong—a small degree of dominance will have a large effect on the equilibrium frequency of deleterious mutations.

Inbreeding will also have substantial effects on deleterious mutations. In populations subject to inbreeding, the equilibrium frequencies of deleterious mutations are reduced because mating among related individuals elevates the frequency of homozygous genotypes above the HW expectation (see Chapter 4). As inbreeding ensues, a larger proportion of the copies of the deleterious mutation are exposed to the effects of selection, which facilitates their elimination from the population (*Purging*; see the Purging Genetic Load section). Even a small degree of inbreeding ($f > 0.05$) is adequate to result in an asymptote approaching very close to zero as selection intensity increases (Hedrick 2009). The net effect of inbreeding results in a reduction of $\hat{q}$ across the range of selection intensities.

In general, the expected equilibrium frequencies of deleterious alleles are quite low and depend on both the fitness effects of the mutation (s) and dominance (h; Fig. 5.16). Mutations with high levels of expression in heterozygotes can persist in populations only when their fitness effects are extremely small. Hence, dominance will reduce the frequency of deleterious alleles due to expression in the heterozygous condition, and their frequencies can be reduced even further when the frequency of homozygotes is elevated due to inbreeding. At such low frequencies, these deleterious mutations will be more susceptible to loss by genetic drift. In the mutation-selection balance model described here, none of the mutations ever go to fixation because the population is infinitely large, but in finite populations, genetic drift will contribute to both the fixation and extinction of mutations.

Recall from Chapter 4 that the probability of fixation of a mutation in a finite population is $1/2N$; for any new mutation the chance of loss from the population is very close to unity (i.e. $1-[1/2N]$]). Under these conditions, we might predict that the large majority of mutations would quickly be lost, so very few deleterious mutations would be segregating in the population. In contrast to this prediction, outcrossing populations tend to harbor large amounts of genetic load, which is primarily due to the accumulation of deleterious mutations that display low levels of dominance (Charlesworth and Willis 2009). Later in this chapter, we will discuss the accumulation of deleterious alleles contributing to fitness declines within populations and the potential for selection to reduce genetic load.

Models of Selection; Synopsis—From the models discussed earlier, we can see the substantial potential for selection to modify the genetic composition of populations and ultimately of entire lineages. These population genetic models indicate that whether selection acts on deleterious or beneficial alleles, as allele frequencies change, the alleles that confer higher fitness increase in frequency. From this observation, we can predict that, as long as the environment remains constant, the net effect of selection will be to increase average fitness of individuals from one generation to the next.

The seventh principle of selection is that populations under selection will increase in average fitness across generations. As selection works to maximize fitness, it generally results in the loss of genetic variation; except in the case of overdominance, only one allele becomes fixed when a stable equilibrium is reached. While these predictions hold true for mathematical models of infinite populations that lack gene flow and mutation, multiple population genetic processes will be acting on populations in nature so the expected outcomes of selection are not *always* realized.

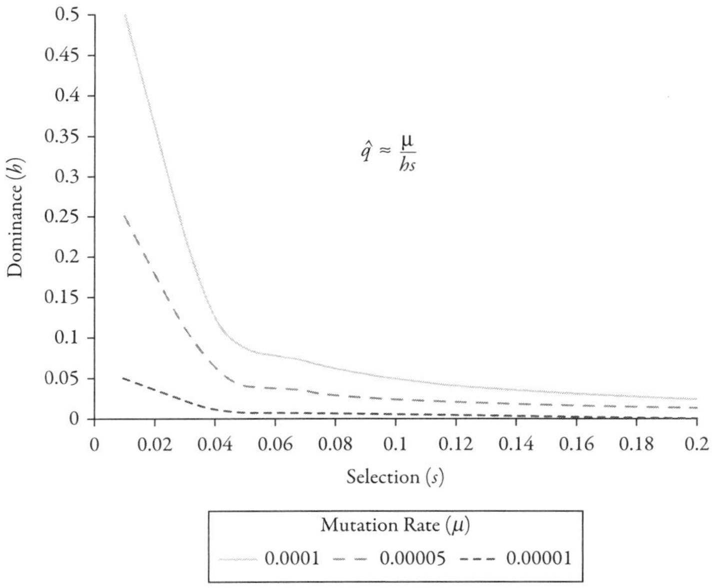

Fig. 5.16. The predicted relationship between the strength of selection (s) and level of dominance (h) for three different mutation rates. Predicted lines are based on an equilibrium allele frequency of 0.02.

5.5 THE BALANCE AMONG POPULATION GENETIC PROCESSES

Selection with Gene Flow and Genetic Drift—Considering our discussions of population genetics in this and the last two chapters, it should clear that evolutionary processes do not operate in isolation. From Chapter 4 we know that, depending on dispersal rates and population size, gene flow may be counteracted by genetic drift or vice versa (Fig. 4.19). Similarly, adaptation through natural selection may be thwarted by the influence of other population genetic processes (Fig. 5.17; refer to color plate). In this metaphorical model, each population genetic process may counteract the others. Allele frequency change is represented as a deviation in the angle of the levers; the dominating force will depress its lever to overpower the other two. If the three forces were perfectly balanced then the net result would be no change in the lever angle, which indicates no modification of the allele frequency (i.e., the population will remain in HW equilibrium). Note that in this case, HW is maintained in spite of the fact that not all the assumptions are met. Even though the balancing of opposing processes should happen only very rarely, it reminds us that the degree of change in allele frequency from one generation to the next is not a perfect indicator of the action of an individual evolutionary process.

Drift, selection, and gene flow probably affect nearly all populations, so allele frequency change between generations reflects the net outcome of all three processes acting simultaneously. We could add a fourth lever to represent the effects of mutation; however, since we are discussing allele frequency change over a short timescale this lever would be very small and difficult to detect relative to the other three. This model of a three-way teeter-totter provides a convenient visualization of the counterbalancing forces of evolution acting on allele frequencies in populations.

Population Genetic Diversity—In the model of evolution depicted in Fig. 5.17 (refer to color plate), the total level of genetic diversity is represented by the height of the pyramid—if genetic variation were lacking then the levers would be flat on the ground and evolution would not be possible (i.e., the teeter-totter would not be able to tilt). The total genetic diversity present in populations is as dynamic as allele

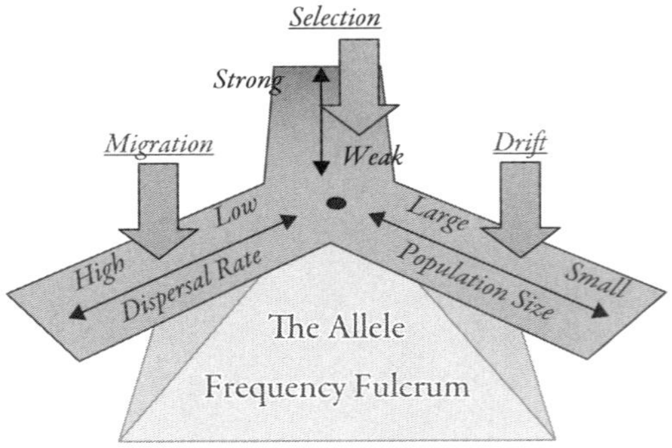

Fig. 5.17. The balancing forces of selection, gene flow, and genetic drift expanding on the metaphorical model for allele frequency change in populations from Fig. 5.19. Selection can be counteracted by migration or drift.

Process	Diversity Within	Differentiation Among
Selection	Decrease	Increase
Gene Flow	Increase	Decrease
Genetic Drift	Decrease	Increase
Mutation	Increase	Increase

Fig. 5.18. Consequences of population genetic processes for levels of variation within, and differentiation among, populations.

frequencies, and it changes as evolutionary forces affect populations. Even as population processes act to change allele frequencies in populations, they also have consequences for levels of genetic diversity and differentiation among populations (Fig. 5.18). Selection reduces genetic diversity within populations as it eliminates some alleles in favor of others. If populations occur in habitats that differ to various degrees, then selection will also increase the level of genetic differentiation among populations. The reduction in genetic diversity will be direct for those genomic regions subject to selection, but selection also has the effect of reducing the effective population size, because it increases the variance in reproductive success. Consequently, the effects of genetic drift will be enhanced in populations subject to selection. Genetic drift also leads to the loss of diversity and increased differentiation, but in this case, allele frequency change is random, so drift will always increase differences among populations regardless of the environmental conditions (see Chapter 4).

While selection and genetic drift reduce genetic variation, gene flow and mutation will elevate levels of diversity within populations (Fig. 5.18). If populations differ in allelic composition, then gene flow will tend to increase genetic diversity within populations and simultaneously will homogenize allele frequencies among populations (see Chapter 4). Mutation contributes to diversity in all populations, but we expect the new alleles generated will be unique, so it also causes higher divergence among populations over longer timescales. The sum result of these processes is to enable evolution by natural selection; as selection and drift deplete genetic variation within populations, mutation and gene flow replenish it. In the next section, we will explore the generation and fate of molecular genetic variation under the influences of drift and selection.

5.6 GENETIC LOAD

The interplay among the population genetic processes of selection, genetic drift, and gene flow can be seen in the progression of the accumulation of deleterious mutations (genetic load). In small populations, selection is less effective due to genetic drift, so we expect the average of effects of mutations fixing in small populations to be greater than mutations accumulating in larger populations. While gene flow can counteract the effects of selection and genetic drift, it can also introduce additional genetic variation that may facilitate the removal of fixed deleterious mutations. In this section, we will discuss the processes affecting the level of genetic load maintained in populations.

Accumulation of Genetic Load—Genetic load results from any variation that reduces fitness below that of a theoretically optimal genotype (Hedrick 2009). From this perspective, any expressed genetic variation segregating in populations has the potential to contribute to genetic load. We can define the total genetic load accumulating in finite populations as the proportion due to the deleterious mutations that have been fixed by genetic drift (*Drift Load*), and the proportion due to mutations that are segregating at intermediate frequencies (*Inbreeding Load*). The latter is evident from the reduction in fitness of offspring from consanguineous matings (inbreeding—including selfing), which is generally referred to as inbreeding depression (see Chapter 4). While the majority of genetic load is due to mutations that are mostly or completely recessive, there may be some fraction that is due to overdominance at individual loci (Charlesworth and Willis 2009), which is sometimes referred to as *Segregation Load*. An additional fraction of genetic load may be due to epistasis—allelic combinations across loci that affect fitness (*Recombination Load*), which may be particularly important for the fitness of hybrid offspring from matings between individuals in divergent lineages (*Hybrid Breakdown*; Chapter 10). Lastly, gene flow between populations occupying different habitats may introduce genetic variation that reduces fitness if the offspring produced are not adapted to the local environment (*Migration Load; Outbreeding Depression*). We will discuss inbreeding load in the next section along with the effects of selection acting to eliminate deleterious mutations (purging). Here, we focus on the processes and consequences of the accumulation of genetic load in finite populations that is due to deleterious mutations that are completely or mostly recessive.

Earlier in this chapter, we discussed equilibrium conditions of mutation-selection balance in infinite populations (Fig. 5.16), but the dynamics of mutation accumulation is quite different in populations of limited size. In finite populations, the number of deleterious mutations will increase over time as genetic drift overwhelms the influence of selection. In small populations, selection is ineffective when its strength is less than $\sim 1/2N$, so mildly deleterious mutations are less likely to be affected by selection and will accumulate across generations (Hedrick 2009). Note that this *Drift Barrier* for the effectiveness of selection depends on population size and dominance; mutations with more substantial fitness effects and higher expression in heterozygotes will tend to be eliminated from populations, resulting in a hyperbolic distribution similar to Fig. 5.16. We expect selection will be more effective in larger populations, so only mutations with very mild combined effects of selection and dominance will accumulate, while more deleterious mutations will escape the effects of selection in smaller populations (i.e., the drift barrier is raised).

As discussed in Chapter 3, the large majority of new mutations will be lost within a few generations; however, some mutations will persist for varying amounts of time and a small fraction will become fixed. Moreover, the rate of accumulation of mutations is independent of population size because the number of mutations produced each generation ($\mu 2N$) and the probability of fixation ($1/2N$) cancel out, leaving only the mutation rate (μ; see Chapter 3). Since the rate of accumulation of mutations depends only on the mutation rate, the primary effect of population size is on the severity of deleterious alleles; small populations will accumulate mutations that are more detrimental compared to the average effects of mutations accumulating in large populations (Lanfear et al. 2014).

Effects of Genetic Load Accumulation—Although individual mutations may have only very minor effects on growth and reproduction, the accumulation of large numbers of deleterious alleles across generations will contribute to a decline in average fitness. From the earlier discussion, we can expect that finite populations will be segregating for substantial numbers of mildly deleterious mutations. There will be a constant turnover of deleterious alleles, as the number of mutations going to extinction each generation will match the fraction of new mutations entering the population. Note that these mildly deleterious alleles will be effectively neutral if their individual fitness effects are lower than the drift barrier, but their collective impacts will take a toll on the average fitness of individuals in the population. The fraction of segregating mutations that become fixed will contribute to the inevitable deterioration of fitness, contributing to drift load and ultimately leading to population decline—a process referred to as *Mutational Meltdown* (Lynch et al. 1995). Fitness decay as a consequence of mutation accumulation is of particular concern for the conservation of small and isolated populations of rare and widespread species (e.g., Busch 2006; Oakley 2013).

The proportion of fixed mutations accumulating in populations due to genetic drift (drift load) results in the reduction of fitness over time, leading to mutational meltdown. The effect of drift load is evident when the fitness of offspring from crosses among individuals within populations is compared to offspring from crosses between populations (Crow 1948; Busch 2006). Heterosis after between-population crosses is common across a wide range of plant species. As discussed earlier, the heterosis from inter-populations crosses is most likely due to the fact that different populations are fixed for different complements of recessive deleterious alleles rather than overdominance: offspring from among-populations crosses will be heterozygous at loci fixed for different deleterious mutations in each population. For isolated populations the accumulated drift load will lead to a reduction in the population size as the average fitness diminishes, and larger proportions of segregating deleterious alleles will become fixed due to the increased effects of genetic drift as the effective population size declines. The compounding effects of mutations accumulating and contributing to drift load will ultimately lead to accelerated fitness declines and possible extinction of populations that are isolated from the effects of gene flow.

This process of mutational meltdown due to drift load can be counteracted by gene flow from neighboring populations that are fixed for different sets of deleterious alleles (Tallmon et al. 2004). We expect that pollen or seed dispersal among populations will provide opportunities for the production of offspring that are more heterozygous, and consequently more vigorous. Inter-population gene flow will infuse populations with genetic variation and will create opportunities for selection to eliminate deleterious alleles as the more heterotic individuals from inter-population crosses produce more offspring in subsequent generations. In a group of populations subject to genetic drift and low levels of gene flow, the accumulation of drift load will be counteracted by heterosis of offspring resulting from inter-population gene flow.

Purging Genetic Load—Fitness loss due to increases in homozygosity after consanguineous matings (inbreeding depression) reflects the proportion of genetic load due to deleterious alleles segregating in the population. The fitness difference between inbred and outbred offspring from crosses within populations is due to deleterious mutations segregating at intermediate frequencies (inbreeding load) and is not affected by the proportion of alleles that have already become fixed (drift load). Consequently, the degree of inbreeding depression detected in selfed progeny reflects only a proportion of the total genetic load. Evidence from crossing experiments indicates that the large majority of the inbreeding load in plants is due to deleterious alleles that are completely recessive or have a low level of expression in heterozygotes (Charlesworth and Willis 2009). In animals, a substantial proportion of inbreeding load is due to low-frequency lethal mutations that are completely recessive, but such mutations appear to be rarer in plants. The reason that mutations of large effect are uncommon in plant populations remains unclear. It could be the case that lethal mutations are difficult to detect in early stages of plant embryo development, or that mutations of large effect are filtered by cell lineage and gametophytic selection. In either case, we can at least say that the inbreeding load affecting fitness after seed germination appears to be mostly due to the collective effects of mildly deleterious mutations (Charlesworth and Willis 2009).

The effects of inbreeding can be severe, possibly leading to decreased fitness of individuals, population decline, and the extinction of inbred lines (Charlesworth and Willis 2009). Furthermore, inbred populations have reduced genetic diversity and may display limited capacity to adapt to challenging conditions (e.g., Oakley 2013). These issues are of particular concern for endemic species with small populations and limited distributions, which may be threatened by inbreeding and an inability to respond to environmental challenges due to climate change. In these cases, conservation biologists may consider human-mediated genetic exchange among populations (*Genetic Rescue*) to infuse additional genetic variation, which could elevate fitness due to heterosis and increase the population growth rate (Tallmon et al. 2004). The effectiveness of genetic rescue efforts depends on having adequate sources of genetic variation that are differentiated but not so divergent that inter-population mating leads to hybrid breakdown (Chapter 10).

The proportion of genetic load that is due to deleterious alleles at intermediate frequencies is sensitive to the effects of selection, effective population size, and the inbreeding history of the population. As small populations inevitably undergo inbreeding, larger proportions of deleterious alleles will occur in the homozygous state and consequently, will be more exposed to the effects of selection. Thus, inbreeding will facilitate the purging of deleterious mutations that are recessive or only partially dominant. Under inbreeding, rare alleles of large effect are quickly eliminated, but the larger number of mildly deleterious mutations are much less susceptible to purging

(Charlesworth and Willis 2009). The difficulty in removing mutations of small effect is exacerbated, because inbreeding reduces the effective population size, so the effectiveness of selection is diminished. Under these conditions, we expect that populations segregating for large numbers of mutations will decline in fitness, as larger proportions of deleterious alleles become fixed and contribute to the drift load. Evidence from inbreeding experiments in plants and animals is consistent with the expectation that inbreeding alone is often insufficient to substantially reduce the inbreeding load (Byers and Waller 1999; Crnokrak and Barrett 2002). Under selfing and biparental inbreeding, a proportion of deleterious mutations may be purged from populations, but for most populations a much larger number are likely to escape the effects of selection and become fixed.

One of the difficulties for interpreting results from purging experiments is the variation in the methods and criteria used to estimate genetic load (Byers and Waller 1999; Crnokrak and Barrett 2002). These experiments typically start with a large number of selfing lines derived from crossing among individuals from one or more populations. For selfing lines that are maintained by single descendants, each line will become almost completely homozygous within a few generations; the proportion of heterozygous loci declines by 50% for each round of selfing, so after five generations the proportion that are still heterozygous is around 3% (i.e. $0.5^5 = 0.031$). As inbreeding ensues, there is typically a high rate of extinction of lines. For example, in an attempt to purge genetic load in a population of common monkey flower (*Erythranthe guttata*; formerly *Mimulus guttatus*), only 335 of the original 1,200 lines persisted after five generations (Willis 1999). Purging may be evident as a small proportion of lines increase in fitness—plants that Darwin referred to as "heroes" (Darwin 1876). On the other hand, this "rebound" in fitness may be due to some degree of adaptation to the experimental conditions rather than the elimination of deleterious alleles contributing to genetic load (Crnokrak and Barrett 2002). In Willis's *Erythranthe* experiment, there was only minor evidence of lessened inbreeding depression after five generations of selfing; progeny from crosses among inbred lines displayed similar levels of inbreeding depression as the original plants did at the beginning of the experiment, which suggests that little or no purging of inbreeding load had taken place. The results of this experiment seem to be typical—most attempts to reduce genetic load in plant populations through selfing over successive generations have had little success.

The lack of substantial reduction in genetic load in crossing experiments is at odds with the observation that many plants routinely self and display low levels of inbreeding depression, which suggests that there must be a mechanism to effectively eliminate genetic load in natural populations. As noted above, we expect that inbred populations would rapidly become fixed for substantial numbers of mildly deleterious mutations. Gene flow among inbred populations fixed for alternative sets of deleterious alleles may contribute to the genetic rescue of inbred populations and facilitate purging through the introduction of "normal" allelic variants at loci fixed for deleterious alleles. This expectation is supported by models that examine the benefits of gene flow in metapopulations (Whitlock 2002), which allows the disassociation of deleterious and beneficial alleles at different loci that would otherwise stall the process of purging (*Selective Interference*; Lande et al. 1994; Winn et al. 2011).

Purging in response to inbreeding and gene flow may be occurring during the range expansion of some invasive species (e.g., Marchini et al. 2016; Fig. 5.19; refer to color plate). In this example of purging in nature, homozygous populations toward the periphery of the expanding range of the *Brachypodium sylvaticum* invasion in North America display high fitness for both selfed and outcrossed progeny. The effectiveness of purging in this case may be due to the unique history of the *B. sylvaticum* invasion; high levels of genetic diversity derived from multiple introductions combined with inbreeding and gene flow in an expanding metapopulation may have provided opportunities for selection to eliminate genetic load. Invasive species that are derived from multiple introductions from divergent regions of the native range may provide ideal conditions for purging and the evolution of high-fitness populations.

The effectiveness of purging genetic load can be seen clearly if we contrast the effects of selfing and outcrossing in metapopulations (Marchini et al. 2016; Fig. 5.20). In these simulations, selfing lines become fixed for deleterious alleles at more than 40% of loci within ten generations. Outcrossing within a single large population (N = 100), or the same number of individuals divided into ten small populations

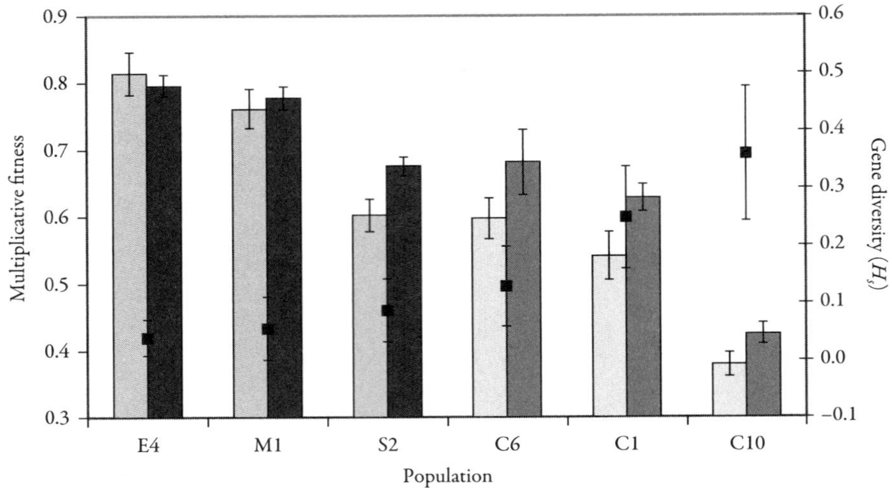

Fig. 5.19. Evidence of purging in newly-colonized populations of *Brachypodium sylvaticum*. The fitness is shown for selfed (light-colored bars) and outcrossed (dark bars) progeny from low genetic diversity (purple bars) and high genetic diversity (green bars) populations. Red symbols indicate levels of genetic diversity. Modified from Marchini et al. (2016).

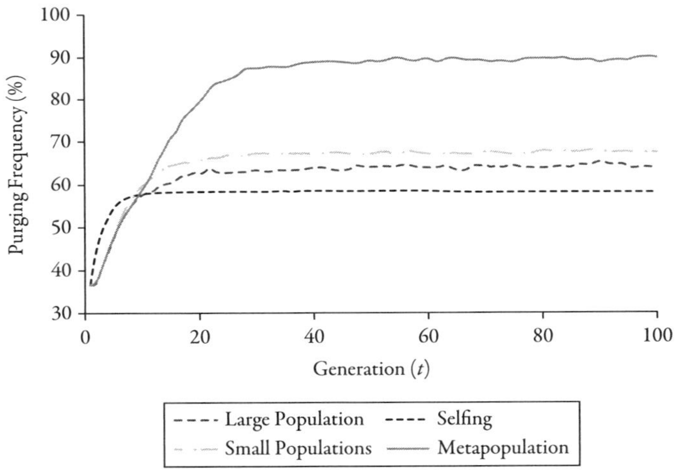

Fig. 5.20. Results from simulations of purging in selfing lines (short dash), a single large population (N = 100; long dash), ten small populations with no gene flow (N = 10 each; dash-dot), and a metapopulation of ten small populations (N = 10) with gene flow (m = 1) every fifth generation (solid). All individuals started with 100 unlinked loci heterozygous for a normal and a deleterious allele. Mutation rate was U = 1 deleterious allele per individual per generation. Selection was imposed by choosing the single most-fit progeny from 200 seeds produced per line each generation. Simulation results generated using programs described in Marchini et al. (2016).

(N = 10) with no gene flow, take longer to purge, but the proportion of deleterious alleles eliminated is not much higher. The highest level of purging (~90%) is obtained when small populations regularly experience gene flow (Fig. 5.20 solid line), which confirms our prediction that regular introduction of genetic variation into inbred populations facilitates the elimination of deleterious recessive alleles. These populations do not become completely purged, but only because mutations in each generation maintain a low frequency of deleterious alleles (Marchini et al. 2016).

The effects of inbreeding and gene flow on the potential for selection to reduce genetic load can be seen for simulated cases of range expansion (Marchini et al. 2016; Fig. 5.21; refer to color plate). After introduction, small populations undergo some initial purging as they rapidly become fixed for different sets of deleterious alleles (Fig. 5.21 top panels; refer to color plate). As range expansion ensues from each purged population, contact between independent phalanxes of expansion results in gene flow among populations fixed for different deleterious alleles, which facilitates additional rounds of purging (darker red regions in the lower panels of Fig. 5.21; refer to color plate). This process is heavily dependent on the initial genetic diversity, so invasions initiated by multiple introductions from several source regions are predicted to have much greater potential for purging and more rapid range expansion (Marchini et al. 2016).

The elimination of inbreeding and drift load in these simulations seems to occur relatively quickly

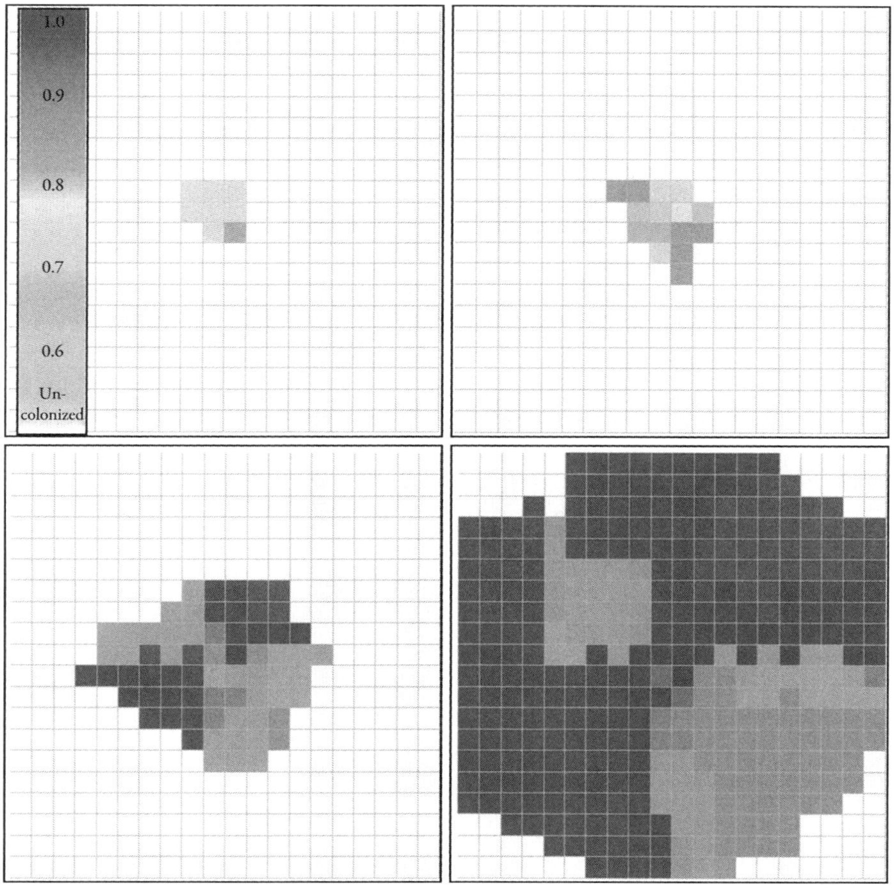

Fig. 5.21. Stages of range expansion after establishment of an outbreeding invasive species; upper left panel—one of a few satellite populations displays higher fitness after colonization by more than one of the original five populations (dark blue outline); upper right panel—invasion continues as new sites are colonized primarily from the first high-fitness population. A second purged population appears on the opposite side of the range; lower right panel—rapid range expansion as satellite populations from the two original purging events undergo admixture. Another round of purging produces higher fitness populations; lower left panel—high-fitness genotypes spread and range expansion is rapid, purging is nearly complete. From Marchini et al. (2016).

under circumstances that should be widespread. In contrast, inbreeding depression is prevalent for the majority of plants, which indicates that they harbor high levels of inbreeding load. This suggests that the basis of inbreeding load is complex and is probably based on a large number of recessive alleles of small effect (Charlesworth and Willis 2009). It is clear that there is much to learn about the architecture of genetic load, including the circumstances that favor the maintenance of inbreeding load and the processes that promote purging. The loss of genetic load is particularly important for establishing conditions that favor the evolution of self-fertilization—a topic we will explore further in Chapter 9.

5.7 GAMETOPHYTIC SELECTION

The Haploid Life Stage—The retention of a multicellular haploid life stage from algal ancestors distinguishes plants from animals, which lack post-meiotic expression of their gamete genomes (Chapter 2). While all of the land plants (embryophytes) have haplodiplontic life cycles (Box 1.3), the degree of dominance and independence of the haploid life stage differs among major groups. The gametophyte stage is dominant and sporophytes parasitic across the bryophytes, but all of the vascular plants have a dominant sporophytic stage. For the seedless tracheophytes, including lycophytes and ferns, the gametophytes are small, independent plants that are exposed to the selection pressures of the local environment as they grow and generate gametes that fuse to form the next diploid generation. Angiosperms and gymnosperms retain their megagametophytes within sporophytic tissues and after fertilization, continue to provide resources to the developing seeds. These seed-bearing plants also protect their microgametophytes by encasing them within pollen grain walls during dispersal and then provide nourishment for pollen-tube growth after germination. Consequently, the gametophytes of the embryophytes are effectively protected from external environmental conditions during their growth and development.

The earliest land plants possessed many of the characteristics of modern bryophytes and seedless tracheophytes including independent gametophytes. During the early stages of embryophyte evolution, the expression of haploid genomes may have been a liability that slowed diversification; selection on independent gametophytes in the early land plants would have limited genetic diversity across co-expressed regions of the genome and may have reduced the capacity of lineages to respond to novel environmental challenges. Exposing haploid genomes to selection by environmental conditions each generation would have limited levels of standing genetic variation and reduced the potential for adaptation to novel habitats.

Some plant lineages apparently developed mechanisms to escape the limitations imposed by environmental selection on the haploid life stage. While most plant groups with independent gametophytes have limited diversity, the modern ferns and peat mosses (*Sphagnum*) underwent accelerated diversification during the Cretaceous. In both cases the increase in rates of diversification were associated with a series of whole-genome duplication events (Chapter 2) that effectively sequestered genetic variation from the effects of selection on the gametophytic life stage (Otto and Whitton 2000; Ricca and Shaw 2010). While each episode of polyploidy may have provided increased opportunities for diversification, this relief from the negative consequences of gametophytic selection was only temporary because gametophytes eventually became functionally haploid again as lineages underwent diploidization. Furthermore, increases in genome size may have slowed rates of plant growth (Knight et al. 2005), which would ultimately limit the pace of adaptation and lineage diversification.

A more effective solution to the limitations imposed by post-meiotic expression of large proportions of the genome in gametophytes developed in the earliest angiosperms as a side effect of the protection of developing seeds, which had been exposed to damage from herbivores in their seed fern ancestors (Mulcahy et al. 1996; Chapter 2). Flowering plants effectively turned haploid expression from a liability into an advantage as their enclosed carpels buffered gametophytes from selection pressures imposed by the external environment and simultaneously created a forum for the filtering of deleterious mutations that may have accumulated during vegetative growth (Chapter 3). For the flowering plants, gametophytic selection as a result of pollen competition may have been a key innovation that helped promoted the rapid diversification and dominance of angiosperms across most terrestrial environments (Walbot and Evans 2003).

Pollen Competition—The potential consequences of selection on gametophytes in flowering plants—and pollen in particular—has been a topic of speculation and discussion since the early twentieth century (Jones 1928). Genes expressed in the haploid state are not sheltered by dominance; deleterious mutations are completely exposed and consequently, the effects of selection on gametophytes can be much greater than on sporophytes. Since there is a large degree of overlap in the genetic determination of traits between life-stages (Borg et al. 2009), selection on the gametophytic stage may have repercussions for sporophytes in the next generation (Mulcahy et al. 1996; Walbot and Evans 2003; Armbruster and Rogers 2004; Seoighe et al. 2005). Haldane was highlighting the predicament of the haplodiplontic life cycle and the potential consequences of selection on gametophytes for the sporophytic life stage when he quipped that "A higher plant species is at the mercy of its pollen" (Haldane 1932).

The earliest reports of gametophytic selection as a consequence of pollen competition came in the first years of the twentieth century from studies of the inheritance of morphological traits in corn (*Zea mays*), evening primrose (*Oenothera*), sorrel (*Rumex*), and other plants (reviewed in Jones 1928). The effects of pollen competition on segregation ratios became evident as studies of inheritance became more common soon after Mendel's work became more widely known. Studies of Mendelian inheritance using plants sometimes found that the expected ratios were distorted to various degrees; more detailed studies attributed this bias to the differential growth rate of pollen tubes carrying alternative alleles segregating at a locus (selective fertilization; *Certation*).

During the early twentieth century, there was no reason to consider the consequences of selective fertilization beyond its occasional effects on the inheritance of a few polymorphic traits. This perspective began to change after 1932 when Haldane pointed out that it would be unlikely for the gametophytic life stage of plants to have their own separate genomic components controlling their growth and physiology (Mulcahy et al. 1996). In spite of the logic of Haldane's assertion, it was generally assumed that gene expression in pollen was limited. This view began to change in the 1960s when agronomists working with corn and other crop plants began to document patterns of non-random fertilization that depended on both the pollen donor and the recipient (e.g., Pfahler 1967). During the late twentieth century, evidence was accumulating that there was substantial overlap in the expression of genes during gametophytic and sporophytic generations, and there was growing interest in the potential of gametophytic selection as a mechanism for crop improvement (Ottaviano et al. 1980).

The first evidence that selection on the gametophytic life stage could affect the vigor of the next generation of sporophytes came in the late 1960s and early 1970s from the work of the Italian agronomists Ercole Ottaviano and Mirella Sari-Gorla, and the American biologists David and Gabriel Mulcahy (Ottaviano et al. 1982; Mulcahy and Mulcahy 1987). In these experiments, investigators manipulated the intensity of pollen competition by increasing the size of pollen loads on stigmas or by increasing the distance that pollen tubes were required to traverse through the style to reach the ovary.

The finding that seedlings were more vigorous with increased selection on pollen led to a proliferation of similar experiments during the last decades of the twentieth century and continued into the early twenty-first century (e.g., Winsor et al. 2000; Pasonen et al. 2001; Lankinen et al. 2009). Many of these studies found that gametophytic selection increased the average, or decreased the variance, in the vigor of offspring in the next generation. On the other hand, the effects of pollen competition on offspring vigor was not consistent across all studies, and some authors cast doubt on the hypothesis that pollen competition could result in significant genetic modifications for offspring in the next generation (Walsh and Charlesworth 1992). In particular, strong directional selection across diploid and haploid life stages should minimize genetic variance for growth rates, in which case there would only be minor potential for the effects of intense pollen competition in a single generation. Furthermore, while gametophytic selection could cull deleterious mutations that accumulate each generation, the magnitude of this effect is expected to be minimal and cannot account for the differences in vigor observed from the effects of pollen competition (Charlesworth and Charlesworth 1992). It is also possible that selection on the haploid stage could have negative fitness consequences for the sporophytic generation (*Antagonistic Pleiotropy*). Critics of the pollen competition hypothesis for increased offspring vigor suggested that the observed effects were more likely due to maternal environment effects on seedlings, or that if there were genetic effects, they would be limited to a single generation.

In spite of the continued controversy over the effects of pollen competition on offspring vigor, the potential for gametophytic selection as a potent force for the evolution of plants has not been diminished. There is no question that the exposure of genetic variants to selection in the haploid state could elicit substantial changes in allele frequencies within a single generation. Many of these studies are problematic because plant reproductive systems themselves limit the manipulations that would adequately control for the conditions present during pollen-tube growth and seed maturation, as well as the study of the effects of those conditions. For example, under conditions of strong directional selection, any variance in fitness (i.e., growth rate) is likely to be due to non-heritable genetic variation that results from dominance and epistasis (Chapter 7) and consequently, is much more likely to be expressed in the diploid state. Hence, selective ovule abortion is more likely to be responsible for observed differences in the vigor of offspring from some pollen competition experiments, but the contributions of this reproductive stage are rarely considered (e.g., Cruzan and Thomson 1997).

The majority of investigations on the effects of pollen competition that have been conducted so far have focused on phenotypic effects that can be measured within a generation or two. A more productive approach may be to examine evidence for historical effects of selection by comparing genomic regions with tissue-specific expression (e.g., Arunkumar et al. 2013; Gossmann et al. 2016). These studies make comparisons among panels of genes whose expression is restricted to one life stage (i.e., pollen vs. seedling) and consistently find evidence for stronger purifying selection and higher K_A / K_S (dN / dS; Chapter 3) ratios for transcribed regions expressed in gametophytes. These results suggest that many of the gene regions expressed in pollen have diverged more recently and evolve faster than comparative regions that are only expressed in sporophytes. The genomic analyses conducted thus far have provided strong evidence for selection on haploid genomes during the gametophytic life stage that is more effective than selection on diploid genomes and suggest that gametophytic selection has considerable potential to contribute to plant evolution and diversification.

From what we have learned so far, it appears that gametophytic selection provides an efficient mechanism to favor the spread and fixation of beneficial mutations. It is also apparent from theoretical analyses that overlapping expression between the gametophytic and sporophytic life stages should be favored in flowering plants for its potential to reduce genetic load (Otto et al. 2015). This point is particularly important because of the potential contributions of somatic mutation during vegetative growth to the mutational load of plant populations (Chapter 3). In spite of the large volume of work that has been completed, it appears that we have only scratched the surface; there is much to learn about the effects of selection on the haploid genome and its consequences for genomic regions with overlapping expression between life stages. Ultimately, a more thorough understanding of plant evolution will require an appreciation of the effects of gametophytic selection on purging of genetic load and the spread of advantageous mutations.

5.8 QUANTIFYING SELECTION

The availability of molecular genetic data has created a spectrum of opportunities for testing hypotheses concerning the importance of evolutionary processes at an unprecedented level of detail. The paradigm shift that occurred—first with allozymes and later with DNA sequence data—was due more to the type of information that became available than the resolution of detail that the new technologies provided. In particular, the availability of allelic data has allowed applications of predictive mathematical models to evaluate the influences of evolutionary processes on patterns of variation. In this section, we summarize a range of neutrality tests that have the common aim of detecting different forms of selection on one or a few sampled genomic regions.

Coalescence Theory—We have already discussed, earlier in this chapter and in the previous chapter, methods for analyzing the distribution of neutral genetic variation developed during the modern synthesis by Fisher, Wright, and Haldane. Population genetic models based on the Fisher-Wright approaches are prospective in nature—they generally aim to predict future distributions of variation given a set of starting conditions. A retrospective approach to population genetics—*Coalescence Theory* (Box 5.1)—finds its roots in the early work of Kimura in the 1950s, but it was not fully developed until the 1980s through the collective work of several mathematicians (e.g., Kingman 1982; Tajima 1983; Hudson 1990; Ewens 2004).

The coalescence approach differs from Fisher-Wright models of population genetics because it informs current distributions of genetic variation by inferring the history of alleles back through time (Fig. 5.22). Moreover, these *Gene Genealogies* can be modeled to test hypotheses concerning historical demographic events and the influence of genetic drift and selection on contemporary distributions of genetic variation (Siol et al. 2010; Pannell and Fields 2014). These methods were particularly applicable to early molecular population genetic studies, because they did not require large samples; a gene genealogy estimated from even a small set of sequences can be used to make inferences concerning historical processes.

In this section, we are particularly interested in the application of predicted distributions of variation due to genetic drift as null expectations to infer the effects of selection. First, we will discuss the predictions of coalescence models under genetic drift. We will then compare sampled variation to patterns generated by neutral coalescence to detect genomic regions under the influence of selection using two general approaches; the first methods utilize comparisons of allele frequency distributions (*Neutral Diversity* and *Site Frequency Spectrum* methods); the second set of methods compare patterns of divergence (*Neutral Divergence* methods). We end this section with an evaluation of the effectiveness of different genomic approaches in identifying adaptive genetic variation.

Predictions under Neutral Coalescence—Coalescence theory provides us with robust neutral expectations for the distribution of genetic variation and patterns of divergence. To predict levels of genetic diversity under genetic drift, we assume conditions provided by the *Infinite Alleles Model,* where each new mutation generates a new allele. The assumptions of this approach are reasonable for most sequence data sets, because the number of substitutions detected tends to be small relative to the number of bases sampled (Hedrick 2009). In other words, even a limited genomic region of a few thousand bases may have enough capacity for mutation accumulation that each base change would result in a new allele. We can use homologous sequences from multiple individuals sampled from the same or different populations or lineages to identify multiple alleles (*Haplotypes* defined by combinations of several SNPs; see Box 7.1).

Using a few simple assumptions, we can make predictions of the number and distribution of alleles in a population subject only to genetic drift. In the absence of selection and gene flow, the amount of genetic variation that is maintained in a population is affected only by the mutation rate (μ) and effective population size (N_e). The expected level of genetic diversity is given by the population mutation parameter, which is defined as

$$\theta = 4N_e\mu.$$

Theta (θ) reflects the equilibrium level of genetic diversity expected for a constant population size and mutation rate. This measure of diversity is related to the expected heterozygosity under equilibrium conditions (H_{eq}) as

$$H_{eq} = \frac{\theta}{1+\theta}.$$

H_{eq} represents the *predicted* level of heterozygosity under a specific set of conditions, so it differs from the *expected* heterozygosity (H_e) discussed in Chapter 3, which is a measure of genetic diversity based on a sample of individuals. For detecting the effects of selection, we will be using comparisons of the equilibrium level of homozygosity instead of heterozygosity in a sample of sequences. We can express the predicted homozygosity as

$$f_{eq} = \frac{1}{\theta+1}.$$

As already mentioned, the level of diversity maintained in a population will depend on the effective population size and the mutation rate. Theta is a convenient parameter for discussing these effects because it summarizes the influences of both processes. For example, θ will increase (and f_{eq} will decline) along with increases in either the population size or the mutation

Box 5.1 Coalescence Theory

In our previous discussions of population genetic models, we have focused on *expected* distributions of genotypes that were determined from current allele frequencies (i.e., according to HW expectations). These analyses focus on transient variation that is the result of processes acting over a few generations; the effects of inbreeding, selection, or gene flow on genotype frequencies can be erased by a single generation of random mating. Coalescence theory, on the other hand, addresses processes acting over hundreds to thousands of generations. In this case, we are concentrating on *predicted* distributions of genetic variation under models that assume specific roles for genetic drift and selection. In these analyses, distributions predicted from the effects of genetic drift are compared to sampled variation to make inferences concerning the influence of selection on distributions of allele frequencies.

To detect the influences of selection and other processes on genetic diversity, we first need a basis for comparison, so we will predict the expected distributions of genetic variation for a population subject only to drift and mutation. We can appreciate the generation of allele frequency distributions more clearly by examining the history of a genomic region using a gene genealogy (Fig. 5.22). Each terminal branch at the bottom of this tree represents an extant variant (i.e., an allele—a DNA sequence made unique by single mutation events). Moving back in time (upward) along the branches that define the history of each allele we see that they join periodically (coalesce) at points (nodes) that represent the mutation events that define each new lineage. For any set of alleles, the chance of a new mutation is the same in each generation, so as the number of lineages increases through time there is a shorter time between coalescence events simply because there are more lineages. Consequently, recent coalescence times are shorter and the amount of time for coalescence of the first two lineages is the longest (i.e. $2N$ generations where N is the population size; Fig. 5.22).

For any set of n alleles, we can predict that the time to their most recent common ancestor (MRCA), as represented by the node at the top of the gene tree in Fig. 5.22, for all those alleles (T_{MRCA})

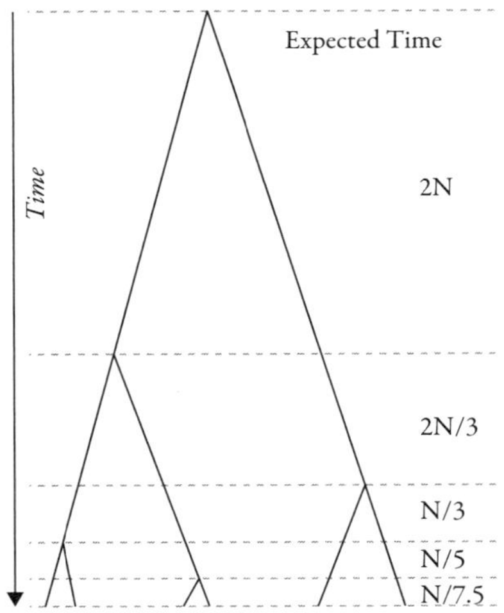

Fig. 5.22. The relative time to coalescence events for six lineages looking back through time. Number of generations to each coalescent event is expressed in terms of the effective population size (N).

will be the sum of expected times for all coalescence events ($T_{MRCA} = T_n + T_{n-1} + T_{n-2} + \cdots + T_2$). The last term ($T_2$) represents the predicted time for the first two alleles to coalesce. The expected time to coalescence for any set of n alleles will be

$$E(T_{MRCA}) = 2\left(1 - \frac{1}{n}\right).$$

Note that we can ignore the mutation rate for this calculation because the timescale used is relative; a range of mutation rates will result in the same distribution of branch lengths even as the absolute times differ. The expected time to the MRCA ranges from 1.0 for two lineages to an asymptote of 2.0 as the number of lineages approaches infinity. We can convert this relative scale to the number of generations by multiplying $E(T_{MRCA})$ by $2N$ (Fig 5.22). For two lineages, the expected number of generations to coalescence is 1.0 ($2N$). For three lineages the total expected time is 1.3333, so the time between three and two lineages is 1 / 3 of the time expected for coalescence from two lineages to one lineage; $1.0 - 1.3333 = 0.3333$, which becomes $2N / 3$ after we multiply by $2N$ (Fig. 5.22). It is important to note that the gene tree in Fig. 5.22 is an idealized representation of coalescence events for the effect of genetic drift acting alone—actual coalescence events approximate this distribution and will range around each of the expected times shown. These neutral distributions provide a basis for comparison with alleles sampled from populations to allow detection of the effects of balancing and purifying selection.

rate. For any given mutation rate, we expect that larger populations will have reduced effects of genetic drift, so they will contain more alleles at equilibrium, and consequently, the homozygosity will be lower. When θ is close to zero the number of alleles that can be maintained in the population is around 1.0. Increasing the population size or the mutation rate will rapidly increase the equilibrium number of alleles. Since we are typically working with sequence data, the mutation rate is summed across all the base positions; estimates of θ can range well above 1.0, and expected numbers of alleles greater than ten are not uncommon.

For population genetic data, we can use comparisons of estimates of θ to infer the historical influence of selection or demographic events. It is important to note that the reliability of inferences we make from population samples is heavily dependent on sample sizes. The accuracy of estimates of θ will depend on the number of individuals sampled; the number of alleles detected will asymptotically approach the neutral coalescence expectation with increasing sample size (Hedrick 2009). In upcoming sections, we will discuss applications of different estimates of θ to infer the historical influence of selection (*Neutrality Tests*) on the levels and distribution of genetic variation.

Neutrality Tests; Diversity Distributions—The expected allele frequency distribution due to genetic drift provides a reasonable "null expectation" for testing population genetic hypotheses. The *Ewens-Watterson (EW) Test* compares sampled distributions to those predicted under drift to detect the influence of selection on the distribution of allele frequencies. Significant departures from expected distributions may be attributed to the action of directional or balancing selection if the assumptions of this test have been met (e.g., constant population size). To make tests for departures from neutrality we need to sample homologous sequences from a number of individuals to generate a frequency distribution of alleles (Fig. 5.23). We will assume that recombination rates across our sampled region are low enough to be ignored, which allows us to use each combination of base differences across the sequence to define each allele (haplotype). Alternatively, we can use other types of multiallelic genetic markers such as microsatellites or allozymes (Appendix). Individual SNPs are not appropriate for this test because they do not meet the assumptions of the infinite alleles model.

By sampling variation across a specific genomic region, we obtain information on the frequencies of

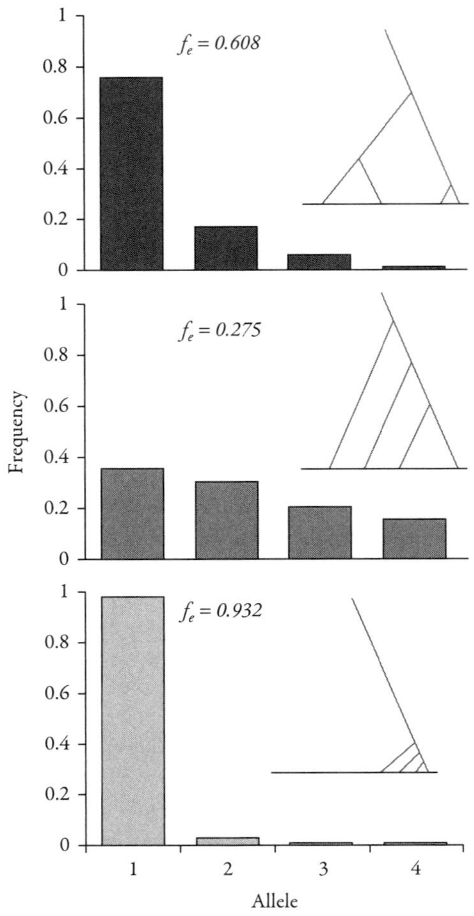

Fig. 5.23. The effects of balancing and purifying selection on the frequency distribution of alleles. The top panel shows the expected distribution with only genetic drift, the middle panel is for balancing selection, and the bottom panel represents purifying selection. The expected homozygosity (f_e) and the corresponding coalescence gene tree is shown for each set of conditions.

alleles, but we do not have direct knowledge of their histories (as depicted in the gene trees in Fig. 5.23). Allele frequency distributions allow inferences of allele history under the assumption of constant population size; in a population subject only to genetic drift the abundance of each allele will directly correspond to its age. Older alleles will have had more time to proliferate and consequently will be carried by more individuals in the population. Note that the tallest bar in each of the three frequency distributions in Fig. 5.23 corresponds to the longest (root) branch, with the shorter bars corresponding to each of the shorter branches of each tree. By assuming that allelic abundances directly reflect their histories, we can use frequency distributions to test for the effects of genetic drift and selection.

Making comparisons to assess the influence of selection on genetic diversity is more convenient if we can use a single parameter. The distribution of allele frequencies is conveniently summarized by the expected homozygosity ($f_e = \Sigma p_i^2$), and this value can be compared directly to the predicted homozygosity parameter described above (f_{eq}) to assess departures from neutral expectations. Notice that f_e and f_{eq} range from zero to one and reflect the evenness of the allele frequency distribution. In the absence of selection, genetic drift leads to a characteristic distribution with the oldest allele being the most common and the youngest occurring at the lowest frequency (Fig. 5.23 top panel). The relative abundances of alleles correspond to the distribution of times to coalescence (Box 5.1). The distribution of allele frequencies in the top panel of Fig. 5.23 results in an estimate of f_e that matches our expectation under neutral coalescence (i.e., for four alleles and a sample size of 200 alleles $f_{eq} = 0.608$; Ewens 2004).

When all the alleles occur at similar frequencies, f_e will be closer to zero, which indicates that the average allele age is greater than expected under genetic drift (Fig. 5.23 middle panel). In comparison to neutral coalescence (Fig. 5.23 top panel) the distribution in the middle panel Fig. 5.23 displays more even frequencies due to the fact that alleles have been retained in the population for relatively long periods of time; branch lengths in the associated gene tree are longer than we expect compared to a population subject only to genetic drift. This pattern can be interpreted as an outcome of balancing selection where multiple alleles are favored because of overdominance or because each allele confers advantages in an alternative environment. As discussed earlier, cases of balancing selection in plants have been difficult to identify, but examples may be more common in selfing species such as *Arabidopsis* (Wright and Gaut 2005). On the other hand, these putative cases are more likely to be due to associative overdominance because of low recombination rates and the cumulative effects of directional selection across large regions of the genome. A much stronger case for the effect of balancing selection can be made for self-incompatibility alleles that have frequency-dependent advantages when they are rare (Kohn 2008; see Chapter 9).

A more common pattern of genetic variation generated by selection is represented by strongly skewed distributions with one abundant allele and all the others at very low frequencies, which produces a value of f_e much closer to 1.0 (Fig. 5.23 bottom panel). This pattern suggests that all but one of the alleles have low persistence times in the population—alleles are much younger than expected when compared to expected ages under neutral coalescence. The prevalence of a single allele is typically interpreted as evidence for purifying selection where the large majority of non-synonymous substitutions are detrimental (Wright and Gaut 2005). Surveys suggest that around 20% of the genome of plants may be affected by purifying selection, and a good proportion of these regions are outside of reading frames (functional locations within intergenic regions; Hough et al. 2013). Based on the limited sample of species available, it appears that lower proportions of the genomes of plants are constrained by purifying selection compared to most animals.

Demographic Instability—Our interpretations of the evidence for selection from patterns of allelic diversity are heavily dependent on a number of assumptions including a lack of gene flow and constant population size. The distribution of allele frequencies is particularly sensitive to variation in demographic conditions. For example, as we discussed in the last chapter, during population bottlenecks of short duration, rare alleles are lost at a higher rate, so a recent bottleneck of short duration could generate the same distribution of allele frequencies that we expect with balancing selection (Fig. 5.24; mild bottleneck). If a low variance in allele frequencies was detected at one locus, we could discriminate between the alternative hypotheses of balancing selection or a recent bottleneck by sampling more genomic regions. Since a bottleneck should have equal effects across the genome, a hypothesis of balancing selection would be supported if allele frequencies at one locus were much more equal than the distributions found at a number of other genomic regions.

A number of other processes can also generate the strongly skewed distribution of allele frequencies expected under purifying selection. For example, if a population bottleneck had been prolonged or severe (e.g., if a new population had been established by just a single individual) then we expect to see a loss of common as well as rare alleles. In this case, new mutations appearing since the bottleneck would have short coalescence times (Fig. 5.24; severe bottleneck)—a pattern that is also consistent with purifying selection. Again, testing of this hypothesis could be improved by broader genomic sampling; however, we still run the risk of misidentifying the cause because positive selection favoring a single allele (selective sweep; Fig. 3.4; refer to color plate) will also generate low regional genomic diversity and may be difficult to discriminate from purifying selection. The appearance of a pattern consistent with purifying selection at some loci could also be generated by the introduction of novel alleles through low levels of gene flow. In general, there is a wide variety of demographic and population genetic processes that can affect distributions of allele frequencies, so interpretations of outcomes from the EW test should be interpreted with caution.

Site Frequency Spectrum (SFS) Analysis—Using a single estimate of genetic diversity to infer the influence of selection has a number of drawbacks, and consequently the application of the EW test has been limited. Tajima's D is a more widely applied method that utilizes information on the relatedness among alleles and compares this to the frequency of polymorphic sites (Hedrick 2009). This analysis uses an estimate of θ based on the nucleotide diversity (π). Recall from Chapter 4 that π quantifies both allele frequencies and the number of nucleotide differences between pairs of alleles, so θ_π represents an estimate of diversity based on the average relatedness. For Tajima's test this estimate of diversity is compared to an estimate of θ based on the frequency of polymorphic sites:

$$\theta_S = \frac{S}{a_1}.$$

Where S is the number of polymorphic sites and a_1 is based on the number of alleles sampled ($a_1 = \Sigma 1/i$; for $i = 1$ to $n-1$ alleles). The estimates of theta based on relatedness (θ_π) and site diversity (θ_S) are expected to be the same under neutral conditions, so the difference between them ($d = \theta_\pi - \theta_S$) will be zero.

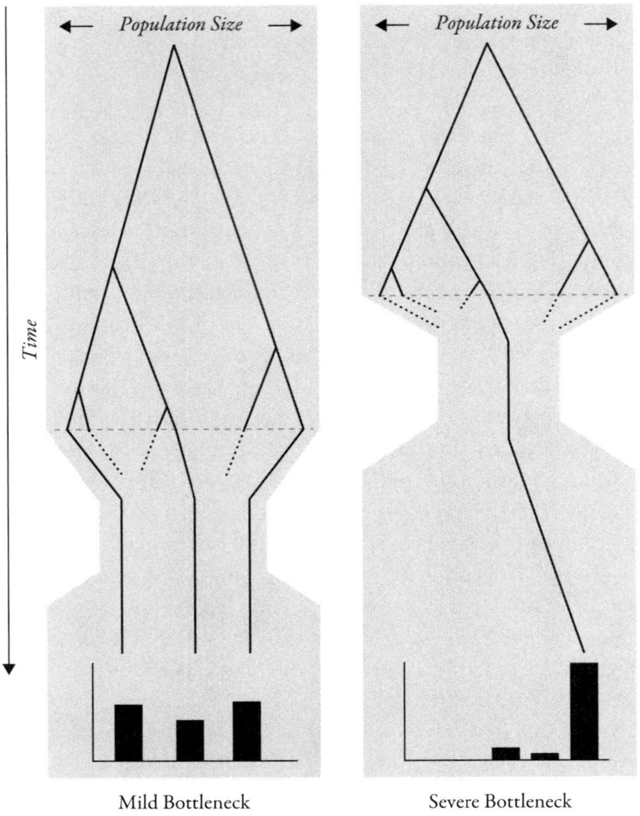

Fig. 5.24. Effects of population bottlenecks on coalescence times. A recent mild bottleneck results in retention of high-frequency alleles and loss of rare alleles. A severe bottleneck leaves only a single allele that begins to diversify.

Tajima's D statistic standardizes this difference by the square root of its variance (the standard deviation):

$$D = \frac{d}{\sqrt{V(d)}}.$$

The two estimates of θ used in Tajima's test differ primarily with respect to their sensitivity to the presence of rare alleles. The estimate of theta based on relatedness among alleles (θ_π) is more sensitive to the presence of rare alleles, because it incorporates both the frequencies ($p_i p_j$) of alleles and their relatedness (π_{ij}—the number of nucleotide differences between alleles i and j). This estimate of theta reflects allele frequency evenness, so in this respect it is more similar to the expected heterozygosity (H_e). The estimate of theta based on site frequencies (θ_S), on the other hand, quantifies the total number of SNPs so it is not affected by differences in allele frequency. Comparison of these two estimates of theta provides an evaluation of genetic diversity that is more responsive to the influences of demographic and population genetic processes than the EW test (Hedrick 2009).

We can use Tajima's D to infer the effects of balancing and purifying selection from population samples. The greater sensitivity of θ_π to allele frequencies results in estimates that are smaller than θ_S and consequently, a negative value of D, when larger numbers of recently derived rare alleles are present. Similar to our outcomes from the EW test, we can infer that such a pattern is expected when a large proportion of sites are subject to purifying selection (Fig. 5.23 bottom panel), the region has been affected by a selective sweep, or the population has experienced a severe or sustained bottleneck (Fig.

5.24; severe bottleneck). Positive values of D will result from cases where relatedness among alleles is lower than expected based on the number of polymorphic sites, which is a pattern predicted under balancing selection (Fig. 5.23; middle panel), or due to a recent bottleneck of short duration that resulted in the elimination of rare alleles (Fig. 5.24; mild bottleneck). As discussed earlier, we can discriminate between demographic and selection hypotheses by sampling a large number of regions across the genome to identify outlier loci, which are candidates for targets of selection.

Tajima's D can be expanded by applying Fay and Wu's H statistic if samples representing multiple lineages are available. The H statistic identifies cases of positive selection by detecting recently derived alleles that occur at higher frequencies than expected under neutral coalescence (Hedrick 2009) and statistic illustrates an alternative approach to the evaluation of cases of selection where allelic diversity is compared in a phylogenetic context. In the next section, we discuss more methods that use comparison among lineages to infer the effects of population genetic and demographic processes.

Linkage Disequilibrium—An alternative to the evaluation of distributions of genetic diversity and site frequencies spectra are tests of *Linkage Disequilibrium (LD)* for pairs of loci within regions affected by strong positive selection (Siol et al. 2010). These tests are based on the expectation that high levels of LD will be generated in genomic regions closely associated with an allele that is on its way to fixation (selective sweep). Under a hard selective sweep (Fig. 3.4; refer to color plate), a region of high homozygosity and LD will be maintained until an adequate number of mutations accumulate. Linkage disequilibria generated by soft sweeps, on the other hand, will be much more ephemeral due to recombination among the collection of alleles affected by selection. In general, lower diversity and higher levels of linkage disequilibria will characterize genomic regions that have recently been affected by selective sweeps.

Neutrality Tests Based on Divergence—Sampling multiple closely related lineages allows for the application of comparative methods that can detect the influence of selection on patterns of genetic diversity. The first of these, the *Hudson-Kreitman-Aguade (HKA) Test*, requires at least two lineages and two genomic regions that differ in their nucleotide diversity (Hedrick 2009). Recall from Chapter 3 that there is generally variation in mutation rates across the genome. Under neutral coalescence, we expect that regions with higher mutation rates will display elevated levels of within-lineage diversity and greater rates of divergence among lineages. The HKA test takes advantage of this expectation to make predictions of levels of divergence under neutral coalescence from levels of diversity within lineages.

Application of the HKA test is illustrated by the example of genome evolution during domestication of maize (*Zea mays mays*) from teosinte, which is a complex of several subspecies (Fig. 5.3; refer to color plate). The development of a large and compact ear from the much smaller and looser inflorescence of teosinte is primarily due to the action of the *teosinte branched 1 (tb 1)* gene (Fig. 5.25). To conduct the HKA analysis a 300 bp "sliding window" is moved across the gene region and the nucleotide diversity (π) is calculated at each 50 bp step (Wang et al. 1999). This analysis reveals that estimates of π in maize track diversity in the ancestral lineages over the exon and intron regions, but the level of diversity is substantially lower for the non-transcribed control region (5' NTR). Note that the levels of polymorphism across the transcribed region of the gene varies substantially, but the nucleotide diversity is similar for both the ancestral and derived lineages, so we infer that these regions have been subject to similar levels of purifying selection. For the NTR region, on the other

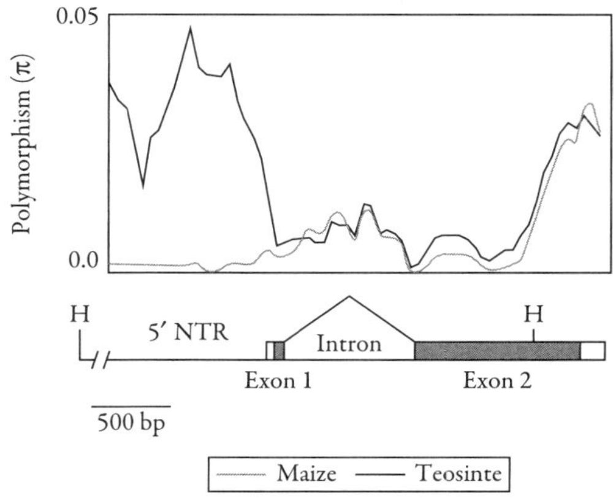

Fig. 5.25. A sliding-window analysis of nucleotide diversity (π) for the *teosinte branched 1* gene in teosinte and maize. From Wang et al. (1999).

hand, the level of polymorphism is substantially lower in maize—a pattern that is consistent with a recent selective sweep. Application of the HKA test indicates a significant departure from the neutral expectation for the NTR region, which matches our expectation that there was strong selection on inflorescence architecture during domestication.

The *McDonald-Kreitman (MK) Test* focuses on comparing synonymous and non-synonymous substitutions among lineages to detect cases of positive selection among closely related lineages (Hedrick 2009). Non-synonymous sites are more likely to be subject to purifying selection so they should generally display lower levels of polymorphism. Variation at both types of sites is a function of the mutation rate, so we expect that the ratio of non-synonymous (N_P) to synonymous (S_P) polymorphisms within lineages should be the same as the ratio of non-synonymous (N_F) to synonymous (S_F) fixed differences (substitutions) between a pair of lineages. Under conditions of strong purifying selection on non-synonymous mutations and constant population size, the ratio of N_P / S_P should predict the ratio of N_F / S_F in lineages derived from a recent common ancestor. However, if any non-synonymous mutations are advantageous in derived lineages then we expect they should fix at a higher rate so the N_F / S_F ratio would be higher than expected, based on the N_P / S_P ratio. The MK method summarizes all of the substitutions across a transcribed region so it may be less sensitive to cases where a small number of base changes are affected by positive selection. If a high enough number of non-synonymous sites is affected by positive selection then the MK test can provide an effective test for adaptive evolution.

Performance of Neutrality Tests—The application of neutrality tests for detecting the effects of selection will depend largely on the type of genomic information that is available. Studies evaluating the effects of selection on molecular genetic variation range from samples obtained from single populations to regional metapopulations and multiple lineages representing intra- and inter-specific variation. In population genetics, we generally focus on genomic variation sampled from one or a few closely related lineages. In cases where we are evaluating the distribution of variation for a few genomic regions sampled from single populations, we can use tests based on diversity distributions such as EW, or we can apply a site frequency spectrum test (Tajima's *D*). For detecting the effects of positive selection, the diversity and linkage disequilibrium tests are most effective if alleles under selection are at intermediate frequencies or have become fixed very recently (Zhai et al. 2009). In cases where samples from a few closely related lineages are available, then either Fay and Wu's *H*, HKA, or MK tests may be applied. Simulations suggest that among these three, the HKA test has the most power for detecting the effects of selection under a range of scenarios (Zhai et al. 2009).

Neutrality Tests with Demographic Instability—Neutrality tests for detecting the effects of selection assume ideal demographic conditions that are unlikely to be met for most populations. Coalescence approaches for the identification of genomic regions under selection will work best if population sizes are large and constant over extended periods of time; in contrast, most plant species have relatively small populations that most likely have experienced one or more bottlenecks. As discussed earlier in this chapter, the effectiveness of selection can be compromised by genetic drift, so neutrality tests in plants may generally be biased to detect only sites that are subject to more intense selection. This is potentially a problem with the MK test, which assumes that non-synonymous mutations are consistently subjected to strong purifying selection (Siol et al. 2010). The expected frequencies of synonymous and non-synonymous substitutions can be refined to account for alleles displaying a wider range of deleterious effects, but this approach does not address the broader problem of demographic instability in plant populations.

One approach to accounting for demographic effects on genetic variation when conducting neutrality tests is to conduct demographic simulations to generate distributions expected under genetic drift (Siol et al. 2010). This *Coalescence Simulation* (*CS*) approach provides a method to produce a refined distribution of expected values for more robust tests for the influence of selection on genetic variation. This approach has been applied to identify gene regions subjected to selection during the domestication of corn from teosinte ancestors (Yamasaki et al. 2007). We expect that the distribution of genetic variation in corn has been affected by a population bottleneck that occurred during its domestication by indigenous peoples of Mesoamerica and a second bottleneck that occurred as modern agriculturalists selected for desirable traits from the indigenous crop lineages. Levels of diversity across the genome were used to guide simulations that varied the size of the initial population of teosinte and the severity of the population bottleneck. These analyses determined that the initial population size was on the order of < 10% of the total teosinte individuals across Central America and that the bottleneck's effective duration was less than 2,800 years. The distribution of genetic diversity generated from these simulations was then used to determine that around 2% of the genes surveyed had been affected by artificial selection during domestication.

In a similar study, De Mita et al. (2007) used twenty-four genes to validate coalescence simulations and determined that *Medicago truncatula* had recently undergone population expansion. They then used expected distributions of Tajima's D and Fay and Wu's H under a population expansion model to identify a single gene out of five candidates involved in root nodulation and signaling for mutualism with nitrogen fixing bacteria that had been subjected to positive selection. Using the CS approach in these studies allowed the elimination of "false-positive" candidates for selection that were actually the consequence of demographic processes.

While the CS approach can provide the basis for more robust tests to detect genomic regions that have been subject to selection, it is important to note that the outcomes of these analyses will only be as good as the accuracy of the demographic model developed. In general, it would be prudent to utilize a large number of neutral regions that are distributed across the genome to ensure the genetic distribution generated is sufficiently accurate. These studies may also be most successful when ancillary information can be used to verify the functional roles of candidate genes, focusing, for example, on gene regions known to influence traits that were likely to be affected by selection as in the cases of maize and *Medicago* described here. Other aspects of the history of populations such as regional substructure, range expansion, and population colonization after local extinction can also affect coalescence processes and should be accounted for in CS models; these remain as an active areas of research (Siol et al. 2010). Appropriate application of coalescence simulations has been shown to have substantial potential to account for the effects of population history and to provide more accurate identification of the targets of selection.

5.9 ADAPTATION WITH GENE FLOW

Variation in the environmental conditions across a species' geographic range may favor local adaptation and the evolution of *Ecotypes* that are associated with specific habitats (see Chapter 7). Locally adapted populations that do not possess reproductive barriers will be subject to migration load due to gene flow from populations in different environments, so selection favoring local adaptation may be counteracted by gene flow. This migration load may be prevalent because *Coarse-Grained* environments consisting of mosaics of contrasting habitats are quite common. For example, watercourses often intersect drier regions and create corridors of riparian habitat bordered by sharp moisture gradients. Plants distributed along these gradients will be under strong selection, but local adaptation may be disrupted by gene flow from adjacent regions. The potential for adaptation in heterogeneous environments will depend on the balance between the strength of selection and the level of gene flow among populations in different habitats.

We can illustrate the effects of selection and gene flow on adaptation more clearly if we focus on a single locus (Fig. 5.26). In this scenario there are two habitats connected by varying levels of gene flow (m), which is equal in both directions. The A allele is favored by selection in one habitat (purple; $1 + s$ for the AA genotype) and is hindered in a second habitat (red) by the degree of $1-\alpha s$ (Lenormand 2002). In this model the maintenance of a polymorphism across the two different habitats depends on the strength of selection, the amount of gene flow, and the selection differential (a may vary from 0 to 1). The lower panel in Fig. 5.26 shows the conditions for the maintenance of a polymorphism (both alleles at intermediate frequency) versus swamping (fixation when $m/s>\alpha/(1-\alpha)$) by the A allele in both habitats.

Across different levels of selection, the maintenance of a polymorphism depends on both m and a (Fig. 5.26 lower panel). When a is zero the fitness of the A allele is $1-0=1$ in the red habitat, and $1+s$ in the purple habitat, so the alternative (A) allele is eliminated (swamping). As a increases,

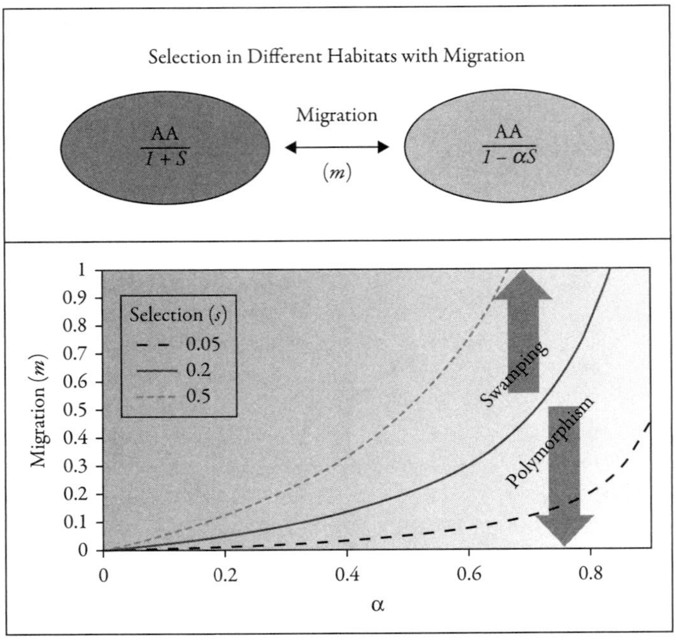

Fig. 5.26. The balance between migration and selection. An allele is beneficial in one habitat (left) and detrimental in a second (right) by the degree *as*. The fitness of the aa genotype is 1 in both habitats and the two alleles are codominant. The level of migration needed to balance selection for different values of selection across a range of values for *a* is shown. For each value of selection, migration rates above the line results in swamping (fixation of A) and below the line a polymorphism of both alleles is maintained (shown only for the line at $s = 0.2$).

the effectiveness of selection (*as*) is greater, so higher levels of gene flow are necessary to maintain a polymorphism across both habitats. In the polymorphic parameter space below the blue line ($s = 0.2$) the frequency of the A allele will decline when *a* is high and *m* is low (whiter region in the lower right-hand corner of the graph). Above the blue line migration overpowers selection in the red habitat and the A allele goes to fixation (upper-left region of the graph). In Fig. 5.26 this balance is illustrated for $s = 0.2$ (blue line) only, so regions of swamping and polymorphism are maintained above and below each line that represents a different strength of selection (i.e., for $s = 0.05$ and 0.5). For all levels of selection, the strongest difference in allele frequencies between the two habitats is maintained when effectiveness of selection (the product of *as*) is greatest and the migration rate is very low (Fig. 5.26 lower panel, lower right-hand corner).

There are a number of examples from the literature where selection has maintained genetic differentiation between populations occupying adjacent habitats (Lenormand 2002; Nosil et al. 2009). Classic among these are the populations of the grasses *Agrostis tenuis* and *Anthoxanthum odoratum* that occupy pastures and adjacent mine tailing soils (piles of materials left from the process of ore removal) that were contaminated with heavy metals (McNeilly and Antonovics 1968; Antonovics and Bradshaw 1970). Colonization of the contaminated soil presumably took place around the edge of the tailing piles as selection favored alleles conferring heavy metal tolerance. As discussed in Chapter 10, evolution of mine tailing ecotypes was probably facilitated by the simple genetic basis of metal tolerance in these plants. Selection on plants occurring in the contaminated soils is very strong, as seeds from the pasture populations and F_1 hybrids from crosses between the two populations suffer high rates of mortality on the tailing piles. In this system, high levels of genetic differentiation and strong *Clines* for traits associated with metal tolerance are maintained between the pasture and tailing populations by strong selection that counterbalances the high rates of gene flow from pastures (Antonovics 2006). For plants, such cases of strong ecological differentiation over short spatial scales are most often associated with edaphic conditions that represent harsh soil environments, such as the serpentine soils of western North America (e.g., Harrison and Rajakaruna 2011).

Isolation by Adaptation—Evidence for divergence due to selection is not always as clear as the well-documented cases of plants adapted to the challenging soil conditions just described. On the other hand, more subtle patterns of variation in environmental conditions across local and regional spatial scales are quite common. Habitat differentiation creates opportunities for the development of patterns of genetic differentiation that are mediated by ecological variation, but ecotones separating adjacent habitats may also facilitate adaptation and gene flow. As seeds and pollen disperse locally, successive generations may allow incremental gene flow across environmental gradients. While we expect selection to limit gene flow at specific genomic regions that confer adaptation, it is less clear whether selection mediated by ecological conditions can lead to genetic differentiation across neutral regions of the genome.

Our expectations for the development and maintenance of genetic differentiation may be quite different for genes subject to selection and for unlinked genomic regions that are neutral. Recall from Chapter 4 that low rates of gene flow and small effective population sizes will lead to divergence through genetic drift. While we expect some level of differentiation among all populations due to isolation by distance (IBDist), selection in different habitats may further reduce gene flow due to *Isolation by Adaptation* (*IBA*), which may allow higher levels of genetic divergence to develop at loci across the genome (Nosil et al. 2009; Fig. 5.27; refer to color plate). Theory predicts that the effects of isolation will be the same whether reduced gene flow is due to geographic distance or to the effects of selection in different habitats—neutral genetic divergence should be higher when gene flow is reduced.

Our expectation of higher divergence at neutral genetic loci between different habitats is borne out in some studies, but many of these appear to be instances where a lack of recombination or historical divergence may be responsible for patterns of genetic differentiation across the genome. The clearest examples

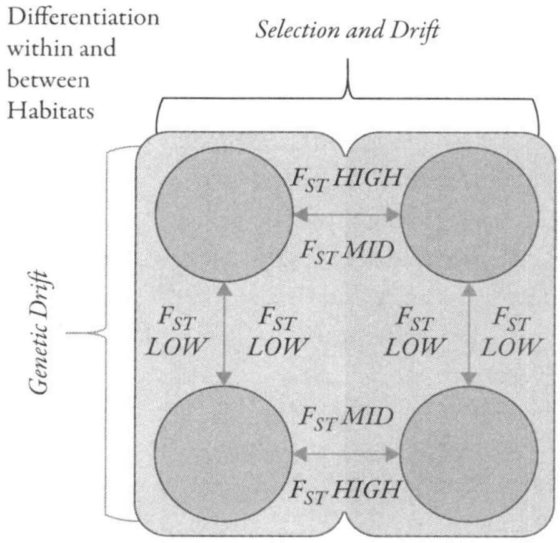

Fig. 5.27. Expected patterns of genetic differentiation within and between habitats in response to selection. Two habitats are indicated by regions of purple (left side) and red (right side). Expected levels of differentiation (F_{ST} —LOW, MID, or HIGH) are given for a locus under selection (red) and an unlinked neutral locus (blue).

of strong genetic differentiation between habitats for genome-wide genetic markers all appear to be cases of plant species that are highly selfing (e.g., Hamrick and Allard 1972; Bedada et al. 2014; Chapter 4). In these plants, selection presumably acts on specific genomic locations and selfing maintains gametic disequilibrium (Box 10.1) across the entire genome—alleles at neutral loci are associated with habitats because the distribution of selfing lineages is determined by selection on a small number of loci.

For outcrossing species on the other hand, convincing examples of associations of neutral genetic variation with divergent habitats in close proximity to each other are more difficult to find. For example, in a study of the marine angiosperm *Zostera marina* (eel grass) in intertidal habitats that were either submerged or exposed at low tide, divergence for most microsatellite markers was similar for pairs of populations located within and between habitats (Oetjen and Reusch 2007). In this system, neutral markers did not display higher differentiation between habitats in spite of contrasting selection regimes, and the few markers that did display stronger differences were shown to be closely linked to gene regions that were under selection. A similar study on a forest herb that primarily outcrosses but is capable of selfing (*Geum urbanum*) failed to find neutral genetic divergence across adjacent habitats (Vandepitte et al. 2007). While neutral genetic divergence associated with habitat differences can develop in species with high selfing rates, it is more difficult to identify strong cases of isolation by adaptation over short spatial scales in outcrossing species.

Over broader geographic scales, there may be more opportunity for IBA to contribute to neutral genetic differentiation. For example, associations between neutral genetic variation and habitats may be prevalent for some groups of closely related taxa occurring across environmental gradients. Lexer et al. (2014), for example, contrasted divergent lineages within the *Restio capensis* species complex over a large geographic region in South Africa. Several reviews have also concluded that there is ample evidence for the role of selection as a barrier to gene flow (Nosil et al. 2009; Orsini et al. 2013). *Meta-Analyses* of the literature have found evidence for IBA (also termed *Isolation by Environment* or *Isolation by Ecology, IBE*). These studies controlled for IBDist and concluded that up to 5% of the variation in neutral markers was explained by ecological distance (Shafer and Wolf 2013; Sexton et al. 2014).

Many of the studies supporting neutral divergence based on IBA were conducted over much broader geographic scales than the investigations of *Zostera* and *Geum* described earlier. Since we are viewing the distribution of these species at one point in time, it is difficult to know whether historical distributions may have facilitated isolation and ecological divergence among groups of populations. Under such a scenario, geographically isolated populations may have undergone neutral divergence, so contemporary patterns of variation may reflect historical processes rather than genome-wide divergence solely due to the effects of selection in contrasting habitats. Of course, a smaller number of genomic locations will be the targets of selection for adaptation to local environments, but we do not expect selection on a limited number of locations to have genome-wide effects on levels of neutral divergence. In general, clear cases of the effects of IBA on unlinked neutral genetic variation that has developed without isolation by distance may be uncommon, but further studies are necessary before firm conclusions on the role of IBA for local-scale population differentiation can be made.

The lack of clear examples of the effects of IBA on neutral loci over short spatial and temporal scales in outcrossing species suggest that populations in different habitats are generally not experiencing strong isolation that is imposed by local selection. We expect that selection may limit dispersal between habitats and consequently should restrict gene flow, which would lead to higher levels of divergence across the genome; however, in outcrossing species it appears that rates of recombination are adequate to allow different genomic regions to respond independently to the influences of selection and genetic drift.

Clues concerning the conditions necessary for IBA to contribute to divergence at neutral loci unlinked to the targets of selection can be gleaned from theoretical analyses. Models of IBA indicate that selection is most effective at preventing neutral introgression when habitats are discrete—they are separated by abrupt *Ecotones* that prevent gradual gene flow across generations. For example, simulations show neutral divergence between these discrete habitats under moderate to strong selection (Thibert-Plante and Hendry 2010), but very strong viability selection is necessary to maintain allele frequency differences at unlinked neural loci along more gradual environmental gradients (Barton and Bengtsson 1986). Natural environments tend to change gradually with distance and such sharp distinctions between adjacent habitats are uncommon. In other models, the inclusion of additional processes such as assortative mating was necessary to maintain high levels of differentiation at unlinked neutral loci (Gavrilets and Cruzan 1998). These analyses suggest that the conditions necessary for selection alone to restrict gene flow to a degree that is necessary for neutral divergence are rather restrictive and may be uncommon.

To better appreciate the difficulty of obtaining significant levels of differentiation at neutral loci that is due to IBA (or IBE), it is instructive to illustrate the processes of ecological divergence (Fig. 5.28; refer to color plate). As we noted earlier in this chapter, adaptation is unlikely to occur by large changes, so we expect environmental gradients (ecotones) will facilitate selection for the exploitation of a novel habitat (Fig. 5.28; Adaptation along an Environmental Gradient; refer to color plate). Populations distributed across an ecotone may display higher levels of divergence for loci under selection, but stepwise gene flow should maintain low differentiation for neutral regions even between populations at the extreme ends of the gradient. Neutral divergence is more likely to occur with isolation by distance (Fig. 5.28; Adaptation with Isolation by Distance; refer to color plate). In this scenario, some degree of neutral differentiation is maintained after changes in geographic ranges create parapatric distributions. In this case, the maintenance of neutral divergence will depend on the level of reproductive isolation; if hybrid populations are formed then introgression may proceed leading to the decay of the neutral differentiation over time. As mentioned earlier, either strong selection or a combination of selection and assortative mating will be necessary to maintain high levels of divergence across neutral regions of the genome.

From these discussions, it appears that patterns of genetic variation consistent with IBA are more likely to be common in studies conducted over broader geographic scales or with groups of populations that have undergone divergence with geographic isolation. At small spatial scales, neutral divergence between contrasting habitats will require very strong viability selection, as in the case of grass ecotypes on mine tailings, or additional factors, such as habitat-specific assortative mating. We also note that the potential for neutral divergence is greater when levels of genome-wide recombination are reduced by high rates of selfing, which suggests that rates of recombination as well as the strength of selection are important determinants of the efficacy of IBA. In the next section, we will take a closer look at patterns of divergence across the genome in response to different levels of selection and regional rates of recombination.

Genomic Islands of Divergence—As advances in genetic marker technologies have allowed higher resolution views of genomic variation, our understanding of the influence of selection on the distribution of genetic variation has become more refined. Improvements in sequencing efficiency have allowed analysis of larger numbers of samples, which has enabled more comprehensive studies in population genomics. To conduct these investigations multiple individuals are sampled from two or more populations and sequenced for multiple regions across the genome. From these sequence alignments, levels of differentiation are quantified based on single-nucleotide polymorphisms (SNPs; Fig. 5.29).

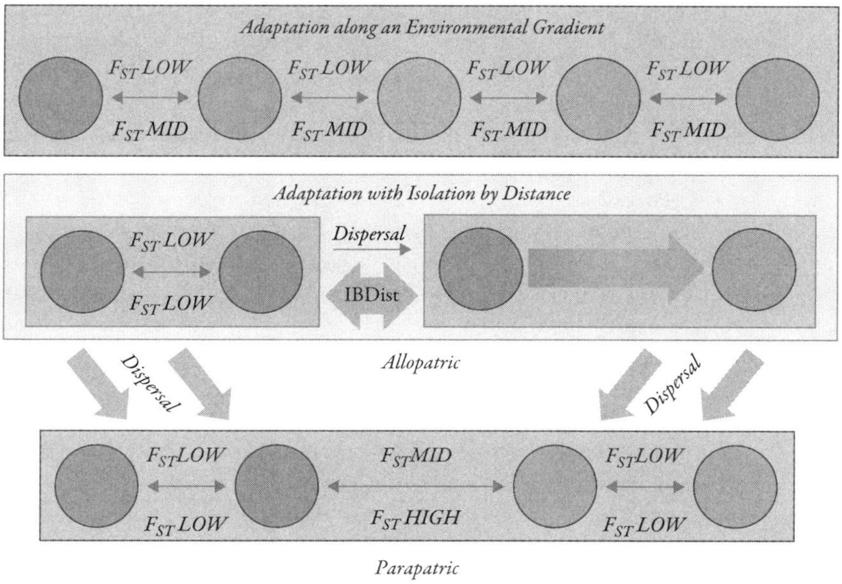

Fig. 5.28. Scenarios for the origin and distribution of ecotypes. In the top panel adaptation occurs along an environmental gradient. Loci under selection differentiate (Red) while low differentiation (F_{ST}) is maintained across neutral regions (Blue). In the second scenario adaptation and neutral divergence occurs with isolation by distance after dispersal (Allopatric Divergence; middle panel). Some level of neutral differentiation persists after dispersal creates parapatric distributions (lowest panel).

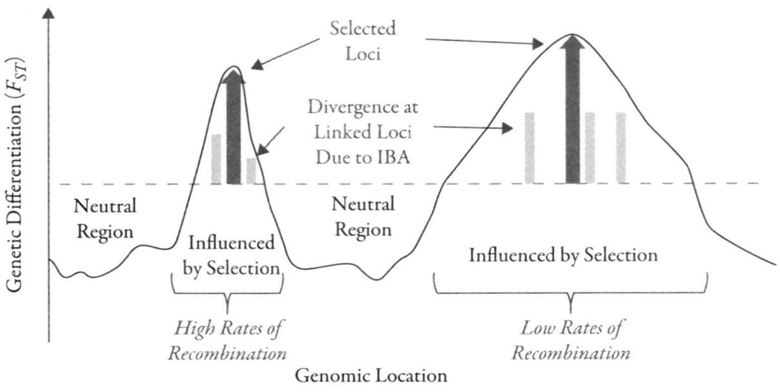

Fig. 5.29. The hypothesis that islands of genomic divergence maintained by selection at specific loci. Levels of divergence along a chromosome (solid line) reveal neutral regions and regions of high divergence. The horizontal dashed line indicates the neutral threshold. Regions of high and low recombination surrounding selected locations are shown.

Lessons we have learned from these analyses largely confirm the information gleaned from low-density genetic marker studies such as those described in the previous section. Sampling variation across the genome from populations occurring in different habitats often reveals large stretches of low genomic divergence among populations with a number of small regions of high differentiation (*Islands of Divergence; Speciation Islands*; F_{ST} *Outliers*; Nosil et al. 2009). It was thought that higher levels of divergence at specific genomic locations are maintained by selection on individual loci, but more recent analyses suggest that this may not always be the case (Pannell and Fields 2014). The characterization of these genomic divergence islands, the consequences of selection on specific loci for linked regions, and the evaluation of processes maintaining islands are active areas of investigation.

We expect that most genomic regions will not be subject to selection, so they will display low levels of differentiation (F_{ST}) that are the outcome of the balancing effects of gene flow and genetic drift (Fig. 5.29). Genes that confer adaptation to local environmental conditions are expected to display elevated differentiation, since alternative alleles will occur at higher frequencies in contrasting habitats. The regions surrounding loci under selection will also show higher levels of differentiation; linkage is expected to shelter neighboring stretches of DNA from the influence of gene flow.

Genomic regions near regions conferring adaptation are thought to be more susceptible to the accumulation of neutral and selected variation due to *Divergence Hitchhiking* (Feder and Nosil 2010). Since regions near selected loci are less affected by gene flow, there is more potential for selection to favor alleles that contribute to local adaptation, so islands of divergence are expected develop around selected loci and to grow over time. Under stable environmental conditions, islands will continue to grow and eventually coalesce, leading to ever larger islands and eventually a completely divergent genome—a process that is thought to contribute to ecological speciation (Nosil et al. 2009). The divergence hitchhiking hypothesis predicts that genomic islands should vary in size, that island size will be positively correlated with age, and that more isolated populations should be characterized by larger islands because they will experience lower rates of gene flow (Feder and Nosil 2010).

It is important to note that the generation of genetic differences under the divergence hitchhiking hypothesis differs from the process of genetic hitchhiking. Divergence hitchhiking is thought to be due to the presence of an allele at one locus that is maintained by selection in the current environment. In this case, the closely linked regions are sheltered from the effects of gene flow, which leads to the accumulation of neutral and adaptive variation. Genetic hitchhiking, on the other hand, occurs during allele frequency change at a locus under selection; alleles at nearby loci that happen to be in linkage disequilibrium with an allele affected by selection will be carried along as it goes to fixation (selective sweep) or extinction (background selection). Genetic hitchhiking will create regions of low genetic diversity in the region of the selected locus, whereas divergence hitchhiking is thought to generate regions of higher diversity. In both cases, the "hitchhiking" effect decays rapidly with decreasing linkage.

Predictions under divergence hitchhiking are a refinement of our expectations under IBA. In both scenarios, local selection reduces the effects of gene flow, which increases the potential for genomic differentiation among populations occupying different habitats. In the previous section, we concluded that this was unlikely to occur in genomic regions that were more distant from loci under selection in outcrossing species, unless the success of immigrants and hybrids was strongly restricted by selection and assortative mating. It seems more likely that levels of differentiation could be enhanced in regions that are closely linked to genes conferring adaptation. On the other hand, theoretical analyses indicate that the conditions necessary to allow genomic island growth may be restrictive (i.e., very small population sizes and very low rates of gene flow; Feder and Nosil 2010), suggesting that the circumstances necessary for divergence hitchhiking may be uncommon.

A critical test of the divergence hitchhiking hypothesis examined the influence of gene flow by evaluating IBDist for sets of sunflower (*Helianthus*) populations (Renaut et al. 2013). In these relatively high-resolution analyses, many small islands of divergence were identified, but no association was found between average island size and levels of gene flow as would be expected under divergence hitchhiking. On the other hand, there was a strong association between island size and recombination rates such that large islands were more likely to occur in genomic regions with low rates of recombination (Fig. 5.29). Moreover, a separate set of analyses of five pairs of closely related species revealed that larger islands harbored lower levels of genetic diversity (Cruickshank and Hahn 2014), which is also consistent with the hypothesis that these genomic regions have lower rates of recombination (see Chapter 3).

The observation that regions of high divergence are not associated with lower rates of gene flow raises questions about the processes responsible for the development of genomic islands (Pannell and Fields 2014). In the study on *Helianthus,* Renaut et al. (2013) did not assess genetic associations with environmental variables so they were not able make more direct tests for the role of selection. The authors hypothesized that low recombination rates resulted in reduced effective population size for these genomic locations, so the higher F_{ST} is probably the result of the enhanced effects of genetic drift.

The hypothesis that genomic islands do not necessarily correspond to targets of selection has been supported by several other studies where the genes associated with selection were not consistently located in regions of high divergence among populations (Pannell and Fields 2014). For example, studies of loblolly pine (*Pinus taeda*) have found that genomic regions harboring allelic variation associated with environmental variation did not correspond to the locations of islands of genomic divergence (Eckert et al. 2010; Frichot et al. 2013). These investigations into the genomic basis of adaptation and divergence are significant because they highlight the importance of quantifying ecological variation, which is a necessary component of information needed to test adaptive hypotheses for the effects of selection on the distribution of genetic variation.

Studies conducted so far have found little evidence supporting the hypothesis that larger genomic island size is associated with age or lower levels of gene flow, so it has been difficult to identify clear cases that support the divergence hitchhiking hypothesis. The role of local adaptation for the origin and maintenance of divergence islands also seems unlikely; instead, these genomic islands may be the consequence of low regional rates of recombination (Renaut et al. 2013). As the resolution of our view of the genome has improved, it has become evident that islands of divergence remain small and do not undergo accretion due to the effects of selection or reduced gene flow. It is also apparent that genes that are associated with environmental variables and are also candidates for selection are not consistently associated with sites of higher divergence.

The one thing that is clear from these analyses is that our understanding of the genomic basis of adaptation is still in its early stages of development. The regions of higher divergence that characterize populations occupying different habitats are sometimes implicated in local adaptation, but the role they play in the process of adaptation and speciation will require further analyses that combine the quantification of ecological and genomic variation.

Identifying Adaptive Genetic Variation—In many ways the focus on isolation by adaptation (IBA or IBE) and divergence hitchhiking in the early 2000s is reminiscent of the selectionist views of allozyme variation that were prevalent in the early 1970s (Chapter 4). In both instances, the assumption of a broad role for selection shaping genetic variation within and among populations eventually became more refined to acknowledge greater roles for other evolutionary processes. For example, a paper by Hamrick and Holden (1979) effectively recanted statements made in an earlier paper (Hamrick and Allard 1972) on the roles of allozyme loci for local adaptation in *Avena barbata*. Similarly, early propositions that IBA could have genome-wide effects on neutral genetic variation have been restated to include only linked regions (divergence hitchhiking). As more theoretical and empirical analyses have become available it is becoming increasingly clear that exceptional conditions are necessary for genomic island accretion and more generally for IBA (Feder and Nosil 2010; Renaut et al. 2013).

It is now apparent that divergence hitchhiking is unlikely to play an important role for adaptation and genetic diversification. Indeed, divergence islands may have been red herrings that attracted undue attention for their assumed role in ecological divergence. Higher levels of divergence at specific genomic locations may be due to greater effects of background selection, which would reduce regional effective population sizes. Thus, genomic islands of divergence may more often be associated with low rates of recombination and represent hot spots for genetic drift rather than regions that typically harbor adaptive genetic variation.

It is clear from this discussion that identifying genomic locations contributing to adaptation requires careful quantification of environmental variation as well as intense genomic surveys. It is also evident that recombination plays an important role in genomic diversification. Depending on frequencies of genome-wide (i.e., due to outcrossing) and regional rates of recombination, we expect varying degrees of independence, so the genome may be effectively more "coarse-grained" (high selfing frequencies) or "fine-grained" (high outcrossing frequencies). For outcrossing species, it may be best to view the genome as being more fluid; for most of the genome, recombination rates are typically high enough to allow different regions to respond independently to population genetic processes. Ultimately, the genomic composition of populations will reflect the balancing forces of genetic drift, gene flow, recombination, and selection resulting from migration load, inbreeding load, and environmentally induced pressure favoring allelic variants that contribute to local adaptation.

The emerging view of a more limited role for genomic islands of divergence for adaptation acknowledges the complexity of processes affecting levels of gene flow, adaptation, and diversification among populations. Genomic islands are of considerable interest as they are candidates for the targets of selection and represent variation that is important for differentiation among populations. The realization that islands tend to remain small while gene flow affects the large majority of the genome is perhaps not surprising when we take a broader view of genetic exchange. Recall from our earlier discussions that most populations are likely to possess varying levels of drift load so that inter-population crossing will generally result in heterotic offspring. If genetic load is generally due to a large number of alleles of small effect, then we would expect that selection pressure would favor introgression of allelic variation across most of the genome, just as it was being opposed at specific locations because reduced of recombination rates (islands) or the presence of alleles favored by the local environment. In the next section we will shift our attention away from adaptive genetic variation to focus on the effects of the vast majority of expressed mutations, which are mostly detrimental and contribute to the genetic load of populations.

5.10 ECOLOGICAL GENOMICS

The immensity of biological diversity that we encounter on a daily basis, and that has been cataloged from the far reaches of our planet, provides ample evidence for the consequences of selection and its promotion of adaptive evolution. In spite of the spectacular range of phenotypic traits, advances in our understanding of the molecular genetic basis of adaptation have been modest. Most of what we have learned of the adaptive evolution in plants comes from a handful of examples of genes of large effect in domesticated crops and in model systems (Siol et al. 2010). Much of our discussion of selection on adaptive molecular genetic variation to this point has considered genomic locations that have large effects on fitness. But, both theoretical and genetic mapping studies tell us that genes with small or moderate effects on fitness more often control adaptation (Ehrenreich and Purugganan 2006). Recently (since around 2010), advancements in the efficiency of sequencing technologies and methods for the analysis of genomic variation across natural environments have begun to yield information on potentially adaptive genetic variation in a broader range of species (Pannell and Fields 2014). Ecological genomic studies on these "non-model" species hold considerable promise to improve our understanding of the effects of selection and the process of adaptive evolution.

Variation in biotic and abiotic environments defines the selective forces that promote adaptation; the field of ecological genetics endeavors to study this interface, which translates ecological variation into changes in the genetic composition of populations. Historically, ecological genetic studies have focused on quantitative and polymorphic phenotypic traits, but there has always been a prevailing desire to drill down into the underlying genetic variation to understand the processes of adaptive evolution at the genomic level. This strong desire to uncover ecologically relevant genetic information was evident from the early enthusiasm for allozymes as markers for adaptive variation. The desire to study allelic variation that is directly subject to selection has been frustrated by a lack of appropriate genetic markers for several decades; however, in recent years this technological barrier has been breached, so we may expect a profusion of ecological genomic studies in the near future.

Detecting adaptive variation requires conducting genetic surveys within an appropriate ecological context. Such endeavors would also be more productive if studies were conducted on species that have relatively broad ecological ranges. Both of these criteria have been met in several studies that have quantified environmental variation across the ranges of non-model species. For example, investigations of expressed genetic variation have revealed a number of genomic locations harboring allelic variation that is associated with environmental variables in *Pinus taeda* (loblolly pine; Eckert et al. 2010) and *Populus balsamifera* (balsam poplar; Keller et al. 2012). Both of these studies utilized population modeling to delineate demographic effects on genetic diversity and to resolve allelic differences that corresponded to ecological variation. In the loblolly pine study, the five loci found to be associated with ecological variation correspond to genes found in *Arabidopsis thaliana* that are known to be important for stress responses, which provides further confirmation of the veracity of this approach. Genomic investigations that integrate habitat information are most likely to provide significant progress in our understanding of the process of adaptive evolution.

The majority of ecological genomic studies have focused on transcribed regions, but phenotypic traits are affected by both variation in protein structure and their levels and patterns of expression during development. As discussed in Chapter 3, large proportions of intergenic DNA are functional and contains various types of control regions affecting gene expression, including locations coding for interference and micro-RNAs. Consequently, we expect that a large proportion of genetic variation affecting traits to be located in regions outside of transcribed DNA, so comprehensive genomic surveys will be required to obtain a more complete picture of the effects of selection on adaptive genetic variation.

Ecological genetic investigations are complicated by the potential for individual plants to alter phenotypic trait expression in response to environmental variation during growth and development. Obtaining a more comprehensive understanding of the genetic and environmental determination of trait variation and the consequences for fitness will require consideration of phenotypic traits as well as the patterns of gene expression and DNA sequence variation that control them. In the next chapter, we continue our discussion of ecological genetics as we explore variation in phenotypes and methods to assess the strength and consequences of selection on phenotypic traits.

5.11 QUESTIONS FOR CONTEMPLATION AND DISCUSSION

1. What do you think would have happened if Charles Darwin had never lived? Would we still have a theory of evolution similar to the one that exists today? What might have been different?
2. Describe how hard and soft selection might apply more appropriately to different types of

selection pressures. Why is there generally a trade-off between the intensity of hard and soft selection?

3. What is the difference between viability and performance selection? How would you design an experiment that measured both?
4. How does the more complex life cycle of plants affect the influence of selection on populations? Why might an independent gametophyte be a liability? How did angiosperms take advantage of the gametophyte stage to elevate their rates of diversification?
5. Are Fisher's models of evolution by natural selection useful for understanding how selection works in real populations? What are the most important points that you learned from these models?
6. Describe in your own words the consequences of dominance for the effects of selection. In particular, what are the dynamics of a recessive deleterious or beneficial mutation when it occurs at a low frequency? Why would a dominant deleterious mutation starting at high frequency be slow to go to extinction?
7. Why do you think clear cases of heterosis (overdominance) are rare?
8. What is the drift barrier and how does it explain the inability of selection to be effective in small populations? How does selection interact with gene flow and drift to influence allele frequency change?
9. What is the difference among the concepts of drift load, inbreeding load, recombination load, and migration load? Which is most important for determining inbreeding depression? Why is it generally difficult to purge genetic load from populations?
10. Describe the distribution of selection and dominance for deleterious mutations segregating in populations. Do you think this reflects to distribution of mutations when they first arise? How is this distribution generated?
11. Describe each of the neutral diversity, site frequency spectrum tests, and neutrality tests based on divergence. For each one list the type of data needed, the assumptions required, and the strengths and weaknesses of the test.
12. Can selection limit gene flow? How do the strength of selection, the presence of ecotones, and recombination act to limit the effects of gene flow across the genome?
13. Why is it so difficult to identify adaptive genetic variation? How could you distinguish F_{ST} outliers that are due to low rates of recombination from those that are due to selection? What type of data could be used to make a stronger case for the role of selection?

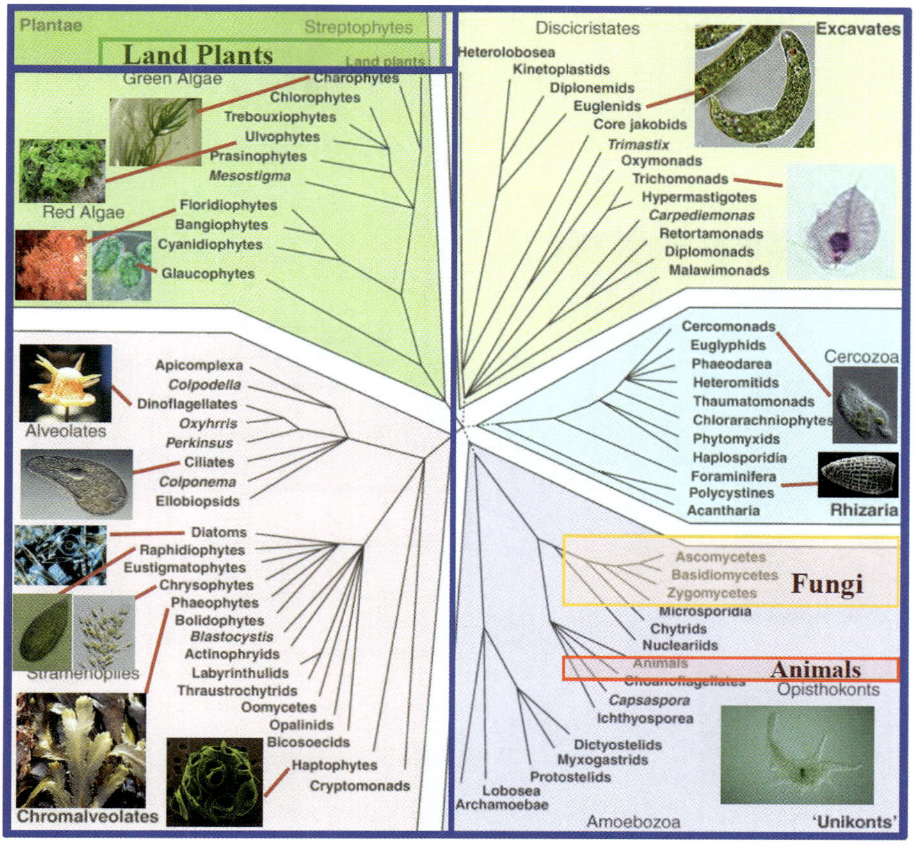
Plantae
Streptophytes
Land Plants
Land plants
Green Algae
Charophytes
Chlorophytes
Trebouxiophytes
Ulvophytes
Prasinophytes
Mesostigma
Red Algae
Floridiophytes
Bangiophytes
Cyanidiophytes
Glaucophytes
Discicristates
Excavates
Heterolobosea
Kinetoplastids
Diplonemids
Euglenids
Core jakobids
Trimastix
Oxymonads
Trichomonads
Hypermastigotes
Carpediemonas
Retortamonads
Diplomonads
Malawimonads
Cercomonads
Euglyphids
Phaeodarea
Heteromitids
Thaumatomonads
Chlorarachniophytes
Phytomyxids
Haplosporidia
Foraminifera
Polycystines
Acantharia
Cercozoa
Rhizaria
Apicomplexa
Colpodella
Dinoflagellates
Oxyhrris
Perkinsus
Ciliates
Colponema
Ellobiopsids
Alveolates
Diatoms
Raphidiophytes
Eustigmatophytes
Chrysophytes
Phaeophytes
Bolidophytes
Blastocystis
Actinophryids
Labyrinthulids
Thraustrochytrids
Oomycetes
Opalinids
Bicosoecids
Haptophytes
Cryptomonads
Chromalveolates
Ascomycetes
Basidiomycetes
Zygomycetes
Fungi
Microsporidia
Chytrids
Nucleariids
Animals
Animals
Capsaspora
Ichthyosporea
Opisthokonts
Dictyostelids
Myxogastrids
Protostelids
Lobosea
Archamoebae
Amoebozoa
'Unikonts'

Fig. 1.1.

Fig. 1.5.

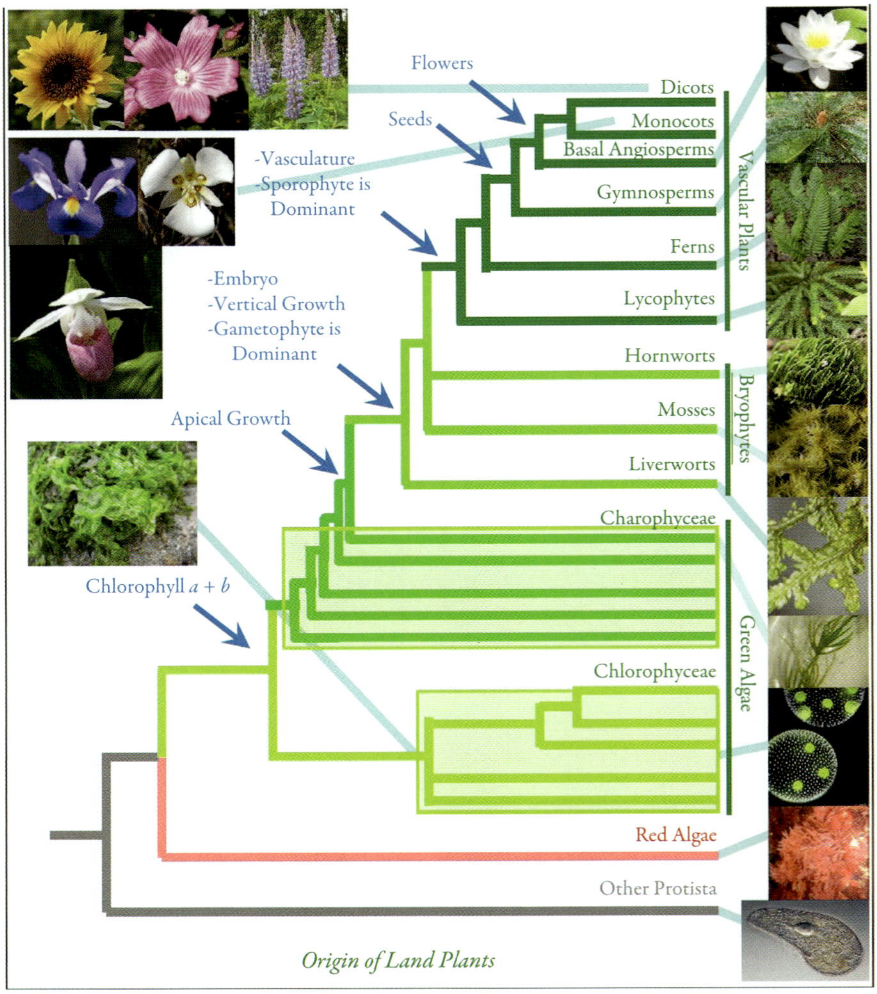

Flowers
Seeds
-Vasculature
-Sporophyte is
Dominant
-Embryo
-Vertical Growth
-Gametophyte is
Dominant
Apical Growth
Chlorophyll $a + b$
Dicots
Monocots
Basal Angiosperms
Gymnosperms
Ferns
Lycophytes
Hornworts
Mosses
Liverworts
Charophyceae
Chlorophyceae
Red Algae
Other Protista
Vascular Plants
Bryophytes
Green Algae
Origin of Land Plants

Fig. 2.1.

Fig. 2.11.

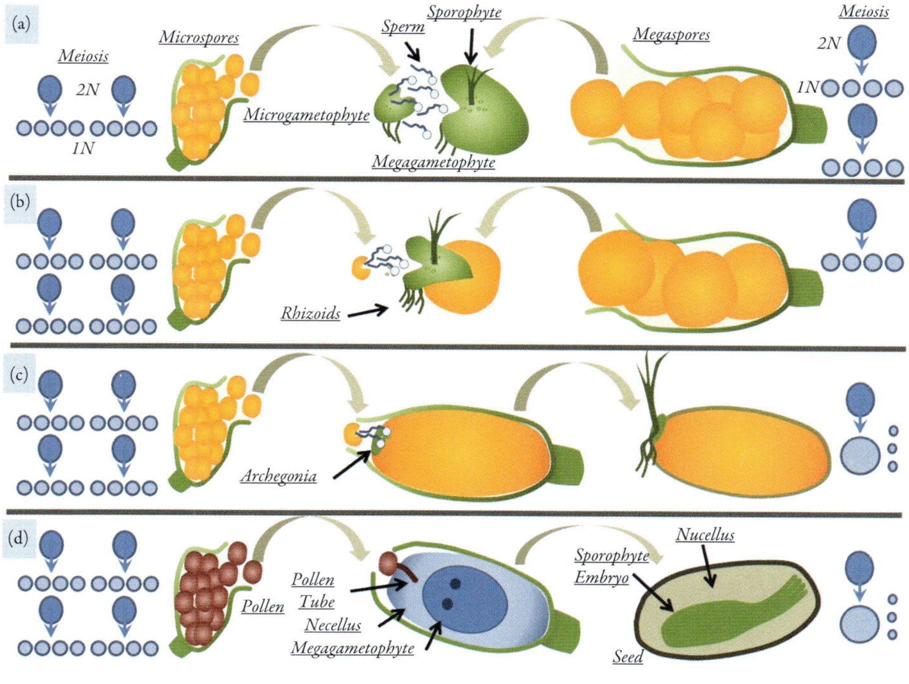

Fig. 2.15.

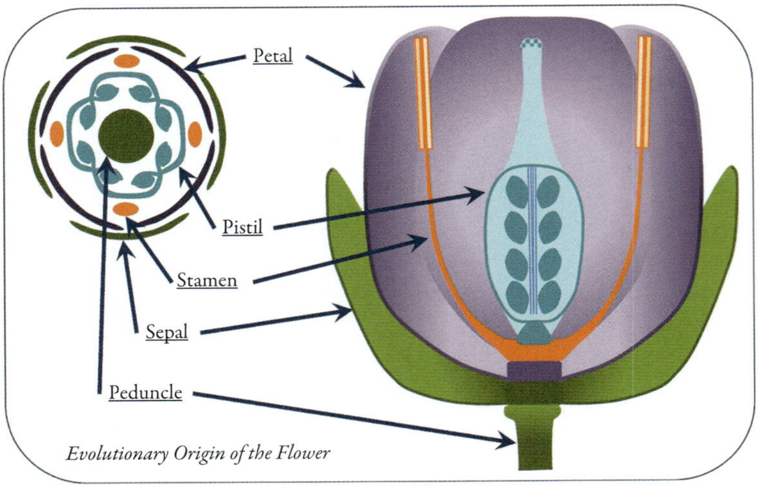

Fig. 2.18.

Fig. 2.22.

Fig. 2.23.

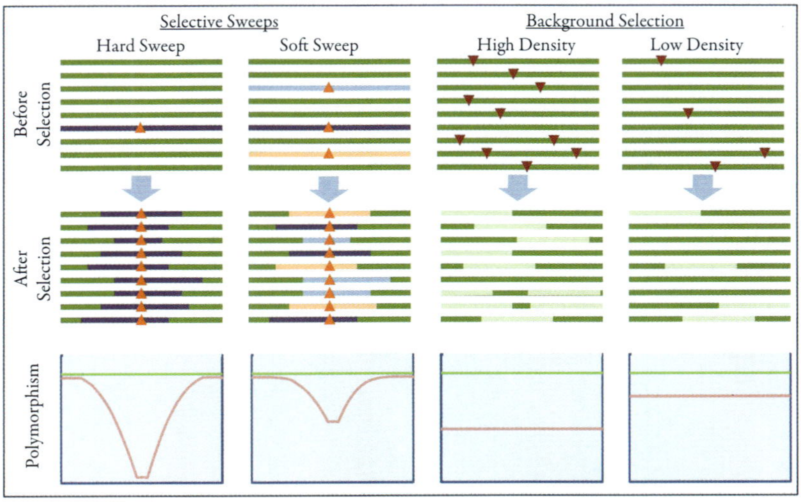

Fig. 3.4.

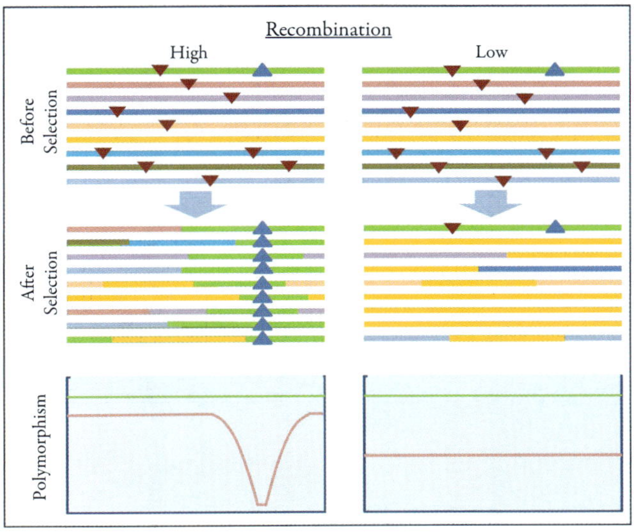

Fig. 3.5.

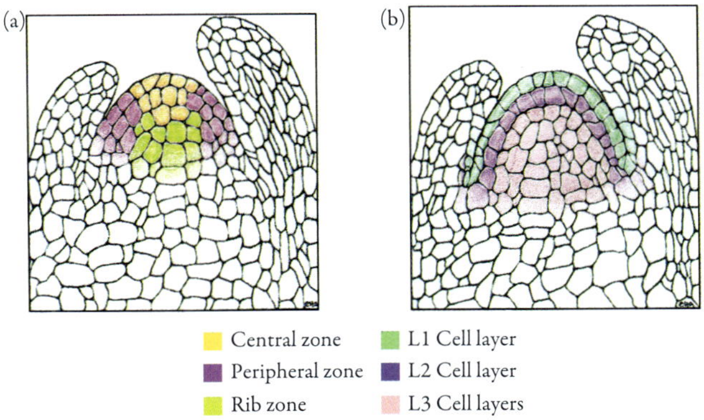

Fig. 3.13.

Fig. 3.15.

Fig. 3.16.

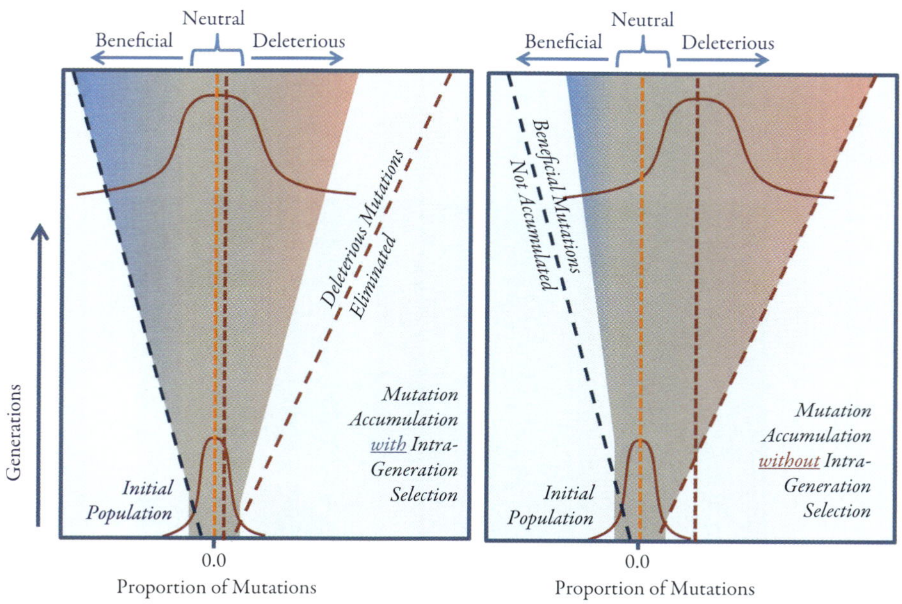

Fig. 3.20.

Fig. 4.5.

Fig. 4.11.

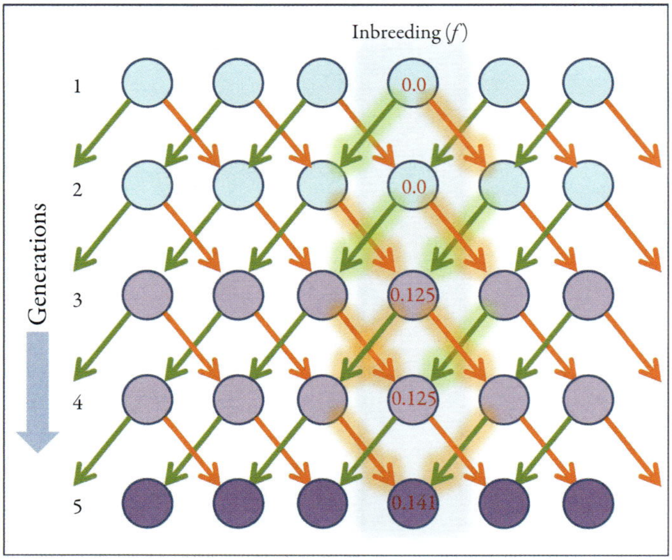

Fig. 4.14.

Fig. 5.3.

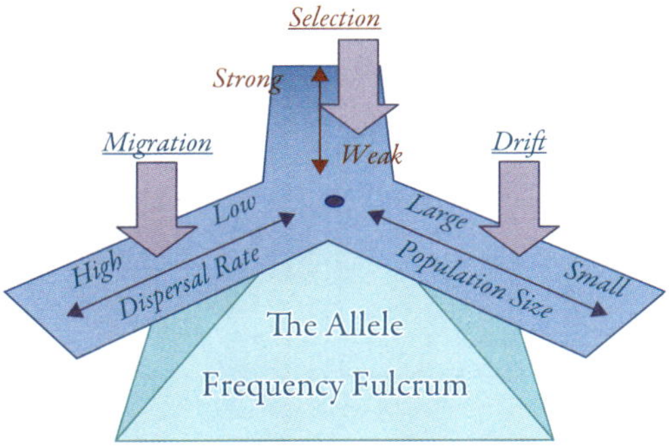

Fig. 5.17.

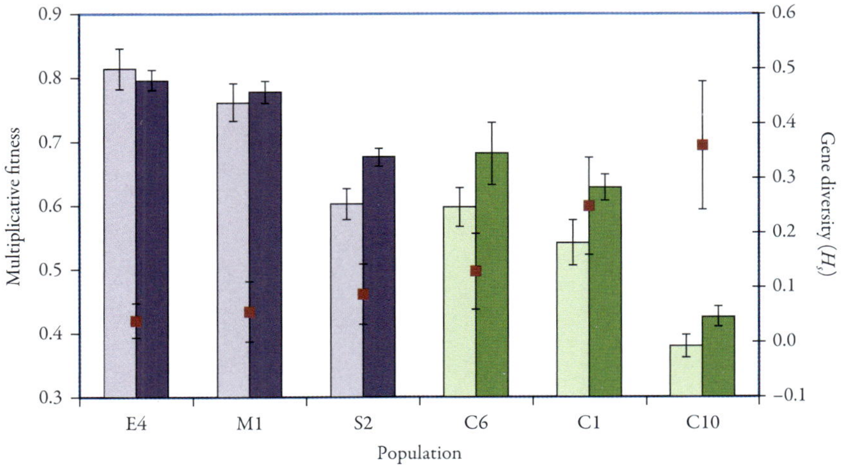

Fig. 5.19.

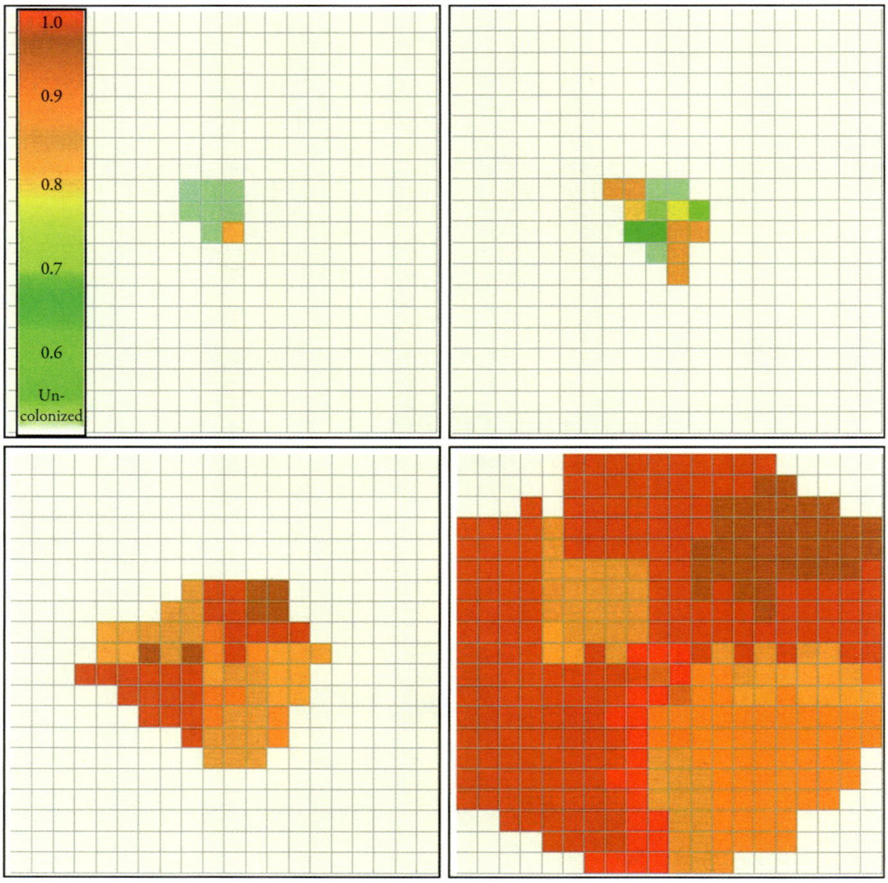

Fig. 5.21.

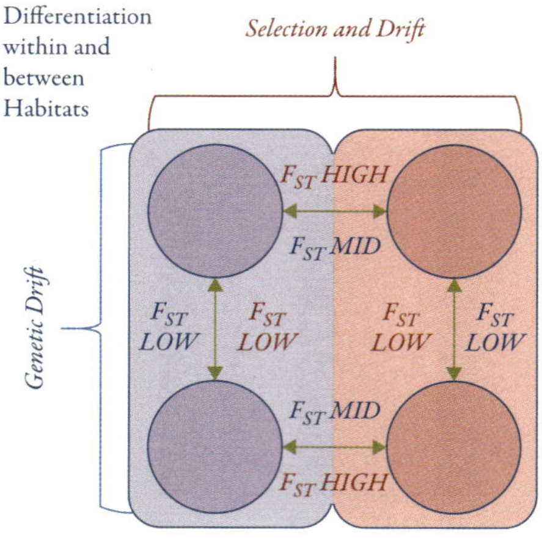

Fig. 5.27.

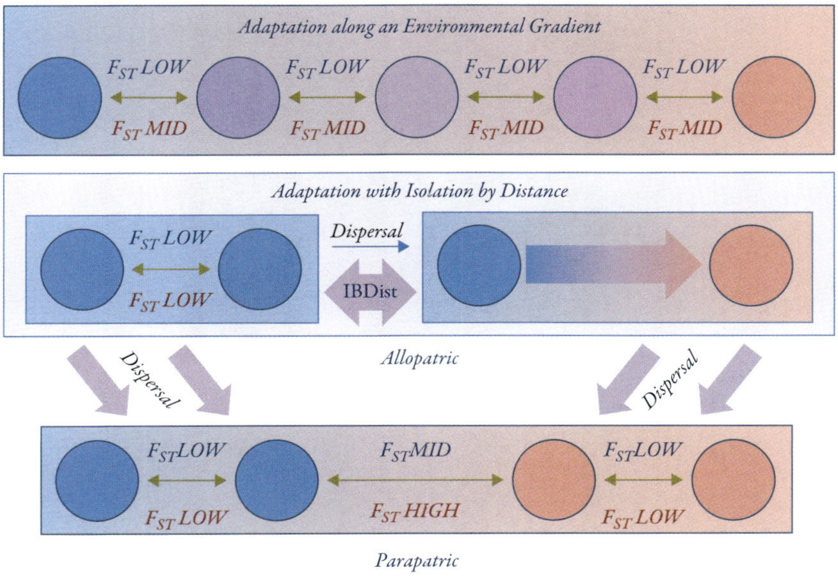

Fig. 5.28.

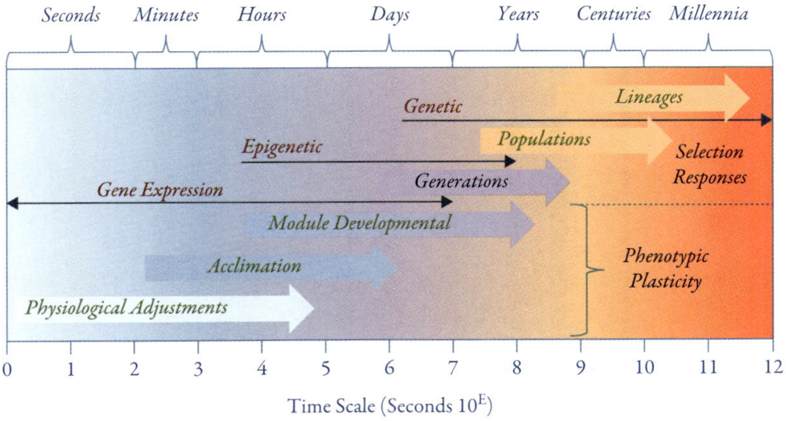

Fig. 6.1.

Fig. 6.3.

Fig. 6.4.

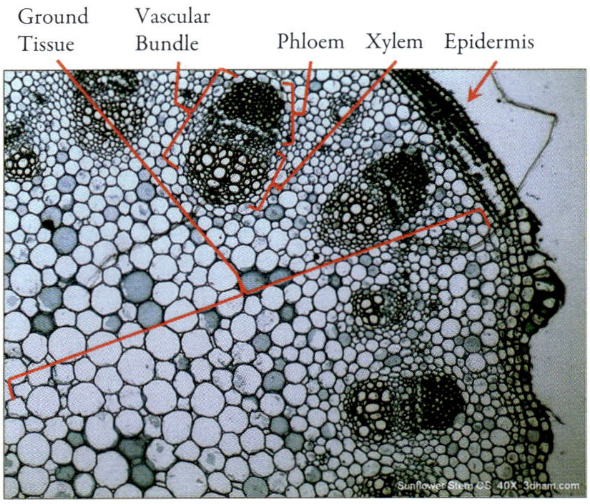

Fig. 6.5.

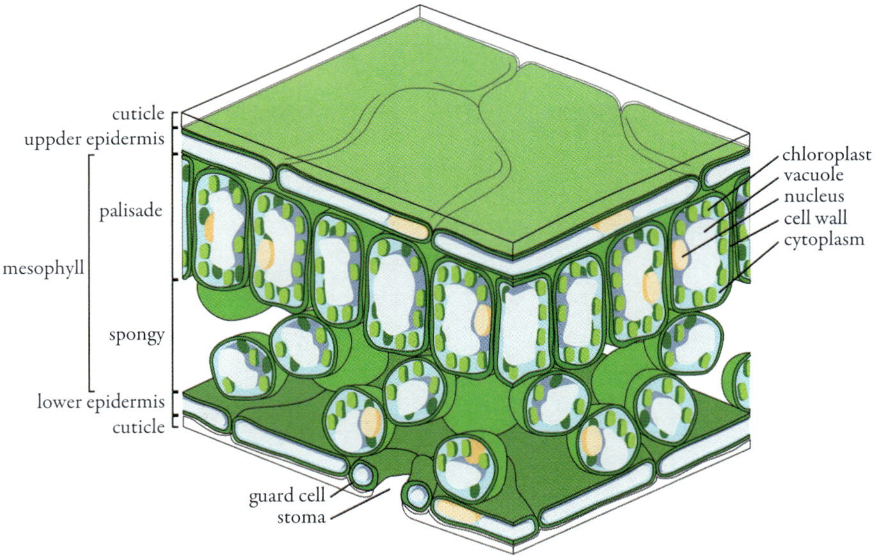

Fig. 6.6.

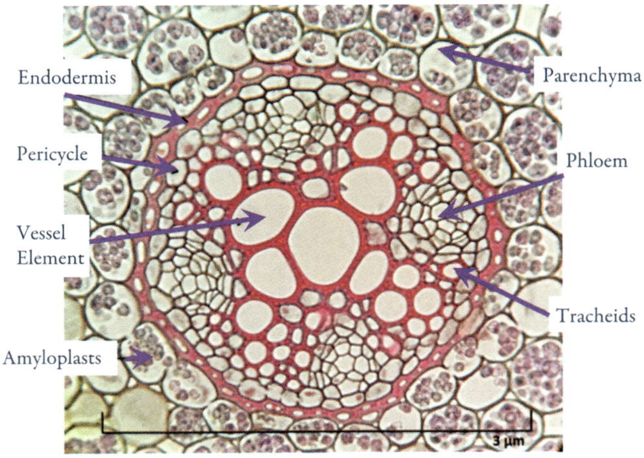

Fig. 6.7.

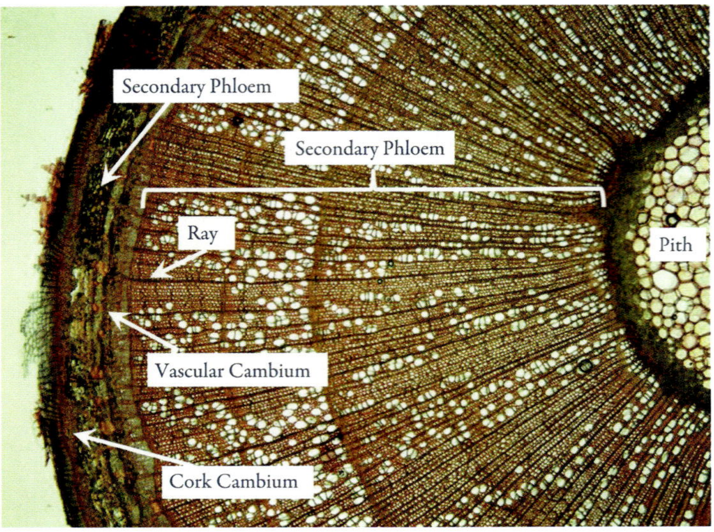

Fig. 6.8.

Fig. 6.9.

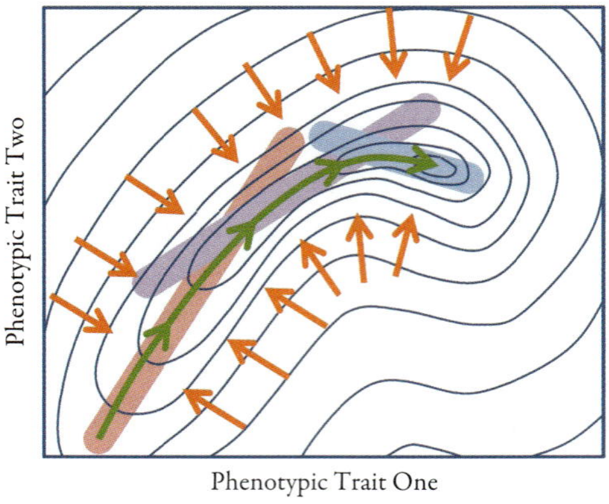

Fig. 6.12.

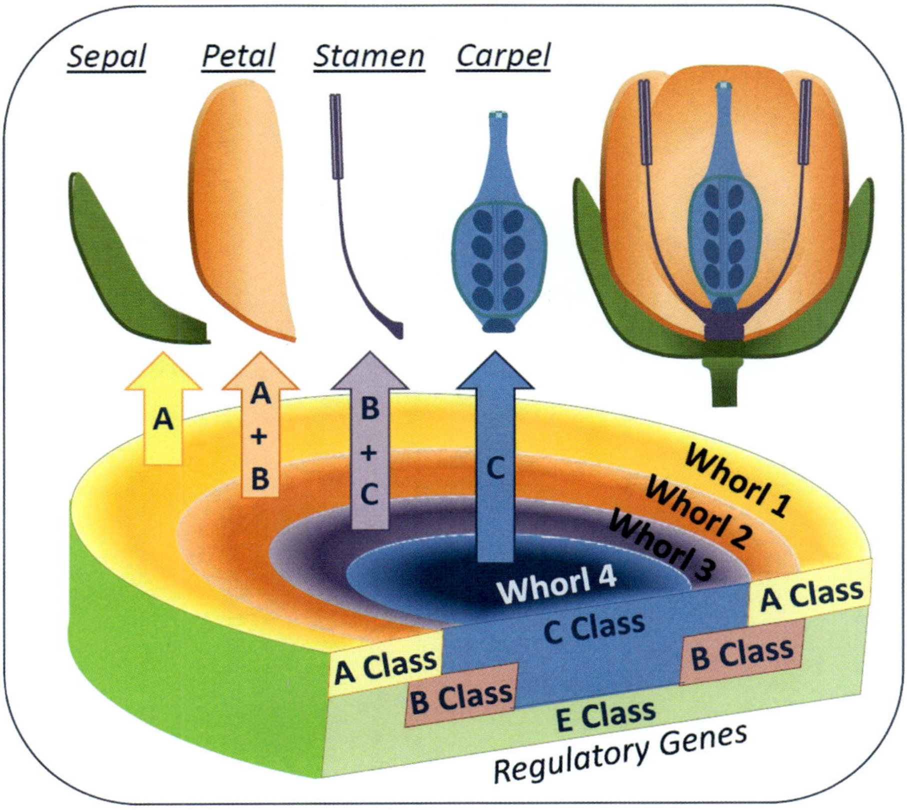

Fig. 6.13.

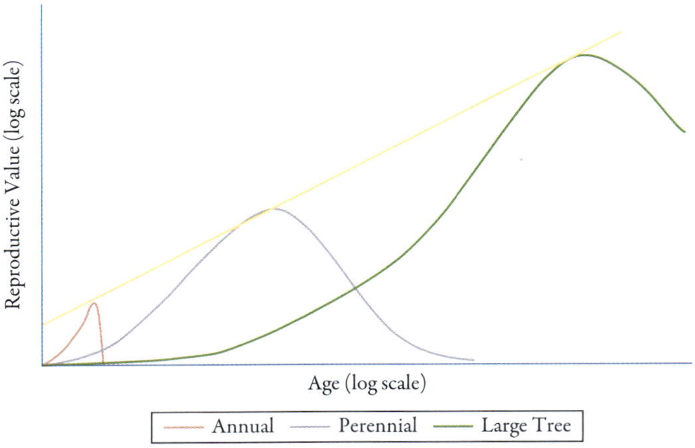

Fig. 6.14.

Fig. 6.18.

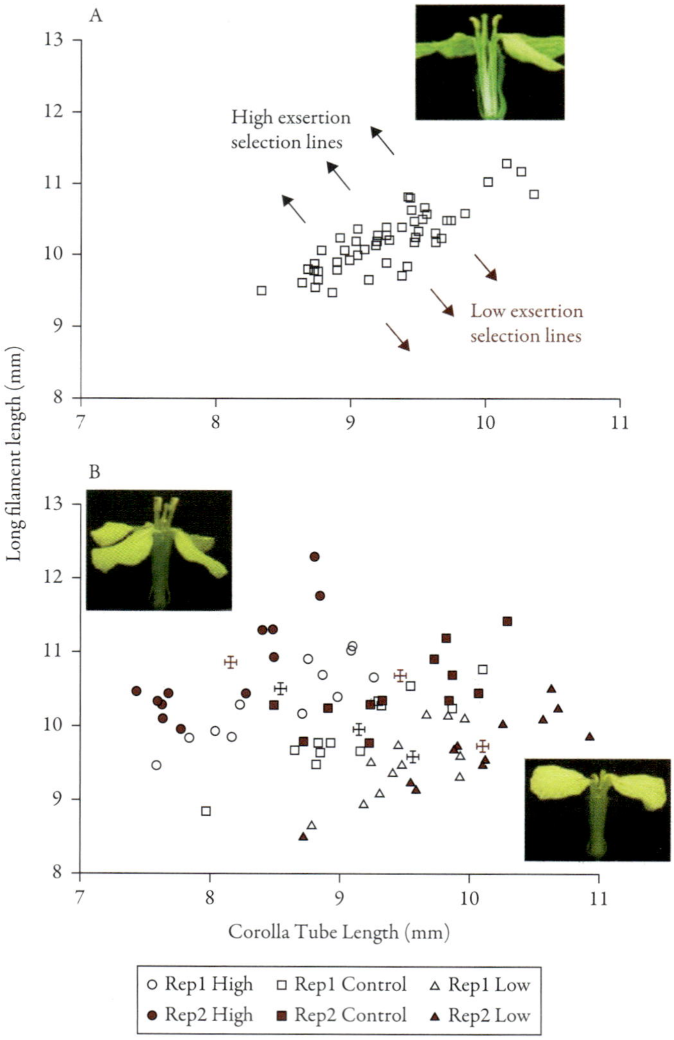

Fig. 6.21.

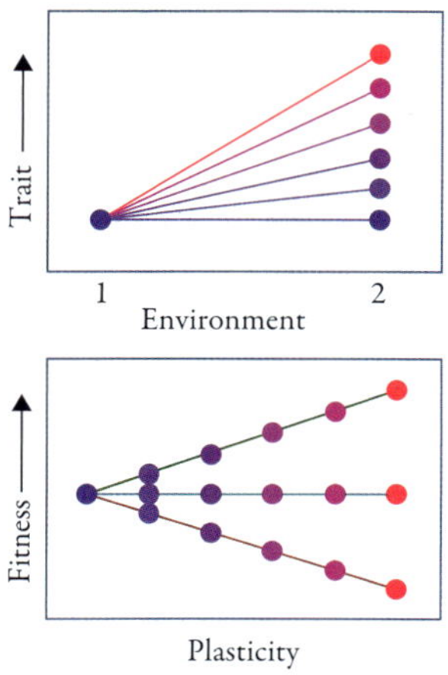

Fig. 6.23.

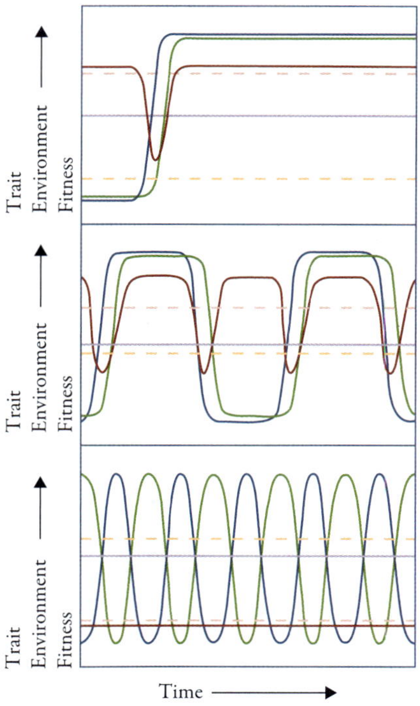

Fig. 6.24.

Fig. 6.25.

Fig. 6.26.

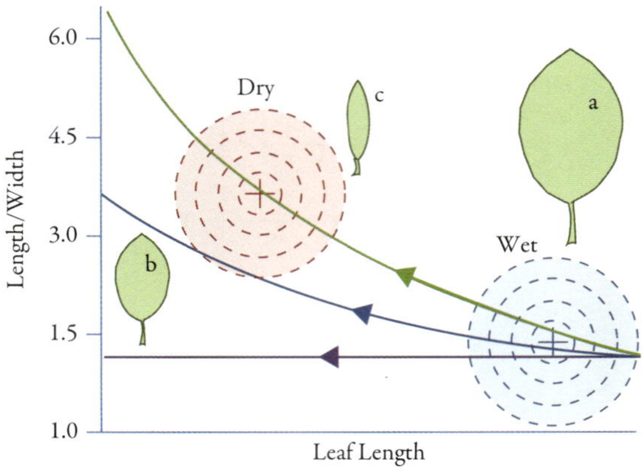

Fig. 6.28.

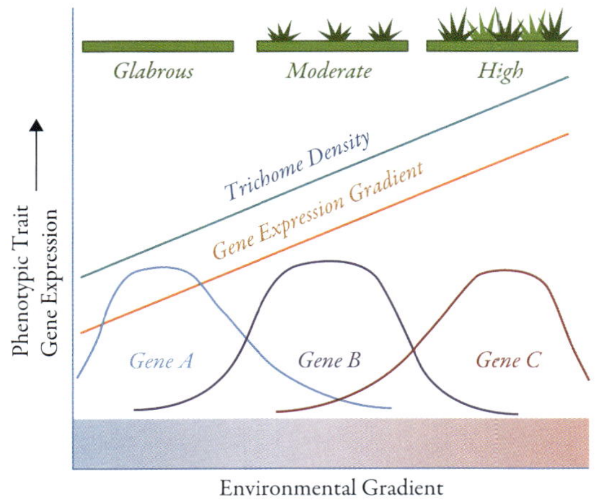

Fig. 6.29.

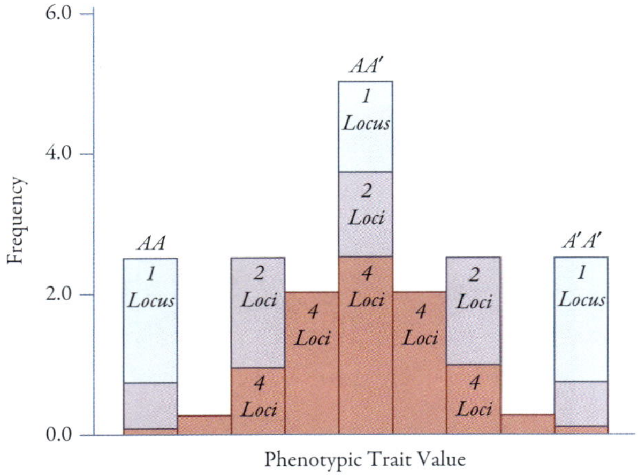

Fig. 7.2.

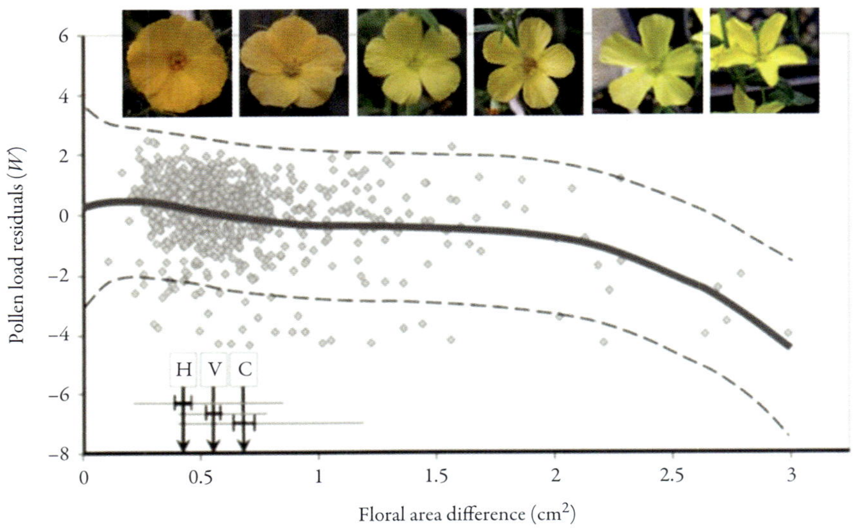

Fig. 7.11.

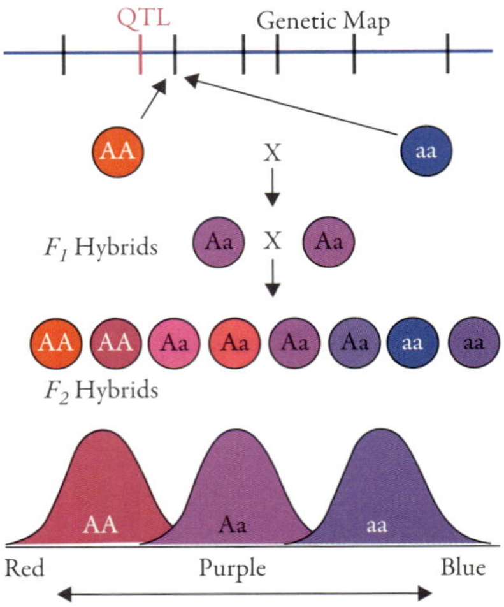

Fig. 7.14.

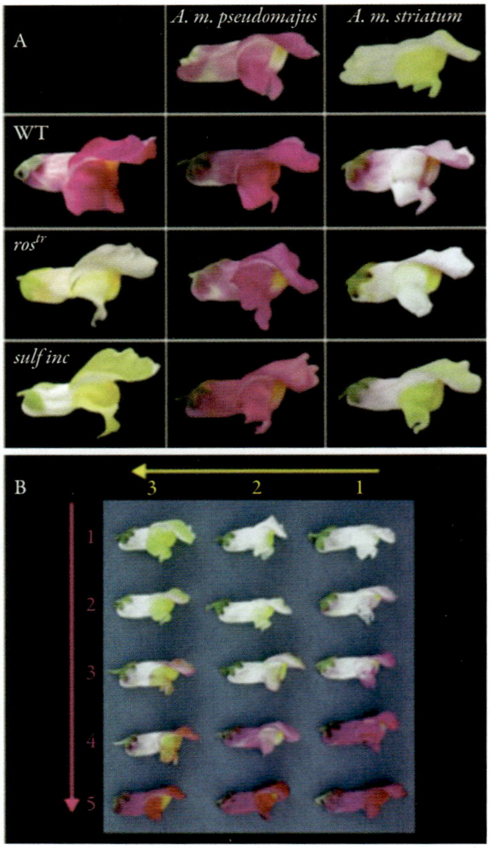

Fig. 7.16.

a
Corolla
(open)
Stigma
(inserted)
Anthers
(inserted)
Nectar
guides
(brushy)
b
Nectar
guides
(flat)
c
Anthers
(exserted)
Corolla
(tubular)
Stigma
(exserted)
d
e
f
g
h
i
j
k
l

Fig. 7.17.

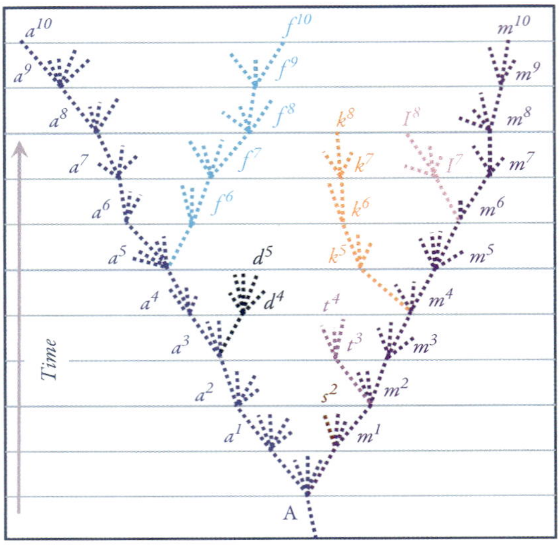
a^{10}
f^{10}
m^{10}
a^9
f^9
m^9
a^8
f^8
k^8
I^8
m^8
a^7
f^7
k^7
I^7
m^7
a^6
f^6
k^6
m^6
a^5
d^5
k^5
m^5
a^4
d^4
t^4
m^4
a^3
t^3
m^3
Time
a^2
s^2
m^2
a^1
m^1
A

Fig. 8.1.

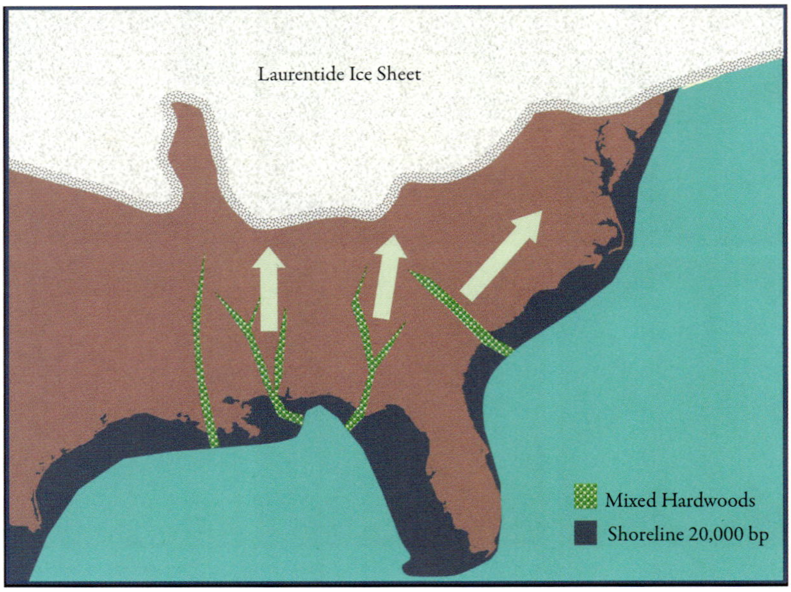
Laurentide Ice Sheet
Mixed Hardwoods
Shoreline 20,000 bp

Fig. 8.2.

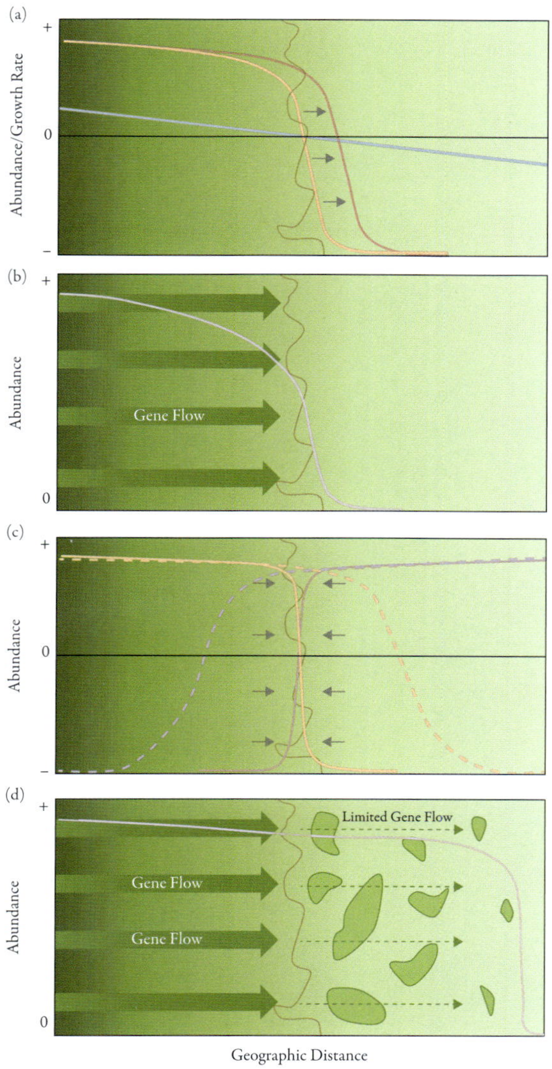
(a)
Abundance/Growth Rate
+
0
–
(b)
Abundance
+
Gene Flow
0
(c)
Abundance
+
0
–
(d)
Abundance
+
Limited Gene Flow
Gene Flow
Gene Flow
0
Geographic Distance

Fig. 8.4.

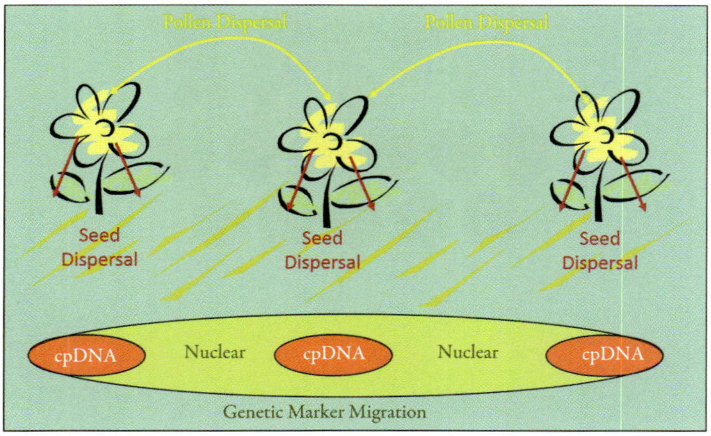
Pollen Dispersal
Pollen Dispersal
Seed Dispersal
Seed Dispersal
Seed Dispersal
cpDNA
Nuclear
cpDNA
Nuclear
cpDNA
Genetic Marker Migration

Fig. 8.5.

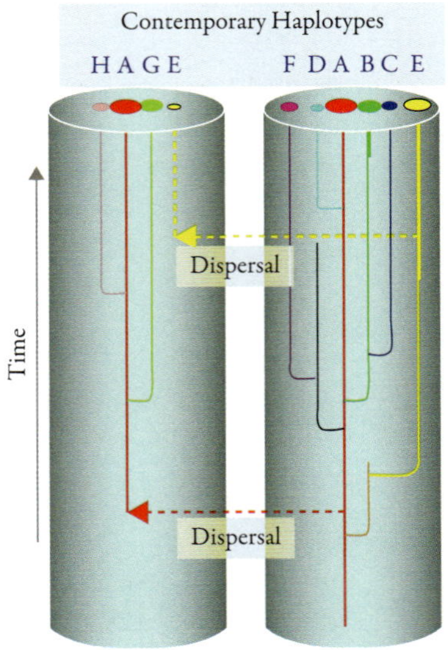

Fig. 8.11.

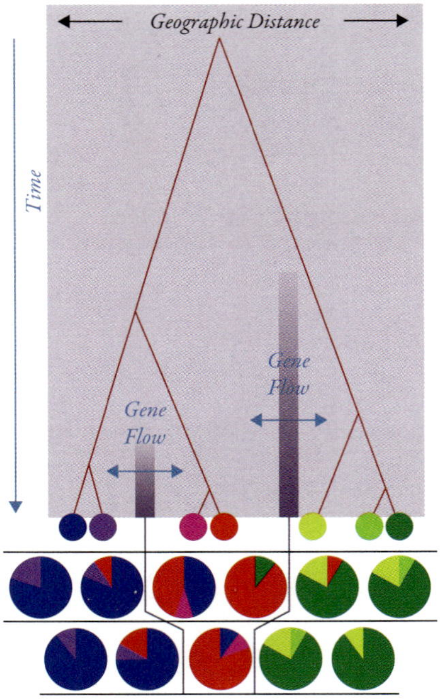

Fig. 8.12.

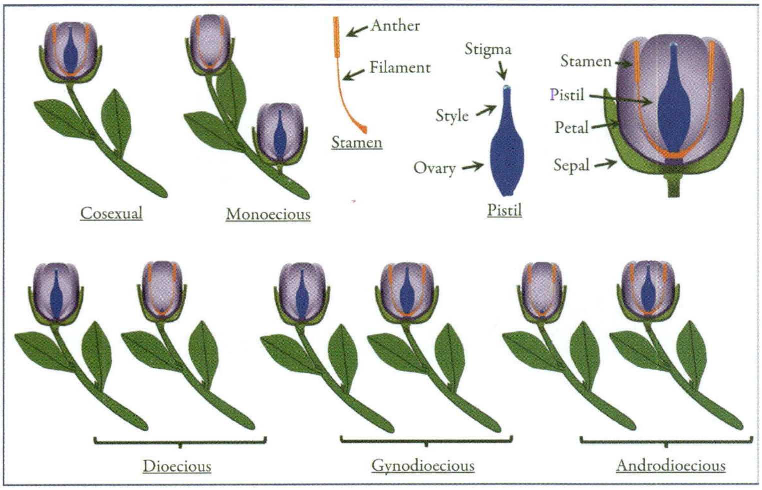

Fig. 9.1.

Fig. 9.2.

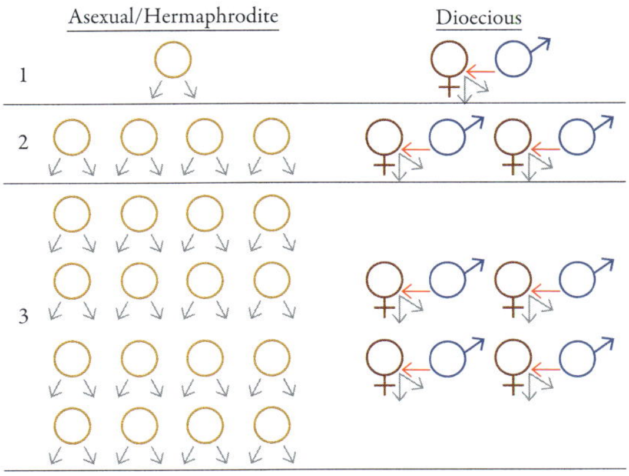

Fig. 9.4.

Fig. 9.8.

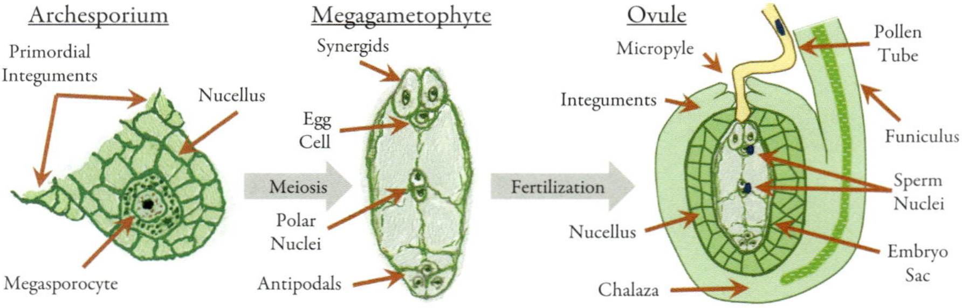

Fig. 9.9.

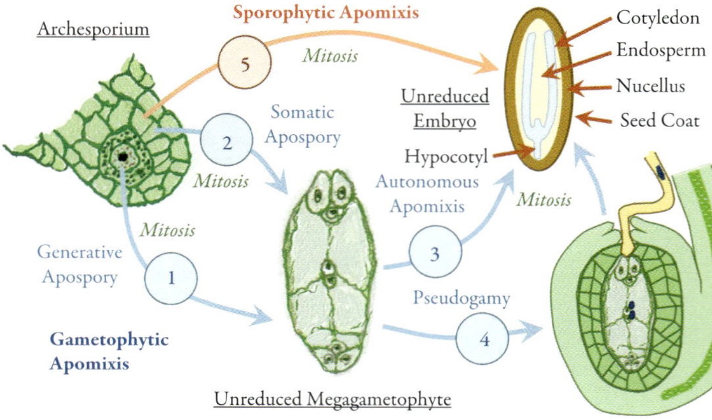

Fig. 9.10.

Fig. 9.11.

Fig. 9.12.

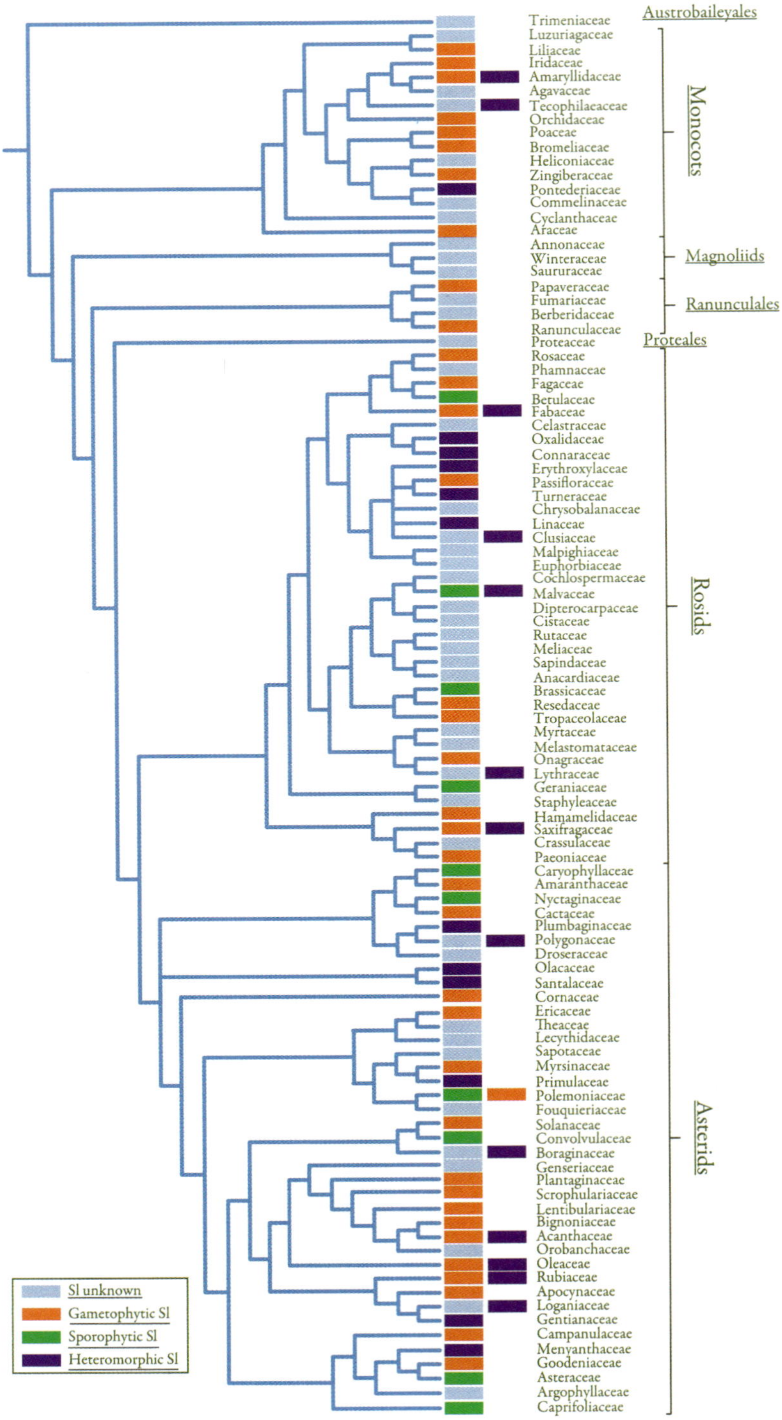
Trimeniaceae
Luzuriagaceae
Liliaceae
Iridaceae
Amaryllidaceae
Agavaceae
Tecophilaeaceae
Orchidaceae
Poaceae
Bromeliaceae
Heliconiaceae
Zingiberaceae
Pontederiaceae
Commelinaceae
Cyclanthaceae
Araceae
Annonaceae
Winteraceae
Sauruaceae
Papaveraceae
Fumariaceae
Berberidaceae
Ranunculaceae
Proteaceae
Rosaceae
Phamnaceae
Fagaceae
Betulaceae
Fabaceae
Celastraceae
Oxalidaceae
Connaraceae
Erythroxylaceae
Passifloraceae
Turneraceae
Chrysobalanaceae
Linaceae
Clusiaceae
Malpighiaceae
Euphorbiaceae
Cochlospermaceae
Malvaceae
Dipterocarpaceae
Cistaceae
Rutaceae
Meliaceae
Sapindaceae
Anacardiaceae
Brassicaceae
Resedaceae
Tropaceolaceae
Myrtaceae
Melastomataceae
Onagraceae
Lythraceae
Geraniaceae
Staphyleaceae
Hamamelidaceae
Saxifragaceae
Crassulaceae
Paeoniaceae
Caryophyllaceae
Amaranthaceae
Nyctaginaceae
Cactaceae
Plumbaginaceae
Polygonaceae
Droseraceae
Olacaceae
Santalaceae
Cornaceae
Ericaceae
Theaceae
Lecythidaceae
Sapotaceae
Myrsinaceae
Primulaceae
Polemoniaceae
Fouquieriaceae
Solanaceae
Convolvulaceae
Boraginaceae
Genseriaceae
Plantaginaceae
Scrophulariaceae
Lentibulariaceae
Bignoniaceae
Acanthaceae
Orobanchaceae
Oleaceae
Rubiaceae
Apocynaceae
Loganiaceae
Gentianaceae
Campanulaceae
Menyanthaceae
Goodeniaceae
Asteraceae
Argophyllaceae
Caprifoliaceae
Austrobaileyales
Monocots
Magnoliids
Ranunculales
Proteales
Rosids
Asterids
Sl unknown
Gametophytic Sl
Sporophytic Sl
Heteromorphic Sl

Fig. 9.14.

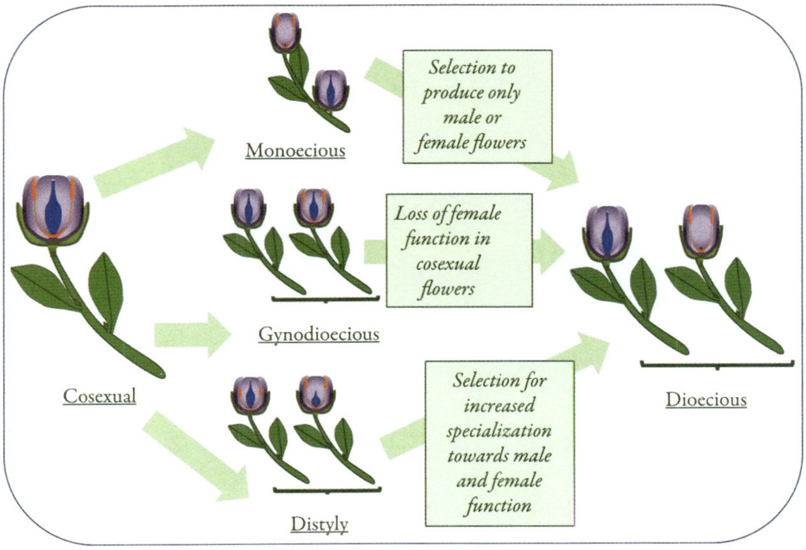

Fig. 9.27.

Fig. 9.28.

Fig. 9.29.

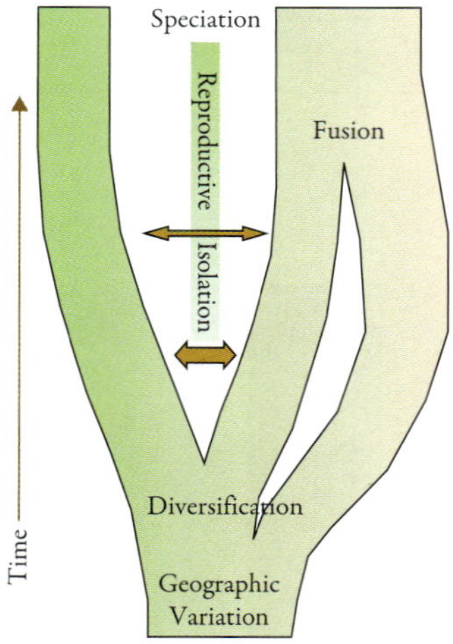

Fig. 10.1.

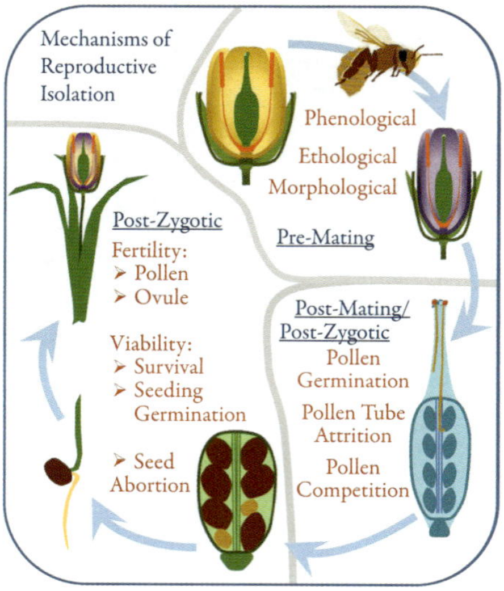

Fig. 10.4.

Fig. 10.6.

Fig. 10.7.

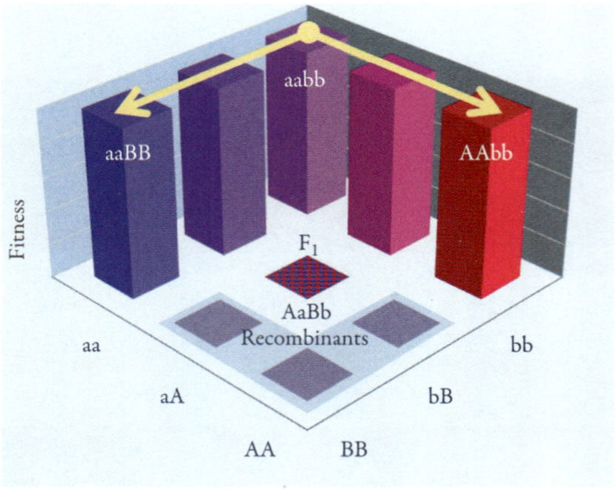

Fig. 10.10.

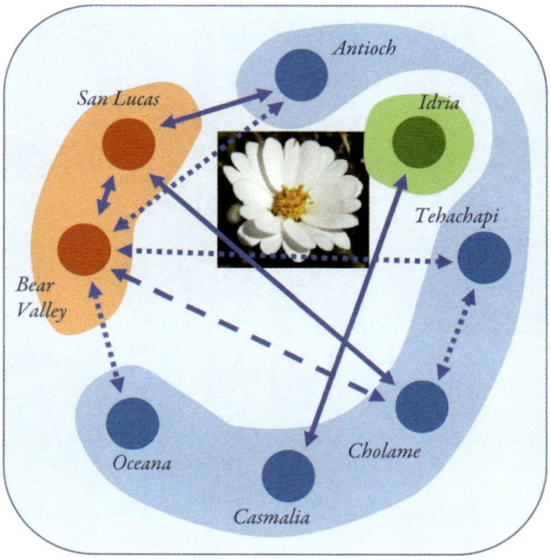

Fig. 10.12.

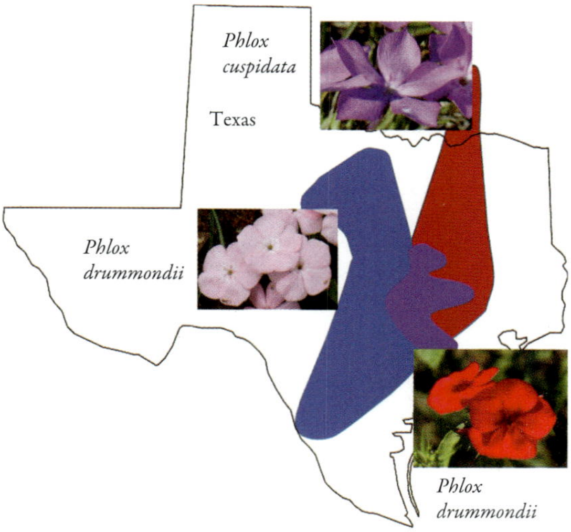

Fig. 10.13.

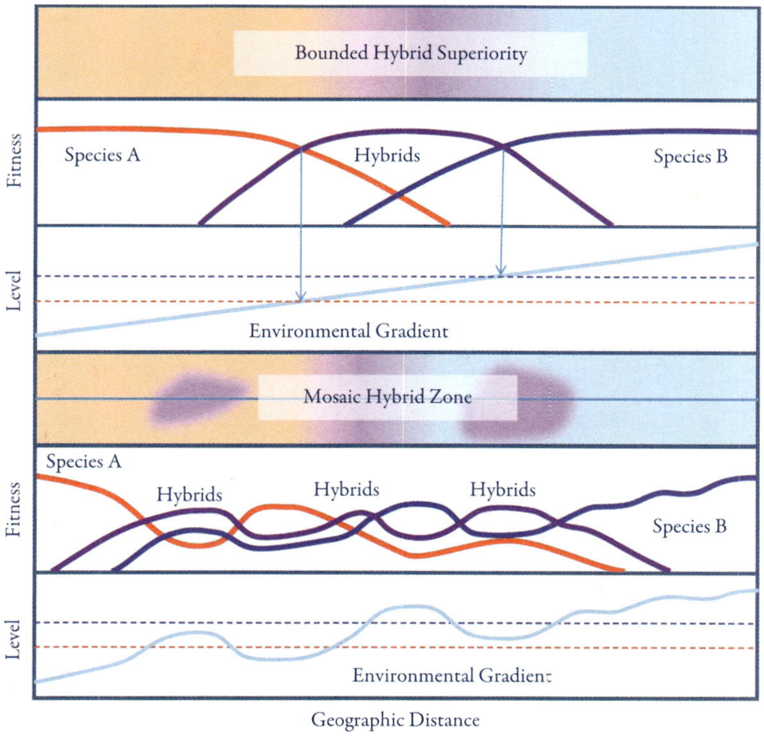

Fig. 10.20.

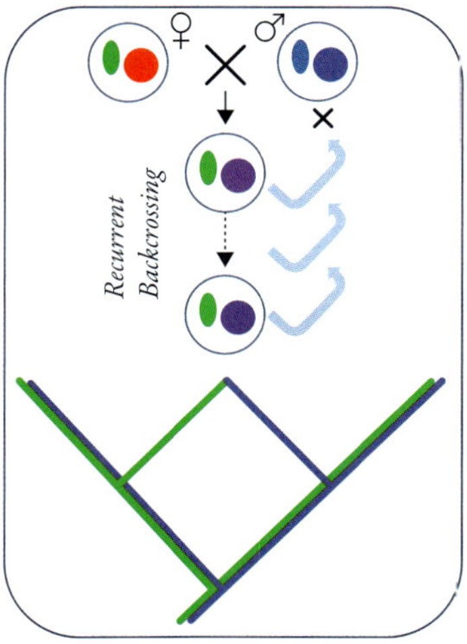

Fig. 10.21.

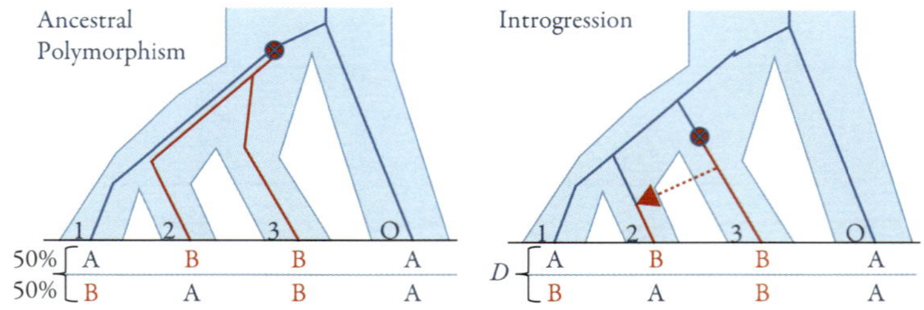

Fig. 10.23.

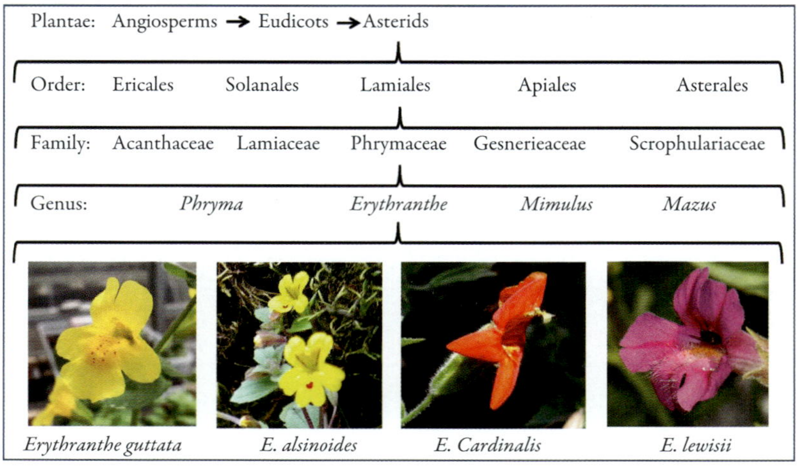

Fig. 11.1.

Fig. 11.2.

Fig. 11.3.

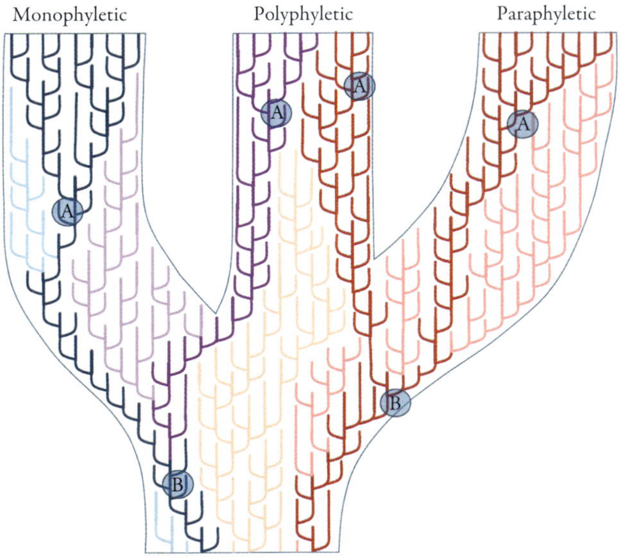

Fig. 11.4.

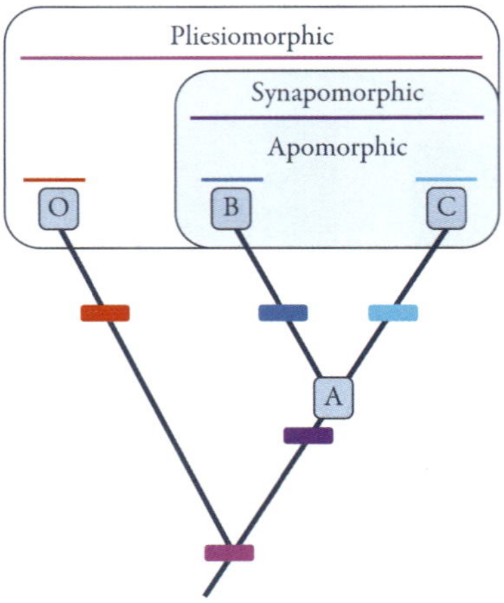

Fig. 11.5.

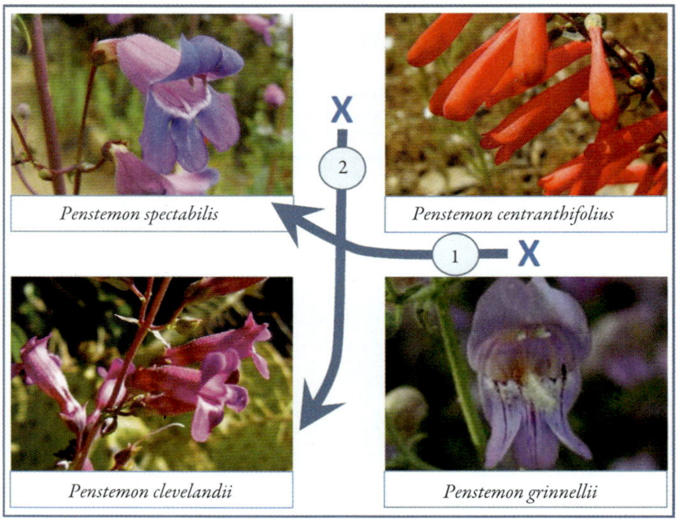

Fig. 11.7.

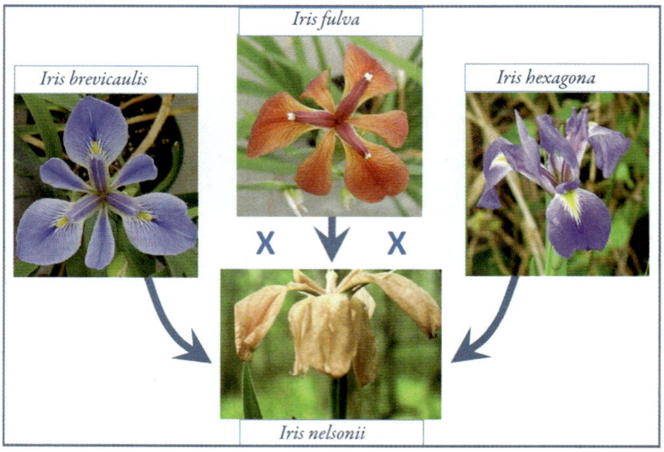

Fig. 11.8.

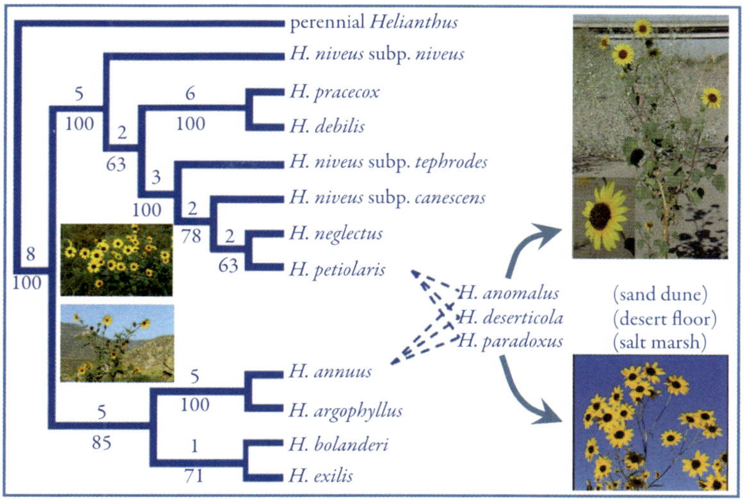

Fig. 11.9.

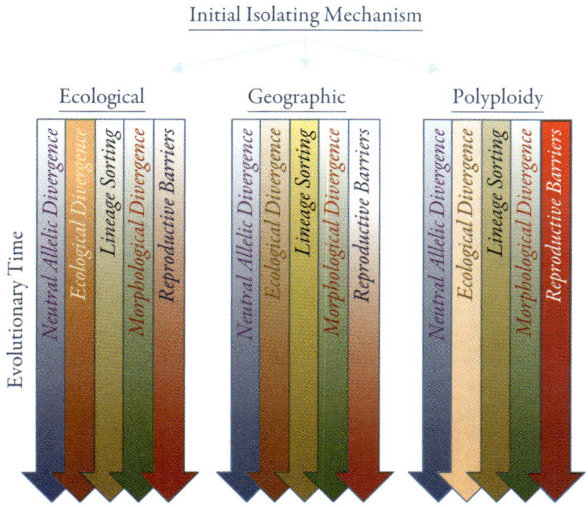

Fig. 11.10.

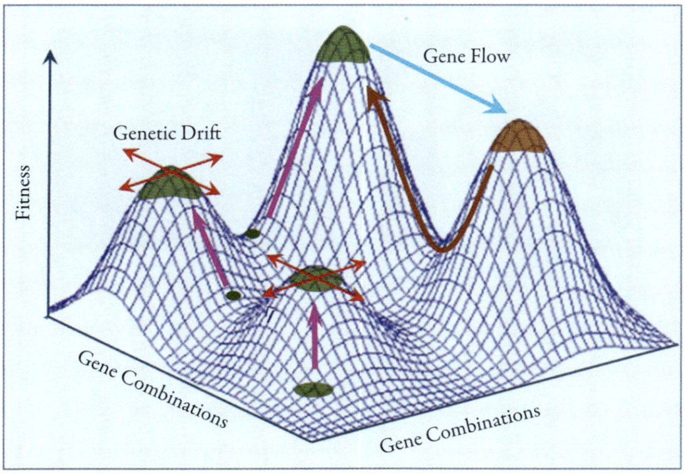

Fig. 11.11.

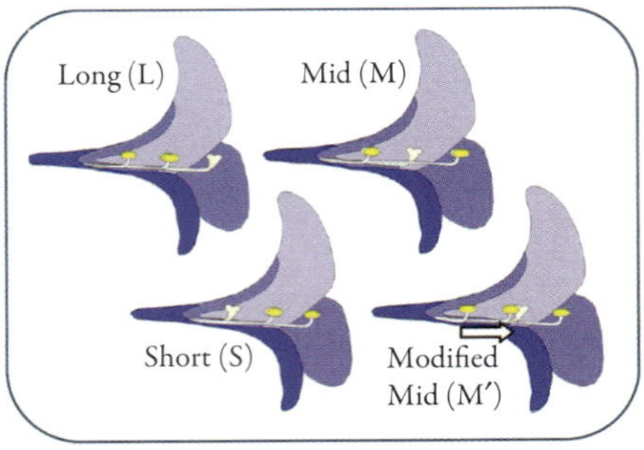

Fig. 11.12.

Fig. 11.17.

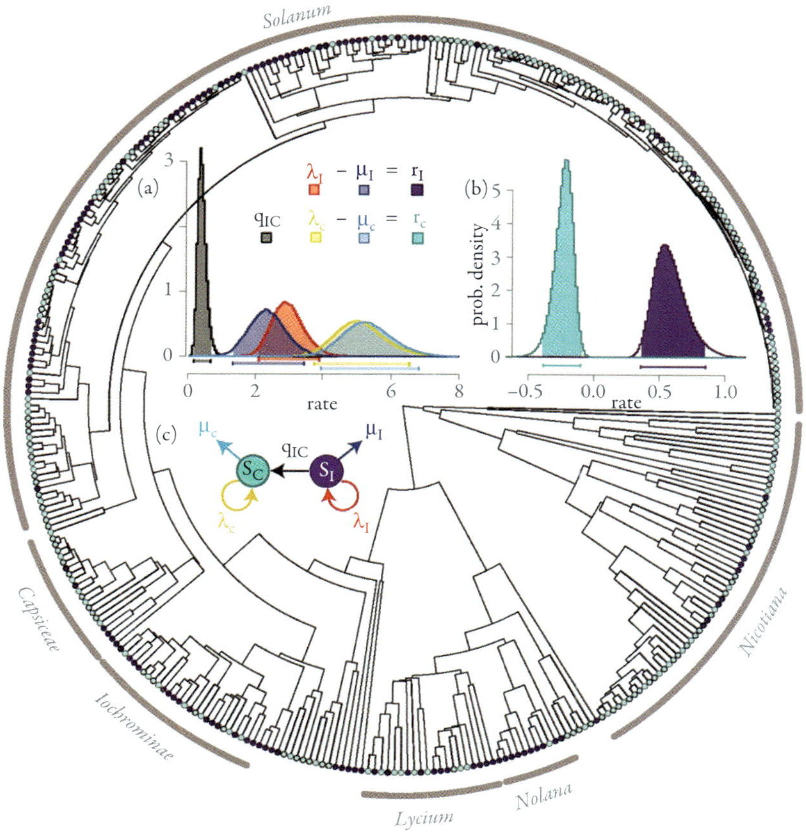

Fig. 11.18.

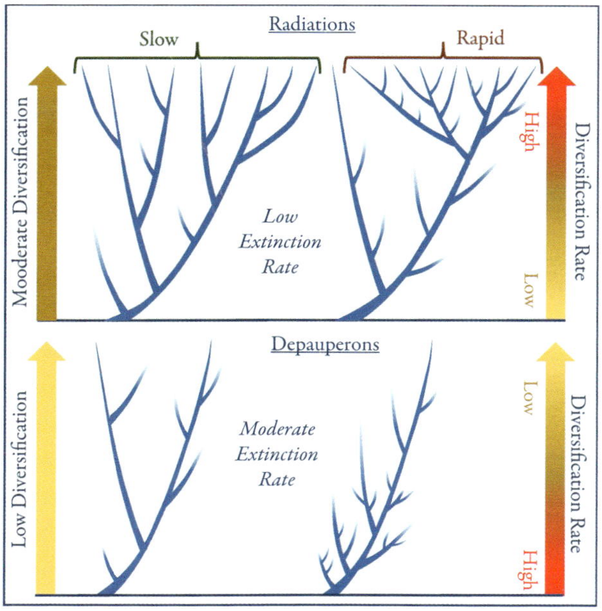

Fig. 11.20.

Fig. 11.21.

Fig. 11.22.

PART III

PATTERNS OF DIVERSITY

6

PHENOTYPIC VARIATION

IF A BUD be torn from the branch of a tree and cut out and planted in the earth . . . . it will grow, and become a plant in every respect like its parent. This evinces that every bud of a tree is an individual vegetable being; and that a tree therefore is a family or swarm of individual plants . . . . [and] the shoot is a succession of individual vegetable members.

Erasmus Darwin (1800) in Phytologia, page 2.

6.1 PHENOTYPIC VARIATION

The genetic information that has accumulated over the last several billion years has become organized into congregations of genes that work in concert to generate a myriad of phenotypes. The coding sequences of genes determine protein structure and function, but phenotypes are as much an outcome of the patterns of gene expression as they are of gene content. The correspondence between the genetic code and the phenotype is a complex network of interactions and collaborations among functional genetic elements. The resulting phenotypes are a product of variation in gene sequences and the tempo and mode of their expression; functional genomic regions can be viewed as players in a performance (phenotype) that is determined by the choreography that genes follow during growth and development. The importance of patterns of gene participation for variation in phenotypes is evident from the observation that much of the functional genetic code occurs in intergenic regions that act to regulate transcription and post-transcription patterns of gene expression. Understanding this complex foundation of the phenotype and the genetic basis of adaptation represents a major challenge for plant evolutionary biologists.

The form and function of plants are primarily governed by the complements of genes inherited from parents, but phenotypes may also be influenced by the prevailing environmental conditions. Organisms may display *Phenotypic Plasticity*—the expression of different phenotypes by a single genotype—that allows them to cope with changes in their environment by adjusting their physiology and morphology as they grow. This is particularly important for plants compared to many other eukaryotes, as they are sedentary and may be subjected to environments that fluctuate over a range of spatial and temporal scales. Abiotic factors such as light intensity, temperature, and moisture availability are important for plant growth and may vary within a single day or across seasons. The ability of plants to make appropriate responses to changes in environmental conditions is critical for the acquisition of resources and consequently, for growth, survival, and reproductive success.

The growth form of most animals is *Determinate* and *Unimodal* so there is little potential for morphological modifications once an individual has reached maturity. Animals compensate for their inflexible morphologies through physiological adjustments (*Acclimation*) and behavioral responses to their environments. As noted in *Phytologia* by Erasmus Darwin, the first comprehensive text on botany (Darwin 1800), most plants have *Indeterminate* growth and consist of repeated units (i.e., they are *Multimodal* and consist of *Metamers*; White 1979). Metamer is a general term that refers to repeated units that are fundamentally similar in structure; it can be applied to describe the structure of both plants and animals. The term *Module* is sometimes used synonymously with metamer (Harper 1977), but this term more commonly refers to groups of traits or genes (modules) that contribute to the same function (see the section below on plant growth and development).

Plant phenotypes are much more flexible than animals because plants can adjust their morphology in response to the environment as each metamer develops. Plants have the advantage of being more plastic in their morphology than animals and retaining the ability to display morphological responses to fluctuating environmental conditions throughout their life. The capacity for physiological and developmental responses to environmental conditions extends ecological tolerances, which has consequences for geographic ranges and responses to selection as novel conditions are encountered.

Phenotypic responses to environmental fluctuations may occur at timescales ranging from less than a minute to many hours, days, or years and may extend across multiple generations (Fig. 6.1; refer to color plate). Plants are most likely to respond to environments that change within the life of an individual through phenotypic plasticity, which can be subdivided into reversible physiological adjustments and irreversible morphological changes that occur during the development of new metamers. Established metamers may make reversible plasticity responses that include short-term physiological adjustments or longer-term acclimation to prevailing conditions. Rapid physiological changes may occur in response to a number of environmental variables including light intensity, temperature, water potential, and CO_2 concentration. Such changes can be detected as changes in stomatal conductance and photosynthetic rates that may occur within less than a minute (Nobel 2005). If an environmental challenge persists for many hours or days, then plants may make physiological changes that are less labile. For example, under sustained drought plants will increase the solute concentrations of leaf cells to improve their ability to extract moisture from the soil. These responses may also be accompanied by the production of heat-shock proteins (*Stress Proteins*) that serve to stabilize other proteins associated with membrane integrity and metabolic pathways (Aroca 2012). Such physiological adjustments are critical for optimizing growth responses to environmental changes, which

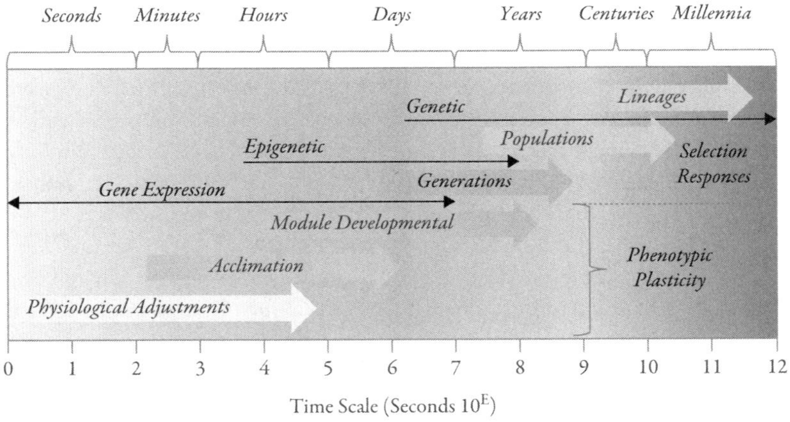

Fig. 6.1. The time scale of phenotypic and genetic responses to environmental challenges. Short-term changes in patterns of gene expression affect physiological responses that may persist for less than a minute or number of days. Acclimation can occur over several days in response to seasonal changes. Epigenetic changes affect gene expression during growth and across generations. Longer-term responses generally occur as selection affects the genetic composition of populations and lineages.

maximizes the contributions of individual metamers to plant fitness.

Short- and medium-term physiological responses and acclimation are reversible and may allow individual leaves to maintain positive contributions to plant fitness across a range of environmental conditions. Permanent modifications to anatomy and morphology may occur during development. If environmental conditions persist for several days, then newly developing leaves may undergo irreversible responses resulting in heterogeneous morphologies among leaves within an individual. For example, individual metamers typically respond to local light conditions by producing larger and thinner leaves when they are shaded by foliage. Similarly, seasonal changes in response to increasing drought stress are also possible and may include reductions in leaf size as well as increases in cell wall rigidity, trichome density, and leaf and leaf cuticle thickness (Nobel 2005; Aroca 2012). This flexibility in the expression of phenotypic traits allows plants to make plasticity responses as they grow. As each metamer (including a leaf) develops in response to local microenvironments and to changes in prevailing conditions, the individual may come to consist of a diverse complement of metamers.

More permanent acclimation responses that affect the whole plant may occur with seasonal changes. For example, in response to cooler temperatures in the fall, perennial plants will undergo "cold-hardening" that includes translocation of carbohydrates to roots, the production of "anti-freeze" proteins, and increases in cell membrane permeability, all of which render plant tissue less susceptible to frost damage. These more persistent seasonal modifications in response to changes in temperature or water availability may involve epigenetic modifications (Chapter 3 including changes in methylation patterns that may be reversed later in development or may be transferred to the next generation (Becker and Weigel 2012; Sahu et al. 2013).

If challenging environmental conditions persist for many days, or if there are shorter periods of environmental extremes, then less-tolerant individuals—those that are not able to make the appropriate plasticity responses—may succumb and will be replaced by more tolerant individuals. Thus, selection pressure on traits is conferred by the environment and may precipitate evolutionary responses if traits are heritable. Persistent environmental pressures may result in longer timescale and directional changes in the genetic composition of populations and eventually of entire lineages (Fig. 6.1; refer to color plate). These genetic and transgenerational epigenetic changes in populations are the foundation of adaptive evolution—a major challenge for plant evolutionary biologists is to obtain a more complete characterization of the separate and interactive roles of plasticity and genetic modifications for adaptive responses to challenging environmental conditions.

While we expect that changes in phenotypic traits in response to selection will be determined by modifications of the genetic code, mapping the connections between phenotype and genotype has proven to be a difficult task. As discussed in the previous chapter, there are several examples where evidence of past selection is found at the genome level. We also have an abundance of cases where responses to selection on phenotypic traits have been demonstrated. While there are many lines of model and crop plants possessing well-annotated mutations that are maintained in germplasm collections and laboratories, these are not typically found in the wild, so they have limited utility for understanding processes of adaptation. Except for cases of mutations of large effect, the numbers of examples where we can identify the specific changes that result in natural variation in phenotypic traits are very few. Nonetheless, progress is being made in this area as more studies incorporate high-density genome scans with studies of phenotypic variation across natural environments.

In this chapter, we will focus on the phenotype: its genetic and environmental determination, the genomic distribution of genetic variation affecting trait expression, and the evaluation of the effects of on quantitative traits that comprise the phenotype. The phenotype is the primary interface for selection, because it is the portion of an organism that interacts with biotic and abiotic factors to determine fitness. The idea that phenotype is the boundary between the genotype and its environment was first proposed by Ford (1971), who is the founder of the field of ecological genetics. An assessment of the factors affecting phenotypic trait expression—both environmental and genetic—is critical to our understanding of the process of adaptive evolution. At the same time, we appreciate that traits are often controlled by underlying allelic variation, so all of the population genetic processes discussed in the last three chapters, including mutation, gene flow, genetic drift, and selection, are potentially important determinants of variation in phenotypic traits.

With this balanced view of phenotypic trait variation in mind, we will begin our discussion of phenotypic variation by reviewing the development of plants and their basic anatomy and morphology to establish a foundation of understanding. We conclude this chapter with a discussion of phenotypic plasticity, including variation in gene expression (*Transcriptomics*). We will reserve for the next chapter discussions of methods to evaluate the strength and form of selection acting on phenotypic traits (*Quantitative Genetics*) and examples of adaptive trait variation in natural populations.

6.2 ORIGINS OF PLANT GROWTH MODES

Our understanding of the major phenotypic features of modern plants may be improved by examining their development in the context of the origin and emergence of the first terrestrial embryophytes. One of the most outstanding features common across land plants is their indeterminate development: as plants grow, they are constantly rejuvenated through the turnover of living tissues. Continuous apical growth is a major feature that distinguishes plants from animals, which are unimodal and discontinue development at maturity. Apical growth was a critical *Preadaptation* that may have been inherited from aquatic ancestors that enabled early embryophytes to invade and dominate terrestrial habitats (Chapter 2). Aquatic habitats differ from terrestrial ones because resource availability is not spatially constrained; aquatic algae are continuously bathed in a relatively benign environment where inorganic nutrients can be replenished through the flow of currents. During the time leading up to the origin of terrestrial plants, much of the algal diversity in freshwater systems was planktonic, but algal macrophytes may have been favored in shallow habitats where anchorage and apical growth would have ensured good exposure to light (e.g., perhaps similar to the modern charophytes). As mentioned in Chapter 2, the primary selective pressure for the invasion of land was probably access to the meager inorganic nutrients present in aquatic habitats. Apical growth would have facilitated access to nutrients, as these algae became emergent and extended their thalli across the soil surfaces in quests for inorganic nutrients. The moist banks surrounding ponds and streams may have provided generous amounts of nitrogen, phosphorous, and other resources necessary for plant growth.

While terrestrial habitats were relatively rich in resources necessary for plant life, they presented significant challenges to would-be invaders. Plants emerging above the water were exposed to extreme dryness and the effects of gravity, so the earliest invaders probably remained closely appressed to moist soil to maintain their water balance. Emergent environments also differ from aquatic ones, because they are effectively two-dimensional: individual plants may monopolize a portion of the surface and the nutrients contained in the underlying soil. Successful exploitation of resources on land required an efficient method of spreading the plant body over and through the soil. The presence of persistent *Meristems* that enabled directional growth by the continuous addition of plant tissue proved to be a *Key Adaptation* that facilitated invasion of terrestrial habitats.

Submerged and emergent plants do not require extensive root systems, so it is not surprising that the growth of the earliest land plants was *Monopolar* (growing from one end of a single axis), as can be seen in some of the more common early land plants such as *Cooksonia* and the zosterophylls (Figs. 2.3 and 2.9). Monopolar growth enabled individuals to migrate across the soil surface as new tissue was constantly added to the apical meristem. This mode of growth can be highly successful as is evident from the abundance and diversity of modern bryophytes and lycophytes, which are monopolar and lack true root systems. In addition, many angiosperms growing from *Rhizomes*, such irises, grasses, and many herbaceous dicots, are effectively monopolar when mature, although they begin life as *Bipolar* seedlings (growing from both ends of the axis—stem and root). During monopolar growth, the older portions of the plant body are discarded as new tissue is added at the growing tips of stems. Consequently, the collection of living cells constituting an individual is constantly in flux as new tissue is added by meristems and the older portions die off. This turnover of living cells within individual plants occurs because of *Programed Cell Death* (*PCD; Apoptosis*) and catabolic cell degradation (*Autophagy*) of tissues and whole organs and is a recurring theme in plants that distinguishes them from most animals.

Bipolar growth was probably favored for the establishment of well-developed root systems, which were necessary to support the above-ground portions of larger and taller plants that were exposed to the effects of gravity and wind. Superior height became advantageous as the moister areas capable of supporting early plant life became crowded, and competition for light and nutrients became more intense. Extensive root systems proved advantageous, as they provided both support for the aerial portion of the plant body as well as access to a larger volume of soil for the extraction of water and nutrients. The root systems of many modern bipolar plants can be massive—sometimes exceeding the biomass of the upper portions of the plant. In monopolar species, roots are *Adventitious* (arising from stem tissue) and are numerous and small in diameter in comparison to the rhizomes and stems from which they arise. Bipolar growth may also have been a necessary innovation for colonization of more challenging habitats—the establishment of large root systems early in development may have provided access to soil water that was necessary for survival in regions with intermittent precipitation. Although there are many notable exceptions, the majority of gymnosperms and angiosperms are characterized by bipolar growth—with well-developed stem and root systems—throughout their lives.

Early land plants such as *Cooksonia, Rhynia* and the zosterophylls were not much more than green stems that served both as a surface for photosynthesis and as a support system for reproductive organs (Figs. 2.3 and 2.9). While small *Enations* and microphylls were present in some species, the development of more specialized organs for the capture of light energy did not appear until the late Devonian—more than fifty million years after the appearance of vascular plants. The evolution of leaves with larger surface areas (megaphylls) may have been constrained by the lack of adequate vasculature and the absence of strong selective pressure (Beerling 2005). Since the first appearance of land plants in the early Paleozoic, the atmospheric concentration of CO_2 had been very high, but it began to decline sharply in the late Devonian until it approached contemporary levels around the end of the Carboniferous. Such a high level of CO_2 in the atmosphere would have enabled higher rates of carbon fixation even with relatively small surface areas and low densities of stomata for gas exchange. We can see examples of low leaf surface areas in dominant species of the mid Devonian such as scale trees (*Lepidodendron*; Fig. 2.11) and giant horsetails (*Calamites*; Chapter 2). It was not until the last third of the Devonian that we see the appearance megaphylls in early trees such as the *Wattieza* and the progymnosperms (e.g., *Archaeopteris*; Fig. 2.12); species that possessed woody trunks with large areas of secondary xylem for the efficient transport of water. The evolution of large leaves was only possible once the transpirational demand of higher numbers of stomata could be supported by more efficient vascular systems. The fact that the stomatal density suddenly began to increase in the late Devonian—and at a much faster pace than the decline in CO_2—suggests that the evolution of large leaves was constrained more by the lack of adequate vasculature than by high CO_2 concentrations. Hence, the decline in CO_2 levels from the late Devonian and leading up to the Carboniferous rainforest collapse was probably a consequence of rapid increases in the accumulation of plant biomass, facilitated by the appearance of more efficient vascular systems that enabled the production of megaphylls.

In the early Carboniferous, we see the origin of the seed fern lineages that ultimately led to the evolution of modern gymnosperms and angiosperms (Chapter 2). By this time, the basic architecture, morphology, and anatomy of the vegetative portions of seed plants had been established. In the next section, we will focus primarily on the flowering plants, as we briefly review their development, morphology, and anatomy. This overview of the fundamentals of morphology will provide a foundation for our understanding of plant phenotypic variation.

6.3 PLANT DEVELOPMENT

Seed Germination—Following dispersal, seeds may germinate if they are exposed to the appropriate environmental conditions and if any physical or endogenous barriers that enforce dormancy are overcome. Germination usually proceeds with the emergence of the *Radicle* (embryonic root), which quickly develops root hairs to anchor the seedling in place. In many large-seeded species such as oaks, the *Hypocotyl* (radicle and shoot extending below the *Cotyledons*) will cease to grow, so the seed will remain at or beneath the soil surface as the *Epicotyl* (shoot above the cotyledons) emerges to form the embryonic shoot in *Hypogeal Germination* (Fig. 6.2). In hypogeal species, the seed remains in place until its resources are depleted and it eventually decays as the plant matures. In plants with *Epigeal Germination,* the hypocotyl continues to extend upward forming a hook that pulls the seed out of the soil. As the hypocotyl straightens, it sheds the seed coat to expose the two cotyledons, which spread open and expand into seed leaves. The seed leaves are generally photosynthetic and may persist for some time before the epicotyl extends from the apical meristem to form the stem and the first true leaves (Fig. 6.2). Germination in monocots may be hypogeal or epigeal, but it differs from dicots in that seeds produce a protective sheath (*Coleoptile*) that emerges from the seed prior to the epicotyl. The epicotyl extends through and then above the coleoptile to unfurl the first leaves. Similarly, the radicle in monocots is enveloped in a *Coleorhiza* when it first emerges from the seed coat.

Stem Development—Seedlings continue to grow from their shoot apical meristems by adding new metamers (internodes and nodes with leaves and axillary buds; Fig. 6.3; refer to color plate). In many herbaceous plants, and in the early growth stages of woody species, the addition of new leaves is accompanied by stem growth through the elongation of internode regions. Internodes continue to extend as new metamers are produced, so the plant stem will grow upward or, as with rhizomes or stolons, horizontally, just above or beneath the soil surface. In plants with aerial stems, the *Cauline* (stem) leaves typically develop at each node and are characteristically alternate (one leaf per node), opposite (two leaves per node), or more rarely whorled (multiple leaves at each node). Leaf arrangements alternate between sides of the stem in species with opposite leaves (e.g., plants in the mint family, Lamiaceae), or are spirally arranged in a Fibonacci series in plants with alternate leaves, to produce a balanced distribution along the stem. As stems grow taller, older leaves may abscise as they become shaded, in response to seasonal climate changes in fall-deciduous species, or due to leaf senescence in evergreen species, or leaves may be shed as conditions become more stressful due to limited water availability or high temperatures.

In some herbaceous annuals and perennials such as *Arabidopsis, Plantago* (plantain), *Primula* (primrose), *Viola* (violets), *Fragaria* (strawberry), *Sedum* (stonecrops), and many others, internode tissue is

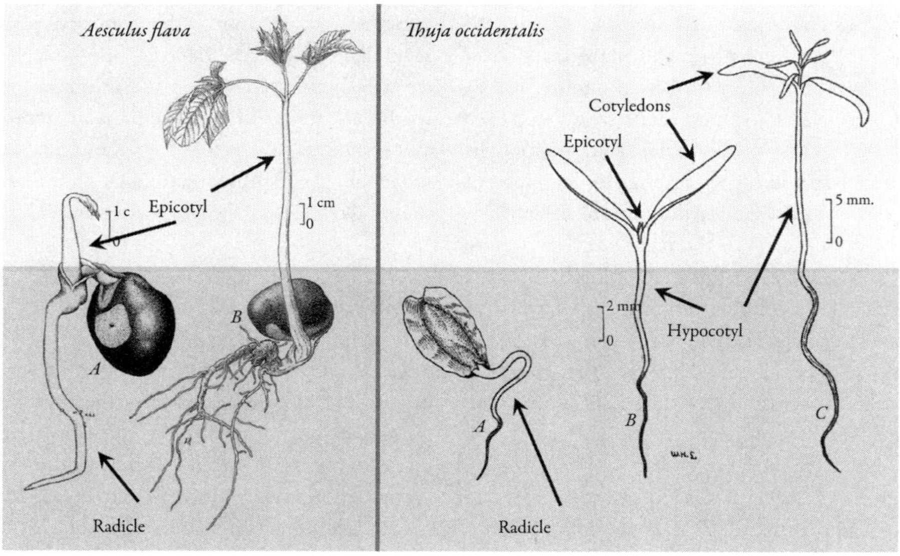

Fig. 6.2. Epigeal (A) and hypogeal (B) and seedling germination. In epigeal seedlings the cotyledons are often pigmented with chlorophyll and serve as seed leaves. In hypgeal species the cotyledons remain below ground or at the soil surface and provide stored carbohydrates for seedling establishment.

Fig. 6.3. Plant stem morphology showing two metamers with their corresponding nodes, internodes, and leaves. Photo by Knulclunk.

minimal so the leaves are closely stacked together to form flattened *Rosettes* or *Acaulescent* (stemless) clusters of leaves that remain close to the ground (Fig. 6.4; refer to color plate). Rosettes may persist for most of the season (annuals) or for several years (biennials and perennials) before the apical meristem elongates to produce a flowering stem (*Bolting*). Rosette plants may die after flowering (annuals and biennials). Perennial rosettes and acaulescent plants may continue to produce flowering stems from axillary buds over many years.

Herbaceous Stem Anatomy—The herbaceous plant stem consists of relatively few tissues and organs (Fig. 6.5; refer to color plate). Stems grow from their tips by the addition of cells that differentiate into tissues in the region just below the meristems. Cells divide in central zones to produce foundation cells that continue to divide and to differentiate into separate tissues depending largely on their spatial position (Box 3.3). Epidermal cells become tightly interlocked and exude waxy cuticles on the cell surface. They may also develop other indumentum characteristics such as trichomes and glaucous waxy coatings. Beneath the epidermis, cells develop into ground and vascular tissue. Ground tissue is distributed throughout the stem and in dicots is divided into a cortical region beneath the epidermis and a pith region that occurs

Fig. 6.4. Basal rosettes of *Dodecatheon pulchellum* var. *cusickii* (Cusick's pretty shooting star) with leafless flowering stems.

within the ring of vascular bundles. Much of the ground tissue consists of elongated parenchyma and collenchyma cells, the latter of which have thickened cell walls and serve to provide structural support for the stem. Both of these cell types retain cytoplasm at maturity, and parenchyma cells near the stem surface may possess high concentrations of chloroplasts. Other cells in the stem undergo apoptosis, so they lose their cytoplasm and are not metabolically active at maturity (Beers 1997). Sclerenchyma consists of long narrow cells (sclereids and fibers) that are nearly filled with lignin deposits and lack cytoplasm. Sclereids and fiber bundles may be distributed in the ground tissue but are most commonly associated with vascular bundles and are important for increasing stem strength. In dicots, there are characteristically four or five vascular bundles distributed in a ring, while vascular bundles are smaller and evenly distributed throughout the stem in many monocots. Vascular bundles include xylem tissue consisting of tracheids

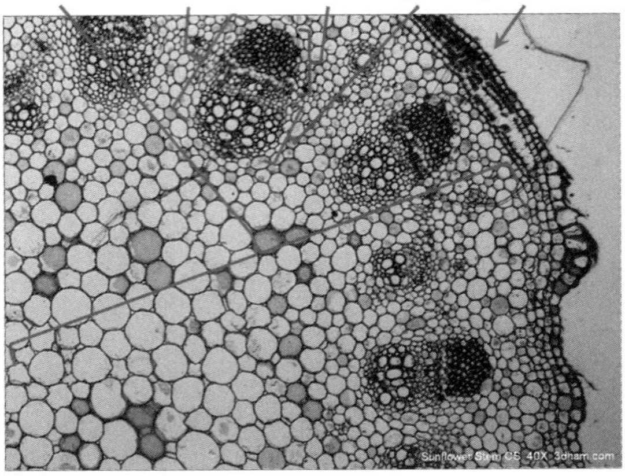

Fig. 6.5. Cross section showing the internal anatomy of a typical dicot stem (*Helianthus annuus*; sunflower). Photo by John Alan Elson.

and vessel elements and phloem that consists of sieve elements and associated companion cells. Xylem cells undergo apoptosis and so are void of cytoplasm at maturity. These cells serve primarily for the transport of water and nutrients to support photosynthesizing tissues in the leaves and stems. Phloem transports photosynthates (sugars) from sources (leaves) to regions of active growth (sinks) including roots, meristems, flowers, and fruits.

Leaf Anatomy—As the stem grows, leaves are produced at nodes from primordia that first appear as small enations along the edge of the meristem dome (Box 3.3). New leaves toward the stem tip are small and continue to expand via marginal meristems as they age. Mature leaves represent resource-generating organs and are the plant's primary interface with the aerial environment. Gas exchange occurs through numerous stomata that occur on the lower leaf surface in most species (Fig. 6.6; refer to color plate). Stoma opening is controlled by a pair of guard cells that respond to demands for CO_2 uptake and to leaf water potential. Transpiration occurs primarily through the stomata and allows for gas exchange as well as the transport of nutrients from the roots as evaporative loss establishes a negative water potential gradient (i.e., suction). CO_2 is taken up and O_2 lost in an exchange across the membranes of parenchyma cells in stomatal cavities; gas exchange through the stomata occurs passively by diffusion.

In many species, the mesophyll cells toward the surface of the leaf are columnar and arranged in rows that are perpendicular to the leaf surface (palisade cells). Vascular strands extend throughout the leaf in parallel to each other (in most monocots) or in a branching network pattern (in dicots) and form the veins of the leaf. As carbon is fixed, photosynthates may be stored locally in leaf cells as starch or converted to sugars and actively loaded into the phloem for transport to growing regions of the plant.

Root Development—Plants simultaneously interact with two contrasting and uniquely demanding environments (atmosphere and soil) to obtain the inorganic nutrients and energy necessary to generate organic compounds. Aerial portions of the plant are specialized for the acquisition of light energy and CO_2. Root systems are specialized to absorb water and nutrients (primary nutrients N, P, and K; secondary nutrients Ca, S, and Mg; micronutrients B, Cl, Mn, Fe, Zn, Cu, Mo, and Ni), act as storage organs, and to interact with microbial mutualists, which may enhance access to water and nutrients. The primary interfaces for the uptake of water and nutrients are the growing root tips where root hairs develop as cells differentiate. There is constant turnover of root hair regions as roots continue to grow and penetrate into the soil. Similarly, mutualists such as *Arbuscular Mycorrhizal Fungi (AMF)* constantly reform connections to the growing root tips, though nitrogen-fixing nodules found in some groups persist on older roots for longer

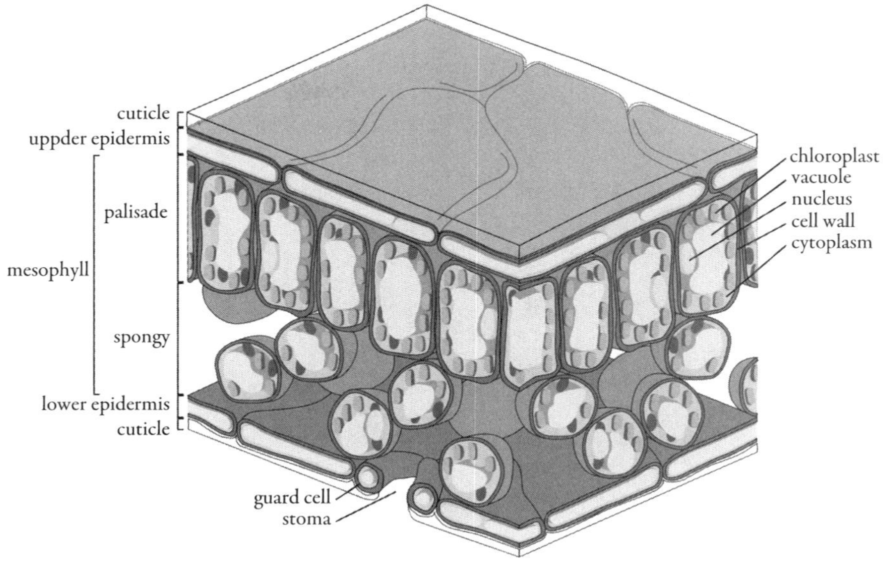

Fig. 6.6. Cross section showing the internal anatomy of a typical dicot leaf. Created by Zephyris.

periods of time. Older roots thicken, serve for the storage of carbohydrates, and provide support for the aerial portions of the plant body.

Most dicots develop primary root systems that originate from the radicle and hypocotyl of the seedling. The primary root branches to produce an array of fine roots that effectively penetrate and occupy the soil region underlying the aerial portion of the plant. Roots differ from stems in lacking nodes and axillary buds. While the growth of stems is primarily determined by phototropic responses, roots respond positively to gravity (gravitropism) and to higher concentrations of inorganic nutrients and water. Associations with fungal mutualists also affect root growth, as heavily colonized root systems tend to be larger.

Roots of perennial herbaceous species serve for the storage of carbohydrates and may become thickened into *Taproots* (carrots, radishes, etc.) or *Tubers* to allow plants to survive through harsh conditions. Not all underground storage organs are derived from roots, as tubers can be derived from stem tissue such as stolons or rhizomes (potatoes, yams, etc.) or from roots (e.g., sweet potatoes, cassava, etc.). Stem tubers can be distinguished, because they usually possess axillary buds (e.g., potato "eyes") that can give rise to new stems. Other underground storage organs derived from stems include *Bulbs* and *Corms*. Bulbs, such as onions, have shortened internodes, and consist of layers of leaves that are modified for storage. Corms consist of one or more thickened internodes and are found in a variety of monocot and dicot species. The tubers of some herbaceous vine species of morning glory (*Ipomoea*), wild cucumber (*Marah*), and kudzu (*Pueraria*) are sometimes referred to colloquially as manroots, as they can become quite large—often exceeding several meters in length and weighing over one hundred kilograms.

In rhizomatous dicots and most monocots, the *Primary Root* senesces early in life and adventitious roots arise from the base of the stem or from rhizomes, stolons, or bulbs. Many perennial bunch and rhizomatous grasses possess *Fibrous Root* systems that consist of numerous adventitious roots that are small in diameter. In larger monocots, such as palms and corn, thicker adventitious roots develop toward the outside portions of the stem base to form *Prop Roots* that serve to support the plant. Adventitious roots persist for the life of the plant in palms and other large monocots or may be replaced annually in species growing from rhizomes or bulbs.

Primary Root Anatomy—The anatomy of herbaceous roots differs from stems as the vascular tissue is typically organized into a single cylinder (*Stele*) located at the center (Fig. 6.7; refer to color plate).

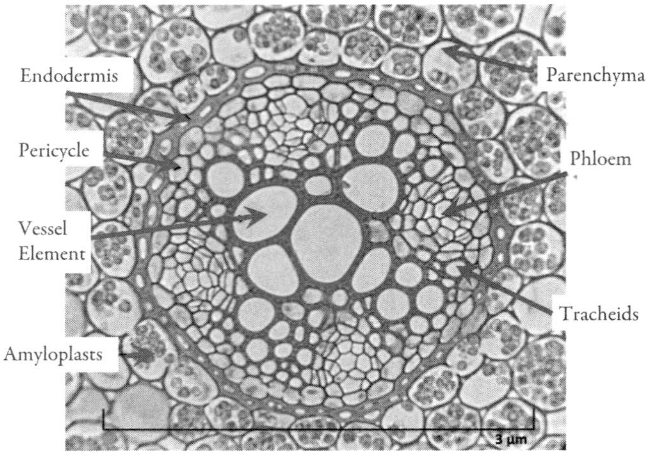

Fig. 6.7. Cross section of a *Ranunculus* (buttercup) root. The endodermis and pericycle surround the central stele containing vascular tissue. Cortical tissue consists of parenchyma cells containing amyloplasts for storage of starch. Photo by Sadierath.

The stele is surrounded by an endodermis (*Pericycle*) and consists of both ground and vascular tissues. Outside the stele, there is a thick region of ground tissue (cortex) that is covered by a single layer of cells comprising the epidermis. Tissues of the root develop in the zone of differentiation from foundation cells that are produced by cell divisions in the quiescent center of the meristem (Box 3.3). Cells produced toward the growing tip of the root undergo apoptosis as they develop thickened cell walls to form the root cap. This tissue serves to protect the growing root tip. Root cap cells are constantly replaced from the quiescent center as they are shed during growth through the soil.

The epidermis of fine roots near apical meristems lacks a cuticle and is densely covered with root hairs. Each hair is an extension of a single epidermal cell that penetrates among soil particles for access to water and nutrients. Inorganic nutrients may be acquired passively through mass flow and diffusion, or actively as root hairs produce hydrogen ions that release nitrogen and phosphorous in the form of cations that are associated with negatively charged soil particles. Water and nutrients may be passed through the *Symplast* by entering root hairs and diffusing through the cytoplasm of adjacent cells via the plasmodesmata that connect adjacent cells. Alternatively, water and nutrients may flow through cell walls (*Apoplast*) and through the root cortex. Water and nutrients can only enter the stele by passing through a membrane to enter the symplast, because a cylindrical sheath of cells (the pericycle) prevents apoplastic movement of materials into the stele. A waterproof material (*Casparian Strip*) surrounds each cell of the pericycle, like a gasket, to seal off the stele, so water and nutrients must pass through cell membranes before being loaded into the xylem. Symplastic regulation of the stele helps maintain higher water potential within the root system and prevents entry of harmful compounds and soil contaminants.

Secondary Growth—In woody plants and some herbaceous species, the stem and roots begin to thicken during the second year of growth. While primary growth refers to increase in length, secondary growth is any accretion in the diameter of the stem or root. In woody plants, two meristem regions (cambium) develop as cylinders just beneath the epidermal surface after the first year of growth (Fig. 6.8; refer to color plate). The more exterior region (*Cork Cambium*) begins to produce cork cells toward the exterior of the stem. These cells undergo apoptosis as their cell walls thicken with suberin to develop an impervious bark layer on the outside of the stem (*Periderm*). New cork cells are constantly produced by the cork cambium as the outside layers of bark split and are shed as the stem increases in diameter. Woody stems retain the ability for gas exchange through groups of living cells that

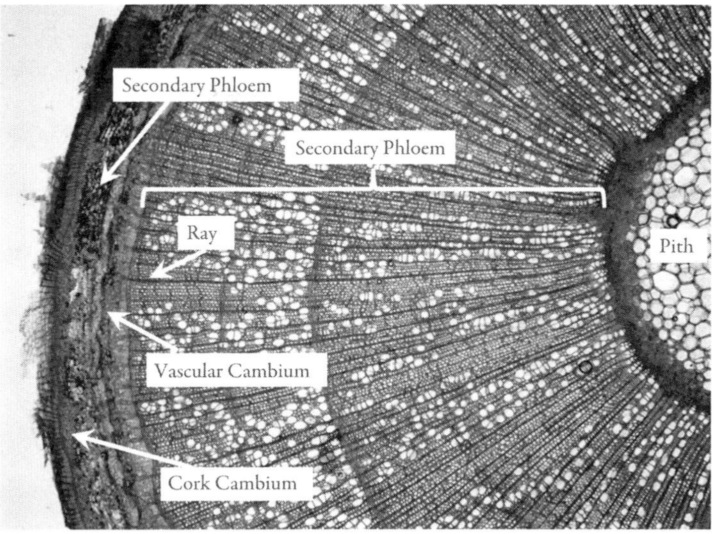

Fig. 6.8. Cross section of a woody stem of *Corylus avellana* (common hazel). Annual rings of secondary xylem can be distinguished. Narrow rays of parenchyma connect the central pith region and the vascular cambium. Photo by annukar1.

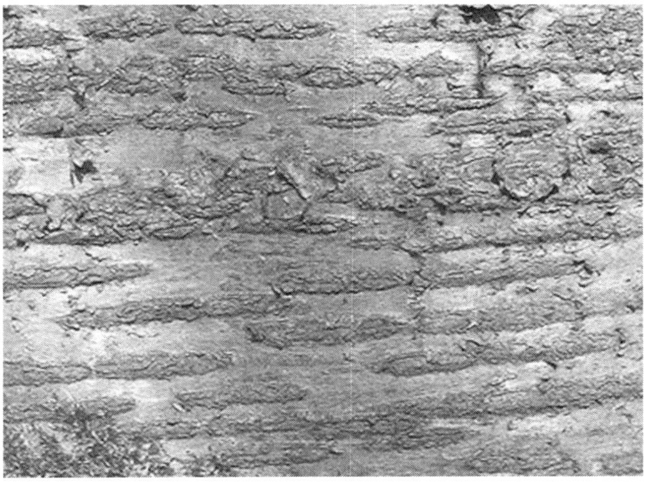

Fig. 6.9. Lenticels appear as horizontal raised regions of the thin bark of this cherry tree (*Prunus* sp.). Photo by Rosser1954.

occupy cracks in the periderm (*Lenticels*). Lenticels are generally hidden deep in crevices in trees with thick bark but are visible in species with thin periderms such as cherry trees (*Prunus* sp.; Fig. 6.9; refer to color plate).

Increase in the diameter of woody stems is primarily due to the addition of secondary xylem and phloem by the *Vascular Cambium*. The vascular cambium is bifacial—producing layers of secondary xylem toward the interior of the stem and secondary phloem in a thin layer between the vascular and cork cambium. Secondary xylem is added each year to form a strong woody central column within the trunks of trees. Secondary growth in roots occurs in pace with the stems and results in a similar anatomical structure. In temperate or seasonally dry regions, the annual layers of xylem can be identified by differences in the color of the wood due to changes in the size of cells toward the end of the growing season (*Growth Rings*). In contrast, trees in wet tropical regions often lack discernible growth rings. Smaller sections of the vascular cambium produce *Medullary Rays* that extend through the functional layers of secondary xylem (*Sapwood*) to the center of the stem. These rays contain parenchyma responsible for depositing resins that occlude the xylem in the central portion of the trunk (*Tylosis*) to form a darker region of wood (*Heart Wood*). These terpene-containing compounds are thought to help protect the wood from attack by microbes that promote decay. Other than the medullary rays, the stem inside the cylinder of the vascular cambium consists of cells lacking cytoplasm; living portions of a mature tree are the thin layers of cambium and phloem just beneath the bark, the medullary rays, and regions of primary growth at the stem tips.

Not all plants undergoing secondary growth possess organized cambial layers or develop woody tissue. For example, the production of tubers and thick taproots occurs through secondary growth by addition of ground tissue. Monocots such as palms undergo secondary growth in a more disorganized manner that is sometimes referred to as "diffuse" secondary growth. These forms of "anomalous" secondary growth proceed by the addition of vascular or ground tissue, by the thickening of cell walls, or by the production of cork and fiber. Some form of secondary growth in aerial or subterranean portions of the plant is a characteristic of all perennial plants.

6.4 PLANT LIFE FORMS

The flowering plants vary widely in their life form (*Physiogamy*) and life history (*Phenology*). The oldest classification system (the Raunkiær system) dates from the early twentieth century and is based solely on the location of dormant meristems (Gurevitch et al. 2006; Fig. 6.10). Annual plants (*Therophytes*) persist for less than a single year and survive during inhospitable conditions as seeds. These plants typically occupy ephemeral habitats that are created by disturbance or by the temporary availability of conditions conducive to growth in otherwise harsh habitats such as deserts. Herbaceous plants and low-growing woody perennials are classified as either *Chamaephytes* (perennating buds close to the soil surface), *Hemicryptophytes* (buds at the surface), or *Cryptophytes* (buds beneath the surface of the soil or water). Chamaephytes include herbaceous and woody plants that produce dormant buds that are close to the ground, such as low-growing shrubs and creeping and matt-forming species. The hemicryptophytes and cryptophytes are herbaceous species that die back to roots, rhizomes, tubers, bulbs, and corms each year. Hemicryptophytes are represented by perennial acaulescent and rosette species that may overwinter as green plants or produce buds at the soil surface. The above-ground portions of cryptophytes typically senesce annually and meristems persist underground (*Geophytes*) or underwater (*Hydrophytes*). Shrubs and trees form overwintering buds at the tips of their branches, which are typically well above the soil surface (*Phanerophytes*). According to the original definition, the overwintering buds need to be at least 25 cm above the soil surface for a plant to be considered a phanerophyte. This delineation would not include many low-growing plants, such as the arctic willow (*Salix arctica*), kinnikinnick (*Arctostaphylos uva-ursi*), heather (*Calluna vulgaris*), and other low-growing shrubs of bogs, heath, rocky outcrops, and arctic and alpine habitats. Plants with woody secondary growth share many of the same characteristics of their life form and phenology, so it may be best to consider phanerophytes to be inclusive of all woody species regardless of stature.

Plant life forms differ primarily with respect to the lifespan of individuals and the number of mitotic cell divisions separating episodes of sexual reproduction (meiosis and recombination). If we consider that the number of cell divisions per centimeter of stem length may be similar for annual and long-lived woody species, then it becomes apparent that the primary differences between these two life forms are the persistence of individuals and the frequency of recombination. Annuals complete their life cycles within just a few months and often grow to just a few decimeters (e.g., *Arabidopsis*) or up to a few meters in height (e.g., *Zea mays*). At the other extreme, trees reach maturity after one to twenty years, often grow to more than 30 meters, and are capable of accumulating large amounts of biomass (Petit and Hampe 2006). The living tissues of trees persist as thin layers just beneath

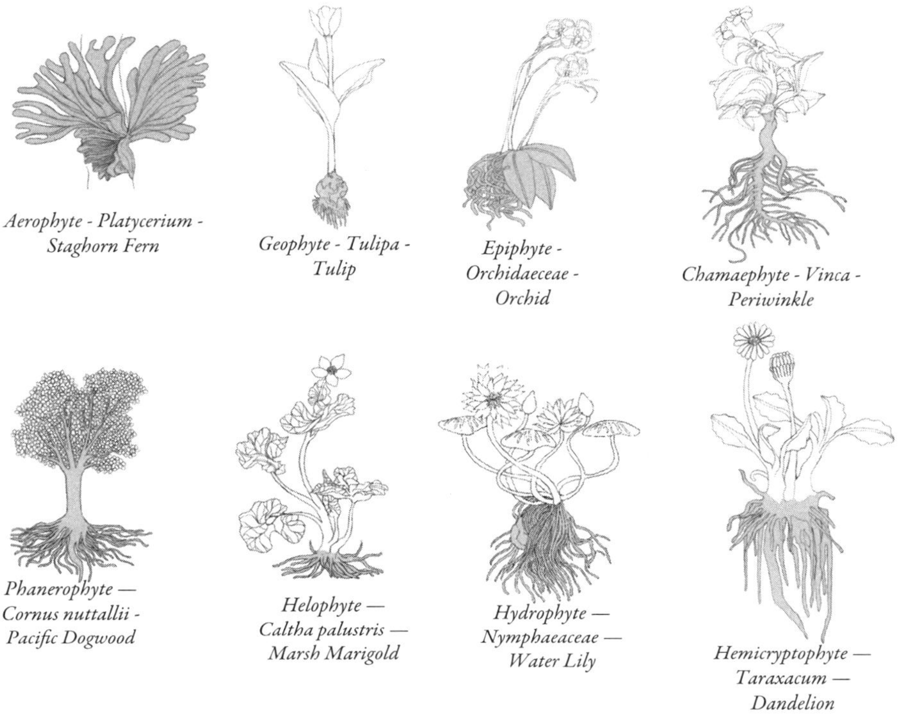

Fig. 6.10. The Raunkiær system of plant life form classification is based on the position of overwintering buds. Persistent portions of each life form are shaded green. Original art by Michelle Nicole Williamson.

the periderm and at stem tips in regions of primary growth. The time between recombination events—from germination to flowering—for large trees may be equivalent to ten, hundreds, or even thousands of generations of an annual species. As discussed in Chapter 4, the frequency of recombination among plant life forms has implications for mutation accumulation and substitution rates over phylogenetic timescales.

Among the embryophytes, woody secondary growth is limited to the seed plants (gymnosperms and angiosperms) and appears to have a single evolutionary origin (Spicer and Groover 2010). The anatomical similarity in wood structure between these groups suggests that both are derived from woody lineages that occupied coal swamps in the late Devonian and early Carboniferous (Chapter 2). Extant gymnosperms are represented by woody shrubs, vines, and large trees. Basal lineages of the flowering plants (i.e., *Amborella*, Nymphaeales, and Austrobaileyales) do not contain any trees as they only include species of herbs, vines, and shrubs. This lack of large trees among extant basal representatives is consistent with the proposal that the angiosperms originated in arid regions or ephemeral habitats where annual and shrub life forms were favored (Chapter 2).

Flowering trees that dominate contemporary tropical and temperate forests did not appear until much later with the diversification of the eudicots during the mid to late Cretaceous (Magallón and Castillo 2009). The transition from woody to herbaceous growth is reflected in the genetic composition of modern herbs, which retain the ability to undergo secondary growth (Petit and Hampe 2006; Spicer and Groover 2010). Transitions between woody and herbaceous growth habits (in both directions) are common across eudicots, with many origins of herbaceous clades from woody ancestors and instances where woody growth developed in recently derived island species. The relatively simple genetic control of the development of vascular cambium reflects the phylogenetic labiality of woody stems; manipulating expression of a few genes can transform annual *Arabidopsis* plants into perennials with secondary growth (Melzer et al. 2008). Moreover, the same genes controlling transition to flowering also affect meristem activity, which reflects a direct trade-off between sexual reproduction and continued vegetative growth. We will continue our discussion of the sometimes subtle differences between annual and perennial growth later in this chapter, when we explore processes of senescence in plants.

The appearance of large perennials with woody secondary growth in the late Devonian represented a major innovation that remains highly successful in contemporary plant communities. Plants with woody secondary growth had advantages over the other early, swamp-forest species such as scale trees, which were effectively gigantic herbaceous plants and were probably *Monocarpic* (*Semelparous*), so they died after reproducing. The tree life form allows individuals to reproduce annually (they are *Polycarpic—Iteroparous*) and to dominate resources in one location over a long period of time (Petit and Hampe 2006). Large trees are capable of effectively modifying and maintaining their local environment—their massive size prevents intrusion by other large individuals and creates understory habitat for shrubs and herbaceous species. Such large and persistent life forms have advantages over short-lived species, as it is not necessary for them to reproduce every season and their protective bark, covering stems, and large size can buffer them from disturbances such as fire and inhospitable or severe conditions. Trees also gain advantages for dispersal and mating as a function of their stature, which contributes to the high levels of genetic diversity in populations of trees compared to herbaceous species (Petit and Hampe 2006). Individual trees may produce millions of seeds over their lifetimes. On the average, only one seed will successfully grow to maturity, so there are many opportunities for selection to favor superior offspring. Trees have low diversification rates but individual lineages tend to endure for long periods. The large size, long lifespan, and high reproductive potential of trees constitute a resilient life form that is a dominant component of the majority of biomes.

6.5 PLANT GROWTH AND DEVELOPMENT

Plant development is iterative, as internodes, nodes, and leaves are repeatedly generated during stem elongation, and as a consequence, plants consist of collections of many interdependent metamers. This metameric construction and the relatively simple composition of organs render plants resilient to genetic and environmental perturbations. Recall that, as discussed in Chapter 4, chromosomal rearrangements and changes in ploidy are common in plants and typically result in perfectly functional phenotypes—perhaps as a consequence of their relatively simple architecture. Similarly, physical damage can easily be overcome as leaves, branches, and even the entire above-ground portions of individuals can be replaced through resprouting and growth of new stems. The metameric structure of plants has provided a flexible and robust architecture that has accommodated the evolution of a diversity of growth forms that occur in a wide range of environments.

As plants grow, they obtain and transport resources and reproduce through the coordination of multiple tissues and morphological features that constitute plant organs (i.e., stems, roots, leaves, flowers, etc.). The efficient function of organs is often dependent on the close organization of multiple phenotypic traits (anatomical or morphological features). Groups of traits that collectively contribute to the same function constitute *Modules* (Wagner 1996; Diggle 2014), and the coordination of functionally associated traits has been referred to as *Phenotypic Integration* (Ordano et al. 2008). Modules may include traits within single metamers or may require the coordination of traits across multiple metamers. It is generally thought that the efficient functioning of modules—stronger phenotypic integration—is closely associated with fitness, so we expect that selection has favored the coordination among traits that contribute to the same function through pleiotropy and shared gene regulatory networks. In this section, we discuss plant architecture and the consequences of shared gene regulation for the development of individual phenotypic traits, metamers, and organs.

Metameric Structure—The repetition of organs and potential for independent growth of metamers was a dominant theme in Erasmus Darwin's depiction of plant biology (Darwin 1800). Indeed, Erasmus considered plants to be collections of separate entities that "are each an individual vegetable being [that] in many circumstances resemble individual animals" (Darwin 1800, page 5). This idea was not new at the time of publication of his *Phytologia*, but traces back to the writings of students of Linnaeus in the mid-eighteenth century (White 1979). While Erasmus Darwin and others may have taken the analogy to animals beyond useful comparison in some instances, the general view of plants as collections of repeated and partially independent units continues to be the general paradigm for plant biologists working in a wide range of fields, from ecology to morphology, physiology, development, and genomics.

The concept of plants as collections of metamers was refined and applied to the field of plant population ecology in the 1970s (Harper 1977; White 1979). Harper considered plant architecture to be effectively hierarchical, with individual genets divided into branches (ramets—whether they remain physically connected or not), which are further divided into what he referred to as modules (i.e., metamers). Harper's primary interests were in the demographics and population ecology of clonal species. White (1979) objected to Harper's use of the term module and proposed terminology that had been developed by plant morphologists in the nineteenth and early twentieth centuries. He referred to plants as metapopulations of phytons (a segment of stem including one internode, node, and one or more leaves arising from the same node), which he considered to be one of several types of metamers. While these terms, and others proposed by White, never came into wide usage, both metamer and module are commonly applied to describe plant morphological structure and development. Metamer has a broader definition, as it applies to animal segmentation as well as the iterative structure of plants. Module is more often used to describe groups of interdependent phenotypic traits that function together and display correlated expression (Wagner 1996; Diggle 2014). In this text, metamer is used to describe the repetitious structure of the vegetative and reproductive portions of the plant, and module is used to describe groups of developmentally and functionally integrated traits. Note that structures comprising metamers are spatially restricted to a single morphological unit (e.g., internode, node, and leaf), while traits included in the same module may be distributed across metamers (e.g., sepals, petals, stamens, and pistils represent separate metamers and are all members of a flower module).

Watson and Casper (1984) extended the concept of plant metameric structure to resource allocation and transport as they considered metamers as *Independent Physiological Units (IPUs)*. These authors used examples from crop species such as *Phaseolus* (bean), *Glycine* (soybean), *Gossypium* (cotton), and *Solanum* (tomatoes and potatoes) to demonstrate that the distribution of photosynthates is localized. For example, the carbon needs of developing fruits are primarily served by leaves on the same or adjacent metamers. Sugars within the plant are not always allocated to the closest developing fruit, as distribution depends on the plant's vascular anatomy. Individual vascular bundles may serve leaves and fruits on the same side of the stem, such that the distribution of resources within the plant is sectoral (along one side of the stem). This sectoral patterning of resource distribution is maintained as plants grow, such that separate groups of branches on a tree may be served by different sections of the trunk, which in turn are connected to separate groups of roots. Similarly, physiological interdependence based on proximity extends to higher levels of organization, as whole stems and collections of branches may display different levels of independence. For example, the fate of fruits can be strongly affected by shading or defoliation of leaves on the same branch while other portions of the tree remain unaffected (Watson and Casper 1984). From these examples, it becomes evident that plants may be best viewed as hierarchically structured collections of physiologically interdependent metamers.

Modularity—The concept of modularity has been applied across a number of biological disciplines to describe groups of closely associated units working in concert to perform a particular task, such as groups of enzymes or phenotypic traits that contribute to the same process. Most often, the term "module" has been used to identify groups of phenotypic traits that collectively operate to achieve a particular function (Wagner 1996; Diggle 2014). Such traits typically covary in their expression due to some degree of common genetic determination (i.e., the collection of pleiotropic effects and shared regulatory elements that comprise *Gene Networks*) that is important for maintaining the function of complex structures (e.g., stamens and pistils) and organs (e.g., flowers). Interactions among traits within a module are important for maintaining module function, as the expression of individual traits changes during development and across generations. For example, the size of the flowers may vary along the stems of an inflorescence, but the relative positions of reproductive structures (anthers and stigma) within flowers may be maintained for a range of corolla sizes to ensure efficient pollen removal and receipt. As selection acts on modules, the individual traits tend to evolve in concert such that functionality is maintained as lineages diverge and adapt to novel conditions.

A complex reproductive structure such as a flower could be defined as a module, but within a flower individual structures such as stamens and carpels also represent collections of interdependent traits that can also be considered to represent modules. Hence, our delineation of modules will depend on the level of functionality that we are concerned with (e.g., pollen presentation versus whole-flower pollen export and receipt) and modules can be defined hierarchically from larger organs (whole flowers) to structures that function within them (e.g., stamens within flowers, pollen within stamens, etc.). This view of plant structure recognizes the importance of trait interdependence and assumes that selection has favored suites of codependent traits that respond to environmental variation and selection pressures in a coordinated fashion to maintain the functionality of complex organs across developmental and evolutionary timescales.

Functional dependence among individual traits and the gene networks that control them results in collective responses to selection pressure that provides stability of complex processes as they evolve. Wagner (1996) suggested that selection should work to increase integration within modules while simultaneously favoring independence among separate modules (Fig. 6.11). He expected that pleiotropy across two functional groups of traits would be reduced (*Parcellation*) so selection would favor independence between the two modules. In the second scenario depicted in Fig. 6.11 selection favors increased coordination within modules, and two modules emerge from a collection of independent traits as a consequence of increased pleiotropy within functionally related groups (*Integration*). We expect that both parcellation and integration could occur simultaneously to increase integration within and delineation among modules during the evolution of complex structures. Thus, pleiotropy and shared regulation among genes allows suites of traits to covary while providing a means for separate modules to change along independent trajectories.

Wagner (1996) portrayed the concerted evolution of traits within modules as an outcome of the simultaneous effects of directional and stabilizing

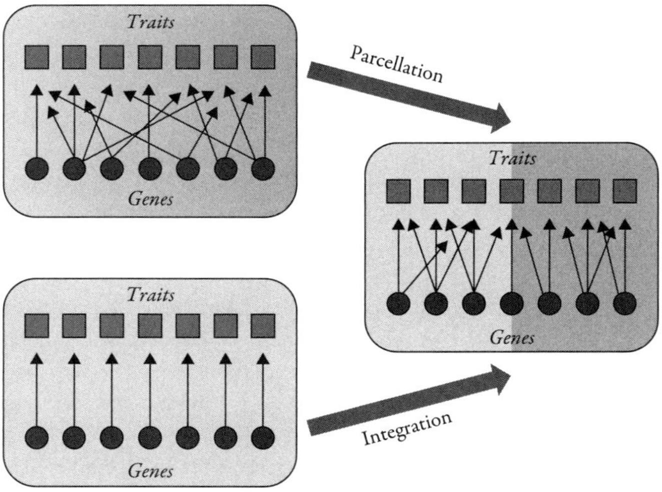

Fig. 6.11. Wagner's (1996) model depicts the evolution of modularity as integration increases within and decreases among modules through changes in the patterns of pleiotropy.

selection. We can visualize this process on a fitness surface as the evolution of trait combinations along a ridge of high fitness (Fig. 6.12; refer to color plate). In this scenario directional selection acts on two traits simultaneously to increase the module's contribution to whole-organism fitness, which is represented as movement up the sloping ridge. Stabilizing selection is imposed as trait combinations off the ridge have lower fitness. As the module evolves along the ridge, selection begins to affect the traits differentially as the trajectory curves toward a peak of high fitness. We could imagine such a pattern of coordinated trait changes for the evolution of flower size: selection favored larger petals, while reproductive structures may have increased in size at a slower rate. This modular view of phenotypic evolution emphasizes the fitness outcomes of combinations of traits and provides a general paradigm for understanding the processes and constraints affecting the evolution of complex organs.

Recognizing the potential for integration among traits that are functionally interdependent provides a more sophisticated and more accurate view of the plant phenotype. We expect selection to favor strong integration among traits that represent functional groups, while parcellation will allow the separation of modules, so they may follow independent evolutionary trajectories (Figs. 6.11 and 6.12; refer to color plate). In general, there should be strong integration within, and parcellation between, vegetative and reproductive characters, but multiple modules may also exist within flowers and suites of vegetative traits (Diggle 2014). Positive correlations are often evident among floral traits such that the entire flower might represent a single module. On the other hand, we expect that parcellation of floral traits would be favored in cases where selection favored the separation of groups of traits that serve different reproductive functions. For example, selection for pollinator attraction may favor larger petal and sepal sizes, while the placement of reproductive organs within the flower would be influenced by selection for efficient pollen transfer. One such example is the red columbine (*Aquilegia canadensis*); in this species, traits associated with pollen transfer, including nectar spur length and the positions of anthers and stigmas, remain the same across a wide range of petal sizes (Kliber and Eckert 2004). This type of parcellation of functions may be more common in flowers that are specialized for particular types of pollinators (e.g., hummingbird pollination in the case of red columbine) compared to flat or bowl-shaped flowers such as buttercups (*Ranunculus* sp.) and many plants in the sunflower family (Asteraceae). Indeed, studies of variation of floral traits in species with specialized flowers such as tobacco (*Nicotiana*) and larkspur (*Delphinium*) indicate that separate pollen-transfer efficiency and pollinator attraction modules may be common (Diggle 2014).

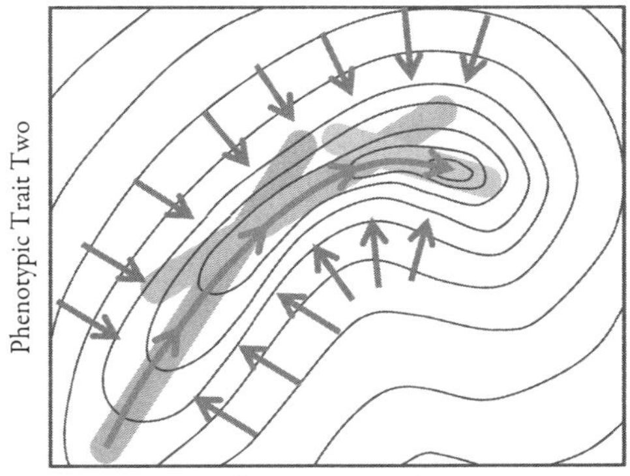

Fig. 6.12. Evolutionary trajectory of two traits that comprise a module (green arrows). Thin blue lines represent fitness isoclines. Directional selection pushes traits along a fitness ridge while stabilizing selection (orange arrows perpendicular to isoclines) maintain module functionality. Red, purple, and blue rods represent constraints imposed by gene networks controlling trait expression.

Genetic Determination of Flower Development—While there are many examples of trait associations and the delineation of modules in plants, our understanding of the underlying genetic architecture of modules and determinants of morphological variation remains rudimentary. Studies of modules in plants have primarily focused on flowers, because they are relatively complex organs that consist of easily quantified structures. As described elsewhere in this chapter, correlations among floral traits and the development of flowers have been studied extensively. We also have a basic understanding of the suites of regulatory genes responsible for the differentiation and development of separate groups of floral structures (i.e., sepals, petals, stamens, and carpels; Lawton-Rauh et al. 2000; Soltis et al. 2007a; Fig. 6.13; refer to color plate). The analysis of phenotypic traits and the characterization of the genetic determination of flower development have been conducted as independent fields with little overlap, but our understanding of the origin and development of plant structures is rapidly improving and will likely facilitate the development of more integrative studies.

The major regulatory genes responsible for flower development display a history of gene duplication accompanied by neofunctionalization and subfunctionalization (Chapter 4) dating back to the diversification of land plants (Lawton-Rauh et al. 2000; Soltis et al. 2007). Orthologs of the major classes of the MADS-Box genes that control flower development can be found in bryophytes, ferns, and gymnosperms indicating that they arose well before the appearance of spermatophytes. The ABCE pattern of expression and determination of floral organs (Fig. 6.13; refer to color plate) has been developed primarily using *Arabidopsis* and *Antirrhinum*, but this developmental model appears to be consistent across the core eudicots.

As reproductive organs of early angiosperm ancestors became more complex, additional gene duplication events allowed specialization of individual structures. Several floral gene duplications occurred over the early history of land plant diversification, with one more extensive episode during the time leading up to the origin of flowering plants. Regulatory genes with B-class functions that are associated with petal and stamen development arose through duplication events that occurred between 230 and 290 MYA (Permian or early Triassic). These dates fall well before the appearance of flowers in the fossil record, in

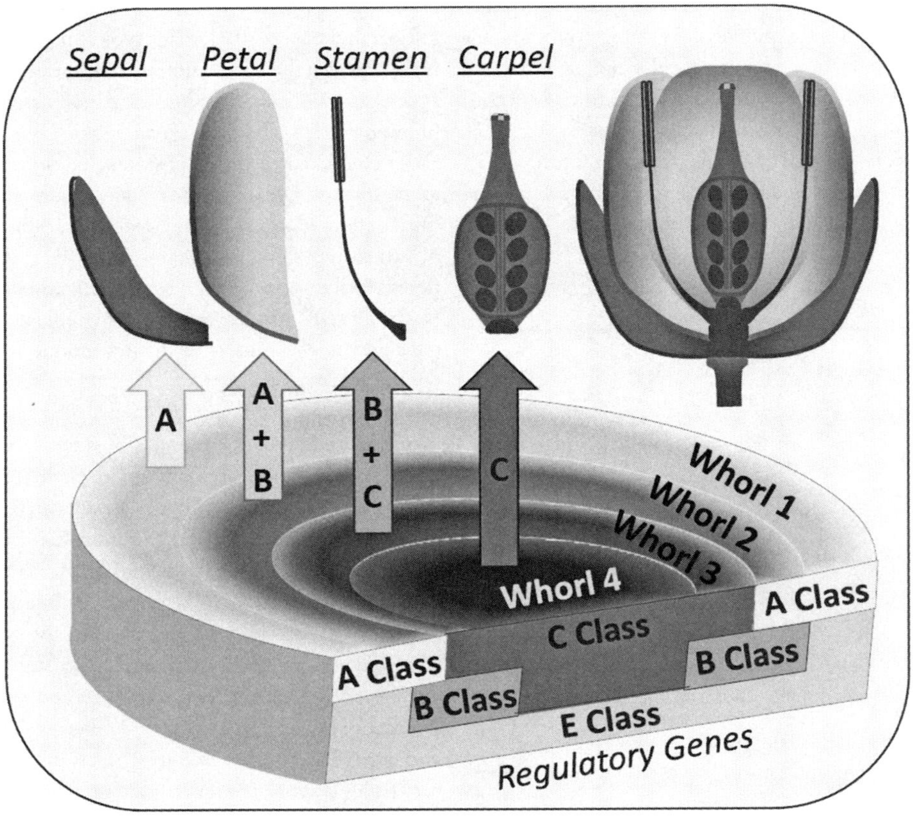

Fig. 6.13. The major classes of regulatory genes that determine the development of floral structures. Each class is expressed in overlapping fields within the floral primordium and the floral structure produced within each whorl is determined by specific combinations of genes.

the early Cretaceous, but the timing corresponds fairly well with a hypothesized Triassic origin of flowering plants, as described in Chapter 2. It's possible that these B-class genes arose along with microsporophylls (i.e., pollen-bearing leaves) that served the combined functions of attraction and pollen presentation (Soltis et al. 2007). These gene duplication events suggest that ancestors of flowering plants may have possessed reproductive structures consisting of sporophylls that attracted animal pollinators—most likely insects in the order Coleoptera (beetles), which coincidentally underwent rapid diversification during the Triassic (Powell 2009).

The ABCE model of gene expression is fairly consistent across the core eudicots, but greater variation in homeotic gene expression patterns during flower development is seen in the basal angiosperms. For example, gene expression in basal dicots such as avocado (*Persea americana*) and water lilies (*Nuphar*) is more broadly overlapping across the regulatory classes, which presumably results in the low differentiation across floral structures present in these species (Yoo et al. 2010). Among core eudicots with lower numbers of petals and stamens, homeotic gene expression during flower development is more spatially discrete. Greater specificity of the timing and locations of regulatory gene expression enabled the evolution of more specialized flower morphologies, which was particularly important with the increasing dependence on animal pollinators during the diversification of flowering plants that started in the Cretaceous.

The characterization of regulatory genes responsible for flower development provides a foundation for more detailed studies of the regulatory basis of differences among floral morphologies within and among species. We have made significant advances in our knowledge of some aspects of floral determination such as differences in flower symmetry (i.e., bilatterally versus radially symmetrical flowers; Preston and Hilernan 2009) and the regulatory networks determining flowering time (Pajoro et al. 2014); however, our understanding of the genetic basis of integration and parcellation within and among modules that extend across multiple types of floral structures remains largely undeveloped. From the phenotypic analyses described above it is clear that different degrees of correlated development can exist across separate floral structures and we expect that both the development and evolution of these functional modules are controlled by pleiotropy and shared regulation. Our understanding of gene regulatory networks continues to improve and to reveal more subtle aspects of the genetic determination of both integration and developmental associations among phenotypic traits (Pajoro et al. 2014).

6.6 SENESCENCE

Death is a critical component of adaptive evolution and diversification, because the rate of genetic change in stable populations is a function of the turnover of individuals. While animals fit well with our expectations of the roles of aging and death in populations, plants challenge our perceptions of senescence and mortality; individual plants may persist for thousands of years even as cell degradation and death play important roles in their growth, development, and the continuous regeneration of new organs and tissues. In animals, we expect senescence—the deterioration of cells and organism function—to contribute to the mortality of individuals as they age, but this is not necessarily the case for plants.

Animals have determinate growth, and older cells are not replaced but age with the individual. As animal cells and organs get older, they suffer diminished function and elevated susceptibility to attack from pathogens and predators. Consequently, the rate of senescence, along with the frequency of death due to stochastic events, will determine average lifespan. Senescence is distinguished from programmed cell death (apoptosis) and catabolic cell degradation (autophagy) because apoptosis and autophagy may occur in juvenile or mature tissues and can be important components of growth and development (Thomas et al. 2009). Senescence, on the other hand, may lead to catabolic degradation of cells, but has no developmental role; senescence is the loss of function that accompanies aging and ultimately contributes to the death of organs and individuals. In this section, we review the major hypotheses that have been proposed for the evolution of senescence. We then discuss how models of senescence may apply to different aspects of plant growth and development.

Aging in Plants—Explanations for the evolution of senescence have been developed to account for the functional decline of unimodal organisms that have definitive mortalities—that is, the entire individual senesces as it ages and eventually dies. Plants do not fit into this paradigm, because individuals do not succumb to senescence; organs or metamers senesce and may be shed without injury to the remaining, living portions of the plant body, so individuals persist. Most cells of a mature animal undergo few replications and are maintained for the life of the individual. In contrast, living portions of all plants are ephemeral, as cells are constantly replaced from meristems and stem cambia. Animal lifespans limited by constraints on the maintenance of aging cells, whereas the maximum age of plants may be limited by accumulation of DNA replication errors during extensive somatic growth. For example, as discussed in Chapter 3, pollen fertility declines in extremely old quaking aspen (*Populus tremuloides*) clones, apparently due to the accumulation of somatic mutations (Ally et al. 2010). In this species, mutation accumulation is predicted to ultimately lead to the loss of sexual function, but there is no obvious process that limits the life expectancy of the vegetative portion of the clone.

The life forms of animals differ significantly from plants with respect to growth, aging, and lifespan; however, exploring the ideas that have been developed to explain senescence in animals can be instructive for understanding the senescence of individual tissues and organs of plants. Examining the various hypotheses that have been proposed to explain senescence provides us with some insights into plant growth forms and serves to illustrate the contrasts between plants and unimodal organisms with determinate growth, such as most animals.

Metabolic Rate Hypothesis—Some of the earliest ideas on the upper limits of longevity and the onset of senescence were based on the assumption that there are biophysical constraints on the amount of energy expended over an individual's lifespan. The *Metabolic Rate Hypothesis,* as suggested by R. Pearl in the early twentieth century, assumes that there is an upper limit on the number of heartbeats over the lifespan of an animal, so those species with higher basal metabolic rates consequently will have shorter lives (Speakman 2005). This idea applies generally across extreme cases of vertebrates (e.g., mice and shrews versus elephants and whales), but there are many exceptions as well. For example, reptiles and birds with similar metabolic rates vary in lifespan from just a few years (chameleons and some passerine birds) to more than one hundred (tortoises and parrots).

In some instances, the metabolic rate hypothesis is consistent with life expectancy in plants. For example, we may conclude that the growth rates of annual plants are higher than long-lived perennials, but there are many exceptions to this generality. A number of long-lived perennials such as bamboo (Poaceae; Bambuseae), kudzu (*Pueraria* sp.), poplars (*Populus* sp.), and acacia (*Acacia* sp.) can have extremely fast growth rates—up to several decimeters of stem length per day. Moreover, photosynthetic rates are more often limited by rates of consumption within the plant (sink strength), temperature, and the availability of light, water, CO_2, and nutrients, across plant life forms (Nobel 2005; Kirschbaum 2011). On the other hand, leaves of deciduous trees with longer lifespans have lower rates of photosynthesis (Matsuki and Koike 2006). This may be due to the greater investment in anti-herbivore compounds in long-lived leaves, but it is also consistent with the metabolic rate hypothesis. For example, there is a linkage between energy expenditure and free radical production, which is associated with rates of senescence in animals (Speakman 2005). Similarly in plants, older tissues have reduced capacity to detoxify *Reactive Oxygen Species* (*ROS*), which are a consequence of free radical production and are implicated in the onset of autophagy (Khanna-Chopra 2012). For plants, the metabolic rate hypothesis may apply to individual organs but may have limited applicability at the whole-organism level, because rates of growth and photosynthesis vary widely and do not display clear relationships with plant lifespan.

The Telomere Hypothesis—Some cytological characteristics of cells have been found to be associated with lifespan and rates of senescence. One of the most frequently cited examples is the association between the deterioration of telomeres (Guanine-rich DNA/protein caps on the ends of chromosomes) and the onset of senescence (McKnight and Shippen 2004). This idea arose from the observed attrition of telomeres that leads to the cessation of mitosis and senescence of animal cells in culture, which has been referred to as *Mitotic Senescence* or *Replicative Senescence*. Telomere degradation has received extensive attention in animal biology because it tends to be associated with a number of diseases including cancer, and variation in the length of telomeres can be predictive of whole-organism longevity.

Telomeres are maintained by high levels of telomerase expression in some animal tissues such as the epidermis and endothelium of the intestine where cell divisions continue at maturity, whereas most somatic tissues do not express telomerase, and cells have shortened telomeres. If a homologous pattern were to occur in plants then we would expect meristems and cambial tissues to have a high expression of telomerase while leaves and fruits should not. Studies of telomerase expression in *Arabidopsis* is consistent with this expectation as expression of *atTERT*, a homologue to human telomerase, is high in meristem cells but not in other tissues (McKnight and Shippen 2004).

To test the role of *atTERT* for the maintenance of telomeres and effects on longevity, plants lacking *atTERT* expression (knockout lines) were created. These plants displayed telomere attrition over several generations, but telomere shortening progressed at a much slower pace than in mouse knockout lines. After several generations, these plants began expressing developmental abnormalities, which were apparently due to chromosome breakage and rearrangements. In contrast to expectations, these telomerase knockout plants exhibited extensive somatic growth and were effectively transformed into perennials, so they had extended lifespans compared to non-mutant lines. Based on the information available it appears that, while the attrition of telomeres can lead to chromosomal abnormalities, it is not predictive of higher rates of senescence and reduced longevity in plants.

Evolution of Senescence—From this discussion, it becomes apparent that telomere shortening and the senescence of cell lines is not an inevitable outcome of multiple rounds of mitosis or the aging of cells. Recognizing that cell lines of microbes and some complex eukaryotes apparently have the potential for immortality prompts us to ask why senescence has evolved, or as George Williams put it, "It is remarkable that after a seemingly miraculous feat of morphogenesis, a complex metazoan should be unable to perform the much simpler task of merely maintaining what is already formed" (Williams 1957). The same question can be asked of individual organs of plants: Why should a leaf, which requires significant energy expenditure for development ever be discarded? Similarly, we can ask why annual plants even exist if perennials can have the advantage of precedence for access to soil and light resources without the risk of establishment. As mentioned earlier, the processes responsible for senescence in plant organisms and organs may be quite different from those operating in animals. Nonetheless, it is useful to review evolutionary explanations for the evolution of senescence in unimodal organisms, as the contrast is informative, and some of these principles may also apply to plants.

Evolutionary theory predicts that senescence is likely to be due to the expression of deleterious alleles late in life. We already have discussed the fact that deleterious mutations are likely to be segregating in populations and will affect the growth and survival of individuals (Chapters 4 and 5). Recessive mutations with low expression in heterozygotes will tend to be more abundant, while those with higher dominance are more likely to be eliminated because they have greater exposure to selection. The effectiveness of selection is also determined by the timing of expression; younger individuals often have a higher *Reproductive Value* (Fig. 6.14; refer to color plate), so selection will be stronger on alleles affecting reproduction early in life. Conversely, deleterious mutations that are expressed late in life have diminished effects on fitness, because reproductive capacity is waning; mutations expressed late in life will be more likely to escape the effects of selection and will tend to accumulate in populations. This principle—the *Mutation Accumulation Hypothesis*—was first outlined by Peter Medawar in 1946 as an explanation for senescence and increased susceptibility to disease in individuals living beyond their natural lifespan (Medawar 1946). For example, improved medical practices and protection from sources of disease and predation has extended the lives of humans and their domesticated pets well beyond the lifespans of their ancestors. According to Medawar's mutation accumulation hypothesis, the physical decline and deterioration of function in associated with aging in modern societies is a consequence of a large number of mildly deleterious mutations that are only expressed in older individuals.

Theory also suggests that there may be tradeoffs between performance early in life and longevity. Selection on mutations is often complicated because a single gene may affect the expression of multiple phenotypic traits (*Pleiotropy*). In some cases, a non-synonymous sequence change may simultaneously increase fitness through its effect on one trait while

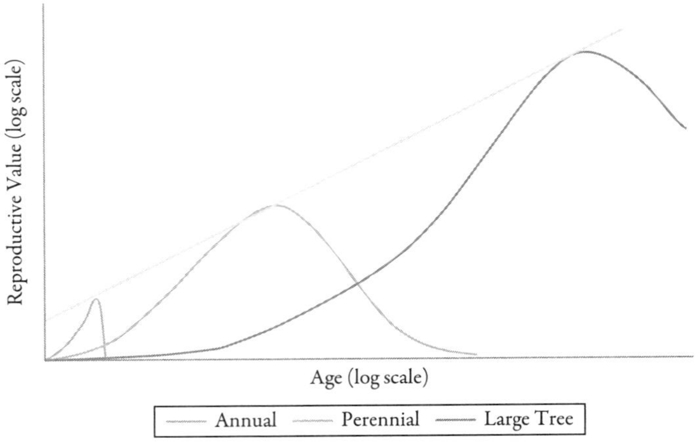

Fig. 6.14. Reproductive value curves for an annual/biennial plant (red line), herbaceous perennial or short-lived shrub (purple), and a long-lived tree (green). The reproductive value at age x integrates the probability of survival and fecundity for all ages $\geq x$. The orange line shows an increase in the maximum reproductive value with increasing lifespan. Both axes are on log-scales.

also causing a decrease in fitness due to effects on a second trait (antagonistic pleiotropy). If the first trait occurs in younger individuals when reproductive value is high, and the second trait occurs only late in life, then the mutation may be pushed to fixation by selection in spite of its detrimental effects on older individuals. Hence, antagonistic pleiotropy can contribute to increases in rates of senescence, if mutations have positive effects early in life (Williams 1957). Mutations that are advantageous early will be favored even as they increase rates of senescence and susceptibility to disease later in life when reproductive values are lower. Antagonistic pleiotropy has the potential to supplement the accumulation of mutations contributing to senescence, but it also generates a tradeoff where selection favoring high rates of reproduction early in life may come at the expense of longevity.

The potential for the accumulation of mutations contributing to senescence depends heavily on the distribution of reproductive values over the lifespan of individuals. The reproductive value is the expected contribution to fitness across the lifespan based on the chance of survival and fecundity (Fig. 6.14; refer to color plate). As the accumulation of deleterious mutations acting at different ages is a consequence of the shape of the reproductive value curves, we should view senescence as an outcome of selection pressure due to differences in life-history strategies among species. For example, if survival past a certain age were unlikely due to the lack of resources or high rates of predation, then we would expect selection to favor high reproductive output early in life, and such a species would probably have a shortened lifespan even under controlled conditions. In other species with higher reproductive values late in life, we expect that selection would prevent the accumulation of mutations contributing to senescence, and the lifespan of individuals could be greatly extended. These cases of whole-organism senescence and elevated susceptibility to disease late in life are commonly observed in animals, but as we will discuss in the next section, the consequences of aging and the factors affecting lifespan may be quite different in plants.

Determinants of Plant Lifespans—Plants do not age in the same way as animals, because their cells are constantly replaced as they grow. The cycle of growth is repeated annually in plants, as they go through the same process of primary growth, leaf and flower production, and the formation of dormant buds. Within the first year or two, most perennial species undergo secondary growth, which increases stem girth, most often by the production of woody tissue. Living tissues are maintained and replaced by meristems and cambial regions, where cells continue to divide over the life of the individual. Consequently, most of the living cells may be less than a few years old even in very old plants, and with the exception of meristem regions, individual cells are not maintained over the life of an individual as they are in animals. The rejuvenation of plant tissue is evident: the newest growth of the oldest plants in a population is characteristically the same as the youngest individuals. In contrast, the tissues and organs of animals age with the whole organism and eventually senesce, leading to death (McKnight and Shippen 2004; Speakman 2005). Since plant tissues remain youthful, it is unclear whether they ever senesce, or at least it is probable that the decline of older plants is not caused by the same processes that occur in aging animals.

The timing of availability and abundance of resources primarily determine the lifespans and phenologies of plants. Annual plants often live in habitats that are ephemeral—they take advantage of resources that are accessible for only a short period of time, so they have a low chance of surviving past a few months. According to the evolutionary theories discussed earlier, we would expect that mutation accumulation would lead to the senescence and decline of annuals if we artificially extended their lives beyond one year. Similarly, we can explain the long lives of trees because the reproductive value is predicted to increase as individuals grow larger (Fig 6.14; refer to color plate). The somatic mutation accumulation theory predicts that older plants should suffer from senescence and degeneration with old age, but this does not appear to be the case at all. For example, disabling genes that control determinacy in *Arabidopsis* results in perennial plants that do not exhibit evidence of senescent decline (Melzer et al. 2008), which we would expect under the mutation accumulation and antagonistic pleiotropy hypotheses. Furthermore, the death of annual plants with determinate growth such as *Arabidopsis* is not caused be senescence in the strict sense, but instead is due to a programed process of autophagy that results in the translocation of the soluble products of catabolic pathways to developing seeds and fruits (Lim et al. 2007). As mentioned earlier, evolutionary transitions from annual to perennial life forms are common in plants, even though we predict this would be extremely difficult if annuals accumulated mutations causing senescence in older individuals. Apparently, mutation accumulation and other processes accounting for senescence in animals do not operate in plants, perhaps because cells and tissues are constantly rejuvenated for as long as individuals persist.

If lifespans of plants are not determined by senescence, then why do they ever die? The answer to this question may be that often plant deaths are due to external forces rather than the erosion of the viability of individuals over time. The large majority of plants are polycarpic (iteroparous) and will continue grow and reproduce until conditions become inhospitable, or they are destroyed by pathogens, herbivores, or disturbances such as soil disruption by animals, windstorms, or fire. Under stable conditions, some plants can become very old. For example, clones of quaking aspen (*Populus tremuloides*) in the Rocky Mountains have been aged at greater than 10,000 years (Ally et al. 2010), and clones of creosote bush (*Larrea tridentata*) in the Mojave Desert can be more than 11,000 years old (Vasek 1980). The fact that very old individuals of these species are uncommon suggests that they persist merely by good fortune, as they have escaped the numerous catastrophes that have culled the other members of their cohort over the years. Under this view, we expect that any individual plant may have the potential to be immortal; in fact, much of the agricultural industry is based on this possibility, as varieties of many ornamentals and of fruits, including wine grapes, strawberries, cherries, apples, pears, peaches, and bananas, are maintained strictly through clonal propagation. On the other hand, we know that plants accumulate somatic mutations during growth (Chapter 4), so the decline of extremely old individuals may occur because of mutational load and the loss of sexual function (Ally et al. 2010).

Determinants of Leaf Lifespan—At the level of the whole organism, plants may have the potential for long-term growth without experiencing the effects of senescence, but individual organs such as leaves appear to decline in function as they age and are regularly replaced. Furthermore, aging in leaves is more similar to senescence in animals, as the function of older individuals declines with age (Kitajima et al. 2002). In temperate regions with harsh winters, the majority of angiosperms have short-lived leaves that undergo autophagy and are discarded in the fall (Harper 1977; Lim et al. 2007). In the same habitats conifers, some angiosperm shrubs (e.g., *Rhododendron*, *Vaccinium*, etc.), and some ferns (e.g., *Polystichum*), retain their leaves year-round. Most gymnosperms are considered evergreen; they annually produce new needles, leaves, or fronds that persist over the winter and may last from two to seven years. The notable exceptions are a few deciduous species such as the bald cypress (*Taxodium distichium*), larches (*Larix*), and the dawn redwood (*Metasequoia glyptostroboides*). In general, the rate of photosynthesis is lower for long-lived leaves and this pattern is consistent across the tracheophytes (Kitajima et al. 2002; Osborne and Beerling 2003; Matsuki and Koike 2006; Karst and Lechowicz 2007).

Leaf longevity may be a flexible trait that is an outcome of historical selection, but the tradeoffs among leaf retention, the rate of photosynthesis, and leaf senescence are not well understood. Differences in leaf longevity may be viewed from the perspective of returns on investment. For example, we could expect that leaves capable of high rates of photosynthesis would provide adequate energy returns within one growing season. Additional investment in overwintering capacity of these leaves may not pay off, because their higher rates of photosynthesis appears to accelerate senescence (Lim et al. 2007)—perhaps as a consequence of higher production of free radicals and ROS (Khanna-Chopra 2012). In the case of evergreen conifers with lower rates of photosynthesis, it may take several years to achieve the same returns as deciduous leaves, and the rate of decline in photosynthesis capacity with age is correspondingly slower. Apparently, investment in herbivory resistance and characteristics allowing persistence through harsh winters comes at a cost in terms of the rate of photosynthesis (Matsuki and Koike 2006). At the same time, the slower rate of photosynthesis results in lower rates of senescence, which allows leaves to remain functional over a longer period of time. Ultimately, the abscission of perennial leaves may be a function of the cost of maintenance relative to the rate of return in photosynthates produced. It appears that associations between leaf longevity, rates of photosynthesis, and declines in photosynthesis with leaf age are dictated by functional constraints. In other words, the tradeoff between leaf longevity and senescence may not be amendable to selection pressure in the same way as the tradeoff between longevity and senescence is in animals. Since both deciduous and evergreen leafing strategies are distributed widely across the tracheophytes, we can conclude that leaf longevity is an evolutionarily labile trait, but additional research is needed to identify the conditions favoring different leaf lifespans.

Evolution of Monocarpy—While plants with indeterminate growth may have long lives, monocarpic (semelparous) species, including some annuals, all biennials, and a few perennials, die immediately after they reproduce (Harper 1977; Fig. 6.14; refer to color plate). The mortality of these plants is effectively programmed, as vegetative tissues undergo autophagy and the recovered resources are translocated to reproductive structures. Selection has favored this type of plant "suicide" under two different circumstances. The first are annual plants that occupy habitats conducive to growth for only a short period. In these cases, plants may have a slim chance for repeated episodes of reproduction, so the annual life history is favored (Young and Augspurger 1991). Annuals with determinate growth such as *Arabidopsis, Brassica, Cardamine*, and many other mustards typically flower early in the season in habitats where favorable conditions are available for a brief period of time. Apparently, the ephemeral nature of these habitats has been consistent enough for selection to favor plants that have limited root systems and no investment in overwintering structures; they are able to maximize their reproductive output by devoting the majority of their accumulated resources into a single episode of reproduction.

Circumstances favoring monocarpy in perennials (including biennials) may be a consequence of exponentially high returns on investment in reproductive structures (Young and Augspurger 1991). Monocarpic perennials that grow as basal rosettes and produce a single stem upon flowering include the biennial scarlet gilia (*Ipomopsis aggregata*), monument plant (*Fasera speciosa*), some terrestrial bromeliads (*Puya* and *Tillandsia*), Hawaiian silversword (*Dubautia* and *Wilkesia*), and many plants in the Asparagaceae such as species of *Agave, Yucca*, and *Nolina*, plus a scattering of species in the Crassulaceae, Asteraceae, and other families. Monument plants, agaves, and yuccas can live for many years before they reproduce, generating a very large flowering stalk that bears numerous flowers and fruits. This type of "Big Bang" reproduction has been predicted to be favored when fitness returns on investment in reproduction are exponential (Fig. 6.15). For these plants, it is thought that producing smaller inflorescences would result in very low reproductive success either because of few pollinator

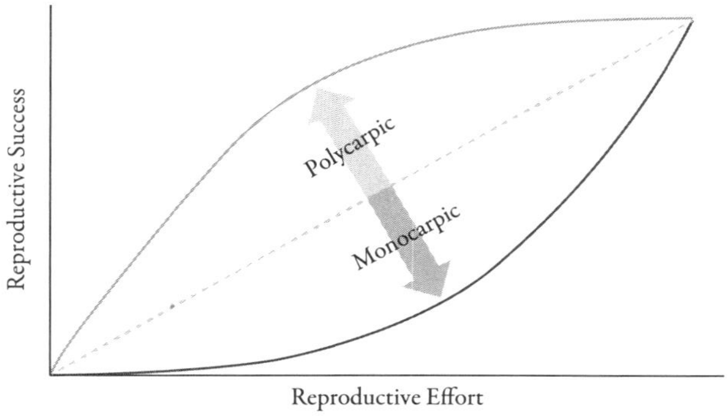

Fig. 6.15. Optimizing returns on investment in reproduction. Iteroparity (polycarpic) is predicted with diminishing returns as resources are allocated to sexual reproduction (upper line). Semelparity (monocarpic) is predicted with exponential returns (lower line).

visits (Schaffer and Schaffer 1979) or high rates of pre- or post-dispersal seed predation (e.g., Tsvuura et al. 2011). Both of these mechanisms may be operating in species of yucca, which must produce large inflorescences to attract their obligate mutualist moth pollinators and also satiate the seed-eating larvae, so large numbers of undamaged seeds will reach maturity (Addicott 1986). In some species of *Agave*, on the other hand, very large inflorescences are required to satisfy the energy demands of their bat pollinators (Good-Avila et al. 2006). While exponential returns on investment in reproduction are not always clearly evident in empirical studies of reproductive success of long-lived monocarpic species (e.g., Udovic 1981), exponential fitness returns from investments in flowering remains our best explanation for this life-history strategy.

For monocarpic species, we would expect that all available resources would be allocated to flower production to maximize reproductive success, but this may not always be the case. Damage to the flowering stem in some annual and biennial species releases axillary buds, which results in multiple flowering and larger numbers of seeds than in undamaged plants. This type of *Overcompensation* in reproductive effort has been observed in populations of *Ipomopsis aggregata* in response to herbivory by mammals (Paige and Whitham 1987). This phenomenon may be due to a developmental constraint in undamaged plants, as apical dominance and flowering are associated with the onset of autophagy in *Arabidopsis* (Melzer et al. 2008). Overcompensation has been associated with increased production of specific gene products and endoreplication resulting in endopolyploidy (Scholes and Paige 2014; Chapter 3), but it is unclear whether these effects are causes or consequences of release from apical dominance. In either case, it appears that reduced reproductive output in undamaged monocarpic plants may be a consequence of the developmental constraints imposed by the strong apical dominance of a single inflorescence.

Life History Evolution—From these discussions, it is evident that the factors determining the timing of reproduction and lifespan in plants are different from those governing the life histories of animals. While senescence may occur for leaves, because cells age along with the entire organ, at the level of the whole organism there is a constant turnover of leaves, flowers, and living tissues in stems, so plants have the potential for extremely long lifespans. For monocarpic plants and individual leaves, these tissues remain highly functional even as autophagy is initiated—plants do not die because of senescence. Leaf turnover and monocarpy, then, are a consequence of selection to maximize reproduction; the processes of degeneration, decline, and increased susceptibility to disease due to aging that are common in animals are apparently rare or absent in plants.

Lifespan in plants is more often a consequence of selection for high reproductive output (monocarpic/semelparous plants) or of external factors that cause accidental death (polycarpic/iteroparous plants).

Note that changes in the reproductive value over the lifetime of individuals are due to both reproductive output and the chances of survival (Fig. 6.14; refer to color plate). The sharp decline in the reproductive value of annuals within a few months of germination often reflects high mortality as environmental conditions become inhospitable. Although perennials may continue to grow and reproduce over many years, mortality due to external factors eventually leads to a decline in reproductive value in older individuals. Extremely long-lived plants, such as large trees and highly clonal woody species, may continue to increase their reproductive output for many thousands of years, but we predict a decline in reproductive value will eventually become evident because of somatic mutation accumulation (Fig. 6.14; refer to color plate). Since there is little or no evidence for the degeneration of tissues and organs in older plants, we can conclude that the differences among life histories is consequence of internal and external factors; plants are not susceptible to the same processes of decline due to late-life expression of mutations (e.g., due to antagonistic pleiotropy) that lead to senescence in animals.

As the growth and longevity of plants are manifestations of constant replication and replacement of cells, they retain many of the characteristics of their microbial ancestors. Just as there is no reason for algal lineages to ever senesce, we should have no expectation that the meristem cell lineages of plants will decline and suffer from degraded function over time. At the same time, plants differ from algae by having more complex structure and organization of cell types into tissues and organs. The size and complexity of plants results in some degree of isolation and perhaps independence among different ramets and metamers of the same individual. Plants face additional challenges as they grow, because branches of larger individuals may be exposed to a range of environmental conditions. In the next few sections, we explore the coordinated growth of complex organs, consequences of the iterative organization, and the responses of individual stems and metamers to variation in the environment.

6.7 DEVELOPMENTAL CONSTRAINTS

Shared regulatory networks may impose constraints on the developmental trajectories of individual structures and growth fields among structures that constitute modules. The ontogenetic relationships among traits are not necessarily linear so that increasing the overall size of a module may result in disproportionate size changes among its component structures. Such *Allometric* relationships among structures may have implications for the functionality of organs. Size differences among structures may come about by either differences in growth rate or variation in the timing of development (*Heterochrony*). Both allometry and heterochrony during development will generate *Heteroblasty*—ontogenetic variation in the size and shape of organs within individuals. Generally, heteroblasty is represented as regular patterns of variation among plant organs along stems or inflorescences (see section on Heteroblasty). In this section, we will discuss the contributions of allometry and heterochrony to size differences among structures within modules and ontogenetic patterns that produce architectural variation along inflorescences and stems.

Allometry—Pleiotropy and shared regulatory elements that control the expression of traits within modules may impose constraints on the growth responses of individual structures. The characterization of the scaling relationships among traits is generally referred to as *Allometry*. This term should not be confused with "allometric," which specifically refers to a differential (exponential) relationship between two traits (Fig. 6.16). Relationships between traits may also be *Isometric*, in which case they change proportionally (linearly) to one another. We expect to find allometry (allometric or isometric relationships) among traits within the same module, and a wide range of relative growth rates among individual structures could be evident.

We can characterize the relationship between any pair of traits (x and y) using a generalized formula for allometry, as $y = kx^a$; where k represents a constant (the y-axis intercept), and a is a scaling exponent that can range from 1.0 (isometric) upward for increasingly allometric relationships. For example, differential growth rates in the length and width of developing leaves result in different shapes; equal rates (isometric growth) would produce a circular (orbicular) leaf regardless of size (leaf a in Fig. 6.16). Increasing growth rate in length relative to width (allometric growth) will produce increasingly elongated (elliptic to lanceolate; leaves b, c, and d) and ultimately linear (acicular or narrowly linear; leaf; e) leaves (Fig. 6.16).

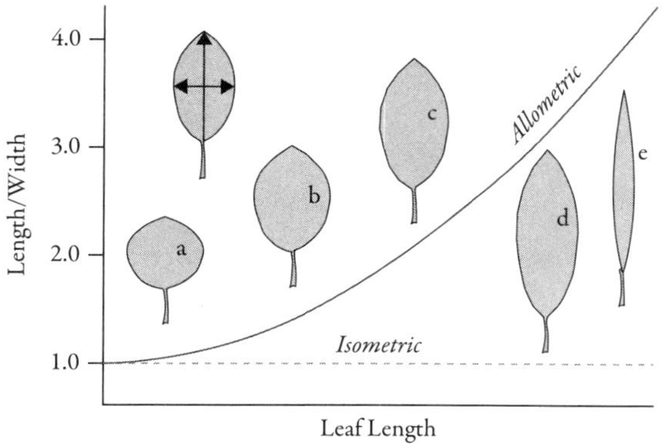

Fig. 6.16. Growth differences in length and width affect the shape of leaves as they grow. Isometric growth, as indicated by the dashed line, produces orbicular leaves (a). More rapid elongation results in allometric growth to produce increasingly narrow leaves from broadly elliptic (b) to elliptic (c), lanceolate (d), and acicular (e).

While ontogenetic processes affect the phenotype of the whole organism in most animals, the same is not true of plants. For example, growing stems often possess leaves at different developmental stages and consequently, of different sizes and shapes due to allometry among different fields of growth across expanding leaf blades (Bar and Ori 2014). *Ontogeny* in plants generally refers to the development of organs (flowers and leaves) beginning from the initiation of primordia on the domes of meristems (Box 3.3). In many plants, mature leaves display allometric growth that results in changes in shape with size; if leaves produced on different portions of the same stem mature at different sizes, then they may also differ in the lengths of dentations or lobes and overall shape (heteroblasty; see section on Heteroblasty). For example, in snapdragons (*Antirrhinum majus*), smaller leaves occur on the lower portion of the stem and are always rounder, while leaves on the upper part of the stem are larger and more elongated (Costa et al. 2012).

Heterochrony—In our discussion of allometry so far, we have made the implicit assumption that morphologies were governed strictly by differential growth rates, but it is also possible that the time of initiation or cessation of growth could generate size differences. During organ development individual structures may grow continually at different rates, or the timing of growth may differ to achieve differences in final size (Li and Johnston 2000). Heterochrony expands on the concept of allometry by recognizing that different morphologies can be attained via both the relative timing and the rate of development.

The term heterochrony was first used by Ernst Haeckel in his 1875 monograph, *Die Gastrula and die Eifurchung der Thiere*. It has been used in a number of different ways since then, but most recent authors adhere to the definition proposed by Gould (1977): "changes in the relative time of appearance and rate of development for characters already present in ancestors" (Gould 1977, page 2). This is a particularly useful definition because it explicitly places developmental processes in an evolutionary and comparative context. Note that heterochrony differs from allometry in that heterochrony refers to how single structures among related taxa develop differently (i.e., comparison to the probable ancestral states) while allometry refers to patterns of development among structures within the organs of a single lineage. In essence, heterochrony provides us with a vernacular of ontogenetic processes that generate morphological diversification.

Heterochrony has been classified into two general categories that refer to the relative sizes of mature structures compared to ancestors (*Paedomorphosis* and *Peramorphosis*; Fig. 6.17). Paedomorphosis refers to the "juvenilization," reduction, or subtraction of structures. Reduction in the final size of structures compared to ancestors may be manifest through reduced growth rate (*Neoteny*),

Process	Rate	Initiation	Cessation
Paedomorphosis			
Neoteny	**Slower**	Normal	Normal
Progenesis	Normal	Normal	**Early**
Postdisplacement	Normal	**Late**	Normal
Peramorphosis			
Acceleration	**Faster**	Normal	Normal
Hypermorphosis	Normal	Normal	**Late**
Predisplacement	Normal	**Early**	Normal

Fig. 6.17. Delineation of heterochrony leading to Paedomorphosis and Peramorphosis. Processes differ based on rate of growth, or the timing of the initiation or cessation of the growth period.

early cessation (*Progenesis*), or late initiation (*Postdisplacement*) of growth. Similarly, the exaggeration of traits (peramorphosis) may come about as a result of an increased growth rate (*Acceleration*), late cessation (*Hypermorphosis*), or early initiation (*Predisplacement*; Fig. 6.17) of growth. Note that identification of paedomorphic or peramorphic structures is only possible if the taxon being studied can be placed in a phylogenetic context, so comparisons to sister species can be made. Determining whether the size differences of structures are due to growth rates or timing of initiation and cessation of growth requires the establishment of temporal landmarks and detailed measurements throughout development. In plants, studies of heterochrony generally start with the initiation of flower or leaf primordia during early stages of meristem development and differentiation.

Studies of heterochrony have primarily focused on animals, but a fairly large literature on the application of these principles to plant development has been established (Li and Johnston 2000). While most studies have compared closely related species and have focused on floral structures, trends over longer evolutionary timescales can also be placed in the context of heterochrony. The reduction of laminar structures in seed ferns to form angiosperm floral components such as sepals, petals, stamens, and carpels can be viewed as a form of progenesis. Similarly, the origin of the much-reduced angiosperm gametophytes in ancestors with gametophytes consisting of thousands of cells came about through both progenesis of development and acceleration (peramorphosis) of gamete production (Friedman and Williams 2004). The application of the principles of heterochrony has the potential to elucidate the mechanisms responsible for the generation of the broad range of morphological diversity displayed among plants.

A number of ontogenetic studies have made comparisons among floral morphologies in closely related species (Li and Johnston 2000; Friedman and Williams 2004). For example, the smaller selfing flowers of *Clarkia xantiana* ssp. *parviflora* compared to ssp. *xantiana* is an outcome of both progenesis and acceleration of organ development (Runions and Geber 2000). Similarly, the development of smaller and closed *Cleistogamous* flowers on outcrossing (chasmogamous) plants (e.g., in *Viola, Lamium, Astragalus, Collomia* and others) is often an outcome of progenesis of petals combined with acceleration or predisplacement of sexual function (Li and Johnston 2000). In other cases, the developmental processes responsible for transitions between pollination syndromes have been elucidated; the tubular flower of *Delphinium nudicaule* was apparently derived from the more open and bee-pollinated flower of *D. decorum* by neoteny of petals and sepals combined with acceleration and hypermorphosis of nectaries (Guerrant 1982). While these studies have been informative in many respects, it has also become clear that the categorical approach to describing ontogenies may be inadequate, as these terms do not always capture the full dynamics of the growth trajectories seen for individual structures during development.

Modifications to the ontogenetic trajectories of plant structures must be governed by underlying gene regulatory networks, so we expect that patterns of gene expression would be coordinated with the timing and rate of development. Antagonistic microRNAs that target gene families controlling flower development may be responsible for heterochronic modifications of floral morphology of *Antirrhinum* species (Geuten and Coenen 2013). Flowers are outstanding models for studies of ontogenetic processes and provide opportunities for integrative studies of gene expression that may elucidate the genetic architecture of heterochronic processes.

Heteroblasty—Consistent patterns of change in the size or shape of flowers or leaves based on order of initiation or spatial position on stems can be referred to as *Architectural Effects* (Diggle 2014). Such architectural effects are evident when developmental processes that depend on the position of each metamer govern the morphologies of organs, rather than environmental conditions and are often due to gradients in microRNAs or other regulatory elements along stems or inflorescences that result in heteroblasty. Heteroblasty for leaf morphology is common in many annuals and herbaceous perennials that die back to overwintering structures below or at the soil surface (e.g., many hemicryptophytes, cryptophytes, and geophytes). Heteroblasty for flower size is common in plants that produce determinate inflorescences such as some genera in the lily family (Liliaceae; including species of *Clintonia, Erythronium, Fritillaria,* and *Streptopus*; Fig. 6.18; refer to color plate).

Regular patterns of variation in the size and shape of flowers and leaves, or the arrangement of organs along stems (*Phyllotaxy*), has been referred to as heteroblasty (Diggle 2014), but some authors reserve this term for cases of abrupt changes in morphology among different stages of plant growth (Zotz et al. 2011). In this text, we use the term heteroblasty in a general context to include ontogenetic differences that produce morphological variation among mature plant organs (flowers, leaves, etc.) that may be gradual or abrupt. We reserve the term *Heterophylly* for cases of abrupt changes in leaf morphology that are induced by environmental conditions (included as cases of heteroblasty by Zotz et al. 2011), which will be discussed later in this chapter in the section on phenotypic plasticity.

Flowers may vary in size within and among individuals, so coordination among structures is critical for proper functioning at maturity. Individual flowers are more analogous to unimodal animals since ontogenetic changes occur as flower buds mature and cease at maturity (*Anthesis*—blooming). In some instances, such as the red columbine mentioned

Fig. 6.18. Archtectural effects on leaf and flower size in *Streptopus lanceolatus* var. *roseus*. The size of both leaves and flowers decrease towards the tips of the stems. Photos by Jason Hollinger and Superior National Forest.

earlier, some floral structures are constrained to a consistent size to ensure functionality (Kliber and Eckert 2004). In other cases, allometric growth among reproductive structures will change the relative positions within flowers, which may have consequences for functionality. For example, the small-flowered varieties of many species such as the blue-eyed Mary (*Collinsia parviflora*), small-flowered gotetia (*Clarkia xantiana* ssp. *parviflora*), and some populations of monkey flowers (*Erythranthe*), display less stigma–anther separation and consequently higher rates of self-pollination (Ornduff 1969; Runions and Geber 2000; Kennedy and Elle 2008; Fig. 11.17; refer to color plate). Such allometric relationships among reproductive structures within flowers may provide evolutionary pathways toward reproductive assurance via selfing in lineages developing smaller flowers (Chapter 9.

6.8 EVOLUTIONARY CONSTRAINTS

Gene regulatory networks may impose interdependencies that generate correlated responses to selection among traits that are members of the same module. Traits that are developmentally linked may not evolve independently such that selection on one may result in responses in a second trait that shares the same regulation. In general, developmental trajectories of individual structures and organs represent pathways for responses to selection, so developmentally linked traits tend to evolve in concert. The underlying genetic architecture of modules may produce *Evolutionary Constraints,* which are defined as anything that slows the approach to, or prevents the attainment of, a phenotype that confers higher fitness. Such constraints affect evolutionary trajectories and the process of adaptive evolution. Because of evolutionary constraints, we expect that the underlying genetic architecture of traits may guide their responses to selection and render some phenotypes difficult or impossible to attain.

Any number of variables may impose constraints on the evolutionary potential of organisms, including simply a lack of heritable genetic variation for selection to act on. For example, in some plants ovule numbers are characteristic of the family (e.g., four seeds in Boraginaceae and Lamiaceae) or genus (e.g., one seed in *Prunus* and ten seeds in *Malus* in the Rosaceae) and remain constant across a range of flower sizes—in these species a response to selection to increase ovule number per flower would not be possible. In other cases, physical constraints may render certain trait combinations impossible to achieve (e.g., a very large fruit supported by a very thin stem). Selection on grain crops for higher yield results in weak stems that cannot support the large seed heads (Reynolds et al. 2009). Physical limitations and lack of genetic variation are constraints that in turn delineate possible trait combinations from infeasible regions of phenotypic space.

More often, evolutionary constraints are discussed in the context of linkage, pleiotropy, or shared gene regulation that cause correlated responses in one trait due to selection on a second trait. In general, we expect that phenotypic evolution will follow paths of least resistance so that relationships among traits during development will determine the trajectories followed during adaptive evolution and lineage divergence. These *Genetic Correlations* are thought to be imposed by the underlying genetic architecture of traits and to define paths of combined trait evolution (Gould and Lewontin 1979). Genetic correlations are generally assumed to impose constraints that render phenotypes outside of developmental trajectories difficult or impossible to achieve. In this section, we will explore the developmental processes that generate morphological variation and their potential to impose evolutionary constraints on patterns of phenotypic evolution.

Genetic Correlations—Genetic correlations transfer the effects of selection on one trait to other traits, due to shared genetic control. Genetic correlations among traits are generally thought to be due to the effects of pleiotropy (including shared gene regulation), but strong linkage can also generate concerted responses. Over time, we expect that recombination will erode associations between alleles at different loci, so genetic correlations due to linkage will generally be transient, while those due to pleiotropy are more permanent. Genetic correlations that are governed by shared genetic determination are thought to impose constraints on the range of phenotypes that can be easily attained through selection.

Genetic correlations among traits fall into two broad categories. First, as described in the previous sections, traits that are members of the same module are expected to share gene regulatory networks and consequently will generally display positive genetic correlations. As selection acts on one trait, we expect to see correlated responses on other traits contributing to the same function due to allele substitutions at pleiotropic loci. For example, selection to increase petal size will often result in longer anthers such that these traits will tend to evolve in concert during evolutionary divergence. We expect that these positive genetic correlations will be maintained for suites of traits that are members of the same module and contribute to the same function (e.g., sexual reproduction).

Genetic correlations may also be generated among traits that contribute to different fitness-related functions. In particular, the available resources will limit growth responses, so investment in one function may come at a cost to others. Plants may increase their investment in vegetative growth, but this reallocation of resources may reduce flower and fruit production. Such tradeoffs in plant function generate negative genetic correlations that are common among life-history characteristics closely associated with fitness (Fig. 6.19). In this figure, note that the phenotypic correlation between allocation to sexual and asexual reproduction is positive as the availability of resources increases. The negative genetic correlation is an outcome of the finite pool of resources available; selection to increase one function results in an automatic decrease in other fitness-related functions. For example, in domestic wheat (*Triticum*), selection for increased allocation to seeds has resulted in lower investment in vegetative structures and roots (Reynolds et al. 2009). To quantify this type of negative genetic correlation we would need to express each trait as the proportion of total resources (e.g., biomass). Just as with the case of evolution of senescence described previously, negative genetic correlations among traits are due to antagonistic pleiotropy that results in tradeoffs among life-history traits that are closely tied to fitness.

Negative genetic correlations that are due to allocation represent physical constraints imposed by resource availability; we would not expect that selection could ever increase one function without an equivalent reduction in other functions that rely on the same resource pool. Positive genetic correlations are not governed by physical constraints so we might expect that they will differ with respect to the strengths and

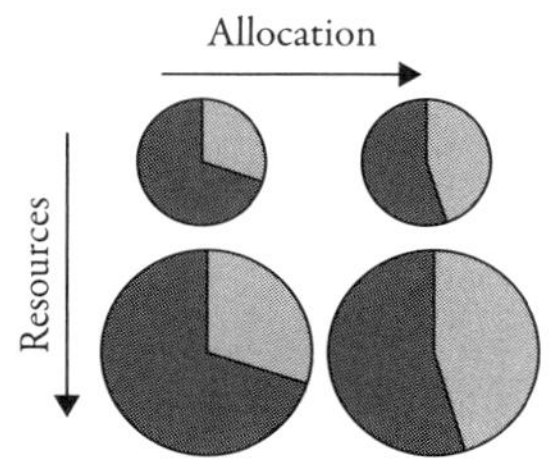

Fig. 6.19. Negative genetic correlations generated among life-history characteristics as a consequence of antagonistic pleiotropy. In this example the purple sections of the pie diagrams represents allocation of resources to sexual reproduction and the blue to asexual reproduction. As resource availability increases both functions increase. The negative genetic correlation is generated as the available pool of resources is allocated to one function at the expense of the other.

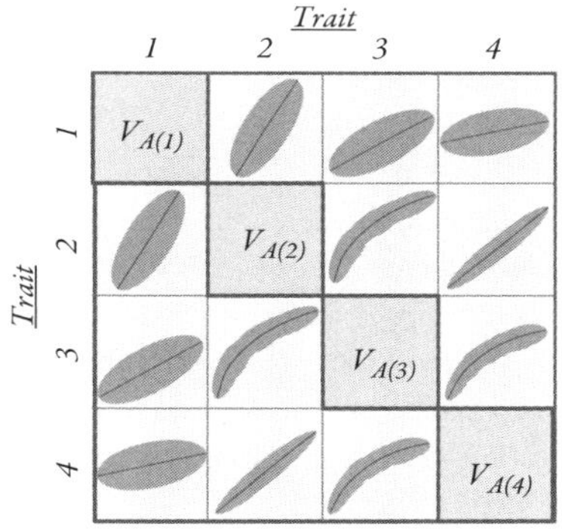

Fig. 6.20. A G-Matrix for four hypothetical traits (1–4). Additive genetic variances are on the diagonal positions and covariances between each pair of traits are illustrated in the off-diagonal cells. Shaded areas indicate the spread of observations around each predicted line.

shapes of their relationships. For any set of phenotypic traits, we can construct a *G-Matrix* of genetic variances and covariances (Arnold et al. 2008; Fig. 6.20). Each pair of traits may have a characteristic relationship that differs with respect to the strength of the correlation (deviations from the line), slope, and linearity (linear of curvilinear). Such G-matrices have been thought to govern evolutionary trajectories and define ranges of phenotypic space (combinations of trait values) that are easily feasible, difficult, or impossible to achieve, and have been proposed as a mechanism to maintain stasis of phenotypes for extended evolutionary timescales (Gould 1966; Arnold et al. 2008).

G-matrices have been considered to represent definitive characteristics of species that govern their evolution and responses to selection (Lande 1979; Arnold 1992), but this may not always be the case. On the one hand we might expect that genetic covariances would define multivariate phenotypic evolution such that past and future trajectories of combined trait expression could easily be predicted, and in some cases, this has proven to be true (e.g., among floral traits of *Brassica campestris*; Dorn and Mitchell-Olds 1991). On the other hand, the majority of attempts to select trait combinations in directions that are not favored by genetic correlations (i.e., perpendicular to the relationship; Fig. 6.21A; refer to color plate) have succeeded to varying degrees (Conner et al. 2011). In this particular example (Fig. 6.21B; refer to color plate), researchers succeeded in manipulating the degree of anther exsertion well beyond and below predicted outcome from the genetic correlation depicted in Fig. 6.21A; refer to color plate).

Numerous studies have found that genetic correlations can be modified by various population genetic processes, including selection, genetic drift, and inbreeding (Arnold et al. 2008), suggesting that these relationships may be much more labile than originally assumed. Changes to the correlation structure among traits may be required to maintain high fitness during adaptive evolution. In the hypothetical example depicted in Fig. 6.12 movement along the ridge of high fitness requires that the slope of the relationship between a pair of traits (represented as the change in orientation of the red to purple and blue rods in Fig. 6.12; refer to color plate) changes to accommodate evolution toward a local adaptive peak.

While we understand that genetic correlations can often be modified, we know much less about how the degree of integration within modules affects the potential for selection to modify these relationships. We might expect, for example, that genetic correlations may be stronger when more precise coordination among traits is critical for maintaining organ function (e.g., the positioning of anther and stigma placement in flowers with specialized pollinators; Fenster et al. 2004). Furthermore, even in cases where the restrictions of genetic correlations can apparently be alleviated by selection, the underlying genetic architecture still imposes constraints such that evolution toward some trait combinations may be slowed considerably (Conner et al. 2011). At this point our understanding of the role of gene regulatory networks for module integration and genetic correlation structures (G-matrices) remains rudimentary, but the potential for more integrative approaches to the study of genomic and phenotypic variation to elucidate the genetic architecture of ontogenetic and evolutionary trajectories continues to grow (Long et al. 2008; Pajoro et al. 2014).

To this point, we have considered the plant phenotype primarily as collections of well-coordinated traits without much discussion of the potential for environmental influences on variation in trait expression. As mentioned previously, we expect the sizes of organs to differ among metamers either because of architectural effects or in response to cues from the local environment. In the next section, we will discuss environmentally induced trait variation, as we explore the causes and consequences of phenotypic plasticity in plants.

6.9 PHENOTYPIC PLASTICITY

As plants grow, leaves developing on each metamer may differ in size and shape as stems encounter different microenvironments and as they respond to seasonal variation in climate and resource availability. Individual metamers often display phenotypic plasticity for morphological traits—particularly those associated with leaf size and shape—in response to environmental cues. For example, leaves shaded by foliage will generally be larger and thinner compared to those exposed to full sun (e.g., Rozendaal et al. 2006), and leaves developing during drought will often be smaller in size and may display increases in trichome (leaf hair) density (e.g., Picotte et al. 2007). While plasticity responses generally result in continual variation in traits, some plants routinely experience starkly contrasting environments and display abrupt changes in leaf morphology (heterophylly—see section 6.9). As discussed in the previous section, variation in leaf size and shape may be due to architectural effects—these systematic changes provide a template for morphological variation that may be further modified by plasticity. Architectural effects and plasticity work in concert, so plants will typically possess heterogeneous collections of leaves because of ontogenetic processes and variation in environmental conditions.

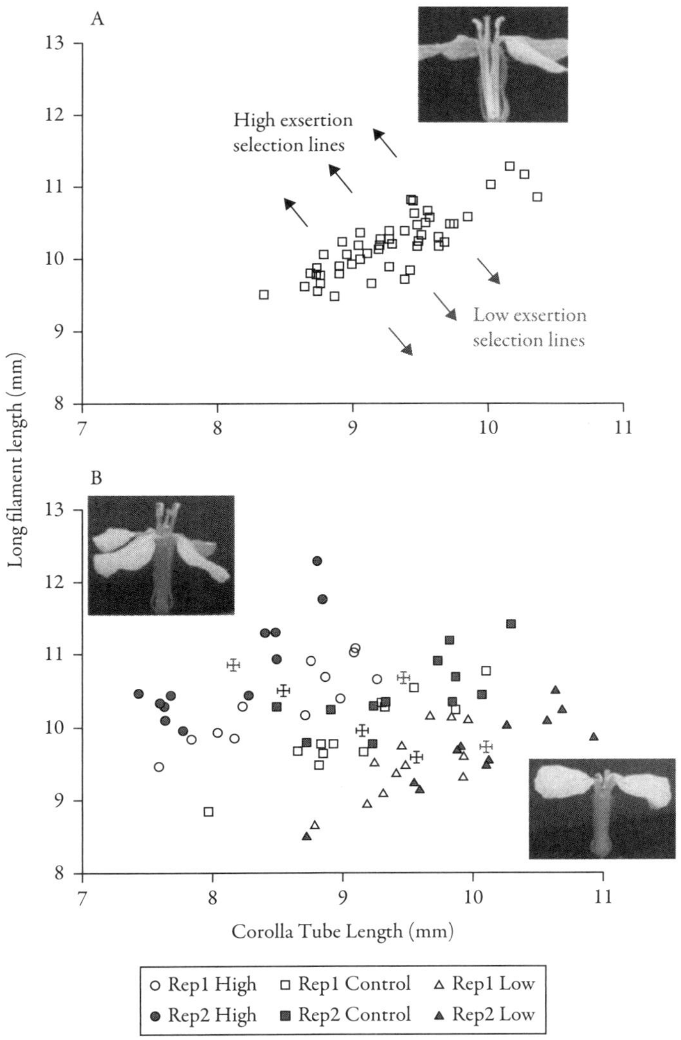

Fig. 6.21. Means for offspring from single crosses showing the genetic correlation between corolla tube length and filament (anther) length in *Raphanus raphanistrum*. The arrows (A) indicate the direction of selection perpendicular to the genetic correlation. The lower panel (B) represents the range of trait combinations displayed in progeny after nine generations of selection. From Conner et al. (2011).

New leaves are continually produced and older leaves may be shed, so the diversity of leaf sizes and shapes may change over the growing season and across plant lifespans. Older leaves may be retained for different periods of time, ranging from days to years, but they eventually senesce or become shaded by new foliage to the point that they represent a liability. Loss of efficient function of individual leaves may also occur if they developed under environmental conditions that are no longer prevalent. For example, leaves developed early in the season in temperate regions will experience cooler and wetter conditions, so their phenotypes may not be optimal for the warmer and drier conditions that may occur in the summer. During growth, many plants are capable of adjusting to changing conditions by shedding older leaves and producing new ones that are better suited to the current environment. This phenotypic flexibility allows plants to optimize growth in temporally variable environments and stands as a hallmark of the success of the embryophytes. In this section, we discuss the importance of phenotypic plasticity as individual metamers respond to prevailing environments.

Defining Plasticity—Plant phenotypes can vary in response to environmental conditions across a broad range of timescales (Fig. 6.1). Phenotypic plasticity is generally defined as environment-dependent phenotypic expression (Bradshaw 1965; Scheiner 1993; Sultan 2003). This definition encompasses all phenotypic modifications, including responses that are strictly physiological, but some authors have taken a more restrictive view by focusing only on morphological characters that are developmentally determined and therefore, are irreversible for an individual metamer (Pigliucci 2001). We should also note that environmentally-dependent variation includes epigenetic modifications that can be inherited by offspring (transgenerational epialleles—Chapter 3; Nicotra et al. 2010; Schlichting and Wund 2014), which blurs the lines between phenotypic plasticity and heritable variation contributing to adaptive processes. We will focus our discussion of plasticity on all aspects of environment-dependent phenotypic variation that is expressed over the lifetime of an individual. This definition encompasses any variation that is influenced by environmental conditions, so it is inclusive of physiological changes, environmentally induced epigenetic modifications, and any changes in gene expression that affect the phenotype. We are including short-term physiological adjustments and acclimation to changes in the prevailing conditions, but our discussion will follow the majority of the literature on this topic (Pigliucci 2001), which focuses on morphological variation that is established during development. Note that our definition includes the environmentally induced changes in gene expression—the conceptual framework of phenotypic plasticity can easily be applied to transcriptomic analyses (Aubin-Horth and Renn 2009).

Studies of phenotypic plasticity typically evaluate changes in single morphological traits in response to contrasting environmental conditions. Plants are particularly suitable for these investigations because they display elevated levels of plasticity in morphology compared to other multicellular organisms. Moreover, identical, or nearly identical, genotypes of plants can often be replicated in multiple environments through vegetative propagation or sowing of seeds from selfing lineages. Usually morphological traits from the same genet or selfing line are compared in two or more environments to produce *Norms of Reaction* (*Reaction Norms*; Fig. 6.22) that illustrate the potential for plasticity responses of individual traits to different environmental conditions (Pigliucci 2001). Most traits evaluated are responsive to environmental variation (i.e., they display phenotypic plasticity), but some traits are unresponsive—they are *Canalized* (fixed)—so phenotypic expression is consistent across a range of environmental conditions (Flatt 2005; Fig. 6.22 top panel). Canalized traits are resilient—the same phenotype is achieved despite environmental variation during development.

We generally assume that the degree of plasticity is a product of past selection, and that phenotypic responses to environmental cues are adaptive (*Adaptive Plasticity*); however, this may not always be the case. Some amount of phenotypic variation within and among individuals may be effectively neutral with no consequences for fitness. Phenotypic plasticity can also result in fitness declines if there are substantial *Costs of Plasticity* including construction costs of traits and if the resulting phenotype proves to be a mismatch for prevailing environmental

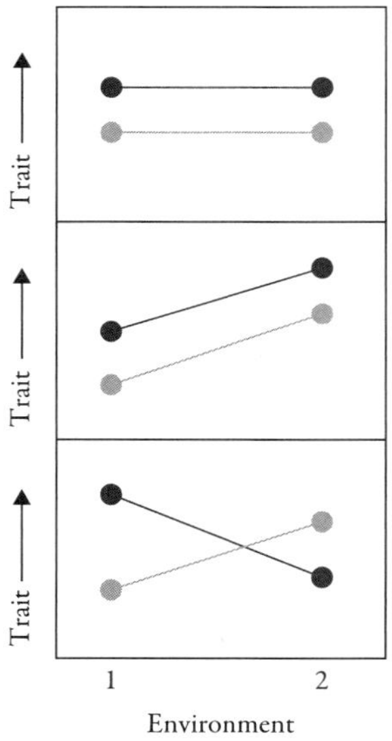

Fig. 6.22. Examples of reactions norms illustrating a canalized trait (top panel), phenotypic plasticity (middle panel), and genotype-by-environment interaction (bottom panel).

conditions. In this section we will explore phenotypic plasticity responses for continuous and threshold traits (heterophylly) to environmental variation, their potential costs, and issues associated with the evaluation of adaptive plasticity responses.

Reaction Norms—Studies of phenotypic plasticity typically focus on the response of traits in contrasting environments. Plasticity is often evaluated under two or more controlled environments where single variables are manipulated and sets of traits are measured on pairs of plants of the same genet or selfing lineage that represent different populations or species (Pigliucci 2001; van Kleunen and Fischer 2005). More rarely, plasticity studies are conducted using field manipulations or by taking advantage of natural environmental variation (e.g., Donohue et al. 2001; Bell and Galloway 2008). The types of environmental variables typically manipulated include temperature, water availability, nutrients, shading by the foliage of other plants, wind abrasion, and herbivore damage. Traits measured include biomass allocation to sexual reproduction and vegetative growth, phenological characteristics such as germination, budding, and flowering times, morphological traits such as differences in above- and below-ground biomass, and the size, shape, and indumentum (surface characteristics such as hairs, bristles, and spines) of leaves. Plasticity studies are also conducted to evaluate plant responses to variation in biotic environments including stem elongation (*Etiolation*) in response to shading from competitors (Schmitt et al. 2003) and the expression of secondary chemicals and morphological changes such as increases in trichome density in response to damage from herbivores and pathogens (Karban et al. 1999; Metlen et al. 2009). Quantification of traits in multiple environments provides information on the range of trait plasticity for individuals and the variance in plasticity responses among genets. By focusing on traits that are understood to be functionally important for fitness, these studies endeavor to characterize plasticity responses for sets of genets, which presumably reflect their potential to succeed under a range of environmental conditions.

Virtually all characteristics of plants vary to some degree within and among individuals. Plasticity studies focus on trait variation that is induced by environmental cues and has the potential to confer environmentally depended increases in fitness. Investigators make a distinction between responses that are "passive" (referred to as "inevitable" by some authors; e.g., Sultan 2003) due to variation in resource availability and those that are "active," as they represent induced responses that have the potential to increase fitness (van Kleunen and Fischer 2005). For example, under conditions of low nutrients or drought, leaves and inflorescences may be smaller, but these responses may be inevitable outcomes of resource limitation and are not necessarily "active" responses. Distinctions between passive and active responses are not always clear, and the magnitude of active responses may be curtailed by size constraints imposed by resource limitations (van Kleunen and Fischer 2005). Even under conditions where resources are limiting, active responses may be evidenced by changes in relative allocation to different functions (e.g., between roots and shoots, or vegetative growth and flower production) or by the production of structures that increase tolerance to stressful conditions (e.g., increases in leaf trichome density in response to drought; Picotte et al. 2007; Nicotra and Davidson 2010). Such active responses require mechanisms that perceive changes in salient environmental conditions and then induce appropriate physiological or developmental responses. The molecular basis of some active responses has been well characterized. For example, signal transduction controls many common plasticity responses, such as phytochrome-induced responses to day length and foliage shading (Schmitt and Wulff 1993) and responses to salt, drought, and temperature stress (Zhu 2002; Shinozaki et al. 2003). We generally assume that active changes in trait expression represent adaptive plasticity responses that are the product of past selection.

Many vegetative traits are responsive to environmental variation, but some—particularly traits associated with sexual reproduction—may be more canalized. Canalization is evident when traits display little or no variation in response to the environment (Fig. 6.22 top panel). In these reaction norms, the lines connect values for the same genet or inbred lineage in two different environments. Canalization is evident because the trait expression for each genet is the same across both environments.

Invariable traits are not necessarily the products of selection for canalization. For example, the consistency of flower architectures (*Merism*)—the number of parts per whorl (e.g., sepals, petals, anthers, etc.)—is often characteristic of plant families (e.g., many core eudicots are quadra- or penta-meris with flower parts in fours or fives, respectively). Such traits are generally considered to be constrained by phylogenetic history rather than recent selection for canalization. Other floral traits that display consistency across a range of flower sizes are more likely to represent outcomes of selection for canalization. Good candidates for recent selection for expression consistency are nectar spur lengths and positions of reproductive structures in red columbine (*Aquilegia canadensis*) that were discussed earlier in this chapter. In this case, these traits display consistency in size across a range of flower sizes (Kliber and Eckert 2004). Moreover, the adaptive value of consistency in these traits is evident as precise positioning is required for effective placement of pollen on the bodies of hummingbird pollinators. Given the lack of covariation with flower size, the consistency of nectar spur length and positions of the anthers and stigma are probably cases of canalization that have been favored by selection.

Phenotypic plasticity is evident whenever differences in trait expression are observed among two or more environments (Fig. 6.22 middle panel). In this example, the reaction norms are parallel so the plasticity responses are the same, but the average level of trait expression across the two environments is greater in one genet compared to the other. It is important to note that reaction norms display only the changes in a particular phenotypic trait between environments and do not necessarily indicate whether plasticity responses are adaptive (van Kleunen and Fischer 2005), for example, in cases where differences in trait expression represent passive responses to resource limitations. Determining whether plasticity responses are adaptive requires careful tracking of the responses of traits that are associated with measures of fitness (e.g., seed production or vegetative growth) as well as those that are putatively of functional importance for the environmental conditions that plants are exposed to—a topic we will explore more thoroughly later in this chapter.

Equal changes in trait expression between environments may be less likely when the reaction norms of genets from multiple populations or species are quantified; more often there is variation in plasticity responses such that the reaction norms are not parallel (Fig. 6.22 bottom panel). Such cases are referred to as *Genotype-by-Environment Interactions (GxE)* and are evident from mixtures of responses, including trait changes in opposite directions as well as reaction norms that display canalization, but any case of non-parallel reaction norms may constitute a GxE response. In these cases, the level and direction of plasticity as well as the mean response across environments may differ among genets. Such diverse plasticity reactions might be expected for collections of genets from different environments where lineages have been subject to contrasting selection regimes, or they may perhaps reflect plasticity responses for traits that have limited effects on fitness so they have not been constrained by selection.

Plasticity Costs and Limits—Fitness gains that may be accrued through plasticity responses of plants to variation in their environments may be compromised for a number of reasons (van Kleunen and Fischer 2005). First, there may be energetic costs associated with sensory mechanisms that promote trait changes and with the construction and maintenance of the induced phenotypes. Since resources are generally limiting, realized plasticity is likely to be compromised compared to optimal responses. Second, phenotypic responses will always lag behind alterations in the environment. Like all organisms, plants are capable of very rapid changes in physiology and cellular metabolism that are evident from alterations in gene expression in response to environmental stimuli (Jones et al. 2012), but phenotypic traits that are established during development are generally not labile, which imposes limitations on phenotypic flexibility and makes optimal phenotypes difficult to achieve.

Costs of plasticity have been categorized as indirect costs associated with sensory and regulatory mechanisms and the costs of construction and maintenance of traits expressed in response to changes in the environment (DeWitt et al. 1998; Callahan et al. 2008). We can refer to these as indirect costs because they may cause fitness reductions due to their association with a plasticity response, but this cost is not a direct consequence of changes in trait expression. Direct costs are defined as fitness reductions that may occur if the phenotype expressed is not a good match for the environment. It is generally not practical to delineate direct and indirect costs; instead, total plasticity costs are usually evaluated as the reduction in fitness associated with variation in trait expression across environments (van Kleunen and Fischer 2005; Fig. 6.23, red line). Using this phenomenological approach, we can compare genets displaying different

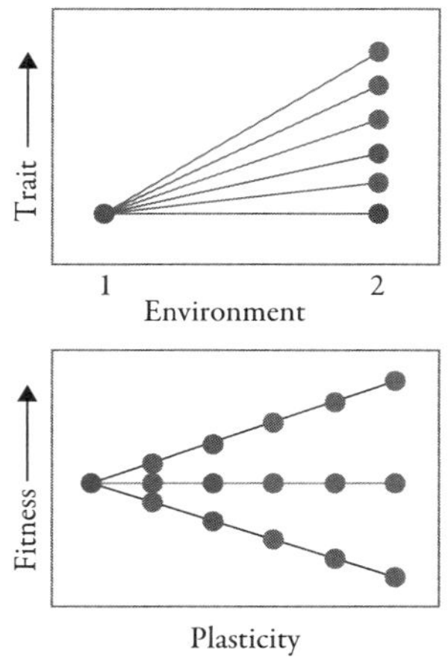

Fig. 6.23. The fitness effects of plasticity to environmental variation. Increasing plasticity (Increasing in environment 2) displayed in the upper panel results in decreases (lower panel, lower line), increases (upper line), or no change in fitness (center line; lower panel).

levels of trait plasticity and measure fitness-associated traits across environments to estimate the costs or benefits of plasticity. Lower fitness observed for the more plastic genets would indicate that there was some cost of trait change, but unless we were able to quantify resource allocation associated with the construction and maintenance of sensory mechanisms and structures, it would not be possible to distinguish between the direct and indirect costs of plasticity.

Spatial Heterogeneity—If a plant were able to perfectly modify its phenotype in response to environmental variation, it would be able to maintain high fitness (i.e., adaptive plasticity) across environments. In essence, plasticity could serve to allow adaptation across spatially heterogeneous environments in place of genetic responses. We could predict that plasticity in this hypothetical "Jack of all trades" species would effectively buffer populations from the effects of selection for local adaptation. Alternatively, with imperfect plasticity, responses to selection could favor local adaptation. We expect that selection is required to maintain plasticity—as long as individuals occurring across a wide range of environmental conditions continued to interbreed the high level of plasticity could be maintained by selection, and genetic adaptation to local conditions would not be favored (Alpert and Simms 2002). While we might imagine that a "Jack of all trades" species would be widespread and successful in a wide range of habitats, such ideal plasticity responses never occur, suggesting that either there are significant costs associated with high levels of plasticity. Most theoretical and empirical analyses of plasticity evaluate responses to spatially heterogeneous environments (Scheiner 1993; Pigliucci 2001), but the implications of variation among habitats for sedentary organisms such as plants requires careful evaluation.

Whether plants generally respond to spatial environmental variation through plasticity or through local genetic adaptation remains as an unsettled question. Depending on the frequency of each habitat and the strength of selection, models of plasticity to heterogeneous environments typically predict selection for the maintenance of plasticity (Scheiner 1993), but plants may typically violate the assumptions of these models in a couple of important ways. First, sedentary individuals experience conditions in a single habitat and are not exposed to conditions across environments. The argument could be made that plasticity would be favored because seeds may be dispersed into multiple habitats; however, even in cases where contrasting habitats occur within close proximity—within the typical dispersal distance of a species—selection may favor local adaptation over plasticity. It is important to recall that most plants generate large numbers of seeds over their lifespan, so individuals could maintain high fitness if a good number of their offspring succeed within the local environment. On the other hand, plasticity could be favored if individuals could increase their fitness by producing more plastic offspring that succeed in a wider range of environments (Pigliucci 2001). While this prediction seems logical, examples of comparative studies of plants demonstrating greater plasticity in response to spatial heterogeneity are difficult to find. For example, the prediction has often been made that plants that are more plastic should have greater success colonizing and invading novel regions and habitats (e.g., Baker 1965), but successful invaders do not appear to display greater capacity for adaptive plasticity (Davidson et al. 2011). On the other hand, spatial heterogeneity could result in selection for alternative alleles in different habitats (balancing selection—Chapter 5). It does appear that more plastic species can sometimes have greater ecological ranges, but this association is not widely consistent

(Davidson et al. 2011). The contribution of spatial heterogeneity to phenotypic plasticity in plants remains an open question.

Temporal Plasticity—The pronounced capacity for phenotypic plasticity in plants reflects their routine exposure to environmental fluctuations. While most animals possess mobility and behavioral mechanisms that enable avoidance of severe environmental challenges, plants are sedentary over much of their life cycle and plasticity enables them to tolerate and often thrive through widely ranging conditions (Bradshaw 1965). For example, individuals living in arid environments often experience diurnal temperature ranges exceeding 30 C and water availability that ranges from flooding to severe drought. The prevalence of temporal environmental fluctuations across biomes and habitats has necessitated the capacity for plants to express a broad range of plasticity responses.

The wide-ranging plasticity displayed by plants could represent outcomes of selection in response to either spatial or temporal variation in environmental conditions (Alpert and Simms 2002). Plasticity would be advantageous for plants that spread vegetatively over large areas because they are likely to experience spatial variation in physical conditions. It has also been suggested that when seed dispersal distance exceeds the environmental grain (patchiness), plasticity will be favored because offspring of single individuals will be exposed to different habitats. Even though we expect spatially heterogeneous environments to favor plasticity, there is little evidence supporting this prediction (e.g., Baythavong 2011). On the other hand, all plants experience temporal changes in their environments at varying timescales, so it is more likely that the pronounced plasticity observed in plants is generally a consequence of environmental variation occurring across days and over seasons (Bradshaw 1965). In spite of the prevalence of temporal variation, the large majority of investigations of phenotypic plasticity have focused on contrasts among static environmental conditions (Pigliucci 2001; van Kleunen and Fischer 2005). This focus on constant conditions may be partly a consequence of the organisms chosen for plasticity studies; the large majority of investigations in plant phenotypic plasticity have used annuals such as *Arabidopsis* and *Impatiens* that may not persist long enough to experience substantial temporal variation. While these studies have garnered a tremendous amount of information on the ecological and evolutionary implications of phenotypic plasticity, plant responses to constant conditions do not capture the particular challenges associated with temporal changes in the environment.

The large majority of plants are perennial and continue growth under seasonal changes in temperature, soil water availability, and the prevalence of pathogens and herbivores, so consideration of temporal environmental variation is important for our understanding of the origins and maintenance of phenotypic plasticity (de Kroon et al. 2005). In most biomes, weather conditions change over seasons in predictable patterns, but variation in the timing of shifts in temperature and the occurrence of precipitation events can be substantial. We expect the phenotypes of new leaves to be influenced by the conditions they are exposed to during development, but under changing conditions, phenotypic responses may lag behind environmental changes for a couple of reasons (Fig. 6.24; refer to color plate). First, short-term environmental fluctuations are common and the cost of producing leaves that are poor matches to the prevailing conditions can be high—particularly in resource-limited environments. The cost of leaf production may not be recovered if a new leaf proves to be a liability for water loss or carbon fixation. To avoid costs invoked by brief environmental fluctuations we expect plant responses to be dampened to some degree such that phenotypic responses are only induced after new conditions have persisted for some period of time (Alpert and Simms 2002; De Kroon et al. 2009).

In general, we expect delayed induction of plasticity when the phenotypic response time is long relative to the typical duration of environmental fluctuations. Hence, physiological responses may be induced quickly as the environment changes, but irreversible traits that are established during organ development will be slower to respond. Such developmental delays slow the total phenotypic response because of the time required for leaves to reach maturity. If such traits are induced too quickly and environmental fluctuations are of a short duration, then mature leaves may be exposed to conditions that are quite different from those that promoted their plasticity responses (Fig. 6.24 bottom panel; refer to color plate). In environments that are highly unpredictable, we would expect a canalized phenotype to have higher

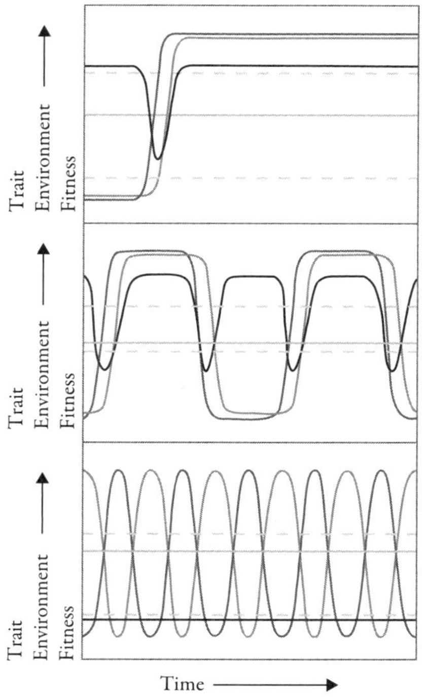

Fig. 6.24. Fitness responses for plastic (red line) and fixed (dashed orange line) genet in environments with different levels of variability (blue line). The trait expression of the plastic genotype lags behind environmental changes (green line). The light-red dashed line indicates average fitness for the plastic genet.

fitness then a plastic one. As the frequency of environmental fluctuations declines, the relative advantage of plasticity increases (middle panel), and we expect the greatest advantage to plasticity when conditions remain constant for long periods relative to the phenotypic response time and the life span of individual leaves (top panel). Whether plasticity or canalization is favored will depend on the average frequency of environmental fluctuations and the time-lag required for phenotypic responses; canalized phenotypes will have higher fitness under rapid fluctuations while plasticity will be advantageous when changes are infrequent and constant conditions are sustained for longer periods (Alpert and Simms 2002; Scheiner and Holt 2012).

Metameric Growth and Plasticity—The metameric structure of plants provides both advantages and limitations to fitness gains that might be accrued through plasticity responses to variation in the environment. On the one hand, plants can more accurately track seasonal changes as plasticity is expressed in new developing metamers during plant growth. This provides a greater level of flexibility in the phenotype as developmentally dependent traits can be modified as stems elongate. On the other hand, under changing conditions older leaves are likely to be poorly matched to the current environment and may represent a burden to total plant fitness. While leaves are likely to provide positive carbon gains soon after they are produced, their fitness benefits will decline as they age. As discussed earlier in this chapter, some of the loss of photosynthetic capacity is a consequence of senescence, but the decline in function would be amplified if shifts in environmental conditions exceeded the compensatory capacity of plasticity in physiological traits. This would be particularly true when conditions become more stressful later in the growing season due to reduced soil water availability and higher temperatures in regions subject to summer drought, or increased shading resulting from leafing-out of the canopy for plants growing in understory habitats. Under altered environmental conditions, older leaves may not be able to maintain positive carbon balances and would become a liability, in which case it may prove more advantageous to the plant to abscise them.

While the ability of individual metamers to respond to conditions as they develop enables flexibility, under changing conditions older leaves may impose a burden. If environmental conditions remain constant for a long enough period of time, then it may be possible for an individual plant to replace all of its leaves with ones that are a better match for the current environment, but note that this effectively extends the phenotypic response time of the entire plant to a large degree. Currently we have little information on the costs and benefits of leaf retention under temporal environmental variation, so this area of study represents a challenge for future plant biologists.

Heterophylly—Heterophylly occurs in some species in which abrupt changes in organ morphology are triggered by the local environment and experienced by newly developing metamers; it includes some of the most dramatic examples of phenotypic plasticity in plants. Since these changes generally occur as plants age, they are effectively ontogenetic, but if the plant remained in its initial environment it would continue to produce only "juvenile" leaves;

consequently, there is not always a clear distinction between heteroblasty because of ontogeny and changes in morphology due to plasticity responses to environmental conditions (heterophylly).

Heterophylly is a polymorphism in leaf morphology within individuals that is induced by the environment (referred to as heteroblasty in Zotz et al. 2011) and is common in plants that are routinely exposed to starkly contrasting physical conditions within their lifespan. For example, the emergent aquatic buttercups (e.g., *Ranunculus aquatilis*) produce finely divided leaves on submerged metamers and entire leaves on emergent portions of the stem (Fig. 6.25; refer to color plate). Submerged leaves differ from emergent leaves by lacking a cuticle and stomata. Other examples are lianas in the genus *Monstera* (Araceae) from tropical rainforests (Fig. 6.26; refer to color plate). These plants begin life in the darkened understory and grow upward along tree trunks until they reach the canopy, where they are exposed to high levels of sunlight. In the understory, *Monstera* plants have small, oval-shaped juvenile leaves that remain closely appressed to tree trunks. Once they reach the upper canopy, these plants produce very large leaves that are divided into lobes and are often marked with characteristic fenestrations (naturally occurring holes). Even though the change in leaf morphology in these heterophyllous plants is typically associated with plant age, the transition from understory to canopy leaves is more dependent on environmental conditions then it is ontogenetic; if young plans are exposed to high light levels, they will transition to adult leaves much more quickly. Since changes in leaf morphology in heterophyllous plants are induced by the environment rather than controlled by ontogeny, we can consider these cases of phenotypic plasticity.

Adaptive plasticity—We generally expect that active plasticity responses are the products of past selection, so they should confer increases in fitness. While this prediction seems reasonable, there are few studies that have demonstrated the adaptive value of plasticity (van Kleunen and Fischer 2005; Schlichting and Wund 2014). Demonstrating fitness increases due to phenotypic plasticity responses of individual traits has been difficult, primarily because any maladaptive plasticity responses should have been eliminated by past selection. In other words, our leverage for demonstrating adaptive plasticity depends on the levels of variation available; if selection on plasticity is

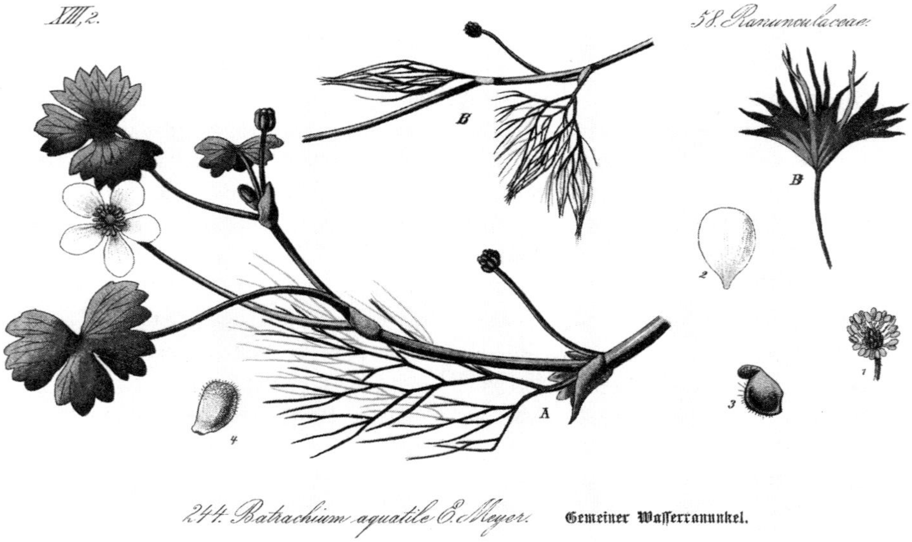

Fig. 6.25. Botanical illustration of heterophylly in *Ranunculus aquatilis*. Submerged leaves are finely divided while emergent leaves are divided into broader lobes. Illustration originally appeared in Otto Wilhelm Thomé *Flora von Deutschland Österreich und der Schweiz* (1885).

Fig. 6.26. Adult (left) and juvenile (right) life stages of *Monstera*. Photo by the author.

strong, then genets with inadequate plasticity will not be present and cannot be contrasted with the genets that have appropriate responses. With a reduced range of variation, it will be challenging to demonstrate that plasticity is associated with fitness, simply because all the genets we test may display similar plasticity responses and consequently will all display relatively high fitness. Conversely, if there is a high level of variation in plasticity (i.e., strong GxE) for genets derived from the same habitat, then it is likely that variation in the trait we are observing is effectively neutral to selection. We expect this because all traits—including the level of plasticity for a trait—that have substantial effects on fitness should display very low variability because of historical selection.

The investigations that have successfully demonstrated adaptive plasticity typically have focused on functional traits for which the implications of expression are evident. One such trait is stem elongation in response to shading (van Kleunen and Fischer 2005). We expect that seedlings competing with other plants for access to light would gain an advantage by increasing their internode length. Note that this etiolation response to shading is induced through the well-known phytochrome signal transduction system and results in plants that are taller (Schmitt and Wulff 1993). Phytochrome responses have been the targets of breeding and gene manipulation in agricultural species to prevent plants from increasing their investment in stem tissue under the crowded conditions typical of crop monocultures (Sawers et al. 2005). Under natural conditions, the increased investment in stems that results from crowding may not represent a liability, as the improved access to light is likely to result in elevated resources and higher levels of reproduction. For example, plants of *Impatiens capensis* from a habitat with more heterogeneous shading demonstrated greater adaptive plasticity across shade conditions in natural environments (Donohue et al. 2001). In this case, access to light outweighed the additional resources that might have been necessary for stem construction—the most plastic genotypes accrued greater average fitness across environments.

Comparing Plasticity in Divergent Lineages—Identifying cases of adaptive plasticity becomes more complicated when we are comparing genets from divergent populations or lineages. In this case, we may find differences in plasticity responses of individual traits that are associated with fitness, but genets derived from different habitats are likely to differ for a number of characteristics (i.e., many traits may be responding simultaneously), so it will be difficult to ascribe fitness effects to the plasticity of a single trait. As we discussed earlier in this chapter, many traits covary because of developmental constraints and genetic correlations. Moreover, if two lineages have been separated for some time then they may differ for large numbers of physiological and morphological traits. Comparisons of divergent genets will be complicated by the difficulty of delineating independent fitness effects of individual trait plasticities.

One solution to the problem of trait covariation is to scramble genomes such that traits are no longer associated with each other—this can be achieved by hybridizing phenotypically divergent lineages for several generations to produce artificial populations of recombinant hybrids (see Chapter 11). This approach also has the advantage of generating a broader range of variation in traits and their plasticities, which increases leverage for assessing the effects of individual traits on plant fitness. Experiments testing the consequences of plasticity in recombinant populations have rarely been conducted. In one such study, a number of leaf traits associated with drought tolerance were tracked over the flowering season, and their cumulative effects on plant growth were evaluated for a population of recombinant hybrids derived from divergent taxa associated with wet and dry habitats (Picotte et al. 2007). In this experiment, individual traits of recombinant plants growing under field conditions displayed a wide range of plasticity responses (Fig. 6.27). Moreover, the plasticities of traits associated with drought tolerance had overriding effects on growth rates. During the wet season (values along the Y-axis) leaves were broader (Fig. 6.27.A), larger (Fig. 6.27.B), and often nearly hairless (*Glabrous*; Fig. 6.27.B), while leaves developing during the dry season were smaller and narrower with increased trichome (leaf hair) densities. Those plants that were able to decrease leaf size and width and increase hair density to a greater extent (i.e., had the greatest plasticity) attained larger sizes by the end of the season (Picotte et al. 2007; Fig. 6.27). Note that some of these recombinant hybrids displayed maladaptive plasticity responses (i.e., increases in leaf width and area and decreases in trichome density in drought). Such inappropriate plasticity responses are not typically observed, and we expect these will be rapidly removed by selection. This study highlights the adaptive value of phenotypic plasticity in temporally variable environments where appropriate responses over the growing season are crucial for growth, survival, and reproduction.

We generally view organisms as products of historical selection so that most of the characteristics we observe—including plasticity responses—are adaptive. This will generally be true for functional traits that are strongly associated with growth and survival under more stressful conditions, but variation in some traits may have little or no consequence for fitness. In general, wide deviations in the direction and amplitude of plasticity among genets collected from the same habitat may be interpreted as cases of variation that are (nearly) neutral to selection. On the other hand, such cases of wide variation in plasticity responses could be due to recent gene flow from divergent populations, so caution needs to be taken when interpreting results from population samples. Experiments evaluating the consequences of trait variation under field conditions will be more likely to provide accurate assessments of the fitness consequences of trait plasticities.

Developmental Processes and Plasticity—Production of adaptive phenotypes through plasticity requires concerted responses among large numbers of traits. While plasticity studies have usually focused on individual features, we recognize that manifestation of optimal phenotypes necessitates coordinated expression of suites of traits. This problem becomes more complicated when the traits involved are members of modules—shared gene regulation may impose constraints that prevent the expression of optimal trait combinations. In the example of adaptive plasticity in *Piriqueta* described earlier, the small narrow leaves that were favored under dry conditions may not be have been attainable unless the relationship between leaf length and width was allometric (Fig. 6.28). In the hypothetical example shown in Fig. 6.28, an isometric relationship results in leaves that are too wide and is maladaptive (leaf b; purple line). An optimal trait combination can be obtained if leaf width decreases faster than leaf length as overall leaf size declines (Fig. 6.28 leaf c; green line). It is important to note that the optimal phenotype for dry conditions in this example can only be achieved through the coordinated responses of both leaf traits; selection would need to act on the underlying shared genetic regulation of development to produce an appropriate response. This multi-trait approach to investigations of phenotypic plasticity emphasizes codependency among plant characteristics and highlights the importance of quantifying collections of traits for understanding the adaptive phenotypic responses to environmental variation.

Environmentally-Induced Changes in Gene Expression—A plethora of studies have demonstrated environment-dependent gene expression in plants (Aubin-Horth and Renn 2009; Nicotra et al. 2010). Such *Transcriptomic* analyses are common and represent an important endeavor in the genomic analysis of plant biology (Box 6.1). While the connection between environmental variation and transcript abundance is well established, it has been more difficult to connect gene expression to specific phenotypic traits. Nonetheless, we can make inferences concerning the patterns and consequences of gene expression that might be associated with phenotypic plasticity. From the gene expression profiles that have been found we can envision two different types of responses (Fig. 6.29; refer to color plate). In the first model, the trait (trichome density in this example) becomes more exaggerated in response to increased expression (*Linear Expression*—orange line). Linear

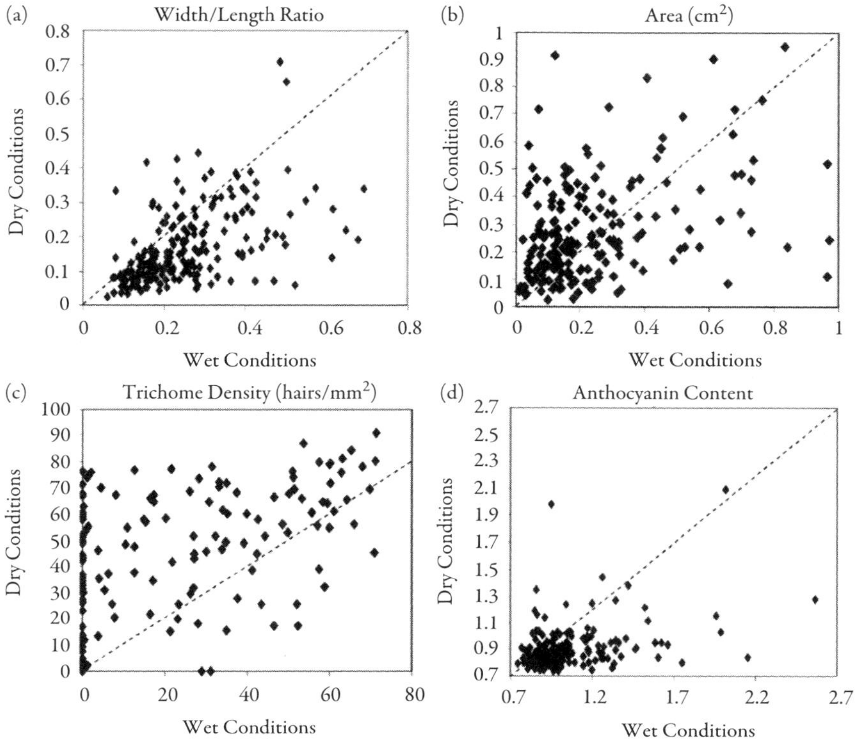

Fig. 6.27. Plasticity responses of leaf shape (A), area (B), trichome density (C), and anthocyanin content (D) for recombinant hybrids of *Piriqueta cistoides* ssp. *caroliniana* in wet and dry environments. Each dot represents the response of an individual genet for a transition from a wet to a dry environment over the growing season. Genets that decreased leaf width and leaf area (A and B—values below the diagonal lines), and increases in trichome density (C—values above the line) had the greatest growth. From Picotte et al. (2007).

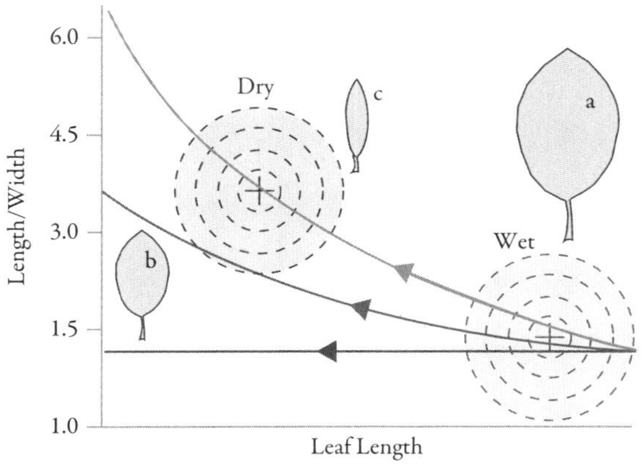

Fig. 6.28. Fitness consequences of allometric plasticity responses for a pair of traits (leaf length and width). The optimal phenotype in the wet environment (blue fitness isoclines) is large and relatively wide (a). An isometric response (purple line) results in a small wide leaf (b) that is non-optimal for the dry environment (red fitness isoclines). The ideal response is allometric resulting in a leaf that is both small and narrow (c).

Box 6.1. Transcriptomics

The transcriptome comprises the complement of messenger RNA (mRNA) molecules that is present within the tissues of an organism at any particular point in time. The composition of the transcriptome is dynamic through development processes and responds to challenges presented by the physical and biotic environments. Changes in the transcriptome can occur over minutes, so charactering the transcriptome requires freezing tissues quickly after collection. After total RNA has been isolated the mRNA component is isolated from other RNAs (transfer and ribosomal RNA) and is then subjected to reverse transcription to produce cDNA libraries. The large majority of mRNA transcripts present in plant cells are typically products of a few genes (e.g., *RuBisCO*), so libraries may be normalized before sequencing to equalize representation of different gene products. The normalized cDNA libraries are then either hybridized to a known set of DNA probes affixed to a glass slide (DNA Chip; Microarray) or are directly sequenced using next-generation technologies (RNA-Seq). Microarrays are analyzed by scanning the DNA chip for fluorescently-labelled cDNA and abundance is estimated from the intensity of the signal (Aubin-Horth and Renn 2009).

Sequencing analysis requires the quantification of the number of sequence reads that correspond to each gene to estimate transcript abundance using bioinformatics pipelines such as RNA-Seq (Aubin-Horth and Renn 2009). In this case, the cDNA libraries are fragmented and used to construct libraries for high-throughput sequencing, such as with Illumna HiSeq instruments. The resulting sequence reads are aligned to a reference transcriptome. Levels of expression can be quantified from the number of reads aligning with different transcribed regions. For studies of gene expression with RNA-Seq it is better not to normalize transcript abundance prior to sequencing. RNA-Seq analysis has an advantage over microarray hybridization because it has greater potential to discover novel gene products.

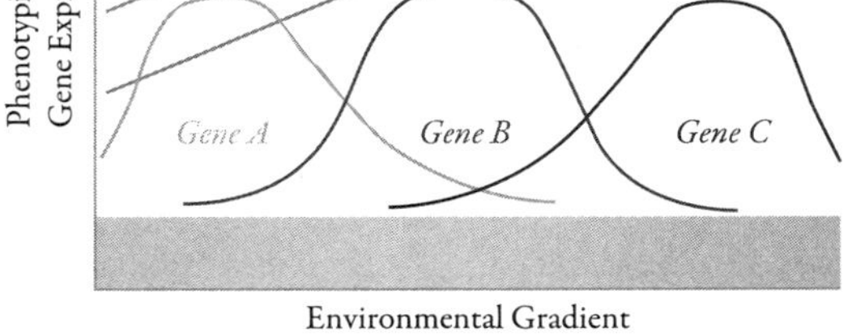

Fig. 6.29. Models for gene and phenotypic trait expression in response to an environmental gradient. In this example leaves are glabrous (hairless) under wet conditions and trichome density increase linearly (from moderate to high) as soil water availability declines. With linear expression (orange line) trichome density is a consequence of increased gene expression. With differential expression trichome density increases as a sequence of genes (A, B, and C) are expressed in response to increasing drought.

responses to gene expression may be a common pattern for many types of traits. Phenotypic variation could also be associated with differential expression among genes such that individual or groups of genes are up or down regulated to achieve a range of phenotypic trait expression (*Differential Expression*—blue, purple, and red lines for the expression of genes A, B, and C). Differential gene effects may also occur, as expression of each gene remains high as additional transcripts become more abundant (i.e., continued high expression for genes A and C as trichome density increases). These scenarios are hypothetical but provide a starting point for envisioning the relationships among genes and phenotypic traits.

Any individual quantitative trait is typically controlled by a number of loci that have varying degrees of influence on levels of trait expression (Mackay et al. 2009), and we expect the underlying genetic architecture of traits to include combinations of linear and differential expression patterns. Even in cases where the functions of genes are fairly well known, clear associations between genes and traits are not always evident. For example, Baker et al. (2012) found that branch number in *Erythranthe guttata* was negatively associated with the expression of three orthologs of *MAX* genes (*MORE AXILLARY GROWTH*), which are known to inhibit branch growth in *Arabidopsis* and crop species. In this study, there was a high level of variation in *MAX* gene expression so a direct connection between the expression of individual genes and the number of branches produced could not be clearly defined.

The lack of direct correspondence between gene and trait expression is common for a number of reasons. For one, these genes do not act in isolation, so the expression of branching is an outcome of interaction among a number of different genes (epistasis). In this case, phenotypic traits may depend on changing expression in more than one gene, which makes it difficult to establish the direct effects of single transcripts. Second, individual genes typically function in more than one role so their expression may be associated with other phenotypic traits (pleiotropy). Pleiotropy results in redundancy such that different genes could have the same phenotypic effect, or the high abundance of a transcript may be due to the multiple roles that it serves across a range of environmental conditions. Third, the timing of expression during development may be critical, in which case it would be difficult to sample tissues within the appropriate time window to detect the salient genes determining specific phenotypic outcomes. It is also possible that traits are induced only above a certain level of gene expression to produce two-state *Threshold Traits* or traits with multiple states (*Meristic Traits*) such as the number of petals on the disk of a flower in the composite family (sunflower; Asteraceae). In this case, subtle changes in the levels of transcript abundance could have large effects on the phenotype. The collective phenotypic influences of the expression of many genes contribute to the complexity of the underlying genetic architecture of traits.

Transcriptomic responses of plants to environmental challenges generally involve changes in the expression of large numbers of genes. While it may be possible to characterize the reaction norms of individual genes (e.g., Aubin-Horth and Renn 2009), such an approach may not be profitable; suites of genes often appear to operate as modules, as they display coordinated changes in expression. Such changes are generally translated into a range of physiological and morphological responses at the phenotypic level. Many of these changes in gene expression are governed by micro RNAs and other regulatory elements, or by changes in methylation and other epigenetic modifications to the genome (Schlichting and Wund 2014). Changes in gene expression that are due to gene regulatory elements may be short-lived, but epigenetic changes in plants can be sustained for longer periods—even after the environmental stress that originally induced the response is removed. In many cases, such epigenetic changes have been shown to be heritable and to be maintained across several generations (Chapter 4).

Facilitation of Genetic Responses—Plasticity is generally viewed as a temporary change in traits, but phenotypic responses may also facilitate genetic adaptation (Schlichting and Wund 2014). As plants respond to environmental challenges, transcript expression patterns change and may expose different genes to the effects of selection. The potential for plasticity to allow colonization of a novel habitat, which then results in the promotion of heritable changes, was first proposed in 1896 (Baldwin 1896). Baldwin framed his arguments in the context of the adaptability of animal behavior to novel conditions, but the definition of the *Baldwin Effect* has been expanded to include a broader range of phenotypic responses. Under this scenario, improved fitness in a new environment is initially due to phenotypic plasticity. Selection then elevates fitness by acting on genes that are only expressed under novel conditions and by enhancing the expression of genes with linear responses to environmental gradients. This type of "evolution facilitated by plasticity" falls under the general category of *Genetic Accommodation*—a process by which environmentally induced phenotypic variants are selected to become genetically determined (West-Eberhard 2005). Note that in plants, genetic accommodation could occur through genetic or heritable epigenetic changes in gene expression and consequently may be quite common—particularly in cases of stress responses (e.g., Grativol et al. 2012).

If conditions are sustained over a number of generations then the capacity for plasticity may ultimately be lost as the phenotype becomes canalized for optimal expression in the new environment—a type of genetic accommodation referred to as *Genetic Assimilation*. Genetic assimilation can be detected by comparison among closely related lineages. For example, invasive populations of *Prunella vulgaris* that have colonized forest habitats have lost the capacity for plasticity (Godoy et al. 2011). Similarly, advanced-generation hybrids of *Piriqueta* display a more canalized leaf phenotype compared to early-generation hybrids (Picotte et al. 2007). In this case the leaves of advanced-generation hybrids are intermediate in width and have high trichome densities, and these plants display higher fitness than the parental taxa within their range in central Florida (Cruzan 2005). Genetic assimilation and other cases of genetic accommodation may be quite common in plants as they often have the capacity for high levels of plasticity and sustained phenotypic responses through the inheritance of epigenetic modifications acquired during vegetative growth.

6.10 THE DIVERSITY OF PLANT PHENOTYPES

The impressive diversity of plant life forms and their reproductive and vegetative morphologies are the products of developmental processes and plastic responses to environmental variation, shaped by hundreds of millions of years of evolution. It is astonishing to realize that the coordinated development and environmental responses of an enormous number of individual characteristics is required to generate functional phenotypes. Much of this choreography is a function of foundational developmental programs that were probably established early in the evolution of the land plants. As plants met new challenges, these programs were elaborated or coopted for new roles to generate the diversity of plant life present today. The persistence of the current wide range of plant life forms into the future will depend heavily on their continued respond to environmental challenges through plasticity and ultimately adaptive genetic modifications.

From the preceding discussions of gene expression and plasticity, it becomes evident that progress in developing models of the genetic architecture of phenotypic traits may be thwarted by a high level of transcriptomic complexity. While we may have information on the functional roles of some gene products from studies with model organisms, we also recognize that large numbers of genes typically affect quantitative traits, and the sum effects of more subtle influences of many individual genes of small effect can be substantial. An alternative to detailed gene expression analyses is to take a phenomenological approach—to partition the observed phenotypic variation into its genetically and environmentally determined components and to evaluate the genetic architecture of traits by mapping the genomic locations of the controlling loci. In the next chapter, we will discuss the field of quantitative genetics, the genomic mapping of loci controlling quantitative traits, and the application of methods for the assessment of the heritability and genetic architecture of phenotypic traits.

6.11 QUESTIONS FOR CONTEMPLATION AND DISCUSSION

1. How has indeterminate growth contributed to the success of the embryophytes and their phenotypic diversification? What are the advantages of this strategy over determinate growth?
2. In what ways are apoptosis and autophagy important for plant function, growth, and development?

3. How is it that some life forms persist in environments where others effectively disappear? How does the combination of environmental conditions and growth forms contribute to plant diversity? Identify habitats where ephemeral growth forms such as therophytes and geophytes are favored by biotic or by abiotic constraints.
4. What conditions favor monocarpy and big bang reproduction?
5. How has metameric structure contributed to the success and diversification of land plants?
6. What are the advantages of parcellation and integration of modules? How does shared genetic determination contribute to adaptive plasticity responses? How might it constrain adaptive genetic responses?
7. What are the differences between allometry and heterochrony? How do these developmental processes contribute to phenotypic diversity?
8. How do architectural effects differ from phenotypic plasticity responses? How do both contribute to heteroblasty and heterophylly? Can you imagine an experiment that could partition the contribution of each to changes in leaf morphology as plants age?
9. What are the possible relationships among module integration, genetic correlations, and evolutionary constraints? Under what situations do you expect plant traits to display positive versus negative genetic correlations?
10. What are the differences between canalization, phenotypic plasticity, and genotype-by-environment interactions? When might you expect canalization versus plasticity to be favored by selection?
11. What are the differences between passive and active plasticity responses? Could we consider both of these types of plasticity to be adaptive?
12. Why is it often difficult to demonstrate adaptive phenotypic plasticity?
13. What are the relationships between gene expression, epigenetic modifications, and phenotypic plasticity?
14. Are epigenetic modifications a form of phenotypic plasticity, or are they more similar to genetic modifications? What about transgenerational epialleles—is that still a form of plasticity?

7

QUANTIFYING PHENOTYPES

IT CANNOT BE too strongly urged that the problem of ... evolution is essentially a statistical problem ... when we know the numerical answers to these questions for a number of species, we shall know the deviation and rate of change in these species at the present day—a knowledge which is the only legitimate basis for speculation as to their past history, and future fate.

Walter Frank Raphael Weldon, 1893, Proceedings of the Royal Society

7.1 QUANTIFYING PHENOTYPES

Plant characteristics vary among individuals and populations and provide the raw material for evolution by natural selection. We learned in the last chapter that the combined effects of genetically determined developmental processes and environmental influences generate morphological variation. Similarly, physiological and biochemical traits are affected by differences in genotypes and environments. Understanding the potential for traits to respond to selection requires the quantification of the proportion of variation among phenotypes that is determined by genetic differences. As suggested by the epigraph quote, early biometricians such as Raphael Weldon, Francis Galton, and Karl Pearson could see the potential for statistical analyses to provide insights into evolutionary processes, even though they did not have the tools to quantify the proportion of phenotypic variation that was genetically determined. In this chapter, we discuss the quantification of phenotypes and the advantages and limitations of statistical and genetic decomposition of trait variation to provide insights into their potential to undergo adaptive evolution in novel and changing environments.

It is fairly easy to observe variation in traits among individual plants growing in the field and under cultivation. It is quite another thing to realize that, by measuring many individuals of known relatedness, we can partition the proportion of variation that is genetically determined and more important, the degree to which trait differences will be reliably transmitted from one generation to the next (i.e., *Heritability*). The tendency for quantitative characteristics of plants and animals to be heritable has been well known from observations of the outcomes of breeding efforts over the last few thousand years (Lush 1937). Even though the transmission of continuously varying traits was widely recognized, this tendency was not quantified until the early twentieth century when heritabilities were estimated in cultivated plants such as tobacco and corn (e.g., East 1915). The mathematical treatments provided by Fisher (1918) further demonstrated that variation in these traits could be accounted for by multiple Mendelian loci (i.e., they are *Polygenic*) and that the variation in traits could be partitioned into genetic variance components.

Methods for the decomposition of the variance in trait expression into components representing heritability, dominance, and epistasis were further developed in the mid-twentieth century (Cockerham

1954; Kempthorne 1954). Statistical evaluation of trait heritability has been particularly important for selective breeding of domestic plants and animals (Mather 1949). The application of these methods in evolutionary ecology became more widespread after 1960 with the first publication of Falconer's *Introduction to Quantitative Genetics*, which has been updated and continues to be a valuable introductory text (Falconer and MacKay 1996; see Lynch and Walsh 1998 for a more comprehsive treatment of the topic). While some have suggested that quantitative genetics will rapidly become supplanted by genomic approaches, this prediction has yet to be realized, and methods for the analysis of quantitative traits remain as valuable tools in evolutionary studies (Hill 2010). In this chapter, we explore the quantitative genetic analysis of plant traits. We discuss approaches for the identification of genomic regions and of individual genes that account for the variation of ecologically relevant traits.

7.2 QUANTITATIVE GENETICS

As we learned in the last chapter, complex networks that include large numbers of genes and regulatory elements typically determine phenotypic traits. Even in the case where traits are discrete (present or absent) or can be categorized into several expression levels (meristic), the underlying genetic architecture may be quite complex as there are typically genetic modifiers of small effect in addition to loci (genes) that govern the majority of variation in polymorphic traits. Furthermore, phenotypic traits are often affected by the environment, so the total variation that we observe generally represents the sum effects of environmental and genetic influences. The partitioning of total phenotypic variation into its genetic and environmental components is the focus of the field of *Quantitative Genetics* (Falconer and MacKay 1996; Hill 2010).

Mathematical evaluation of phenotypic traits began in earnest in the late nineteenth century with the work of Raphael Weldon, Francis Galton, Karl Pearson, and other members of the biometrician movement. Pearson worked on the analysis of quantitative biological data and developed a number of statistical methods for describing phenotypic variation. He focused on distributions of variation and covariation among traits (e.g., Pearson's correlation coefficient), but these analyses were not explicitly placed in the context of the genetic determination of traits. It was perhaps Person's denial of the relevance of Mendelian genetics for quantitative traits that impeded his ability to advance understanding of the genetic basis of quantitative traits. Meanwhile, his colleague and collaborator Ronald Fisher demonstrated the Mendelian basis of continuously varying traits in his 1918 paper (Fisher 1918), which effectively established the field of quantitative genetics.

Fisher evaluated the effects of alleles segregating at a single locus on a phenotypic trait as deviations from a central value that is defined as the midpoint between the phenotypes of the homozygous genotypes (Fig. 7.1; $h_1 = 0.5$). Variation in the effects of alleles that are segregating at a locus affecting the expression of a trait (*QTL—Quantitative Trait Locus*) is determined by their relative expression (*Dominance*) in the heterozygous state. The dominance of each allele (h; $h_2 = 1-h_1$ with two alleles) will determine the phenotype of the heterozygous genotype. When two alleles have the same expression in the heterozygote the level of dominance is equal ($h = 0.5$) for both, and their expression is considered to be *Additive* (each allele contributes equally to trait expression). The dominance of each allele can vary from zero to one; perfectly additive expression is probably rare, but the concept of additivity becomes important when considering the heritability of traits across sexual generations.

Note that a number of terms are used to describe the lack of complete dominance by one allele in the

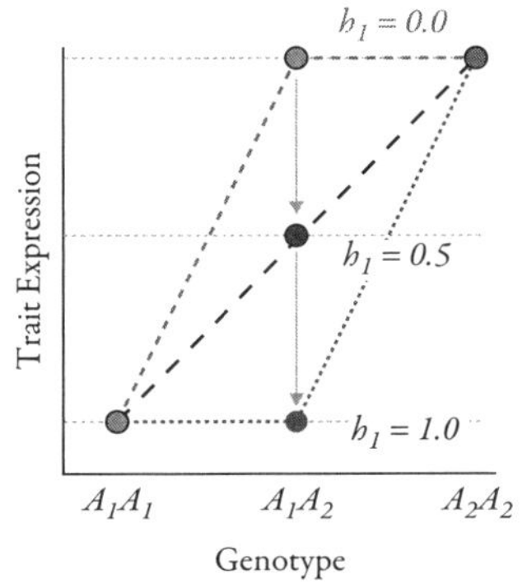

Fig. 7.1. The effects of alleles at a single locos on the expression of a phenotypic trait.

heterozygote (*Partial* and *Incomplete Dominance*, and *Codominance*), and are sometimes distinguished based on the effect of the two alleles on the phenotype (blended expression versus coexpression, respectively). Cases of codominance (coexpression of alleles to produce a composite phenotype) are rare so this classification is not particularly useful. When h falls between zero and one, we can describe dominance as partial or incomplete, and if dominance is either greater than one or less than zero then this would be a case of overdominance or underdominance, respectively.

To apply quantitative genetics methods, we assume that a large number of QTLs are contributing to variation in each phenotypic trait (i.e., the trait is polygenic). As suggested by Fisher, increasing the number of loci controlling a trait to more than two generates a distribution of values that becomes more continuous and is normally distributed when allele frequencies are equal (i.e., produces a bell-shaped curve; Fig. 7.2; refer to color plate). In this simple example, we assume that alleles at each locus occur at equal frequencies, and there are up to four loci (one locus, A; two loci, *A* and *B*; or four loci, *A, B, C,* and *D*). The normal allele at each locus has no effect on the trait and the alternative alleles (*A′, B′*, etc.) all have equivalent effects—allele effects are additive within and among loci and there is no epistasis. Differences in dominance and interactions among loci will affect the distribution of values and will rapidly increase the complexity of translation between genotype and phenotype. The large majority of traits we may choose to measure for their ecological and evolutionary importance will be controlled by many QTLs and will display a continuous—although not always a normal—distribution. A primary focus of quantitative genetics is to determine the average effects of alleles segregating at loci that influence phenotypic traits and in particular, to delineate the proportion of genetic variation that is additive (see the Genetic Determination of Traits section).

Quantitative genetic methods assume that loci controlling variation in traits have equal effects. This is probably never the case, as studies generally show individual locus effects follow a skewed distribution, with few loci having large effects, and the frequencies of loci increasing as their effects diminish (Fig. 7.3). In the example depicted in Fig. 7.3, 333 loci affecting flowering time in maize were detected, but only 7 of them changed this trait by more than one day (Buckler et al. 2009). Specific features of organisms such as flowering time, petal length, or leaf area may typically be controlled by a few hundred loci, and traits that are tied closely to fitness, such as plant size or growth rate, could be affected by thousands of loci. At the other extreme, defense responses to herbivores or

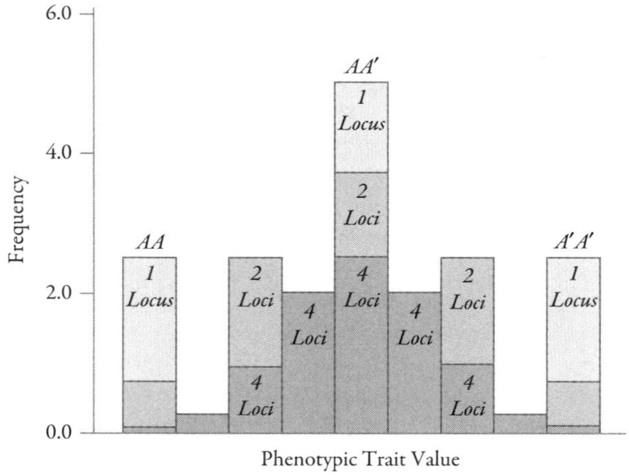

Fig. 7.2. The distribution of phenotypic trait expression with an increasing number of QTLs. With one locus (blue bars) there are three categories; this increases to five categories with two loci (purple bars), and nine levels of trait expression with four loci (red bars). Allele effects are assumed to be additive within and among loci.

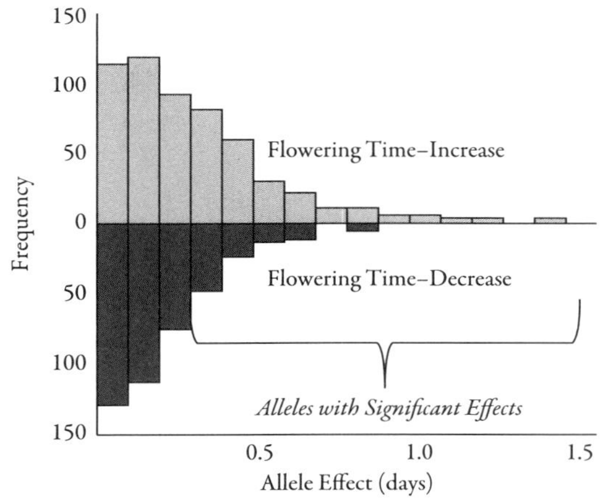

Fig. 7.3. The magnitude of effects of QTLs for flowering time in maize. Alleles caused increases (bars above zero) or decreases bars below zero) in flowering time from zero to 1.5 days. Only alleles with larger effects were significantly different from zero. Modified from Buckler et al. (2009).

pathogens may involve relatively few genes (Holeski et al. 2012). In all of these examples locus effects are not equal, but the methods for estimating genetic components of trait variation that we discuss in the next two sections are robust to departures from equal locus effects and also perform well when the number of loci affecting traits is relatively small (Falconer and MacKay 1996; Hedrick 2009).

In the next two sections, we discuss the relative genetic determination as manifested by phenotypic consistency within and among clonal replicates of genets (*Broad-Sense Heritability*), and in the subsequent section, we will explore the consistency in trait expression across sexual generations (*Narrow-Sense Heritability*). Our discussion will focus on both the theoretical underpinnings of heritability estimates and on the practical application of methods to estimate heritability. To accurately estimate phenotypic variance components, we need to take into account a number of environmental and genetic influences that may bias the values we obtain. In particular, male (pollen donor) and female (pollen recipient) parents differ in their potential to affect offspring phenotypes, so we will end this section with a discussion of *Maternal Effects.*

The Genetic Determination of Traits—The focus of quantitative genetics—and evolution in general—is on variation among individuals rather than on mean trait values, because the range of trait expression sets an upper limit on the potential for change across generations. When we were discussing phenotypic plasticity in Chapter 6, we were making comparisons among trait means for each genet across environments. Focusing on the distribution of deviations from the mean (*Variance*) allows us to evaluate the opportunities for selection and other evolutionary processes to precipitate changes in traits over generations. In studies of trait evolution, we evaluate the phenotypic variance (V_P), which is calculated as the sum of squared deviations from the mean ($V_P = \Sigma(\bar{P} - P_i)^2$; where $\bar{P}$ is the trait mean and P_i is the trait value for each individual). Variances represent convenient summaries of the level of trait variation and hence, the potential effects of selection on trait evolution.

The potential for trait evolution depends first on the proportion of variation that is determined by the complement of alleles (genotype) that each individual is carrying relative to the influences of the environment. We can define the contributions of the genotypes (V_g) and environments (V_E) to the total phenotypic variance simply as their sum:

$$V_P = V_G + V_E.$$

Variances have the convenient property of being additive in the same way that we can sum the sides of a right triangle to determine the length of the hypotenuse (i.e., according to the Pythagorean Theorem $a^2 + b^2 = c^2$, which is analogous to $V_G + V_E = V_P$;

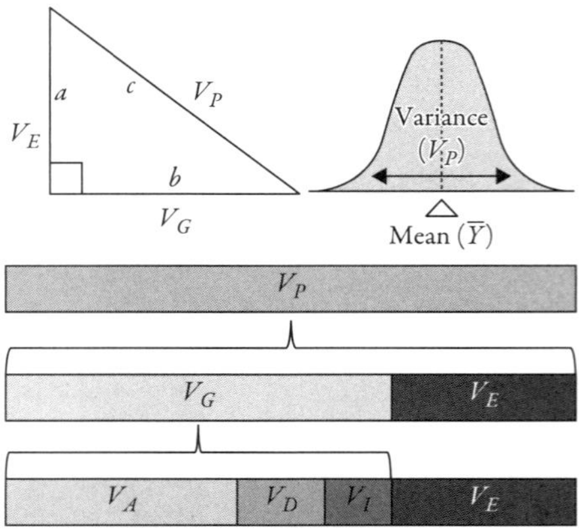

Fig. 7.4. Decomposition of the phenotypic variance (V_P) into its genetic and environmental components. The phenotypic variance is represented as the distribution of values around the mean ($\bar{Y}$). The phenotypic variance is decomposed into V_G and V_E according to relationships among orthogonal (uncorrelated) variances. The genetic variance can be further decomposed into its additive (V_A), dominance (V_D), and interaction (epistasis; V_I) components.

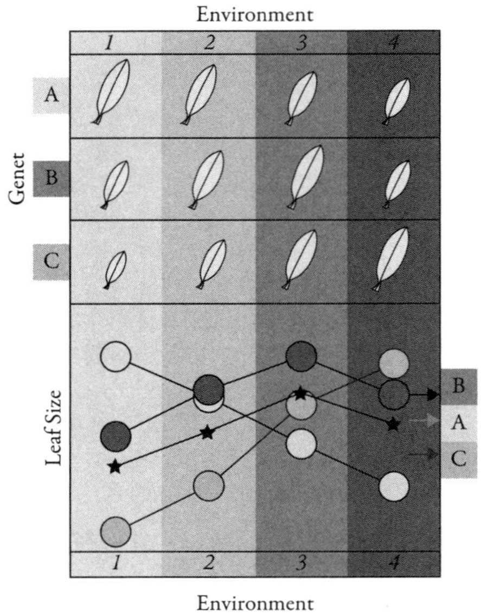

Fig. 7.5. A common garden experiment for measuring broad-sense heritability. Top panel; three genets (A, B, and C) are replicated in four environments (1–4) and leaf sizes are measured. Bottom panel; reaction norms for leaf size are plotted for each genet. Means for each genet are indicated on the right axis (arrows and corresponding letters). Average leaf sizes for all genets in each environment are indicated by the location of stars.

Fig. 7.4). Note that this relationship holds true only if the effects of genotypes and environments are completely independent—that they are orthogonal (perpendicular) to each other ($COV_{GE} = 0$). This will generally not be the case, as we discovered during our discussion of phenotypic plasticity in Chapter 6; genets respond to environmental variation with correlated trait expression (plasticity) and genotype-by-environment (GxE) interactions, which represent lack of independence between genotypes and environments for trait expression.

The COV_{GE} represents the degree to which genotypes respond to environments in different ways, so it is more closely related to GxE than it is to plasticity (correlated responses to the environment). For example, the COV_{GE} will be greater than zero, as genotypes tend to demonstrate independent plasticity in response to resource availability (Fig. 7.5 lower panel). In this example, the average leaf size across genotypes differs among environments as indicated by the location of the stars—if these averages were the same across environments then the V_E would be equal to zero, but the COV_{GE} would depend on the degree to which genotypes change order across environments (Fig. 7.6 lower panel). Independent responses to environmental variation reduce the independent effects of V_G and V_E on phenotypic variation. We account for this interdependence by explicitly defining the contribution of COV_{GE} to the total phenotypic variance:

$$V_P = V_G + V_E + COV_{GE}.$$

When genotypes maintain exactly the same order across environments, COV_{GE} is equal to zero, and V_P is the sum of V_G and V_E (Fig. 7.6 top and middle panels).

The relationships just described allow us to decompose the total phenotypic variance into its genetic and environmental components, and to evaluate the relative genetic determination of phenotypic traits. The genotypic variance is difficult to determine directly, unless environments are carefully controlled. Even then, our ability to keep conditions perfectly constant across an experiment is limited; variation in plant growth and trait expression within growth chambers and across greenhouse benches is common.

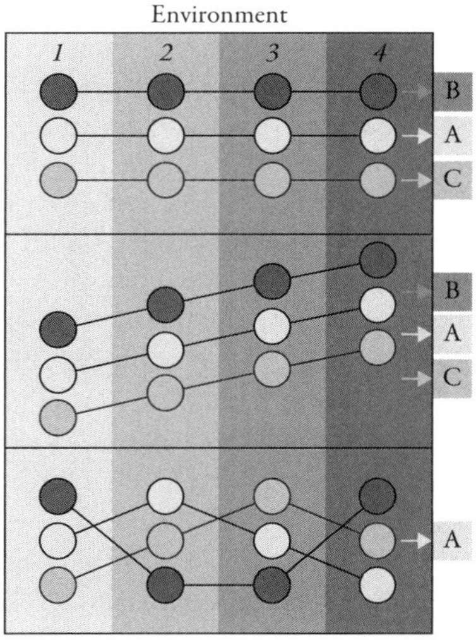

Fig. 7.6. Three scenarios for the responses of individual genets (A, B, and C) to four different environments. In the upper panel each genet is canalized ($V_E = 0$), but there is variation among genets ($V_G > 0$). In the middle panel there is phenotypic plasticity (V_E and $V_G > 0$). In the bottom panel there is strong genotype-by-environment interaction (GxE) so the $COV_{GE} > 0$ and all three genets have the same mean ($V_G = 0$).

Consequently, it is more convenient to evaluate genetic determination as the proportion of variation due to differences among genotypes, which is an estimate of the broad-sense heritability:

$$H^2 = \frac{V_G}{V_P}.$$

The broad-sense heritability can be estimated by placing replicates of multiple genotypes in different environments and measuring trait expression (Fig. 7.5). This is relatively easy to do with many plant species, as they can be cloned by vegetative propagation of rhizomes, tubers, stem cuttings, or tillers. Traits are then measured for each genotype across all the environments and the means calculated—the variance among these means represents an estimate of V_G (Fig. 7.5). The phenotypic variance is calculated as the variance among all plants including all genotypes and environments, and comparing the genotypic (variance among genet means) to phenotypic variance provides an estimate of the broad-sense heritability. In the example in Fig. 7.5 the genet means are not identical, indicating that V_G, and consequently, the broad-sense heritability is greater than zero.

Estimates of H^2 are sensitive to the diversity of genotypes included in the experiment and are dependent on the range of variation represented by the environments used. It is important to note that variation in H^2 can be difficult to interpret, because this value is a ratio of two variances—increasing or decreasing the numerator or the denominator will have large effects on heritability estimates, so interpretation of this parameter needs to be approached with caution (Visscher et al. 2008). For example, V_G for a trait will be greater if we were to collect genets from across multiple populations occurring in different habitats, compared to sampling only from within a single population. Similarly, H^2 will be sensitive to the variation in environmental conditions—it will be smaller if we test our collection of genets across a broader range of environments and larger if we test variation among genets in a single environment. The design of such *Common Garden* experiments will depend on the goal of the study and the degree to which the experimenter wants the estimate of broad-sense heritability to reflect levels of variation typically found in natural populations. Testing heritability in controlled environments such as growth chambers will produce inflated estimates, and if we wish to evaluate the typical genetic determination of traits in nature, then we will want to place our common gardens in the full range of conditions where the species we are studying typically occurs.

Estimates of broad-sense heritability are primarily a function of levels of variation among genets and relative trait consistency across environments. Highly canalized traits will tend to display higher H^2 when genets differ in trait expression (Fig. 7.6 top panel). Similarly, plastic traits may display high broad-sense heritability if the relative ranking of trait expression remains consistent for genets across environments (Fig. 7.6 middle panel). Low variation in trait expression among genets and high levels of GxE across environments will result in low estimates of heritability (Fig. 7.6 lower panel). Broad-sense heritability reflects relative variation in trait expression among a collection of genets grown in a particular set of environments and does not reflect aspects of genetic architecture such as the distribution of QTL effects or the number of QTLs influencing a trait (Visscher et al. 2008).

Broad-sense heritability is useful in evaluating the consistency of trait expression for members of the same genet (clone) and has applications for many agricultural species. Trait consistency within genets across environments is important for the quality of products from fruit trees, wine grapes, and vegetable crops that are propagated clonally. On the other hand, broad-sense heritability does not accurately predict the similarity between parents and their sexually produced offspring. To understand heritability of traits across generations, we will need to decompose the genotypic variance into components, due to the average additive and non-additive effects of QTLs—a topic we will explore in the next section.

Heritability across Sexual Generations—The degree of resemblance between parents and offspring is critical for the transmission of phenotypic variation across generations and for facilitating phenotypic responses to selection. The total genetic variation described earlier (V_G) does not adequately predict the similarity between parents and offspring because genetically determined differences among genets will include the dominance effects of alleles segregating at each locus and the effects of interactions among alleles at different loci (epistasis). Dominance effects reduce the heritability of trait expression, because heritability represents deviations from the midparent value (i.e. $h = 0.5$; Fig. .1). A perfectly heritable trait would have additivity across all QTLs such that all offspring would display trait expression that was exactly midway between the parental phenotypes. This scenario is similar to the hypothetical model of *Blending Inheritance* that was favored by many scholars during the nineteenth century. The fallacy of blending inheritance and perfect additivity is that all siblings would be identical, and over generations, all individuals in the population would converge on the same phenotype, which would be the average of the variation expressed across all individuals when the population was first established. Of course, this never occurs, because both dominance and epistasis contribute to variation among siblings and families.

Darwin never accepted the blending inheritance model, but instead favored *Pangenesis* (from the roots "pan"—encompassing, and "genos"—origin). This model assumed that particles of inheritance (*Gemmules*) were transferred to offspring in different mixtures, which explained the variation among siblings that was commonly observed in plant breeding and animal husbandry. Mendel's astonishing insight was to realize that the idea of particulate inheritance could be reduced to just two particles (alleles), and that the alleles segregating at a locus interacted (i.e., by dominance) in a way that explained the patterns of variation that he observed between parent and offspring generations of his pea plants. The complexity of the translation between the genotype and phenotype—including the pervasive role of epistasis—was not fully realized until the advent of modern genomics, which has allowed a more thorough characterization of gene regulatory networks and the genetic architectures of traits.

It may seem counterintuitive that the same processes that generate variation—dominance and epistasis—also reduce the heritability of traits. On the other hand, it is important to note that deviations from the midparent value contribute both additive and dominance variance components; in other words, the average dominance (h) across QTLs is not equivalent to V_D. To understand this, we can imagine a best-fit regression line for each of the three dominance scenarios depicted in Fig. 7.1; the slope of this line and its fit to the observed phenotypes indicates the degree to which a QTL contributes to V_A. For example, when the A_1 allele is completely recessive ($h_i = 0$) the homozygous A_1A_1 genotype has the lowest trait expression, and substituting one A_2 allele will increase the trait value; consequently, we expect that selection could increase the frequency of the A_2 allele even though there is no difference in trait expression between the heterozygous and homozygous A_2A_2 genotype. QTLs segregating for alleles with unequal dominance still contribute to additive genetic variation because the average phenotype of one homozygote plus the heterozygote differs from the alternative homozygote (i.e., in this example the average phenotypes of $A_2A_2 + A_1A_2 > A_1A_1$). In the case where $h_i = 0.5$ there is perfect correspondence between the regression line and the observed levels of trait expression, and such a QTL would contribute only to V_A. In all cases of dominance, some proportion of the variance contributed by the QTL will be additive; the only time there is no contribution to V_A is for cases overdominance and underdominance where both homozygotes exhibit equal trait expression (i.e., the regression line is flat so the slope is zero).

Epistasis among QTLs may contribute to additive genetic variation in much the same way as dominance among alleles segregating at one locus. To explain this, we can apply the same principle just described to explain the contribution of dominance to the additive variance by comparing the average trait expression for different combinations of alleles at both loci. Differences among averages of different combinations of alleles at pairs of loci will contribute to V_A, unless alternative double-homozygous genotypes display equivalent phenotypic expression—similar to the effects of overdominance and underdominance described earlier. As long as there is a consistent directional change in trait expression with allele substitution at one or both of the interacting loci, there will be some contribution of epistasis to the additive genetic variance.

The interaction genetic variance may also be converted to additive genetic variation if one of the interacting QTLs becomes fixed for one allele (Hedrick 2009). In this case, a proportion of V_I is effectively converted to V_A and V_D because, without the obfuscation due to variation at an interacting locus, alleles segregating at the remaining QTLs will have more direct effects on trait expression. The loss of interaction genetic variation and its conversion to additive variation can be particularly important in small populations where increases in V_A are commonly observed after some inbreeding (Goodnight 1988). Persistent inbreeding will likely result in the loss of genetic variation as the population becomes more homozygous (Visscher et al. 2008), but this will also depend on the mating structure; populations consisting of selfing lineages that rarely outcross may maintain high levels of genetic diversity for extended periods of time, while biparental inbreeding will more quickly result in a population that is completely homozygous. The additive genetic variance within a population could be elevated by gene flow, and in general, we recognize that our estimates of variance components are subject to population history including levels of inbreeding, gene flow, and—as we will discuss further later—the strength of selection on traits.

Narrow-Sense Heritability—From this discussion, it becomes clear that the similarity between parents and offspring will be primarily determined by the proportion of genetic variation that is additive, so estimating V_A will be critical for predicting the heritability of traits across generations and for predicting responses to selection. We can decompose the total genetic variation (V_G) into its constituent components using the same principle that we invoked to define the genetic and environmental variances (Fig. 7.4). Applying the principle of additivity of variances, we can delineate the additive (V_A), dominance (V_D), and interaction (V_I) components of genetic variation"

$$V_G = V_A + V_D + V_I.$$

By definition, these variance components are independent (orthogonal) so they are uncorrelated and covariance terms do not need to be considered. The total phenotypic variation can then be decomposed into its genetic and environmental components:

$$V_P = V_A + V_D + V_I + V_E + COV_{GE}.$$

It is generally not feasible or of much utility to estimate the individual effects of the dominance and interaction variance components, so these are generally pooled into "non-additive" genetic variation. The additive variance component, on the other hand, is extremely important, as it defines the proportion of the genetic variance that is transmissible from one generation to the next. This parameter defines the narrow-sense heritability (h^2), which we define as the proportion of the total phenotypic variation that is additive:

$$h^2 = \frac{V_A}{V_P}.$$

This narrow-sense heritability is often referred to as the *Breeder's Value*, as it provides an estimate of the potential response to selection—in agricultural applications it is used to predict probable outcomes of breeding efforts for crop improvement. Estimation of h^2 has value for studies in evolutionary ecology, where it can be used to indicate the potential for traits to respond to natural selection. Just as with broad-sense heritability, narrow-sense heritability is a ratio of values that will be sensitive to the variation among genotypes and environments used for its estimation. In the next section, we discuss different methods for estimating h^2 and some of the caveats associated with interpretations of the outcomes of these experiments.

Estimates of Narrow-Sense Heritability—Estimating the additive variance and narrow-sense heritability requires crossing experiments where trait expression is compared between parents and offspring (*Parent-Offspring Regression*) or among groups of offspring from different sets of parents (*Sibling Comparisons*). In this section, we discuss the basis and application of these prospective methods, which aim to predict future responses to selection. In the next section, we will explore an alternative approach that is retrospective—the *Response to Selection* is measured for offspring after selection has been applied to a population.

The similarity between parents and offspring is the most intuitive method for estimating h^2. Under ideal conditions, traits are measured for both generations growing together in the same environment, but this is rarely feasible (e.g., as in the case of annual species—discussed later in this section). If the trait is heritable, then we expect to see a correlation between the mid-parent (mean for each offspring's parents) and average offspring trait values (Fig. 7.7 top panel). As the strength of this relationship increases the slope (b) approaches a one-to-one correspondence between the mid-parent and offspring trait values, which would represent a case where all of the variance is additive ($h^2 = 1.0$). Note that in this experiment we are comparing phenotypic variances, so even under tightly controlled conditions there will be environmental effects that will result in deviations away from a perfect regression. If we are using the mid-parent value, then the slope is equivalent to the heritability ($h^2 = b$), and if we are using trait expression for one parent from each cross, then $h^2 = 2b$ (Falconer and MacKay 1996).

When we consider designing experiments to estimate narrow-sense heritability using parent-offspring regression, the difficulties of this approach quickly become apparent. Heritability estimation based on the similarity between parents and offspring is inconvenient, because both generations typically cannot be grown in the same environment. This is a particularly difficult problem for field estimates of heritability, because no two growing seasons are exactly

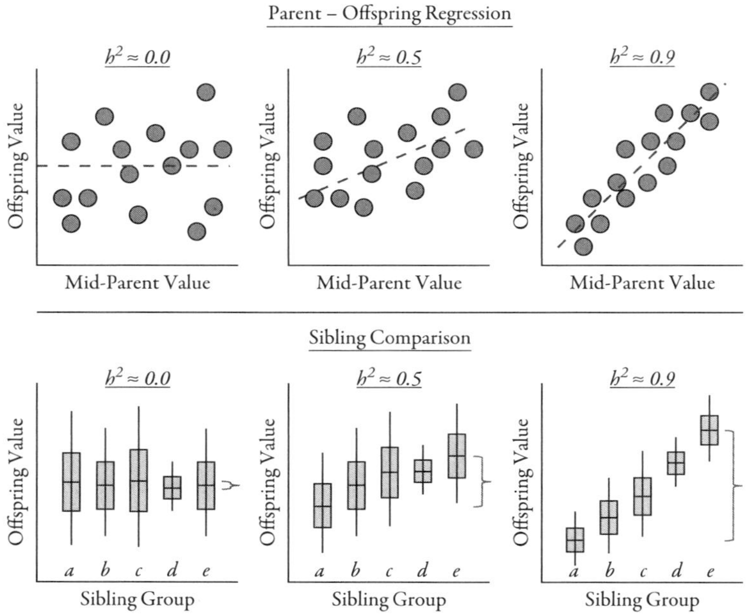

Fig. 7.7. Estimates of narrow-sense heritability (h^2) from parent-offspring regressions (top panel) and sibling comparisons (bottom panel). For parent-offspring regression the heritability is based on the slope of the regression line (b), and for sibling comparisons it is based on the variance among family means compared to the average within-family variance.

alike, and trait expression compared between years may be strongly influenced by the environment. Even if we were to grow plants under tightly controlled conditions such as a growth chamber, a bias could be introduced if parents and offspring were grown at different times. Such differences between the parent and offspring environments would result in a perfect heritability that does not correspond to the 1:1 relationship for mid-parent and offspring trait expression described here and would be likely to over- or underestimate the narrow sense heritability.

The large majority of studies estimating narrow-sense heritability utilize various modifications of sibling comparisons (Fig. 7.7 lower panel). In this type of experiment we can obtain estimates of V_{Total} (the total variance among all individuals), V_{AF} (the variance among family means), and V_{AP} (the average variance among progeny within families). We do not need to compare parents to offspring, and we can make comparisons among seed families to estimate the narrow-sense heritability. For a full-sib design (progeny in each family have the same parents), we can estimate narrow-sense heritability as:

$$h^2 = \frac{2V_{AF}}{V_{AF} + V_{AP}} = \frac{2V_{AF}}{V_{Total}}.$$

Note that, V_{AF} provides an estimate of the additive variance (but see further explanation later in this section) and V_{AP} provides an estimate of V_E. Just as with broad-sense heritability, estimates of these variance components will be affected by the diversity of genets and environments included in the experimental design. This type of experiment is typically conducted in a single environment, but trait expression in multiple environments can be evaluated if plants can be replicated by vegetative propagation. In this case, we can estimate an additional variance component among genets within families (V_{AG}), which is equivalent to the total genetic variance (V_G). The variance among genets should always be larger than V_{AF}, because it should include both the additive and non-additive genetic variance components.

The primary difficulty with the full-sib design is that each plant is crossed to only one other individual, so there is the potential for V_{AF} to be inflated by the effects of dominance and epistasis, in which case it would not accurately reflect the additive genetic variance. This bias is due to the fact that the effects of interactions between genomes in paired crosses is only tested against single individuals, so a more robust design crosses every individual reciprocally with a number of others (Fig. 7.8). In this *Diallele Cross*, we can obtain independent estimates of components for the variances among pollen donors (V_{PD}) and recipients (V_{PR}; also referred to as sires and dams, respectively). These provide independent estimates of the additive genetic variance for calculating heritability based on the among-donor and among-recipient variance components, respectively:

$$h^2 = \frac{4V_{PD}}{V_{PD} + V_{PR} + V_{AP}}, \text{ and } h^2 = \frac{4V_{PR}}{V_{PD} + V_{PR} + V_{AP}}.$$

Pollen Recipient

Pollen Donor	P_1	P_2	P_3	P_4	
P_5	X_{15}	X_{25}	X_{35}	X_{45}	$\overline{X}_5$
P_6	X_{16}	X_{26}	X_{36}	X_{46}	$\overline{X}_6$
P_7	X_{17}	X_{27}	X_{37}	X_{47}	$\overline{X}_7$
P_8	X_{18}	X_{28}	X_{38}	X_{48}	$\overline{X}_8$
	$\overline{X}_1$	$\overline{X}_2$	$\overline{X}_3$	$\overline{X}_4$	

Fig. 7.8. Diallele cross design using different sets of pollen recipients (P_1–P_4) and donors (P_5–P_8). Cell values (X_{ii}) represent seed family mean trait expression for each cross. The variance among recipients (V_{PR}) and donors (V_{PD}) are calculated from the marginal means for each group of parents ($\overline{X}_1$–$\overline{X}_4$ and$\overline{X}_5$ –$\overline{X}_8$, respectively). In a full diallele design the same set of individuals are crossed reciprocally so they are used as both pollen donors and recipients (see Maternal Effects).

We can average the pollen donor and recipient variances to obtain a more robust estimate of narrow-sense heritability:

$$h^2 = \frac{2(V_{PD} + V_{PR})}{V_{PD} + V_{PR} + V_{AP}}.$$

Diallele cross experiments are applied widely in plant breeding programs and studies of plant evolutionary ecology to evaluate the evolutionary potential of quantitative genetic traits (Falconer and MacKay 1996; Visscher et al. 2008). These experiments can be made more robust by including larger numbers of plants, but the size of such experiments increases exponentially with the number of donors and recipients (i.e., 3 x 3 requires 9 crosses, which increases to 16, 25, 36 for 4, 5, and 6 plants, respectively). The amount of effort required to conduct diallele cross experiments is substantial, so such studies typically test offspring in only one environment. In an example of one such study, the heritabilities of scarlet gilia (*Ipomopsis aggregata*) floral traits were tested in a diallele cross design using a common garden in its mountain meadow natural habitat (Campbell 1996). Since information was available on the strength of pollinator-mediated selection on floral traits in this species (Campbell et al. 1996), the authors were able to make substantive inferences concerning the evolution of floral morphology. Such thorough analyses of potential trait evolution in response to selection in nature are uncommon, but in cases where the primary selective forces and target traits can be identified, they provide important insights into the potential for evolutionary responses to environmental challenges.

Sib-comparison crossing designs have advantages over parent-offspring regression, as they allow for the estimation of the genetic covariances of the G-matrix, if multiple traits are analyzed in the same experiment (Fig. 6.20). Note that these genetic correlations are measured among traits using family means from a diallele or other sibling comparison design. Understanding the genetic covariance structure of traits allows prediction of correlated selection responses and the potential for evolutionary constraints to render some trait combinations difficult to achieve. On the other hand, as discussed in Chapter 6, the predictive power of G-matrices may only be useful for short-term projections, as they are not necessarily stable and may vary among populations and over time.

Response to Selection—An alternative approach for estimating narrow-sense heritability examines the translation of selection on the parental to the next offspring generation. Taking a retrospective approach to estimate narrow-sense heritability avoids the problems associated with limited samples of parents and complicated crossing designs that may produce estimates of additive variance that are confounded with other genetic and non-genetic influences (e.g., see the discussion on maternal effects later in this chapter). To conduct such an experiment, we would first measure trait expression for a population of plants (base population) and then select individuals with the most extreme trait values for crossing (selected population; Fig. 7.9). The difference between the trait means of the base and selected populations represents the strength of selection (S, the phenotypic selection coefficient). We then grow the offspring and measure their trait expression (offspring selected population). The narrow-sense heritability is calculated as the response (R; difference between the base and offspring selected population means; Fig. 7.9) relative to the strength of selection:

$$h^2 = \frac{R}{S}.$$

This is often referred to as the *Breeder's Equation* because of its application for the improvement of domesticated plants and animals (Lush 1937). The advantage of this approach is that an experimenter can potentially examine heritability under more realistic conditions compared to the prospective approaches described above. The primary limitation is that selection response can be evaluated on only one trait at a time—especially considering the amount of effort involved. On the other hand, selection experiments provide an opportunity to measure indirect responses in non-target traits due to genetic correlations (Fig. 6.20).

There are difficulties with selection response experiments, including the bias that can be introduced if the offspring and parental generations are grown at different times. To avoid this problem, a control treatment that consists of random crosses among plants in the base population (offspring random population) can be included in the experiment. With this

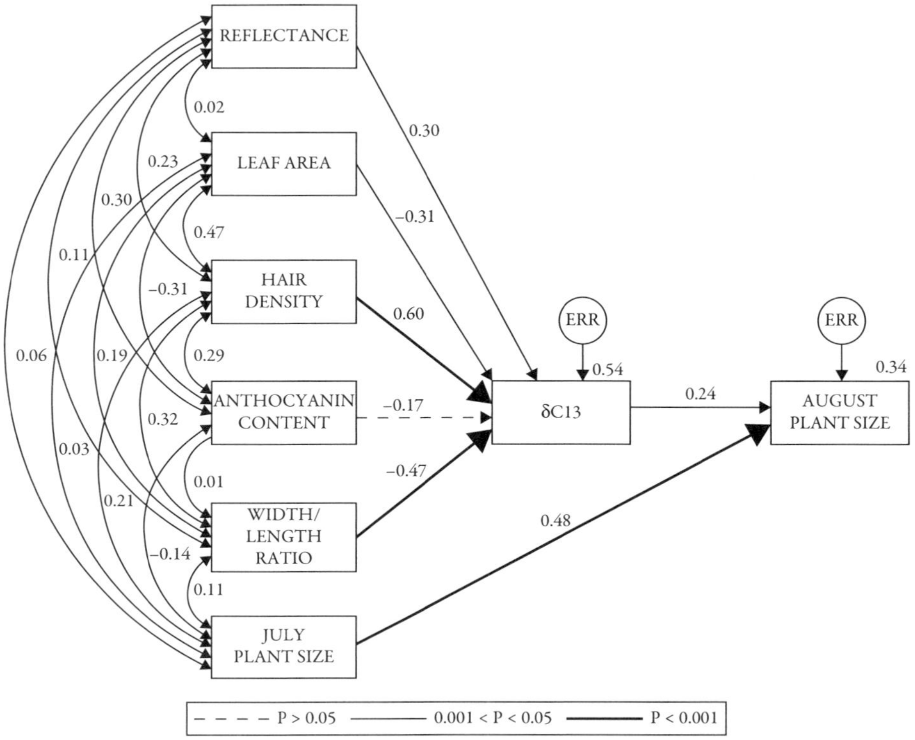

Fig. 7.9. Path analysis for the effects of leaf morphological traits on proximate (δC13—water use efficiency; WUE) and ultimate (August Plant Size) fitness estimates in *Piriqueta cistoides* ssp. *caroliniana* grown under drought conditions in the field. Single-ended arrows represent the effects of each leaf trait on WUE and plant size. Double-ended arrows indicate correlations among leaf traits. From Picotte et al. (2007).

design, we expect the offspring random population mean to be the same as the base population mean under identical conditions, so including this control allows robust estimates of the response to selection in the offspring generation. Note that we are not able to test the selected parents under the same growth conditions as the offspring, so we are assuming that there is no genotype-by-environment interaction that is based on the level of trait expression. We can also strengthen this experimental design by replicating several base populations and by growing the offspring in multiple common gardens.

Heritability Estimates for Long-Lived Species—Application of methods for the estimation of heritability is generally limited to species that have faster growth rates and shorter generation times. Evolutionary ecology studies of heritability in plants have generally focused on annual species. Or vegetative propagation of perennial species has been used to estimate broad-sense heritability as an upper bound for h^2 (Geber and Griffen 2003; Visscher et al. 2008). The latter strategy is not particularly productive since narrow- and broad-sense heritabilities are often not strongly associated. In some economically important forest trees, plantations have been established to facilitate the estimation of narrow-sense heritability (Cornelius 1994). Sibling comparison studies for such long-lived species require careful planning and patience, as it often takes a decade or more before traits of mature individuals can be measured. In plantations, and in other circumstances where the pedigrees of plants are known, the sibling comparison approach described here can be applied, or estimation procedures based on similarities between parents and offspring living in the same environment can be used (i.e., the animal model; Sillanpaa 2011). Natural plant populations generally consist of individuals representing multiple generations that cannot be distinguished, so pedigree-based methods cannot be applied.

An alternative approach for estimating heritability in long-lived species is to use algorithms that use information from neutral genetic markers to estimate relatedness among individuals and combine these with measurements of trait expression (Ritland 1996; Ritland 2000). To apply this method, individuals in a natural population are assayed for genetic markers that are used to estimate relatedness. This approach has not been widely used because it is often difficult to find adequate levels of population structure; high levels of variation in relatedness over a relatively small area are required to obtain robust heritability estimates. Another difficulty with this approach is that environments are likely to be spatially correlated. In species with limited seed dispersal, the genetic structure of populations, and hence relatedness among individuals, will be confounded with spatial environmental variation. Such spatial autocorrelation of environments would tend to inflate heritability estimates; individuals that are more closely related are also likely to be growing in similar environments. The potential effects of the environment on heritability estimates can be controlled for, to some degree, by measuring features of the habitat that are likely to affect trait expression (e.g., light and moisture availability, intensity of competition, etc.), but the effectiveness of this approach depends heavily on the ability of the experimenter to obtain accurate measurements of salient environmental characteristics.

The other difficulty with using genetic marker, is that errors associated with the relatedness estimates can be quite high, so these methods may result in heritabilities that are out of bounds (i.e., less than zero or greater than one). For example, in a study of European columbine, Clara Castellanos et al. (2011) found several non-significant heritabilities that were negative and significant estimates that were as high as ~4.0. This wide range of estimates may be due to the relatively low mean number of alleles (<4) across the ten microsatellite loci assayed. Higher levels of genetic marker polymorphism may provide more reliable relatedness estimates. In a study of populations of a long-lived cycad (*Zamia fairchildiana*) using highly polymorphic markers (averaging 16 alleles per locus), all of the heritability estimates fell between zero and one, even though only six loci were assayed (Lopez-Gallego and O'Neil 2014). Utilization of molecular assays that generate large numbers of polymorphic markers (e.g., genome-wide SNP assays) holds promise for more accurate relatedness estimates and broader application of population-based heritability studies in long-lived species.

Maternal Effects—The goal of crossing experiments is to combine genetic material from pairs of individuals so phenotypic outcomes of different gene combinations in offspring can be assessed. Efforts to estimate heritability are often complicated by the differential influences of pollen and seed parents on offspring phenotypes (Roach and Wulff 1987; Donohue 2009). In particular, in flowering plants, pollen recipients (maternal parents) contribute more material to the developing ovule, as all the contents of the megagametophyte including the cytoplasm and organelles of the egg cell are passed on to the developing ovule. By contrast, microgametophytes (pollen) typically contribute only nuclear DNA and do not pass on organelles or any other constituents of their cytoplasm. As seeds develop, they are provisioned by the maternal parent and may be affected by the environmental conditions that the seed parent is exposed to. These *Maternal Effects* represent the sum of the extra-nuclear genetic influences that seed parents have on offspring phenotypes and that need to be taken into account for heritability estimates and other quantitative genetic studies.

Maternal effects on offspring phenotypes can be broken down into several categories: 1) cytoplasmic genetic effects due to differential contributions of plastids, mitochondria, and other cytoplasmic constituents; 2) differential genetic contributions to the endosperm (usually 2x maternal and 1x paternal resulting in a triploid endosperm; Chapter 9); 3) differences in seed provisioning; 4) maternally determined seed dormancy; and, 5) influences of the maternal plant's environment on developing seeds that translate to seedling phenotypes.

Maternal effects are detected by comparing offspring from reciprocal crosses in which the same set of plants are used both as pollen donors and recipients (i.e., diallele cross design; Fig. 7.8). In this type of experiment, we can identify maternal effects as any difference in trait expression between two groups of offspring from reciprocal crosses between the same two parents. The individual that served as the pollen parent for one group serves as the seed parent for the second group (i.e., $P_1 \times P_2$ and $P_2 \times P_1$, for pollen donor x recipient). In this case, the two groups of offspring are identical in terms of the nuclear genetic contributions from both parents but will differ with respect to maternal influences on trait expression (i.e., traits in offspring from $P_1 \times P_2 \neq P_2 \times P_1$). Note that this reciprocal contrast differs from variation among pollen or seed parents (trait mean differences among columns or rows, respectively; Fig. 7.8), which are estimates of the additive genetic variation. Differences in offspring from reciprocal crosses represent differential contributions to offspring by parents and are usually assumed to be due to maternal rather than paternal influences (Roach and Wulff 1987).

It is generally difficult to delineate the individual sources of maternally determined variation in quantitative traits, but we can group them into *Transient Maternal Effects* and *Heritable Maternal Effects*. We expect that differences in seed provisioning (seed size), endosperm influences, seed dormancy, and the maternal environment (listed as items 2, 3, 4, and 5 earlier) will be transient as they will tend to have stronger effects when seedlings are young but will be overwhelmed by the influence of the offspring's own genetic composition as the plant grows (Roach and Wulff 1987). In contrast, heritable genetic effects may continue to affect the phenotype over the lifespan of individuals and will be passed on to offspring through seeds.

As we analyze the potential influences of maternal parents on offspring phenotypes, we can see many cases where our distinctions among different sources of variation become less clear. For example, the maternal environment can affect seed provisioning (e.g., due to nutrients available to the seed parent) along with other transient influences on early seedling growth. Heritable maternal influences on offspring phenotypes include cytoplasmic effects but also the effects of the maternal-environment that result in heritable epigenetic modifications (Boyko and Kovalchuk 2011; Chapter 3), so maternal-environment effects can be transient or heritable. Such long-term effects on offspring can be induced by seed parents—particularly in response to stress (Sahu et al. 2013)—through heritable epialleles that may improve the fitness of offspring in the local environment. As these transgenerational epialleles can be sustained for several generations, they provide a flexible mechanism for long-lived species to produce offspring that are better adapted to local conditions and to respond to environments that may change over the scale of several growing seasons (Herman et al. 2014).

Evolvability—To this point, we have been discussing the narrow-sense heritability of traits as a measure that is predictive of evolutionary potential. Heritability is the basis of the breeder's equation described earlier, and is widely used in agriculture and evolutionary ecology to predict evolutionary trajectories in response to selection; however, the ratio of the additive genetic variation (V_A) with the phenotypic variation (V_P) may not be the best measure for predicting the evolvability of traits. This measure of evolvability has been criticized, because it is based on the total phenotypic variance, which tends to display correlations with other variance components (Hansen et al. 2011). The lack of independence between the additive and phenotypic variances will have consequences for predicted responses to selection in two different ways. First, because additive and non-additive genetic variance components are not independent and tend to display complex relationships (e.g., V_I can be converted to V_A as described earlier). This reduces the reliability of heritability as a predictor of evolvability, because it is based on scaling the additive variance by the total phenotypic variance (i.e. $h^2 = V_A / V_P$). Second, the environmental variance (V_E) will tend to affect the heritability and the selection gradient in opposite directions, which will have consequences for predictions of responses to selection based on the breeder's equation ($R = h^2 / S$). We can illustrate this problem more clearly if we imagine a case where plants with higher resource availability grow to a larger size; phenotypic variance is elevated because V_E is larger. As traits subject to selection also may covary with plant size, the higher environmental variance will tend to increase the slope of the selection gradient. A large V_E also inflates the total phenotypic variance, while V_A remains constant, so the estimate of heritability is lower, and consequently, the heritability and the selection gradient will tend to be negatively correlated.

As an alternative method to predict evolvability, Hansen et al. (2011) suggest that the additive genetic variance should be scaled by the mean of the trait (m) rather than the phenotypic variance ($e_\mu = V_A / m^2$). They argue that this estimate will provide a more reliable prediction of selection responses, because the additive variance and the trait mean should be independent, and consequently, e_μ will more reliably reflect variation in V_A. Because of differences in the scaling factors used in these estimates (i.e., V_P versus m^2), heritability and evolvability estimates are not strongly correlated. For example, traits that are closely associated with fitness often have low heritabilities, but scaling the additive genetic variance by the mean suggests that these traits have high potential to evolve in response to selection. At this point, comparative studies are not available, so whether estimates of evolvability based on scaling by the phenotypic variance or the trait mean turn out to be better predictors of the responses to selection remains an open question.

Prospects for Heritability Estimation—The ultimate goal of heritability experiments in evolutionary ecology is to understand the potential for traits to respond to selection. From the discussions above, it becomes clear that heritability estimates are broadly affected by the experimental design. Heritability can be affected by the collection of genets chosen for crossing, the environment or range of environments that their progeny are exposed to, and whether genetic variances are calculated for maternal or paternal parents.

Heritability studies on plant species have generally found that experiments conducted in controlled environments or from a limited range of field environments will overestimate the heritability of traits (Geber and Griffen 2003). Including a greater diversity of genets from different habitats is likely to inflate the estimate of V_A and consequently, result in a higher heritability estimate. Conversely, including a broader range of environments will increase the estimate of V_E, which contributes to a larger phenotypic variance and results in lower heritability. This difficulty stems from the fact that heritabilities are calculated as ratios of genetic and phenotypic variance, which is necessary because these parameters are proportions—estimates of V_A and V_G need to be scaled by the total phenotypic variance. As responses to selection in nature are generally manifest within populations, experiments endeavoring to estimate heritabilities will be most fruitful when the ranges of genetic and environmental variation take into account the dispersal ability of the organism under study and the spatial scale of populations.

Maternal effects may inflate the additive genetic variance by elevating similarities among siblings. Transient maternal effects will artificially inflate heritability estimates and can be avoided to some degree by measuring offspring traits in mature individuals. Heritable epigenetic modifications are heritable, as are beneficial somatic mutations, and both may contribute to selection responses. The ability to acquire beneficial genetic and epigenetic modifications during vegetative growth may have been favored by the sedentary life form of plants and may have been instrumental in their early success and continued diversification.

A major limitation for heritability studies—and quantitative genetics in general—is that researchers are generally limited in the range of traits that can be measured conveniently or accurately. When designing experiments for the estimation of heritability, traits are often chosen based on breeding goals or on the experimenters' perception of the likely targets of natural selection. In studies of evolutionary ecology, the ultimate goal may be to understand adaptive processes, and one limitation of such studies is the large number of physiological, developmental, and morphological traits that may be important for adaptation. Comprehensive phenotypic assessments are not feasible, so we rely on experimenter experience and intuition for choosing smaller sets of salient traits that can be accurately measured. In the next section, we discuss the evaluation of the strengths and characteristics of selection affecting phenotypic traits.

7.3 SELECTION ON PHENOTYPIC TRAITS

Understanding the processes leading to adaptation to novel, challenging, and changing environments has been a major goal of researchers working in the fields of evolutionary ecology and ecological genetics for many decades. From our discussion of the outcomes of selection in Chapter 5, we understand how allele frequency change can be manifest at the genomic level. However, selection acts directly on variation in phenotypic traits, so it is the collective expression of these characteristics—the sum effects of the genome and the environment on the phenotype—that determines fitness. The study of phenotypic traits for understanding adaptive evolution is inherently attractive, because the value of traits in particular ecological circumstances is often evident. This apparent fit between organisms and their environments motivated the widespread resistance by evolutionary biologists of the mid-twentieth century to the acceptance of genetic drift as an important evolutionary process. While we now recognize that neutral evolution is prevalent at the genomic level, it is reasonable to expect that much of the variation we observe in phenotypic traits represents the outcomes of past selection. In this section, we focus on methods to evaluate the strength and pattern of selection currently acting on populations and analyses aimed at detecting the effects of historical selection on the distribution of variation in phenotypic traits.

Selection in Natural Populations—The effect of contemporary selection acting on phenotypes (i.e., in contrast to historical selection) is typically evaluated based on associations between vegetative or reproductive traits and estimates of fitness (Fig. 5.6, upper panel). The goal of these analyses is to determine whether traits not directly associated with fitness (target traits) are correlated with some aspect of reproductive success or vegetative growth that more directly reflects total plant fitness (see Chapter 5 for a discussion of plant fitness measures).

In plants, investigations estimating selection coefficients fall into two general categories: those focusing on 1) the effects of vegetative traits on plant growth under particular environmental conditions (e.g., habitat-associated selection on traits); or 2) the effects of traits relevant to pollinator attraction or effectiveness on sexual reproductive success (e.g., pollinator-mediated selection on floral traits). Studies examining selection on vegetative traits include a wide variety of traits reflecting variation in the physiology and morphology of leaves as well as aspects of plant architecture (e.g., branching frequency) and the timing of phenological events (Ackerly et al. 2000). Studies of pollinator-mediated selection typically focus on aspects of flower morphology, color, odor, and pollen and nectar rewards that are relevant to the effectiveness of pollinator attraction and rates of pollen receipt and removal (Harder and Johnson 2009; Sletvold and Ågren 2014). Studies estimating the strength and pattern of selection typically focus on sets of related traits and analyze their effects using multiple regression models (see the Multivariate Selection section).

Choosing sets of traits that are likely to be under selection and pairing them with appropriate estimates of fitness requires knowledge of the biology and ecology of plants. For example, to evaluate selection on leaf traits (e.g., morphology, anatomy, and physiology) appropriate surrogates for fitness would be traits associated with plant growth. These could be traits that are more directly affected by the target traits (proximate fitness estimates; e.g., rate of photosynthesis for leaf traits), or they could be a more indirect response that is closely associated with plant fitness (ultimate fitness estimates; e.g., plant growth or seed production). For selection on floral displays, we could assess rates of pollinator visitation or pollen removal from anthers and capture by stigmas as proximate fitness estimates and seed set and siring success as ultimate estimates of fitness.

When evaluating the effects of selection on quantitative traits, it is best to assess both proximate and ultimate fitness estimates and to evaluate the relationship between the two. For example, in a study of the effects of leaf traits on plant fitness in drought conditions, Picotte et al. (2007) estimated fitness with a proximate trait (δC13; WUE) and an ultimate trait (plant growth; Fig. 7.9). In this case, the authors found that WUE was strongly associated with plant growth under drought conditions, even though none of the leaf traits measured had strong direct effects on plant size. Proximate fitness traits may not always be good predictors of plant fitness, because ultimate fitness can be affected by other plant traits. For example, above-ground biomass may be affected by competition with other plants or by differences in allocation of resources to root and shoot growth. Similarly, fruit and seed set can be affected by pollen receipt and by the availability of resources for seed production. Proximate estimates of fitness may also be affected by characteristics that are extraneous to the target traits (e.g., flowers may self-pollinate as they age), but in general, proximate fitness estimates will more reliably reflect variation in the target traits.

Multivariate Selection—From our discussion on developmental modules in Chapter 6, we can predict that quantitative traits will often be associated with each other, and in general, we expect that shared gene regulatory networks will generate genetic correlations among traits. Consequently, traits with overlapping genetic determination will also display correlated responses to selection, because of changes in allele frequencies at pleiotropic loci. Selection on one trait may engender indirect responses in other traits due to genetic correlations caused by shared genetic determination. Because of the potential for correlated responses to selection, it may be difficult to determine which traits are the direct targets of selection.

We can account for genetic correlations while evaluating the effects of selection on traits by utilizing multiple regression models (Lande 1979). We assume that selection may be acting on several traits simultaneously and that the sum effects of direct and indirect selection will predict the amount of change in each trait from one generation to the next. These models are based on the prediction that knowledge of the additive genetic variances (G_{ii}) and covariances (G_{ik}; for traits I and k) for a set of n traits will allow us to predict the change in a trait (ΔZ_i) in response to selection:

$$\Delta Z_1 = B_1 G_{11} + B_2 G_{12} + B_3 G_{13} + \cdots + B_n G_{1n}.$$

Where B_1, B_2, through B_n represent selection coefficients for trait 1, 2, through n, respectively. The simultaneous effects of selection can be predicted for a set of n traits if all of their variances and covariances can be estimated (i.e., the G-Matrix described in Chapter 6; Fig. 6.20).

It is generally not practical to estimate the G-Matrix for a set of traits, because it would require large crossing designs and comprehensive evaluations of traits for large numbers of individuals under controlled conditions. The G-Matrix can provide information about the genetic architecture and relationships among traits, and it has been used to predict changes in traits and constraints on selection responses. On the other hand, the G-Matrix does not tell us whether the traits measured are targets of selection. The covariance structure offers predictions of the expected patterns of phenotypic changes if selection were applied to individual traits, but it may not be particularly useful for understanding the selection pressures that organisms are subject to in their natural environments.

To understand the selection pressures acting on traits, we can account for phenotypic variances and covariances, using multiple regression models to estimate the strength and pattern of selection (Lande and Arnold 1983). These models are constructed by measuring a number of target traits and regressing them on a proximate or ultimate estimate of fitness (w):

$$w = b_1 P_1 + b_2 P_2 + b_3 P_3 + \cdots + b_n P_n.$$

Where b_i represents the slope of the regression (*Linear Selection Coefficient*) for trait P_i on fitness (w). Multiple regression analysis provides estimates of the partial sums of squares, so this analysis indicates the independent effects of each trait included in the model. Even in cases where traits covary to some degree, this approach provides estimates of the independent relationship between each target trait and a trait that provides an estimate of fitness. Note that the genetic covariance structure is not explicitly included in this analysis, so we assume that each trait in the model is directly affected by selection.

We can expand the multivariate selection analysis to include estimates of non-linear relationships between each target trait and fitness by including the quadratic term (P_i^2) for each trait, along with its associated regression coefficient (*Quadratic Selection Coefficient*). In these models, a significant negative quadratic term is indicative of stabilizing selection, and a positive quadratic term is consistent with disruptive selection (Fig. 7.10). Applying multiple regression analysis in natural populations allows simultaneous estimation of the strength and form of selection on an estimate of fitness.

Regression models have been widely used to understand selection forces acting on plant traits in natural populations (Geber and Griffen 2003). Recall that characterizing selection alone will not predict evolutionary outcomes in the next generation, because this will also depend on the narrow-sense heritability. On the other hand, knowledge of the patterns of selection acting on plant traits allows predictions of changes in traits in the next generation, given that additive genetic variation is present. Understanding whether traits are subjected to directional, stabilizing, or disruptive selection allows us to predict whether we would expect to see changes in traits in the near future. This is particularly important, as many populations are subject to selection pressure due to climate change.

A considerable number of selection studies have been conducted in natural plant populations, and those including multiple years of data have often found that the strength and even the direction of selection on traits can vary from one year to the next (Geber and Griffen 2003). Such fluctuating patterns of selection may not be surprising when we consider that the trait values we observe represent the outcomes of many decades of historical selection and may reflect the average effects of varying patterns of

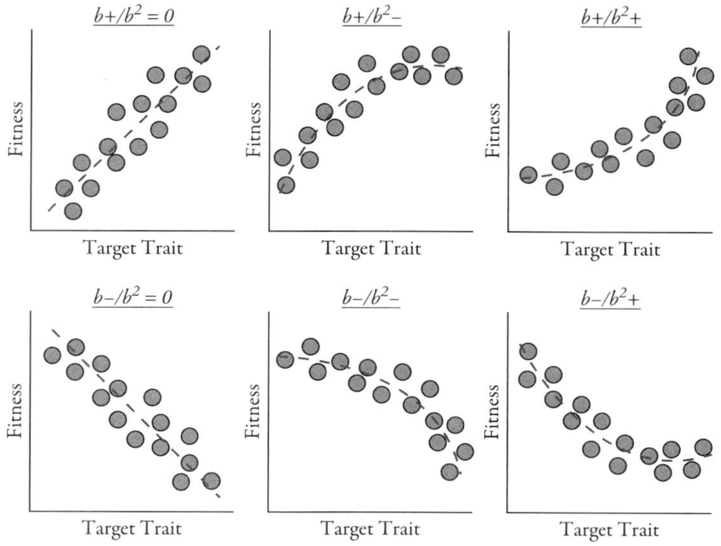

Fig. 7.10. Possible outcomes of quadratic selection analysis for an individual trait. The linear selection coefficient can be positive (upper panels) or negative (lower panels). When the quadratic term is negative (center panels) it may result in stabilizing selection if the linear slope is not too steep (i.e. when it is close to zero). Similarly, a positive quadratic term can indicate disruptive selection.

selection across many growing seasons. While there could be longer-term directional trends in selection and the corresponding changes in traits, these may be difficult to detect from studies of selection that span only a few years.

Selection Gradient—Quadratic selection analysis can provide estimates of the general pattern and strength of selection, but the relationship between target traits and fitness is often more complex (Schluter 1988). While we could add more terms in a polynomial regression analysis, exponents greater than the quadratic (i.e., P_i^3, P_i^4, etc.) are not easily interpreted, and these complex regression models may not fully capture subtle changes in the slope across different values of a trait. An alternative approach—*Selection Gradient Analysis*—utilizes a "moving window average" (spline estimation) to visualize the selection gradient across a range of trait values (Schluter 1988; Fig. 7.11; refer to color plate). In the study by Anton et al. (2013), the effects of the floral traits on the amount of pollen deposited on stigmas was analyzed. The plants studied differ for the size and shape of their bowl-shaped flowers as indicated by the total petal area and the separation between petals. The floral area was positively associated with pollen load size, and the quadratic regression term was not significantly different from zero, indicating directional selection for increased flower size. Flower shape was measured as the difference in the area of the flower compared to a circle that just touched the tips of the petals (Fig. 7.11; refer to color plate), so lower values of the proportional area difference indicate rounder flowers (smaller clefts between petals). Multiple regression analysis indicated stabilizing selection for the degree of petal separation with negative linear and quadratic coefficients, but note that the shape of the selection gradient in Fig. 7.11; refer to color plate) is more complex than a simple quadratic curve (see Fig. 7.10 for comparison). Selection gradient analysis may provide more detailed information on the strength and direction of selection across different ranges of trait values.

Selection in Recombinant Populations—While multiple regression and spline analyses are powerful approaches for the evaluation of the strength and pattern of selection in natural populations, there are limitations to this approach. In particular, estimates of regression parameters are sensitive to the variables that are chosen; excluding traits that are the targets of selection will inflate estimates of the selection coefficients

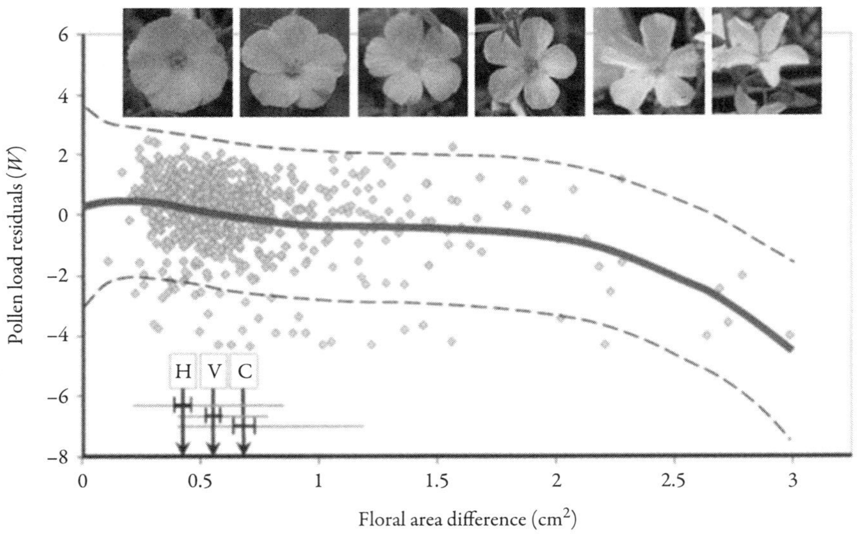

Fig. 7.11. Selection gradient analysis for the effects of floral area difference (area of the flower compared to a perfectly circular flower) on the size of the pollen load in an artificial recombinant hybrid population of *Piriqueta cistoides* ssp. *caroliniana*. Higher values indicate flowers with less petal separation (flowers representing different shapes are arranged across the top). The best fit line (solid) by spline analysis is bounded by upper and lower 95% confidence intervals (dashed lines). The average flower shapes for two morphotypes (C and V) and a derived hybrid taxon (H) are indicated by the arrows associated with each letter. Horizontal gray lines represent ranges and capped lines are standard errors of the means. Modified from Anton et al. (2013).

of correlated traits included in the model. We expect that large numbers of traits will covary, so the ideal quadratic analysis should include all characteristics possibly subject to selection. This is generally not possible, because our knowledge of the complexity of interactions among traits and environments is incomplete, and because we cannot measure all salient traits accurately. For example, a large number of traits may contribute to drought tolerance in plants—from the morphology and anatomy of leaves and stems to a plethora of traits associated with cell physiology and whole-plant hydrology—and it is generally not feasible to measure all of them. This is particularly evident when we consider the effort involved in measuring the large number of individual plants required for robust estimates of selection coefficients—generally at least two hundred (Lande and Arnold 1983).

The potential effects of unmeasured traits on estimates of selection coefficients will always be a concern for studies of selection in natural populations, because historical selection will tend to generate covariances among traits even if they are not developmentally linked. Selection will always favor the combination of traits that achieves the maximum fitness, which will generate linkage disequilibria among QTLs and consequently, will inflate genetic correlations. Sampling individuals across heterogeneous environments will result in correlations among traits due to developmental constraints and the historical effects of selection. The large numbers of correlations generated among traits will hamper efforts to estimate the independent effects of traits on fitness. Moreover, because selection will tend to remove non-optimal phenotypes we expect the range of trait values found in natural populations will be limited. The lack of more extreme trait values reduces the leverage for the estimation of selection coefficients and diminishes the robustness of the analysis.

An alternative approach to obtain estimates of the independent contribution of traits to fitness is to generate artificial recombinant populations by making crosses between phenotypically divergent lineages (e.g., hybridization between subspecies; Lexer et al. 2003b). Recombinant genotypes can be produced by crossing among F_1 individuals to form an F_2 generation or by crossing them with either parental lineage to generate backcross (B_1) hybrids. Continued crossing or inbreeding within lines for several generations will tend to reduce phenotypic trait correlations, as recombination breaks-up associations among QTLs and allows the fitness effects of traits to be assessed in a broader range of genetic backgrounds. Furthermore, interbreeding between more divergent lineages (i.e., subspecies) will often result in a wider range of trait expression in recombinant hybrids than is found between the two parental lineages, due to *Transgressive Segregation* (Rieseberg et al. 1999; Chapter 10). The wider range of trait values plus the increased independence among traits that can be found among recombinant hybrids provides for more substantive analyses of the strength of and pattern of selection on individual traits.

The two studies, described earlier, by Picotte et al. (2007) and Anton et al. (2013) were conducted using artificial recombinant hybrids from crosses between two ecologically divergent morphotypes of *Piriqueta cistoides* ssp. *caroliniana*. Note the relatively weak association among traits (double-ended arrows in Fig. 7.9) and transgressive trait expression (the range of trait expression is well beyond the parental morphotypes in Fig. 7.11; refer to color plate) exhibited by the plants used in these studies—both of which are patterns expected in recombinant hybrid populations. Depending on the life-history characteristics and variation available for the plant under study, constructing artificial hybrid populations may be a viable option to provide robust estimates of the independent effects of selection on individual traits.

Predictive Selection Analysis—Up to this point, we have discussed experiments designed to make predictions about future changes in traits solely based on contemporary patterns of selection. When the biological function of traits is well known, we may be able to make predictions about responses to selection. Such "functional" traits include physiological and morphological characteristics that have well-documented roles for plant water status and photosynthesis. For example, in a study of populations of a species of sea rocket (*Cakile edentula* var. *lacustris*) that occupy habitats differing in water availability, the functions of traits were used to predict their observed changes in response to drought (Dudley 1996). In this case, the roles of functional traits were confirmed by estimating selection gradients for plants grown in wet and dry common gardens in the field. In other cases, researchers have been able to recreate historical scenarios of selection and compare the predicted selection regimes to the distributions of trait expression in contemporary populations. This "retrospective" approach can be quite powerful for understanding the long-term consequences of different selection regimes, but reconstructing realistic historical circumstances is feasible only under specific circumstances. In particular, information on the historical distribution of plants can be used to predict the expected outcomes of selection as they colonize novel habitats. In the investigation of variation in flower morphology in *Piriqueta* described earlier, trait transitions for flower size and shape observed in contemporary hybrid populations were predicted from selection gradients measured in artificial recombinant populations in field common gardens; note that, as predicted by the selection gradient, the mean for the hybrids is lower than either parental morphotype (Fig. 7.11; refer to color plate). Such retrospective studies of selection are particularly robust, as they simultaneously document predictions and outcomes of selection by reconstructing historical scenarios.

The large number of cases of the establishment and spread of invasive plants in novel environments provides unprecedented opportunities for conducting predictive selection analyses. As discussed in the next chapter, care needs to be taken to accurately identify the source regions and to discriminate between the effects of selection and genetic drift on trait changes that may occur during invasion. In one such study, the source regions for an invasive bunchgrass (slender false brome—*Brachypodium sylvaticum*) were identified using genetic markers, and this information was used to make focused analyses of trait transitions between the native and invasive ranges (Marchini et al. 2018; Fig. 7.12). Predictions for functional trait responses were made based on the more severe drought conditions present in the introduced range. Making comparisons to the specific source regions, rather than the entire native range of introduced species, provides a much more robust framework for the testing of adaptive hypotheses.

Accounting for Genetic Drift—Gene flow among populations affects the distribution of genetic diversity and the potential for populations to adapt to local environmental conditions. Recall from Chapter 5 that both selection and genetic drift

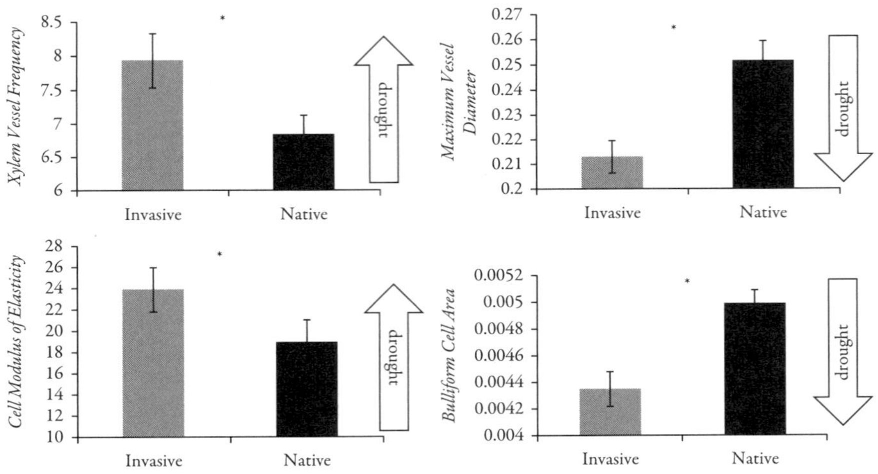

Fig. 7.12. Trait transitions after introduction to a drier climate in the invasive range of slender false brome (*Brachypodium sylvaticum*). Predictions for trait changes in response to drought (arrows) were made based on the known function of the traits measured. Modified from Marchini and Cruzan (2018).

can contribute to genetic differentiation among populations. With moderate to large population sizes, we expect that the effects of genetic drift would be reduced, so that gene flow would tend to homogenize allele frequencies for neutral loci. In contrast, differentiation among populations could be maintained at loci that are subject to selection. We predict that the effects of selection on the distribution of genetic variation would be particularly evident in heterogeneous environments, since selection will promote allele frequency differences among populations for QTLs contributing to local adaptation. The contrasting expectations for neutral variation and loci contributing to local adaptation provides a framework for detecting the effects of historical selection on the distribution of genetic variation (Whitlock 2008).

We generally cannot directly measure levels of variation at QTLs that may be subject to selection, but we can estimate differentiation among populations for the traits that they control. Quantitative trait variation among populations is estimated as Q_{ST}, which is based on the level of genetic variation among populations (V_B), and the additive genetic variation within populations (V_W):

$$Q_{ST} = \frac{V_B}{V_B + 2V_W}.$$

Note that this estimate of Q_{ST} assumes that V_E is effectively zero, so genetic variance estimates need to be made using plants from controlled crosses that are grown in uniform environments (e.g., common gardens). Typically, this is done using the same types of sibling comparisons we previously discussed for estimates of narrow-sense heritability. The average variance in trait expression among sibling groups from the same population provides an estimate of V_W, and the variance among populations can be used to estimate V_B. Estimates of Q_{ST} can be used to infer the historical effects of genetic drift and selection on phenotypic traits by comparing the distribution of variation in phenotypic trait expression across populations to variation based on neutral genetic markers (F_{ST}; Chapter 4).

Since estimating Q_{ST} is dependent on growing offspring from multiple plants from several populations under controlled conditions, this type of experiment will require considerable effort and will be more appropriate for faster growing species. For longer lived plants, phenotypic traits can be measured in natural populations to obtain an analogous parameter that estimates total phenotypic variation among populations (P_{ST}), but note that these estimates will tend to be inflated due to environmental effects on trait expression (i.e., due phenotypic plasticity; Leinonen et al. 2013). Even for short-lived plants, care needs to be taken to obtain accurate estimates of Q_{ST}. For example, just as with heritability estimates, seed collections made from plants growing under field conditions may produce biased estimates due to maternal environment effects. Obtaining robust estimates of Q_{ST} requires carefully designed experiments that include multiple groups of siblings representing several populations grown in controlled environments.

Estimates of Q_{ST} can be compared to F_{ST} from neutral genetic markers to infer the effects of historical selection on phenotypic traits. In cases were populations occupy different environments, we expect that traits conferring local adaptation will have been affected by selection as well as genetic drift, due to limited population sizes. In particular, we expect that Q_{ST} will be greater than F_{ST} ($Q_{ST} > F_{ST}$) when directional selection has led to trait divergence among populations (*Diversifying Selection*). Conversely, if selection regimes are similar across environments, then phenotypic trait expression may be more consistent than expected based on neutral diversification ($Q_{ST} < F_{ST}$; *Homogenizing Selection*). If a trait is not subject to selection, or if selection is not strong enough to overcome the effects of gene flow and genetic drift, then Q_{ST} will be approximately equal to F_{ST} ($Q_{ST} \approx F_{ST}$). Comparisons of Q_{ST} and F_{ST} across populations provides a powerful method for detecting the historical effects of selection on distributions of phenotypic trait expression within and among environments.

There are a number of caveats to be aware of when interpreting comparisons of Q_{ST} and F_{ST}. Whitlock (2008) cautions that Q_{ST} estimates need to be compared to a number of independent estimates of F_{ST} rather than the average across loci and that probability distributions need to be generated for robust tests of departures from neutral expectations. Potential biases for estimates of Q_{ST} may be introduced due to non-additive genetic variation affecting the traits under study, so investigations taking into account the effects of selection on local adaptation will be more robust if ancillary information can be used to support predictions for the outcomes of selection. For example, in the study described earlier on invasive slender false brome, the authors were able to make predictions about the effects of directional selection on invasive populations, and the outcomes of diversifying selection were confirmed after comparison of Q_{ST} estimates to neutral divergence (Marchini et al. 2018). Across a number of studies, the majority of $Q_{ST} - F_{ST}$ comparisons have found evidence for diversifying selection on phenotypic traits (Leinonen et al. 2013), but this pattern may reflect the fact that researchers are probably more likely to focus on traits that are associated with adaptation to heterogeneous environmental conditions.

Outcomes of Selection—Up to this point, we have discussed phenotypic traits as if they were always malleable, but lack of variation and constraints may sometimes limit responses to selection. Phenotypic traits may be canalized in the sense that they do not respond to environmental influences, so they are not phenotypically plastic. Note that canalization does not necessarily refer to the lack of heritability of a trait—individuals may display intransient expression of traits, but heritable variation may be present within and among populations. Whether a canalized trait is susceptible to selection may be a function of the trait history—characteristics that are more recently canalized in response to prevailing environmental conditions may be more responsive to selection. On the other hand, traits that have remained constant over longer phylogenetic periods (e.g., flower merism) may be buffered from genetic and environmental perturbations to the degree that they are completely unresponsive to selection. We expect that canalized traits will be evident as cases where $Q_{ST} < F_{ST}$, which could be due to either recent homogenizing selection across populations or to constraints imposed by phylogenetic history (Lamy et al. 2012).

As discussed earlier in this chapter, we expect that responses to selection will be directly proportional to the narrow-sense heritability—the proportion of phenotypic variation that is due to the additive genetic effects of the QTLs controlling a trait. On the other hand, some traits may be constrained from responding to selection due to developmental associations that generate genetic correlations; if selection opposes changes in developmentally linked traits then selection responses in correlated traits may be difficult to obtain unless the underlying genetic architecture can be modified (Chapter 6).

More often we might expect that traits' responses to selection will be limited because recurrent directional selection has depleted allelic variation, contributing to additive genetic variation (Fig. 7.13). For example, traits that are closely tied to fitness

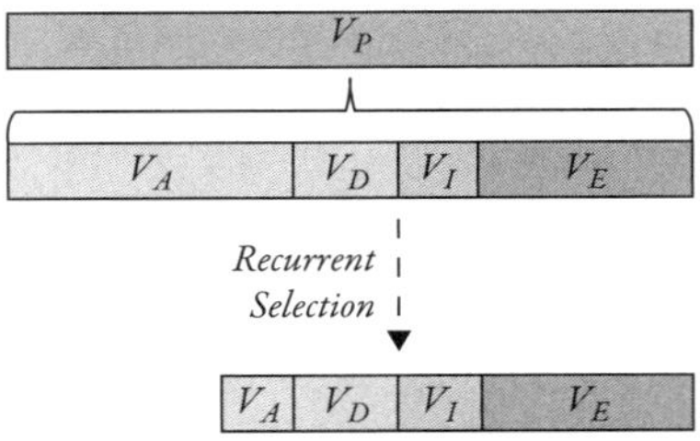

Fig. 7.13. The consequences of recurrent directional selection for the level of additive genetic variation present.

(e.g., growth rate and seed production) will be subject to directional selection across environmental conditions such that any heritable genetic variation will rapidly become fixed. This is a prediction of *Fisher's Fundamental Theorem,* which states that the rate of change in fitness is a function of the level of heritable genetic variation available. Similarly, we would expect that the heritability of traits that are the targets of breeding programs would decline as they are subject to directional selection over many generations.

While lower additive genetic variation for traits subject to recurrent selection seems intuitive, and may generally be true for plants (Mazer and LeBuhn 1999), the depletion of additive genetic variation for traits under selection may not be so straightforward. For example, this prediction holds for some traits that have been subject to artificial selection, but in other cases heritability seems to remain high even after one hundred generations of selection (e.g., for the oil content of corn; Hill 2010). The potential for long-term selection responses may be directly related to the genetic architecture of traits—corn oil content is controlled by a large number of QTLs of small effect—while other traits that cease to respond to selection after a few generations are usually determined by a smaller number of loci of large effect. Of course, we expect that even when very large numbers of QTLs contribute to trait variation, continued directional selection will eventually deplete heritable variation, and the response to selection will plateau. It is clear from this observation and our previous discussions that the constraints that govern the range of trait expression and the potential for responses to selection may be a function of genetic determination, which is a topic we will explore further in the next section.

7.4 GENETIC DETERMINATION OF PHENOTYPIC TRAITS

Complete nuclear genome sequences have been completed for dozens of plant species, and the availability of sequence data is increasing at a dramatic pace. The promise of genomics lies in the comprehensive information that becomes available as more genomes are sequenced and as more information is gathered on the functions of expressed regions. On the other hand, variation in the genome is not easily linked to the expression of phenotypic traits, and major-effect genes are known for only a handful of traits (Ehrenreich and Purugganan 2006; Rockman 2012; Travisano and Shaw 2013). Moreover, the large majority of traits that confer adaptation are polygenic and may be affected by a large number of genes of small effect, in which case the characterization of variation at the genomic level would be unlikely to reveal clear insights into processes responsible for adaptive evolution (Rausher and Delph 2015). It is important to keep in mind that selection operates directly on phenotypic traits; by favoring particular trait states over others, selection indirectly influences the frequencies of alleles segregating at one or more QTLs that determine trait expression. From this point of view, characterization of variation at the phenotypic level may be adequate for our understanding of adaptive evolutionary processes. On the other hand, quantitative genetic methods provide only generalized information on the sum effects of allelic variation across many loci. Even while selection is mediated by ecological processes, at its core, evolution is a genetic phenomenon; to gain insights into adaptive processes we should strive to understand the genetic determination of phenotypic variation, including the number, relative influence, and genomic distribution of genes governing traits that affect fitness in changing and challenging environments. In this section, we focus on the potential for methods that integrate genetic and phenotypic information to characterize the genetic architecture of quantitative traits and to serve as windows into the process of adaptive evolution at the genomic level.

QTL Mapping—One approach to understanding the genetic determination of quantitative traits is to search for variation at neutral genetic markers that is associated with variation in phenotypic traits. This type of analysis can be conducted by making crosses among individuals that differ in trait expression and then examining associations between trait expression and the mapped locations of genetic markers across the genome (*QTL Mapping*; Fig. 7.14; refer to color plate). Mapping the genomic distribution of QTLs allows for characterization of the location and the phenotypic effects of specific genomic regions—note that information on the connections between variation at the genomic and phenotypic levels cannot be gleaned from examination of sequence variation alone. QTL mapping has been used widely in plant breeding and studies in evolutionary ecology—mostly using plants that have substantial genomic resources available (Mauricio 2001; Erickson et al. 2004; Ehrenreich and Purugganan 2006).

To identify the likely genomic locations of QTLs, a detailed genetic map needs to be developed for the species being studied. A large number of variable genetic markers are identified, and their map locations are estimated using segregation distortion analysis in the F_2 progeny. Putative QTLs are located on the map by searching for non-random associations between marker allele segregation and phenotypic

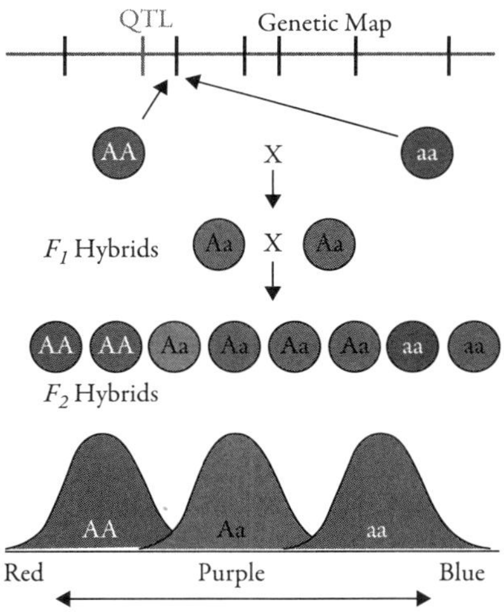

Fig. 7.14. Mapping of quantitative trait loci (QTLs) that control the expression of phenotypic traits. Colors represent trait expression in the parents (red and blue) and in the offspring (shades of purple). In this example the parents display alternative homozygous genotypes (AA and aa) at a marker locus that is close to the location of a QTL. Linkage between the marker and QTL causes a non-random distribution of phenotypes (color) among the marker genotypes.

trait expression (Fig. 7.14; refer to color plate). In practice, this mapping is conducted using software applications that calculate the Log-Odds score (*LOD Score*) for different positions along each chromosome (Fig. 7.15). LOD scores reflect the probability of the observed level of segregation distortion relative to random segregation. Scores exceeding 3.0 indicate that the odds of the mapping interval not containing a QTL are 1000 to 1, which is generally considered an appropriate threshold for identifying the position of QTLs. Significant LOD scores indicate that one or more QTLs are probably located within the mapped genomic region.

When using codominant markers, the dominance relationships between the alleles segregating at the putative QTL can be discerned by comparing the phenotypic expression of the three marker genotypes (Fig. 7.14; refer to color plate). In this example the phenotype of the heterozygote is intermediate between the homozygotes, so we would infer that the sum of allelic effects of QTLs in this region is codominant ($h = 0.5$; Fig. 7.1). If the heterozygote phenotype were closer to one homozygote or the other than this would indicate dominance ($h \neq 0.5$); or the heterozygote phenotype could exceed the averages for the homozygotes, which would suggest overdominance or underdominance. The QTL map provides a summary of the number, relative influence, and genomic distribution of regions associated with variation in the expression of a quantitative trait.

QTL mapping is a powerful and useful method for the genomic characterization of phenotypic traits, but there are several limitations to this approach. For one, a large number of genetic markers need to be screened for a substantial sample of F_2 offspring (usually > 300 offspring). The ability to detect a particular QTL is directly related to its relative influence on trait expression and its recombination distance to the nearest genetic marker; QTLs that are closer to marker locations and those with larger effects are more likely to be detected. With larger numbers of genetic markers—particularly with SNP detection methods available using NGS technologies—the resolution of genetic maps can be dramatically improved (Bergelson and Roux 2010). Even with large samples of markers and individuals, the resolution of most genetic maps is not high enough to identify individual genes; a region as small as five cM (centimorgans—map units) may contain hundreds of genes (Erickson et al. 2004).

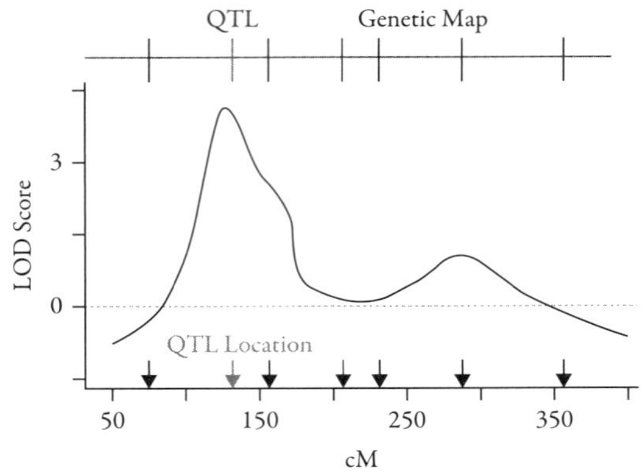

Fig. 7.15. Hypothetical LOD score analysis for the association between quantitative trait variation and genetic map location. Arrows indicate the location along the chromosome. The scale is in centimorgans (cM).

Furthermore, genetic mapping approaches are more likely to detect genes of large effect (i.e., those explaining > 15% of the variation in a trait), so even with high-density maps, the large majority of QTLs affecting a trait will be missed, because they have small effects. This problem is highlighted by the fact that the sum of the effects of the identified QTLs typically explains only a minority of the variation found in phenotypic traits (Brachi et al. 2011).

There is another major limitation for QTL mapping: because a small number of individuals are crossed to generate the mapping population, these experimental designs only include a small fraction of the phenotypic variation found in nature. This problem can be partially addressed with more complicated crossing designs that include larger numbers of accessions (Erickson et al. 2004), but there are practical limitations to the number of parents that can be crossed in the first generation. Moreover, the requirement for multiple generations of plants limits the range of life histories that are appropriate for QTL mapping experiments, so this approach may only be feasible with short-lived plants.

One approach to QTL mapping is to develop artificial populations of F_2 individuals that have been inbred as separate lines for several generations (Dalziel et al. 2009). The resulting phenotypically diverse mapping populations of *Recombinant Inbred Lines* (*RILs*) or *Near Isogenic Lines* (*NILs*) are powerful tools for inferring the genetic determination of phenotypic traits. RIL mapping requires a large amount of effort, so it has been conducted with only a few models (e.g., *Arabidopsis*) and annual crops and is not feasible for the majority of plant species. Even with high densities of genetic markers, the mapping resolution for QTLs is limited by recombination, because a few generations are inadequate for disassociating alleles at different loci—relatively large genomic blocks that may contain several marker loci and many hundreds of expressed and functional genomic regions.

Association Mapping—Genome-Wide Association Studies (*GWAS*) uses mature individuals growing under controlled or natural conditions and can serve as a more widely applicable alternative to QTL mapping. To apply this method, a high-density of SNPs can be generated using NGS technologies (Korte and Farlow 2013; Rellstab et al. 2015). High densities of markers are not necessarily useful in most QTL studies, because even after six to ten generations, recombination has not had time to break-up allelic associations at smaller genomic scales, and groups of markers on the same genomic blocks will segregate together. The genomic resolution can be much greater for GWAS, since plants with a wide range of divergent pedigrees can be included, recombination has had more time to break up associations, and so much finer mapping resolutions can be attained. Sampling across a wider range of genotypes also increases the chances of detecting rare alleles that may contribute to phenotypic variation.

GWAS has been successfully applied across a number of crop, model, and non-model species, but there are a number of limitations to its application.

In particular, population structure will increase the chances for "false positives"—cases where the association between marker and phenotypic variation is generated by differences in the relatedness among individuals sampled rather than linkage (Brachi et al. 2011). For example, local groups of related individuals may become fixed for marker alleles and may appear to be divergent for phenotypic traits due to the effects of genetic drift and local inbreeding. Such coincidental occurrences may generate associations between marker alleles and QTL alleles at unlinked loci; high LOD scores close to the marker locus will be falsely interpreted as a strong phenotypic influence by the nearby genomic region. While population structure can be accounted for to some degree by correcting for the relatedness among plants sampled, these methods are unlikely to completely eliminate the possibility of detecting false positives.

Another difficulty for the application of GWAS stems from the fact that recombination rates are not even across the genome, so linkage associations may be maintained over large genomic blocks. Such regions would be more likely to contain QTLs affecting traits in opposite directions, so phenotypic influences may be obscured. Moreover, without a linkage map, individual effects of QTLs cannot be determined, and it would not be possible to infer the distance between markers and putative QTLs. In general, genetic background effects—associations due to relatedness and uneven recombination—will lead to the appearance of "ghost" QTLs (marker associations with non-causative genomic regions—false positives) and "invisible" QTLs that are overlooked because of lack of recombination and weak effects on phenotypes. As we discussed in Chapter 4, selfing and local inbreeding can generate strong population structure; even with corrections for relatedness among individuals, large sample sizes, and high densities of markers, both false and invisible QTLs are likely to continue to be a problem for genome-wide association analyses conducted across local genetic neighborhoods and populations. On the other hand, putative QTLs identified using GWAS can be further evaluated using controlled crosses and QTL mapping.

Finding the Genes that Matter for Adaptation—In spite of the potential drawbacks, QTL mapping and GWAS have generated valuable insights into the genetic determination of ecologically relevant traits (Bergelson and Roux 2010; Brachi et al. 2011; Korte and Farlow 2013). A good example is flowering time in *Arabidopsis thaliana*. Variation in flowering time across latitudes in this species appears to be under strong selection, as it is closely tied to fitness: flowering too early leaves plants subject to frost damage while late-flowering plants may be unsuccessful because of lower water availability and competition with other plants. Mapping studies in *A. thaliana* have successfully identified candidate genes affecting flowering time. For example, the *FRIGIDA* locus has been shown to affect flowering time, and the frequency of putatively functional alleles segregating at this locus are associated with flowering times across a latitudinal gradient (Stinchcombe and Hoekstra 2008).

One of the most successful mapping studies was conducted on oil content in maize. Mapping across lineages that had been selected for seventy generations revealed fifty small genomic regions (2–3 cM) of roughly equal effect (Laurie et al. 2004). While the majority of mapping efforts have been conducted using crops and *A. thaliana*, there is a growing number of cases where these methodologies have been successfully applied to analyze the genetic basis of ecologically relevant traits in non-model species (Stinchcombe and Hoekstra 2008).

Some of the most successful studies evaluating the genetic basis of adaptive traits in non-model plants have focused on aspects of flower morphology. These analyses have been facilitated by the fact that genes of major effect often control flower color. In the snapdragons of the *Antirrhinum majus* species complex, flower color is controlled primarily by three genes: the *ROSEA* and *ELUTA* loci control the intensity and patterning of magenta pigments while *SULPHUREA* controls the distribution of a yellow pigment (Whibley et al. 2006). Two subspecies of *A. majus* with contrasting flower colors form narrow hybrid zones in the Pyrenees (Fig. 7.16; refer to color plate). These hybrid zones appear to be stable, while post-pollination reproductive isolation barriers are weak (Andalo et al. 2011), which suggests that the hybrid zone is maintained by pollinator-mediated selection on flower color.

Other studies of the genetic architecture of flower morphology have found relatively small numbers QTLs for petal shape in *A. thaliana*, flower color in Louisiana irises, nectar spur length in columbines, and flower color and shape in two species of *Erythranthe* (*Mimulus*; Mauricio 2001; Fig. 7.17; refer

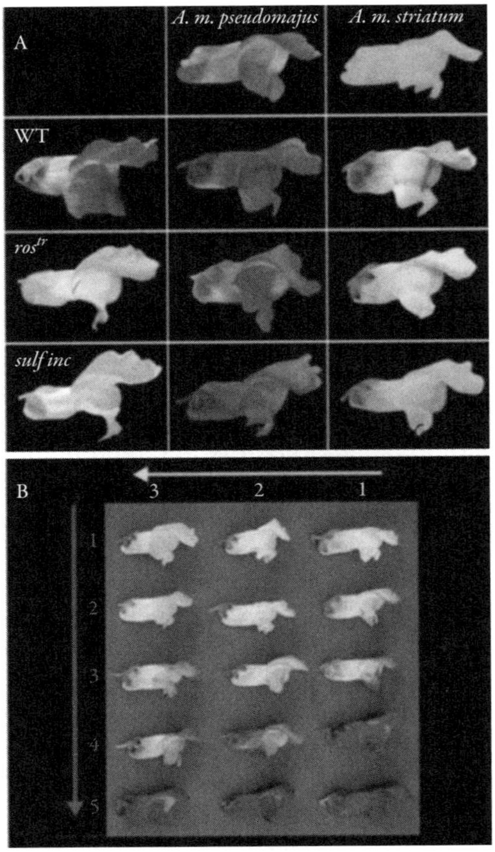

Fig. 7.16. Flower color variation in subspecies of *Antirrhinum major.* Panel A—First-generation hybrid flowers from crosses between *A. m. pseudomajus* and *A. m. striatum* (top row) and cultivated *A. majus* genotypes possessing different flower color genes (left column). Panel B—The range of flower color variation found among plants in the hybrid zone in the Pyrenees. From Whibley et al. (2006).

to color plate). In the latter case, the authors analyzed eight morphological traits separating two species of monkey flower (*E. cardinalis* and *E. lewisii*) and found that each was determined by at least one QTL of major effect (Fig. 7.17; refer to color plate). These two species are often found growing in close proximity in their native ranges in the Pacific Northwest region of North America. In spite of being interfertile, hybrids are only rarely found in nature. Different pollinators serve these two species; hummingbirds commonly visit *E. cardinalis* while *E. lewisii* is normally visited by bumble bees. The reproductive barriers between these species are presumably maintained by pollinator preference and selection on flower morphological traits (Chapters 9 and 10).

Prospects for Association Mapping—While mapping and association analyses have often been successfully used to determine the genetic basis of phenotypic traits, the application of these methods remains problematic. In addition to the issues with false positives and the genetic background effects mentioned earlier, the success of these studies often depends on the environmental conditions that plants are exposed to. For example, studies of flowering time in *A. thaliana* that were conducted in the greenhouse and in the field detected largely non-overlapping sets of QTLs (Bergelson and Roux 2010). These cases suggest that controlled environments may produce misleading results, as only two of the previously identified candidate genes for flowering time were included in the twenty-five QTLs detected by GWAS conducted under field conditions. Such outcomes suggest that the expression of many QTLs is subject to genotype-by-environment interactions, and their detection will depend on the environments that plants are exposed to as well as the genetic composition of the screening population.

Therefore, the question remains: Have we made much progress finding the genes that matter for adaptation? The discrepancies between QTLs found for plants grown under controlled and field conditions, the bias toward genes of large effect, and the large "invisible fraction" of QTLs that remain undetected in almost every association study has led some to make disparaging remarks concerning our progress in understanding the genetic basis of ecologically relevant traits. In particular, Rockman (2012) suggested that the few genes of large effect that have been detected are not representative of the large majority of genes that are important for adaptation. This argument stems from the observation that QTLs of large effect are rare and generally do not explain large proportions of the heritable variation in phenotypic traits. Furthermore, many of these genes are identified based on their presence in domesticated organisms, and many are due to mutations of large effect that are not found in natural populations. In contrast, most quantitative traits are polygenic and appear to be controlled by large numbers of genes of small effect. Some of the largest scale GWAS

Fig. 7.17. The range of flower morphologies for *Erythranthe cardinalis* (*Mimulus cardinalis*; upper right) and *E. lewisii* (*M. lewisii*; upper left) and their hybrids. From Mauricio (2001).

analyses that have been conducted on human traits confirm this; schizophrenia, for example, appears to be controlled by tens of thousands of QTLs (Purcell et al. 2009). Furthermore, the suggestion that QTLs of small effect are important for adaptive evolution supports Fisher's geometric model of adaptation, which predicts that the approach to fitness optima would be more likely to occur through mutations of small effect (Rockman 2012; Chapter 5). Even while the idea that QTLs of small effect appear to be the norm may be supported by theoretical expectations, it does not quell our concerns that evaluation of the genetic basis of phenotypic traits responsible for adaptation will be difficult. On the other hand, even though QTLs of large effect are rare, they may well be relevant for phenotypic evolution, and it may be true that evolutionary processes acting on them are similar to processes affecting genes with weaker influences on phenotypic variation.

The fact that large numbers of QTLs with very small effects may explain the large majority of variation in ecologically relevant traits may seem to discourage attempts to characterize adaptive genomic variation, because these regions are difficult to detect and to study in detail. On the other hand, alternative approaches may be more productive for understanding adaptation at the genomic level (Rockman 2012). For example, statistical genetics can be used to track the average effects of large numbers of genomic regions associated with SNPs. Assaying large panels of genetic markers for many samples continues to become more feasible, and these methods can be combined with experimental manipulations—including selection responses—to provide insights into the genomics of adaptation. These efforts can be complemented with continuing efforts to characterize QTLs of large effect, but the question whether evolutionary processes acting on such loci are broadly characteristic of the large majority of genes of small effect remains unanswered.

It is a little startling to realize that the whole-genome sequencing efforts of the last few decades have brought us only incrementally closer to understanding the genetic basis of adaptation. While we have learned much about a handful of genes with large effects, we know little of the many genes that have infinitesimal influences, but which collectively explain the largest proportion of heritable variation in phenotypic traits. This realization does not diminish the tremendous advances that have been made in understanding the phenotypic effects of a number of plant genes. These efforts have established a rich foundation of knowledge that will provide a basis for comparison as our understanding of the genetic basis of ecologically and agronomically important traits expands. These advances are providing important tools for crop improvement, including the use of marker-QTL associations to focus breeding efforts (*Marker-Assisted Selection*; Collard and Mackill 2008). The challenge for evolutionary biologists in the future will be to integrate a number of approaches—including quantitative evaluation of phenotypic variation, association mapping, statistical genetics, and genomic characterization of candidate genes—to gain insights into the genomic basis of adaptation and our understanding of the translation of selection at the phenotypic level to the genome.

7.5 QUANTIFYING GENETIC VARIATION

One of the primary goals of evolutionary ecology is to understand the potential for organisms to respond to challenging and changing environments. As we discussed earlier in this chapter and in Chapter 6, a large proportion of phenotypic variation is manifested as responses to environmental variation. While plants may withstand environmental fluctuations through plasticity responses, genetic modifications may be necessary to ensure the long-term persistence of populations. There is general acknowledgement that assessing genetic diversity is important for predicting the evolutionary potential of populations, but there is much less agreement over the most appropriate methods for measuring levels of genetic variation. While genetic markers have been established as the mainstay for the quantification of levels of genetic diversity (Hartl and Clark 2006; Hedrick 2009), some have questioned whether neutral markers are useful for assessing levels of variation at QTLs that may be important for adaptation—particularly for rare, threatened, and endemic species (Edwards 2015). This point of view is supported by the observation that neutral genetic markers are poor predictors of the levels of quantitative genetic variation (Reed and Frankham 2001). On the other hand, estimating genetic variance components for phenotypic traits requires a considerable amount of effort, is not feasible for a wide range of organisms, and may sometimes provide misleading information concerning evolutionary potential. In this section, we will explore the possible reasons for the discrepancy between genetic diversity estimates obtained from genetic markers and quantitative genetic analyses and evaluate the potential for neutral genetic markers to provide information that is relevant for evaluating the evolutionary potential of populations and species.

Why Genetic Diversity Estimates Differ—To understand why there may be differences in the levels of genetic variation measured using markers and quantitative traits, we need to review how evolutionary processes affect the distribution of genetic variation within and among populations (Chapters 4 and 5). Recall that the effective population size will determine the tendency for variation to be lost by random processes, so populations that have historically been smaller will tend to contain less genetic diversity. Gene flow can counteract the effects of genetic drift, as it infuses genetic variation into populations and equalizes levels of variation among them. As pollen, seeds, and other propagules disperse among populations, they carry with them entire haploid or diploid genomes; in other words, alleles at neutral markers move coincidentally with alleles at QTLs. Similarly, loss of genetic variation through drift proceeds through the failure of individuals—entire genomes—to survive or produce offspring in the next generation. As drift is a random process with respect to genotype, there is no reason it should differentially affect neutral and functional regions of the genome. If the entire genome were only affected by gene flow and genetic drift, then we would expect that levels of genetic variation at neutral markers and QTLs would be strongly correlated, which is the foundation of the F_{ST}–Q_{ST} comparisons that we discussed earlier in this chapter.

While we expect that drift and gene flow will have equal effects on markers and quantitative traits, the same is not true for selection (McKay and Latta 2002). From our previous discussions, recall that selection will tend to reduce genetic variation within populations, and may elevate differentiation among populations when environments are heterogeneous. Selection acts to reduce the survival and reproductive success of individuals expressing particular phenotypes, so unlike drift and gene flow, it has the potential to differentially affect levels of variation among different genomic regions. Moreover, from our discussion of Fisher's fundamental theorem earlier in this chapter, we generally expect those quantitative traits that are most closely tied to fitness will display the lowest levels of heritable genetic variation.

It is important to note that neutral genomic regions are not completely immune to the effects of selection. For example, as selection leads to the death or reproductive failure of some individuals it also reduces the effective population size, which enhances the effects of genetic drift equally across the genome. Individual neutral markers can also be affected more directly by selection if they are linked to loci subject to directional selection. As we discussed in Chapter 3, the surrounding genomic region will coincidentally be subject to loss of diversity due to genetic hitchhiking, which can be viewed as a regional genomic bottleneck. Overall, we expect that the large majority of neutral genetic markers will not be closely linked to genomic regions subject to selection, so they will display higher genetic diversity within populations. Given similar mutation rates, QTLs affected by selection will always harbor lower levels of genetic diversity than neutral genomic regions.

Predicting Differences in Diversity Estimates—From the above considerations, we can begin to make some predictions concerning the relationship between the level of allelic diversity at quantitative traits and for genetic marker loci. First, the average level of genetic variation should be lower across QTLs because all genomic regions are equally affected by drift and gene flow, while quantitative traits may also be subject to selection. Second, assuming that selection strength varies across quantitative traits, the range of allelic variation across QTLs should be much greater than for neutral genetic markers. This may also depend on the history of selection and the genetic architecture of traits; as we discussed previously in this chapter, those traits that are controlled by larger numbers of QTLs may retain heritable variation even after many generations of selection. Third, for those quantitative traits subject to spatially varying selection, we expect that levels of genetic diversity will be greater among than within populations. This prediction assumes that environments are relatively homogeneous within populations but differ among them, which generates higher diversity in trait expression among populations relative to the neutral expectation (i.e., $Q_{ST} > F_{ST}$). Conversely, if habitats are homogeneous then quantitative trait variation among populations should be lower than for neutral markers. Across all of the selection scenarios that we might imagine—whether it is directional or stabilizing within populations or diversifying or homogenizing among populations—we can predict that the average allelic diversity at QTLs will always be lower within populations than for neutral genetic marker. Because they are not subject to selection, neutral loci provide a reliable set of expectations for the effects of gene flow and genetic drift on the distribution of genetic variation within and among populations.

Using Neutral Genetic Markers to Understand Evolutionary Potential—So are neutral genetic markers useful for understanding the adaptive potential of populations? This question is being asked with increasing frequency concerning species that are challenged by global climate change. If populations are subject to changing environmental conditions, then we might expect that some quantitative traits that have not been subject to selection in the past may become important for maintaining high rates of survival and reproduction in the near future. At the same time, there may be a number of traits that have been subject to selection in the past and will continue to be important for fitness in the future. This reasoning provides us with two categories of traits—those that have been subject to historical selection and those that may become important for survival in the future. Although we expect that there is a full range of historical and future selection scenarios, these categories represent the extreme ends of a continuum and are useful for thinking about the type of information that neutral markers might provide to gain insights into adaptive potential.

If we view genetic markers as a convenient way to estimate the levels of genetic diversity in the absence of selection, then we can see that this information can be quite valuable for predicting the responses of traits that will become important for adaptation in the future. In this scenario, changes in the environment leads to increasing directional selection on a subset of phenotypic traits. If the QTLs controlling these traits have not been subject to strong selection in the past, then their allelic diversity should be quite high—perhaps close to the neutral expectation if selection has been very weak. We expect that the evolutionary responses of such traits will depend on two things; 1) the levels of standing genetic variation present within populations, and 2) the amount of allelic diversity that is distributed among populations. In species with high rates of gene flow, the within-population diversity will be high, and selection responses may be quite rapid. With low rates of gene flow, most of the genetic variation would be distributed among populations, and adaptive responses will be more dependent on population connectivity and dispersal rates.

Neutral genetic markers will also be very useful for predicting the selection responses of traits that are continually subject to directional selection as climatic conditions change. In this case, we expect that much of the QTL variation within populations has been depleted. Future responses to selection would depend on new mutations appearing within populations, or novel alleles being introduced via gene flow or hybridization with divergent lineages. In either case, the rate of gene flow among populations, which is a function of connectivity and dispersal, is of critical importance because it will determine the rate of spread of advantageous mutations, which will ultimately determine the rate of response to selection in the face of changing environmental conditions.

Using Quantitative Traits to Estimate Evolutionary Potential—Is it worthwhile to use quantitative genetic variation to predict the adaptive potential of populations? Perhaps, but there are a number of issues that require consideration before proceeding with experiments to estimate quantitative genetic parameters. First, one has to decide which traits to analyze. On the one hand, this may be obvious; for example, if models predict more severe drought for a particular geographic range, then it would seem reasonable to focus on physiological and morphological traits associated with drought tolerance. On the other hand, there is a large number of traits to choose from and it may not be practical to measure them all. A practical solution may be to choose a "representative" set of traits that may span the range of the types of characteristics that may become important for adaptive responses to future selection regimes.

Measures of quantitative genetic variation for phenotypic traits will chronically underestimate evolutionary potential. Depending on the genetic architecture, the amount of genetic variation estimated for each trait may vary widely and will tend to reflect the effects of historical selection. In cases where the levels of genetic diversity at the QTLs controlling a trait are low, the evolutionary potential will be a function of the distribution of diversity among populations and the rates of gene flow among them. In this case, it would be prudent to estimate genetic parameters for a number of populations, but this approach would not be sufficient because combinations of alleles that are distributed among populations may determine the full evolutionary potential. We could include crosses among individuals from different populations, but such experimental designs will quickly become unmanageable as larger numbers of populations are included. Moreover, it is generally thought that adaptation to novel conditions is often due to alleles segregating at low frequencies in populations (Morjan and Rieseberg 2004), which would be difficult to detect unless extremely large numbers of individuals were included in the experiment. Even with large crossing designs that include multiple populations, some amount of genetic variation that may be important for adaptation will almost certainly be missed, so estimates of evolutionary potential based on quantitative genetic analysis of traits should generally be viewed as conservative.

Predicting Evolutionary Potential—From the previous discussions we can conclude that estimates of genetic diversity from neutral genetic markers and quantitative genetic analyses may provide different types of information on the evolutionary potential of populations. On the one hand, we may be skeptical about the utility of neutral markers, because within-population genetic diversity estimates from these two approaches are only weakly correlated (Mittell et al. 2015). While it is clear that different evolutionary processes are affecting neutral markers and QTLs, the lack of strong correlations between these diversity estimates is not surprising (McKay and Latta 2002). The most critical question is whether allelic variants that will confer adaptation are currently present in populations. Clearly, neither of these methods is adequate to thoroughly address this question, but they may provide information that could be useful for making generalized predictions for the evolutionary potential of populations and species.

One of the most difficult challenges for understanding evolutionary potential is that the genetic variation that may be critical for adaptation to changing environments is more likely to be sequestered among a minority of populations across the species' range. Since most studies of quantitative genetic variation sample a limited number of individuals and populations (Hill 2010), they are likely to miss large proportions of genetic variation. Moreover, population differentiation raises the potential for combinations of alleles from different populations to be important for adaptive responses. While larger numbers of populations are sometimes surveyed for quantitative variation (e.g., Savolainen et al. 2011; Stock et al. 2014; Oakley 2015), rarely, if ever, are crosses among populations included in these experiments. Comparing levels of quantitative trait variation within populations to crosses made among individuals from different populations will provide estimates of variation that more closely approximate the evolutionary potential of species challenged by climate change.

Since a substantial amount of genetic variation is likely to be distributed among populations, it becomes clear that understanding dispersal and gene flow is a crucial component of predicting evolutionary potential. Neutral genetic markers can provide valuable information on rates of gene flow, but these estimates also need to be interpreted carefully (see Chapter 4). Furthermore, it is becoming increasingly clear that analyses of gene flow should incorporate landscape features, as rates of seed movement are likely to depend heavily on the behavioral characteristics of dispersal vectors (Holderegger et al. 2010). Analyses of gene flow and landscape resistance to dispersal will provide important information on connectivity among populations, which is critical for the spread of advantageous alleles (Morjan and Rieseberg 2004). Understanding the potential for levels of quantitative genetic variation of populations to be elevated through gene flow will be important for management decisions for species of concern.

The evaluation of quantitative genetic parameters will continue to be important for our understanding of evolutionary potential; however, these estimates should be viewed in the context of the probable effects of historical selection. Varying levels of selection will result in lower additive genetic variation, and population estimates will not reflect the full evolutionary potential of a wide range of phenotypic traits, because the majority of genetic variation may be present among populations. With the exception of severely range-limited species, it will generally not be possible to make comprehensive surveys of quantitative variation within and among populations. With these limitations in mind, it may be best to combine quantitative trait analyses with neutral genetic marker studies of gene flow and landscape resistance. The combined information from these approaches will provide assessments of the adaptive potential of populations and will serve to inform management decisions for the preservation of species that are subject to selective pressure due to climate change.

7.6 QUANTITATIVE GENETICS GOING FORWARD

The increased application of genomics in studies of evolution and ecology in the last few decades has yet to supplant quantitative genetics as a productive field of study, and probably never will. The realization that large numbers of QTLs of small effect determine most phenotypic traits has diminished hopes that genomic approaches alone will provide large amounts of novel information on adaptive genetic processes. It is becoming increasingly clear that improvements in our understanding of the genetic architecture of ecologically relevant traits and processes of adaptive evolution will require several approaches, including association studies with high-density markers, characterization of candidate genes, and traditional quantitative genetic analyses of phenotypic traits. It is also clear from the earlier discussion that traits within populations do not evolve in isolation; we expect that gene flow among populations has been and will continue to be a critical process for the spread of beneficial mutations that confer adaptation to changing environments. In the next chapter, we discuss the processes and consequences of gene flow in more detail, as we explore dispersal and the development of geographic and ecological variation within species.

7.7 QUESTIONS FOR CONTEMPLATION AND DISCUSSION

1. What is the relationship between V_E and COV_{GXE} and phenotypic plasticity? Describe how variance among genotypes can be detected when there are strong plasticity responses with and without COV_{GXE}.
2. Is the concept of broad-sense heritability useful? Under what circumstances would it be important?
3. Describe an experiment that would provide estimates of V_G and levels of phenotypic plasticity. Do you think that testing your genets across multiple environments will lead to an underestimate of V_G?
4. How are estimates of V_A biased by the selection of genetic material and environments? What do you think would be the most appropriate approach for understanding the potential for traits to respond to selection?
5. Think about a comparison between prospective and retrospective approaches to the estimation of heritability. Do you think one is more reliable than the other? Does one of them tend to provide more information?
6. Is it feasible to estimate heritability in long-lived species? What are the caveats that need to be overcome for accurate estimates of V_A?
7. How do maternal effects potentially bias estimates of heritability? Do you think this is a real bias, or could it be that heritable maternal effects could contribute to evolvability of populations?
8. What are the caveats associated with estimating selection gradients on phenotypic traits? Can the estimation of selection in contemporary populations be a useful endeavor?
9. Design an experiment to infer the presence of the historical effects of diversifying or homogenizing selection on one or more phenotypic traits. What type of information will you need? What would be the best approach for measuring the phenotypic traits?
10. What are the difficulties associated with QTL mapping and GWAS? Is the information gleaned from these studies useful for understanding the potential for adaptive evolution?
11. What is the best approach for inferring the genetic determination of adaptive phenotypic variation? Design a research program that would identify variation at the genomic level that contributes to adaptive responses to environmental challenges.
12. What is the difference between retrospective and prospective analyses of selection? Do you think that retrospective analyses have value for

understanding how plants might respond to future climate change?

13. Are neutral genetic markers useful for understanding adaptive potential of populations? Present arguments for and against the use of quantitative genetics and population genetics with neutral markers to predict adaptive potential.

8

INTRASPECIFIC VARIATION

AFTER A THOUSAND generations, species (A) is supposed to have produced two fairly well-marked varieties, namely a[1] and m[1]. These two varieties will generally continue to be exposed to the same conditions which made their parents variable, and the tendency to variability is in itself hereditary, consequently they will tend to vary, and generally to vary in nearly the same manner as their parents varied.

Charles Darwin (1859) On the Origin of Species by Means of Natural Selection, page 109.

8.1 DIVERGENCE AMONG POPULATIONS

The majority of species are distributed over large geographic areas, so it is likely that groups of populations will be isolated to various degrees and will have the potential to respond to selection for adaptation to regional environments. Hence, we expect that most species will be heterogeneous and will consist of mosaics of divergent varieties. This idea is reflected in Darwin's (1859) description of intraspecific diversification, where he depicts divergence among lineages as a crucial step in the process of speciation (Fig. 8.1; refer to color plate; Chapter 11). Both Darwin and Wallace (1865) assumed that the varieties they observed within species were based on heritable differences, but in the late nineteenth century this view was not widely held. During this time, most naturalists viewed intraspecific variations as induced by the environment and assumed that they were reversible (i.e., they were due to phenotypic plasticity; Chapter 6). This hypothesis was supported by some of the earliest *Transplant Experiments* conducted by researchers such as Bonnier and Kerner in the 1890s (reviewed in Turesson 1922). In these studies, and others conducted in the first two decades of the twentieth century, experimenters repeatedly demonstrated that environmentally induced phenotypic changes could be reversed by exposure to alternate conditions. These results bolstered the Mendelians, who viewed minor variations as inconsequential for evolution and assumed that speciation was due to genes of large effect (Chapter 3).

Views of intraspecific variation began to change in the 1920s with the publication of a series of transplant experiments by Göte Turesson (Turesson 1922). Turesson fully acknowledged the presence and importance of environmentally induced phenotypes, but he also saw consistent evidence of sets of traits that were characteristic of populations and appeared to confer adaptation to the local habitat. He envisioned species as consisting of many ecologically distinct lineages that he came to refer to as ecotypes. Turesson's work was the first to demonstrate that ecotypic differences were genetically determined and were maintained under different environmental conditions (e.g., plants from lowland and alpine habitats grown together in a common garden).

Turesson's pioneering work was not widely appreciated until several decades later when a trio of plant biologists—geneticist and ecologist Jens

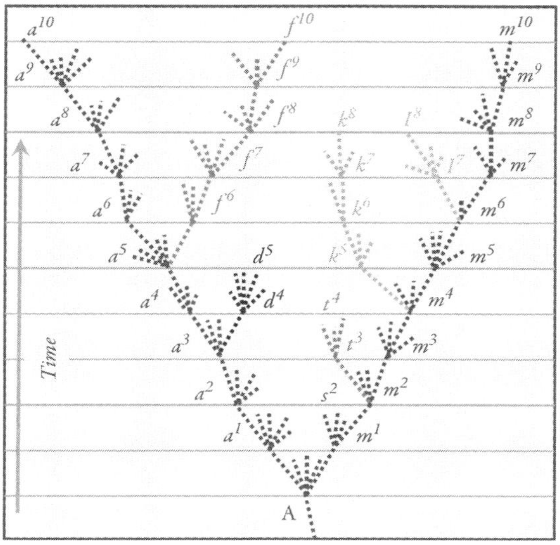

Fig. 8.1. A rendering of a portion of the only figure to appear in Darwin's (1859) *Origin of Species by Means of Natural Selection*. Distance between horizontal lines represent a thousand generations. Dotted lines represent varieties within species. When dotted lines persist to the next horizontal line, they "have formed a fairly well-marked variety, such as would be thought worthy of record in a systematic work."

Clausen, taxonomist David D. Keck, and physiologist William Hiesey—working at the Carnegie Institute of Washington at Stanford University, published a series of five monographs between 1940 and 1948. These authors confirmed Turesson's conclusions about ecotypic variation by documenting genetically determined differences among coastal and alpine ecotypes of several plant species including *Drymocallis* (*Potentilla*) *glandulosa* (sticky cinquefoil) and *Achillea millefolium* (yarrow). By this time, the controversy between the Mendelians and biometricians had been nearly forgotten, and the modern synthesis of evolutionary biology was well underway, so there was wider acceptance of the evolutionary importance of variations within species. Consequently, Clausen, Keck, and Hiesey's results fit well with developing views on adaptation and the importance of natural selection for phenotypic evolution.

Responses to selection for local adaptation and ecologically induced variation within species continue to be active areas of investigation, but in the late twentieth century, there was also growing recognition that neutral processes can affect differentiation within and among populations across much of the genome. While we expect that environments will vary across landscapes and will impose selection for adaptation to local conditions resulting in the ecological races, we also understand that random processes and historical contingencies can contribute to genetic variation among lineages. As we saw in previous chapters, levels of gene flow will be reduced over longer geographic distances, when suitable habitat is sparsely distributed, or if landscape features limit the movements of pollen and seed vectors. With less exchange among populations, increased isolation leads to divergence, as allele frequencies are subject to the effects of genetic drift and selection.

Gene flow and genetic drift are expected to affect the distribution of genetic variation over recent history (i.e., within the last ten to one hundred generations; Selkoe and Toonen 2006), but longer timescale processes may produce distinct patterns of divergence that cannot be accounted for by contemporary processes. In particular, genetic discontinuities can occur across a relatively homogeneous environment, such as in cases where genetic divergence among groups of populations has been generated by historical changes in the geographic range (Soltis et al. 1997; Nieto Feliner 2014). In these cases, populations in different portions of a species' range may be characterized by unique histories. For example, colder climates during the last glaciation caused many species to be confined to small geographic regions with warmer climatic conditions (*Glacial Refugia*). Range expansion out of separate refugia after the glaciers retreated resulted in a patchwork of genetically distinct lineages across geographic ranges.

In this chapter, we explore how historical contingencies and the process of selection have produced divergent groups of populations across geographic ranges of plants. We first examine how historical climatic changes have affected geographic distributions of plant species and communities. We discuss the processes of range expansion and the biotic and abiotic influences on range limits and describe how the resulting patterns of variation can be evaluated using population genetic and phylogeographic analyses. In the last section of the chapter, we examine the origin and maintenance of ecotypic variation within species.

8.2 CHANGES IN GEOGRAPHIC RANGES—GLACIAL CYCLES

Changes in geographic distribution have been common throughout the history of plant evolution. Periods of warmer or cooler climates and changes in ocean currents have altered seasonal temperatures and precipitation across continents. The most dramatic changes occurred because of periodic glaciations. The most recent glacial period of repeated glaciations started around 2.5 MYA during the Pleistocene when glaciation cycles lasting 41,000–100,000 years affected the distribution of species worldwide. Over the past 20,000 years—since the *Last Glacial Maximum* (*LGM*)—the distribution of vegetation has changed as climates warmed and the ice sheets retreated northward. During this time, many plant species shifted their distributions by thousands of kilometers, and the effects of these range expansions on patterns of genetic variation are still apparent today.

The distribution of plants and composition of communities are still in flux, as species continue to respond to changed climatic conditions, taking advantage of opportunities by dispersal into unoccupied habitats. Different modes of range expansion—in particular the number of individuals dispersing into new areas—affect the distribution of genetic variation across the expanding range and offer the potential for adaptation to novel habitats. In this section, we will explore the consequences of historical (post-glacial) and contemporary range expansion for the distribution of genetic variation.

Patterns of Plant Migration after the Pleistocene—With the end of the last ice age (the end of the *Pleistocene*, and the beginning of the *Holocene*), around 11,000 years ago, the distributions of organisms were shifting as climates changed. During the last glaciation (*Wisconsonian*, 110,000–11,000 years ago), plant species that are currently widely distributed were confined to the glacial refugia in the southernmost regions of continents of the northern hemisphere where the climates were more favorable. In North America, there were refugia located in northern Mexico, southern Florida, and areas close to major rivers that drained into the Gulf of Mexico and the Atlantic Ocean (Fig. 8.2; refer to color plate). Similarly, glacial refugia for temperate species existed in Europe in the southernmost regions of Spain, Italy, and Greece (Hewitt 1999) and along the southern coastal regions of Asia, including coastal islands such as Taiwan (Qiu et al. 2011). Refugia also existed in the ice-dominated northern portions of continents in areas of exposed land, including mountain peaks that protruded above the ice (*Nunataks*) and areas along the coast where mountains shielded the land from glaciers flowing outward from steep valleys toward the ocean (e.g., along the coast of British Columbia; Soltis et al. 1997).

In North America, much of Alaska and the adjacent region in Asia (collectively referred to as *Beringia*) remained ice-free and were connected by a broad land bridge that allowed dispersal of a number of species, including humans, into the American continents. Fragmentation of species distribution on nunataks and refugia to the north and south of the ice sheets for extended periods of time allowed for divergence through the accumulation of novel mutations and the actions of selection and genetic drift.

Once the glaciers began to melt, plants and animals migrated northward and southward from refugial areas. Range expansions of temperate species advanced along predictable paths (*Migration Corridors*), such as low-lying regions and valleys. As the climate continued to warm, the migration fronts from multiple refugia came into contact to form *Suture Zones* between groups of divergent populations (Remington 1968). The outcome of contact between pairs of

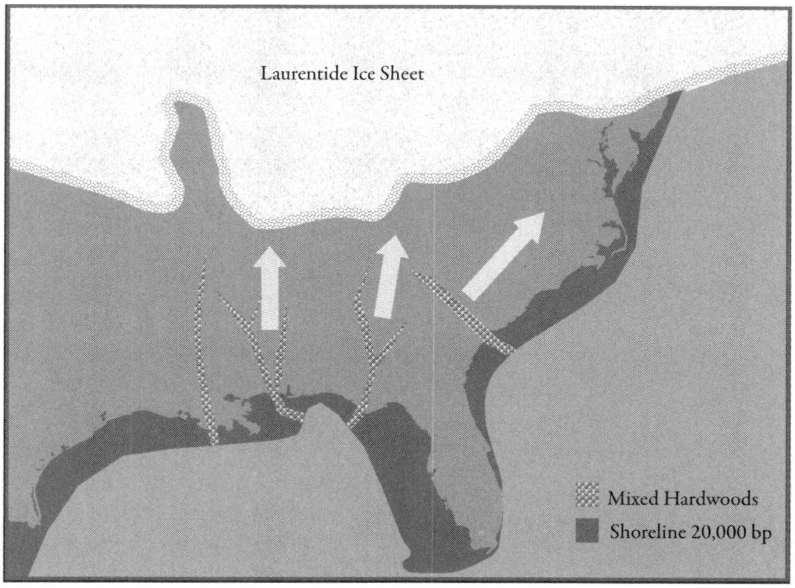

Fig. 8.2. An example of the effects of the furthest extent of Pleistocene glaciers on the distribution of plant communities. Hypothetical locations of glacial refugia of temperate species (mixed hardwoods) are indicated by green-checkered regions.

divergent lineages (*Contact Zones*, synonymous with suture zones) ranged from extensive hybridization and *Introgression*, to *Parapatric* distributions of reproductively isolated species (Chapter 10). These migration events had the effect of creating continuous distributions of species that consisted of large mosaics of genetically divergent populations with non-overlapping geographic distributions. Such patchworks of divergent lineages are often phenotypically uniform, so geographic heterogeneity can only be revealed using molecular genetic markers (see section 8.7).

During range expansions, plants from warmer regions invaded established plant communities that were closer to the ice sheets. Broad expanses of coniferous forests and a relatively narrow region of tundra vegetation occupied the landscape between the glaciers and collections of temperate species in the south. As the climate warmed, plant distributions began to change as populations toward the southern ends of ranges declined, and colonization extended ranges northward. With the exception of bare ground exposed near the edges of the melting ice sheets, the range expansion in response to climate warming required that species to the south invade established plant communities. This process of range expansion must have occurred in much the same way as the spread of invasive plants is occurring today; species from the south invaded intact communities by taking advantage of openings created by disturbance. This process would have been facilitated by natural disturbances, such as forest fires and broad-scale death of trees due to outbreaks of insects and diseases, which would have become more common as populations became weakened due to changing climatic conditions. The increases in frequency and scale of wildland fires and insect outbreaks that we are experiencing today are probably characteristic of the type of vegetation shifts that occurred at the end of each glaciation cycle. On the other hand, contemporary climate change is proceeding at a much faster pace, so it is unclear how this will affect the distribution, composition, and, patterns of extinction for plant communities in the future.

8.3 RANGE EXPANSION

Whether we are discussing post-glaciation changes in distribution of plant species or the spread of invasive plants in contemporary times, there are a number of common denominators that govern the pattern and process of range expansion. The movement of *Propagules* (seeds or vegetative fragments) from one place to another (*Dispersal*) will affect the genetic composition of populations if the migrant successfully reproduces. Dispersal of spores (e.g., pollen) will potentially affect the genetic composition of populations but does not contribute to changes in the range of species, so our discussion here focuses on the dispersal of propagules. Most migrants are probably not dispersed to suitable sites, fail to establish, or do not reproduce because of a lack of compatible mates (i.e., in self-incompatible or dioecious species), so their movement has no consequences for the distribution of genetic variation.

Dispersal Modes—When new sites are successfully colonized, we expect that bottlenecks will affect genetic variation in newly established populations unless the migrant pool is large (e.g., when a large number of individuals disperse to a site simultaneously or within a few generations). If suitable sites are abundant and widely available for colonization, then range expansion may proceed primarily by short-distance dispersal (*Contiguous Range Expansion; Phalanx Range Expansion*). With contiguous range expansion, *Migrant Pools* are likely to be large and populations with relatively high levels of genetic diversity will be established. This mode of range expansion may be more likely after the destruction of large tracts of existing vegetation by widespread disturbances such as wildland fires.

The distribution of suitable habitat is rarely continuous enough to allow for contiguous range expansion; more commonly, small disturbances open up sites for colonization within existing vegetation, so range expansion often proceeds in a more disconnected manner by *Long-Distance Dispersal* (*Non-Contiguous Range Expansion; Extra-Range Dispersal*; Ramakrishnan et al. 2010). In this mode of range expansion, new populations are established by a few individuals, so bottlenecks, inbreeding, and loss of genetic variation are expected to accompany the spread of plant populations into new regions (Excoffier et al. 2009). These "outpost" populations may eventually spread and coalesce to produce a more continuous distribution across the landscape, but population bottlenecks as a consequence of non-contiguous range expansion will establish a gradient of genetic diversity that declines from source regions to the most recently colonized populations at the leading edge of the expanding range (Le Corre and Kremer 2012). Examining patterns of genetic diversity across ranges provides insight into the contributions of each of these modes of range expansion to historical and contemporary changes in the geographic distribution of species (see section 8.7).

Several lines of evidence indicate that non-contiguous expansion is much more common than contiguous range expansion and is probably responsible for the large majority of invasions by alien species. One line of evidence for the importance of non-contiguous expansion comes from the discordance between observed seed dispersal distances and the historical rates of range expansion that occurred following the last glaciation (*Reid's Paradox*; Clark et al. 1998). For many plant species, the large majority of seeds fall within a few meters of the parent plant, and rare long-distance dispersal events are extremely difficult to detect (Cain et al. 2000). If we make projections of the rate of range expansion based on measures of seed dispersal, we quickly realize the paradox: based on direct estimates of seed dispersal, not nearly enough time has elapsed for most species to have spread from refugia to reach their contemporary distributions. The widely accepted solution is that long-distance dispersal was primarily responsible for the recolonization of regions once covered by glaciers. Although long-distance dispersal events are extremely difficult to detect by direct observation, they appear to have had a disproportionate effect on post-glacial range expansions.

Non-contiguous range expansion has had consequences for the latitudinal distribution of genetic variation. Temperate and arboreal plant species of the northern hemisphere generally display gradients of declining genetic diversity from southern regions to the northern limits of their distributions (Cain et al. 1998; Hewitt 2000; Excoffier et al. 2009). This pattern would be expected for species that had experienced serial population bottlenecks—due to non-contiguous range expansion—during their spread northward as they invaded and displaced plant communities in the wake of the retreating ice sheets. While a number of species display patterns that are consistent with this expectation (e.g., Cwynar and MacDonald 1987; Allen et al. 1996; Vik et al. 2010), there are some notable exceptions. For example, for a wide range of species in Europe, the confluence of migration corridors from multiple glacial refugia produced *Admixture* (hybridization; Chapter 10) and an increase in the genetic diversity at intermediate distances between the northern and southern limits of geographic distributions (Petit et al. 2003). Farther north there is a decrease in genetic diversity and an increase in differentiation among regions, as would be expected from recurrent long-distance dispersal along migration corridors. With a sparse distribution of available sites, few migrants will establish isolated populations, so range expansion will proceed through serial bottlenecks that will deplete genetic diversity toward the margins of the geographic distribution.

Range Expansion During Invasion—Further evidence for the primary importance of long-distance dispersal for plant range expansion comes from observations of invasions by introduced species (Hastings et al. 2005; Pauchard and Shea 2006). As was the case with range expansion after glaciation, the rate of spread of invasive plants cannot be explained without significant contributions from long-distance dispersal (Cain et al. 2000; Nathan et al. 2008). On the other hand, contemporary range expansions differ from post-Pleistocene expansions, as humans are implicated in the introduction and spread of invasive species around the world (Von der Lippe and Kowarik 2007; Wichmann et al. 2009). Humans differ from other dispersal vectors because movement is generally limited to established transportation routes, which often results in larger numbers of migrants arriving at specific foci of human activity (Wilson et al. 2009). While human-mediated dispersal has the potential to increase the size of migrant pools, the potential for establishment at a new site is still limited by the invasibility of the existing plant community, which in turn is generally a product of disturbance (Theoharides and Dukes 2007).

Habitual movements of humans generally result in predictable patterns of dispersal and increased levels of propagule pressure at specific sites (i.e., *Directed Dispersal*; Howe and Smallwood 1982; Wenny 2001). For example, sites that consistently experience higher levels of human activity, such as roadside pullouts, recreation sites, and areas subject to natural resource harvesting, are likely to be subject to higher levels of disturbance and increased rates of seed deposition. Genetic marker analyses indicate that populations established by human activities often have relatively high genetic diversity that reflects dispersal from multiple source populations (e.g., Ramakrishnan et al. 2010; D'Hondt et al. 2012) because of movement along transportation routes and habitual visitation to these sites. Therefore, in many ways, humans are ideal vectors for invading plants; they act as efficient agents of long-distance dispersal, both moving propagules and creating open habitat by the destruction of native vegetation for the establishment of invasive populations.

Population Genetic Consequences of Range Expansion—Long-distance dispersal generally results in bottlenecks and inbreeding, because few individuals are likely to establish new populations. Colonization results in loss of genetic diversity because uncommon alleles are less likely to be dispersed from source populations. Moreover, with few colonists, inbreeding will ensue within a few generations because offspring will tend to share recent common ancestors (Chapter 4). In normally outbreeding species, the effects of inbreeding are expected to be more detrimental; higher genetic load in outbreeding populations will result in higher levels of inbreeding depression, which reduces individual plant vigor and the potential for population growth. Inbred populations at the leading edge of a range expansion can be "rescued" by additional gene flow. Such a pattern was found by Ramakrishnan et al. (2010)—newly established populations were generally smaller and highly homozygous, and populations farther from the edge of the expanding range were larger with higher levels of genetic diversity.

In species where small migrant pools are responsible for range expansion, we expect the rate of geographic spread will be slowed because of repeated population bottlenecks. Recent expansions are typified by declining genetic diversity toward the limits of the expanding range (Pannell and Fields 2014). As described previously, colonizing populations are expected to suffer from reduced genetic diversity and inbreeding, which may result in reduced population growth rates due to inbreeding depression. While slowed range expansion due to inbreeding may generally be the case, there is also evidence that repeated bottlenecks during range expansion may result in purging of genetic load (Pujol et al. 2009; Chapter 5). In particular, simulations and empirical evidence suggest that repeated rounds of inbreeding followed by gene flow during range expansion can result in efficient purging, and this is particularly true when genetically divergent lineages come into contact (Marchini et al. 2016; Fig. 5.21; refer to color plate).

Purging of genetic load during range expansion could precipitate the evolution of self-fertilization (Chapter 9). Once the fitness of selfed offspring exceeds at least half the fitness of outcrossed offspring, there will be a genetic transmission advantage of self-fertilization, and we expect that selection will favor any mutation that increases the frequency of self-fertilization (Lande and Schemske 1985). While there is not yet evidence for evolution of increased selfing frequencies in invasive plants, range expansions associated with the retreat of glaciers after the last ice age may have favored selfing in northern latitudes. For example, northern populations of *Campanulastrum americanum* (American bellflower) display substantially reduced levels of inbreeding depression and higher selfing frequencies than those further south (Barringer et al. 2012). Both selfing and asexual reproduction tend to occur at higher frequencies in northern latitudes, but it is not yet clear whether these strategies are an outcome of selection via genetic transmission or ecological circumstances that favor reproductive assurance (Chapter 9).

While purging during range expansion has potential to contribute to the evolution of selfing, we currently have limited information on how frequent loss of genetic load due to repeated bottlenecks promotes increased selfing frequencies. Between 10% and 15% of plant species are considered highly selfing (i.e., less than 20% outcrossing; Goodwillie et al. 2005), and this proportion is maintained even though selfing lineages tend to be short lived and to experience higher rates of extinction than outcrossing lineages (Goldberg et al. 2010). In spite of the prediction that the emergence of new selfing lineages must be relatively frequent, we currently have no examples of populations that are clearly at intermediate stages of transition to a high selfing frequency. It is possible that the evolution of selfing is episodic; in other words, perhaps range expansions that are due to major climatic changes would promote higher frequencies of purged lineages and the origin of multiple selfing species over a relatively short span of time. If this were the case, it would explain why species in transition toward selfing are currently difficult to identify. This situation may change in the coming centuries as anthropogenic climate change is expected to precipitate a new round of range shifts, which will perhaps be accompanied by the origin of a new cohort of selfing lineages.

8.4 RANGE LIMITS

At any particular period during changes in geographic distributions, we expect the range occupied by a species would be somewhat heterogeneous. Environments vary across landscapes, so as individuals move farther from one location, they are more likely to encounter different biotic and abiotic conditions. Physical factors affecting plant growth (e.g., climate and soils), as well as biotic factors (e.g., plant, soil microbe, pathogen, pollinator, and herbivore communities), all change with geographic distance, as well as latitude, elevation, and proximity to geographic features (e.g., water bodies and mountain ranges). In general, we expect environmental change across the landscape to be gradual, as conditions become more different with distance, but variation in elevation, soils, or water availability may create habitat patches of greater similarity at intermediate distances (Fig. 8.3). For example, in habitats created by streams that support riparian communities, plants would be more similar to other plants within patches than to plants in the intervening vegetation. These discontinuous communities are expected to become more different with distance, due to changes in other physical and biotic factors, so the peaks of similarity seen in Fig. 8.3 decline at greater distances. Whether the environment is patchy (*Coarse-Grained*) or gradually changing (*Fine-Grained*), conditions will change until the limits of the distribution of individual species are eventually reached, as population growth and the density of individuals decline. In this section, we explore the biotic and abiotic factors affecting range limits, and the influences of gene flow and genetic drift on the potential for populations to adapt to local conditions.

Abundance and Range Limits—The idea that individuals of a species will occur at higher density toward the center of their distribution—because that is where the habitat is most "ideal"—and will tend to decline toward range limits (*Abundant Center Hypothesis, ACH*) has been widely accepted in the fields of ecology and biogeography (Lawton 1993).

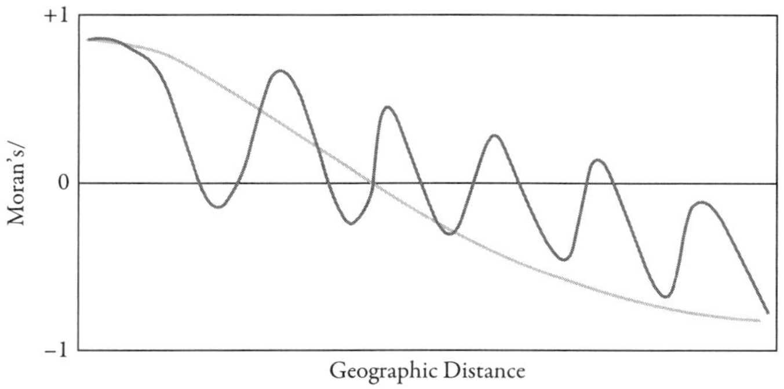

Fig. 8.3. Spatial autocorrelation for habitat similarity across landscapes. Values of Moran's *I* are equivalent to correlation coefficients for pairs of locations at different distances. Environmental gradients will produce gradual changes in the habitat (smooth line), while patchy environments may produce undulations in similarity that decline with distance (wavy line).

On the other hand, surveys of the literature suggest that only about half of the species that have been analyzed have distributions that are consistent with the abundant center hypothesis (Sagarin and Gaines 2002; Sexton et al. 2009). For plant species in particular, the majority of studies do not detect a pattern of declining abundance toward range boundaries (e.g., Ribeiro and Fernandes 2000; Kluth and Bruelheide 2005; Samis and Eckert 2007). To understand the possible reasons abundance distributions may not follow the expectations of the abundant center hypothesis, we need to examine the underlying assumptions of this idea and the factors that may contribute to range limits.

From an ecological standpoint, we may expect the boundaries of a species range to be abrupt if population growth declines as environmental conditions become less suitable toward the distributional limits. Abrupt range limits occur in response to environmental gradients if the fitness of individuals—and consequently population growth rate—declines gradually until the growth rate becomes negative; at this point a threshold is reached beyond which populations can no longer persist (Fig. 8.4A; refer to color plate). This boundary can be extended if expanding populations are within close enough proximity to maintain declining populations through dispersal (red line in Fig. 8.4A; refer to color plate). From a metapopulation perspective, these marginal populations act as "sinks" that can only be maintained by dispersal from "source" populations (Holt and Keitt 2000). These population growth rate/habitat suitability models predict a pattern consistent with the abundant center hypothesis; population size should decline from the center of the range to the edge, as long as environmental gradients are smooth and gradual.

The prediction of declining abundance toward the range limits just described assumes that populations are not able to adapt to local conditions. If populations were able to respond to local selection, then growth rates would remain high across the environmental gradient, and the abundant center hypothesis would not hold true; however, a number of processes limit responses to selection. For one, we expect that gene flow from adjacent populations will counterbalance the influence of selection and prevent local adaptation (Moeller et al. 2011). Populations toward the center of the range may have higher population growth and consequently, would have greater potential to export seeds and pollen. At the same time, we expect much less gene flow among populations toward the range limits, because populations near the range limits would have lower growth rates, so gene flow from the center of the range will tend to swamp the effects of local selection. Adaptation to local conditions would be limited further by small population size, which would increase the effects of drift and further curtail the influence of selection.

If habitat conditions deteriorate toward the range limits of a species, then both the effective population size and rates of gene flow among populations would decline. Hence, for a species with a stable range (i.e., one that is not undergoing range expansion), we expect

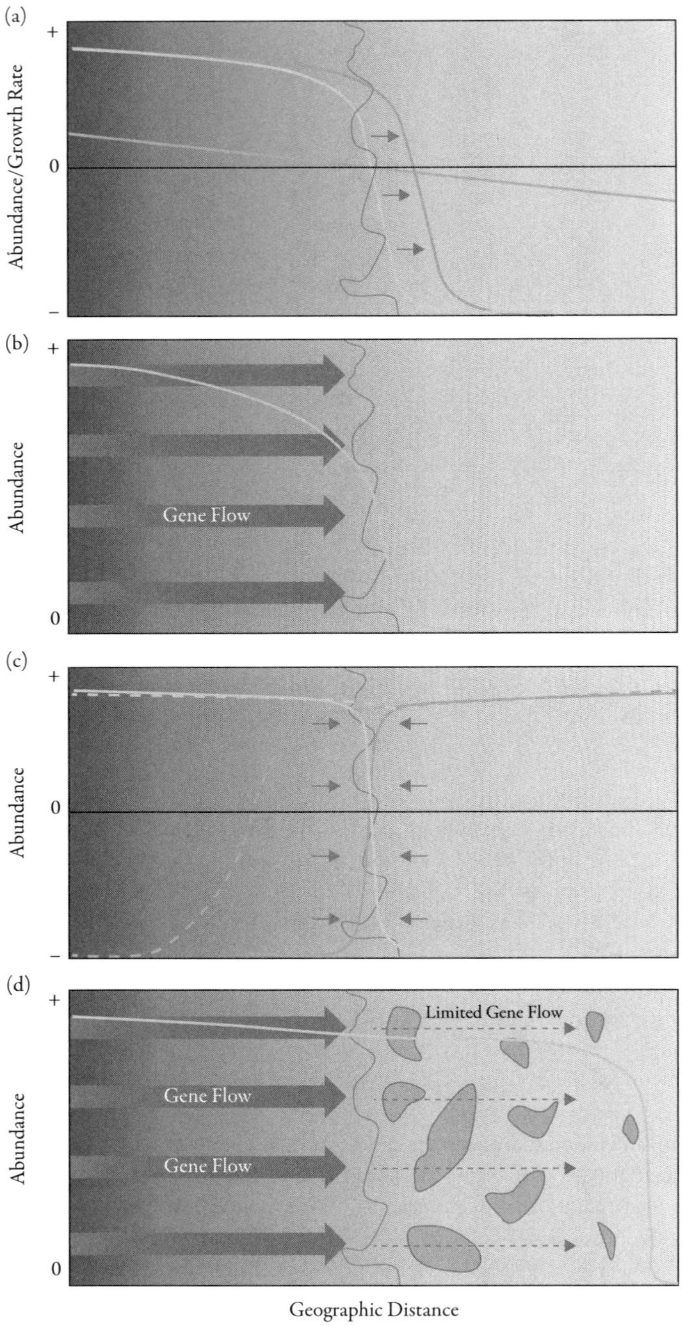

Fig. 8.4. Mechanisms affecting the distributional abundance of individuals within a species: A. Population growth rate (blue line) declines linearly until it becomes negative, creating a threshold for abundance (orange line). Sink populations in regions of negative growth may be maintained by dispersal (red line); B. Populations toward the range margin are prevented from adapting to local conditions by gene flow and genetic drift; C. Competition between two species creates parapatric distributions with sharp boundaries. Orange and purple solid lines are realized abundance, dashed lines represent the potential distribution without competition; D. Populations in habitat islands outside the main range adapt to local conditions when gene flow from the primary range is limited. Abundance within populations is indicated by the gray line.

decreasing genetic diversity within populations and increasing differentiation among populations toward range boundaries—a pattern that has been found in a wide range of studies (Eckert et al. 2008). By considering the effects of population genetic processes in marginal populations, we can predict a pattern that is consistent with the abundant center hypothesis, but in this case smaller population sizes near the range limits may be maintained, because gene flow and drift thwart local adaptation (Fig. 8.4B; refer to color plate).

Both the ecological and population genetic scenarios just described predict patterns that are consistent with the abundant center hypothesis; however, only a minority of studies surveyed have found that the abundance of individuals and average fitness decline toward the range margins (Sexton et al. 2009). There are a number of reasons the abundant center hypothesis may not always predict patterns of abundance across the geographic range. For one, the distribution of a species may not be in equilibrium, because climatic conditions are changing and the species has recently been affected by a range contraction. In this case, populations may be larger than can be maintained by current conditions, and small populations that were located toward range margins may have already disappeared. If the species being studied were a long-lived perennial, then there may not have been enough time for population size to decline to match recent climatic changes; a situation that is exacerbated by the sedentary nature of plants since individuals tend to occupy a site until removed because they become susceptible to disease or are destroyed by disturbance. Under this scenario, we may expect that fecundity in marginal populations would be lower, even if density is high and individual plants are large (e.g., Samis and Eckert 2007).

Another reason that population size may not decline toward distribution boundaries is that the range margin could be defined by sharp environmental gradients (e.g., when a species' distribution abuts a major geographic feature such as a mountain range or large body of water). Elevation or water availability may be relatively constant across a broad area and then change rapidly as the feature is approached. Abundance will remain high across most of the range but drop off sharply in regions where the environment changes abruptly.

Competition may also affect the abundance of individuals near range margins. The presence of a competitor that has a somewhat different range of tolerance to abiotic conditions may produce a parapatric distribution of two species with high abundance in adjacent geographic ranges (Fig. 8.4C; refer to color plate). If the environmental tolerances (*Fundamental Niche*) of the two species overlap, then each may maintain a high population density on either side of a boundary that reflects the threshold of competitive superiority. The sharp boundary is maintained by competitive interactions, which effectively define the *Realized Niche*, and produce abundance distributions that remain high up to the range edges. Competition and other types of biotic interactions (e.g., mutualisms) may produce patterns of distribution that do not match the expectations of the abundant center hypothesis.

Another scenario where abundances do not decline near range limits can occur if increased isolation allows for local adaptation and the maintenance of high-density populations near range margins. If ecological factors reduce rates of dispersal and consequently gene flow, then marginal populations may be able to respond to local selection. This could occur if the distribution of suitable habitat patches becomes sparse near the range limits or if pollen or seed vector densities are lower in regions close to the geographic boundary of the plant species. Undulations in environmental conditions, such as small changes in elevation or water availability, will tend to produce habitat patches of decreasing size and density as the average conditions change over geographic distance (Fig. 8.4D; refer to color plate). With a sparse distribution of populations near the range edge, low rates of gene flow may introduce adequate genetic variation to allow for adaptation to local conditions without swamping the effects of selection (see Chapter 5—Adaptation with Gene Flow). The ultimate limit to the ability of populations to adapt to conditions farther along the gradient may be reached due to lack of adequate genetic variation within the species or to evolutionary constraints that produce tradeoffs among traits conferring adaptation to different environments (e.g., Angert et al. 2008).

Tests of the Abundant Center Hypothesis— Ultimately, the distribution of most species is probably affected by a combination of factors, including abiotic conditions, population genetic processes, and interactions with other species. As is the case for many simplistic models, we generally learn more about the ecological and evolutionary forces that determine distribution and abundance from species that do not meet the expectations of the abundant center hypothesis. For example, in contrast to the predictions of the abundant center hypothesis, sometimes the lowest abundance is found toward the center of the range. In the case of an endemic shrub in Brazil (*Coccoloba cereifera*), lower density toward the center of the range is associated with higher fire frequency (Ribeiro and Fernandes 2000). In another study, of the annual mustard *Hornungia petraea* in Europe, lower density in the range center was found to be governed by seed bank dynamics (Kluth and Bruelheide 2005). It is also important to keep in mind that a species' distribution could match predictions of the abundant center hypothesis for extraneous reasons. For example, a species undergoing range expansion is likely to have smaller populations toward the expanding edge because of the effects of recent colonization.

8.5 LONG-DISTANCE DISPERSAL

While the processes governing abundance across a species' range can be complex, more challenging still is understanding the intersection of rare circumstances that have allowed some species to persist and flourish well outside their native ranges after *Long-Distance Dispersal* (Baker and Stebbins 1965; Williamson 1996). The chance that a seed traveling some distance beyond the normal geographic range would happen to land in a location offering suitable conditions for growth and survival to flowering seems astronomically small. Yet, such large shifts in geographic distribution have frequently occurred by natural processes, as evidenced by the existence of disjunct populations (e.g., island populations established without human intervention) and ecologically specialized populations (ecotypes) of many species. Instances of successful colonization and establishment after long-distance dispersal have increased in recent history as a direct consequence of human activities. We have facilitated the success of numerous invasive species by accidentally and intentionally transporting organisms across broad geographic distances.

After long-distance dispersal, newly established populations are likely to encounter novel environments in the invasive range. To some degree, long-distance dispersal may create conditions that are conducive to adaptation because newly established populations are released from the constraints imposed by gene flow, as discussed in the previous section. Movement of individuals or propagules to locations well outside the native range creates opportunities for adaptation to novel habitats (Keller and Taylor 2008; Prentis et al. 2008). At the same time, population bottlenecks commonly occur during colonization and establishment (Bock et al. 2015), so reduced genetic variation may limit adaptive potential (Chapter 4). In spite of the small chance of landing in a location with suitable conditions and the limits imposed by low levels of genetic diversity, there are many examples of successful colonization after long-distance dispersal.

Examples of successful colonization after long-distance dispersal can be seen in numerous cases where divergent populations and species have evolved on islands (Baker and Stebbins 1965; Stuessy 2008). Similarly, long-distance dispersal may have been responsible for the diversification of some species to produce complexes consisting of many ecotypes or subspecies (Clausen 1962; Lowry 2012). While unique island taxa and ecotypes represent examples of historical diversification, in the last few centuries we have seen human activities dramatically increase the frequency of long-distance dispersal and hence, opportunities for adaptation to novel geographic ranges (Olivieri 2009; Wilson et al. 2009). Some of these displaced populations have undergone rapid evolution to flourish in their invaded ranges (Colautti and Lau 2015). While a lot of attention has been given to the ecological and economic damage inflicted by invasive species, these "unnatural and uncontrolled experiments" also present us with unique opportunities to investigate the processes promoting diversification and adaptation in novel habitats.

Adaptation During Invasion—A number of studies have documented phenotypic changes after human-mediated invasion by comparing individuals from native and invaded ranges (Bossdorf et al. 2005). While it might be assumed that the changes observed in invasive populations are adaptive, unequivocally demonstrating adaptation to novel conditions can be difficult (Keller and Taylor 2008). Identifying sources of introduction is crucial, because traits often vary widely across the native range, so comparison to non-source populations may lead to the false conclusion that phenotypic changes have occurred in the invasive range. Reliable methods for identifying sources of invasive populations have been proposed (Keller and Taylor 2008; Estoup and Guillemaud 2010) and successfully applied in a number of cases (e.g., Rosenthal et al. 2008; Gaudeul et al. 2011).

Detecting the effects of selection after introduction is complicated by the potential for genetic drift; phenotypic shifts in introduced populations may be due to population bottlenecks that occurred during colonization and establishment. On the other hand, genetic marker studies indicate that most successful introductions have only small reductions in genetic diversity, indicating that bottlenecks were minor. More often, it appears that successful invasions have been initiated by large migrant pools, multiple introductions, and hybridization among divergent lineages from different parts of the species' range (e.g., Lavergne and Molofsky 2007; Rosenthal et al. 2008), which would create more opportunity for selection to act on colonizing populations. Nonetheless, the hypothesis that adaptive phenotypic shifts have occurred during introduction and establishment in a novel range requires careful evaluation, and nonadaptive hypotheses may sometimes be difficult to exclude.

High genetic diversity invasions are common, and successful invasions often display little or no loss of genetic diversity in the introduced range, which suggests that there may have been many opportunities for selection to promote the adaptation of invasive genetic lineages to their new environments. Some of the most notorious cases of alien plant species that have wreaked economic and ecological havoc apparently stemmed from multiple introductions, and colonizing propagules often originated from multiple geographic regions of the native range (Olivieri 2009; Estoup and Guillemaud 2010). Examples of species with high genetic diversity include some of the most aggressively invasive species known, including *Triadica sebifera* (Chinese tallow tree; DeWalt et al. 2011), *Pueraria lobata* (kudzu; Pappert et al. 2000; Sun et al. 2005), and *Phalaris arundinacea* (reed canary grass; Lavergne and Molofsky 2007) in North America and the *Opuntia robusta* (wheel cactus; Baker et al. 2008) in Australia. It is also the case that common garden studies have demonstrated increased vigor for individuals derived from introduced compared to native ranges for a number of species (e.g., Lavergne and Molofsky 2007; Dlugosch and Parker 2008; Xu et al. 2010; Hodgins and Rieseberg 2011), which suggests that adaptation after invasion might be common.

While it is tempting to interpret higher fitness in common gardens as evidence of adaption to novel conditions in the invasive range, it is striking that invasive plants are often generally more vigorous regardless of the environment they were exposed to. These results may not be surprising, considering the history of native and invasive plants; native populations typically have been stable for long periods of time, so mating within populations would promote inbreeding and consequently a loss of vigor. In contrast, the process of introduction during early stages of invasion may have afforded opportunities for crossing among individuals derived from different geographic regions in the native range. Such admixture would produce more heterozygous offspring, which would alleviate any history of inbreeding and infuse invasive populations with greater genetic variation (Chapter 10). The observation of increased vigor of offspring from crosses among inbred lines—often termed heterosis—is common in plant breeding and studies of populations (Chapter 4). Native populations, on the other hand, may be suffering from inbreeding to a greater degree than invasive populations, so we expect they would have lower vigor in all environments. The contrasting histories of native and invasive populations confound comparisons, making it difficult to distinguish the effects of heterosis from adaptation to the invaded range.

Even though it is difficult to make comparisons between native and introduced populations, there are some convincing examples of adaptation to local conditions by invaders. For example, invasive ragweed (*Ambrosia artemisiifolia*) displayed greater growth than plants from the native range under specific light conditions (Hodgins and Rieseberg 2011). Similarly, flowering times of ragweed and Canary Island St John's wort (*Hypericum canariense*) growing in common gardens were associated with latitude (Dlugosch and Parker 2008), suggesting that invasive populations have responded to selection to optimize the chance of successful reproduction. Studies that focus on critical life history traits (e.g., Xu et al. 2010;

Chun et al. 2011) or abiotic factors that have severe effects on the growth and survival of species (e.g., Hodgins and Rieseberg 2011) are more likely to demonstrate adaptive responses in invasive populations.

Invasion Without Genetic Adaptation—While there has been much focus on admixture and adaptation in the invasive species literature, there are a number of cases of invasion success with little or no genetic modification. The most dramatic of these is the invasion of wide regions of Europe and North America by a few clones of Japanese knotweed (*Fallopia japonica*). Genetic analyses revealed that single genotypes were widespread across continents constituting large and widespread clones. By using methylation-sensitive restriction enzymes, Richards et al. (2012) demonstrated that epigenetic variation—differences in DNA methylation patterns—was associated with ecological variation. The authors suggested that differences in patterns of gene expression controlled by methylation patterns may be critical for local adaptation.

In other cases, uniform habitat characteristics across continental regions may have facilitated wide-scale invasion by a few genotypes. For example, water hyacinth (*Eichhornia crassipes*), a floating aquatic plant (Fig. 9.2; refer to color plate), is represented by just a handful of clones in its invasive range in tropical and subtropical Asia and North America (Zhang et al. 2010). Aquatic habitats in tropical regions generally provide relatively consistent abiotic conditions, so invading water hyacinth may have met few challenges as it colonized new regions.

Another case of dramatic invasion success by cheat grass (*Bromus tectorum*) suggests that preadaptation may also facilitate success. Populations of cheat grass consist of selfing lineages that display non-random associations with ecological conditions across arid western North America (Ramakrishnan et al. 2006; Merrill et al. 2012). Lineages are homozygous across genetic marker loci and outcrossing is extremely rare, so the potential for this species to adapt to habitats beyond those for which they are preadapted may be limited by the lack of recombination.

Invasive Species under Climate Change—In the future, we can expect that human activities will continue to spread species across the globe. We already know that the introduction of alien species can be extremely disruptive as these plants flourish outside their native ranges. In many cases—particularly on islands—invaders can provoke irreparable damage, including the extinction of many endemic species. On the other hand, increased rates of dispersal may alleviate loss of species, as plants and animals are dispersed into regions made more favorable by rapidly changing climates. We can be hopeful that human activities will serve to promote the preservation of diversity in the face of climatic shifts, as human-mediated dispersal increases opportunities for outcrossing and admixture and may promote adaptation to changing conditions.

8.6 GENETIC CONSEQUENCES OF SEED AND POLLEN DISPERSAL

Long-distance dispersal has been responsible for most historical and contemporary range expansions of plant species, but it is extremely difficult to study the rare long-distance movement of propagules by direct observation (Nathan et al. 2003). More insights into the processes contributing to range expansion have been gained through the application of genetic marker techniques. In particular, the contrasting inheritance modes of cytoplasmic (i.e., chloroplast and mitochondrial) and nuclear genomes can provide tools to separate the effects of colonization by seeds from pollen-mediated gene flow on the distribution of genetic variation.

For most Angiosperms, chloroplast (cpDNA) and mitochondrial (mtDNA) genomes are inherited solely from the maternal parent (Lynch 1997; White et al. 2007), while nuclear markers are transmitted via seeds and pollen. This occurs because organelles and plastids are generally not transferred from the pollen tube into the egg sack of the ovule where fertilization takes place (Box 9.1). For species with limited seed dispersal, we expect that the movement of cpDNA markers will be restricted relative to nuclear markers (Fig. 8.5; refer to color plate), so comparison of genetic differentiation among populations for nuclear and cytoplasmic markers can provide separate estimates of gene flow via seeds and pollen (Ouborg et al. 1999). The chloroplast genome does not recombine, so sets of cpDNA markers are inherited together as haplotypes. For most species, the level of differentiation among populations is greater for cpDNA markers because average seed-dispersal distances are much less than for pollen dispersal.

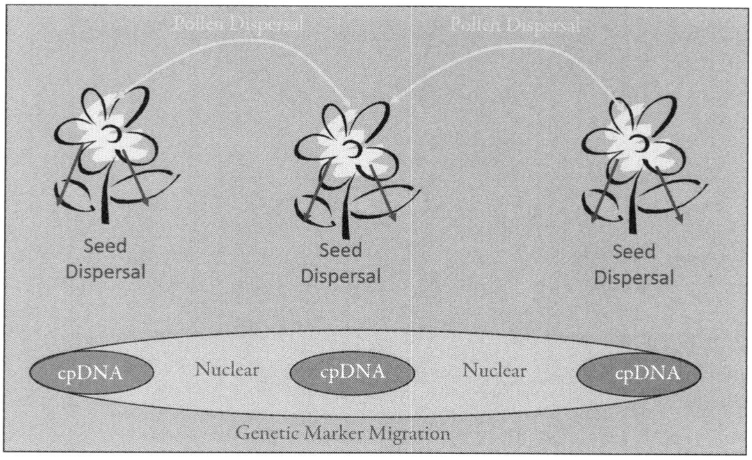

Fig. 8.5. The effects of pollen and seed dispersal on the migration of genetic markers. With limited seed dispersal, maternally-inherited chloroplast (cpDNA) or mitochondrial (mtDNA) markers are expected to produce much lower rates of migration. Nuclear markers migrate via pollen and seeds so "shadows" are much larger.

The consequences of lower rates of gene flow may be compounded by the smaller effective population size (N_e; Chapter 4) for cpDNA markers; chloroplast genomes are haploid, so N_e is expected to be half that of a diploid (Ennos 1994). The probability of losing cpDNA haplotypes each generation is greater because each individual transmits only one allele to its offspring, so rare cpDNA haplotypes are more likely to be lost, as fewer individuals carry fewer copies. This transmission disadvantage results in larger effects of genetic drift and consequently, stronger differentiation among populations for chloroplast markers compared to diploid (nuclear) markers.

We expect that patterns of seed dispersal will have strong effects on the distribution of maternally inherited cpDNA haplotypes. The work of Petit et al. (1997) provides an excellent example of the application of chloroplast genome variation to characterize the historical effects of long-distance seed dispersal. These authors found a patchwork of cpDNA markers distributed across populations of oaks in France, where each "patch" was nearly fixed for one genetic variant (Fig. 8.6). Neighboring trees tended to possess the same marker, so the entire population consisted of a mosaic, with individual patches covering areas averaging around 30 to 40 kilometers in diameter. They interpreted this pattern as evidence of long-distance dispersal events that occurred during post-glacial range expansion. Presumably, trees became established initially as openings in the existing vegetation became available (the region was occupied by conifers when the climate was cooler), and then the space in between was filled in through local seed dispersal and contiguous range expansion until the expanding patches coalesced to form a continuous distribution. Wind dispersal distributed pollen among colonizing trees, so individual nuclear alleles are widespread over a much broader area covering multiple cpDNA patches. The resulting pattern is one of low differentiation across the range for nuclear markers and a patchwork of regions that are strongly differentiated for chloroplast markers.

Mutation Surfing during Range Expansion—The distribution of genetic marker variation is affected by the mode of inheritance and characteristics of seed and pollen dispersal. During range expansion by long-distance seed dispersal, colonists will become established well beyond the primary range and will be sparsely distributed. These isolated individuals may then exchange pollen over relatively long distances, because their nearest neighbors are some distance away. Particularly in species that can only outcross, any seeds produced by isolated individuals would be the result of long-distance pollen dispersal. The tendency for sparse distributions to facilitate long-distance pollen dispersal has been shown for isolated individual trees and remote forest fragments (Young et al. 1993; Nason and Hamrick 1997; Sork

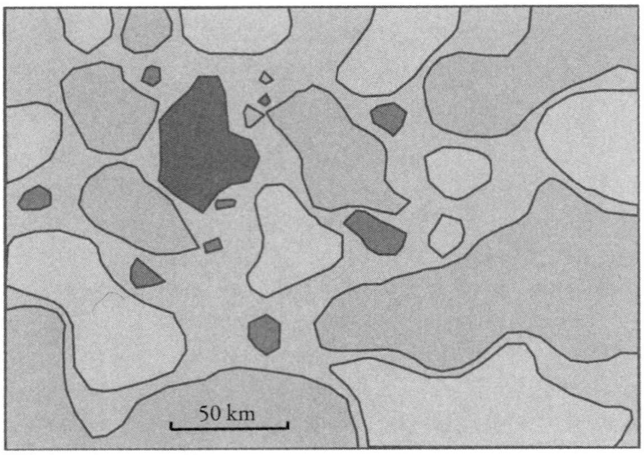

Fig. 8.6. The distribution of regions of oak forest with high probabilities of finding one of four cpDNA haplotypes. Modified from Petit et al. 1997.

and Smouse 2006; Broquet and Petit 2009). Over its lifespan, an isolated individual would be likely to receive pollen from a number of donors, so the amount of genetic variation contributed to a new outpost population through pollen dispersal could be quite large. In other words, range expansion by long-distance seed dispersal can facilitate higher levels of gene flow via pollen and consequently, larger migrant pools for nuclear genomes. The relatively high levels of genetic diversity introduced into isolated populations by pollen dispersal during range expansion increases the chance that most of the genetic diversity from ancestral regions will be distributed into the new range.

Large pollen migrant pools will have implications for the distribution of genetic variation across the expanding range of a species, and in particular, they can facilitate the migration and spread of rare alleles. Simulations predict that rare alleles can be swept along with the leading edge of the range, as they are caught up and effectively "surf" the wave of expansion (*Mutation Surfing*; Excoffier et al. 2009; Fig. 8.7). Rare alleles have a much better chance of traveling with the expanding range when migrant pools are relatively large (i.e., greater than 5% of the source populations; Klopfstein et al. 2006), so mutation surfing is much more likely to occur by pollen transfer than by seed dispersal.

To see the effects of large migrant pools on genetic variation and the possible effects of mutation surfing, we can examine the distribution of markers that are inherited solely via pollen in conifers. While maternal inheritance of cpDNA markers holds true for most flowering plant species, the mode of inheritance for cytoplasmic and nuclear markers is quite different for conifers. Cone-bearing species typically have maternal inheritance for mtDNA (seed dispersal), paternal inheritance for cpDNA (pollen dispersal), and biparental transmission for the nuclear genome (seed and pollen dispersal; White et al. 2007). Paternal inheritance of cpDNA genomes will produce higher levels of gene flow for cpDNA markers in conifers because of the high rate of pollen dispersal. These higher rates of pollen dispersal may produce conditions that facilitate mutation surfing during range expansion.

A possible case of mutation surfing is described for silver fir (*Abies alba*) in Europe by Vendramin et al. (1999). In silver fir, the level of cpDNA diversity is much higher in northern regions than in areas that were thought to have acted as glacial refugia, suggesting that mutations arising during post-glacial range expansion were retained and carried along by long-distance pollen dispersal as the species spread northward. Comparison of the distribution of maternally inherited (seed dispersed) mtDNA markers with paternally inherited (pollen dispersed) cpDNA markers in silver fir indicates a much higher rate of gene flow across long distances for the latter (Liepelt et al. 2002), a pattern that is consistent with our expectations for pollen-dispersed markers in larger migrant pools. This contrast in the distributions of pollen-dispersed (cpDNA) and seed-dispersed (mtDNA) markers probably reflects both the number

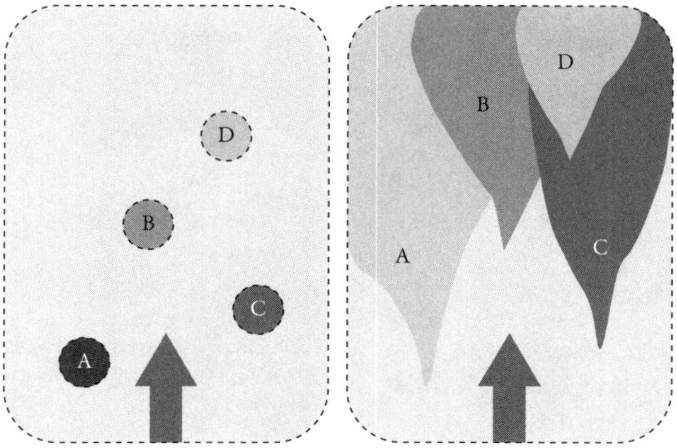

Fig. 8.7. The effects of migrant pool size during range expansion on the distribution of new mutations. Limited number of migrants results in restricted distributions of new mutations (A-D; left). Large migrant pools promotes mutation surfing and the spread of new mutations with the advancing front (right).

of migrants involved (i.e., because of the large amount of pollen exchange among widely spaced trees) and the low frequency of long-distance seed dispersal (i.e., because the establishment of trees by long-distance dispersal is probably rare).

The Effects of Dispersal Vectors on the Distribution of Genetic Variation—Migrant pool size often depends on the behavioral characteristics of dispersal vectors. As described earlier in this chapter, directed seed dispersal (Howe and Smallwood 1982; Wenny 2001) can increase propagule pressure to sites where chances of survival are higher and consequently, increase the size of migrant pools. Directed seed dispersal is found most often in cases of animal dispersal, especially dispersal by humans and frugivorous birds. For example, the habitual visitation of huckleberry plants (*Vaccinium membranaceum*) by birds in recently colonized regions following volcanic activity has led to newly established populations that possess high levels of genetic diversity (Yang et al. 2008). Dispersal of seeds by birds may also have had strong effects on the distribution of genetic diversity over larger geographic scales. In another species of *Vaccinium*, high levels of genetic diversity were found toward the margins of its geographic distribution (Yakimowski and Eckert 2008), which is in contrast to the majority of species, which display declines in genetic diversity toward range limits (Eckert et al. 2008). In both of these cases, the regular visitation of berry patches by birds has probably increased rates of gene flow and consequently, facilitated the maintenance of high levels of genetic variation across the ranges of these species.

The distribution of genetic variation generated by range expansion, and then modified by subsequent gene flow among established populations, will depend both on the frequency of long-distance migration events and the size of the migrant pools involved. Much of the dynamics of gene flow via seeds is determined by the frequency of disturbance in existing vegetation, which allows for the establishment of dispersed propagules. The level of genetic diversity present in colonizing populations is contingent on the mode of dispersal; different plant species have evolved fruits that take advantage of a variety of vectors, and it is the behavior of these dispersal agents that ultimately determines migrant pool sizes. The behavior of seed vectors will affect the frequency of long-distance dispersal to open sites and ultimately, the distribution of genetic variation for maternally inherited genomes. Once outpost populations are established, pollen dispersal among them will facilitate the long-distance movement of larger migrant pools, which will promote the spread of genetic variation across the expanding geographic range.

Possible examples of the effects of long-distance pollen dispersal on the distribution of genetic variation during range expansion are revealed in surveys of populations in recently colonized regions (Klopfstein et al. 2006). In these cases, genetic variation in newly established populations at the northern limits of range expansion were higher than expected based on the assumption of severe population bottlenecks due to limited seed dispersal. Understanding dispersal processes helps solve the paradox of why populations of plants are not completely homozygous at their northern range limits; pollen dispersal among sparsely distributed populations during range expansion may have promoted the spread of alleles to the far reaches of the species' range. Uncovering the processes maintaining diversity and cohesion among populations of species will require detailed analyses of the distribution of genetic variation, along with an understanding of the underlying biological and ecological processes affecting gene flow. In the next section, we will discuss in more detail how genetic relatedness among individuals and populations can be used to infer historical processes of migration and range expansion.

8.7 PHYLOGEOGRAPHY

From our discussions earlier in this chapter, it is clear that the geographic distributions of plants and animals are in flux, as they are responding to long-term shifts in global climates. It is a little shocking to realize that the plant communities that we are familiar with—and that are seemingly permanent features of the landscape—have only recently been established. Individual populations and groups of populations have unique histories, and consequently, regional variants of species are characterized by genetic signatures that reflect the pathways to their establishment. While these differences are often evidenced by phenotypic traits that distinguish ecotypes, more commonly, species seem to be homogeneous across broad regions, and the divergent lineages present are not obvious.

Unraveling the histories of populations and lineages through the analysis of phylogenetic relationships within and among populations is the primary goal of *Phylogeography* (Avise 2004). In particular, phylogeographic investigations focus on the history of lineages since the last glacial maximum. These analyses depend on the estimation of *Network Phylogenies* (Fig. 8.8) that display relationships among haplotypes (a region of DNA sequence having a unique complement of mutations), which represent independent lineages constituting an individual species or spanning two or more closely related species (Box 8.1). Network phylogenies differ from species phylogenies (Chapter 11) because the branches are defined by single mutations (substitutions or indels), and internal *Nodes* represent potentially extant haplotypes.

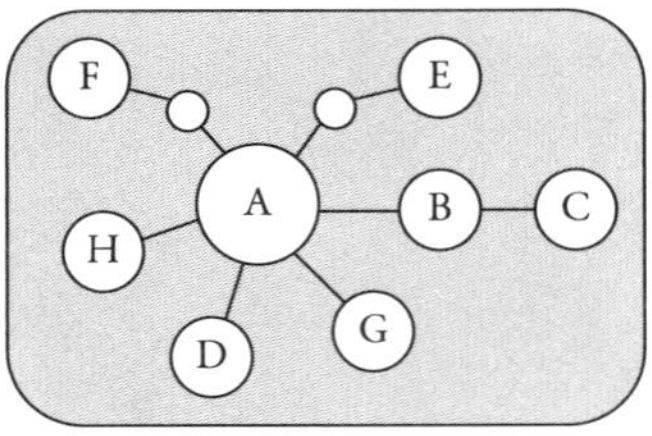

Fig. 8.8. Haplotype network based on a sample of individuals from one or more populations. Each circle represents a haplotype and lines indicate single mutational steps. Empty circles indicate haplotypes that were not sampled.

Reconstructing intraspecific phylogenies requires data on sequence variation from orthologous regions of nuclear or cytoplasmic genomes for individuals sampled from across the geographic range of a species. In plants, chloroplast sequence variation is often used for phylogeographic studies, because this genome is not subject to recombination, which can confound phylogenetic relationships among nuclear haplotypes (Fig. 8.10). Unlinked genetic markers that are subject to recombination, such as nuclear SNPs and microsatellites, are not suitable for reconstructing network phylogenies, because haplotype integrity is not maintained from one generation to the next, and haplotypes cannot be ordered by ancestor-descendent relationships. Network phylogenies represent a snapshot in time of the genealogical diversification of lineages within species (Box 8.2).

The haplotype composition of a population or region may represent a combination of in situ diversification by the accumulation of novel mutations and the immigration of haplotypes from neighboring regions by dispersal (Fig. 8.12; refer to color plate). Understanding the distribution and abundance of haplotypes within and among populations and regions allows us to infer phylogeographic relationships, which reflect the historical processes of migration and dispersal that have led to contemporary distributions (Avise 2000). Inferring historical processes that have

Box 8.1 Network Phylogenies

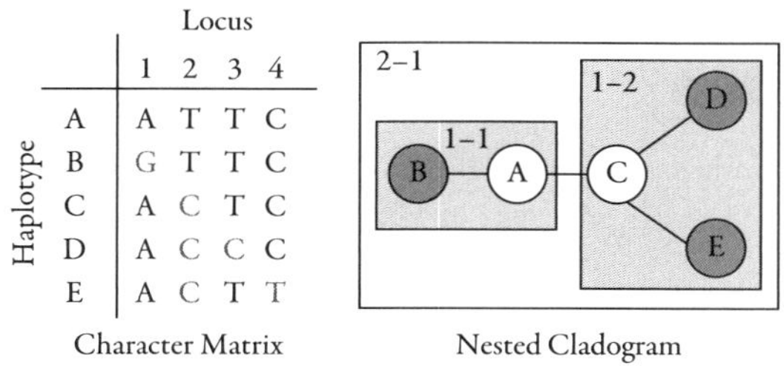

Fig. 8.9. A character matrix for five haplotypes (A–E) of nucleotide base substitutions and the resulting network phylogeny. Red bases represent mutation difference relative to the A haplotype. Labeled boxes represent groupings for nested clade analysis. Modified from Cruzan and Templeton (2000).

Relationships among haplotypes can be represented by unrooted network phylogenies if the number of mutations in the dataset is relatively small. Networks can be constructed by hand from character matrices (Fig. 8.9), or using software applications (e.g., Clement et al. 2000). Each mutation (substitution or indel) in the character matrix is represented by a single step between haplotypes in the network phylogeny. Haplotypes with descendants will occupy interior positions (interior haplotypes—A and C in Fig. 8.9) while those without descendants will occur at exterior positions (tip haplotypes—B, D, and E). Unsampled haplotypes may be included in the network (Fig. 8.8). Difference in the relative abundance of haplotypes is often indicated by differences size of haplotype circles (Fig. 8.8).

Rules for construction of networks are dictated by parsimony, so the minimal number of steps possible links haplotypes together. Networks can include ambiguities (loops) where individual haplotypes are connected to more than one ancestor by equal numbers of steps (e.g., Dobes et al. 2004). Loops can often be resolved by the geographic distribution and abundance of haplotypes, or by the relative frequency of different types of base substitutions (i.e., transitions versus transversions; Templeton et al. 1992; Crandall and Templeton 1993). One-step clades can be grouped by starting with the tip haplotypes and working inward. Information on the geographic distributions of haplotypes and clades can be used to develop hypotheses concerning the historical processes that have led to contemporary distributions (Box 8.3).

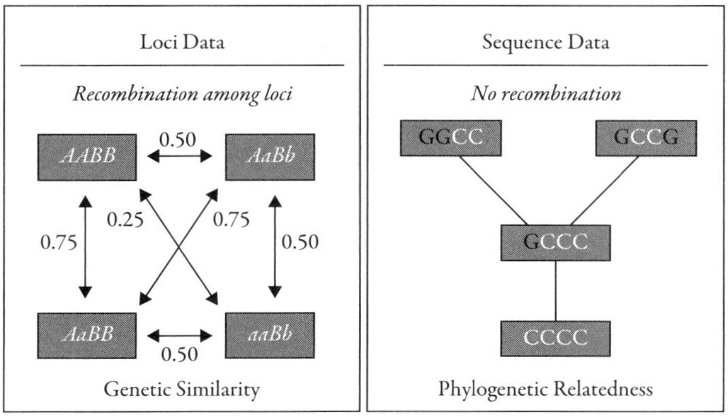

Fig. 8.10. Contrasting types of genetic marker data. Loci data are unlinked so they can recombine, and provide measures of genetic similarity. Sequence data indicates filial relationships among haplotypes.

Box 8.2 Inferring Historical Events

We can use the principles from coalescence theory and population genetics to make interpretations of the distribution of genetic variation within and among populations to infer historical processes. The haplotypes in the network depicted in Fig. 8.8 are distributed between two populations that are connected by dispersal events in Fig. 8.11 (refer to color plate). In the present, the population to the left has four haplotypes (H, A, G, and E) and the one to the right has six haplotypes (F, D, A, B, C, and E). The thickness of the lines and size of the circles at the top of the gene trees in Fig. 8.11 (refer to color plate) indicate the relative abundance of each haplotype. This depiction allows us to see the mutation and dispersal events that have led to the contemporary distribution of these haplotypes. The first dispersal event results in the colonization of a new site (the population on the left) by the red haplotype (A), and this new population begins to accumulate novel mutations. Each branch event represents a single mutation; for example, the green haplotype (B) is a single mutational step from the red haplotype, and the yellow haplotype (E) is two mutational steps away (Figs. 8.8 and 8.11; refer to color plate). There is a second dispersal event, so now both the red and yellow haplotypes are shared between the two populations.

It is important to remember that the only information we have when we sample these two populations is the abundance and genealogical relationships of extant haplotypes (Fig. 8.8); our challenge is to use this information to infer historical events. We note that two of the haplotypes (A and E) were found in both populations, and the remaining haplotypes were found in only one population. Applying the principles outlined above, we can make the following inferences:

1. The A haplotype in Fig.s 8.8 and 8.11 (refer to color plate)is ancestral (i.e., the oldest) in this sample. Two lines of evidence support this conclusion: 1) All of the other haplotypes are derived from the

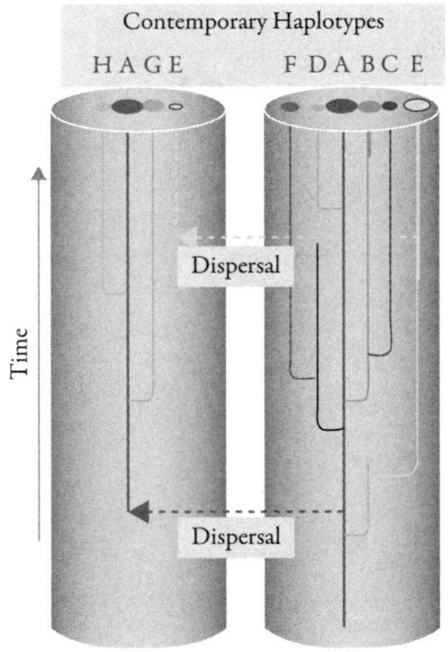

Fig. 8.11. Historical processes leading to contemporary distributions of haplotypes. Each cylinder represents a site that may support a population. The relationship among these haplotypes is depicted in the network phylogeny in Fig. 8.8.

A haplotype, and 2) A is the most abundant in both populations. Having been around longer would allow the A haplotype to be represented in a larger number of individuals, which also increases the chance that any one of them could be carrying a new mutation. The most recently derived haplotypes (tip haplotypes) are much less abundant compared to central haplotypes and may be carried by only one or a few individuals.

2. The population to the right in Fig. 8.11 (refer to color plate) is older than the one to the left. We can infer this because there is evidence of more mutations in the right population, based both on the number of haplotypes and on the number of mutational steps from the ancestral haplotype (A). An alternative explanation would be that the left population was recently subjected to a bottleneck. Under this scenario, only the A haplotype would have survived, so it would have been difficult to eliminate this hypothesis without ancillary information. For example, perhaps the site now occupied by the left population was not available until recently because of changes in sea levels or the presence of glaciers, and the right population was located in a region that could have served as a glacial refugium. With multiple lines of evidence from independent sources and corroborative phylogeographic patterns from other species, the inferences that we can draw from the patterns found in a single species become more robust.
3. The dispersal event or population bottleneck that established haplotype A in the left population in Fig. 8.11 (refer to color plate) was not recent. This inference comes from the observation that enough time has passed for several other haplotypes to accumulate in the population.
4. The E haplotype in the left population in Fig. 8.11 (refer to color plate) represents a second dispersal event. It would also be possible that the presence of this haplotype in both populations is an example of convergent evolution (i.e., an independent mutation event in each population). If the mutation rate is very low, then the chance of parallel mutations is highly unlikely because the probability of observing the same mutation twice is exceedingly small. We still might argue that the dispersal event could have gone in the opposite direction; however, the E haplotype is much more abundant in the right population, indicating that it has been present for a much longer period of time. The lack of daughter haplotypes and low abundance for haplotype E in the left population indicates that the dispersal event was relatively recent.

This description of how inferences are made from the network phylogeny (Fig. 8.8) and distribution of haplotypes between populations (Fig. 8.11; refer to color plate) highlights both the strengths and weaknesses of phylogeographic analyses. The conclusions that can be drawn from this analysis are tentative and should be considered as hypotheses that require further testing. As discussed in the text, conclusions from phylogeographic studies can be made more robust using multiple methods of analysis (Box 8.3), independent lines of evidence from other disciplines, and by analyzing multiple species from the same region.

led to contemporary patterns of intraspecific diversification requires knowledge of the relatedness among lineages and their relative abundances across the geographic range of study. Consequently, studies of intraspecific diversification require samples of multiple individuals from within populations and from across the entire range, or a contiguous region occupied by the species. In this respect, studies of intraspecific diversification examine longer time frames than population genetics and shorter periods than phylogenetics.

The idea that phylogenetic studies could be conducted in a geographic context to reveal the history of populations was first suggested by John Avise and colleagues in a seminal paper on phylogeography (Avise et al. 1987). In general, phylogeographic analyses focus on the history of populations since the last ice age for the identification of contact zones, migration corridors, and glacial refugia as discussed at the beginning of this chapter, but these methods can also be applied to infer more recent cases of

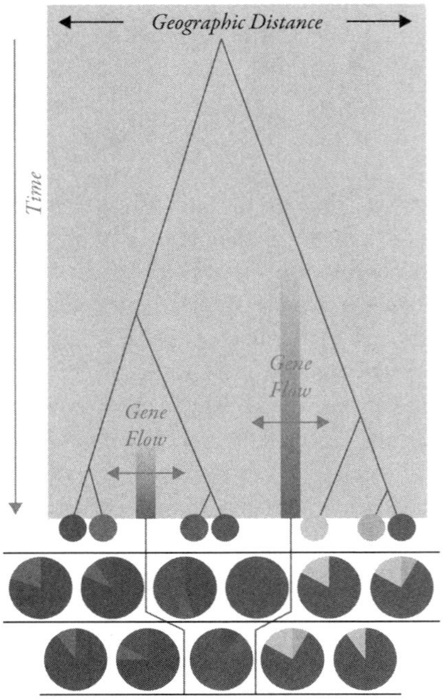

Fig. 8.12. The development of geographic differentiation within a species through the accumulation of novel mutations and gene flow. Each region is characterized primarily by regional lineages, but there is some lineage sharing between neighboring regions due to gene flow. Each pie chart represents a population.

long-distance dispersal and range expansion. For example, it is possible to identify cases of long-distance dispersal and the source regions responsible for the establishment of introduced and invasive populations (e.g., Rosenthal et al. 2008; Gaudeul et al. 2011).

While human-mediated introductions are often well documented, this is not always the case, so the history of an apparent introduction can be ambiguous. For example, fragrant water lily (*Nymphaea odorata*) is common in lakes of eastern North America and in a smaller disjunct region in the Pacific Northwest. Herbarium records indicate that this is a case of recent long-distance dispersal to the lakes of Washington, Oregon, and British Columbia, that was probably mediated by humans in the early twentieth century. Although unlikely, it is possible that early botanists overlooked this large white-flowered plant and that it has been present in the Pacific Northwest for some time (i.e., since before the last ice age). If that were the case, then we would expect that populations in the Pacific Northwest would have had time to undergo diversification through the accumulation of mutations, making them unique to some degree compared to populations from the east. Alternatively, recent introduction would be evident if all the haplotypes found in the introduced range were identical to those from the native region.

In this section, we will explore the applications and procedures used in phylogeographic analyses. We will start by developing a basic understanding of the theoretical underpinnings for the analysis of gene genealogies (coalescence theory; Chapter 5) and then apply these principles to interpret the geographic distributions of closely related lineages. We will discuss the advantages and limitations of different methods of phylogeographic analysis and provide examples of how they have been used in conjunction with *Palynology, Paleoecology*, and *Biogeography* to develop our understanding of changes in the distribution of plant species over the last twenty thousand years. Phylogeography provides historical information on the distribution of genetic variation and complements methods from population genetics and landscape genetics that provide analyses of contemporary processes (Chapter 4).

Gene Trees—Our ability to make inferences about the origins of populations and patterns of migration and dispersal depends not only on genetic profiling but also on methods that allow us to link individuals together in an order that reflects ancestor-descendent relationships. If we compare DNA sequences from a specific genomic region among individuals of the same species, we are likely to find variation. For any particular genomic region—whether transcribed or not—we will often find that populations within species can be grouped into a number of *Subtaxa* (i.e., subspecies, varieties, ecotypes, morphotypes, etc.) that are related to different degrees. Within each subtaxon, there may be a number of genetic lineages, and relatedness among lineages can be determined by comparing homologous sequences to construct gene trees, which are also referred to as *Gene Genealogies*. Each genetic lineage (branch in the genealogy) is represented as a haplotype. In any particular generation, some individuals will carry a novel mutation, which is represented as a new branch in Fig. 8.11; refer to color plate. In an ideal world, we would be able to trace each descendent generation with unique mutations, but mutation rates are not high enough to make that possible, so a number of closely related individuals will share the same haplotype. If these individuals survive to reproduce, their haplotypes will be passed on to the next generation, but within any given time period, a number of individuals may fail to reproduce, and some lineages will go extinct.

The first thing that becomes apparent when we examine gene genealogies is that, at any particular time, there are similar numbers of lineages present (Fig. 11.4; refer to color plate). If population sizes and numbers are constant over long periods, then there will be continual turnover of lineages as some are lost due to genetic drift, and new ones are formed by mutation. The number of lineages that can be maintained within a population will depend on the effective population size (N_e; Chapter 4), and the ultimate fate of all lineages is extinction. Some lineages will diversify more than others, but at any point in time, all extant individuals will be descended from a single common ancestor (the *Most Recent Common Ancestor; MRCA*; marked by the circled letters A in Fig. 11.4. With constant population size, the time back to the MRCA will remain about the same; the last common ancestor will progress forward in time, as lineage turnover changes the composition of extant lineages over time. The timespan back to the MRCA can be shortened if there has been a recent bottleneck or extended if the total number of individuals—all populations of the taxon combined—has been increasing (Chapter 5; Box 5.1). Similarly, if the historic effective population size has been small, the time back to the MRCA will be shorter as lineages will be lost at a higher rate due to genetic drift. Conversely, the time back to the MRCA will be longer when the historic effective population size has been large (Nielsen and Beaumont 2009).

The depiction of genealogical variation in Fig. 11.4 (refer to color plate) includes events that split the distribution into three isolated groups. When the species first split, and for some time afterward, the isolated groups of populations share a number of closely related lineages. It is not until many generations later that the diverging species will be characterized by sets of lineages that are all derived from common ancestors that originated after the split (e.g., they would all appear like the monophyletic case in Fig. 11.4 (refer to color plate). At this point, the isolated groups would have different MRCAs and would have achieved a state of reciprocal monophyly. Consequently, their relationships can be characterized using techniques for reconstructing relationships among species (*Phylogenetics* as described in Chapter 11). However, for a considerable amount of time after initial separation among groups of lineages representing each species, there is considerable haplotype sharing, which diminishes as the relatedness among haplotypes within each incipient species increases over time. This process of *Lineage Sorting* can take a large number of generations if effective population sizes are large, so it may be some time before a pair of diverging taxa will be characterized by reciprocal monophyly.

Relationships among haplotype lineages cannot be accurately described using techniques developed to infer phylogenetic relationships among species (Chapter 11). In a species phylogeny, the branch tips represent living taxa and the nodes represent hypothetical ancestors that are extinct. A number of mutations may define tree topologies and branch lengths in a species phylogeny. In a gene genealogy, single mutational steps define nodes, and nodes as well as tip haplotypes may be represented by extant lineages. These intraspecific genealogies can be accurately depicted using unrooted network phylogenies (Fig. 8.8; Box 8.1) and analyzed using methods based on coalescence theory and principles from population genetics. In the remainder of this section, we will explore the principles of coalescence theory, genetic drift, and migration and discuss how they can be applied to infer historical processes that have produced the contemporary distribution of genetic variation within species.

Phylogeographic Principles—The transfer of genetic material from one generation to the next produces a traceable history (genealogy) that can be recovered by comparing homologous DNA sequences among individuals of the same species. People use similar methods to uncover the history of their family from historical records and comparison of surnames. In many families, ancestors occasionally made spelling mistakes or intentionally changed their surname; sometimes those errors were reliably transferred on to all their descendants (e.g., Smythe becomes Smithe and then Smith, etc.). This is effectively what happens as DNA sequence mutations accumulate within a lineage and are transferred from one generation to the next. If we look at this process in reverse—looking back down the genealogy from the branch tip toward the root—we will notice that each branch point is defined by a novel mutation occurring in the gene copy of one member of the lineage (Box 8.2; Fig. 8.11; refer to color plate). One individual passing a new mutation on to their offspring will initiate a new lineage (branch).

As we look down along the branches, we notice that for each time interval (generation) there is an equal chance of a new mutation appearing in any individual, but since mutations are relatively rare, branches split off only occasionally at random intervals. We also notice that not all branches make it to the top of the "tree"—some may die off after shorter or longer time periods, so that the number of lineages currently represented is a subset of all the lineages that have existed over time. Since we are looking at this process backward through time, we refer to branch points as "coalescence events"—literally, merging or comingling of lineages. In our analysis, we are not privy to the historical events that led to the current genetic composition, but we can infer those processes using the abundance and relatedness of genetic lineages that are distributed within and among populations.

Under conditions of constant population size, we expect new lineages to accumulate through mutation, and the total number of lineages present in a population will approach an equilibrium between mutation and extinction that depends on the effective population size. Because the genomic region under study is selectively neutral, we expect that the abundance or extinction of lineages will be governed by genetic drift. When a new mutation—creating a new haplotype—appears in a population, it has a very high chance of being lost within a few generations (Chapter 4). For those mutations that persist, we expect that the number of individuals carrying the new haplotype will gradually increase, so that on average, a higher proportion of individuals in the population will represent older haplotypes (Fig. 8.11; refer to color plate). The first prediction we can make from considerations of coalescence theory and drift dynamics is that (1) *we expect the oldest haplotypes (lineages) to also be the most abundant.*

The spread of haplotypes among populations will depend on their abundance within populations and hence, their relative age. In any single generation, there is a small chance that any individual will successfully migrate to a new location. If migration success is random, then the probability of a haplotype being carried to a new site is directly proportional to its relative abundance in the source population. As dispersal occurs among populations, lineages will spread, and if migrant pools are small, those that are more common within populations will also occupy larger geographic ranges. The second prediction we can make is that (2) *the oldest lineages will be the most geographically widespread.* We can also conclude that new haplotypes will probably be restricted to their population of origin.

In any single generation, each individual recruited into a population has an equal chance of carrying a new mutation that would define a new haplotype (lineage). Since older lineages are more likely to be represented by more individuals, they will also tend to generate novel lineages. So, the third prediction we can make is that (3) *the oldest lineages will have more daughter lineages,* and consequently they will tend to be located at interior positions within the haplotype network (e.g., haplotype A in Figs. 8.8 and 8.11; refer to color plate).

During the time a species (population) occupies a particular site or region, novel mutations will appear and new lineages will tend to accumulate. This process can be relatively slow if mutation rates are low; as long as population size is stable or growing, the number of lineages maintained will be proportional to the age of the population. We expect that regions of a species' distribution that have been occupied for longer periods will be characterized by higher haplotype diversity, and conversely, if migrant pools are small then more recently colonized sites or regions will contain relatively few lineages. So, the fourth prediction we can make is that (4) *regions that have been occupied for longer periods of time will tend to be the most diverse.*

Applying Phylogeographic Principles—Identifying the processes contributing to contemporary distributions of haplotype lineages by applying the principles just outlined depends on whether the species and genomic regions being studied meet some underlying assumptions. First, the development of phylogeographic structure for genetic markers depends on limited dispersal among populations. A high migration rate would disperse all haplotypes across populations so any structure that developed during range expansion and colonization would rapidly erode. We can test for the effects of limited dispersal—whether haplotypes have non-random geographic distributions—by comparing the geographic range (*Dispersion*) of tip and interior haplotypes; restricted dispersal will result in lower levels of dispersion for tip haplotypes (Templeton et al. 1992). Demonstrating restricted migration is a necessary step before proceeding with phylogeographic analyses.

Successfully conducting phylogeographic analyses depends on assaying genomic regions with adequate levels of genetic variation within and among populations. The mutation rate of the genomic region being analyzed needs to be high enough to generate a number of novel haplotypes within the time frame being studied (i.e. within twenty thousand years), but not so high that identical mutations will appear frequently in more than one population (*Homoplasy*). If migration is limited, then using genomic regions with lower mutation rates to analyze geographic patterns provides us with a view of historical processes. For example, the mosaic pattern of cpDNA variation in European oaks that developed during post-glacial range expansion (Petit et al. 1997) may not have persisted with high migration rates. While low mutation rates can contribute to the maintenance of patterns that were created by historic processes, geographic variation in these genomic regions will not reflect the effects of contemporary processes of migration and drift, because local populations will tend to share the same haplotypes.

Reconstruction of genealogical relationships depends on the reliable transmission of haplotypes from one generation to the next (Fig. 8.10). For this reason, genomic regions that do not recombine (e.g., cpDNA and mtDNA) are often used for phylogeographic analyses (e.g., Schaal et al. 1998; Maskas and Cruzan 2000; Dobes et al. 2004; Fehlberg and Ranker 2009; Diaz-Perez et al. 2012; Sede et al. 2012). For phylogeographic analyses of flowering plants, chloroplast genomes have the additional advantage of usually being maternally inherited, so patterns of variation in cpDNA markers will reflect seed dispersal and colonization processes. On the other hand, mutation rates for the chloroplast genome are very low (White et al. 2007), but this limitation can be overcome by using intergenic regions that have higher levels of variation (Lynch 1997) and by sequencing a larger proportion of the chloroplast genome (Cronn et al. 2012).

Finding consistency among distributions of haplotypes from multiple genomic regions will provide more robust phylogeographic inferences (Templeton 2004; Knowles 2009). Nuclear regions can be analyzed to supplement inferences based on cpDNA, but network phylogenies based on nuclear regions will typically have ambiguities due to recombination events (loops; Box 8.1). Loops may be resolved if the two ancestor haplotypes display a strong difference in abundance or if the geographic distribution of haplotypes makes one connection more probable (Crandall and Templeton 1993). Information on the geographic distribution of networks can be analyzed to infer different historical processes using traditional phylogeographic methods of inference such as *Nested Clade Phylogeographic Analysis* (NCPA; Templeton 1998), and specific hypotheses can be tested using methods from *Statistical Phylogeography* (Nielsen and Beaumont 2009; Box 8.3).

Coalescence Modeling—This collection of procedures utilizes likelihood algorithms to assess alternative hypotheses for phylogeographic patterns and is sometimes referred to as statistical phylogeography (Hey and Machado 2003; Nielsen and Beaumont 2009); however, this is really a misnomer as NCPA also utilizes statistical approaches, so *Coalescence Modeling* may be a better descriptor. Coalescence-based methods provide a set of robust tools for testing phylogeographic hypotheses and can provide insights into the demographic history of populations and the estimation of divergence times for different lineages and populations (Arbogast et al. 2002; Fig. 8.14). On the other hand, these algorithms are extremely complex, and including multiple parameters in models of population history (e.g., population size and time since divergence) can produce multiple combinations of conditions that generate the same pattern of divergence (Nielsen and Beaumont 2009; Templeton 2009a; Fig. 8.15). In the pair of scenarios depicted in Fig. 8.15, larger numbers of genetic lineages are retained by divergent populations when the ancestral population is large, so diversification and development of unique sets of lineages occurs over a shorter timespan compared to a smaller ancestral population size. Multiple solutions from coalescence modeling that are equivalent can be resolved through comparison to results from NCPA and by using ancillary information on known historical events affecting the distribution of genetic lineages.

Models of population history are relatively limited in scope and cannot include the entire range of

Box 8.3 Methods of Phylogeographic Analysis

When the field of phylogeography was first developed by Avise and colleagues in the 1980s (Avise et al. 1987), inferences of historical processes were made through the interpretation of the pattern of geographic distribution of haplotypes. With some notable exceptions (e.g., Soltis et al. 1991; Dumolin-Lapègue et al. 1997a; Soltis et al. 1997; Comes and Abbott 1998), the large majority of early phylogeographic investigations were conducted with animal species—probably because levels of variation in animal cytoplasmic genomes were much higher than in most plant species. Identification of hypervariable intergenic cpDNA regions in the mid-1990s (Demesure et al. 1995; Dumolin-Lapègue et al. 1997b) allowed analysis of a wider range of plant species. Advances in sequencing techniques in the last few decades (e.g., Cronn et al. 2008; Kohrn et al. 2017) has increased the feasibility of sequencing more of the chloroplast genome for larger numbers of samples. Since the entire genome is inherited as a single non-recombining sequence, the information from multiple cpDNA regions can be combined to construct composite haplotypes (e.g., Dobes et al. 2004).

Nested Clade Phylogeographic Analysis (NCPA)—The idea that statistical methods could be employed for hypothesis testing in phylogeography was first introduced by Templeton and colleagues in the early 1990s (Templeton et al. 1992; Templeton 1998). Since NCPA was first applied to analyze geographic and phenotypic distributions of cpDNA variation in 2000 (Maskas and Cruzan 2000; Templeton et al. 2000), it has become widely used for phylogeographic analyses of plants (Dobes et al. 2004; Fehlberg and Ranker 2009; Garrick et al. 2009; Sosa et al. 2009; Diaz-Perez et al. 2012; Sede et al. 2012). The widespread use of NCPA has continued in spite of a number of criticisms (Knowles and Maddison 2002; Knowles 2009; Nielsen and Beaumont 2009; Templeton 2009a, 2009b). To understand the strengths and potential pitfalls of this approach we need to review the methods used in NCPA for phylogenetic inference.

Collections of samples taken from multiple populations across the range of a species are sequenced for one or more genomic regions, and haplotypes are arranged into nested phylogenies as described in Box 8.1. The geographic dispersion (D_c) and displacement (D_n) of each haplotype and nested clade level is evaluated by comparison to random distributions (Fig. 8.13). Patterns of significant departures

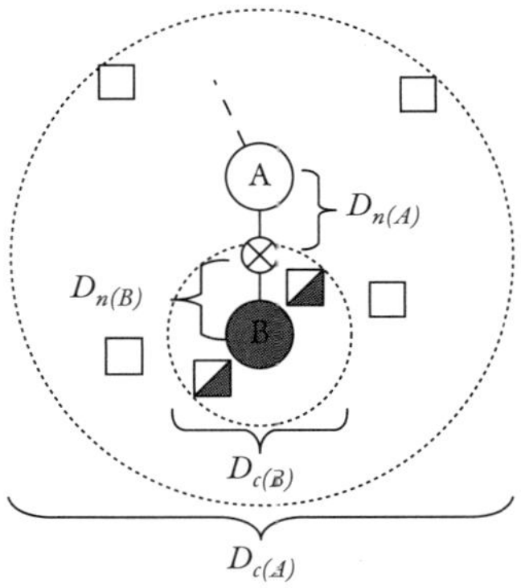

Fig. 8.13. Parameters used for nested clade analysis of the geographic distribution of haplotypes and clades. Two haplotypes are depicted; the A haplotype is located at an interior position relative to the B haplotype, which is at a tip position. Squares indicate the locations of populations and are shaded to show the presence of each haplotype. Haplotypes are placed at their geographic centers; the X indicates the geographic center of the clade; D_n values indicate the displacement of each haplotype center from the clade center; and D_c values indicate the geographic dispersion of each haplotype and clade (not shown). Modified from Cruzan and Templeton (2000).

in dispersion and displacement are interpreted using published dichotomous keys (Templeton 1998, 2004). The types of geographic patterns that can be detected using this approach include:

- Contiguous range expansion—newer haplotypes/clades have more restricted distributions and have displaced geographic distributions relative to the parent clade (Fig. 8.16D).
- Long-distance dispersal—a haplotype/clade displays a high level of displacement from the geographic center of the parent clade (Fig 8.16C).
- Multiple refugia—haplotypes/clades display non-overlapping geographic distributions (both are significantly displaced from the parent clade center; Fig. 8.16A and B).
- Allopatric divergence—geographically displaced clades display high levels of haplotype diversification (Fig. 8.16B).

Nested clade phylogeographic analysis has been criticized for having a high false-positive error rate when used with a single genomic region, and it has been shown to perform poorly when analyzing simulated data sets. On the other hand, when applied to real data, it almost always corroborates conclusions obtained with coalescence-based methods (Nielsen and Beaumont 2009). This may reflect a failure of simulations to generate realistic scenarios, but it has also been suggested that use of the dichotomous key encourages biased interpretations based on the subjective impressions of the researcher. Templeton (2009a, 2009b) counters that the high false-positive error rate problem can be solved by analyzing multiple genomic regions. He also contends that NCPA is the only method available that allows discovery of unknown historical processes because coalescence-based simulations require a priori models for testing (see the Coalescence-Based Methods section). Hypotheses generated using NCPA can then be tested using Bayesian modeling of populations and validated with independent evidence from biogeography, geology, paleontology.

Coalescence-Based Methods—These programs simulate population history to identify the historical processes that best explain abundance and distribution of haplotypes. Coalescence theory provides a means to mathematically connect demographic processes with the network phylogeny, so the probability of obtaining the observed network can be calculated for a particular set of demographic parameters (e.g., migration rate, population size, growth rate, etc.). The probability of obtaining a particular tree cannot be calculated directly, so instead varieties of simulation methods have been developed. Many of these rely on Bayesian mathematics, which provides estimates of the probability of obtaining the observed data given a particular set of initial parameter values (*Priors*). A reasonable range of priors are iterated and posterior probabilities (i.e., chance that the priors explain the data) for each combination are compared. The combination of parameters that best explains the data is then accepted as the best model of population history.

information typically collected in a phylogeographic study. Other approaches, such as NCPA, observation of the geographic distribution of lineages, and inferences from known biogeographic patterns, can provide information about geographic processes affecting the distributions of lineages within species. For example, phylogeographic analyses can aid in the development of hypotheses concerning the location of Pleistocene refugia, patterns of historical range expansion, and the origin of parapatric distributions of distinct lineages (e.g., Soltis et al. 1997; Tremblay and Schoen 1999; Maskas and Cruzan 2000). Moreover, in many of these cases, populations were morphologically uniform over broad geographic regions, and the existence of divergent lineages within the range was not evident until phylogeographic analyses had been conducted.

Coalescence modeling can then be applied to address specific hypotheses within the broader phylogeographic context. For example, Garrick et al. (2009) tested multiple hypotheses for the timing and location of *Vicariance* events for the desert

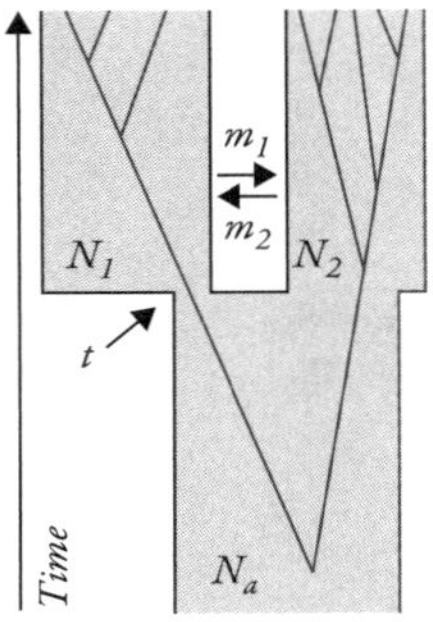

Fig. 8.14. An example of coalescence modeling of a vicariance event. Parameters included in simulations include the population sizes for ancestral (N_a) and derived populations (N_1 and N_2), timing of the vicariance event (t), sizes of population bottlenecks, and rates of migration between diverging lineages (m_1 and m_2).

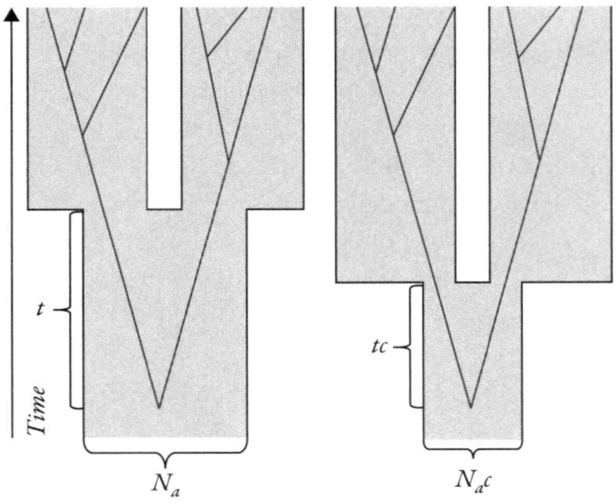

Fig. 8.15. Effects of the timing of a vicariance event and ancestral population size of patterns of diversification. The same level of diversification can be obtained if the ancestral population size and divergence time change by the same proportional amount as indicated by a constant (c).

plant *Euphorbia lomelii* (Euphorbiaceae) along the Baja Peninsula that had been developed based on phylogeographic patterns for several animal species. In this case, these authors were able to reject alternative scenarios for multiple, older events that might have affected the distribution of genetic variation within this species in favor of a single, younger vicariance event. In this example, the pattern of range expansion along the peninsula was inferred from a separate analysis of patterns of diversity for the nuclear genomic regions. Corroboration among multiple methods, including NCPA and coalescence modeling, provided a clear picture of historical events that led to contemporary distributions of genetic lineages along the Baja Peninsula.

Phylogeographic Patterns—Sampling plant populations for variation in cytoplasmic and nuclear genomes will generally yield geographically structured distributions of haplotypes that were generated by demographic and population genetic processes in response to historic climate change. The types of patterns that might be uncovered will often fall into a set of general categories that reflect the major processes affecting the distribution of plants (Fig. 8.16). By applying the principles derived from consideration of the effects of coalescence theory, genetic drift, and migration outlined above we can make interpretations of these patterns to provide insights into post-glacial histories of a wide range of species.

Isolation in geographically separated Pleistocene refugia can produce patterns of parallel range expansions along migration corridors (Fig. 8.16A). Geographic features that present dispersal barriers, such as mountain ranges and major rivers, may define these pathways of range expansion, but genetic patterns reflecting parallel migration can also be found across broad areas that lack major geographic obstacles. Descendants from each glacial refuge may colonize adjacent regions to produce continuous distributions with a contact zone separating areas occupied by different sets of haplotypes. For example, Fehlberg and Ranker (2009) found evidence for separate migration corridors across the continuous range of brittlebush (*Encelia farinosa*) in desert regions of southwestern North America. In this case, populations from central and eastern Arizona were derived from a refuge in central Mexico, while those in California were descendants from a refuge on the Baja Peninsula.

In most cases, species with similar habitat requirements will tend to follow the same migration corridors. For example, the contact zone between migration fronts found for the distribution of *Piriqueta* (Maskas and Cruzan 2000) across northern Florida corresponds to contact zones found for several animal species (Walker and Avise 1998), indicating that they were associated with the same glacial

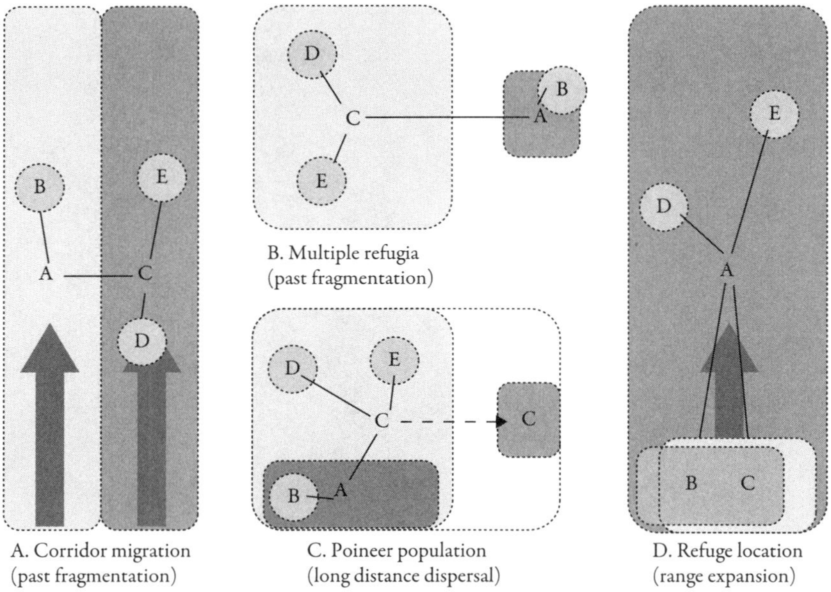

Fig. 8.16. Hypotheses for historical processes responsible for contemporary distributions of haplotypes. Containing circles or rectangles indicate the geographic range of each haplotype, solid lines indicate placement on the network phylogeny, and block arrows indicate the direction of range expansion. A. Range expansion along separate migration corridors leads to a contact zone between the A and C haplotypes; B. A disjunct distribution due to multiple refugia that are characterized by different sets of haplotypes; C. A disjunct distribution that is due to long-distance dispersal (dashed arrow); D. location of a Pleistocene refuge based on the high diversity of haplotypes. Modified from Cruzan and Templeton (2000).

refugia in southern Florida and along the gulf coast. Concordant post-glacial changes in geographic distribution are also evident for a number of plant species in the Pacific Northwest region of North America (Soltis et al. 1997). In Europe, migration corridors extending northward from refugia in Iberia, Italy, and Greece are well characterized for a number of plant and animal species (Fig. 8.17; Dumolin-Lapègue et al. 1997a; Vendramin et al. 1999; Hewitt 2000; Sharbel et al. 2000; Petit et al. 2003).

For the examples just described, range expansion from multiple glacial refugia resulted in continuous distributions, but in other cases migration corridors remained geographically separated, resulting in disjunct distributions. Divergent groups of populations can be constrained to isolated regions because of the distribution of suitable habitat or geographic barriers that prevent admixture and allow for haplotype diversification in each region—a pattern referred to as allopatric fragmentation (Fig. 8.16B). For example, separated regions of suitable habitat led to geographically isolated lineages of Douglas fir in southwestern North America and northern Central America (Gugger et al. 2011). After the last glaciation, range expansion of Douglas fir followed migration corridors along the parallel ranges of the Sierra Madre Mountains. Populations on each mountain range maintained separate geographic distributions that are characterized by unique sets of haplotypes.

Cases of allopatric fragmentation require isolation over relatively long time frames, but the nature of dispersal barriers creating conditions necessary for independent haplotype diversification are not always clear, for example, for tulip poppy along different portions of a single mountain range in Mexico (Sosa et al. 2009). The distribution of these tulip poppies has not changed substantially as post-glacial changes in distribution occurred along elevation gradients within valleys separated by ridges, so populations retained their genetic coherence with little mixing. In other cases, the dispersal barrier is more obvious, such as between island and continent

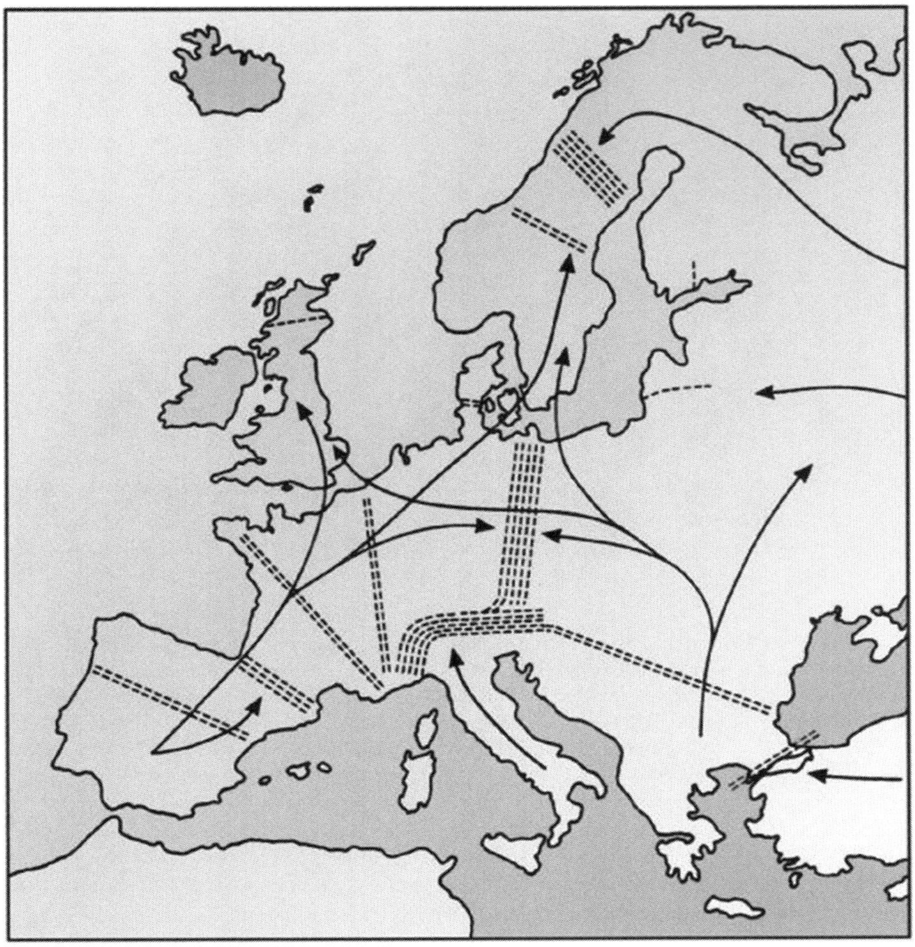

Fig. 8.17. Major migration corridors (arrows) and contact zones (dashed lines) in Europe. From Hewitt 2000.

populations (Maskas and Cruzan 2000) and for tropical dry forest regions that stradle the Amazon basin (Collevatti et al. 2009)

Disjunct distributions can also originate by colonization after long-distance dispersal (Fig. 8.16C). Cases of establishment in a novel geographic range will be distinguished from allopatric fragmentation by low levels of haplotype diversity and lack of novel haplotypes in the introduced range. Examples of recent long-distance colonization are common for introduced and invasive plants (e.g., Dlugosch and Parker 2008; Rosenthal et al. 2008; Richards et al. 2012), but documented instances of more ancient long-distance colonization events are less frequent (e.g., Maskas and Cruzan 2000; Dixon et al. 2009). The distinction between long-distance migration and allopatric fragmentation is primarily a matter of time scale; more recent cases will have low diversity and high levels of haplotype affinity with source regions, while more ancient long-distance dispersal events will be marked by higher levels of diversity and larger numbers of endemic haplotypes (e.g., Sede et al. 2012).

Post-glacial changes in geographic ranges of many temperate and boreal species meant that populations in refugial regions went extinct as the climate warmed, so the genetic diversity they had accumulated during the previous glaciation was mostly lost. In other situations, climate changes have not been so dramatic, and conditions have allowed for the persistence of populations near the regions they occupied during the last glaciation; in these cases, we expect to see regions of relatively high haplotype diversity that correspond to refugia locations (Fig. 8.16D). Evidence for glacial refugia have been found in alpine plant species such as *Dryas integrifolia* (Tremblay and Schoen 1999). These authors were able to find correspondence between high levels of haplotype variation and fossil evidence of persistence during the last glaciation in Beringia and northern Arctic regions of North America. Molecular evidence for similar locations of arctic refugia were found for the alpine plant, *Arabis holboellii*, as well as refugial locations further south in the Rocky Mountains (Dobes et al. 2004). Not surprisingly, correspondence in refugial locations for multiple species is common and in many cases is supported by fossil evidence as well as molecular data (Cruzan and Templeton 2000; Hewitt 2000; de Heredia et al. 2007). Other cases of molecular evidence for locations of glacial refugia include *Encelia farinosa* in North American deserts (Fehlberg and Ranker 2009), *Trilium grandiflorum* in eastern North America (Griffin and Barrett 2004), for a wide range of European species in Iberia, Italy, Greece, and the Balkans (Dumolin-Lapègue et al. 1997a; Vendramin et al. 1999; Hewitt 2000; Sharbel et al. 2000; Petit et al. 2003), and for a number of species in Asia (Qiu et al. 2011; Liu et al. 2012).

Phylogeographic Inference—Collection of nuclear or cytoplasmic sequence data for multiple individuals from populations across the contemporary range of a species allows application of statistical procedures of phylogeographic analysis (Box 8.3), which produce inferences of historical processes contributing to contemporary distributions. For any particular sample of data, some inferences from any single analysis procedure are likely to be weak, but conclusions can be strengthened by analyzing data with multiple methods and by obtaining ancillary supporting information (Cruzan and Templeton 2000; Nielsen and Beaumont 2009; Garrick et al. 2010). Corroborative support for inferences from different methods of data analysis and from independent lines of evidence provides a powerful set of tools for uncovering the history of species.

Additional corroboration of phylogeographic inferences can be obtained by analyzing multiple species from the same geographic region, which allows for the characterization of general biogeographic patterns. A large number of studies have contributed to a comprehensive view of the history of plant and animal communities in Europe (Hewitt 2000). Parallel patterns have also been found for groups of species in the northwestern (Soltis et al. 1997), southwestern (Garrick et al. 2009), and eastern (Walker and Avise 1998; Maskas and Cruzan 2000; Griffin and Barrett 2004) regions of North America, and for southern regions of Asia (Qiu et al. 2011; Liu et al. 2012). These efforts have provided a much more detailed picture of the history of plant and animal communities than can be obtained using paleontological techniques.

As phylogeographic investigations have become more common they have integrated complementary methodologies by including environmental information (*Ecological Niche Modeling; Bioclimatic Envelope Modeling*) and projections of habitat distributions under different climate change scenarios (e.g., Jakob et al. 2009; Gugger et al. 2011). These studies provide historical insights into ecological, genetic, and evolutionary investigations of contemporary populations, efforts to conserve endemic, rare, and endangered species, the control of invasive species, and our understanding of the implications of climate change for species and plant and animal communities in the future (Scoble and Lowe 2010). Phylogeographic studies of plants, in particular, will be central to improving our understanding of the history of organisms and communities, as their sedentary nature and close association with their physical and biotic environments render plants as sensitive proctors of ecological and evolutionary processes.

Application of phylogeographic analyses, along with ecological experiments and analyses of morphological and physiological traits, gene expression, and sequence variation, holds much promise for elucidating such paradoxes as the origin of ecotypic variation and the evolution of range limits. While phylogeographic analyses do not account for adaptation as populations and lineages encounter novel environments, they also do not exclude the possibility that lineages can diversify into contrasting habitats. In the next section, we will first discuss the analysis of ecotypic variation within species and then apply an historical perspective to understand the origin and diversification of ecotypes.

8.8 ECOTYPIC VARIATION

The idea that varieties within a species can become adapted to unique habitats dates back at least to the work of Darwin (1859) and Wallace (1865), but the concept of ecotypic variation within species was not fully developed until the early part of the twentieth century (Turesson 1922; Clausen et al. 1940, 1948). When first introduced, this idea was somewhat revolutionary because it recognized that small variations within species could represent responses to selection for local adaptation. After the introduction of the term ecotype by Turesson in 1922, the concept received some criticism because it was interpreted as delineating discrete ecological "types" (Lowry 2012). In spite of this misreading by some (e.g., Mayr 1963), Turesson, Clausen, and others clearly viewed ecotypic variation as continuous and evolving, rather than categorical and static. During this time, the focus of plant biologists on adaptation via quantitative variation distinguished their view of evolution from that of many animal biologists, who remained focused on discrete polymorphisms (e.g., Dobzhansky 1970; Ford 1971). The other major point of contention over the ecotype concept was whether the unique varieties represent intermediate stages in the process of speciation (Chapter 10). In the early part of the twenty-first century, there has been a resurgence in interest in ecotypes—the genomic analysis of local adaptation in particular—and divergence of locally adapted lineages continues to be an active topic of discussion (Nunez-Farfan and Schlichting 2001; Lowry 2012). In this chapter, we will discuss analyses of phenotypic variation within species and the conditions that favor ecotypic divergence.

Experimental Analysis—Increased recognition of the validity of the ecotype concept was a direct consequence of the development of methodologies to identify cases of genetically determined phenotypic divergence. Plant biologists of the early twentieth century were certainly aware that morphological variation within species was common, but it was often assumed that such variants were distributed randomly with respect to environmental variation and perhaps were due to phenotypic plasticity (Fig. 8.18A) rather than to genetic differences (Turesson 1922). Up through the early twentieth century, much of the study of plant evolution and ecology had been conducted through observations of plants in nature, and carefully controlled experimental manipulations were uncommon. Turesson's revolutionary development was the utilization of experimental manipulations to determine that such variation was due to genetic differences that reflected adaptation to local habitats. While transplant experiments were not new—they date from at least the experiments by Bonnier and Kerner in the 1890s (reviewed in Turesson 1922)—the earlier studies were not as thorough or carefully designed as later experiments conducted by Turesson, and later Clausen, Keck, and Heisey.

Turesson (1922) noted that different varieties of *Hieracium umbellatum* were associated with different habitats, and by growing varieties together under the same environmental conditions (common garden), he was able to determine that these phenotypic differences had a genetic basis. Clausen et al. (1940, 1948) improved on this design by growing plants in common gardens reciprocally in the different habitats that were associated with the putative ecotypes of *Achillea millefolium* (*lanulosa*) and *Drymocallis* (*Potentilla*) *glandulosa* (*Reciprocal Transplant Experiment*; Fig. 8.18). This experimental design allowed these researchers to test for the maintenance of phenotypic differences in different environments and for habitat-associated responses in plant growth and reproduction. Their results provided some of the clearest evidence to date that genetically determined phenotypic differences can confer local adaptation (Nunez-Farfan and Schlichting 2001) and also provided a template for an experimental design that has been replicated numerous times since.

The concept of ecotypes has persisted in the literature and remains a topic of discussion in spite of past predictions of its loss of relevance and ultimate demise (Lowry 2012). Its staying power probably has much to do with its appeal to our sense that organisms should be well adapted to their environments, but this is an expectation that does not always hold up to scrutiny. For example, Galloway and Fenster (1999) found mixed support for local adaptation between native sites in Maryland and Illinois (USA) in reciprocal transplant experiments with the annual legume *Chamaecrista fasciculata*. In these experiments, there appeared to be a strong nuclear contribution to local adaption in Maryland; but Maryland genotypes also appeared to be superior to native genotypes at Illinois sites (*Universal Genotype*; Fig. 8.18B). Similarly, the

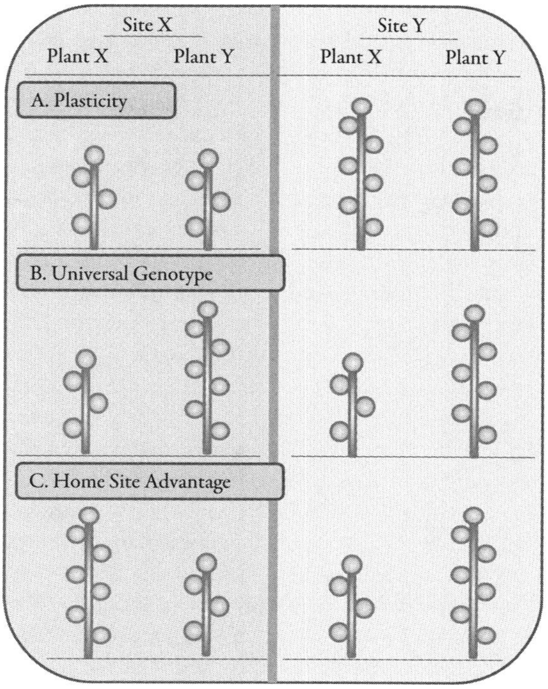

Fig. 8.18. Possible outcomes from reciprocal transplant experiments: A. Plant phenotypes are determined by plasticity; B. One genotype is superior in both environments; and, C. Home site advantage indicates local adaptation.

Home-Site Advantage (Fig. 8.18C) is not consistently found and has been shown to be weak or nonexistent in a number of other cases. For example, between plants associated with shade and open sites for *Arabidopsis thaliana* (Callahan and Pigliucci 2002); between arctic fellfield and snowfield habitats for a long-term study of *Eriophorum vaginatum* (Bennington et al. 2012); and for dry tropical habitats that differ with respect to soils and precipitation for *Ruellia nudiflora* (Ortegon-Campos et al. 2012). These observations suggest that local selection can be overcome by genetic drift, gene flow, and historical contingencies that may have an overriding effect on the potential for populations to respond and adapt to local conditions.

On the other hand, there are a number of dramatic examples of adaptation to local conditions that have been documented for cases of responses to strong selection, which sometimes occur over very short time scales. Many of these examples are associated with sharp discontinuities in environmental conditions and especially with changes in soil (edaphic) environments. For example, serpentine-derived soils in California and Oregon (USA) are characterized by high concentrations of magnesium relative to calcium, low availability of nitrogen and phosphorous, and low water-retention ability. These regions support unique floras of serpentine-specific endemic species and ecotypes of more widely distributed species (Harrison and Rajakaruna 2011). Similarly, chalk soils in Britain, which have high concentrations of calcium and are nutrient poor, are characterized by floras consisting of endemic ecotypes (Snaydon and Bradshaw 1961; Fischer and Stocklin 1997). In another study Lowry et al. (2009), used recombinant inbred lines (RILs; Chapter 7) to identify QTLs associated with salt-spray tolerance between coastal and inland ecotypes of *Erythranthe gutatta* (*Mimulus guttatus*). In these examples, the serpentine and chalk soils and salt-spray habitats are inhospitable to the degree that transplanted individuals of the alternative ecotypes have high rates of mortality.

Some of the most impressive examples of ecotypic divergence are cases where locally adapted populations have evolved in very recent times. Ecotypes of some plant species have become adapted to mine tailings—piles of debris from mines that have high concentrations of heavy metals (McNeilly and Antonovics 1968; Wu et al. 1975; Al-Hiyaly et al. 1988; Antonovics 2006). The new habitat was created adjacent to established plant populations, so we can infer that evolution of these unique ecotypes occurred in situ—without large changes in distribution—and probably under pressure of gene flow from nearby populations not associated with contaminated soil. These evolutionary transitions may have been facilitated by a relatively simple genetic architecture for heavy metal tolerance (Bratteler et al. 2006; Richau and Schat 2009; Burrell et al. 2012) and similarities between the challenges imposed by serpentine and mine tailing soils (Gailing et al. 2004; Brady et al. 2005). As discussed in Chapter 5, stringent edaphic conditions can create a selection regime that overcomes high levels of gene flow by preventing the establishment of hybrid offspring. While environments modified by human activities provide clear evidence of recent evolution of ecotypes, in other cases where ecotypes are geographically separated, their history is not so clear.

Origin of Ecotypes—The observation that ecotypes of a given plant species can occupy different geographic regions suggests that they may have diverged over a relatively long time frame. It is possible that the genetic races that we view as relatively static—consisting of lineages with strong ecological associations—may have been allopatric for extended periods and only recently achieved their contemporary distributions (Fig. 8.19). Even for classic cases of ecotypic variation discussed earlier (e.g., Clausen et al. 1940, 1948), we often do not have information on phylogenetic relationships among lineages, which raises the possibility they were originally derived in regions that were far removed from their current geographic locations. For *Achillea millefolium*, it appears that all of the eight ecotypes occupying divergent habitats across western North America were all derived since the last glacial maximum, and that many populations classified as the same ecotype likely have independent evolutionary origins (Ramsey et al. 2008). In this example, the ecologically divergent lineages appear to have sorted into different habitats during the post-glacial range expansions (*Ecotype Sorting; Multiple Origins*), but it would also be possible for ecologically similar populations to all have been derived from the same lineage (Fig. 8.19).

We can see another possible example of ecotype sorting during range expansion in the recently introduced *Bromus tectorum* (cheat grass) in North America, where divergent genotypes have become associated with different ecological conditions (Ramakrishnan et al. 2006; Merrill et al. 2012). It is impressive that this species' distribution has been established within such a short timeframe (i.e., within the last two centuries). On the other hand, it is unlikely that there has been much opportunity for local adaptation of these recently established populations, because these populations consist of separate selfing lineages and recombination among them is rare (Ramakrishnan et al. 2006). This view of ecotypic divergence provides insight into the varied results from reciprocal transplant experiments; our expectation of consistent local adaptation of plants is dampened by the realization that populations have not evolved for very long in their current habitats. As discussed earlier, drift and gene flow may limit the effects of selection, so it may generally be the case that regional ecotypes are insufficiently adapted to their local environments. Given the relatively short time frame for adaptation since the last glacial maximum, it is perhaps better to view local varieties as the "best solution available" rather than "finely tuned" ecotypes that are closely adapted to local conditions (Savolainen et al. 2007; Hall et al. 2010; Siol et al. 2010).

Incorporating the history of populations into our understanding of adaptive processes provides a more sophisticated view of the ecotype concept. Given the large changes in species' distributions and continuing changes in climate, it is not surprising that adaptation to local conditions is often far from ideal. Fluctuations in biotic and abiotic conditions on different time scales presents a constantly moving target for selection as it pushes populations toward fitness optima that continually shift. On the other hand, it is also clear that local adaptation can be relatively rapid (e.g., Ramsey et al. 2008; examples of mine tailing ecotypes described previously). A better appreciation of the dynamics of local adaptation and ecotypic divergence can be gained by considering phylogenetic history (e.g., Banta et al. 2007; Oberle and Schaal 2011; Robakowski et al. 2012). Understanding the conditions that facilitate or stall rapid adaptation to novel conditions represents a challenge for plant

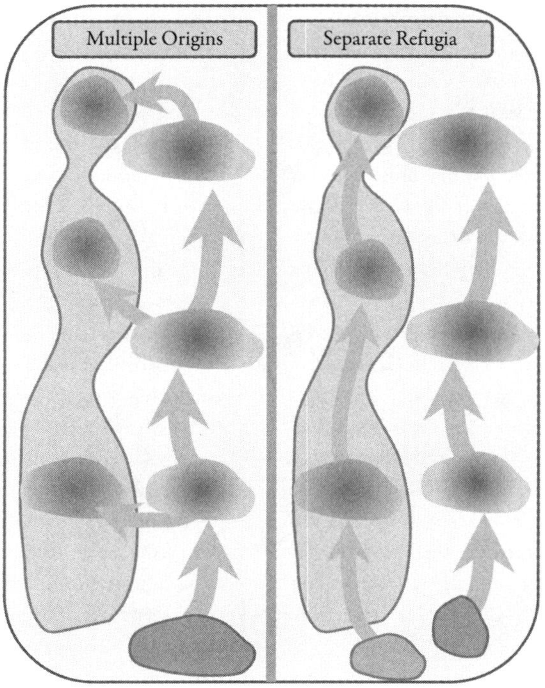

Fig. 8.19. Two scenarios for the origin of ecotypes. Derivation of ecotypes within each region during post-glacial range expansion would lead to multiple origins. Alternatively, divergence may have taken place in separate refugia, as lineages colonized separate habitat corridors during post-glacial range expansion.

evolutionary ecologists; future studies that incorporate information on genetic relatedness and utilize genomic approaches in combination with transplant experiments are likely to provide unique insights into the processes that promote and limit adaptation (Gonzalez-Martinez et al. 2006; Hohenlohe et al. 2010).

8.9 INTRASPECIFIC VARIATION IN PLANTS

Limited dispersal ability and close associations with physical and biotic environments make plants ideal subjects for phylogeographic investigations. The well characterized and predictable responses of plant species to physical conditions makes them reliable tools for revealing historical changes in climates. Furthermore, for many continental regions there is a substantial amount of palynological and paleontological data on changes in the distribution of plants following glaciation (Delcourt and Delcourt 1991; Whitlock and Bartlein 1997), which can be used to corroborate inferences from the distribution of genetic variation (Cruzan and Templeton 2000). In spite of the advantages of plants as subjects for investigations into the distribution history of species, the number of published studies on plants has lagged behind the burgeoning literature on animal phylogeography. In part, the slower rate of increase in phylogeographic studies on plants is due to lower levels of variation available in chloroplast genomes, but this limitation has been overcome by the identification of more variable regions and by improvements in the introduction of targeted capture and methods for more efficient approaches to whole-genome chloroplast sequencing (e.g., Kohrn et al. 2017).

The collective information from phylogeographic studies on plants has provided us with detailed views and a deeper understanding of historical processes that have culminated in contemporary distributions. Limited seed dispersal combined with uniparental inheritance of plastid and mitochondrial genomes contribute to our ability to evaluate fine-scale ecological and demographic processes that affected the distribution of genetic variation during post-glacial changes in geographical ranges. We have learned much, but there is still tremendous potential for improving our understanding of how ecological processes contribute to the development of intraspecific variation, by integrating ecological data, niche modeling, landscape genetics, and genomic approaches with phylogeographic methods.

The distribution of intraspecific genetic variation across landscapes is largely governed by the reproductive characteristics of plants. Levels of gene flow among populations is determined by the dispersal of pollen and seeds, and genetic differentiation can be elevated depending on the degree of selfing and local inbreeding. In the next chapter, we examine plant reproductive biology in general, and aspects of plant reproduction that contribute to genetic differentiation.

8.10 QUESTIONS FOR CONTEMPLATION AND DISCUSSION

1. In Darwin's graphical model of evolution, at what point do you think he envisioned speciation to occur? What aspects of this model are unrealistic?
2. What were the effects of the Pleistocene ice ages on vegetation in the region where you live? What was the vegetation like during the last glacial maximum?
3. What are the effects of range expansion on the distribution of genetic diversity? How do contiguous and long-distance dispersal, the size of the migrant pool size, and directed dispersal affect the distribution of genetic diversity in a recently expanded range?
4. What patterns of genetic diversity do you expect from a stable versus an expanding geographic range?
5. How do population bottlenecks during range expansion affect genetic diversity and the adaptive potential of populations toward the leading edge of the expanding range?
6. What does the abundant center hypothesis predict about patterns of abundance and genetic diversity across the range of a species? How do dispersal, competition with other species, and the distribution of habitat patches affect range limits?
7. Design an experiment that will successfully characterize adaptation during invasion. How can you distinguish the effects of genetic drift versus selection on traits? Can you distinguish the effects of heterosis after admixture from improved adaptation to the invaded range?
8. How do dispersal vectors affect the distribution of genetic variation? Why does the distribution of cpDNA markers often differ from nuclear markers?
9. What is mutation surfing? What conditions are necessary for mutation surfing to occur?
10. Assuming a constant population size, what are the predictions of coalescence theory for the abundance and distribution of central versus tip haplotypes?
11. What are the advantages of nested clade phylogeographic analysis and coalescence modeling for inferring historical patterns of vicariance and migration?
12. What are the patterns expected for glacial refugia, range expansion, and long-distance dispersal?
13. Design an experiment that will distinguish phenotypic plasticity from ecotypic variation. Can you think of a way to distinguish genetic from epigenetic adaptation?
14. How can phylogeographic analyses inform studies of adaptation? How could you determine the evolutionary origin of multiple ecotypes? Are the genomic bases of adaptation the same?
15. What is the relationship between landscape genetics and phylogeography? Can landscape genetic analyses be used to inform phylogeographic patterns?

PART IV

PROCESS OF DIVERSIFICATION

9

REPRODUCTION

*S*EXUAL REPRODUCTION IS *the chef d'oeuvre, the master-piece of nature.... From the sexual, or amatorial, generation of plants new varieties, or improvements, are frequently obtained; as many of the young plants from seeds are dissimilar to the parent, and some of them superior to the parent....*

Erasmus Darwin (1800) Phytologia, 103–104

9.1 MODES OF REPRODUCTION

The transmission of genetic information across generations is integral to the continuity of lineages—genetic inheritance enables adaptation to novel conditions and promotes the diversification of the multitude of species that constitute life on Earth. Reproduction is an intrinsic property of life, but organisms differ in the ways in which they generate progeny. The two modes of propagation available to eukaryotes—sexual and asexual reproduction—differ primarily with respect to the potential for the reshuffling of genomic information and the genetic diversity among offspring. Early in the history of life on Earth, prokaryotes and the earliest eukaryotes were only capable of asexual reproduction, and opportunities for recombination among lineages were limited. The origin of sexual reproduction in diploid organisms—and most importantly recombination during meiosis –supported the evolution of larger and more complex organisms (Butterfield 2000; Chapter 1). As organisms increased in size, their generation times became longer, and the potential for rapid population growth rate declined. Sexual reproduction probably emerged as a mechanism to allow more rapid evolution to overcome the deficits of slower population growth and lower reproductive potential in early multicellular eukaryotes. Hence, sex utilizes recombination to leverage the higher genetic diversity potential of diploidy to generate variable offspring and to overcome the evolutionary constraints imposed by mutation accumulation in asexual lineages.

As early sexual eukaryotes increased in size, selection favored gametes that contained sufficient resources to provide a start for the newly formed offspring (Bulmer and Parker 2002). Successful fertilization in aquatic environments may have occurred by chance encounters, but larger gametes are also less mobile and therefore, were less likely to be fertilized. Consequently, there was simultaneous selection for large numbers of small and motile gametes that increased the likelihood of successful fertilization. These opposing forces amounted to disruptive selection for two different types of gametes: large gametes that contained enough resources to produce viable progeny and large numbers of small motile gametes that increased the frequency of fertilization. Disruptive selection may have resulted in a compromise in gamete size until the establishment of two different mating types (+ and -) that could respond to selection independently (Togashi et al. 2012). The outcome was the emergence of two different types of diploid individuals that produced different sizes of gametes (*Anisogamy*); those individuals considered "female" produced large gametes containing adequate resources for the establishment of offspring while "male" individuals produced large numbers of small and motile gametes. In contrast, *Isogamous* species produce only a single type of gamete and consequently do not have two separate sexes.

It's a little astonishing to realize that all eukaryotic life—from the "protists" to plants, fungi, and animals—are descended from a single sexually reproducing lineage that first arose in the early Proterozoic (Fig. 1.1; refer to color plate). From this common origin, these separate groups diversified to express a variety of sexual systems. The foundation of sexual reproduction is the presence of two mating types that, with the exception of isogamous fungi and algae (Ni et al. 2011), correspond to male and female roles based on the size and dispersal ability of their gametes. Among plants, the seedless tracheophytes, bryophytes, and some gymnosperms (e.g., gingko) are much like animals in producing flagellated sperm and immotile ova contained within archegonia. Gymnosperms and flowering plants (spermatophytes—the seed plants) produce two different types of gametes that correspond to male and female function (i.e., individual plants may act as *Pollen Donors* and *Pollen Recipients*, respectively). Unlike other eukaryotes, the pollen and sperm nuclei of most spermatophytes are immotile. Pollen dispersal to receptive surfaces—micropyles of ovules in the gymnosperms and stigmas in the angiosperms—is achieved by coopting biotic and abiotic vectors. Once pollen grains arrive on a receptive surface they germinate, and with the exception of some gymnosperms such as gingko, sperm nuclei are delivered to the megagametophyte for fertilization via a pollen tube.

In this chapter, we explore the wide variety of sexual expression, mating systems, and asexual reproduction in plants. We start by reviewing the process of reproduction and discussing variation in the distribution of sexual function within and among individuals. We review the modes of reproduction in plants and examine the conditions favoring sexual and asexual reproduction. This section is followed by overviews of sexual and asexual reproduction, which sets up more detailed discussions on the evolution of self-fertilization (selfing) and *Outcrossing*, processes affecting the non-random transmission of genetic material including gametophytic selection because of pollen competition (Chapter 5) and *Selective Ovule Abortion*, and the evolution of sex expression within and among individuals. Much of our discussion applies to all plants, but we focus primarily on flowering plants and the wide range of mating strategies that they represent.

9.2 PLANT LIFE CYCLES

During the Proterozoic, separation of the consumer lineages that gave rise to the metazoans (animals) from the producer lineages (autotrophs) that diversified into the photosynthetic algae and plants occurred early in the evolution of the eukaryotes. The lineage responsible for the origin of all animals reduced its haploid life stage to gametes (diplontic life cycle; Fig. 1.3), while photosynthetic eukaryotes retained a multicellular haploid stage (the gametophyte stage of the haplodiplontic life cycle) throughout their history—from algae to the earliest land plants and including contemporary bryophytes, ferns, lycophytes, gymnosperms, and angiosperms.

The loss of a multicellular haploid stage in the metazoans may reflect the intensity of competition and predation in the aquatic communities of the early Proterozoic (Chapter 1). Under limiting resources, the maintenance of a haploid stage may have represented a liability, both as a target of predation and for gathering the resources necessary to maintain additional individuals. Moreover, producing complex haploids that could survive in these environments would have necessitated the expression of many of the same genes responsible for diploids, which would effectively have erased the advantages of diploidy for sequestering genetic variation responsible for generating variable offspring. Emergence of a diplontic life cycle in early metazoans created a more resilient and evolutionarily adept lineage that is responsible for the tremendous of diversity of animals that followed.

If the diplontic life cycle was such an advantage for consumers, then why not for the producers? In modern plants, there is a range of variation in the dominance and independence of the gametophytic stage. The dominant life stage in modern bryophytes is haploid with a parasitic sporophyte, and the seedless tracheophytes retain an independent gametophyte stage along with the dominant sporophyte (Chapter 2). For seed plants, the gametophytes—contained within pollen and ovules—are severely reduced in complexity and have limited exposure to the environment, as they are effectively parasitic on the sporophyte. For the flowering plants, the gametophytic life stage functions not only as a mechanism to pass genetic material to offspring but may serve as an arena for selection to filter mutations originating during pollen-tube growth (Chapter 5). For all plants, the process of sexual reproduction may be best viewed as a relay alternating between haploid and diploid populations that are subject to independent rounds of selection as their genomes are passed along from one generation to the next.

9.3 PLANT SEXUAL REPRODUCTION

The plant life form confers both advantages and disadvantages for successful reproduction. The stationary nature of the sporophytic life stage facilitates asexual reproduction through indeterminate growth of stems and vegetative spread. On the other hand, being sedentary presents challenges for successful sexual reproduction. Since the gametophytes and spores of seed plants lack mechanisms for auto-mobility, transfer of pollen containing microgametophytes from anthers to receptive stigmas by other agents is required for fertilization of the megagametophytes contained within ovules (Box 9.1). Furthermore, the seeds produced may be more likely to encounter sites with available resources if they can be moved away from the parent plant, so dispersal vectors are generally required to facilitate the transport and success of seeds. As we discussed in Chapter 2, the continued spread and diversification of plants has been enabled through their ability to coopt biotic and abiotic vectors for the transport of seeds and pollen.

Once pollen successfully arrives on receptive stigmas, the next phase in the reproductive odyssey commences (Fig. 10.4; refer to color plate). Pollen germination is initiated as the grains adhere to the stigma and become hydrated, but their success is not guaranteed. For one, self-incompatibility mechanisms may prevent germination or may halt the progress of pollen tubes during their growth to the ovary. Moreover, many more grains often arrive on stigmas than there are ovules to fertilize, so microgametophytes are subjected to pollen competition, as only a subset of the population of pollen tubes successfully fertilize ovules. Competition among pollen grains for access to ovules results in gametophytic selection, which may have effects on phenotypes in the sporophytic stage (Chapter 5).

Successful pollen tubes extend through the *Transmitting Tissue* of the style to the ovary, where they initiate fertilization by entering an ovule through its micropyle (Box 9.1). The process of fertilization in plants is not a simple affair, as two sperm and usually three nuclei from the megagametophyte are required. This process of double fertilization results in a diploid embryo and usually a triploid endosperm, but there are many variations on the process of fertilization endosperm ploidy across flowering plants (Friedman et al. 2008).

The newly fertilized ovules begin development immediately after fertilization. During seed and fruit maturation, the individual ovules within an ovary and developing fruits on the same plant effectively compete among themselves for limiting maternal resources (Sadras and Denison 2009). Ovule and fruit abortion is common—particularly in long-lived perennial species—and is generally due to the effects of competition and the expression of deleterious alleles in developing embryos, resulting in selective ovule abortion. Consequently, fruits often contain shriveled ovules and inviable seeds that were aborted at various stages of development (Bawa and Webb 1984). In many woody species, excess flowers are produced so fruits are abscised at early stages (*Overproduction of Flowers*—see section 9.10), particularly if they only contain a small number of developing seeds (Stephenson 1981).

The cycle of sexual reproduction is completed with the release and dispersal of seeds. A surprisingly large number of plant species lack a specialized seed dispersal mechanism (i.e., they are considered to be *Autochoric* or *Barochoric*—gravity dispersed), so nearly all of their seeds fall within close proximity to the parent plant. In contrast, many species coop vectors to move seeds to various distances away from the maternal plant (*Allochory*). Even when morphological features that would promote dispersal are lacking, seeds may be moved after their initial dispersal by blowing wind or snow (secondary *Anemochory*), or if they become lodged in the hooves of ungulates or captured in mud that adheres to the feet of birds and other animals (secondary *Ectozoochory*).

Many fruits have morphological features that promote the movement of seeds away from the maternal plant (Howe and Miriti 2004). The impressive diversity of fruit morphologies suggests that dispersal ability has been a key feature of the evolutionary success of many plant lineages. Across the seed plants, there are numerous examples of convergent evolution to promote vector-specific dispersal. Adaptations for anemochory are evident as wing and parachute structures that increase buoyancy. This mode of dispersal is perhaps the most common, as it appears in a large proportion of the composites (Asteraceae) and milkweeds (Asclepiadaceae), in many shrubs and tree genera including the willows and cottonwoods (Salicaceae), and in nearly all the conifers (Pinaceae). Ectozoochory appears in a diversity of annuals and herbaceous perennials as hooks on the surface of fruiting structures and barbed awns (in grasses) that can catch in the fur of mammals. A wide diversity of herbaceous species (especially the Solanaceae), shrubs, and trees produce fleshy fruits that are eaten by birds and mammals (*Endozoochory*) and typically are undamaged after they pass through the animal's digestive track. Other dispersal mechanisms include movement by water (*Hydrochory*), by ants (*Myrmecochory*), or explosive dehiscence (*Ballistic Dispersal*).

The effectiveness of seed dispersal has substantial consequences for gene flow and the genetic structure of populations (Chapter 4). In the next sections, we explore other aspects of plant reproduction that affect evolutionary processes of plant populations, including the expression of sexual functions within and among individuals, modes of sexual and asexual reproduction, and plant traits that promote nonrandom mating within and among populations.

9.4 PLANT SEX EXPRESSION

Plants display a much wider variety of sexual systems than any other group of organisms. Most angiosperms are hermaphroditic and produce only *Cosexual* (*Bisexual; Perfect*) flowers that contain both pollen and ovules, but some hermaphroditic flowering plants and most conifers in the family Pinaceae (pines, firs, spruces, hemlocks, and larches) have separate male and female flowers on the same individual (*Monoecy*; Fig. 9.1; refer to color plate). With the exception of the conifers, gymnosperms such as gingko, cycads, and the gnetophytes (*Welwitschia, Ephedra,* and *Gnetum*) are mostly dioecious, with separate plants producing only pollen or ovules. The gametophytic life stages of bryophytes and seedless tracheophytes are either monoecious or dioecious, as they produce antheridia and archegonia either on the same (homosporous) or on different (heterosporous) haploid *Thalli*. Dioecy also occurs in the flowering plants—particularly on islands (Chapter 11)—but it is not always clear-cut, as some species are *Subdioecious*, a category that includes *Gynodioecy* (plants bearing either cosexual or female flowers) and *Androdioecy* (plants with cosexual or male flowers; Fig. 9.1; refer to color plate). Sex expression in most dioecious flowering plants is generally not controlled by sex chromosomes, and some species such as maples (*Acer*) and Jack-in-the-pulpit (*Arisaema triphyllum*) have *Diphasic Sex Expression*—the ability to change sex over their lifetime (Korpelainen 1998). In contrast, most animals are dioecious and have fixed sex expression that is usually controlled by sex chromosomes.

In addition to the wide range of variation in the distribution of pollen and ovule production within and among individuals, many hermaphroditic plants have the capacity to reproduce through self-fertilization, and some plants—particularly weedy species—automatically self in the bud or during early stages of anthesis without assistance from a pollen vector (i.e., they are autonomously autogamous). Complete selfing is more common in annuals and is rarely found in perennials. Outcrossing is characteristic of some annuals and long-lived perennials such as woody shrubs and trees and is generally enforced by strong physiological *Self Incompatibility* (*SI*). While many species are nearly or completely outcrossing, a substantial proportion of plants routinely produce both selfed and outcrossed offspring to various degrees and are considered to have *Mixed Mating Systems*. In some plants such as violets (*Viola*), impatiens (*Impatiens*), several species in the grass family (Poaceae), and peanuts (*Arachis hypogaea*), mixed mating is achieved by producing mixtures of *Cleistogamous* (closed flowers that autonomously self) and *Chasmogamous* (open, outcrossing) flowers.

Plants are unique among eukaryotes, because a large proportion of them reproduce asexually as well as sexually (Fig. 9.2; refer to color plate). Many flowering plants reproduce through vegetative growth by producing root sprouts, runners, and rhizomes. Some tree species such as the quaking aspen (*Populus*

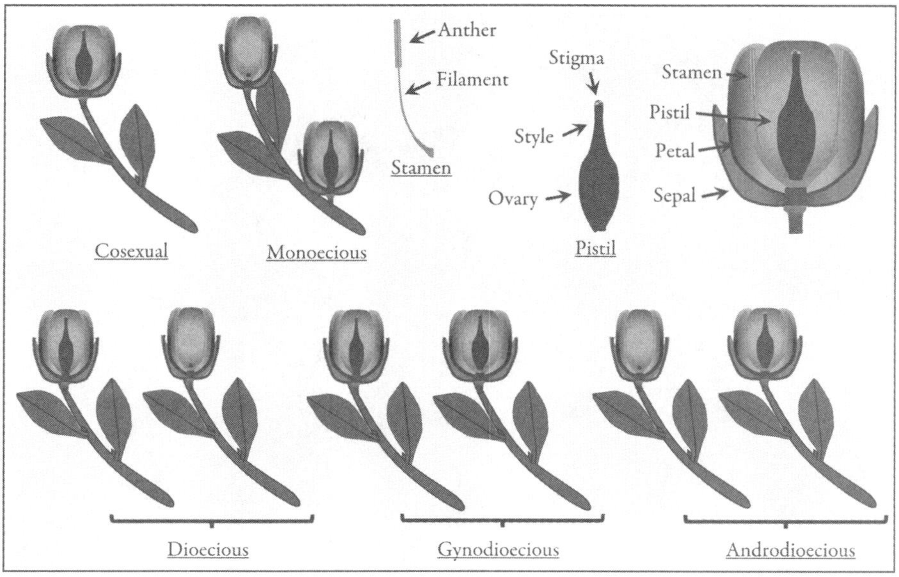

Fig. 9.1. The range of sexual expression in flowering plants.

tremuloides) engage in extensive vegetative spread such that a single individual may occupy several hectares (Lin et al. 1994; Macaya-Sanz et al. 2016). While individual ramets of each aspen genet remain connected, in other species such as the water hyacinth (*Eichhornia crassipes*) individual stems become separated as the clone grows, which allows them to spread and cover large proportions of lakes and ponds in tropical and subtropical regions. Asexual reproduction can also be accomplished through specialized structures such as vegetative propagules (*Vivipary; e.g., Bulbils*—Fig. 9.2; refer to color plate), or asexually produced seeds that are generated by various types of apomixis (Box 9.1).

While asexual reproduction is common in plants, cases of the complete loss of the capacity to reproduce sexually are extremely rare (Sharp and Otto 2016), and virtually all apomictic plants display some degree of sexual reproduction (Hojsgaard 2014). Unlike most other multicellular eukaryotes, many plants have "the best of both worlds"—asexual reproduction to allow superior genets to spread within a local habitat and sexual reproduction to facilitate adaptive responses to fluctuations and challenges in their biotic and abiotic environments. In the next few sections, we will delve deeper into the evolutionary origins of sex and the environmental and genetic conditions favoring sexual and asexual reproduction.

9.5 THE EVOLUTION OF SEX

The transition to sexual reproduction early in the evolution of eukaryotes provided a mechanism for the generation of more variable progeny by recombination among divergent lineages. Recombination, along with diploid inheritance of alleles that displayed varying degrees of dominance and epistasis, created more variable offspring phenotypes than is possible through clonal reproduction. Moreover, recombination facilitates the accumulation of beneficial mutations into single lineages during adaptation (Fig. 9.3). In asexual populations, individuals possessing different beneficial mutations may compete, leading to the loss of some lineages and the beneficial mutations that they carry (*Clonal Interference*). Eventually, beneficial mutations may arise independently in the same clonal lineage, but adaptation will always be more rapid in outcrossing sexual populations because of recombination. Hence, we expect that the introduction of sex early in the Proterozoic provided natural selection with a much greater range of variation to act on. Sexual reproduction effectively speeds up rates of divergence and facilitates more rapid adaptation to novel conditions.

While recombination and the production of variable progeny may accrue advantages for sex

Fig. 9.2. Examples of sexual reproduction in plants: Fragmentation of vegetative ramets in *Eichhornia crassipes* (water hyacinth), clones of quaking aspen (*Populus tremuloides*) display differences in fall leaf color change on a mountain slope in Alaska; plantlets produced on the edge of a leaf of *Bryophyllum daigremontianum* (mother of thousands); bulbils are produced on the lower inflorescence of *Bistorta vivipara* (*Persicaria*; alpine bistort); and apomictic seed production in the flowers of *Pilosella aurantiaca* (hawkweed). Photos by CrazyD, Jorg Hempel, Beeblebrox, Ted Center, and Orchi.

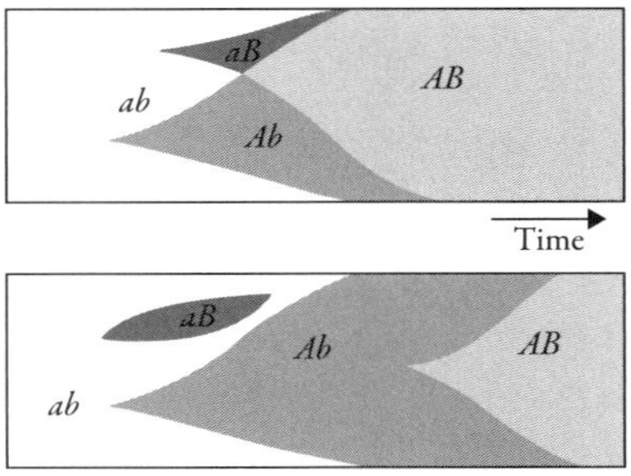

Fig. 9.3. Mutation accumulation and adaptation in sexual (top panel) and asexual lineages (bottom panel). Two beneficial mutations are more rapidly combined into a single lineage in a sexual population. Clonal interference leads to the initial loss of one mutation (B), but the same mutation eventually reappears in an asexual lineage. Graphic created by MykReeve.

by increasing the rate of adaptation, there are also disadvantages associated with sexual reproduction. For example, survival to adulthood is generally very low for most species, so it is a risky endeavor for a successful individual to recombine its genome with others that may not be as fit. Moreover, recombination will break up advantageous combinations of alleles across loci, so the average fitness of offspring is likely to be lower than that of both parents (recombination load—see the Mathematical Models for the Evolution of Sex section). Sexual reproduction also requires the successful transport and receipt of compatible pollen, so there is a risk of lower reproductive success compared to asexual reproduction (i.e., asexuality confers greater reproductive assurance). In spite of these drawbacks and others described later in this chapter, sexual reproduction is prevalent across the land plants, and eukaryotes in general.

Even if we assume that there are no risks due to genomic amalgamation and lower reproductive assurance, there are still intrinsic fitness deficits associated with sexual reproduction that can be very difficult to overcome. First and foremost, there is the demographic detriment that is levied on dioecious sexual lineages, because only daughters are capable of producing offspring (Fig. 9.4; refer to color

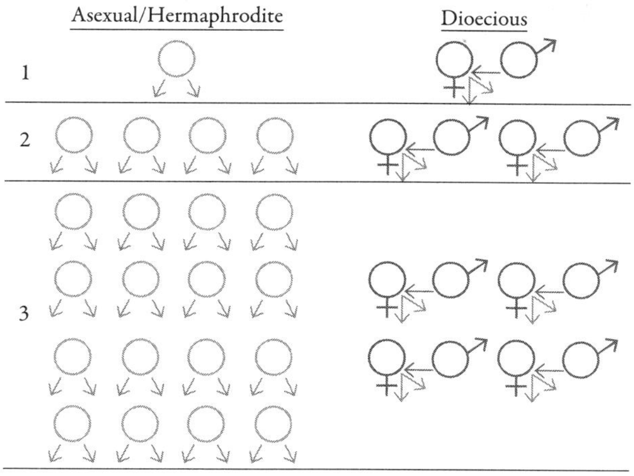

Fig. 9.4. The cost of males. Asexual and hermaphroditic populations grow twice as fast as dioecious populations because only females are capable of producing offspring. In these scenarios each asexual or female individual produces four offspring. There is no cost of males for hermaphrodites because each individual functions as both a male and a female.

plate); females contribute to population growth by producing progeny for the next generation while males only serve the purpose of fertilizing eggs. For dioecious species, this *Cost of Males* may reduce the capacity for population growth by half (Maynard-Smith 1978). The production of male offspring may substantially reduce the potential for population growth compared to asexual lineages, which only produce female offspring.

The second disadvantage for sexual reproduction is the fitness loss that may be imposed by the production of haploid gametes (*Cost of Meiosis*; Williams 1975). For dioecious lineages there appears to be a genetic transmission deficit, because clonal offspring possess two copies of the maternal genome, while only one is transmitted to offspring resulting from sexual reproduction (the other genome copy is donated by the male). On the other hand, this cost of meiosis for dioecious species appears only when we limit our view to a single generation; the predicted transmission deficit for sexual reproduction ignores the fact that male offspring are equally capable of transmitting genetic material (Treisman and Dawkins 1976). If we look across generations, we see that male and female offspring (Fig. 9.4; refer to color plate; generation 2) transmit equal proportions of their mother's genome to the next generation (Fig. 9.4; refer to color plate; generation 3—grandchildren of the original maternal plant). Because each parent passes on two copies of its genome to the next generation—one through female and one through male offspring—there is no cost of meiosis for dioecious lineages separate from the cost of males. On the other hand, there are still half as many copies of the maternal (and paternal) genome(s) transmitted between generations in sexually reproducing populations (i.e., due to the cost of males). Hence, we can say that for dioecious species there is a *Twofold Cost of Sex* (Maynard-Smith 1978) that reduces the rate of population growth and the transmission rate of parental genomes across generations. This twofold cost of sex, which includes both demographic and genetic transmission deficits, is expected to present a severe impediment to the establishment and spread of sexual reproduction.

Note that the expected cost of producing males rests on some important assumptions. First, for isogamous species the sexes are not differentiated, so the cost of males, and therefore the twofold cost of sex, does not apply. While isogamous species generally have two different mating types (e.g., + and – strains in fungi), each is equally capable of producing offspring, so there is no demographic cost for sexual reproduction (Lee et al. 2010). For this reason, the cost of males may be more appropriately referred to as the cost of anisogamy. Second, the potential for population growth is reduced only if the number of propagules is limited such that producing male offspring equally reduces the number of female offspring. For example, the cost of males may be lower if fewer resources are necessary for producing males (i.e., because they are smaller). Third, the twofold cost of sex assumes that each individual expresses only one sex. In some species, sex expression is labile such that an individual may start out life as a male and switch to being female when larger (e.g., as in maples and jack-in-the-pulpit, mentioned above). Because each individual can fulfill both the male and female roles required for sexual reproduction, there would be little or no demographic cost of sex for these sequential hermaphrodites. Since the large majority of plants function as both pollen donors and recipients throughout their reproductive lives, there may be no demographic cost of sex for most plants and other simultaneous hermaphrodites (Fig. 9.4; refer to color plate).

While there may be no demographic cost of sex for hermaphrodites, there may be genetic transmission deficits associated with the production of haploid gametes (Fig. 9.5). In a large outcrossing population of cosexual plants, we expect that, on the average, each individual would succeed in fertilizing one ovule through male function for every seed produced. The same can be said for self-fertilizing species, because each seed receives two genomic contributions from a single parent. On the other hand, a mutation for apomixis in an outcrossing cosexual population would confer an immediate genetic transmission advantage because individuals carrying this mutation would be able to transmit three copies of their genome (i.e., two through seeds and one through pollen) for every two copies transferred through sexual reproduction in a strictly outcrossing population (Fig. 9.5). Hence, there would be a 50% (1.5-fold) cost for sexual reproduction in a hermaphroditic species (Holsinger 2000). Such an advantage for asexual reproduction depends on the availability of unfertilized haploid megagametophytes, so the advantage of apomixis in cosexual populations would decline as sexual genotypes became less common. On the other hand, there would not be an advantage for apomixis in a selfing population since there would not be any ovules available for fertilization by asexual plants that continued producing pollen (i.e., due to a mutation for apomixis in a selfing population). Hence, this 1.5-fold advantage of asexuality is sometimes referred to as the *Cost of Outcrossing* (Holsinger 2000). Note that this same principle would apply to an outcrossing isogamous species if asexual individuals continued to produce gametes. With everything else being equal, a reduced genetic transmission advantage for sexual hermaphrodites would result in the spread of apomixis until the entire population was asexual (Fig. 9.5). In sum, hermaphroditic populations may be subject to a 1.5-fold (50%) cost of meiosis that reduces the advantage of sexual reproduction without having any demographic consequences, but this transmission advantage depends on frequencies of selfing and apomixis in the population.

There may be additional costs of sex for hermaphroditic species because of low reliability of pollen transfer and fertilization. For example, if pollen transfer to stigmas were limited, then male (and female) reproductive success would be compromised and hermaphrodites would have lower fitness compared to asexual lineages. Even if access to ovules were not limited, there is often variation in fertilization success among males so there may be a cost of sex for a good proportion of individuals in outcrossing populations that have lower than average

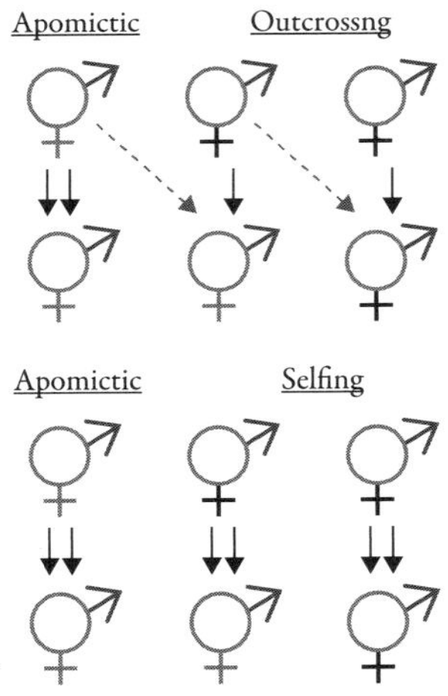

Fig. 9.5. The cost of meiosis in populations of hermaphrodites. Apomictic individuals transmit three copies of their genomes to the next generation; two through their seeds (solid arrows) and one through pollen (dashed arrows). Since sexual individuals only transmit two copies (one through seeds and one through pollen) there is a 1.5-fold cost of sex when apomicts occur at a low frequency. Note that a mutation for apomixis could spread through both seeds and pollen, and would be expected to rapidly increase in frequency.

success through male function. Because the pollen transfer to receptive stigmas can be inefficient, we expect that sexual lineages will have lower reproductive assurance than asexual lineages, and consequently, they may have reduced reproductive success when pollen vectors are limited or absent. Overall, we can predict lower reproductive success for cosexual plants under some circumstances, and combined with the 1.5-fold cost of meiosis just described (Fig. 9.5), we can predict a substantial cost of sexual reproduction for hermaphrodites. On the other hand, the cost of sex for hermaphrodites will always be substantially lower compared to the cost for dioecious species, which are subjected to the twofold cost of sex as well as lower reproductive assurance compared to asexual individuals.

The Paradox of Sex—To understand why sexual reproduction is prevalent for both hermaphroditic and dioecious species, we can examine the hypotheses and models concerned with the evolution of sex. Based on the costs of sex, we can predict that dioecious sexual lineages would need to produce more than twice as many offspring and hermaphrodites 1.5 times as many offspring to outcompete the asexual lineages present in the same population (Fig. 9.4; refer to color plate and 9.5). In spite of the negative demographic and fitness consequences, sexual reproduction is prevalent across the eukaryotes, including the land plants. Asexual lineages have appeared in many groups of metazoans, but these represent a small fraction of the total number of species (Sharp and Otto 2016). Moreover, all of the known asexual eukaryotic species are recently derived, which suggests that these lineages are short-lived (Burt 2000; Otto and Lenormand 2002). Sexual reproduction is widespread in spite of its potential for reduced reproductive assurance and the severe demographic and genetic transmission disadvantages.

The incongruity between the predicted deficits of sexual reproduction and its prevalence represents a paradox that has occupied the attention of biologists since the late 1800s. As is evident from the quote at the beginning of this chapter, Erasmus Darwin was one of the first to recognize the potential for sexual reproduction to produce an exceptional range of variation among offspring (Darwin 1800). While Erasmus did not mention that there might also be deficits associated with sexual reproduction, his grandson Charles suggested that the drawbacks of sex could be overcome through increased fitness due to "complementation" (heterosis). This suggestion came from the younger Darwin's experience with garden experiments in which he compared the vigor of seedlings produced from selfing and outcrossing. Darwin's opinion on this topic did not necessarily reflect the mainstream, since most authors in the 1800s assumed that variation among progeny was adequate to explain the prevalence of sexual reproduction without consideration of the potential disadvantages of sexual reproduction. Careful evaluations of this idea were not forthcoming until August Weismann (1889) developed detailed arguments for the advantages of sex. Consequently, Weismann is generally recognized as the first to present arguments for the advantages of sex, based on variation among offspring (Burt 2000). While genetic variation is probably one of the primary reasons for the dominance of sexual reproduction, models that make simplifying assumptions often find that sex is advantageous only under restrictive conditions (Otto and Lenormand 2002). In the next sections, we discuss models for the evolution of sex that contribute to our understanding of how characteristics of natural systems have led to the prevalence of sexual reproduction across the eukaryotes.

Understanding the Advantages of Sex—Weismann's predictions for the advantages of more variable progeny were analyzed using mathematical approaches in the mid-1900s, but the primary conclusion from these models was that sexual reproduction was favored only under a limited set of circumstances (Otto and Lenormand 2002). In spite of these predictions, sexual reproduction is widespread; in fact, it may be the case that all eukaryotes—even those presumed to be asexual—may engage in some form of recombination among lineages at least occasionally (Hojsgaard 2014; Sharp and Otto 2016). To understand the reasons for the contrast between the theoretical predictions and the observed distribution of sexual reproduction, it is instructive to examine the assumptions of the models used to evaluate the fitness consequences of sexual and asexual reproduction.

Since Darwin's first writings on evolution in the mid-1800s, evolutionary biologists have focused on individual fitness. We expect that heritable characteristics that increase an individual's chances to survive and reproduce will be favored by selection and consequently, will increase in frequency. In contrast, sexual reproduction is a trait that may not have a direct and immediate benefit to an individual, because producing offspring that are more variable does not guarantee success. Much of the paradox of sex comes from the perception that recombination can be disadvantageous in the short-term but may be favored in lineages over many generations.

Consideration of the advantages of sex have been made from several different points of view. Most mathematical treatments have focused on the immediate advantages conferred to individuals in sexual and asexual lineages (i.e., a *Short-Term Advantage* for sex). On the other hand, verbal arguments and simulations have predicted advantages for sex in biotic and abiotic environments that vary spatially and temporally (*Long-Term Advantage*). The discrepancy between the conclusions made from these approaches may be due to the simplifying assumptions of analytical models and their inability to adequately reflect the complexities of structured populations affected by environmental variation (Otto and Lenormand 2002).

When considering long-term advantages of sex, we imagine that selection may act among sexual and asexual lineages; however, much of the perception that the two-fold disadvantage for sex is difficult to overcome may be due to a reluctance to consider selection acting at the levels above the individual (e.g., on populations or lineages). During much of the 1900s there was a strong focus on individual selection as the primary mechanism of adaptive evolution, and most biologists dismissed the notion that selection could be effective above the level of the individual (*Group Selection; Multi-Level Selection*). Group selection had been invoked as a mechanism to explain the evolution of cooperation in animal populations in the early 1900s, but this idea has remained controversial (Goodnight 2013). Understanding the potential for a long-term advantage for sex may require that we examine the success of populations and lineages over a succession of generations and take into account both short- and long-term consequences of recombination and sexual reproduction.

Verbal Models for the Evolution of Sex—One verbal model for the long-term advantage of sex proposes that sexual reproduction arose as a mechanism for larger eukaryotes with longer generation times to keep up with their faster-evolving microbial pathogens and parasites (Chapter 1; Van Valen 1973; Bell 1982). This idea became popularized as the Red Queen hypothesis, which makes an analogy between the Red Queen from Lewis Carroll's *Through the Looking Glass* with multicellular eukaryotes that have to evolve faster just to maintain their place in an evolutionary race with their parasites. The Red Queen hypothesis assumes a long-term advantage for sex: clonal lineages are more likely to succumb to pathogens because they lack adequate genetic variability to evolve resistance. This hypothesis assumes that variation in biotic environments is likely to be more rapid than changes in physical conditions and that this coevolutionary race between species and their pathogens is responsible for the origin and maintenance of sexual reproduction.

In the late 1900s, several verbal arguments for the evolution of sex were developed to account for the short-term advantages of sex. One of the most persuasive lines of reasoning made an analogy between asexual reproduction and lottery tickets (the *Lottery Model*; Williams 1975). In this model, we imagine that the environment is patchy such that there is matching between different genotypes (ticket numbers) and specific combinations of environmental conditions (the winning numbers). Under this model the relative advantages of sexual (each ticket has a different number) and asexual reproduction (all tickets are identical) depends on the distribution of environmental variation; if the home environment of the maternal genotype is rare then producing a wider variety of genotypes is more likely to succeed. A female that produces more variable offspring is more likely to have at least a few of them arrive in habitats that provide a good match for their particlar combination of phenotypic characteristics. On the other hand, if we assume that the maternal plant is a good match for its home environment, then asexual reproduction will be more advantageous if the maternal environment is common. Under the lottery model, sexual reproduction is advantageous only when environments are highly variable and the maternal environment is uncommon.

The lottery model and others that assume variation in environments or differences in resource acquisition among offspring (e.g., the *Tangled Bank Model*; Ghiselin 1974), have been criticised as simplified views that may overstate the importance of minor variations in the environment. Moreover, empirical tests of the advantages of sex conducted in the late twentieth century that were based on environmental and genetic variation largely failed to support short-term benefits for sexual reproduction (e.g., Kelley 1989; Bierzechudek 1990; Ronsheim 1996). More progress was made on our understanding of the advantages of sex in the twentieth and early twenty-first centuries when mathematical models and simulations that incorporated the potential for long-term advantages of sex became more common. In the next section we review the consequences of different assumptions for a family of models that examine predictions for the evolution rates of recombination as a sorrugate for the evolution of sex.

Mathematical Models for the Evolution of Sex—Models for the evolution of sex have generally focused on conditions that would render sexual populations resistant to invasion and replacement by asexual lineages (Otto and Lenormand 2002). Under weak directional selection with large population sizes, the generation of genetic variation among offspring is generally not sufficient to confer short-term advantages for sex. Indeed, recombination among lineages is analogous to a double-edged sword: on the one hand, some offspring may have novel genotypes that have high fitness; on the other hand, recombination breaks up genotypes already proven to have high fitness in the current environment. For example, if fitness depends on alleles across multiple loci, then the best combinations may appear in only a small fraction of the offspring produced by a successful individual. The breakup of adapted genotypes can result in lower average offspring fitness—a phenomenon that is referred to as *The Reduction Principle.* The reduction principle predicts lower fitness in the short term in a constant environment, but there may be short- and long-term advantages to producing variable offspring in heterogeneous environments.

Assessing the outcome of competition between sexual and asexual populations can be analytically difficult, but we can take an alternative approach by examining the conditions that favor an increase or decrease in the rate of recombination between linked loci (Fig. 9.6). These models assume that sex is effectively equivalent to recombination and that the primary purpose of sex is the reshuffling of alleles across loci to break up existing genetic associations and create new ones.

To start our discussion of models examining the evolution of sex, we consider single large populations of haploid individuals that are subjected to weak directional selection (Otto and Lenormand 2002). These models focus on changes in the rate of recombination based on variation at a modifier locus (M) to determine whether sexual reproduction is favored (Fig. 9.6). To understand the potential negative or positive effects of recombination we consider two linked loci (A and B) that are segregating for alleles (a and b) that increase fitness in the current environment. The fitness of the allele combination *a-b* depends on the independent selective advantage of each allele and the sign and level of epistasis between them. These models consider the effects of recombination given the starting conditions for the current level of linkage disequilibrium and the effects of epistasis. We predict a short-term advantage for sex if recombination will increase the average fitness for the next generation (Fig. 9.6; upper left panel) and a long-term advantage for sex if recombination increases the variance in fitness (Fig. 9.6; lower panels). Given the assumption of epistasis among alleles at different loci conferring fitness, we can predict that selection will favor an increase in the rate of recombination under a broad range of conditions.

To understand the potential for the evolution of sex given the assumptions of the basic model just described, we need to assess the conditions occurring in populations after several generations of selection and recombination. The first thing we note is that selection will tend to make the specific allele combinations that are favored by epistasis more common. For example, with positive epistasis, we expect to find an excess of *a-b* genotypes in the population, but these high-fitness genotypes will be broken up by recombination leading to lower fitness in the next generation—a phenomenon referred to as recombination load. In general, selection each generation will promote linkage disequilibrium in the adult population that is the same sign as epistasis (Otto and Lenormand 2002). When these individuals mate, the lower fitness genotypes will be generated through recombination and fertilization, which means the short-term cost of sex is generally high. Under this set of conditions, we predict that increased recombination (sexual reproduction) will be favored only when the long-term advantage outweighs the short-term cost of sex (i.e., when epistasis is weakly negative and

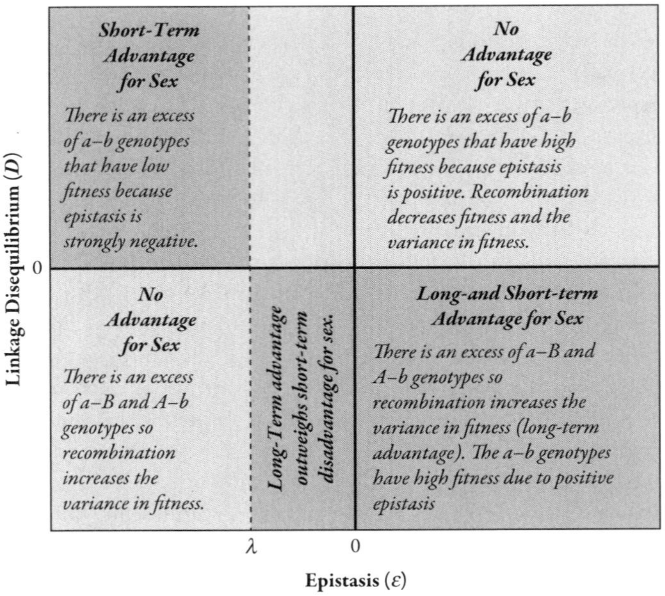

Fig. 9.6. Results from modifier models for the evolution of sex. In a large haploid population the recombination rate between two loci (A and B) is controlled by a single locus (M). Fitness is determined by the presence of alleles a and b plus an interaction term (epistasis) that can be positive or negative. Each panel explains the consequences of recombination given the starting level of linkage disequilibrium (D) and the effects of epistasis (ε).

gametic disequilibrium is also negative; Fig. 9.6). On the other hand, the weak negative epistasis required for the evolution of increased recombination appears to be uncommon (Otto and Lenormand 2002), which suggests that this class of models is too unrealistic to understand the conditions favoring the evolution of sex and recombination.

The model described in Fig. 9.6 could be more useful if we could apply it to identify ecological conditions that are more likely to favor the evolution of sex. In the scenario just described, we assumed a constant environment. Under constant conditions, sexual reproduction is not favored, because the linkage disequilibrium has the same sign as epistasis, and recombination breaks up favorable allele combinations to produce high proportions of low-fitness offspring. Variable environments, on the other hand, could generate patterns of disequilibria that produced a short-term advantage of sex. When environmental conditions change every two to five generations, the sign of linkage disequilibrium would be opposite the sign of epistasis just after a transition (Otto and Lenormand 2002). This model assumes that a coupling genotype (*a-b*) is favored under one set of conditions and the repulsion genotypes (*a-B* and *A-b*) are favored under the other. Hence, in fluctuating environmental conditions, breaking up genotypes that were favored under previous conditions is more likely to generate recombinant genotypes that have high fitness under the current conditions. This model also predicts a short-term advantage of sex if we assume spatial heterogeneity and moderate levels of gene flow among populations occurring in these two types of environments.

So far, the models we have discussed have assumed very large populations of haploid individuals. With haploid populations, we assume that the alleles at every locus are expressed and that fitness is a product of their individual effects and their interactions. When we consider diploid populations, the predicted outcomes change because of the potential for the accumulation of deleterious recessive mutations (segregation load). New mutations will be present as single copies in diploid individuals but could be made homozygous through inbreeding. Hence, segregation load is due to sorting of alleles at a single locus, while recombination load (mentioned earlier) is due to the break-up of advantageous combinations of alleles at different loci. Both segregation load and recombination load are components of genetic load.

In diploid populations, the frequency of heterozygotes is expected to increase each generation as individuals homozygous for deleterious mutations are eliminated. If mutations are nearly or completely recessive, then heterozygotes will persist in the population and will produce a proportion of low-fitness homozygous offspring each generation. Hence, in diploid populations we expect that the average fitness of offspring will be lower than that of the adult population. Mutations can also accumulate in diploid asexual lineages, but the chance that the same mutation would appear in both alleles at one locus is vanishingly small, so recessive deleterious mutations can accumulate across asexual generations without negative effects on the vigor and growth of the clone. Consequently, diploidy results in a strong short-term disadvantage for sexual reproduction due to the accumulation of recessive mutations. While there could be a long-term advantage of sex when the effects of mutations are very weak (i.e., because of the increase in the variance in fitness; Otto and Lenormand 2002), the conditions necessary are quite restrictive and do not appear to be common in natural populations.

To understand how sexual reproduction might be advantageous, we need to consider alleles segregating at multiple loci and the rates of recombination among them. For large diploid populations, our predictions are very much the same as for haploids (Fig. 9.6), but the outcomes of these models change when we consider multiple small populations that are subject to selection, genetic drift, and gene flow (Otto and Lenormand 2002). Results from simulations that track many loci indicate that when genetic drift acts alone, it leads to the fixation of the full range of allelic combinations across populations. If we assume that many of these loci are segregating for deleterious recessive alleles with a smaller proportion of alleles that are advantageous in the current environment, we find that the combination of genetic drift and selection generates disequilibrium across loci within each population. In particular, as advantageous alleles become associated with deleterious alleles across the genome the progress of adaptation stalls—selective interference (the Hill-Robertson Effect; Hill and Robertson 1966). Across multiple small populations, we expect that each one would become fixed for different combinations of alleles; as populations become more homozygous, beneficial alleles become associated with detrimental alleles, which limits the potential for responses to selection. Gene flow among populations allows for exchange of different combinations of alleles across loci, and adaptation progresses as recombination breaks up the associations among alleles that have developed in each population. Genetic purging of deleterious alleles and fixation of advantageous alleles in each population are facilitated by recombination with immigrant genomes. In these simulations, there is a short-term advantage for recombination across a broad range of population sizes even if we assume there is a two-fold cost of sexual reproduction (Otto and Lenormand 2002). Hence, the advantage of sex and recombination is most evident when we consider many small populations that are connected by gene flow.

It is important to note that we assumed a constant environment when we predicted a short-term advantage for sex and recombination in this model of small diploid populations subject to selection and gene flow. In essence, genetic drift tends to counteract the effects of selection and halt the processes of adaptation as populations become more homozygous. Progress toward continued adaptation is facilitated by gene flow among populations fixed for different combinations of alleles, but we would expect that adaptation of the whole metapopulation would eventually stall as adaptive variation across populations becomes exhausted (i.e., because all populations become fixed for the same beneficial alleles). Hence, the short-term advantage of recombination is expected to decline in conjunction with loss of genetic differentiation among populations. Under these circumstances, the continued advantages for sexual reproduction become dependent on the appearance of novel mutations that confer adaptive advantages to environments that vary temporally and spatially. Since environments are generally variable, we can conclude that both short- and long-term advantages could have contributed to the prevalence of sexual reproduction in diploid lineages.

Summary of Models—In this section we have learned that there is a two-fold demographic cost of sex for dioecious species due to the production of male offspring (cost of males) and a 1.5-fold genetic transmission cost of sex for hermaphrodites due to the production of haploid gametes (cost of meiosis). In spite of these predicted deficits and other uncertainties associated with mating, sexual reproduction is ubiquitous across the eukaryotes and probably occurs to some degree in all plants. The solution to the paradox of sex is derived from short- and long-term advantages accrued through recombination (Fig. 9.7). Short-term gains for sex can occur in variable environments that fluctuate at a frequency of two to five generations, which is frequent enough for recombination to generate allelic combinations that are favored in the alternative environment (Fig. 9.7, third row below header). Advantages of sex can also develop in diploid populations from recombination after gene flow among populations subject to genetic drift, which alleviates selective interference among alleles at separate loci, facilitates purging of genetic load, and allows adaptation to proceed (Fig. 9.7, fourth row below header). Long-term advantages of sex are likely due to temporal variation in environments, which favors higher phenotypic variance among offspring. Under environmental change or colonization of novel habitats, we expect that sexual lineages will continue to adapt while asexual lineages may not be able to respond rapidly enough and consequently, are more likely to suffer extinction. Apparently, these benefits of sexual reproduction have contributed to the long-term persistence and diversification of sexual lineages across phylogenetic timescales.

Variation and Recombination as Separate Advantages of Sex—While Weismann's hypothesis for the advantage of sexual reproduction through the generation of more variable progeny has been shown to be correct in the broadest sense, this idea misses the mark; the short-term and long-term benefits of sex appear to be primarily due to the potential for recombination among divergent genotypes. In other words, the primary advantage of sex is that it enables incorporation of novel genetic variation into populations and lineages, and it provides for the generation of advantageous combinations of alleles across the genome through recombination. Of course, Weismann and other scholars of the nineteenth century could not have made this connection, simply because there was no understanding of the processes of genetic inheritance. Indeed, direct evidence of the association between the cytological phenomenon of chiasmata (crossing over) and genetic linkage was not uncovered until thirty years after the importance of Gregor Mendel's work became widely recognized (Creighton and McClintock 1931). Even after linkage and recombination were more fully understood, their connections to the advantages of sexual reproduction

Model Assumptions	Short-Term Effects	Long-Term Effects	Predicted Outcome
Haploid Large Population Constant Environment	Loss of Beneficial Allele Combinations	Increases the Variance Among Offspring	Sex Generally not Favored
Diploid Large Population Constant Environment	Homozygosity of Deleterious Alleles in Offspring	Increases the Variance Among Offspring	Sex Generally not Favored
Haploid or Diploid Large Population Variable Environment	Beneficial After Environmental Change	No Benefit with Only Two Environments	Sex Favored with Changes every 2–5 Generations
Diploid Small Populations Variable Environment	Alleviates Selective Interference and Facilitates Purging of Genetic Load	Beneficial with Continued Environmental Change	Sex Favored with Gene Flow and Under Changing Conditions

Fig. 9.7. Expected outcomes for different models of the evolution of sexual reproduction (recombination; Otto and Lenormand 2002). Outcomes for the first three rows are for analytical models, and the last row represents predictions from simulations. Short-term benefits refer to effects expected for offspring from the current generation and long-term effects apply to consequences that may become apparent over a number of generations.

were not recognized. Through most of the twentieth century, nearly all scholars followed the verbal arguments of Fisher (1930) and Muller (1932) who suggested that differences in variation among offspring was adequate to explain the prevalence of sex, but there was not much effort to test this idea.

In the late twentieth century, there was a resurgence in interest in the paradox of sex. During this time, much of the theoretical and empirical effort to find advantages of sex focused solely on variation among offspring (e.g., as in the lottery model), but these efforts largely failed to discover short-term advantages of sexual reproduction. Similarly, the Red Queen hypothesis, while possibly applying to the origins of sex in the earliest eukaryotes (Chapter 1), may have been of less importance for the maintenance of sex, because most multicellular eukaryotes are capable of rapid immune responses that provide good degrees of protection from pathogen attack (e.g., as in the plant immune system; Jones and Dangl 2006). The Red Queen hypothesis and the lottery model both rely on the contrast in genetic variation among sexual and asexual offspring as the primary advantage of sex, but the plurality of empirical evidence does not support the hypothesis that variation alone is responsible for the prevalence of sexual reproduction.

Our understanding of the advantages of sex improved through the late twentieth and early twenty-first centuries, as the focus shifted away from genetic variation per se to the consequences of recombination as a mechanism to facilitate the purging of genetic load and adaptation to novel environments (Fig. 9.7). Analytical models and simulations focusing on the consequences of recombination found short-term advantages of sex by alleviation of selective interference among beneficial and detrimental mutations. Testing the predictions of these models has been more difficult, as analyses of the effects of individual alleles and of different genetic combinations on fitness are required.

Empirical Tests for the Advantages of Sex—Empirical assessments of the consequences of recombination for the maintenance of sex were fairly rare through the early twenty-first century but have become more common with better access to large-scale sequencing technologies (McDonald et al. 2016; Sharp and Otto 2016). In their study of adaptation in sexual and asexual strains of yeast, McDonald et al. (2016) found alleles with substantial detrimental effects were fixed by genetic hitchhiking in asexual lines but not in sexually reproducing populations. Adaptation was generally faster in sexual populations, and the spectrum of alleles going to fixation was substantially different compared to asexual populations. Several other studies published since 1995 have provided clear evidence of faster rates of adaptation in sexual lineages, but the study by McDonald et al. (2016) was the first to demonstrate that changes in allelic associations across loci were responsible for the advantage of sexual reproduction.

The faster rates of adaptation that have been found in the studies of yeast just described represent a consequential advantage of sexual reproduction that is expected to lead to greater chances of persistence and diversification over phylogenetic time scales. In a broad sense, these results support the original ideas of Weismann (1889), Fisher (1930), and Muller (1932)—that the advantage of sex was associated with greater variability among offspring. Specifically, the advantage of sexual reproduction is garnered through opportunities for the mixing of lineages and generation of individuals having novel combinations of characteristics. Note, however, that faster rates of adaptation because of recombination represent a long-term advantage of sex and do not address short-term demographic and genetic transmission costs. Hence, when sexual and asexual lineages are in direct competition, the short-term costs of sex could easily obfuscate such long-term advantages, particularly in environments that are constant or changing very slowly. From the outcomes of the models described earlier, we can infer that short-term costs of sex could be overcome in populations subject to genetic drift, selection, and gene flow, but to this point, there have not been any direct tests of this hypothesis. Consequently, the paradox of sex has not yet been fully solved; while we believe the answer lies in understanding the consequences of recombination for the dissolution of detrimental genetic associations that develop through selection and genetic drift, these hypotheses require validation through further experimentation and empirical evidence. For example, it is not yet clear whether recombination with immigrants in small populations will typically confer large enough advantages to sexual individuals to outcompete their asexual neighbors (but this may be in cases of gene flow among inbred populations; Ingvarsson and Whitlock 2000; Chapter 4). The resolution of this question will improve our understanding of how selection on individuals and lineages in constant and variable environments has contributed to the prevalence of sexual reproduction across the eukaryotes.

9.6 THE EVOLUTION OF ASEXUAL REPRODUCTION

During the early stages of the invasion of terrestrial environments, the horizontal spread of plants across the soil surface was a key innovation that allowed aquatic embryophytes to colonize emergent habitats in continental regions (Chapter 2). These earliest terrestrial plants may have been able to take advantage of their elongated thalli to integrate variation in moisture and nutrient availability along the borders of ponds and wetlands. The propensity for vegetative spread was passed on to the descendants of the first land plants and can be seen in many groups of modern plants as well as the earliest fossil tracheophytes such as *Cooksonia* and the zosterophylls. The tendency for horizontal elongation was lost in many species, as vertical growth became a priority due to competition for light in the swamp forests of the Devonian and Carboniferous. During this time, many plants must have retained the potential for vegetative spread and fragmentation into independent ramets, because this characteristic is present in large proportions of extant lineages of lycophytes and ferns and in many groups of flowering plants.

Asexual reproduction through vegetative growth appears to be an ancestral condition that is generally characteristic of most embryophytes but is lacking across the divergent groups of gymnosperms. Vegetative reproduction is widespread among herbaceous perennial angiosperms and is particularly common in the grasses and the large diversity of aquatic species (Grant 1981). Virtually all perennial grasses reproduce asexually either through tillers (rhizomes) in the bunchgrasses (tussock grasses such as *Deschampsia, Festuca,* and *Stipa*) or through runners in the stoloniferous grasses, such as *Digitaria, Cynodon,* and dune grasses such as *Leymus* and *Ficinia* (Fig. 9.8; refer to color plate). Many aquatic flowering plants have extensive vegetative spread and often form clones that cover large areas. Some examples of aquatic plants capable of extensive clonal growth include: submerged aquatic species of *Zostera* (marine eelgrass), *Thalassia* (marine seagrass), *Elodea* (waterweeds), and *Potamogeton* (pond weeds); free floating plants such as *Eichhornia crassipes* (water hyacinth) and *Lemna* (duckweed); and emergent aquatic species such as *Typha* (cattail), *Bressenia* (water shield), *Spartina* (cordgrass), and *Phragmites* (common reed).

The propensity for vegetative spread is common across herbaceous perennials and can occur by rhizomes, stolons, tubers, and *Adventitious Rooting* of vegetative or woody stems. In monocotyledonous geophytes such as lilies and tulips, asexual reproduction proceeds by the splitting of bulbs and corms. Vegetative spread is absent in some groups of woody species, including many trees of temperate and tropical forests, but is quite common for some groups of woody plants including species in the willow family (Salicaceae, including *Salix* and *Populus*) and is characteristic of many bamboos (Poaceae; several genera in the subfamily Bambusoideae). The capability for vegetative propagation appears to be retained in many woody species; even those that do not usually engage in horizontal spread can often be coaxed into rooting from cut stems or even branches. Indeed, in tropical regions posts made from live branches often take root and resprout to form "living fences."

Apomixis—While reproduction through vegetative growth is widespread and common across the embryophytes, asexual reproduction through seeds (agamospermy) is a much more recent innovation and occurs only in the angiosperms. The term apomixis ("away from-mixing") has sometimes been interpreted broadly to include all forms of asexual reproduction but is more commonly applied to cases of asexual reproduction through normally sexual reproductive structures. Under this definition, apomixis encompasses cases in which flowers are replaced by bulbils, as in *Poa bulbosa* and *Bistorta vivipara* (Fig. 9.2; refer to color plate). Agamospermy is a more specific term, as it refers to instances of asexual reproduction through seeds. Agamospermy differs from *Parthenogenesis* ("virgin-origin") because the production of asexual seeds in the large majority of apomictic plants requires participation of a pollen tube (pseudogamy; pseudogamous apomixis; Box 9.1). In pseudogamous species, the egg cell is unreduced so the embryo originates through parthenogenesis (i.e., mitosis of the egg cell), but development of the endosperm requires fusion of sperm nuclei with the polar nuclei of the megagametophyte. Agamospermous

Fig. 9.8. Examples of vegetative spread including the ancient "King Clone" of *Larrea tridentata* (creosote bush), the tussock grass, *Spinifex* sp., forming a ring where the older tillers in the center have died, sprouting and rooting at the nodes of the stoloniferous dune grass *Desmoschoenus spiralis*, and an illustration of stoloniferous growth in Bermuda grass (*Cynodon dactylon*). Photos by Klokeid, Mark Marathon, and Arne Hückelheim.

embryos develop from single unreduced cells by parthenogenesis, but the entire seed is a complex structure containing several different tissues having independent origins. Agamospermous seeds are not strictly parthenogenic, so the more general term of agamospermy is a more appropriate descriptor of asexual reproduction through seeds.

Across agamospermous species, there are two general pathways for the production of asexual embryos. In gametophytic apomixis the megagametophyte is derived by mitosis of the megasporocyte or other cells in the archesporium, and the embryo develops directly from the unreduced egg cell by parthenogenesis (Box 9.1). In sporophytic apomixis (adventive embryony) embryos develop by mitosis of sporophytic cells that are outside of the megagametophyte, such as the nucellus or integument. Whether the embryo originates through gametophytic or sporophytic apomixis, the remainder of seed development may proceed normally. In most agamospermous species, endosperm development still requires fusion of the sperm and polar nuclei (pseudogamy), but in some plants—particularly in the Asteraceae—the endosperm is produced without participation of the microgametophyte (autonomous apomixis). Seeds of agamospermous species are sometimes heavier, but they typically possess the same dispersal characteristics as the seeds of their sexual relatives. Consequently, agamospermous species often have much greater dispersal potential than other modes of asexual reproduction, and individual clones can be distributed over large geographic areas (Hojsgaard 2014).

Agamospermy has been detected in at least seventy-eight plant families, which is nearly 20% of recognized families (Grant 1981; Hojsgaard 2014). In general, there is an association between the total number of species in a family and the number of agamospermous species. For example, there are greater numbers of agamospermous species in some

Box 9.1 Origin of Seeds through Sexual Reproduction and Agamospermy.

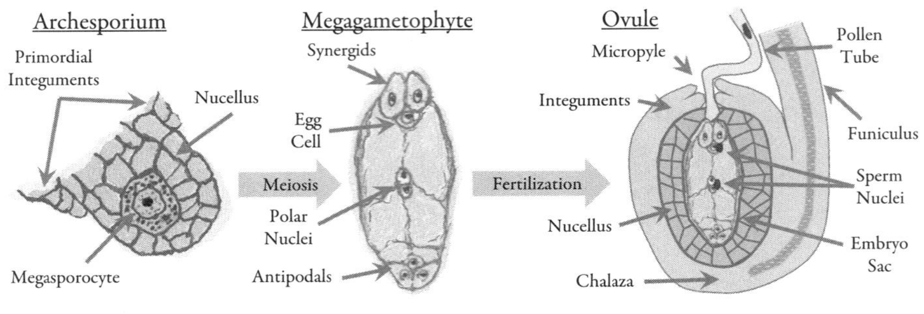

Fig. 9.9. Development of the mature megagametophyte from a megasporocyte, and fertilization by a pollen tube. During double fertilization one sperm nucleus fuses with the egg nucleus, and the second fuses with the two polar nuclei to form the triploid endosperm.

Lineages with sexual reproduction produce megagametophytes by meiosis of *Megasporocytes* within archesporia, which are evident as small protuberances within the developing ovary (Fig. 9.9; refer to color plate). Typically, three of the four products of meiosis degenerate while the fourth undergoes three rounds of mitosis to produce the eight-celled megagametophyte. The megagametophyte is surrounded by *Nucellus* cells of sporophytic origin and enclosed within an ovule consisting of one (gymnosperms) or two (angiosperms) layers of integuments. After pollination, microgametophytes extend their pollen tubes to the ovary to deliver two sperm nuclei to each ovule: one that fuses with the egg cell to form the zygote and the second that joins with the two polar nuclei to form a (typically) triploid endosperm (double fertilization). Carbohydrates for the developing endosperm and embryo are provided via a vascular strand that extends through a funiculus to the chalazal end of the ovule. The integuments eventually develop into the seed coat, and the ovary wall expands to accommodate the developing seeds. In fleshy fruited species, the pericarp develops from the ovary wall, but other portions of the flower may contribute to its development in some groups (e.g., the hypanthium of the flower forms most of the fruit in some members of the Rosaceae such as apples).

Asexual production of seeds (*Agamospermy*) can be broadly divided into cases were the embryo arises from an unfertilized and unreduced egg cell (*Gametophytic Apomixis*; Fig. 9.10; refer to color plate, pathways 1–4), and those where it is derived from the nucellus or other sporophytic cell outside of the megagametophyte (*Sporophytic Apomixis; Adventive Embryony*; Fig. 9.10; refer to color plate, pathway 5). Gametophytic apomixis is further divided into two pathways depending on the origin of the megagametophyte. In *Generative Apospory* (*Diplospory*; 1) the megagametophyte is derived directly by mitosis of a megasporocyte (megaspore mother cell) of the *Archesporium*. Note that production of a haploid megagametophyte in sexual species would normally proceed by meiosis of the megasporocyte. In *Somatic Apospory* (2) the megagametophyte develops from mitosis of a cell of the nucellus. In either case the endosperm may develop without participation of a sperm nucleus (*Autonomous Apomixis*; 3) or more commonly, requires the fusion of one or two sperm with the two polar nuclei (*Pseudogamy*; 4). Depending on the base ploidy of the sporophyte (2X), the resulting endosperm will be either 4X (2X + 2X from fusion of the two polar nuclei for autonomous apomixis), 5X (2X + 2X + 1X for pseudogamy with one sperm nucleus), or 6X (2X + 2X + 1X + 1X for pseudogamy with two sperm nuclei).

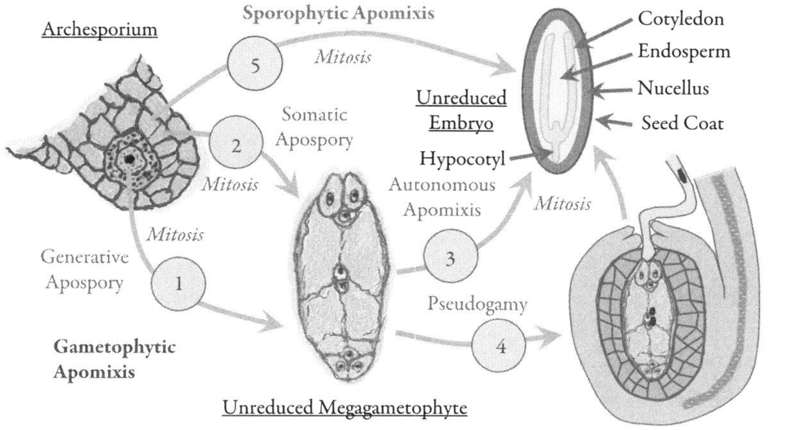

Fig. 9.10. Pathways for the origin of asexual seeds (agamospermy).

particularly large families such as in the grasses (Poaceae), and members of the rose and sunflower families (Roseaceae and Asteraceae). Gametophytic apomixis is typically restricted to species that have higher ploidy levels (Hojsgaard 2014) but is known to occur in diploid species of *Potentilla* (Rosaceae), *Ranunculus* (Ranunculaceae), *Hieracium* (Asteraceae), and *Arabis* (Brassicaceae). Many cases of gametophytic apomixis also appear to be polyploid lineages that arose after hybridization between species that are not normally interfertile. Given the high ploidy levels and hybrid origin, it is possible that many agamospermous lineages would have low fertility through sexual reproduction. Hence, agamospermy may often serve as a mechanism to preserve genomic combinations that have vigorous vegetative growth but because of irregular segregation, may have low success through sexual reproduction.

Mechanisms of asexual reproduction that are independent of the megasporocyte—including somatic apospory and sporophytic apomixis—are more common than development of a clonal embryo by mitosis of the megasporocyte (generative apospory; Hojsgaard 2014; Box 9.1). These alternative pathways for the origin of somatic embryos may be more prevalent because of the complexity of interrupting the normal meiotic origin of the megagametophyte. It may be that the direct development of the megagametophyte or embryo by mitosis of a cell other than the megasporocyte is easier to achieve. Indeed, somatic development of embryos is quite common and sometimes results in the presence of both sexual and asexual embryos in the same ovule. In these cases, the sexually derived embryos are more often subject to failure and replacement by the asexual embryos, which is consistent with the idea that the genomes of apomictic species often suffer from abnormal segregation during meiosis. The propensity for the development of clonal embryos in species with sporophytic apomixis often results in multiple embryos within single seeds (*Polyembryony*). Both sporophytic apomixis and the tendency for polyembryony are particularly common among species in the genus *Citrus* and are widespread among tropical plants. Across the angiosperms, sporophytic apomixis is the most common form of agamospermy (found in 148 genera), followed by somatic apospory (110 genera) and generative apospory, which occurs in only 68 genera (Hojsgaard 2014).

Most agamospermous species are represented by many independently derived asexual lineages, which suggests that these taxa possess characteristics or are subject to conditions that favor frequent transitions from sexual to asexual reproduction (Hörandl et al. 2008). New asexual lineages may arise in three different ways: 1) through clonal diversification by the accumulation of somatic mutations among independent asexual lineages (Chapter 3); 2) by independent origin of agamospermy in homoploid lineages (i.e., without genome duplication) or in hybrid polyploid individuals; or 3) through the spread of genes controlling asexual reproduction by the transfer of pollen from asexual to sexual lineages (*Genome Hijacking*—see the Genome Hijacking section). Genome hijacking is possible because agamospermous lineages retain viable pollen and the potential to fertilize the ovules of their sexual neighbors, resulting in asexual offspring. Hence, agamospermy could persist over long periods of time, as the genes controlling this trait are repeatedly transferred into novel genetic backgrounds (van Dijk 2003; Hojsgaard et al. 2015). All three of these mechanisms may contribute to the high levels of clonal diversity often observed in agamospermous species.

Geographic Distribution—Agamospermous species often differ with respect to the distribution of clonal lineages within and among populations, and at least in temperate regions, there appears to be two general patterns. In the first group, which includes weedy apomictic species such as *Taraxacum* (dandelion), populations are usually highly polyclonal, and individual lineages can be distributed over large geographic regions (van Dijk 2003). The less weedy agamospermous species tend to have a somewhat different distribution of clonal lineages. Like the weedy apomicts, these species often consist of many different clonal lineages, but populations more typically consist of a single clone (Hörandl et al. 2008). This pattern is particularly true for asexual species with distributions that extend further north and to higher elevations and for sexual species with agamospermous populations toward their range edges.

The tendency for agamospermous species to have wider and more northern geographic distributions has been referred to as *Geographic Parthenogenesis* and seems to be a common pattern for both the weedy and non-weedy species with gametophytic apomixis (Hörandl et al. 2008). While species with gametophytic apomixis often have wider geographic distributions than their sexual relatives, the same contrast does not hold for species with sporophytic apomixis, but it is also true that less is known about this latter mode of agamospermy. Within species with gametophytic apomixis, the non-weedy asexual taxa more often have ranges that extend to higher altitudes and further north into arctic regions, while weedy asexual populations tend to be more widespread and more invasive but do not typically display substantially broader ecological ranges (Hörandl et al. 2008). For example, triploid agamospermous populations of the weedy common dandelion (*Taraxacum officinale*) are distributed across multiple continents but always occur in disturbed habitats, while diploid sexual populations are restricted to southern Europe (van Dijk 2003). The agamosporous populations tend to have a more northern distribution in Europe and are generally more widespread, which is consistent with the hypothesis that the higher reproductive assurance conferred by agamospermy contributes to greater colonization success and wider geographic distributions. The observation that a smaller number of clones tend to occur over larger geographic areas has led to the suggestion that they represent *General Purpose Genotypes* that may have higher phenotypic plasticity or other attributes that contribute to their greater success (van Dijk 2003).

Whether the wider geographic and ecological ranges of agamospermous species are due to greater colonization ability or to the superiority of particular asexual genotypes remains unresolved. This is particularly true for species with gametophytic apomixis, because the presence of polyploidy and hybrid origins are generally confounded with agamospermy. In an attempt to disentangle these effects, Hörandl (2006) compared non-apomictic polyploid and diploid species pairs and found there was no difference in the breadth of their geographic distributions, and no single factor of hybridity or reproductive assurance (i.e., comparison to selfing species) completely explained the distributional differences between sexual and asexual populations. In a separate study, comparison of sexual and asexual species in the genus *Oenothera* with the same ploidy levels found that the asexual species had more northern distributions into formerly glaciated regions, suggesting that reproductive assurance was an important contributor to geographic parthenogenesis (Johnson et al. 2010). These studies suggest that reproductive assurance may promote geographic parthenogenesis, but they do not rule out the possibility that ecologically specialized and general purpose genotypes also contribute to the extended ranges of agamospermous lineages.

It is important to note that it is difficult to test the compound effects of factors that may lead to geographic parthenogenesis. The greater colonization ability of particular asexual lineages may be as much a function of their superior genotypic composition as it is a symptom of their mode of reproduction (i.e., because of reproductive assurance). Moreover, it may be that the answer to the dilemma of geographic parthenogenesis differs for weedy and non-weedy apomicts; improved colonization ability may be more critical for weedy plants, while the capture of novel genotypes through genome highjacking as clonal lineages may be an important mechanism for non-weedy species to extend their ranges into more challenging habitats. Further study will continue to improve our understanding of the processes contributing to geographic parthenogenesis and of the balance between sexual and asexual modes of reproduction in general.

Asexual Reproduction in Plants—The prevalence of sexual and asexual reproduction in plants depends on whether clonal expansion is occurring through vegetative growth or apomixis. On the one hand, asexual reproduction through vegetative growth does not impose a cost in terms of a plant's capacity for sexual reproduction, so it is not surprising that many sexually reproducing plants are also capable of extensive vegetative growth. Indeed, these plants have the potential to have the best of both worlds, as they can generate variable progeny through sexual reproduction, and successful genotypes retain the capacity to spread vegetatively to dominate a particular habitat. While there may be a tradeoff between resource allocation to sexual reproduction and vegetative growth, the cost of clonal spread is much less than for apomixis; each new ramet has the capacity to produce reproductive structures, so sexual reproductive success will tend to increase along with clone size, and the models for the cost of sex described earlier do not apply.

Once we recognize that there is not a high cost for clonal reproduction in plants, it is not surprising that the large majority of herbaceous perennials, and many woody shrubs and trees, tend to engage in a mixture of sexual and vegetative reproduction. For asexual reproduction through seeds, on the other hand, individual species and populations are usually completely sexual or reproduce primarily through agamospermy. While we generally do not find populations consisting of mixtures of sexual and agamospermous lineages, there are a few notable exceptions. In particular, both sexual and agamospermous populations of *Antennaria* (pussytoes) are common, and sometimes plants displaying both modes of reproduction occur in the same population (Bierzechudek 1990). It is also possible for individual plants to produce mixtures of sexual and asexual seeds. This is particularly true for species with sporophytic apomixis, since the megagametophyte is retained within the ovule and is available for fertilization. Completely asexual lineages may not exist because virtually all agamospermous plants appear to engage in low levels of sexual reproduction (Hojsgaard 2014).

Agamospermy versus Selfing—In contrast to vegetative reproduction, the conversion of sexual structures to serve as a means for clonal reproduction presents a direct tradeoff between the capacity for sexual and asexual reproduction. In these cases, we would expect that the reduced genetic transmission potential for sexual individuals would lead to the spread of agamospermous lineages, but this expectation differs for selfing and outcrossing populations. From the considerations of the cost of meiosis discussed earlier, we can predict that asexual reproduction would increase in frequency in outcrossing populations until ovules available for fertilization became limiting. Under this scenario, outcrossers would always be at a disadvantage, because asexual plants could fertilize their ovules. For selfing populations on the other hand, ovules would not be available for fertilization, so agamospermous lineages would not have a genetic transmission advantage. In other words, we expect that outcrossing populations would be susceptible to invasion and conversion to asexuality, but transitions to agamospermy from selfing should be uncommon.

The prediction that it would be difficult for agamospermy to spread in populations of selfing species appears to be supported by the distribution of apomixis across plants with different life histories. In particular, agamospermy is common in herbaceous and woody perennials, but rare in annuals. Coincidentally, we see that selfing is relatively common in annuals, less common in herbaceous perennials, and least frequent in long-lived woody species (Barringer 2007). Exceptions to these trends include the agamospermous annuals *Erigeron annuus* (annual fleabane) and *Aphanes arvensis* (parsley-piert; Mogie 1992) and the partially selfing species of *Scalesia* trees and shrubs in the Galapagos (daisy trees; Nielsen et al. 2003). Selfing and agamospermy are often viewed as equivalent strategies for reproductive assurance, and consequently, we would expect that both would be common among annuals, which are often weedy species that rely on colonization of ephemeral and disturbed habitats for persistence. Once an annual is highly selfing, we would predict that it would be resistant to conversion to agamospermy, but this does not address the question of why apomixis is infrequent in annuals.

Agamospermy may be less common in annuals simply because their life histories facilitate frequent transitions to self-fertilization. The shorter lifespans of annual plants correspond to fewer mitotic cell divisions between episodes of sexual reproduction, and consequently less potential for somatic mutation accumulation (Chapter 3). Under this hypothesis we expect that long-lived species would have greater genetic load due to their extensive vegetative growth, which possibly explains the low frequency of selfing in trees (Barrett 2017). Since transition to agamospermy results in *Fixed Heterozygosity*, the larger numbers of recessive deleterious somatic mutations accumulating in perennials would remain unexpressed. Coincidentally, the higher genetic load of long-lived species would present a barrier for the evolution of selfing (see section 9.7). Annuals, on the other hand, display more frequent recombination relative to the number of mitotic cell divisions, which would reduce the occurrence of somatic mutations and facilitate the purging of deleterious alleles. Since genetic load is expected to be generally lower in short-lived species, we can expect that transitions to self-fertilization would be more frequent in annuals. Once the transition is complete, selfing populations would be resistant to the spread of agamospermy because the genetic transmission of both reproductive strategies is equivalent.

Reversion to Sexual Reproduction—The expected lower fitness of dioecious and outcrossing hermaphroditic lineages due to demographic and genetic transmission costs presents an impediment for the spread of sexual reproduction in asexual populations. In a population of asexual individuals that produce functional pollen, it would be difficult for sexual reproduction to become established because of the cost of generating haploid gametes. The cost of meiosis could be overcome if the new sexual lineage was able to fertilize its own ovules; however, the spread of a selfing variant would be thwarted because its offspring are likely to display low fitness because of high genetic load due to the accumulation of somatic mutations. Expression of deleterious mutations in sexual progeny would be a particularly difficult problem to overcome if sexual individuals were derived from lineages that have long histories of asexual reproduction. In this case, the lack of recombination over many generations would have allowed for the accumulation of large numbers of deleterious recessive mutations that would be made homozygous in inbred offspring. A mutation for outcrossing would face similar difficulties due to *Biparental Inbreeding* among closely related individuals. The high bar for overcoming inbreeding depression due to somatic mutation accumulation may explain the rarity of sexual lineages emerging from asexual ancestors (Holsinger 2000). On the other hand, it is possible that transitions to sexuality could be favored if they were accompanied by polyploidy, which could alleviate the deleterious effects of accumulated somatic mutations, as they may be sheltered in offspring after segregation by dominant alleles (Hojsgaard 2014).

Genome Hijacking—While asexuality has often been viewed as an evolutionary dead end (Stebbins 1950; Grant 1981), some agamospermous lineages appear to have continued to diversify over extended periods of time (Holsinger 2000; van Dijk 2003; Hojsgaard 2014). For example, asexual lineages of buttercups in the *Ranunculus auricomus* complex originated during the late Miocene around 11 MYA, and aging based on divergence rates indicates that some modern lineages had hybrid origins around 70,000 years ago (Hojsgaard et al. 2015). One hypothesis for the long-term success of asexuality postulates that the genes controlling agamospermy can be transferred into sexual lineages via pollen. Under this scenario, advantageous genomic combinations are effectively captured by the integration of genes for asexual reproduction. The genes controlling agamospermy could be quite old, because they may have been repeatedly transferred into novel genetic backgrounds that confer adaptation to novel habitat conditions—a process we describe as genome hijacking.

The primary difficulty with the hypothesis that genomes are hijacked from sexual lineages by genes conferring asexuality is that agamospermous populations are often geographically displaced from their sexual relatives (van Dijk 2003; Hojsgaard et al. 2015). For example, in the case of *Taraxacum*, invasive agamospermous populations occur on multiple continents, whereas sexual populations are only found in Europe. Similarly, agamospermous populations of *R. auricomus* are distributed in arctic regions where sexual lineages are absent. In both of these cases, agamospermous plants retain pollen viability and function, so rare outcrossing could occur among asexual lineages. On the other hand, levels of clonal diversity toward range edges may be limited and may not be adequate to explain the broad range of habitats that agamospermous lineages have colonized.

The second difficulty with the genome-hijacking hypothesis is that genes controlling aspects of reproduction do not display patterns of evolution that would indicate that they are significantly older than the rest of the genome. If asexuality genes continually spread by capturing sexually generated genomes, then their patterns of mutation accumulation should reflect a history of prolonged asexuality (i.e., relatively high numbers of synonymous substitutions) compared to the rest of the genome. In contrast to this expectation, genes associated with reproduction in asexual lineages of *R. auricomus* do not have high levels of synonymous substitutions (Pellino et al. 2013). Instead, they are characterized by high K_A/K_S ratios (Hojsgaard et al. 2015; Chapter 3), which indicates that they were subject to selection as they diversified within agamospermous lineages. Similarly, asexual lineages of *Oenothera* have diverged from their sexual relatives by having reduced investment in floral structures (Johnson et al. 2010). Such reductions in investment in reproductive structures may be due to the accumulation of somatic mutations that reduced investment in flowers, which would have been favored by selection as resources were allocated to increase fitness through asexual reproduction. Note that, in both these cases, it is the asexual lineages that appear to display stronger signs of histories of selection and divergence than their sexual cousins. These observations suggest that agamospermous lineages are capable of undergoing some degree of adaptation and diversification, so they should not necessarily be considered as inevitable dead ends. As discussed in Chapter 3, the potential for cell lineage selection to promote clonal evolution and adaptation is a potentially important, but sorely understudied, aspect of plant evolutionary biology.

Summary—Asexual reproduction is an ancestral condition and remains prevalent across the embryophytes. Clonal propagation through vegetative growth and fragmentation is common among herbaceous perennials and some groups of woody plants, but this characteristic has been lost in many groups of trees that grow in forest habitats where horizontal spread would generally not be advantageous. For those species that retain the capacity for clonal growth, vegetative reproduction contributes to overall plant fitness and does not represent a tradeoff with sexual reproduction through pollen and seeds.

In contrast, apomixis coopts normally sexual structures for asexual reproduction through bulbils or agamospermous seeds. While we predict that populations of outcrossing species would always be susceptible to invasion and appropriation by mutations for asexual reproduction, agamospermy remains uncommon across all families of angiosperms. From our discussion of models for the evolution of sex in the previous section, we can conclude that sexual reproduction—and recombination in particular—should be advantageous under a broad range of conditions. Agamospermy appears to be mostly limited to cases of high ploidy and hybridization, or circumstances where reproductive assurance is a priority. Agamospermy is more common in herbaceous and woody perennials and rare in annual species. In many cases, it appears that agamospermy has effectively captured genotypes that are locally advantageous or are general purpose genotypes successful across broader geographic regions. Because of recurrent origins of agamospermy along with genome hijacking and the accumulation of somatic mutations, the genetic diversity of asexual species tends to be quite high. Agamospermous lineages appear to have some capacity for clonal adaptation and diversification over moderate evolutionary timescales but do not display the same potential for long-term persistence as their sexual cousins.

While agamospermy has sometimes been viewed as similar to selfing for the purpose of reproductive assurance, these two modes of reproduction are far from equivalent. Some of the same impediments that apply to the cost of sexual reproduction also apply to the evolution of selfing, but there are additional obstacles due to the expression of genetic load in selfed offspring. In the next section we will explore the advantages and disadvantages of selfing and outcrossing, conditions that favor the evolution of selfing, and the evolutionary consequences of each mating strategy.

9.7 SELFING AND OUTCROSSING

The shift to self-fertilization is considered one of the most frequent evolutionary transitions occurring among flowering plant lineages. This is impressive because the high level of genetic load that is typical of an outbreeding population would represent a substantial impediment that must be overcome for selfing to have an advantage over outcrossing. Moreover, the evolution of a primarily *Autogamous* mating system is usually associated with a number of changes in floral traits including a reduction in flower size and pollen production, and evolution of autogamy is often accompanied by a transition to an annual life-history from perennial ancestors (Wright et al. 2013; Chapters 6 and 10). As discussed in Chapter 11, selfing lineages display higher rates of speciation—presumably because of their higher level of reproductive isolation—but this is counterbalanced by higher rates of extinction. Consequently, the diversity of selfing clades remains low, and extant selfing lineages tend to have recent origins as they occur primarily on short phylogenetic branches.

In spite of the relatively high frequency of evolution toward selfing, we do not have any clear examples of lineages that appear to be at intermediate stages of this transition. Furthermore, attempts to elicit purging of genetic load under controlled conditions often result in only modest reductions in inbreeding depression, and it has been difficult to demonstrate evidence of substantial purging by selfing or inbreeding within populations (Byers and Waller 1999; Crnokrak and Barrett 2002). This may not be surprising given that theoretical treatments have predicted that purging would be extremely difficult if inbreeding depression was caused by large numbers of recessive alleles of small effect (Goodwillie et al. 2005), yet this appears to be likely for most species (Lynch et al. 1999; Charlesworth and Willis 2009). Weakly deleterious mutations are difficult to remove from populations, because inbreeding reduces the effective population size and consequently, lowers the efficacy of selection for removing deleterious alleles (Chapter 5). Hence, populations rapidly become fixed for large numbers of deleterious alleles, and the process of purging becomes stalled (Fig. 5.20). As predicted by the models of purging in a metapopulation (Whitlock 2002), simulations conducted by Marchini et al. (2016) demonstrated that continued progress toward lower genetic load depended on gene flow among populations fixed for different combinations of deleterious alleles. Purging of genetic load through gene flow among inbred populations may be occurring for species invasions after multiple introductions from divergent sources in the case of *Brachypodium sylvaticum* (Fig. 5.19; Marchini et al. 2016), but as of yet there is no evidence of the evolution of increased selfing in this species.

Explanations for the frequent evolution of selfing have been made from both the point of view of the advantages accrued and the stance that selfing avoids disadvantageous circumstances (Uyenoyama 1986). On the positive side, we expect that a selfing variant would have a 1.5-fold (50%) transmission advantage over outcrossing. This is the same argument we discussed for the cost of meiosis (cost of outcrossing) advantage of asexual over sexual reproduction (Fig. 9.5—replacing selfing for asexual). From a more negative point of view, selfing has been proposed as a mechanism for reproductive assurance in unfavorable pollinator environments (Barrett 2017). Under this scenario, we expect mutations that elevate the propensity for selfing to increase in frequency when pollinators are unreliable. Whether plant mating systems are the outcome of the balance between automatic selection based on genetic transmission versus inbreeding depression in selfed offspring or are a product of ecological circumstances, remains as an active topic of discussion. In this section, we discuss models for the evolution of self-fertilization and the circumstances that may favor selfing, outcrossing, and mixed-mating strategies.

Historical Views on Selfing and Outcrossing—Prior to the 1800s it was widely assumed that the cosexuality of flowers was a convenient mechanism to guarantee the purity of each species through self-fertilization (Vogel 1996). During this time, intercrossing and hybridization were viewed as anomalies that resulted in mongrel offspring; a conclusion that was a natural outcome of the commonly held typological concept of species (Chapter 11). Like other naturalists of the late seventeenth century, Christian Konrad Sprengel was well aware that insects often visited flowers, but he, like others of his day, viewed this as superficial to the primary purpose of self-pollination within the flower (autogamy), which was assumed to occur naturally with or without the participation of insects. Nevertheless, Sprengel became intrigued by the intricate morphology of flowers and engaged in a series of studies of flowers and their visitors. These studies led him to conclude that insects served to ensure the reproduction of plants by transferring pollen between flowers. While the sexual role of flowers had been recognized since the time of Linnaeus, it was not until the publication of Sprengel's careful observations that the function of flowers for the attraction of insects to serve as pollen vectors was laid clear by Christian Konrad Sprengel (1793; for an English translation see Sprengel 1996). Unfortunately, Sprengel's monograph on "Discovery of the secret of nature in the structure and fertilization of flowers" and other similar studies from earlier in the eighteenth century were mostly ignored—largely because they were not consistent with the widely held view that the primary function of flowers was for the continuance of the lineage through self-fertilization.

According to Hermann Müller (1883), Sprengel's failure to demonstrate an advantage for cross fertilization diminished the relevance of his observations. In spite of the fact that several scholars had reported results of experiments that indicated the superiority of outbred progeny (e.g., Knight 1799), the significance of cross-pollination was not appreciated until the publication of Charles Darwin's *Origin of Species* (Darwin 1859; Chapter iv. On the Intercrossing of Individuals). As described by Müller (1883, page 5), "Darwin . . . emphasised Knight's law as a general law of nature, placing it on broader and surer foundations and uniting it intimately with his theory of natural selection." Far Darwin, Sprengel's observations took on particular significance, and because of his emphasis of this work, Sprengel's book emerged into the limelight after seventy years of obscurity.

For Darwin, Sprengel's observations served to solidify his conclusion that many of the characteristics he observed in flowers represented adaptations that favor crossing over self-pollination, which he considered to be generally injurious (Darwin 1876). Darwin did not reach this conclusion casually; in fact, he was surprised and somewhat skeptical when he first observed reduced vigor in offspring resulting from self-pollination in a small crossing experiment that he had conducted in his garden. Darwin was intrigued with these results and was motivated to proceed with an impressive series of meticulous experiments and observations that included more than fifty plant species and spanned a period of thirty-seven years. In the large majority of cases, he found that outcross pollinations produced more seeds per fruit and that outcrossed progeny consistently outperformed those from self-pollinations. In some instances, he discovered species to be self-sterile, and he noted that others would preferentially self-fertilize (i.e., they were autogamous). The series of experiments that he summarized in his book, *The Effects of Cross and Self Fertilization in the Vegetable Kingdom* (Darwin 1876), firmly established the conclusion that selfing, and inbreeding in general, usually resulted in progeny with diminished vigor and reduced chances of survival.

Sprengel's work enjoyed a bit of a renaissance in the mid to late nineteenth century, and it inspired Darwin and a number of his contemporaries such as Hermann Müller to engage in studies of the interactions between flowers and their insect visitors (Vogel 1996). Most of these naturalists were advocates for Darwin's new theory of evolution by natural selection, and the floral characters that appeared to promote outcrossing provided fertile ground for gleaning examples of adaptive evolution. Note that this view is a complete about-face from mainstream views in the earlier part of the nineteenth century, when it was assumed that the primary function of cosexual flowers was self-fertilization. Much of this early work is summarized in *The Fertilization of Flowers* (Müller 1883), which presented a thorough analysis of plant reproduction and the advantages of outcrossing, but also recognized that selfing provided some assurance for seed production when pollinators were unreliable (reproductive assurance). In spite of Müller's suggestion of a function for self-fertilization, most studies of pollination biology continued to emphasize the presumed advantages of outcrossing through the late 1900s, but researchers lacked the tools to quantify the efficiency of outcrossing mechanisms. This situation changed in the 1970s when molecular genetic markers (i.e., allozymes; Chapter 4) came into general use and provided a convenient tool for estimates of outcrossing frequencies (Ritland 2002; Barrett 2017).

During the late 1900s, there was a resurgence in interest in pollination biology, and many of these studies were enhanced by new methods for the estimation of outcrossing frequencies with genetic markers. Interest in plant mating systems during this period was propelled by a pair of influential papers published by Russell Lande and Doug Schemske that predicted stable equilibria should exist for nearly complete outcrossing or complete selfing, and mixed mating should be rare (Lande and Schemske 1985; Schemske and Lande 1985). The observation that intermediate outcrossing frequencies were much more common than predicted by Lande and Schemske's model occupied the attention of plant evolutionary biologists for several decades, and explaining the prevalence of mixed mating remains as a major focus of studies in plant reproductive biology. In the next section, we discuss floral mechanisms and modes of selfing and outcrossing in cosexual flowers; in the following sections, we explore the models for the evolution of plant mating systems and the conditions that might serve to explain the commonness of mixed mating.

Modes and Mechanisms of Self- and Outcross-Pollination—The large majority of flowering plants are hermaphrodites with cosexual (bisexual; perfect; *Monoclinous*) flowers (Renner 2014). Cosexuality presents a challenge for species with high genetic loads because the close proximity of anthers and stigmas within the same flower would be likely to promote self-pollination (autogamy). Furthermore, pollen transfer among flowers within the same plant, or among separate ramets of the same genet, is also self-pollination (geitonogamy). Cosexual flowers often display morphological and developmental characteristics that appear to reduce levels of self-pollen deposition, and these same mechanisms may reduce interference between the pollen dispersal and receipt functions of the flower.

A variety of characteristics of flowers that appeared to promote outcrossing caught the attention of Sprengel, Darwin, Müller, and many other biologists of the nineteenth and twentieth centuries. Most of these features represented differences in the timing (*Dichogamy*) or placement (*Herkogamy*) of pollen dispensing and receipt within the flower and appeared to be mechanisms to discourage self-pollination. Dichogamy has been used as a general term referring to the displacement of sex expression over time and is used to describe sequential hermaphroditism over the lifespans of animals. The term originates from Sprengel (1793) who applied it to describe the timing of anther dehiscence and stigma receptivity, and in plants its usage is restricted to the timing of sex expression within individual flowers.

Since flower parts in the androecium and gynoecium represent modified leaves originating from different nodes of a compressed stem (Chapter 2), the natural sequence of maturation proceeds from outside whorls toward the center of the flower (i.e., up the stem). This default pattern of maturation results in the release of pollen from anthers prior to stigma receptivity (*Protandry*; Fig. 9.11; refer to color plate). The natural patter n of maturation of flower parts results in some overlap between pollen dispersal and stigma receptivity, but rarely is the timing of these two functions simultaneous. More often, the degree of protandry is exaggerated such that nearly all the pollen has been dispersed prior to the initiation of stigma receptivity (Stout 1928). In a minority of species, the maturation sequence is reversed so that the gynoecium matures prior to pollen dehiscence (*Protogyny*). Both of these terms fall under the general category of dichogamy, and were originated by the German botanist, Friedrich Hildebrand, in the late 1800s (Stout 1928).

The efficiency of dichogamy as a mechanism to promote the receipt of outcross pollen has been questioned, because most dichogamous plants produce multiple flowers that may open over a time frame of several days or weeks (Lloyd and Webb 1986). As originally noted by Sprengel, because of the potential for geitonogamy, dichogamy would be most effective for the promotion of outcross pollination for those flowers at the very beginning or end of the flowering sequence, or for plants that bear their flowers one at a time (Stout 1928).

The inability of dichogamy to prevent geitonogamy has led to the development of alternative hypotheses for its function. Most notably, using phylogenetic comparisons, Routley et al. (2004) found strong associations between dichogamy and the presence of self-incompatibility. Most species displaying protandry were also self-incompatible, while protogynous plants mostly lacked incompatibility mechanisms. Since most protandrous flowers could prevent self-fertilization through rejection of self-pollen, these authors interpreted this pattern to indicate that the primary function of protandry was to reduce *Sexual Conflict* within a flower and suggested that pollen export would be more effective for flowers that are pre-receptive. Protandry could be particularly effective as a mechanism to reduce interference between functions for pollen dispersal and receipt for species having large or lobbed stigmas (e.g.,

Fig. 9.11. An example of dichogamy (protandry) in *Geranium maculatum*. When flowers first open the stigmas are closed and pollen is dehisced from the anthers (left panel). Stigma lobes separate and curl backwards in older flowers after most of the pollen has been dispersed (right panel). At this stage the stigma lobes may contact anthers resulting in delayed selfing. Photos by Robert Klips (http://bobklips.com).

as in *Geranium maculatum*; Fig. 9.11; refer to color plate). Protogyny, on the other hand, was associated with self-compatibility suggesting that it may be favored as a mechanism to reduce self-pollination. This suggestion seems more feasible when we consider that insect pollinators arriving at a plant will be more likely to deposit large amounts of outcross pollen on the first few flowers that they visit. In this case, protogynous flowers may be more effective at preventing self-pollination because protandrous flowers could have substantial amounts of self-pollen remaining during the time of stigma receptivity, which would promote higher levels of autogamy.

The spatial separation of anthers and stigma within flowers is generally referred to as herkogamy and can occur in two arrangements. *Approach Herkogamy* refers to the placement of the stigma in a position that is further exerted from the receptacle than the anthers, and in *Reverse Herkogamy* the anthers are placed above (further exerted) than the stigma (Fig. 9.12; refer to color plate). Distylous species such as *Primula vulgaris* display *Reciprocal Herkogamy* (Fig. 9.12; refer to color plate), with both approach and reverse herkogamy flower morphs and individual plants producing all of one morph or the other. The large majority of plants are homomorphic (i.e., one flower morph) with approach herkogamy. Approach herkogamy is thought to promote outcrossing, because pollinators will be more likely to contact the stigma before the anthers as they begin to forage on a flower (Webb and Lloyd 1986; Luo and Widmer 2013). Approach herkogamy is particularly common among bird- and bat-pollinated species, where stigmas are often well exerted beyond the opening of the floral tube. Reverse herkogamy is more common in moth- and butterfly-pollinated flowers that have narrow floral tubes.

The effectiveness of herkogamy for the prevention of self-pollination is generally determined by the amount of stigma-anther separation (Luo and Widmer 2013). This appears to be especially true for a wide range of species displaying approach herkogamy, where the physical separation between the stigma and anthers is strongly associated with the proportion of outcrossed seed produced. The effect of stigma-anther separation is much less clear for plants having reverse herkogamy, which generally experience much higher levels of self-pollen deposition on stigmas. In these species, the arrangement of anthers and stigma within the flower may provide other advantages. For example, Webb and Lloyd (1986) point out that herkogamous species are often self-incompatible and suggest that spatial separation of the anthers and stigma might be a mechanism to reduce interference between pollen dispersal and receipt (sexual conflict). This proposal is supported by studies demonstrating that plants with reverse herkogamy (exerted anthers) often export larger amounts of pollen than those with approach herkogamy and consequently, fertilize more ovules on neighboring plants (Kohn and Barrett 1992; Luo and Widmer 2013; Keller et al. 2014). Given these results and those discussed above, it appears that both dichogamy and herkogamy may often function to promote outcrossing as well as to reduce interference between the pollen dispersal and receipt functions of the cosexual flower.

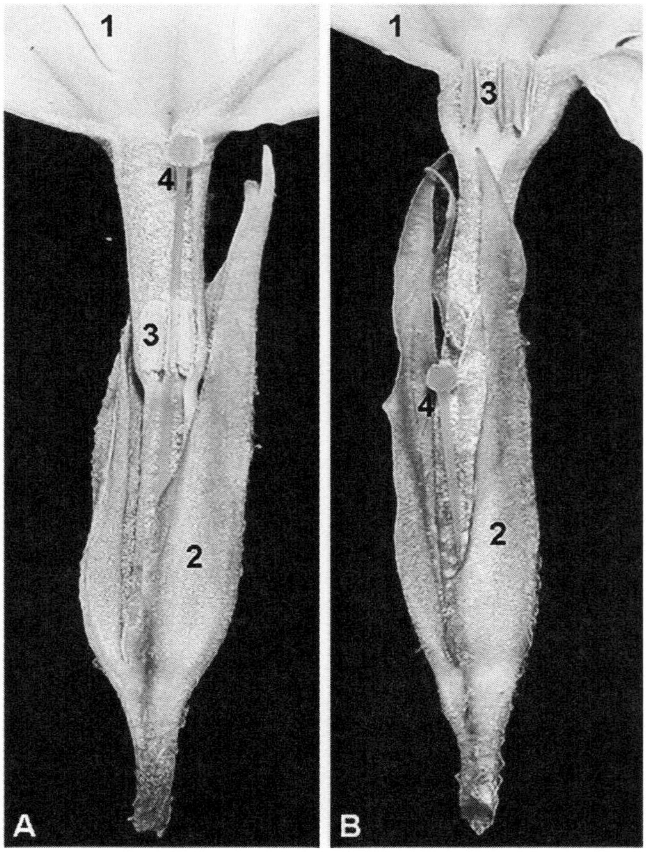

Fig. 9.12. Reciprocal herkogamy in *Primula vulgaris*. Flowers of this species have a narrow floral tube (1) surrounded by a calyx (2). In approach herkogamy the stigma (4) is placed above the anthers (3; left panel; long-styled morph). In reverse herkogamy the anthers are near the opening of the floral tube and the stigma is below. Photos and graphics by Berru.

Self-Incompatibility—Mechanisms for self-incompatibility (SI) are present in about 60% of the flowering plants (Hiscock and Tabah 2003) but are absent in the gymnosperms and other embryophytes (Chapter 2). In species with self-incompatibility, the potential for fertilization depends on the match between the alleles carried by pollen at self-incompatibility loci (S alleles) and the corresponding S alleles expressed in the recipient style (Fig. 9.13). Self-incompatibility is broadly categorized into three groups: homomorphic species that 1) inhibit self-pollen during germination on the stigma (*Sporophytic Self-Incompatibility; SSI*) or 2) during pollen-tube growth (*Gametophytic Self-Incompatibility; GSI*), and 3) heteromorphic species having incompatibility systems associated with distyly or tristyly (*Heteromorphic Self-Incompatibility; HSI*). These classifications are largely artificial, because incompatibility has been independently derived multiple times within each category (Barrett and Cruzan 1994; Allen and Hiscock 2008; Igic et al. 2008). GSI, for example, can be divided into several different groups. RNase-based incompatibility occurs in at least thirty-six families encompassing the large majority of the eudicots, including the Solanaceae, Rosaceae, and Plantaginaceae. Other forms of GSI occur in a few families outside the large Asterid and Rosid clades of the eudicots (e.g., Papaveraceae and Poaceae; Fig. 9.14; refer to color plate; Igic et al. 2008). Both SSI and HSI occur in multiple families (ten and twenty-five, respectively) that are distributed across the angiosperm phylogeny (Fig. 9.14; refer to color plate). The occurrence of SSI and HSI on widely scattered branch tips indicates that these incompatibility systems have multiple

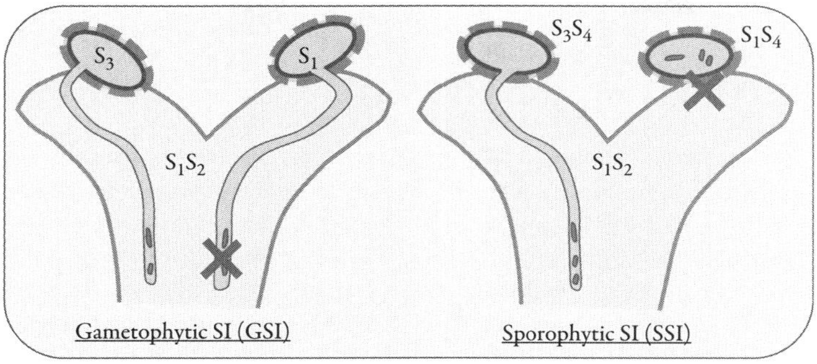

Fig. 9.13. Homomorphic self-incompatibility (SI). In gametophytic SI (GSI) the incompatibility reaction is based on the matching between the haploid S genotype of pollen and the diploid genotype of the style. Microgametophytes having S alleles that match the style are inhibited during pollen-tube growth (X). In sporophytic SI (SSI) the incompatibility reaction is based on proteins deposited in the pollen grain wall by the pollen parent. In this case a match with either of two alleles carried by the pollen will cause an incompatibility reaction and prevent germination (X).

origins and that they have been derived much more recently than RNase-based GSI. In all, some form of self-incompatibility occurs in at least ninety-four families and appears to have evolved independently at least thirty-five times (Igic et al. 2008). Nevertheless, within each broad group (i.e., GSI, SSI, and HSI), incompatibility systems tend to share similar genetic determination of self-incompatibility, and species having HSI share similar morphological characteristics of reproductive structures for each floral morph across divergent species.

Plants having GSI, SSI, and HSI differ with respect to the stage of microgametophyte development in which incompatibility is expressed (pollen germination and growth through the style; Fig. 9.13). For species with SSI, incompatible pollen is inhibited during germination on the stigma, whereas incompatibility is expressed later during pollen-tube growth for species having GSI (Hiscock and Tabah 2003). Incompatibility for plants with SSI is based on the diploid genotypes of the pollen donor and recipient and is controlled by proteins deposited in the exterior of the pollen grain's exine wall by the pollen parent. If either of the two alleles carried by a pollen grain match the recipient genotype, then the stigma papillae adjacent to the grain will typically exhibit callose deposition, and germination is prevented.

Once pollen has germinated, the cytoplasm is evacuated from the grain and migrates along with the pollen-tube tip during growth. Initially, growth through the style relies on stored resources but then transitions to a parasitic stage during which the pollen tube absorbs resources provided by the cells of the transmitting track in the style. For plants in the poppy family (Papaveraceae), which exhibit glycoprotein-based GSI, the incompatibility reaction occurs very soon after germination, so incompatible pollen-tubes will not penetrate very far into the stigma (Franklin-Tong and Franklin 2003). For the large majority of plants with RNase-based GSI, the incompatibility reaction does not occur until pollen tubes have transitioned into the parasitic stage of growth, which generally occurs in the upper portion of the style. For all types of GSI, the incompatibility reaction is based on the genotype of the haploid microgametophyte (Fig. 9.13). If the S allele matches either of the two alleles expressed in the recipient style, the growth of the pollen tube slows as it becomes thicker and exhibits increased callose deposition. Eventually incompatible pollen tubes cease growing and are thereby prevented from reaching the ovary. Compatible pollen tubes carry alleles not matching the stylar genotype, so they are fully capable of fertilizing ovules and will continue to grow toward the ovary. Note that with GSI, the diploid pollen donor could have one of two alleles that match one of the alleles expressed by a recipient, in which case around half of the pollen it produces would be compatible.

Heteromorphic incompatibility systems (HSI) display a much greater diversity of incompatibility reactions, but they also share a number of commonalities. Heteromorphic species are either distylous (two floral morphs; Fig. 9.12; refer to color plate) or tristylous (three floral morphs; Fig. 11.12). As previously described, flower morphs display reciprocal placement of the stigma and anthers within flowers. In distylous species, the expression of floral morphological characters and incompatibility are determined by a single locus that segregates for two alleles (Barrett and Cruzan 1994). One allele is completely dominant to the other and usually determines the short-styled morph (Ss), while two recessive alleles are carried by plants having flowers of the long-styled morph (ss; referred to as the thrum and pin, respectively, by Darwin 1877). When inter-morph incompatibility is strong, all crosses will be heterozygotes (short-styled plants) by homozygous recessives (long-styled plants; Ss x ss) and will naturally segregate into 50% short- and 50% long-styled offspring. Tristyly is determined by two loci (S and M) with one allele dominant at each locus (Barrett and Cruzan 1994). A dominant allele at the M locus determines the mid-styled morph (ssMm), and a dominant allele at the S locus determines the short-styled morph. The dominant S allele is epistatic and dominant to both alleles at the M locus, so any genotype having the dominant S allele will be short-styled (e.g., Ssmm and SsMm). The long-styled morph is homozygous recessive at both loci (ssmm). Just as for distyly, strict inter-morph mating will lead to equal morph ratios among offspring.

In general, pollen morphology in heterostylous plants is associated with anther length. Across heterostylous taxa, shorter anthers generally produce larger numbers of smaller pollen grains while longer anthers produce fewer and larger grains. The degree of stigma-anther separation often corresponds to differences in pollen size and can be associated with the strength of the incompatibility reaction (Barrett and Cruzan 1994). Across species with HSI, incompatibility reactions can occur during pollen germination or pollen-tube growth, but it is generally true that the site of the incompatibility reaction will differ among style morphs within a species. Since the compatibility of pollen depends on the donor genotype, HSI is effectively sporophytic even though the incompatibility reaction may occur during pollen-tube growth (i.e., when the haploid genome of the microgametophyte is expressed). Given the multiple independent origins of HSI (Fig. 9.14; refer to color plate), mechanisms of incompatibility most likely differ among floral morphs and across heteromorphic taxa and has not been well characterized for any individual species.

Hetermorphic and homomorphic incompatibility systems share the same tendency for frequency-dependent selection (balancing selection; Chapter 5), which maintains S alleles at similar frequencies within populations. As long as incompatibility is strong, any S allele that is infrequent will automatically have a selective advantage, because plants carrying rare alleles will be compatible with a larger proportion of recipients in the population. The tendency for S alleles to be resistant to loss from populations by genetic drift renders them persistent within and among lineages over long time scales (Kohn 2008). Additionally, because individuals are only compatible with mates carrying different S alleles, the S locus will always be heterozygous in homomorphic species, and no individuals will be homozygous for the dominant S or M alleles in heteromorphic systems. The maintenance of heterozygosity allows for the accumulation of neutral variation and deleterious recessive mutations in the region that is tightly linked to each S allele (Llaurens et al. 2009). The sheltering of deleterious mutations in the region of incompatibility loci may render species with SI more resistant to the evolution of selfing because of the potentially high level of inbreeding depression associated with homozygosity at the S locus.

Cryptic Self-Incompatibility—A number self-compatible species possess mechanisms that favor outcross- over self-pollen when both are deposited on the same stigma. This phenomenon is referred to as *Cryptic Self-Incompatibility* (*CSI*) because these plants are fully self-compatible and preferential outcrossing can only be detected after pollination with mixtures of self- and outcross-pollen (Bateman 1956; Cruzan and Barrett 1996). The differential success of self- and outcross-pollen can be due to either the failure of pollen during germination, pollen-tube growth, or fertilization of ovules (*Pollen-Tube Attrition*; Cruzan 1989) or to different rates of pollen germination or pollen-tube growth through the style (*Pollen Competition; Selective Fertilization; Certation*; Jones 1928).

Pollen-tube attrition and pollen competition both contribute to CSI, and their effects on outcrossing frequencies depend on both the size and composition of the pollen load on the stigma (Fig. 9.15). The difference in pollen-tube attrition between self- and

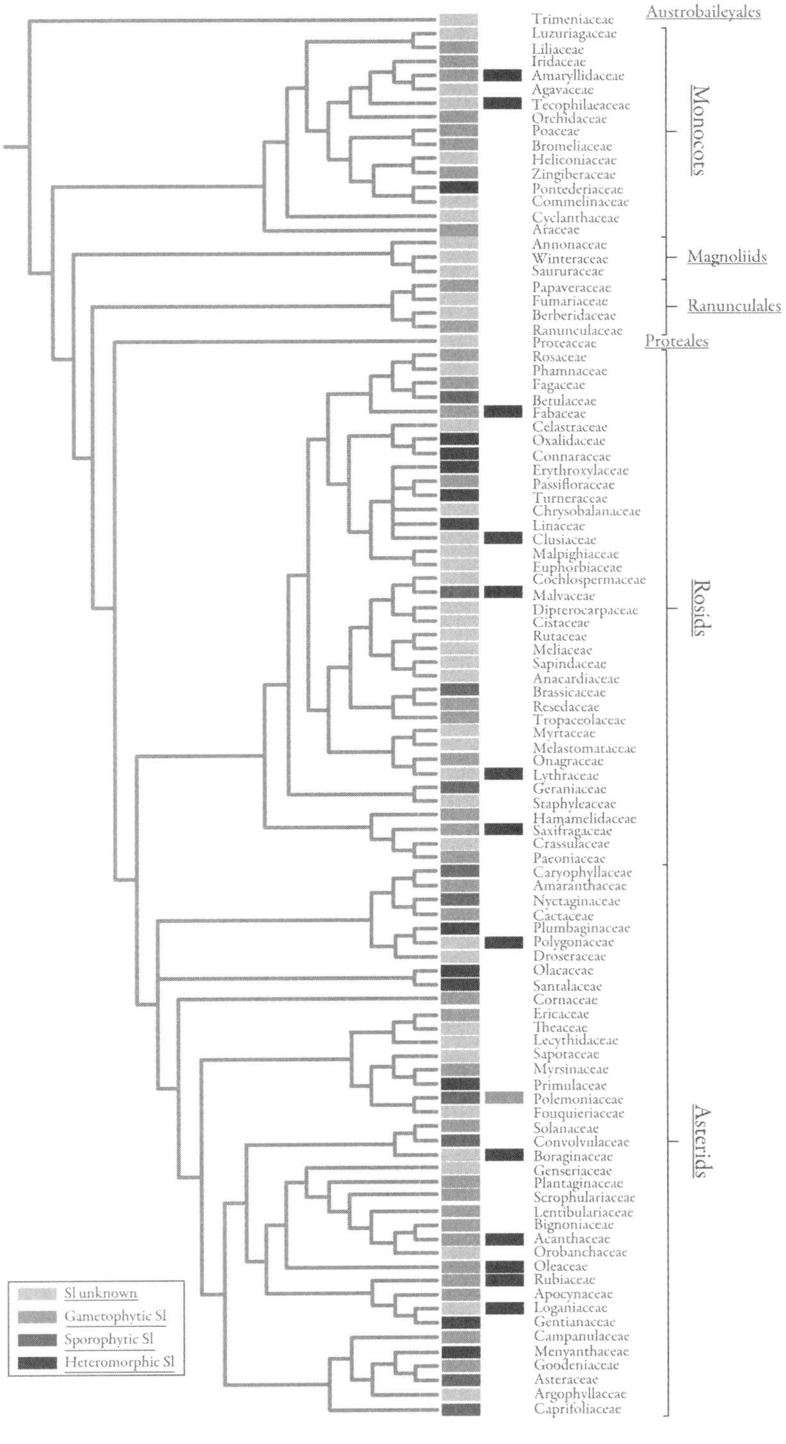

Fig. 9.14. The distribution of self-incompatibility (SI) across 105 angiosperm families. Color codes indicate the modes of SI. Families with more than one form of SI show two codes. Redrawn from Igic et al. (2008).

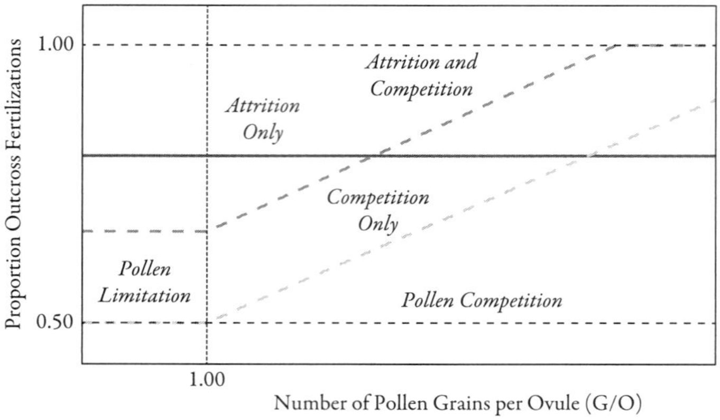

Fig. 9.15. The contributions of pollen competition and pollen-tube attrition to cryptic self-incompatibility (CSI) after pollination with mixtures of self and outcross pollen (50% of each in these scenarios). Under pollen limitation ($G/O < 1.0$) only differences in pollen-tube attrition affect the fertilization success of self- and outcross-pollen, while the effects of pollen competition due to differences in pollen-tube growth rates are seen when $G/O > 1.0$. Three scenarios are shown; 1) only pollen competition affects fertilization success (short dashed line), 2) only attrition affects fertilization success (solid line), and 3) both attrition and competition affect fertilization success (long dashed line). Redrawn from Cruzan and Barrett (1996).

outcross-pollen reflects the rates of failure of each pollen type and will have a consistent effect on the outcrossing frequency, regardless of pollen load size. The effect of pollen competition, on the other hand, depends on pollen-tube growth rates; a small advantage for outcross pollen is exaggerated when pollen load sizes are larger, because the outcross pollen-tubes can secure a larger majority of fertilizations before self-pollen-tubes reach the ovary. Hence, outcrossing frequency will increase with pollen load size, but seed set is still assured when pollen is limiting, because self-pollen is capable of fertilizing ovules. Species having CSI that is due to pollen competition effectively have the best of both worlds: they can produce mostly outcross progeny when pollinator activity is high but are capable of reproductive assurance through selfing when the availability of outcross pollen is limited.

Late-Acting Self-Incompatibility—The potential for self-pollen to fail during pollen-tube growth in the ovary, during the fertilization of ovules, or after fertilization has been collectively referred to as *Late-Acting (LSI)* or *Ovarian Self-Incompatibility* (Seavey and Bawa 1986; Gibbs 2014). The proposal that SI can act after fertilization is based on the observation of higher frequencies of abortion for selfed ovules and fruits. On the other hand, a number of authors have pointed out that higher frequencies of ovule abortion after selfing could be due to the effects of inbreeding depression expressed during seed development (*Early-Acting Inbreeding Depression—EID*; Krebs and Hancock 1990; Husband and Schemske 1996). From this perspective, it would be difficult to distinguish LSI from the effects of deleterious recessive mutations, since both would result in the abortion of self-fertilized ovules. On the other hand, there is evidence that LSI has a simple genetic determination that is similar to incompatibility expressed during pollen germination and growth (SSI and GSI). For example, in the case of poke milkweed (*Asclepias exaltata*), crosses among full sibs revealed pairs that were either fully compatible or incompatible, suggesting that they were segregating for alleles at a single locus that were responsible for LSI (Lipow and Wyatt 2000).

Although patterns of ovule abortion appear to segregate according to expectations for a simple incompatibility system in some cases, it is generally difficult to discriminate between the hypotheses that LSI or EID is responsible for selfed ovule abortion. We note the EID explanation is a general outcome of the accumulation of genetic load, which is a natural outcome of mutation and genetic drift. On the other hand, LSI requires the origin and maintenance of a specialized self-recognition system that is only expressed in the ovary. Furthermore, the phylogenetic distribution of LSI supports the hypothesis that post-fertilization discrimination against self-fertilized ovules is due to genetic load in most cases. Indeed, there is a negative association between taxonomic groups having mechanisms of pre-zygotic SI and those that display the symptoms of LSI (Gibbs 2014). Moreover, the clades having post-zygotic discrimination against selfing are generally trees and shrubs or long-lived geophytes in the Liliaceae, Iridaceae, and Asparagaceae. Furthermore, species in the Fabaceae display both GSI and LSI, but all of those having post-zygotic abortion of selfed ovules are woody, long-lived species. Hence, the lack of self-fertility in species displaying higher frequencies of selfed ovule abortion is consistent with the prediction that long-lived species have greater capacity for the accumulation of deleterious somatic mutations (Chapter 3). While some herbaceous perennial genera such as *Aconitum* (monk's hood) in the Ranunculaceae and *Ipomopsis* (gilia) in the Polemoniaceae apparently possess LSI, mechanisms promoting selfed ovule abortion are completely absent among annual species.

Support for the hypothesis that ovule abortion is more often due to inbreeding depression is evident from the lack of consistency in discrimination against self-fertilized ovules. If putative cases of LSI were due to an incompatibility system analogous to GSI and SSI, then we would expect rejection of selfed embryos regardless of whether outbred ovules are developing in the same fruit. In contrast, many of the species that have been proposed to possess LSI abort fruits from flowers that have been self-pollinated but will maintain mature fruits containing mixtures of selfed and outcrossed seeds (Gibbs 2014). Since we would expect the rejection of selfing to occur for individual embryos, the observation that viable self-fertilized seeds are produced after flowers are pollinated with mixtures of self and outcross pollen is inconsistent with the presence of ovarian incompatibility. The weight of the available data suggests that most, if not all, putative cases of LSI are due to inbreeding depression and selective abortion of fruits containing high proportions of self-fertilized ovules. Presumably, maternal resources that might have been wasted on low-fitness offspring in selfed fruits can be reallocated for other functions contributing to fitness (Wiens et al. 1987).

The Effectiveness of Outcrossing Mechanisms—Transfer of pollen within and among plants by biotic and abiotic vectors creates opportunities for plants to increase reproductive success as pollen donors and recipients but may also induce costs in terms of lost opportunities for outcrossing and reduced fitness due to inbreeding depression of selfed offspring. An ideal pollen vector would be one that methodically visited only single flowers on each plant (i.e., eliminating geitonogamy) as it transferred pollen from one plant to the next, but the reality is that pollen is dispersed among nearby flowers whether they are borne by the same or separate plants. Moreover, as noted in Chapter 10, bees often engage in grooming as they move between flowers, so much pollen can be lost or displaced to portions of the insect's body where it would be unlikely to come into contact with receptive stigmas. The foraging behavior of animal pollinators is generally not ideal for plant reproductive success, and many characteristics of flowers appear to be compromises designed to take the best possible advantage of the inefficient and often unreliable pollination services that are available.

The reproductive strategies of self-compatible plants often appear to be guided by tradeoffs between the advantages and deficits of selfing and outcrossing. Separation of the timing and positioning of pollen dispersal and receipt functions within flowers can promote higher outcrossing frequencies, but optimizing floral characters for pollen dispersal may limit a plant's potential for reproductive assurance through selfing. As the degree of dichogamy or herkogamy increases, the potential for the receipt and export of pollen can increase, but plants with strong separation between these functions may suffer lower seed set when pollinator availability is low. In contrast, synchronizing pollen dehiscence and stigma receptivity may improve reproductive assurance, but improved autogamy may impose fitness costs due to reduced ability to secure fertilization of ovules by neighboring plants (see *Pollen Discounting* discussed in the section on Pollen Discounting) and inbreeding depression of selfed progeny. Many plants avoid the fitness deficits associated with selfing by having strong SI, but again, this comes at the cost of reduced reproductive potential when pollinators are rare.

Tradeoffs between pollen export, reproductive assurance, and fitness gains through the production of selfed and outcrossed progeny can be circumvented through more flexible mechanisms that favor outcrossing. In some species, the degree of herkogamy declines with flower age; the anthers and stigma come into close contact in older flowers resulting in *Delayed Selfing* (Fig. 9.11; refer to color plate). These plants effectively favor outcrossing when pollinator activity is high and have a high degree of reproductive assurance through autonomous autogamy. Some degree of delayed selfing may be present in many species with cosexual flowers through some accidental transfer of pollen within flowers, but seed set through autogamy is generally much lower in these plants. Surprisingly, species with more deliberate mechanisms that promote delayed selfing are relatively uncommon (e.g., *Geranium maculatum*; Fig. 9.11; refer to color plate). Similarly, only a handful of cases of cryptic SI have been documented, so in spite of the potential "best of both worlds" benefits of flexible mating systems, these mechanisms appear to more often be the exception rather than the rule.

To more fully understand the circumstances that may favor higher frequencies of selfing and outcrossing, or mixed mating strategies, it is instructive to consider the selective forces acting on plant mating systems. In the next section, we discuss some of the mathematical approaches that have been developed to understand mating system evolution.

Models for the Evolution of Selfing—After Darwin's pioneering work on the fitness effects of selfing and outcrossing through the late 1900s, most plant biologists held the view that mating systems were the product of ecological circumstances. It was generally assumed that outcrossing would be favored for the large majority of plants, while those that were weedy and persisted through colonization of ephemeral habitats were driven to be self-fertilizing for the purpose of reproductive assurance (Baker 1967; Stebbins 1974). There were no mathematical assessments of the conditions that would favor selfing or outcrossing until Fisher (1941) evaluated the effects of automatic selection for selfing. In this paper, Fisher aimed to clarify some aspects of his models of selection from previous publications (e.g., Fisher 1930) and simultaneously to criticize Wright's mathematical treatments of adaptive evolution. Fisher's main aim was to demonstrate that the mean fitness of the population does not change as an advantageous allele for selfing goes to fixation. This idea is particularly relevant to Wright's shifting balance model of evolution (Chapter 5), because a change in the mean fitness of a population is a condition that would be necessary for evolutionary "hill-climbing" on an adaptive landscape. Although it was not the focus of his paper, Fisher's model was the first to demonstrate that an allele for selfing would rapidly go to fixation in a population because of the genetic transmission advantage (i.e., the cost of outcrossing discussed earlier).

Fisher's model for the evolution of selfing only examined the genetic transmission differences for alleles controlling the mating system, but later analyses incorporated the effects of inbreeding depression and ecological conditions for the evolution of selfing and outcrossing (e.g., Crosby 1949; Lloyd 1979). These models generally predicted that either complete selfing would evolve, or that a balance would be struck between the genetic transmission advantage of selfing and the detrimental effects of genetic load, which would result in stable mixed mating. For example, Lloyd interpreted his results in terms of the quantity and quality of offspring; when pollinators were unreliable, autonomous selfers were predicted to produce larger numbers of seeds, but these offspring would have lower viability due to inbreeding. Outcrossers, on the other hand, produced fewer, high-quality seeds. Differences in seed numbers and qualities counterbalanced each other such that each mode of reproduction resulted in the same number of surviving offspring at some intermediate level of outcrossing.

Following Fisher's pioneering work, the early models and simulations for the evolution of plant mating systems assumed that genetic load and levels of inbreeding depression would not change. The novel insight introduced by Lande and Schemske (1985) was that genetic load could evolve in response to increased levels of selfing. They predicted that outcrossing mating systems would be resistant to the evolution of increased selfing rates because of the lower fitness of selfed progeny due to inbreeding depression (Fig. 9.16 lower region). When inbreeding depression is severe, fitness is reduced because ovules are preempted by self-pollen and are not available for producing high-fitness outcrossed progeny (*Seed Discounting*). In other words, selfed seed abortion reduces the total genetic transmission rate to the next generation (fitness). In contrast, when outcrossing species are subject to unreliable pollination environments or population bottlenecks, selfing and biparental inbreeding may become more prevalent and elevate the potential for purging of deleterious alleles. Once genetic load is reduced to the

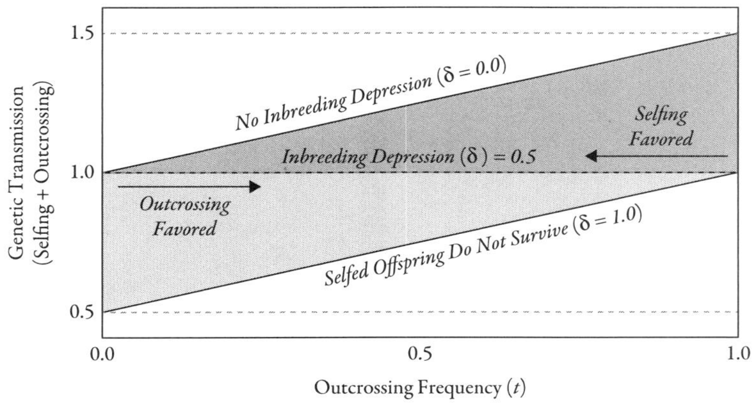

Fig. 9.16. The effect of inbreeding depression ($\delta = 1 - w_s/w_o$, where w_s and w_o are the fitness of selfed and outcrossed progeny, respectively) on mating system evolution. The outcrossing frequency varies from $t = 0.0$ to $t = 1.0$ (i.e. complete selfing to complete outcrossing). The total genetic transmission (fitness) is determined by the proportion of selfed and outcrossed progeny produced and the effect of inbreeding depression on the fitness of selfed progeny. If selfed progeny are not viable then any amount of selfing leads to lower fitness due to seed discounting, in which case complete ourcrossing is favored ($\delta = 1.0$; lower line). When the fitnesses of selfed and outcrossed progeny are equal the genetic transmission advantage results in rapid evolution towards complete selfing ($\delta = 0.0$; upper line). A threshold occurs at $\delta = 0.5$; if inbreeding depression is higher than 0.5 then complete outcrossing will be favored (lower region), and when $\delta < 0.5$ the mating system will evolve towards complete selfing (upper region). Outcrossing is assumed to be random (i.e. no biparental inbreeding—see text for more details).

point where the fitness of selfed offspring rises above 50% that of outcrossed progeny (Fig. 9.16 dashed line), a threshold is reached where selfing is more advantageous than outcrossing (i.e., there is a selection gradient favoring evolution toward complete selfing; upper region in Fig. 9.16). After passing this threshold, the genetic transmission advantage associated with increased selfing, along with the increasing fitness of selfed progeny due to continued purging, drives the mating system until it approaches complete selfing.

The combination of the genetic transmission advantage of selfers and purging of genetic load leads to the prediction that there should be two stable equilibria for plant mating systems, one at complete outcrossing and the other at complete selfing (Lande and Schemske 1985). Based on this prediction, we expect that outcrossing frequencies should remain high in large populations that have adequate pollinator service, and selfing should evolve in populations subject to bottlenecks and unreliable pollinator environments. Consequently, the distribution of mating systems across species should be bimodal (Fig. 9.17); most species should either be selfing ($t <$ 20%) or outcrossing ($t >$ 80%). Under the conditions considered by Lande and Schemske, mixed mating ($t >$ 20% and $<$ 80%) is considered to be temporary, and should be uncommon, because it will only be observed in species that are in the process of evolving toward complete selfing.

The data presented to test Lande and Schemske's model initially appeared to support the prediction that the distribution of outcrossing frequencies would be bimodal (Schemske and Lande 1985; Fig. 9.17). On the other hand, we expect that species with intermediate outcrossing frequencies would be rare, so the number of species displaying mixed mating is somewhat higher than expected. Moreover, additional analyses of outcrossing frequencies showed that a bimodal distribution found by Schemske and Lande was only apparent for wind-pollinated species, and when only animal-pollinated species were considered, the distribution was fairly flat (Fig. 9.17 upper panel; Goodwillie et al. 2005). With the addition of more studies, the bimodality of the distribution for both biotic and abiotic pollination was only vaguely apparent, and large proportions of species displayed mixed mating (Fig. 9.17 lower panel). Such large numbers

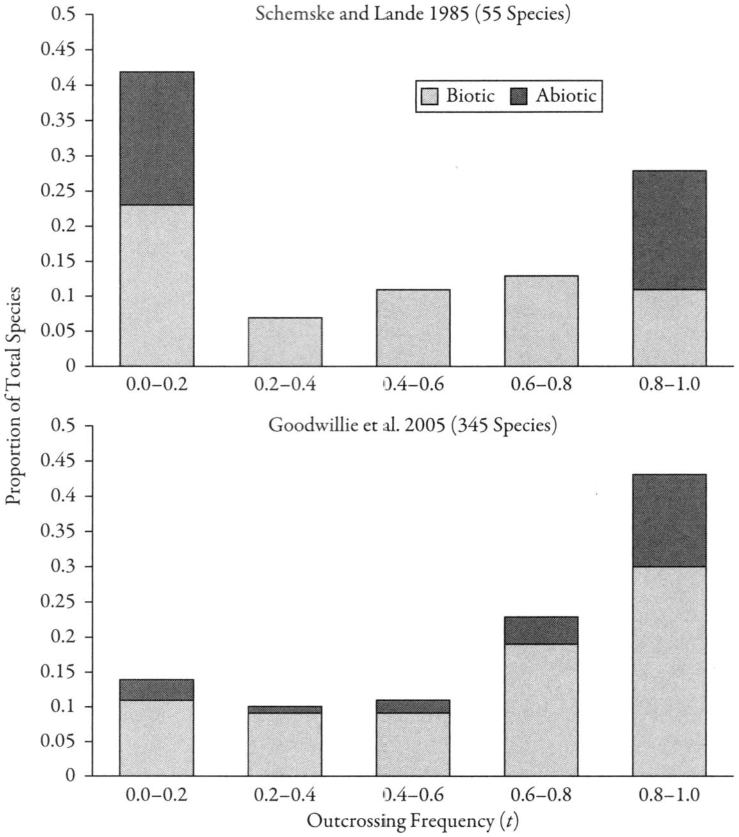

Fig. 9.17. Outcrossing frequencies for biotic and abiotic pollinated species. Under the mixed mating model, the outcrossing frequency (t) is the proportion of seeds sired by other genets, and seeds can only be selfed or outcrossed ($t = 1–s$; where s is the selfing frequency). The upper panel is based on data presented by Schemske and Lande (1985), and data for the lower panel is redrawn from Goodwillie et al. (2005).

of species that were presumably transitional would not be sustainable because outcrossers would rapidly disappear, as virtually all species would evolve toward complete selfing. In contrast to predictions made by Lande and Schemske (1985), the large number of intermediate outcrossing frequencies indicates that mixed mating can be stable.

Models for the Evolution of Stable Mixed Mating—The apparent discrepancy between the predictions of the model proposed by Lande and Schemske (1985) and the observed distributions of outcrossing frequencies led to a proliferation of analyses aimed at explaining the high frequency of species having mixed mating systems. The Lande and Schemske model was compelling because the predicted dynamics of mating system evolution were clear, but it also contained a number of simplifying assumptions that may have limited its predictive value. In particular, their model assumed that pollinators did not typically limit reproduction, populations were very large, outcrossing was random, and inbreeding depression was controlled by relatively few loci of large effect. These assumptions provided fertile ground for the development of more complex analyses that aimed to explain the prevalence of mixed mating (Goodwillie et al. 2005).

Models focusing on conditions that favor the evolution of stable mixed mating fall into several categories: 1) those assuming that the genetic transmission advantage declines as selfing frequencies increase due to pollen discounting; 2) analyses focusing on the potential for the genetic transmission of outcrossers to be elevated by biparental inbreeding; 3) models assuming that the process of genetic purging becomes stalled due to selective interference; 4) analyses incorporating pollen limitation and the benefits of reproductive assurance; and 5) models that account for the effects of *Reproductive Compensation* and variance in progeny fitness to stultify the cost of selfing. To understand how each of these modifications might contribute to stable mixed mating we will examine each one in turn.

Pollen Discounting—Pollen discounting can reduce the genetic transmission advantage and leads to stable mixed mating because it compromises the potential for selfing variants to sire seeds through outcrossing. In the models developed by Fisher (1941) and Lande and Schemske (1985), selfing variants possess a genetic transmission advantage, because they are able to fertilize the ovules of plants that preferentially outcross. The potential for selfing individuals to reproduce through both selfing and outcrossing depends on whether adequate amounts of pollen are available to achieve high seed set through autogamy, as well as through the fertilization of ovules borne by other plants.

We expect that plants transitioning toward selfing may have less pollen available for outcrossing because autogamous species typically have smaller flowers that produce less pollen (i.e., as a symptom of the "selfing syndrome"; Sicard and Lenhard 2011; Fig. 11.17; refer to color plate). Selfing flowers also tend to have anthers adjacent to the stigma, so as described earlier, the positioning of reproductive structures may interfere with pollen export to other plants. The reduced capacity for selfers to fertilize ovules on other plants, either because of lower levels of pollen production or diminished capacity for pollen export (i.e., pollen discounting), effectively curtails their genetic transmission advantage (Goodwillie et al. 2005; Barrett 2017). As pollen discounting increases, the genetic transmission advantage (fitness) of selfers declines, and the fitness threshold for selfed offspring to promote the spread of selfing increases (Fig. 9.18). Hence, pollen discounting can counterbalance the genetic transmission advantage of selfers leading to equivalent fitnesses for selfing and outcrossing at intermediate outcrossing frequencies, which are the conditions necessary to promote stable mixed mating (Harder et al. 2008).

Biparental Inbreeding—The genetic transmission advantages of selfing variants in the Lande and Schemske (1985) model also depends on the assumption that outcrossing is random. In populations structured by limited dispersal, individuals that are in close proximity to each other are more likely to be closely related. Since outcrossing is most likely to occur among neighboring individuals, biparental inbreeding will effectively elevate the genetic transmission advantage of outcrossing, because large proportions of the genomes will be shared between mates (i.e., they will be identical by descent—IBD; Chapter 4). As the frequency of selfing (or inbreeding in general) increases, the level of biparental inbreeding will also increase, and consequently genetic transmission through outcrossing will be higher (Uyenoyama 1986; Goodwillie et al. 2005).

On the surface, it would appear that local inbreeding would benefit genetic transmission through outcrossing and inhibit automatic selection toward selfing; however, it is important to note that biparental inbreeding will have equivalent effects on the genetic transmission across a range of outcrossing. Recall that the genetic transmission advantage of selfers ($t = 0$) is conferred through siring seeds on other plants, so the benefits of biparental inbreeding on genetic transmission will apply equally to selfers and outcrossers (Fig. 9.19). As long as there is no pollen discounting, biparental inbreeding will elevate the genetic transmission regardless of the outcrossing frequency. Consequently, the higher level of genetic transmission due to biparental inbreeding will still favor selfing and is not expected to promote stable mixed mating.

The effects of biparental inbreeding on mating system evolution is confounded by the potential declines in outcrossed offspring fitness due to inbreeding depression. Biparental inbreeding has differential effects on the fitness for outcrossers and selfers because of the opposing effects of genetic transmission and inbreeding depression (Fig. 9.20). This analysis examines the effects of inbreeding depression on the difference in fitness between selfers ($t = 0.0$) and outcrossers ($t = 1.0$) for a range of biparental inbreeding. The fitness of selfers increases more rapidly with biparental inbreeding because of the greater genetic transmission advantage associated with selfing (Fig. 9.19), while there are stronger effects of biparental inbreeding depression on outcrossers because of reductions in the fitness of both the seeds they produce and the seeds they sire. The overall effect of biparental inbreeding is to effectively lower the threshold of inbreeding depression

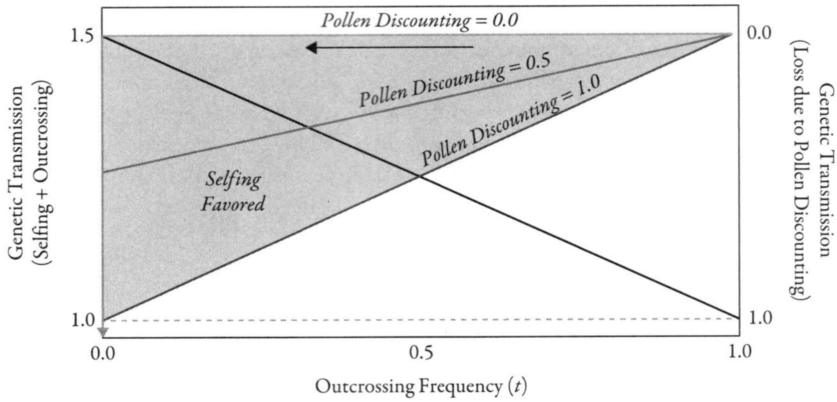

Fig. 9.18. A heuristic model for the effects of pollen discounting on genetic transmission and mating system evolution. When plants completely outcross they transmit one copy of their genome through both seeds and pollen (genetic transmission = 0.5 + 0.5 = 1.0). As selfing increases the genetic transmission through seeds increases while transmission through pollen remains the same, until the total transmission rate reaches 1.5 when outcrossing through seeds = 0 (corresponds to the scenario in Fig. 9.16 with no inbreeding depression). Pollen discounting decreases genetic transmission because less pollen is available for outcrossing as the selfing frequency increases. With no pollen discounting the highest genetic transmission (highest fitness) is achieved when outcrossing = 0. As the rate of pollen discounting increases, the highest genetic transmission occurs at intermediate outcrossing frequencies. When pollen discounting is directly proportional to the selfing frequency (pollen discounting = 1.0) the highest rate of genetic transmission occurs when the outcrossing frequency = 0.5. Hence, when the pollen discounting rate is greater than zero intermediate outcrossing frequencies are stable. Note that this model assumes no inbreeding depression.

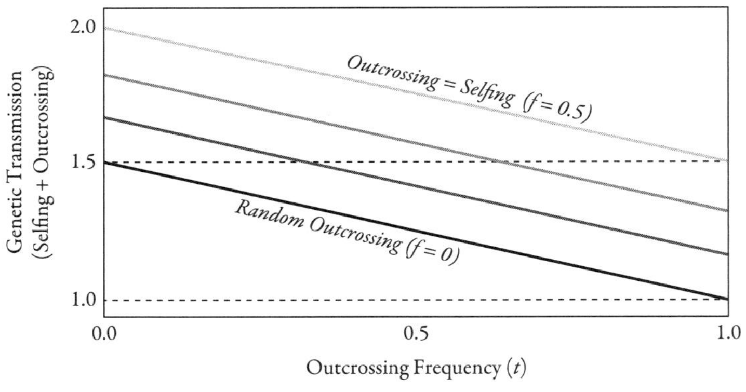

Fig. 9.19. The expected effects of biparental inbreeding (f) on the genetic transmission advantage (through selfing and outcrossing) across outcrossing frequencies. For complete outcrossing (t = 1.0), genetic transmission is expected to be 1.0 (0.5 through seeds and 0.5 through siring seeds on other plants), and the genetic transmission will increase as selfing increases until $t = 0$. Under random outcrossing ($f = 0$), the genetic transmission ranges from 1.0 for complete outcrossing to 1.5 for complete selfing (t = 0.0; corresponds to the scenario in Fig. 9.16 with no inbreeding depression). Under biparental inbreeding, the genetic transmission of selfers will increase to the same degree as for outcrossers if there is no pollen discounting. These predictions assume that the number of seeds sired through outcrossing is equal to the number produced by each plant, and there is no inbreeding depression.

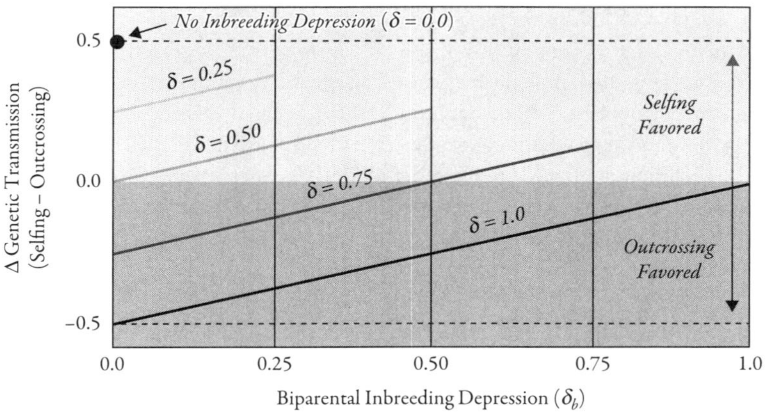

Fig. 9.20. A heuristic model for the effects of inbreeding depression and biparental inbreeding on mating system evolution. Biparental inbreeding depression (δ_b) is expressed as a fraction of inbreeding depression after selfing (δ; ranges from 0.0 to 1.0). The advantage of complete selfing or outcrossing is expressed as the difference in genetic transmission (Δ Genetic Transmission). With no inbreeding depression, selfing is predicted to have a strong genetic transmission advantage (i.e. 1.5 – 1.0 = 0.5; corresponds to the scenario in Fig. 9.16 with no inbreeding depression). For all levels of inbreeding depression biparental inbreeding reduces the fitness of outcrossed progeny and promotes the evolution of selfing. This model examines the loss of genetic transmission (fitness) due to biparental inbreeding depression across the range of outcrossing frequencies. These predictions assume no pollen discounting.

that favors the evolution of increased selfing; even when δ is as high as 0.75, moderately high levels of biparental inbreeding could favor the evolution of increased selfing (Fig. 9.20). This model assumes no reductions in genetic load through inbreeding, but we can predict that it is probable that any degree of purging would elevate the fitness advantage of selfing over outcrossing (Uyenoyama et al. 1993; Goodwillie et al. 2005).

Selective Interference—A critical aspect of the Lande and Schemske (1985) model is the potential for selfing to purge genetic load and alleviate the effects of inbreeding depression. While Lande and Schemske's model incorporated a couple of different scenarios for the determination of genetic load, all of their analyses assumed that inbreeding depression was controlled by few loci and that deleterious alleles displayed some degree of dominance. These conditions may have facilitated reductions in genetic load through selfing, which is a necessary condition for evolution toward complete self-fertilization.

It is more likely that the genetic architecture of inbreeding depression is more complex (Charlesworth and Willis 2009). Laboratory experiments and analytical analyses indicate that purging of genetic load may become stalled due to selective interference among loci affecting fitness and *Drift Barriers* that reduce the effectiveness of selection. Because of the effects of genetic drift, small populations are likely to become fixed for mildly deleterious alleles. These increases in homozygosity will only contribute to differences in fixed deleterious mutations among populations (drift load; Chapter 5) but will not contribute to inbreeding depression for mating within populations. However, due to the reduced effectiveness of selection in small populations, many loci are likely to continue to segregate for large numbers of mildly deleterious alleles (inbreeding load) that will contribute to differences in the fitness of selfed and outcrossed progeny. Whether because of drift barriers to the effectiveness of selection or selective interference among loci, many inbreeding populations continue to segregate for substantial levels of genetic load (Goodwillie et al. 2005; Charlesworth and Willis 2009; Winn et al. 2011). If purging becomes stalled then the fitness difference between selfed and outcrossed progeny will be sustained, which would make it easier for other factors (e.g., pollen discounting or reproductive assurance) to favor stable mixed mating.

Reproductive Assurance—The model considered by Lande and Schemske (1985) differed from previous models in assuming that pollinator service was usually high, and population bottlenecks or loss of pollinators inevitably let to the purging of genetic load and the evolution of selfing. On the other hand, pollinator availability for many species may vary among years and locations, in which case there may be benefits for mechanisms that promote reproductive assurance through self-fertilization (Lloyd 1979). For many species, reproductive assurance occurs by accident when pollinator service is low, and wind, or visitation by insects that are ineffective pollinators, can dislodge pollen from anthers and result in some degree of self-pollination. When pollinators are more common, self-pollination from autogamy and geitonogamy can become costly due to seed discounting when inbreeding depression is substantial and SI or cryptic SI are absent. On the other hand, delayed selfing (e.g., Fig. 9.11; refer to color plate) is much less likely to cause seed discounting as it occurs toward the end of the life of the flower when the receipt of outcross pollen is unlikely. As long as inbreeding depression is not too severe, reproductive assurance through selfing can result in stable mixed mating for species that are subject to variable pollinator environments or that rely on frequent colonization for persistence (Jarne and Auld 2006; Barrett 2017).

Reproductive Compensation—Reproductive compensation occurs as plants produce more ovules than can be matured into seeds or more seeds than are expected to survive to adulthood (Harder et al. 2008). The potential for reproductive compensation to favor the evolution of intermediate outcrossing frequencies can depend on when inbreeding depression is expressed. In cases where selfed embryo abortion is high, mixed mating can be stable if the surviving selfed seeds experience only mild inbreeding depression. In this case, we assume that seed discounting is minimal, so the number of outcrossed progeny generated is not reduced by selfing. Mixed mating is favored because the success of the surviving selfed seeds promotes the maintenance of self-fertility, and inbreeding depression across reproductive stages offsets the genetic transmission advantage of selfing.

Models of mating system evolution generally do not address the variance in fitness of offspring but instead assume that inbreeding depression has equal effects on all selfed progeny. At the same time, we recognize that segregation and recombination will generate variation among offspring, and occasionally, vigorous selfed progeny can be generated—individuals that Darwin (1876) referred to as "heroes." The effect of rare selfed progeny that have high fitness would be negligible if only small numbers of progeny were generated; however, the large majority of plants produce many ovules and seeds—sometimes in the many thousands, or even millions, for long-lived perennials—and only a small fraction of these will ever survive to adulthood. High fecundity for long-lived individuals is a characteristic that distinguishes plants from many other eukaryotes. Among the animals, only insects and aquatic invertebrates produce large numbers of offspring, and reproductive capacity generally declines with lifespan. For plants, it is exactly the opposite as seed production per year usually scales with size, and as individuals live longer and grow larger, their reproductive capacity continues to increase. Even if a large fraction of these seeds fail because they are deposited in inhospitable sites or are consumed by predators, selection can act on the many thousands of survivors to leave the one or few that by good chance and superior vigor grow to replace their parents.

Producing large numbers of offspring that vary in fitness may compensate for the effects of inbreeding depression. Selfed and outcrossed progeny in the tails at the upper ends of the fitness distributions may be more likely to have similar growth and survival. This is especially true for log-normal fitness distributions (Fig. 5.1), because the average of the highest fitness individuals are likely to be more similar than the means of the entire distributions. Furthermore, low-fitness offspring can be culled prior to seed dispersal through selective ovule abortion by competition among developing seeds for limited maternal resources (Rigney et al. 1993; Melser and Klinkhamer 2001). While the effects of inbreeding depression can be compensated by the overproduction of ovules and seeds, empirical and theoretical analyses have only considered the average fitness of selfed and outcrossed progeny (Goodwillie et al. 2005; Charlesworth and Willis 2009), and the potential consequences of the overproduction of flowers and variance in fitness for mating system evolution have not been explored. If the highest fitness offspring from self and outcross matings are more similar then we should expect that the production of many more ovules and seeds than are expected to survive to adulthood would moderate the effects of inbreeding depression and favor the evolution toward increased selfing, and perhaps stable mixed mating.

Evolution of Mixed Mating—The view that inbreeding will inevitably promote purging and the evolution of selfing has been a dominant theme for discussions of mating systems evolution since this idea was first introduced by Lande and Schemske. Our exploration of some of the possible departures from the assumptions made by Lande and Schemske (1985) reveals that both ecological (i.e., pollen discounting and reproductive assurance) and genetic (selective interference) processes could counterbalance automatic selection due to the genetic transmission advantage and result in mixed mating. Selective interference can stall the progression of selection for purging of genetic load and lead to the production of mixtures of selfed and outcrossed progeny. Pollen discounting limits the genetic transmission advantage by reducing the number of progeny produced through outcrossing. Both selective interference and pollen discounting would result in mixed mating when pollinators are abundant.

In contrast, ecological circumstances may have important influences on outcrossing frequencies for species that are routinely subject to low pollinator availability. Maintaining the ability to self-fertilize would be particularly important for early successional species that rely on colonization of new sites for persistence. If there were substantial levels of genetic load then we would expect that such species would have mechanisms that promote outcrossing, such as cryptic incompatibility or delayed selfing, and in any case, variation in outcrossing frequency would track with pollinator abundance.

So far, our discussion of mating system evolution has made the implicit assumption that mixed mating represents a less desirable compromise that is imposed by ecological or genetic forces that oppose the evolution of complete selfing. For example, we assume that plants with reproductive assurance mechanisms will preferentially outcross or that in the absence of pollen discounting or selective interference, the mating system would continue to evolve toward complete selfing. On the other hand, this perspective is inconsistent with our observations of mating systems across species; if we assume that producing mixtures of selfed and outcrossed progeny is unlikely to result in the highest fitness, it becomes difficult to explain the high proportion of species that display mixed mating. Intermediate outcrossing is clearly not always a product of ecological circumstances or forces that stall selection for selfing, as a substantial number of species are known to routinely produce mixtures of cleistogamous and chasmogamous flowers (Oakley et al. 2007). The proportion of each flower type produced by these plants is presumably under genetic control, which suggests that there must be advantages for generating mixtures of selfed and outcrossed progeny.

We can gain some insights into the control of the ecological and genetic influences on mixed mating by examining the effects of cryptic SI on the outcrossing frequency (Cruzan and Barrett 2016; Fig. 9.21). In this study, the strength of stylar discrimination ability (cryptic SI) was measured by applying mixtures of genetically marked pollen to track fertilization success among the progeny produced. This approach allowed the prediction of outcrossing frequencies that would be expected under random mating with no pollen limitation. By comparing the predicted and observed outcrossing frequencies, we can obtain an estimate of the influence of pollinator environments on the mating system (vertical arrows in Fig. 9.21). In contrast to expectations, cryptic SI in this species did not always promote complete outcrossing; instead, the variation in stylar discrimination resulted in the routine production of mixtures of selfed and outcrossed progeny. Variation in the strength of stylar discrimination explained the large majority of variation in outcrossing frequencies (> 80%), indicating that the pollinator environments had only minor influences on mating system variation in this species.

The strong association between stylar discrimination and outcrossing frequencies in *E. paniculata* suggests that the strength of cryptic SI, and consequently the mating system, is primarily under genetic control. These results challenge the assumption that intermediate outcrossing frequencies are the outcome of reproductive assurance or represent mating systems that are prevented from evolving toward complete selfing by pollen discounting or selective interference. The populations surveyed differ in style morph composition and possibly in other aspects of their histories that have favored different outcrossing frequencies (Cruzan and Barrett 2016). Note that genetic control

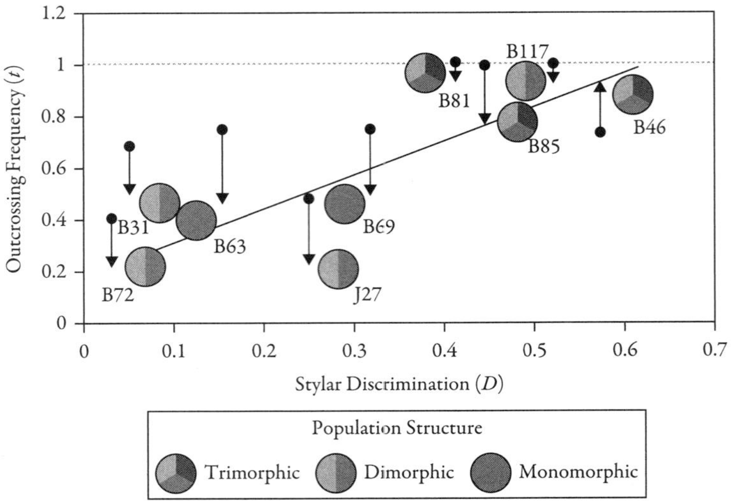

Fig. 9.21. The effects of different strengths of cryptic self-incompatibility (stylar discrimination—D) on outcrossing frequency (t) across populations of the tristylous *Eichhornia paniculata* having different stylar morph compositions (Fig. 10.11). Stylar discrimination was measured a departures in fertilization success compared to frequencies of each pollen type used in pollination for mixtures of pollen. Predicted values of t for each population are indicated by small solid dots, and arrows indicate the departure between predicted and observed outcrossing frequencies, presumably as a consequence of pollination conditions. Modified from Cruzan and Barrett (2016).

of cryptic SI in this species is unknown, but there is no a priori reason to assume that it is associated with genes contributing to genetic load, selective interference, or pollen discounting. The results from analyses of the influence of cryptic SI on mating system variation in *E. paniculata* suggest that producing mixtures of selfed and outcrossed progeny can be an optimal strategy that is favored by selection, but at present it remains unclear what particular forces are responsible for the evolution of intermediate outcrossing frequencies in this and other species.

Summary—The prevalence of mixed mating along with documented cases where selection appears to have favored the production of both selfed and outcrossed progeny underlines the inadequacy of our current models and understanding of mating system evolution in hermaphroditic species. It is also clear that the large majority of species with mixed mating retain significant levels of genetic load (Winn et al. 2011). Rather than take the stance that the production of selfed offspring is a compromise imposed by ecological or genetic circumstances, it might be advantageous to develop approaches that aim to reveal the potential advantages of mixed mating.

There are unexplored avenues of investigation that could provide insights into the factors favoring the production of both selfed and outcrossed progeny. For example, it could be the case that selfed progeny are better adapted to the home microhabitats of their maternal parents, but since most selfed offspring are subject to inbreeding depression, this particular advantage may be evident in only a small minority of recombinant individuals. If we consider the potential for reproductive compensation and overproduction of flowers to filter out the lowest fitness offspring, then we can begin to see the possibility that the surviving selfed offspring might represent those few "heroes" that by chance have circumvented the detrimental effects of genetic load. A better approach to understanding the effects of genetic load on mating system evolution may be to consider the fractions of offspring that display the highest fitness rather than focusing on the larger lot of average individuals out of the multitude of seeds that are produced over the lifespan of an individual.

If selfing could be advantageous for genetic transmission—even when there is substantial inbreeding depression—then what might be the possible benefits of maintaining intermediate or high frequencies of outcrossing? If we consider genetic transmission alone, then it appears that outcrossing is generally not advantageous. On the other hand, the ability to interbreed with other individuals may contribute to long-term viability of lineages and their potential for continued diversification. As we discuss in Chapter 10, hybridization among divergent lineages presents opportunities for the acquisition of beneficial genetic elements. When we consider that the potential advantages of acquiring novel genetic variation through gene flow among populations and lineages, it is easy to conclude that there may be long-term advantages for retaining the ability to outcross. The potential for admixture to alleviate inbreeding depression and promote diversification is an oft-mentioned, but perhaps underappreciated, mechanism that could be favoring the maintenance of outcrossing.

Mixed mating is common for only a subset of flowering plants, and there are large numbers of species that are almost completely selfing or outcrossing (Fig. 9.17). The large swaths of mating-system variation among plants are strongly associated with their life history; plants with intermediate outcrossing are most often herbaceous perennials or short-lived woody species, while the large majority of selfers are annuals, and long-lived trees almost exclusively outcross (Barrett 2017). Selfing is much more prevalent among annuals, and the autonomous selfing syndrome, with its reduced flower size and low levels of dichogamy and herkogamy, is completely restricted to the annuals and short-lived perennials (Sicard and Lenhard 2011). Moreover, autonomous selfers are typically weedy and are common in recently disturbed sites. Hence, evolution toward complete selfing in these species may have been promoted by their demographic and ecological circumstances, rather than being an outcome of the genetic transmission advantage associated with selfing.

At the other extreme, the large majority of long-lived trees and shrubs outcross almost exclusively. Exceptions to this generality seem to be restricted to island species that have only recently evolved woodiness from colonizing ancestors that were typically self-compatible herbaceous annuals or perennials. For example, trees in the genus *Scalesia* (Asteraceae) have radiated in the Galapagos archipelago and have retained the ability to self (Nielsen et al. 2003). For the vast majority of trees, it has been suggested that long-lived species suffer from higher levels of inbreeding depression due to somatic mutation accumulation to the extent that any self-fertilized ovule is likely to abort (Barrett 2017; Chapter 3). If this were true then it would suggest that there were limits to the potential for reproductive compensation to overcome the effects of genetic load even when many thousands (or millions) of seeds are produced over the lifespan of an individual. It is also difficult to explain the maintenance of selfing in woody species such as *Scalesia*; if loss of the ability to self-fertilize were strictly due to somatic mutation load then we would expect that these trees would be incapable of producing selfed seeds. On the other hand, we currently do not have any information on levels of inbreeding depression in species that have only recently evolved woodiness and tall stature. The other possibility is that the high capacity for reproductive compensation in long-lived species obfuscates the necessity for any degree of reproductive assurance, so ovules displaying even a small degree of inbreeding depression are routinely aborted. Understanding the contribution of somatic mutations to genetic load and mating system evolution could provide important insights into mating system evolution and the prevalence of outcrossing in long-lived species.

While self-incompatibility and cryptic SI can be effective mechanisms for the avoidance of self-fertilization, outcrossing is also enforced by the separation of sexual functions among individuals within a population or among flowers on a single individual (Renner 2014). In the next section, we will discuss the evolution of investment strategies associated with reproduction and its consequences for differential sex expression including dioecy and various forms of unequal sex expression within and among individuals.

9.8 INVESTMENT STRATEGIES

The partitioning of resources to different functions over a plant's lifespan defines its life history, which includes its potential for sexual reproduction, vegetative growth, and longevity. At the broadest level, we can see differences in the determination of life histories through the allocation of resources among structures and functions contributing to fitness. The monocarpic life history (including annuals) is determined by investment of all available carbohydrates into a single episode of sexual reproduction that leads to the death of the individual, while polycarpic perennials partition resources between sexual reproduction and vegetative growth throughout their lifespans. We assume that this allocation of resources has been molded by selection to optimize fitness under different conditions: high mortality due to external factors early in life would favor an annual life history, while more stable habitats offer higher expectations for future reproductive success so polycarpic perennials are expected to have higher fitness (Chapter 6). Patterns of resource allocation between sexual and asexual functions are the outcome of selection for optimal investment strategies and are molded to maximize fitness under prevailing environmental conditions.

To the same extent that plants may be selected to allocate resources between growth and sexual reproduction, we can predict that selection will favor the optimization of the size and number of individual structures within major functions. For example, a plant could concentrate all of its sexual reproductive effort into one or a few large flowers, or it could produce large floral displays of many small flowers. Within flowers, resources are distributed to support structures (e.g., pedicels and receptacles), attractive structures (e.g., petals), and rewards (pollen and nectar). Plants also balance their investment among sexual functions contributing to reproductive success. We expect that allocation of resources to the production of seeds and pollen will depend on expected returns as measured by increases in fitness. As described at the beginning of this chapter, plants may gain the highest fitness by serving both sexual functions through individual flowers (cosexual hermaphrodites), separating these functions among flowers on a single individual (monoecious hermaphrodites), or separating reproduction through pollen and seeds among individuals in a population (dioecy; Fig. 9.1). In this section, we explore conditions favoring different allocation strategies for investment in reproduction through pollen and seeds and patterns of deployment of sexual structures and functions that maximize fitness.

Evolutionarily Stable Strategies—New mutations may increase in frequency and continue to spread if they confer some fitness benefit. We often view this process as incremental improvements in growth and reproductive success in a particular environment, but advantageous changes may also be manifest as modification in the allocation of resources among structures and functions that contribute to fitness. If we assume that a limited pool of resources is available, then we can predict that tradeoffs among functions contributing to fitness are common. Changes in allocation patterns may be favored if increased investment in one function over another results in higher fitness.

An *Evolutionarily Stable Strategy* (*ESS*) is one that confers the highest fitness in the current environment and consequently cannot be replaced by alternative strategies. This idea was first applied to *Game Theory*, where it has been used to understand the evolution of interactions among individuals within animal populations (Maynard-Smith 1982). The ESS approach can be employed to analyze most situations where participation in one activity limits the ability to engage others—anytime when tradeoffs limit the maximum expenditure of effort in the endeavors that contribute to fitness. The ESS approach and game theory have been widely used in behavioral ecology and have been utilized in fields outside of biology, such as economics and political science.

Tradeoffs in allocation among plant structures and functions have long been recognized, but an ESS approach was not applied to resource partitioning until the early 1980s (Charlesworth and Charlesworth 1981; Charnov 1982). The ESS approach can be applied to predict the optimal allocation of resources to male and female (pollen and seed) functions under the assumption that there is a limited pool of resources available for sexual reproduction. The optimal strategy for allocation to male and female function will depend on their contributions to fitness, which may not be equal and may change across different levels of investment. The benefits of increasing investment in each sexual function can be depicted as the relationship between resources allocated and returns on fitness—a *Fitness Gain Curve* (Figs. 9.22 and 9.23).

ESS Analysis of Sex Expression—The analysis of evolutionarily stable strategies has been particularly useful for understanding sex allocation and the evolution of different modes of sex expression. These models assume a tradeoff such that investment in one sexual function (i.e., reproduction through seeds or pollen) corresponds to a decreased investment in the other function. If we assume that there is a direct correspondence between the level of investment and returns in fitness, then the fitness gain curves would be linear (Fig. 9.22). Since there is a direct tradeoff between fitness returns on investment in male (r) and female ($1-r$) function, all investment strategies result in the same total fitness (male plus female reproductive success). If fitness returns on investment in pollen and seeds were equally costly then these gain curves would have equal slopes ($k_m = k_f$; dark blue line in Fig. 9.22). On the other hand, we might expect that fitness returns through pollen would be more efficient (i.e., when $k_m > k_f$) because individual pollen grains are smaller than seeds (lighter blue lines in Fig. 9.22).

It is important to note that the three lines in Figure 9.22 represent different scenarios that reflect the intrinsic characteristics of a particular species for fitness returns through male function. Within each scenario, the allocation to male and female function (r and $1-r$, respectively) can change, but coefficients (k_m and k_f) are constants that reflect the efficiency of reproduction; larger values of k_m indicate that reproduction through male function is greater for an equivalent amount of investment. The characteristic k_m for a species can depend on its mating system and the foraging behaviors of its pollinators (see discussion in the next section).

For the linear fitness gain curves depicted in Fig. 9.22, the sum of fitness returns through male and female function ($w = m + f$) remains the same across all values of r, but this situation changes when one or both relationships are non-linear (i.e., when b or c are not equal to 1.0; Fig. 9.23). When one or both fitness curves decelerates with higher levels of resource investment (b or $c < 1.0$), the highest total fitness can only be achieved through partial investment in both functions (blue line and arrow). In contrast, when fitness gains accelerate with investment (b or $c > 1.0$), the highest fitness returns occur with complete allocation to one sexual function or the other (i.e. when $r = 0.0$ or 1.0; brown line and arrow).

The sum of fitness gains through male and female function can be depicted as a fitness set (Fig. 9.24). If fitness gain curves are linear (as in Fig. 9.22), then all allocation strategies are equivalent (dashed lines in Fig. 9.24). Decelerating fitness returns result in the highest total fitness for investment in both male and female function and is expected to favor cosexual flowers as the ESS (convex blue line in Fig. 9.23). Accelerating fitness gains results in a concave fitness functions (b > 1.0) for the tradeoff between investment in male and female function and predicts the highest fitness when all resources are allocated to either one sex or the other in separate individuals

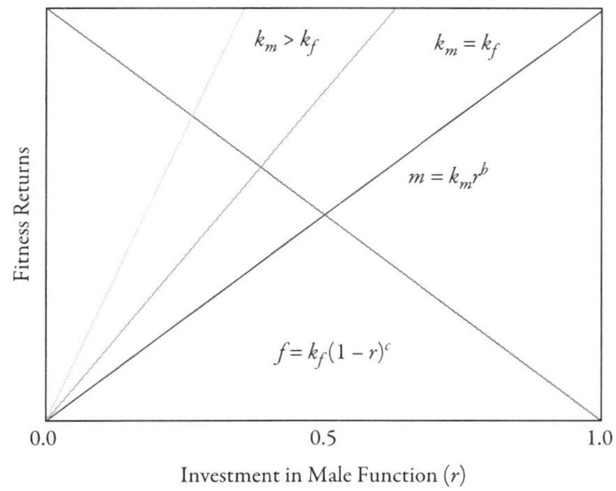

Fig. 9.22. Fitness gain curves for reproductive success (fitness returns) through male (pollen) and female (seed) functions. This model assumes that resources are partitioned between male (r) and female ($1-r$) reproductive effort. The slope of the male gain curve is steeper when returns on investment are higher ($k_m > k_f$). Both male and female fitness returns are linear ($b = c = 1$).

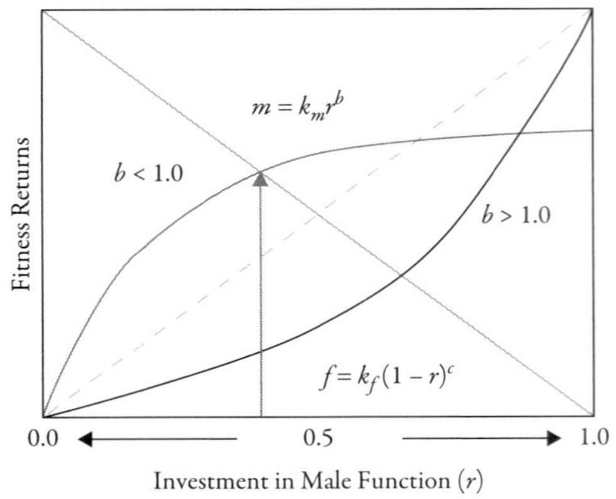

Fig. 9.23. The consequences of non-linear fitness returns on the evolution of cosexual flowers and dioecy. In these scenarios fitness gain curves trough male function are non-linear by being either decelerating ($b < 1.0$) or accelerating ($b > 1.0$). A decelerating function predicts the highest total fitness when resources are partitioned between male and female functions resulting in cosexual flowers decelerating line and vertical arrow). Accelerating returns results in higher fitness for complete investment in either male or female function and predicts dioecy (accelerating line and horizontal arrows).

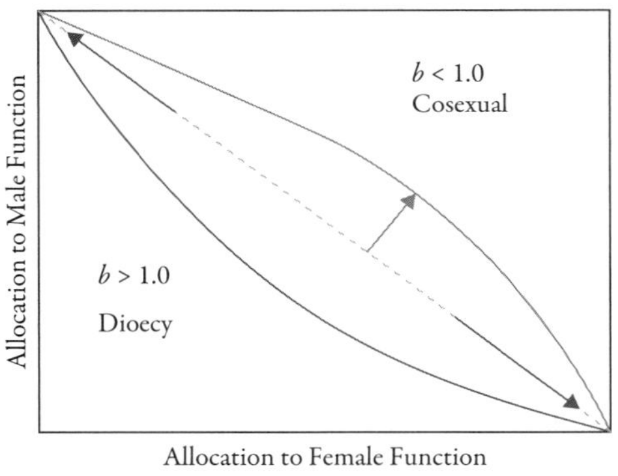

Fig. 9.24. The sum of fitness returns through male and female function under decelerating (convex line and arrow) or accelerating (concave line and arrows) fitness returns as shown in Fig. 9.23. The dashed blue line correspond to linear fitness returns through male function (Fig. 9.22).

(dioecy) or during different phases of a plant's lifespan (diphasic sex expression).

While the separation of sexes is generally viewed as a mechanism to enforce outcrossing, the ESS approach provides us with an alternative perspective on forces that might promote the evolution of monoecy and dioecy (Campbell 2000). By illustrating returns on investment in pollen and seed production with fitness gain curves, we can contrast the outcomes of combining sexual functions within flowers with the separation of functions among individuals. Such ESS analyses can provide a more thorough understanding of the motivations for the evolution of different modes of sex expression.

Critical to our understanding of the expected outcomes from ESS analyses is the shape of fitness gain curves and in particular, whether they are accelerating or decelerating. The shape of these fitness functions depends heavily on a complex set of ecological processes, including the behavioral characteristics of dispersal vectors and the potential for competition among pollen for access to ovules or among seedlings growing in the same habitat. In the next few sections, we discuss the ecological factors affecting the shapes of fitness gain curves for male (pollen) and female (ovule and seed) functions.

Fitness Gains Through Male Function—We expect that returns on investment in male function will generally be decelerating in plants for several reasons. First, if we assume that there are a limited number of receptive stigmas available then we can predict there will always be diminishing returns because of dilution in the pollen pool. We can understand this prediction more clearly if we consider a hypothetical population with random mating where 10 individuals each contribute 100 gametes to a pool of 1,000. If one pollen donor were to double its pollen production, its expected siring success would increase from 10% (100 out of 1,000) to only 18% (200 out of 1,100) instead of the 20% return that would be expected from producing twice as much pollen. As investment in pollen production increases, the proportional return continues to diminish. For example, quadrupling investment would only result in 30% successful fertilizations. In general, dilution in the pollen pool results in a reduction of proportional fertilizations with increased pollen production, and consequently, fitness returns through male function will always be decelerating.

Reproductive success through male function is also complicated by the fact that pollen serves as a food resource for many animal pollinators. As mentioned in Chapter 2, the large diversity of bee species in the Anthophila clade (bees) of the order Hymenoptera (ants, bees, and wasps) are characterized by having branched hairs that cover their bodies. These hairs are effective at picking up large amounts of pollen as bees forage. Stiffer hairs on the hind legs serve as combs to move pollen into baskets located on the legs or abdomen for transfer back to the nest or hive where it is fed to offspring. During grooming, a large amount of pollen is made unavailable for fertilization, as it is moved to pollen baskets or simply falls off the bee's body (Harder and Wilson 1998; Harder and Routley 2006). As bees groom, a small proportion of pollen remains on their body in locations where it can be transferred to stigmas during subsequent visits, while the majority ends up serving as a food resource or is simply lost. The pollen gathering and grooming behaviors of bees effectively cap the proportion of pollen that could ever contribute to reproductive success and is expected to result in flat fitness returns above a particular level of investment in male function.

While bees may disperse only a limited amount of the pollen produced to the receptive stigmas of neighboring individuals, plants can effectively increase their reproductive success through male function by attracting more pollinators (Harder and Thomson 1989; Thomson 2006). Once we recognize that each floral visitor has a limited carrying capacity for pollen that is available for deposition on stigmas, we can predict that the primary factor governing reproductive success through male function is the number of pollinators leaving a plant. This prediction is underscored by the recognition that there is constant turnover of the pollen load carried by bees and that the majority of pollen removed from a flower is deposited on the next few flowers visited (Thomson and Thomson 1989). Overall, the effects of pollinator grooming and the tendency for animal pollinators to visit multiple flowers on a single plant, along with dilution of each plant's contribution to the pollen pool, are likely to contribute to diminishing fitness returns for investment in male function.

Fitness Gains Through Female Function—Reproductive success through female function will depend on the distribution (deployment—see section 9.10) of ovules among flowers and the allocation of resources to developing seeds. As with other resource allocation strategies, we generally assume that there is a limited pool of resources available that results in balances among investment options, including a tradeoff between the size and the number of seeds produced. Such tradeoffs will govern the number of ovules contained within each flower and the total number of flowers produced. Ovule packaging and other characteristics of floral displays will often depend on historical availability of pollinators and their energetic needs. Hence, the allocation of ovules among flowers can be affected by pollinator-mediated selection on characteristics of floral displays, including the size and number of flowers. The consequences of differences in inflorescence size are complex and will affect reproductive success through both male and female function. This topic will be discussed in more detail later in this chapter (see later discussion on the evolution of floral displays). In this section, we focus on strategies concerning the provisioning or abortion of developing seeds and the consequences of seed size and number for fitness returns on resource investments.

Seed size varies substantially across the angiosperms, from the minute "dust seeds" of tropical orchids (~0.0001 mg) to the extremely large water-dispersed seeds of the coconut palm (*Cocos nucifera*; ~1.4 kg) and the cocos de mer (*Lodoicea maldivica*; up to ~17 kg), a palm endemic to the Seychelles and neighboring islands. Seed size—or more specifically the amount of resources stored for seedling establishment—is characteristic for each species and may be an outcome of selection for dispersal capability and seedling success, as well as the constraints imposed by phylogenetic history. Phylogenetic constraints on seed size are often apparent, as fruit types are characteristic of whole families, subfamilies, or tribes, and closely related species tend to be more similar for seed size (Sims 2012). For plants with multi-seeded fruiting structures, both seed number and size often varies, but the range of variation in seed size within species and genera is generally quite limited compared to the wide range observed across the flowering plants.

There is often a tradeoff between dispersal ability and seed size—particularly for wind and gravity dispersed seeds—with smaller, lighter seeds generally dispersing over greater distances. On the other hand, some of the largest seeds, such as coconuts, are water dispersed over long distances, and there was a general trend toward larger seed size as endozoochory became more prevalent in the late Cretaceous and early Tertiary (Eriksson 2008). While selection for dispersal ability has been important for the evolution of seed size, provisioning requirements for seedling survival may override selection for smaller and more numerous seeds (Moles et al. 2005). For example, a good proportion of selection pressure for increased seed size in the early Tertiary was probably due to changes in plant communities when closed canopies prioritized competition for light in shady environments. This trend can be seen even today in the relatively large seeds produced by many canopy species and tropical understory palms and trees growing in closed-canopy communities.

Maternal Provisioning—While seed and fruit characteristics generally reflect phylogenetic history, seed size often varies to some degree within and among plants of the same species and even within individual fruits (Temme 1986). We expect that the constraints imposed by limited resources would select for an optimal level for the provisioning of seeds, but there is likely to be a difference between the levels of investment that maximizes maternal versus offspring fitness (*Maternal-Offspring Conflict*; Smith and Fretwell 1974; Fig. 9.25). In this model, we assume that there is a minimum level of provisioning that is necessary for seedling survival. Above this level, offspring fitness increases rapidly, but the rate of increase slows as resource investment increases and eventually flattens out above a certain level. We predict that the level of provisioning that maximizes maternal fitness would occur at the point of inflection of the offspring fitness gain curve (M^*; Fig. 9.25), above which there are diminishing returns. For higher levels of provisioning, offspring fitness continues to increase because of improvements in the chances of survival under conditions of scarce resources or intense competition. Consequently, there will always be a conflict of interests between the seed parent and its offspring; limiting investment to M^* allows for production of more seeds (higher parental fitness), while each individual offspring would always accrue higher fitness benefits from higher levels of provisioning.

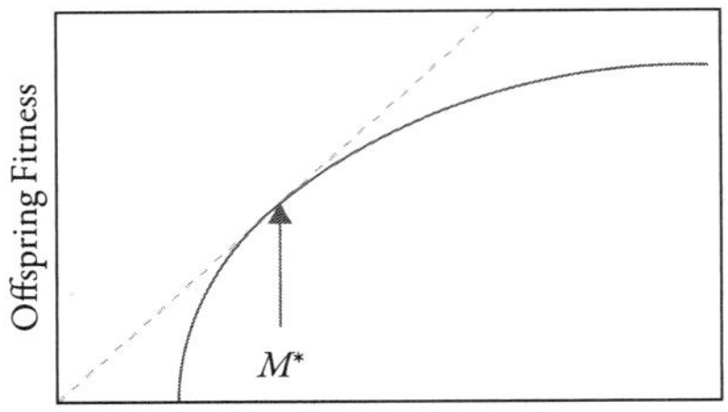

Fig. 9.25. Expected returns in offspring fitness for different levels of maternal investment (seed size). The optimal investment per seed for maximizing the maternal parent's fitness is indicated by M^*. Investments above this level would increase the offspring's fitness, but result in diminishing returns for the maternal plant. Modified from Smith and Fretwell (1974).

From Smith and Fretwell's model, we predict that if provisioning is under maternal control then there should be very little variation in seed size. Indeed, some plant species have been thought to have exceedingly consistent seed weights. For example, carob seeds (*Ceratonia siliqua*) were originally used to establish a metric for the weight of jewelry (the carat; Turnbull et al. 2006), which is consistent with the hypothesis that the range of variation in seed size within species should be limited. Strong maternal control over seed size is suggested by the observation that some plants appear to be capable of setting an upper limit on seed size (Pelabon et al. 2015). It has also been suggested that the higher maternal contribution (2N) to the triploid endosperm compared to the paternal contribution (1N; Box 9.1) provides a mechanism of greater maternal control over seed provisioning (Westoby and Rice 1982; Endress 2011).

While the Smith and Fretwell model suggests that seed size should be consistent, variation in seed weight within species is often substantial, averaging around 25%, and varying by as much as six-fold within a fruit (Stanton 1984; Turnbull et al. 2006). This is true even for the case of the carob seeds mentioned above: the myth of seed mass consistency in this species appears to be based on human perception. Indeed, variation in seed size due to environmental, developmental, and genetic effects within and among fruits is common (Pelabon et al. 2015). Hypotheses for sources of variation in size among seeds developing within the same fruit include: 1) selection for resource sharing among full siblings and competition among embryos having different paternal parents (i.e., due to *Kin Selection; Inclusive Fitness*; Westoby and Rice 1982); 2) competition among developing seeds for limiting maternal resources (Lee 1988); 3) preferential maternal investment in the most vigorous embryos (Temme 1986); and 4) maternal control of seed size variation to improve the chances that at least some will be successful in a variable environment (bet hedging). The first hypothesis can probably be dismissed, since size variation among seeds from the same fruit is often similar after pollinations with pollen from single versus multiple donors (e.g., Cruzan and Thomson 1997; Pelabon et al. 2015). Apparently, lower relatedness among developing seeds does not increase the variance in seed weight as would be predicted under the kin selection hypothesis.

It may be difficult to distinguish between the second and third hypotheses because their predictions are effectively the same: the most vigorous offspring will receive a disproportionate level of resources whether it is due to competition among offspring or preferential provisioning by the maternal parent. On the other hand, the physiological mechanisms responsible for resource allocation and abscission of fruits suggest passive participation by the maternal plant. In particular, fruit and seed provisioning and abortion are thought to be governed by the strengths of "sinks" (the *Pressure Flow Hypothesis*; Lee 1988). Under this model, fruits and ovules compete for maternal resources by establishing osmotic gradients through the uptake of sugars. Auxin production is also implicated in resource provisioning and fruit abortion, as the maintenance of a positive gradient (from fruit to stem) is important for preventing abscission. When multiple fruits are developing on the same stem, they effectively compete by maintaining strong sugar gradients and producing greater amounts of auxin. The same mechanisms may act among seeds developing within the same fruit as sink strength and auxin production are associated with the rate of embryo growth. Under this hypothesis, less vigorous embryos would develop into smaller seeds. The weakest developing seeds may not be able to maintain positive auxin gradients and abortion would be initiated by callose deposition in the chalaza (Fig. 9.9; refer to color plate).

While there will generally be a tradeoff between the size and number of seeds produced, ovule abortion and seed size variation should result in linear returns on maternal investment; the number of seeds produced should generally be a linear function of total investment even when seed size varies. On the other hand, competition among seedlings could result in decelerating fitness returns on investment when suitable germination sites are limited (Dufay et al. 2014). Competition among seedlings may be particularly acute in dense populations of plants that produce large numbers of seeds. In a manner similar to pollen competition for access to ovules, lower proportions of the seeds produced may successfully germinate when only a fixed number of germination sites are available. Such density-dependent responses may set upper limits fitness on fitness returns through pollen and seeds.

The potential for seedling competition to produce decelerating fitness returns depends on the assumptions of limited seed dispersal and limited availability of germination sites. Competition among seedlings could be alleviated by maternal investment strategies that favored the success of offspring across a wider range of environments (McGinley et al. 1987). This hypothesis (maternal control, number 4 in the list) assumes mechanisms that produce variation in seed size that is not necessarily associated with embryo vigor. For example, differences in final seed size may be due to fertilization order, or constraints imposed by the position of ovules within fruits. While most variation in seed size is unordered with respect to position within a fruit, there are some dramatic exceptions to this generality. For example, *Raphanus raphanistrum* produces tapering fruits (siliques) containing seeds that range in size as much as six-fold (Stanton 1984). If variation in mass and other seed characteristics led to variation in dispersal distance of germination success in different environments then competition among siblings could be reduced, and consequently, fitness returns on maternal investment would become more linear.

Summary—It should be clear from these discussions that we generally expect fitness returns on investment in male function to be decelerating and returns on allocation to female function should be linear, or possibly decelerating if post-dispersal sibling competition reduces seedling survival at high densities. Note, however, that these expectations are derived from our understanding of the ecological circumstances affecting reproductive success. While it would be desirable to estimate the shape of fitness gain curves, unfortunately it is difficult to make accurate measures of the parameters needed to estimate fitness returns (Campbell 2000). Our ability to estimate the shape of fitness gain curves depends on the accuracy of assessments of resource investment in each sexual function, the quantification of seeds sired, and estimates of the number of offspring that successfully survive to reproduction. Moreover, estimates of the costs of reproduction through each function in cosexual flowers can be ambiguous because many floral components such as pedicels, receptacles, attractive structures, and pollinator rewards can benefit both sexual functions. Assessment of reproductive success is also intrinsically difficult because of the complexity of tracking the fertilization success of pollen and the survival of seedlings.

While there are a number of impediments for the experimental analysis of sex expression based on fitness returns on investment, the ESS approach provides a conceptual framework that allows insights into the circumstances that might favor the evolution of separate sexes. In the next section, we discuss the distribution of sex expression in plants and evaluate the potential for fitness returns and the avoidance of selfing to explain the evolution of separate sexes.

9.9 SEX EXPRESSION IN PLANTS

While the large majority of plants produce homomorphic cosexual flowers throughout their lives, a small minority of species produce separate male and female flowers that are borne on the same individual (monoecy) or occur separately on individuals in the same population (dioecy, gynodioecy, and androdioecy). In addition to these categories of fixed sex expression, some plants modify investment in male and female function over their lifespans. Changes in investment over a plant's lifespan can be expressed as the production of only male or female flowers depending on plant size or age (diphasic sex expression), but many other plants may continue to produce cosexual flowers while making more subtle changes in the allocation to male and female function. In this section, we explore patterns of sex expression and the evolutionary implications of different sex allocation strategies.

Labile Sex Expression—Perennial plants typically increase in size over their lifespan so the total amount of resources available for sexual reproduction (R) is likely to increase as well. As plants grow, the optimal investment in male and female function may change with R for several reasons (Zhang 2006). For one, the minimal investment necessary for successful reproduction may differ for male and female function (Fig. 9.26). If the minimum investment required for successful reproduction through pollen were lower than for seeds, then smaller plants would accrue higher fitness by investing only in reproduction through male function. Second, there may be a difference in resources necessary for successful parenting of each offspring through pollen and seeds. Even though more pollen grains than ovules may be needed for each seed parented, the total cost of anthers is much less than seeds and fruits in most all species. Consequently, we predict that diphasic reproduction may be advantageous in many perennials, because they are more likely to have higher total fitness by investing only in male function when smaller and then investing only in female function when they are larger (R is higher; Fig. 9.26). This *Size Advantage Hypothesis* was originally applied to animals capable of sex change over their lives (Ghiselin 1969) and has been adapted to predict diphasic sex expression in plants (Lloyd and Bawa 1984).

In spite of our prediction that there may often be fitness advantages for sex change with increasing size, diphasic sex expression in plants is rare (Korpelainen 1998; Vega-Frutis et al. 2014). Diphasic sex expression is known from several plant families including the Araceae (*Arisaema triphyllum*; Jack-in-the-pulpit),

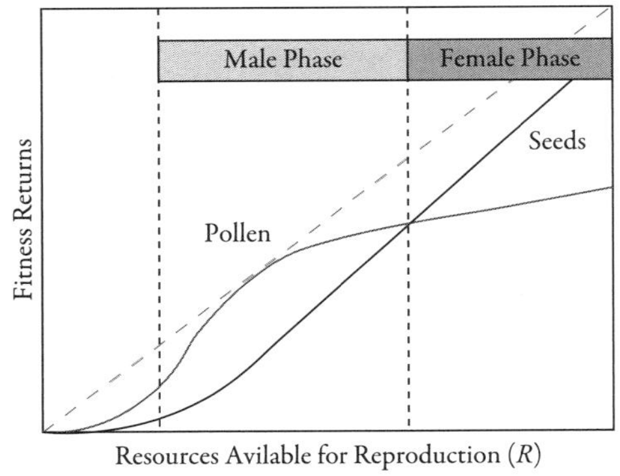

Fig. 9.26. The effects of minimal investment requirements for reproductive success through pollen and seeds on diphasic sex expression. When total resources available for sexual reproduction (R) is lower (small plant size), higher fitness returns can be attained through male function (pollen) than through seeds—small plants should produce only male flowers (male phase). At larger sizes investment is predicted to shift so plants are predicted to produce only female flowers (female phase).

Araliaceae (*Panax ginseng*; ginseng), orchids in the subtribe Catasetinae (*Catasetum viridiflavum*), and trees in the Aceraceae (*Acer* sp.; maples). On the other hand, many plants with cosexual flowers are capable of *Size Dependent Sex Allocation* (Klinkhamer et al. 1997). These species do not display true diphasy (i.e., they typically produce cosexual flowers) but change reproductive efforts through pollen and seeds depending on plant size. Cosexual flowers may be produced over a plant's lifespan, but they may be functionally male when plants are small, while large plants may reproduce through both pollen and seeds. In an extreme case, a species of lily (*Lilium apertum*) produced female sterile flowers when small and larger numbers of cosexual flowers when more resources were available for reproduction (e.g., Zhang et al. 2014).

Diphasic sex expression may be rare because developmental constraints present impediments that prevent the production of unisexual flowers, but it is also possible that fitness gains through the less efficient sexual function are always great enough to favor size-dependent allocation to cosexual flowers. It is important to note that even though fitness returns are typically greater through either male or female function in the scenario depicted in Fig. 9.26, the relative advantage of investing in each sex depends on the slope of the fitness returns. When slopes are similar, it may continue to be advantageous to invest in both sexes as resource availability increases. This is particularly true for larger plants (when R is larger), where fitness returns through male function remain high and may explain the reason for a lack of true diphasy in the example of *L. apertum* mentioned above. Cosexual flowers are also more efficient, since investment in rewards and attraction will benefit both sexual functions. Indeed, it is possible that size-dependent sex allocation is common, but its prevalence is difficult to assess because we do not currently have many detailed studies of sexual function for plants that change size over their lifespan.

While some plants may change expression in each sexual function based on internal resources, external resource availability and habitat quality also appear to be important for sex expression in plants (Vega-Frutis et al. 2014). This may be true for many dioecious plants—from bryophytes to angiosperms—that display *Environmental Sex Determination* where sexes are non-randomly distributed with respect to habitat conditions (Korpelainen 1998; Pannell 2017). In these cases, male plants will typically occur under more stressful conditions or in habitats that have lower nutrient availability, while female plants occupy resource rich locations. In some animals, environmental conditions determine sex expression early in development, but developmental sex determination is rare in plants. For plants capable of producing unisexual flowers, sex expression is generally much more plastic and can depend on both internal (diphasy) and external resource availability. The non-random distribution of sexes with respect to habitat is most often due to labile sex expression, which we can expect to follow the resource-dependent sex expression model described above (Fig. 9.26).

Distribution and Frequency of Dioecy—The large majority of plants are hermaphroditic and produce cosexual flowers, while dioecy occurs in only about 5–6% of species (Renner 2014). Dioecious taxa are widely distributed across the angiosperms, found in 43 plant families with at least 870 independent origins from cosexual ancestors. While generally uncommon, 43% of dioecious species are concentrated in 34 clades that are entirely dioecious. Dioecy tends to be more common in tropical trees, grasses, and wind-pollinated species in general. A good proportion of dioecious species occur on islands, so, as discussed in Chapter 11, we can infer that selection for outcrossing probably played a role in the origin of separate sexes in these cases. This may not be true for the large majority of dioecious species that occur on continents, where it is not always true that dioecy is derived from self-compatible ancestors (Renner 2014). Alternatively, ESS analysis predicts that accelerating fitness gain curves for one or both sexual functions would be adequate to explain the origin of dioecy. While selection for outcrossing and accelerating fitness gain curves present two distinct hypotheses for the origin of dioecy, we currently do not have a clear understanding of how each of these motivations influence the evolution and maintenance of dioecy.

Several pathways have been proposed for the evolution of dioecy (Barrett 2002; Dufay et al. 2014; Käfer et al. 2017). The origin of separate sexes from hermaphroditic ancestors requires the reciprocal loss of one sexual function in different individuals. This is extremely unlikely to occur through simultaneous mutations for each sexual variant, so the evolution of dioecy is generally assumed to occur in a stepwise fashion such that mutations lead to modified flowers displaying complete loss or reduced capacity of one sexual function followed by the origin of flowers specialized for the alternative function (Fig. 9.27; refer to color plate). In all of these scenarios, we assume that plants representing the intermediate steps must accrue higher fitness than the cosexual alternative and will consequently increase in frequency and spread among populations. Since intermediate steps in the evolution of dioecy are predicted to be stable, we expect that clades containing dioecious lineages would also be likely to include species possessing characteristics of the intermediate ancestors. Hence, sexual systems that are viable candidates for intermediate stages that might lead to dioecy include monoecy, gynodioecy, androdioecy, and distyly. All of these pathways may have contributed to the diversity of dioecious species, but some of these sexual systems appear to have more often served as intermediate stages.

The most common pathway for the origin of separate sexes appears to be through an intermediate stage of monoecy (Renner 2014). A relatively high proportion of angiosperm genera (1.4%) include both monoecious and dioecious species, and there is a strong statistical association for the co-occurrence of these two sexual systems. For the transition from monoecy to dioecy to take place, we hypothesize that disruptive selection leads to changes in the proportions of male and female flowers produced by individuals and eventually to populations consisting of plants that produce only one type of flower (Dufay et al. 2014). The monoecy pathway may be particularly common for the large proportion of wind-pollinated dioecious species, as both of these sexual strategies are common in clades of anemophilous species containing both of these strategies (Käfer et al. 2017).

The other possible pathways for the origin of separate sexes include intermediate stages where separate plants produce different flower types: gynodioecy, androdioecy, and distyly. Of these, gynodioecy appears to be the most common, with both dioecious and gynodioecious species occurring in 0.4% of angiosperm genera (Renner 2014), and these two sexual systems occur together more frequently than expected (Dufay et al. 2014). There is much less support for the other proposed pathways. Androdioecy is quite rare, because the conditions

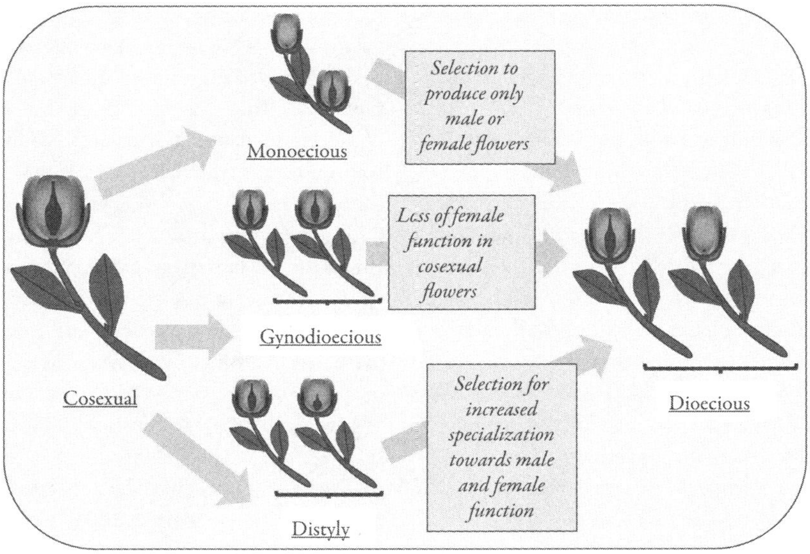

Fig. 9.27. Pathways for the evolution of dioecy. In the monoecious pathway separate sexes is the outcome of selection on the frequency of male and female flowers produced by each plant. When gynodioecy is the intermediate stage there is selection for the loss of female function in cosexual flowers. In the case of distyly, dioecy is the outcome of selection for increased specialization towards male and female flowers from short- and long-styled morphs, respectively.

necessary to maintain male plants, which must compete for access to the ovules of cosexual flowers, are restrictive. Consequently, androdioecy is much less likely serve as an intermediate step for the evolution of separate sexes. Dioecy has also been hypothesized to evolve from distyly through selection for increased specialization toward male and female function in short- and long-styled morphs, respectively (Rosas and Domínguez 2009; Barrett 2010). While this pathway seems plausible, distyly has been independently derived multiple times across the angiosperms, and there are few clades that also include dioecious taxa. The observed lack of associations does not completely rule out the possibility that androdioecy and distyly can serve as pathways to separate sexes, but the rarity of finding these sexual systems together with dioecy suggests that the evolution of separate sexes through androdioecy and distyly is uncommon.

Evolution of Dioecy—Regardless of the pathway leading to the emergence of a dioecious sexual system, the intermediate stages need to present fitness advantages compared to homostylous cosexual flowers. Note that transitional stages may include flowers that reproduce exclusively through either male or female function, so there must be adequate fitness gains for reproduction through one sexual function to exceed the fitness gained through cosexual flowers. As with any other mutation, we would not expect unisexual flower production to spread unless it was advantageous. Since we expect intermediate stages to be evolutionarily stable, it is reasonable to assume that they may persist for extended periods of time. It may also be true that the selection pressure responsible for transition to an intermediate stage may be different from the one leading to dioecy, especially in cases where the transitional step includes two different mating types. In this section, we evaluate evidence for each of these motivations for the evolution of intermediate stages and for the transition to complete dioecy.

From our previous discussions, it is clear that many dioecious taxa fall into a few distinct groups based on their geographic location (i.e., oceanic islands) or life history. General categories of dioecious species include tropical trees that usually bear small, greenish or white flowers (Thomson and Brunet 1990), wind-pollinated plants including grasses and gymnosperms (Renner 2014), and long-lived woody species on islands (Chapter 11). These categories are not comprehensive, as there are examples of dioecy among insect-pollinated herbaceous plants including species in the genus *Antennaria* (pussytoes) and *Anaphalis* (pearly everlasting) in the Asteraceae, *Silene* (catchfly) in the Caryophyllacee, *Solanum* (nightshades) in the Solanaceae, *Humulus* (hops) and *Cannabis* (marijuana) in the Cannabaceae, *Dasiphora* (shrubby cinquefoil) and *Rubus* (blackberry) in the Rosaceae, *Asparagus* in the Asparagaceae, and a number of others. Nearly all dioecious species are herbaceous or woody perennials, and with rare exceptions (e.g., *Mercurialis annua* in the Euphorbiaceae), dioecy is absent among annual plants.

While dioecious species are widely distributed across plant families, this sexual system is still quite rare. The rarity of dioecy has been attributed to higher rates of extinction and lower rates of diversification, but this may not always be the case. For instance, 43% of dioecious species of angiosperms are concentrated in 34 clades that are entirely dioecious (Renner 2014). In all of these instances, it would be reasonable to assume that separate sexes evolved once, followed by diversification from the original dioecious lineage. On the other hand, at least in some species, sex expression is more variable, suggesting that reversals to hermaphroditism with cosexual flowers is feasible (Käfer et al. 2017). It is notable that variable sex expression is generally not present in clades that are completely dioecious, so reversals to hermaphroditism may not account for the general rarity of dioecy. It is also true that chromosomal sex determination is extremely rare in plants, as it occurs in only about 40 species out of 15,600 dioecious plants including species of *Humulus, Cannabis*, and *Silene*. Given that a large proportion of dioecious taxa occur in a small number of clades that are completely dioecious, it would appear that separate sexes could be stable and lead to diversification under some circumstances, which underlines the paradox of the rarity of dioecy among plants.

The rarity of dioecy is also difficult to understand given the broad diversity of dioecious taxa and the number of times that it has evolved (at least 871 independent origins; Renner 2014). Recognizing the large number of independent origins opens up the possibility that different circumstances may have been responsible for the evolution of separate sexes in each case (Käfer et al. 2017). The factors responsible for the origin of dioecy may be difficult to discern from studies of contemporary populations, because ancestral lineages may have evolved separate sexes under different circumstances than those that are responsible for the maintenance of dioecy.

Based on these discussions, we can partition the evolutionary drivers for the origin of dioecy into three categories: 1) the avoidance of selfing in lineages that had previously lost mechanisms of self-incompatibility; 2) accelerating fitness gains leading to investment in a single sexual function (Figs. 9.23 and 9.24); and 3) different levels of resource availability that favor investment in one sexual function (size dependent sex allocation; Fig. 9.26). As discussed in Chapter 11, selfing avoidance is likely to have been the primary motivation for the evolution of dioecy on islands, but the evolution of separate sexes from self-compatible ancestors may also be common in continental regions. For example, monoecious and gynodioecious species—the two most common pathways for dioecy—are typically self-compatible (Dufay et al. 2014), suggesting that selfing avoidance is a likely candidate as the impetus for the evolution of separate sexes (Darwin 1876; Barrett 2002). On the other hand, if separate sexes typically evolve as an alternative to self-incompatibility, then we would expect there to be a strong negative association between self-incompatibility and dioecy across the angiosperms, but the presence of this association has been difficult to resolve (Renner 2014; Käfer et al. 2017).

Part of the difficulty in detecting a strong negative association between self-infertility and dioecy may be the lack of information on the causes of self-infertility. As we have previously discussed, the inability to produce seeds after self-pollination may be caused by either the failure of self-pollen (prezygotic self-incompatibility) or by selective ovule abortion after self-fertilization. Hence, self-infertility may not always be due to self-incompatibility but can be the consequence of early acting inbreeding depression resulting in the abortion of selfed seeds (see the earlier discussion on late-acting self-incompatibility). Abortion of ovules after self-fertilization results in seed discounting, because they are not available for fertilization by outcross pollen. By contrast, self-incompatibility prevents fertilization, so reproductive success through female function is not compromised. Moreover, if the abortion of self-fertilized ovules is due to inbreeding depression, then the impetus for the evolution of dioecy—selection for selfing avoidance—and self-fertility are confounded.

Resolving the confusion over the cause of self-infertility for the evolution of separate sexes may be especially important for dioecious species of long-lived woody shrubs and trees that occur both on oceanic islands and in tropical continental regions (Thomson and Brunet 1990; Käfer et al. 2017). It is notable that nearly all large-statured woody angiosperms are outcrossing, which is probably a consequence of higher genetic load due to their greater capacity for somatic mutation accumulation (Barrett 2017; Chapter 3). It is also notable that large individuals will typically bear large numbers of flowers and consequently, are subject to higher levels of self-pollination through geitonogamy. With the elevated potential for self-pollination and inbreeding depression, we expect that trees lacking the ability to reject self-pollen prior to fertilization would be more likely to be subject to selection for selfing avoidance through other mechanisms such as dioecy. Evaluating the prediction that dioecy is more likely to evolve from ancestors that lack the ability to reject self-pollen will require more detailed studies of mechanisms of self-infertility—and specifically whether it is due to self-incompatibility or to selective ovule abortion—in clades that include both dioecious and cosexual species.

Summary—The evolutionary motivation for the origin of separate sexes in plants has generally been considered to be the avoidance of self-fertilization (Darwin 1876), but Darwin also recognized that there were potential fitness costs associated with dioecy due to lower reproductive assurance and loss of the potential to produce seeds in male plants. The fact that the production of unisexual flowers is not more widespread suggests that the circumstances leading to the origin of diphasy and dioecy are uncommon. The origin of separate sexes is clearest on islands, where the strong association between dioecy and the origin of woodiness from self-compatible ancestors implicates the avoidance of selfing as the primary motivation. For dioecious species that occupy continental regions the situation is more confused. As discussed here, it is possible that the origin of dioecy via selection for selfing avoidance is prevalent, but this hypothesis is difficult to assess without detailed information on the causes of self-infertility.

It is also unclear whether consideration of the shapes of fitness return curves provides much insight into the origin of dioecy. The ESS approach certainly provides a useful conceptual framework to consider for the evolution of separate sexes, but the predictions of these models have proved hard to test in natural systems. On the other hand, it is clear that, for species with labile sex expression, resource availability is important for investment in sexual functions. More detailed analyses of reproductive success for species that have a diversity of sexual systems are needed before we can draw firm conclusions concerning the influence of the shape of fitness return functions for the origin of sex lability and dioecy.

Increases in investment in sexual reproduction generally translate into producing more flowers, but to this point, we have not explored the implications of inflorescence size for reproductive success through male and female functions. In the next section, we evaluate strategies for investment in reproductive functions including the size of floral displays, temporal variation in deployment of reproductive structures on reproductive success, and consequences for the shape of fitness gain curves. We will examine the effects of floral displays, rewards, timing of investment in sexual functions, and returns on allocation to reproduction through pollen and seeds.

9.10 EVOLUTION OF FLORAL DISPLAYS

Floral displays serve the primary purpose of attracting animal pollinators for sexual reproduction through pollen dispersal and receipt (i.e., male and female function, respectively). The production of attractive structures along with nectar and pollen floral rewards emerged as early angiosperms developed stronger associations with their animal pollinators. As discussed in Chapter 2, the behavioral characteristics of the earliest insect pollinators, such as beetles and flies, rendered them ineffective for pollination services, but they were likely an improvement over abiotic alternatives. This would have been particularly true in understory habitats where air movement is limited. Stronger associations with insect pollinators developed in the late Cretaceous with the emergence and diversification of bees (Anthophila). Members of this clade typically engage in methodological foraging behaviors that provide effective pollination services. The parallel diversification of angiosperms and various groups of their animal pollinators is often held up as testimony for the importance of these mutualisms for the success of flowering plants.

While animal pollinators have been important for angiosperm diversification, the relationships between plants and their pollinators are generally compromises, as each partner selfishly attempts to take advantage of the other. Animal pollinators are motivated to acquire floral resources as efficiently as possible regardless of whether their foraging behavior serves the reproductive requirements of plants. Consequently, plants have evolved floral display characteristics designed to manipulate their animal pollinators into foraging patterns that provide more effective pollination services. This conflict of interests results in pollination systems that are not optimal for either the plant or their animal pollinators. In this section, we discuss the evolution of floral display size and its consequences for plant reproductive success through male and female function.

Deployment Strategies—The reproductive success of plants will depend on the total size of the floral display and on how pollen and ovules are distributed (deployed) among flowers. For female reproduction, we could imagine two extremes; one where all ovules are deployed in a single flower and a second where every flower bears only one ovule (Fig. 9.28; refer to color plate). Examples of the former include some aroids (Araceae) such as skunk cabbage (*Symplocarpus* sp.), *Rafflesia* (corpse flowers; Rafflesiaceae), and species in the Melanthiaceae including *Trillium* (trillium) and *Paris* (herb-paris). Plants that produce many flowers with only a single ovule include many fruit trees that produce drupes (stone fruits) in the Roasaceae such as peaches and cherries, species in the Oleaceae including olives (*Olea europaea* and other *Olea* sp.), and trees in the cashew family (Anacardiaceae) such as the mango (*Mangifera indica*). Other examples include the grasses (Poaceae), which produce wind-pollinated flowers that produce a single caryopsis (grain), and trees in the laurel family (Lauraceae) including the avocado (*Percea americana*), which is sometimes considered to be a drupe, but is also referred to as a single-seeded berry because the seed coat is very thin.

The allocation of ovules among flowers may depend on historical levels of pollinator activity or more generally, on the probability of successful pollination. Many species that bear single flowers also have extended periods of anthesis (i.e., flowers remain open

Fig. 9.28. Variation in ovule deployment among flowers and inflorescence size. From upper left to lower right; the corpse flower, *Rafflesia arnoldii*; cherry trees, *Prunus serrulata*; large-flowered trillium, *Trillium grandiflorum*; and an umbel of wild carrot, *Daucus carota*. Photos byIngfbruno, Wilson44691, Christian Fischer, and SofianRafflesia.

longer) and occur in environments where pollinators tend to be scarce. For example, trilliums, herb-paris, and some aroids such as skunk cabbage bloom in early spring when insect abundance is low. Many species of aroids and *Rafflesia* occur in tropical forest understory habitats and are typically pollinated by flies and beetles associated with rotting flesh. The longevity of these carrion flowers may have more to do with the reliability of their insect pollinators then it does with their abundance.

In other cases, plants that are associated with pollinators that are infrequent but more reliable deploy one or a few flowers per day. For example, tropical understory species in the Zingiberaceae such as the spiral gingers in the genera *Costus* and *Heliconias* (e.g., false bird-of-paradise; Fig. 9.29; refer to color plate) typically deploy a single flower each day per inflorescence and are visited by hummingbirds that display consistent foraging patterns (i.e., trap-lining behavior; Ohashi and Thomson 2009). In contrast, plants with large floral displays are generally associated with habitats where insect abundance is high. Species that historically have been in environments with high pollinator densities will generally produce large numbers of flowers and have few ovules per flower.

While historical pollination services may affect patters of ovule deployment, it is important to recognize that the size and number of flowers produced each season can also be determined by architectural constraints. Tree species may be constrained to have large displays, because the number of flowers is associated with the number of apical meristems, so larger trees will also produce larger floral displays. There is much more potential for variation in floral display size among herbaceous species. For example, the species of *Trillium* and *Paris* mentioned above are constrained to produce single flowers each season, but other genera in the same family such as corn lily (*Veratrum* sp.) and death camus (*Zigadenus*) are plants that produce inflorescences with many small flowers and tend to flower later in the season when pollinators are more abundant.

Fig. 9.29. A species of false bird-of-paradise (*Heliconias* sp.) being visited by a green-crowned brilliant hummingbird. Photo taken at Reserva El Copal, Pejibaye, Costa Rica by the author.

While the size and number of flowers produced can vary substantially within many plant families, it is also clear that ovule deployment is sometimes subject to phylogenetic constraints, because consistent ovule numbers per flower sometimes characterize whole plant families or subfamilies. For example, plants in the Boraginaceae (forget-me-not family) and Laminaceae (mint family) always have four ovules per flower, several genera are characterized by drupes that have only a single ovule per flower, members of the Maloidae (Rosaceae) such as apples, pears, crabapples (*Cretaegus* sp.), and shadbush (*Amelanchier* sp.) produce flowers that consistently have ten ovules, and species in the Apiaceae (carrot family) are characterized by fruits consisting of a two-seeded schizocarp. For the larger majority of species that produce dry capsules, berries (e.g., watermelons), aggregate fruits (e.g., raspberries), siliques (mustards), or legumes (peas and beans), ovule number per flower is variable to some degree, with smaller flowers and fruits typically containing fewer ovules and seeds.

For many species having flowers with only one or a few ovules, multiple flowers are often aggregated to form an inflorescence that functions as a single flower. This is generally true of plants in the Apiaceae, which have flowers clustered into umbels that allow pollinators to forage on multiple flowers by crawling, rather than flying, from one flower to the next (Fig. 9.28; refer to color plate). Clustering of flowers is also characteristic of plants in the Asteraceae (sunflower family), where each floret produces a single seed, and many florets are borne on a single receptacle to form a single composite flower.

The reproductive potential of plants is not a simple function of total number of flowers produced but may depend on the timing of anthesis in individual flowers, which will determine the size of the floral display across the flowering season. In general, we expect that the deployment of attractive structures and floral rewards should be the outcome of selection mediated by the most prevalent and effective pollinators (Chapter 10). Hence, the optimal strategy may depend on both the historical abundance of pollinators and their energetic needs. Flowers that are typically pollinated by larger animals such as birds and bats will be larger and sturdier (in the case of bat pollination) and will tend to have high nectar production. As mentioned earlier, plant species associated with trap-lining hummingbirds will tend to dole out flowers one at a time over extended periods. For these and other bird and mammal pollinated species, the energetic cost of flower and nectar production is high, so individual flowers will tend to have greater amounts of pollen and larger numbers of ovules. In contrast, plants pollinated by animals with lower energetic needs, such as bees and butterflies, will generally produce larger numbers of flowers with less pollen, fewer ovules, and less nectar. The set of floral characteristics—including flower size, floral scent, shape, and color of the flower, and nectar composition and volume—that are generally associated with each different type of pollinator are referred to as *Pollination Syndromes* (Chapter 10). Beyond these broad generalities, there is much variation in the characteristics of flowers and floral displays among species associated with each type of pollinator.

Consequences for Reproductive Success—We generally assume that increasing inflorescence size will increase reproductive success, but the contributions to fitness through male and female function may not increase equally. Increasing the number of flowers—and floral display size in general—has often been interpreted as an outcome of selection through male function. A number of studies have found that pollen receipt will generally saturate after a few visits to a flower, while pollen removal, and consequently reproductive success through male function, may continue to increase (Bell 1985; Harder and Johnson 2009). The hypothesis that floral displays may be more strongly associated with male reproductive success is supported by the observation that male plants in dioecious species often have larger flowers (Lankinen and Larsson 2009) and retention of post-pollination flowers that enhance display size may increase pollen removal more than pollen receipt (Cruzan et al. 1988). On the other hand, when pollinators are scarce, female reproductive success is more likely to be limited due to lack of pollen deposition on stigmas, and we expect that larger displays will accrue increased reproductive success more equally through male and female function.

Many perennial plants enhance their floral displays by producing more flowers than they are capable of maturing into fruits (overproduction of flowers; Sutherland 1986). The number of ovules produced each season represents the maximum fitness return that might be accrued through female function, but only if every ovule were fertilized and if there were adequate resources to mature all of them into seeds. Hence, reproductive success not only depends on the potential for maximum seed production, but also on successful delivery of high-quality pollen (i.e., from unrelated pollen donors). As discussed previously, seed set can be lower than the total number of ovules produced because of both inadequate pollination and selective ovule abortion.

The overproduction of flowers followed by high levels of fruit abortion may be a mechanism to increase fitness through both male and female reproductive success. As discussed earlier, larger floral displays generally attract more pollinators and consequently, increase reproductive success through male function. For female reproductive success, it is generally thought that pollen limitation of seed production is common, so increasing pollinator attraction through larger displays should also benefit female success when pollinators are uncommon (Harder and Johnson 2009). Furthermore, even when pollinators are abundant, the presence of larger numbers of developing seeds and fruits may allow for the culling of inferior offspring through selective ovule abortion (Bawa and Webb 1984; Korbecka et al. 2002; Sadras and Denison 2009). For example, self-fertilized seeds are more likely to be aborted when outcrossed seeds are developing in the same fruit (e.g., Rigney et al. 1993), and selective ovule abortion can lead to improved offspring performance (Cruzan and Thomson 1997; Mena-AlÍ and Rocha 2005). By terminating less vigorous offspring early in development, resources may be redirected for allocation to vegetative growth and other functions contributing to fitness. The ability to filter out weaker genotypes may be particularly important for plants with larger inflorescences; even though larger displays may attract more visitors, individual pollinators will tend to visit more flowers on the same plant, and consequently will increase selfing through geitonogamy (Karron and Mitchell 2012). Increasing inflorescence size may result in diminishing returns on investment in female function because as selfing increases, so will seed and fruit abortion, or alternatively, if selfed ovules are not aborted, then increased inflorescence size will lower the quality offspring of offspring produced.

From these discussions, it is clear that pollen dispersal and deposition is often limited by the behavior and abundance of animal pollinators, and we can predict that pollen presentation by plants can be influenced by selection through male reproductive success in several different ways (Thomson 2006). First, plants often extend their flowering period and effectively meter out pollen over time by opening flowers sequentially on inflorescences. As discussed earlier, this can be accomplished by opening individual flowers over extended periods of time. Doling out flowers individually or a few at a time could be advantageous for species in low-density pollinator environments for several reasons. First, it will increase reproductive success through male function, because extending the flowering period will increase the chance of attracting more pollinators. Second, parsing flowers over the flowering season can also increase reproductive success through female function, because it is likely to reduce levels of geitonogamy. Third, low pollinator densities may affect patterns of anther dehiscence within flowers. For example, anthers may dehisce sequentially over the life of the flower, or individual anthers may release pollen slowly. The timing of flower deployment and pollen release appear to depend on the historical pollinator environments; species that are typically exposed to low densities of pollinators (e.g., those flowering in early spring) have slow rates of pollen dehiscence and extended flowering, while those that flower when pollinators are more abundant may release their pollen over a shorter period (Thomson 2006).

9.11 THE DIVERSITY OF PLANT REPRODUCTION

The discussions here provide just a small taste of the diversity of floral forms and strategies for sexual and asexual reproduction across the diversity of flowering plants. It should be clear that a complex array of selective pressures and evolutionary constraints have colluded to generate an impressive range of diversity in flowers, fruits, and seeds. Much of the complexity and diversity of plant reproductive structures originates from pressures associated with a sedentary life form, which necessitates the cooption of biotic and abiotic vectors for successful reproduction and dispersal. While we have mostly focused on selection processes operating within individual populations and species, we recognize that plant communities often consist of many species that may flower simultaneously. Hence, selection may not only serve to improve the efficiency of pollination within a species but may at times operate to reduce interference among species flowering together in the same environment. In the next chapter, we continue our discussion of plant reproductive processes as we explore mechanisms of reproductive isolation among species.

9.12 QUESTIONS FOR CONTEMPLATION AND DISCUSSION

1. What are the advantages and disadvantages of recombination (outbreeding)? In particular, how does recombination accelerate adaptation? Under what circumstances would sexual reproduction be disadvantageous?
2. What are the advantages and disadvantages of asexual reproduction? Under what circumstances would asexual reproduction be more advantageous then sexual reproduction.
3. What is the cost of sex? How do demographic and genetic transmission deficits contribute to the cost of sex?
4. Describe each model for the evolution of sex. List the assumptions and predictions for both short-term and long-term advantages of sexual reproduction.
5. Describe the different components of genetic load including drift, inbreeding, and recombination (segregation) load. What is the relationship of each to the advantages of sexual reproduction?
6. What are the circumstances that might favor the evolution of agamospermy? How do weedy agamospermous species differ from those that colonize range edges?
7. What is genome hitchhiking, and how is it important for the evolution of asexual lineages?
8. Describe how sexual conflict and the avoidance of selfing may have contributed to the evolution of different modes of herkogamy and dichogamy.
9. What are the possible fitness tradeoffs between self-incompatibility and self-compatibility?
10. What are the advantages of flexible mating systems such as delayed selfing and cryptic self-incompatibility?
11. Can you design an experiment that would allow you to discriminate between early-acting inbreeding depression and late-acting self-incompatibility as the mechanism causing selfed ovule abortion?
12. Why do we expect the distribution of outcrossing frequencies to be bimodal?
13. Describe each of the models that have been proposed to explain the prevalence of mixed mating systems. What are the strengths and weaknesses of each explanation?
14. What is the ESS explanation for the evolution of dioecy? Why is this model so difficult to test? How well do ESS and selfing avoidance models predict the occurrence of dioecy?
15. Describe the conditions that can contribute to linear and decelerating fitness gain curves for male and female function.

16. How do resource availability and environmental conditions affect sex expression in plants?
17. What are the circumstances that affect the evolution of seed size? Include descriptions of ecological conditions, maternal-offspring conflict, offspring vigor, bet hedging, and maternal provisioning in your answer.
18. How does the behavior and availability of pollinators affect the deployment of ovules among flowers, the longevity of flowers, the deployment of flowers, and the display size?

10

REPRODUCTIVE ISOLATION AND HYBRIDIZATION

Hence maintenance of life is possible only if the gene patterns whose coherence is tested by natural selection are prevented from disintegration due to unlimited hybridization.
Theodosius Dobzhansky, 1937, Genetic Nature of Species Differences, American Naturalist

The rapidity of evolution in these bursts of creative evolution may well have been due to hybridization. At such times diverse faunas and floras were brought together in the presence of new or greatly disturbed habitats where some hybrid derivatives would have been at a selective advantage.
Edgar Anderson and George Ledyard Stebbins Jr., 1954, Hybridization as an Evolutionary Stimulus, Evolution

10.1 GENETIC ADMIXTURE

Speciation is usually a slow process. As isolated lineages diverge through selection and the accumulation of mutations through genetic drift, they generally retain the capacity to interbreed—often for extended periods of time (Fig. 10.1; refer to color plate). If two divergent lineages (taxa) by chance come into contact, there may be opportunities for inter-lineage mating (*Hybridization*). If these matings are successful, they will result in populations consisting of hybrid offspring of mixed genetic ancestry (admixture). The outcomes of interactions between lineages often depends on the time since divergence. Depending on phenotypic and genetic differences, pollinations between plant lineages may fail at any stage from pollen transfer to stigmas through fertilization of ovules, seed development, or the viability or fertility of hybrid offspring. Even if first-generation hybrids are viable they may fail if the parental taxa occupy different habitats or are associated with different pollinators. At the other end of the spectrum, more closely related lineages may produce viable and fertile offspring that interbreed and backcross to the parental taxa. If hybridization and backcrossing are successful to some degree, then beneficial genetic elements may be transferred from one lineage to the other. It is this type of hybridization that Edgar Anderson and G. Ledyard Stebbins (1954) were focusing on in the quote at the beginning of the chapter; hybridization and inter-lineage gene flow (introgression) may have provided the necessary genetic diversity to promote bursts of rapid evolution and the adaptive radiations that mark the phylogenetic histories of the immense diversity of species that surrounds us today.

The potential for different species to interbreed has been widely recognized for some time. Hybridization has been used for the improvement of domesticated plants and animals for centuries. While there is consensus that hybrids and hybrid populations exist in nature, there has been much less agreement on the evolutionary importance of hybridization. With a few notable exceptions, early scholars viewed hybridization as an aberration or evolutionary noise of little or no consequence and natural hybrids as "mistakes" that are infertile or unlikely to succeed under natural

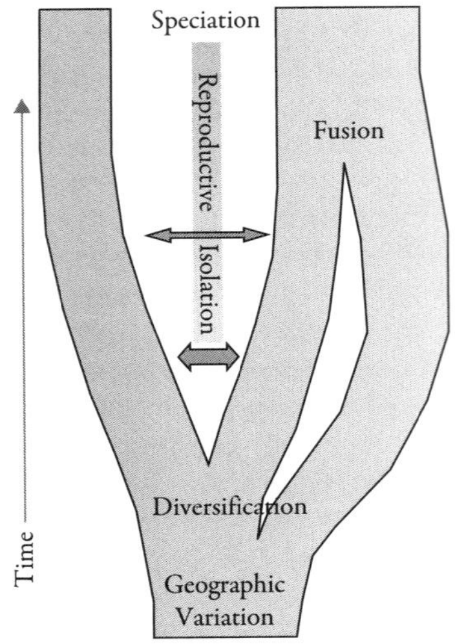

Fig. 10.1. The process of divergence and development of strong reproductive barriers over time. Horizontal arrows indicate hybridization when reproductive isolation is incomplete.

conditions. This view began to change with the introduction of improved genetic marker and sequencing technologies in the late twentieth century, and now, the prevalence of hybridization and its importance for diversification are firmly established. In this chapter, we discuss the history of changing views on the importance of hybridization, mechanisms that isolate lineages and species, the processes of hybridization and introgression, and the conceptual models that have been developed to account for hybridization and its contribution to diversification.

10.2 CHANGING VIEWS ON HYBRIDIZATION

While the importance of hybridization for the evolution of plants and animals is widely accepted today, this was not always the case. The recognition that different pairs of lineages that are classified as species could occasionally interbreed dates back to the work of Carl Linnaeus, who was the first to propose that hybridization could result in new species (Briggs and Walters 1969). These ideas were largely ignored, or considered to be of minor importance for the evolution of species, and today, Linnaeus is much better known for his contributions to the field of systematics. For most of the last several hundred years since the publication of Linnaeus's *Systema Vegetabilium* in 1774, the prevailing attitude has been that hybridization is inconsequential for evolution. During the nineteenth century, this reluctance may have stemmed from the view that species were fixed entities (*Ideal Forms*) that were established by the creator, so any mongrel progeny were most likely doomed to failure. Even as ideas of transmutation, and later, evolution, became more widely accepted, species were viewed as outcomes of natural selection toward fitness optima, so intermediates were likely to be unfit. Darwin seemed to take a more nuanced view: the wide range of fertility between species suggested they fell along a continuum, even as they were inevitably separated by morphological gaps that presumably reflected the existence of separate adaptive optima (Darwin 1859; Mallet 2008).

Hybridization remained largely disregarded through the mid-twentieth century. One exception to the prevailing view was the proposal made by Johannes Lotsy (1916) that the generation of wide varieties of phenotypes in the F_2 generation of hybrid crosses was a mechanism for species diversification. In spite of the fact that Lotsy provided a number of examples of domesticated plants and animals that were derived via hybridization and described cases of viable hybrids persisting in nature, these ideas did not gain acceptance and were largely ignored. Part of the reason may be that, as Heiser (1949, page 646) suggested, Lotsy overstated the case as he "attempted to explain all evolution in terms of crossing."

Lotsy's arguments were apparently not persuasive enough to sway the prevailing opinion that species were carefully shaped by natural selection to match their habitat requirements and consequently, any hybrids generated would inevitably fall between fitness optima. This mind set is reflected in the *Ecological Niche* concepts that ecologists were developing during the early to mid-twentieth century (Schoener 2009). Ecologists, as well as most biologists, viewed species as occupying well-defined niches and assumed that all available niches were filled; by implication, there would be no resources for intermediate individuals (hybrids).

In the 1940s, Edgar Anderson developed an explanation for the persistence of hybrid populations (*Hybridization of the Habitat*) that accounted for fitness optima by proposing that disturbances and ecotones created novel niches where hybrids could persist (Anderson 1948). Anderson supported his arguments with detailed descriptions of hybrid populations of Louisiana irises (*Iris fulva, I. hexagona,* and *I. brevicaulis*) and several other species pairs where hybrid plant populations occupied intermediate or disturbed habitats (Anderson 1949). Anderson was the first to propose that these regions of phenotypic intermediates (*Hybrid Zones*) occurred along ecotones and often consisted of stable hybrid populations. Anderson's "hybridization of the habitat" model immediately attracted the attention of many plant biologists, because it was based on the adaptive norm paradigm and allowed for the integrity of the hybridizing species to be maintained.

Anderson's detailed investigations and models for the persistence of hybrid populations inspired many plant biologists to recognize the elevated importance of hybridization (Heiser 1949; Baker 1951; Grant 1953; Grant 1981). He later expanded the importance of hybridization in an influential paper, written with George Ledyard Stebbins, that caste "hybridization as an evolutionary stimulus" (Anderson and Stebbins 1954). According to Anderson and Stebbins, hybridization—facilitated by intermediate and disturbed habitats—had been an important mechanism of species diversification throughout evolutionary history. These arguments echoed previous proposals for a creative role for hybridization made by Linnaeus and Lotsy. While these ideas were quite popular among plant biologists, most zoologists continued to consider hybrids rare and assumed that hybridization did not contribute in any significant way to the evolution of species—at least for animals. This view was promoted primarily by two influential zoologists—Theodosius Dobzhansky and Ernst Mayr (Dobzhansky 1937; Mayr 1942)—and prevailed until the late twentieth century (Dowling and Secor 1997).

The decades following the publication of Anderson's book *Introgressive Hybridization* (Anderson 1949) saw a deflation of the enthusiasm for the creative role of hybridization in the evolution of plants. Several authors voiced the opinion that "hybrid swarms" found in nature were transitory and likely to remain isolated, so they would have little influence on the evolution of species (Heiser 1973). While it was widely accepted that hybrid populations could occupy disturbed and intermediate habitats, there was increasing skepticism over the role of introgression. This sentiment reflected the opinions of influential authors such as Mayr, who viewed species as the "endpoints" of the process of speciation and introgression as genetic contamination that would most likely reduce the fitness of a species (Mayr 1942, 1963). Hybrids were generally considered to be unfit, so selection would surely favor characters that would strengthen reproductive isolation—an idea that had originated with Alfred Russell Wallace (1889). The idea that selection would strengthen reproductive barriers between hybridizing species was first referred to as the *Wallace Effect*, but was later called *Reinforcement* by Remington (1968), and the latter term is now more widely used. Doubts over the influence of introgression arose primarily from the fact that it was difficult to detect solely from observations of phenotypic traits. Heiser (1973) pointed out that evidence from crossing experiments supported the potential for introgression but lamented that adequate tools for detailed studies of introgression were not available—a situation that would soon change.

Lack of wide acceptance of introgression was consistent with the view that natural selection was the sole force shaping species. This optimistic view of species as near-perfect fits to their adaptive norms dominated the mid-twentieth century and has some similarities to previous depictions of species as "ideal forms," but with the creator replaced by an all-powerful version of natural selection. Randomness was given only a limited role in evolution, as were the potential effects of phylogenetic history and developmental constraints (Chapter 6). This idealized view of natural selection as a pervasive power that molded species to its will became more tempered as more authors found flaws with the "Adaptionist Programme" (Gould and Lewontin 1979). The adaptionist view was criticized as being Panglossian: taking the view that every species was the best of all possible versions. Gould and Lewontin pointed out that traits are often derived from preexisting ones that become co-opted to serve new purposes (*Exaptations*; Gould and Vrba 1982) and are generally not perfect in their new functions. This was not a novel concept but was similar to a proposal made by Darwin for the origin of complex traits from earlier ones that had served different functions in his sixth edition of *On the Origin of Species* (Darwin 1872)—an argument that the adaptionists had apparently overlooked.

The acceptance of the view of species as imperfect or as the "best solutions available" coincided with the accumulation of new information from genomics and biochemistry that supported more prominent roles for randomness and cooption in the evolution of genes and proteins (Kimura 1983; Brosius and Gould 1992). A few short decades toward the end of the twentieth century saw the collapse of the adaptionist program and the rise of the *Neutral Theory* as a more sophisticated view of evolution emerged—one that balanced the roles of random processes and constraints with the formative powers of natural selection.

With the development and application of DNA sequencing technologies for the study of natural variation, the frequency of hybridization became more widely appreciated. Phylogeneticists working with sequence data proposed that cases of *Reticulation*—where cytoplasmic and nuclear genomes indicate contrasting phylogenies—were likely due to historical hybridization (Rieseberg and Wendel 1993; Arnold 1997). Plant biologists had also documented numerous cases of recent and historical allopolyploidy (whole-genome duplications) that most likely originated from instances of hybridization (Chapter 3). A broader appreciation of high frequencies of historical hybridization from cytogenetic and phylogenetic investigations coincided with increased application of genetic markers for the study of contemporary patterns of hybridization and introgression in plants and animals (Arnold 1997; Dowling and Secor 1997). Genetic markers provided tools to study hybrid zones and introgression in much greater detail than had previously been possible, and several influential papers during this period firmly established the evolutionary importance of introgressive hybridization (e.g., Rieseberg et al. 1988; Mallet and Barton 1989; Harrison 1990; Cruzan and Arnold 1993). Advances in genetic marker technologies in the early twenty-first century have allowed for highly detailed analyses of introgression in a wide range of species (Mallet 2008; Abbott et al. 2013)—including humans (Stringer 2012).

The emerging view of hybridization recognizes that introgression during early stages of species divergence may be prevalent (Fig. 10.1; refer to color plate), but discussion continues over the importance of hybridization for adaptation and diversification (Seehausen 2004; Arnold et al. 2012; Abbott et al. 2013; Barton 2013). While many of the same arguments over levels of introgression and its impact on speciation processes continue, our appreciation of the impact of hybridization has become much more nuanced as we recognize that susceptibility to genetic exchange among lineages varies across the genome. Changing views on the creative potential of natural hybridization has begun to influence the perspective of conservation biologists (Jackiw et al. 2015). While hybridization has traditionally been viewed as a threat to the integrity of species, there is a growing realization that in some cases genetic exchange among lineages may pose a viable solution to the challenges imposed by climate change and other threats to sensitive species (Taylor et al. 2015). In the next sections, we discuss the processes associated with introgression and then explore the ecological and genetic processes that promote reproductive isolation among divergent lineages. In the last section, we describe models of hybrid zones and the consequences of hybridization and introgression.

10.3 THE PROCESS OF ADMIXTURE

Anytime that closely related lineages come into close contact, there is potential for genetic exchange. If the lineages are interfertile, the hybrids resulting from these encounters can display a spectacular range variation. These instances of hybridization have fascinated naturalists and modern evolutionary biologists. The ease with which hybrids are formed within many plant groups has inspired breeders to develop numerous novel strains of hybrids that have formed the foundation of the agricultural and horticultural industries. In this section, we define the processes and outcomes of hybridization.

Defining Hybridization—Gradual divergence results in a continuum for the exchange of genetic material, from gene flow within and among populations to introgression via hybridization (admixture) among more divergent groups of populations (lineages or taxa; Fig. 10.1; refer to color plate). There are no natural discontinuities across this range of divergence, so any definitions that we might propose to distinguish gene flow within species from introgression will be somewhat artificial. In spite of this difficulty, a number of definitions of hybridization have been made over the last few centuries, but we will focus on those that have been discussed since the mid-twentieth century.

The use of the term "hybridization" has sometimes been reserved for crosses between species, but Stebbins (1959) points out that the reasoning behind this definition is circular; species are generally considered to be reproductively isolated lineages (biological species concept; Mayr 1942), in which case hybridization would not be possible. To circumvent this problem Stebbins defined hybridization as "crossing between individuals belonging to separate populations which have different adaptive norms" (Stebbins 1959; page 231). Like most evolutionary biologists of his time, Stebbins viewed divergent lineages primarily in an ecological context (i.e., adaptive norm = niche). Harrison (1990; page 70) broadened the definition of hybridization to consider interbreeding between groups of populations that differed by "one or more heritable characters" regardless of whether the lineages were ecologically divergent. Arnold (1997) followed Harrison's definition but was careful to limit the focus to cases of natural hybridization where viable and fertile offspring are produced at some frequency in the F_1 generation. The definitions used by Harrison and Arnold are an improvement over the one proposed by Stebbins, but they do not adequately capture cases of genomic divergence that are not necessarily reflected by phenotypic differences.

In this text, we use a broadened definition of hybridization that includes instances of gene flow between groups of populations (lineages) having unique evolutionary histories. To distinguish hybridization between lineages from gene flow within lineages we consider cases where a period of isolation has allowed for the accumulation of genetic differences. Note that under this definition, lineage divergence does not require fixed alleles, only that the period of isolation has been long enough for distinct genetic profiles to develop in each lineage. This definition allows us to include cases of admixture between individuals from different geographic regions that may or may not be distinguishable based on phenotypic characters. We delineate hybridization from cases of gene flow among populations that are likely to have had some level of genetic exchange within a relatively short time frame (~50 to 100 generations). By categorizing genetic exchange between divergent populations as cases of hybridization, we recognize that the outcomes of genetic admixture are distinguishable from gene flow within lineages, because they have unique evolutionary implications. The goal of this chapter is to explore the processes and outcomes of interbreeding among divergent lineages of plants. We will describe the processes of hybridization and genetic exchange among lineages (introgression) and its impact on evolution and diversification.

Introgression—In natural systems, introgression proceeds as dispersal occurs among divergent populations followed by the production of hybrid offspring (Fig. 10.2). This same process is mimicked as "artificial introgression" for the improvement of horticultural and agricultural species (Fig. 10.3). Crossbreeding among divergent lineages has been used as a tool in agriculture for perhaps thousands of years and accounts for the origin of a number of domesticated plants and animals (Lush 1937; Vavilov and Love 1992). For crop improvement, breeders make crosses to a lineage that possesses one or more desirable traits. The goal is to transfer individual traits from one line to another without compromising the characteristics already established in the crop lineage. This is accomplished by first crossing the parental lines to create an F_1 generation, and then repeatedly backcrossing to the crop lineage. With each backcross generation, offspring with desirable combinations of traits are again crossed back to the original parental

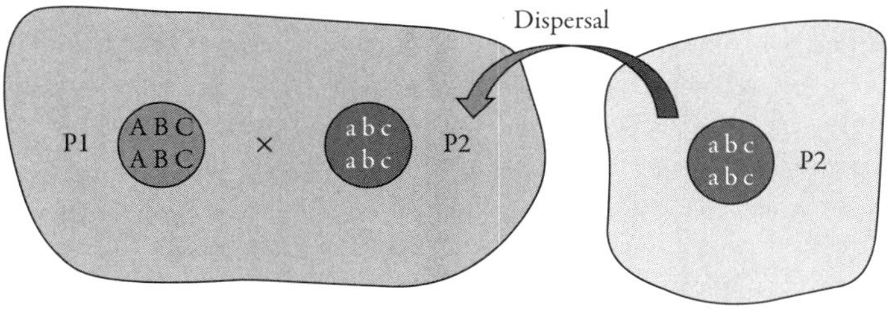

Fig. 10.2. A model for the initiation of hybridization following dispersal.

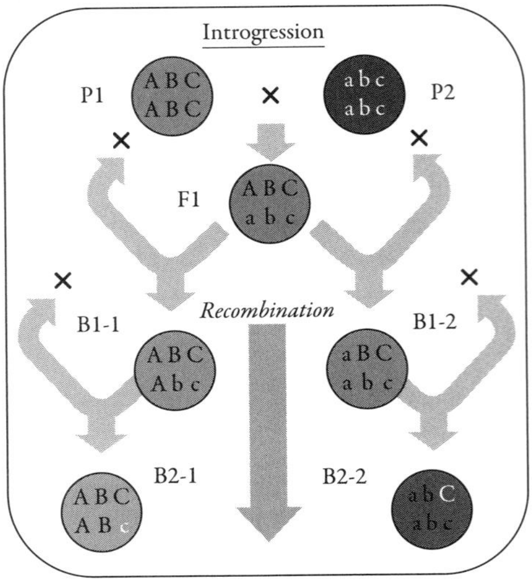

Fig. 10.3. Introgression of genetic material from one lineage to another occurs through repeated backcrossing to each parental line. Hybridization is initiated when the two parental lineages (P1 and P2) cross (indicated by the X) to produce an F1 hybrid. Introgression occurs through repeated backcrossing between hybrids and the parental lineages.

line (Fig. 10.3) until the selected traits are transferred into a new hybrid lineage that also retains the original phenotypic profile.

All of the generations produced by interbreeding divergent lineages are considered hybrids, but each generation is characteristically different from the others. The F_1 hybrids are genetically identical and completely heterozygous across all loci that were fixed for alternative alleles in the two parental lineages. The chromosomes inherited from the parental lineages have not recombined, so they remain intact in first-generation hybrids (Fig. 10.3). Recombination occurs when the F_1 hybrids produce gametes, so the next generation of hybrids and all subsequent generations will inherit chromosomes that are mosaic combinations of chromosomes from each of the parental lineages. Consequently, generations after the F_1 including F_2 to F_n hybrids (i.e., from F_1 x F_1 crosses and their offspring generations) and backcross hybrids (hybrid generations with a history of crossing to a parental line) are considered to be *Recombinant Hybrids* (Fig. 10.3). With each additional hybrid generation, recombination continues to break up associations, so the chromosome blocks inherited from the parental lines are divided into increasingly smaller genomic regions.

After the F_1 generation, each recombinant hybrid offspring inherits a unique combination of alleles from the original parental lines, so the phenotypic variation among individuals is generally very high—at least spanning, and sometimes exceeding (transgressive segregation—see next section) the range represented by the parental phenotypes. Natural hybridization commonly results in highly variable populations consisting of early-generation recombinant hybrids. Natural *Hybrid Swarms* (*Admixed Populations*) occur in numerous plant genera (e.g., Lotsy 1916; Anderson 1948, 1949; Anderson and Stebbins 1954; Stebbins 1959; Arnold 1994, 1997; Cruzan 2005). Once recombinant hybrid populations are formed, the opportunity exists for introgression into the parental lineages (*Introgressive Hybridization*).

Transgressive Segregation—In hybrid generations, trait expression exceeding the range within and between parental lineages is commonly observed after both inter- and intraspecific crosses (Rieseberg et al. 1999). Although transgressive expression had been known for some time (Lotsy 1916; Darlington and Mather 1949), Stebbins (1959) was the first to define the wide variation observed in recombinant hybrid generations as "transgressive variation" and suggested "transgressive segregation" as a possible explanation, but he did not describe any specific genetic mechanisms that might be responsible. Rieseberg et al. (1999) summarized the explanations that have been proposed to account for transgressive segregation and conducted a survey of crossing and QTL mapping studies (see Chapter 7) in plants and animals. These authors concluded that transgressive expression is primarily due to the complementary (additive) effects of alleles derived from the parental lineages, with more minor contributions from epistasis and overdominance. This is an important result, because it suggests that a large proportion of transgressive variation will be heritable, which would facilitate adaptive responses to novel environments. Transgressive variation in hybrid populations has the potential to promote the evolution of novel lineages with trait expression outside the range of parental taxa (*Transgressive Evolution*). Transgressive variation has been noted for crosses between species of *Nicotiana* (Stebbins 1959) and accounts for the origin of novel phenotypes in the hybrid species *Helianthus paradoxus* (Lexer et al. 2003) and a hybrid lineage of *Piriqueta cistoides* ssp. *caroliniana* (Anton et al. 2013; Fig. 7.11; refer to color plate). In the next section, we explore the processes that may facilitate or inhibit the formation and survival of hybrids as we discuss mechanisms and evolution of reproductive isolation between divergent lineages.

10.4 REPRODUCTIVE ISOLATION

Recently diverged lineages often come into close proximity through changes in their geographic ranges. As described in Chapter 8, multiple glacial cycles during the Pleistocene provided opportunities for divergence followed by range expansion, which generated numerous instances of hybridization among divergent lineages of plants. This history is evident from numerous suture zones characterizing regions of contact between the geographic ranges; but not all of these instances of contact resulted in hybridization.

The outcome of interactions between recently diverged taxa will depend on the collective strength of the reproductive isolating mechanisms that separate them. Closely related lineages may come into contact through changes in their geographic ranges, but hybridization would only occur if the pollen of one taxon can successfully fertilize the ovules of the other to produce viable and fertile offspring. If the combined strength of reproductive barriers between sister taxa is great enough, then there may be little or no intermixing, but even a low frequency of fertile hybrids may generate opportunities for introgression (e.g., in Louisiana irises; Arnold et al. 2012). To understand the potential for introgressive hybridization, we need to evaluate the sequence of barriers occurring at each stage of sexual reproduction that may collaborate to prevent the formation and success of hybrid progeny.

Reproductive barriers in plants are best envisioned as a sequence of steps that lead to the production and establishment of offspring (Fig. 10.4; refer to color plate). Successful production of hybrid offspring requires transfer of pollen to stigmas, pollen germination and growth through the style, ovule fertilization, seed development, and germination and survival of seedlings (see Chapter 9 for a more thorough discussion of plant reproduction). Mechanisms of reproductive isolation between taxa are traditionally partitioned into pre- (behavioral and mechanical) and post-zygotic (viability and fertility) stages. In some animals, these two stages are often treated equivalently as pre- and post-mating; however, there is potential for selective fertilization after copulation in plants and animals (Dobzhansky 1937; Howard 1999), so a third division for post-mating processes seems appropriate (Fig. 10.4; refer to color plate). For plants in particular, processes occurring after pollination (i.e., mating) and before fertilization can have substantial effects on gamete transmission and the formation of hybrid zygotes. Failure of interspecific pollen may occur at a number of stages, including pollen germination and pollen-tube growth through the style. Post-pollination processes contribute to *Conspecific Pollen Precedence* (Howard 1999) that may function in collaboration with pre-mating and post-zygotic mechanisms to prevent hybridization.

In the next sections, we discuss the process of reproduction to provide an overview of the prevalence and efficiency of reproductive isolation mechanisms in plants. We will start with *Pre-Mating Isolation* mechanisms that are non-floral (i.e., geographic, ecological, phenological, and autogamous) and then we discuss mechanisms of *Floral Isolation*, which

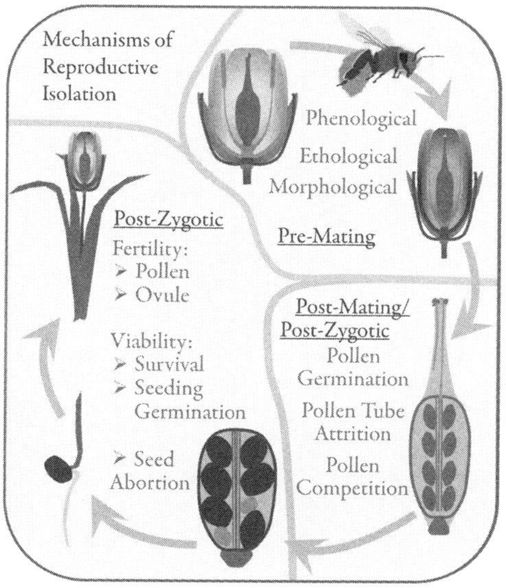

Fig. 10.4. Mechanisms contributing to reproductive isolation. Pre-Zygotic isolation is divided into Pre-Mating (pre-pollination) and Post-Mating (post-pollination).

are pre-mating isolation mechanisms that involve pollinator behavior (*Ethological Isolation*) and morphological isolation based on the form and function of flowers (*Morphological Isolation; Mechanical Isolation*). Finally, post-mating isolation (conspecific pollen precedence) and post-zygotic mechanisms will be discussed in separate sections below.

10.5 PRE-MATING ISOLATION

Physical separation has probably been the most effective and the most common reproductive barrier promoting the formation of divergent lineages (Ramsey et al. 2003; Schemske 2010; Baack et al. 2015). *Geographic Isolation* can effectively limit gene flow among populations and allow further divergence through the accumulation of novel mutations, genetic drift, and selection for local adaptation. As discussed in Chapter 8, geographic separation can occur either by long distance dispersal or by fragmentation of an existing range. Within continental regions, fragmentation in response to climate change—the Pleistocene glacial cycles in particular—has probably been the most common mechanism promoting isolation. Long distance dispersal has promoted the evolution of many endemic lineages on islands but probably has been much less important for the establishment of geographic isolation within continents.

Divergent lineages have frequently been brought into close contact through changes in geographic ranges, but in cases where ranges overlap, other mechanisms may preserve isolation. Opportunities for interbreeding may not exist if divergent lineages are associated with different habitats (*Ecological Isolation*), if flowering times do not overlap (*Phenological Isolation*; Fig. 10.5), or if one or both lineages are primarily selfing (*Autogamous Isolation*). Ecological isolation is common in cases where lineages have become adapted to different habitats, but this mechanism generally does not operate on its own. As an example, the mine-tailing ecotypes of *Anthoxanthum* flower earlier than plants in adjacent grasslands but probably do so because the piles of mine debris tend to become dry earlier in the spring (McNeilly and Antonovics 1968). Furthermore, mine-tailing soil contains high concentrations of heavy metals and is inhospitable to most plants, so any seeds from the grassland ecotypes or any hybrid seeds will have little chance of survival on the mine tailings. Similar patterns of immigrant inviability are found for recently diverged species that are adapted to contrasting soil types in *Erythranthe* and *Senecio* (Moyle et al. 2012; Melo et al. 2014). The reproductive barrier between these ecotypes appears to be due to the combination of ecological, phenological, and post-zygotic isolation barriers that work in concert to inhibit hybridization.

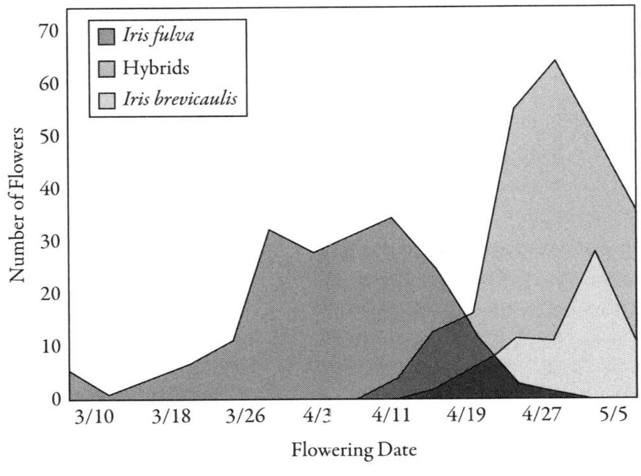

Fig. 10.5. Phenological isolation between *Iris fulva* and *I. brevicaulis* and their hybrids. The period of overlap between the two species is shown in purple. Modified from Cruzan and Arnold (1994).

Flowering time differences are common and contribute to reproductive isolation in a wide variety of species, including the Louisiana irises (Cruzan and Arnold 1994; Tang et al. 2010; Fig. 10.5), coastal and inland ecotypes of *Erythranthe guttata* (Hall et al 2010), mine-tailing and adjacent grassland ecotypes of *Anthoxanthum* (McNeilly and Antonovics 1968), and among species of tropical understory plants in the genus *Heliconia* (Stiles 1975). Flowering phenology may rarely provide complete reproductive isolation, because it tends to be phylogenetically constrained: closely-related species tend to flower at similar times and to respond to the same abiotic cues (Davies et al. 2013). The Louisiana iris species *Iris fulva* and *I. brevicaulis* may be an exception to this trend, as common-garden studies found no flowering overlap over three years (Tang et al. 2010); however, hybridization between these species is fairly common and phenological isolation appears to be incomplete at least in some years (Cruzan and Arnold 1994; Fig. 10.5). Seasonal differences in flowering time overlap, such as those observed among years in irises, are probably common and likely to make phenological barriers incomplete. Furthermore, as mentioned earlier for *Anthoxanthum*, flowering time is also sensitive to local environmental differences among habitats, which may elevate or reduce its effectiveness as an isolating barrier on a local scale. The tendency for closely related lineages to flower at similar times, variation in seasonal patterns of precipitation and temperature, and microhabitat variation may render differences in flowering time an unreliable isolating barrier.

Evolution of selfing in one or both derived lineages may contribute to reproductive isolation, because it reduces the opportunities for inter-lineage pollen transfer. Some of the better-known cases of autogamy isolation include the mine-tailing ecotype of *Anthoxanthum* (Antonovics 1968) and higher selfing rates in contact zones between species of *Phlox* (Levin and Kerster 1967), *Arenaria* (Fishman and Wyatt 1999), and *Erythranthe* (Martin and Willis 2007; Fig. 11.17. A more extreme case is represented by *Glycene falcata*, which produces subterranean cleistogamous (closed) flowers that automatically self (Moyle et al. 2004). It is important to note that selfing taxa often continue to outcross to some degree (e.g., *G. falcata* also produces open-pollinated flowers), so there is still some potential for interspecific pollen exchange.

Floral Isolation—The second group of pre-mating isolating mechanisms involves processes affecting inter-lineage pollen transfer, including those based on the behavior of animal pollinators (ethological isolation) and those based on floral morphologies that favor visitation by specific pollinators or enforce the placement of pollen on limited portions of the pollinator's body (mechanical isolation). Verne Grant (1949) emphasized the importance of floral isolation as a unique characteristic of reproductive barriers in plants and defined the categories of "mechanical" and "ethological" isolation as analogous to mechanisms that had previously been defined for animals (Dobzhansky 1937; Mayr 1942). Grant proposed that the tendency for some classes of animal pollinators such as birds, long-tongued flies, and bees to exhibit some degree of *Flower Constancy* was a major mechanism of species diversification in flowering plants. In Grant's view, these ethological barriers are further enforced by morphological differences in flowers associated with different types of pollinators that prevent contact with reproductive structures or the placement of pollen on different portions of the pollinator's body (Grant and Grant 1964; Grant 1993, 1994). These ideas were corroborated by the observation that flowers could generally be classified into distinct *Pollination Syndromes* (Faegri and van der Pijl 1979; Fig. 10.6; refer to color plate) that were interpreted as being the product of strong associations with different pollinators.

For most of the second half of the twentieth century, the importance of floral isolation was widely accepted (e.g., Stebbins 1981), but toward the end of the century, those who viewed plant-pollinator interactions as much too generalized to provide an effective mating barrier started to challenge this paradigm (e.g., Waser et al. 1996). While the diversity of pollination syndromes among closely related species seems to provide testimony to the prevalence of pollinator-mediated selection and the effectiveness of floral isolation (Fenster et al. 2004), the potential for animal pollinators to mediate reproductive isolation continues to be a topic of active discussion (Ollerton et al. 2009). In the next two sections, we discuss the effectiveness of mechanical and ethological isolation barriers.

Mechanical Isolation—Grant's inferences concerning the effectiveness of mechanical isolation relied primarily on observations of the fit between

Fig. 10.6. Evolution of multiple pollination syndromes. The columns from left to right are examples of bee-, hummingbird-, and moth-pollination syndromes. Each row shows representatives of the same genus. Starting from the top row are the columnines: *Aquilegia bertolonii, A. formosa,* and *A. longissimi* (Ranunculaceae). Second row are beardtongues: *Penstemon spectabilis* and *P. centranthifolius* (Plantagineaceae). Third row are catchfly: *Silene virginica* and *S. latifolia* (Caryophylaceae). The fourth row are morning glories: *Ipomea indica, I. quamoclit,* and *I. alba* (Convolvulaceae). The fifth row are larkspurs: *Delphinium nuttallianum* and *D. cardinale* (Ranunculaceae). The last row are gilias: *Ipomopsis aggregate* and *I. longiflora* (Polemoniaceae). Photos by Stickpen, Walter Siegmund, Cstubben, Lionnel, Derek Ramsey, Jebulon, Dinkum, Forest & Kim Starr, S. Zenner, and Jerry Friedman.

flower and insect morphologies (e.g., Grant and Grant 1964). In this example, two species of bee-pollinated mints (*Salvia mellifera* and *S. apiana*) grow sympatrically in the mountains of southern California and have strikingly different floral morphologies (Fig. 10.7; refer to color plate). The Grants interpreted the differences in positioning of reproductive parts between these two species as a mechanism that would enforce isolation through precise placement of pollen on different portions of the bee's body. The hypothesis that mechanical isolation is effective between these species breaks down when we consider that grooming behavior generally spreads pollen away from the primary locations of deposition to other parts of the bee's body (Mitchell et al. 2009). For this reason, it is unlikely that bee-pollinated species with granular pollen could be effectively isolated through differences in the placement of pollen. On the other

Fig. 10.7. Flowers of *Saliva mellifera* (left) and *S. apiana* (right). The positioning of anthers and stigmas enforces contact with the dorsal surface of the abdomen of bumblebees in *S. mellifera* and with the flanks in *S. apiana*. Photos by Curtis Clark and Stan Shebs.

hand, there are good cases of mechanical isolation for animal pollinators that do not groom (e.g., hummingbirds) and for plant species that package their pollen into *Pollinia* (e.g., orchids and milkweeds in the genus *Asclepias*). For example, Kay (2006) found strong mechanical isolation between a pair of sympatric hummingbird-pollinated tropical plants in the genus *Costus* in Costa Rica. Mechanical isolation also appeared to be strong for several species of *Asclepias* through placement of pollinia on different portions of the bodies of the bee and wasp pollinators when grown in a common garden (Kephart and Theiss 2004). Mechanical isolation may not be perfect in *Asclepias* and *Costus*, as other mechanisms contribute to reproductive isolation between these species. Mechanical isolation depends heavily on animal pollinators performing specific patterns of behavior. While there are some outstanding examples of floral morphological differences that enforce specific placement of pollen on pollinators, variation in pollinator behavior will often erode the effectiveness of mechanical isolation.

Ethological Isolation—The effectiveness of ethological isolation barriers depends on the foraging patterns of animal pollinators. Many types of pollinators, such as butterflies and some types of flies and beetles, are generalists and often visit several species of plant during a single foraging bout. Such behaviors erode isolation barriers, but even those pollinators that are considered to be specialists (e.g., hummingbirds, long-tongued flies, and bees) will often visit a range of flower types and may switch between species within a single foraging bout (Waser et al. 1996; Ollerton et al. 2009). For example, bumblebees will generally forage primarily on one species of flower but will continue to sample other species in the same community (*Majoring and Minoring*; Heinrich 1979). This switching behavior causes regular instances of interspecific pollen transfer (Morales and Traveset 2008) and reduces the effectiveness of floral isolation mechanisms. Geographic differences in pollinator behavior may also contribute to variation in the strength of ethological isolation. For example, two species of *Ipomopsis* are isolated by forging preferences of their hawkmoth and hummingbird pollinators at some locations but hybridize extensively at others (Campbell 2008). Pollinators using flowers for food resources are strongly motivated to engage in behaviors that allow them to meet their energetic needs; consequently, their foraging activities may degrade the effectiveness of ethological isolating barriers among plant lineages.

One exception to the generality that pollinators are behaviorally flexible is the European honeybee (*Apis mellifera*). Individual honeybees tend to display floral constancy throughout their lives, even as the hive will collectively utilize any rewarding plants that are available (von Frisch 1967). It is important to note that this species originally had a limited geographic distribution and has only recently been introduced around the world, so the large majority of plant species have evolved along with pollinators that are much less constant.

In addition to the "sampling" behavior of foraging pollinators just described, other characteristics of animal pollinators can generate opportunities for interspecific pollen transfer (Ollerton et al. 2009; Schiestl and Schluter 2009). Naïve animal pollinators will often explore several potential food sources and consequently, will transfer pollen among species. Most animal pollinators have innate color preferences (e.g., red for hummingbirds; blue or yellow for bees), but most are capable of learning to forage on species that are outside the range of their normal floral syndrome. For example, bees are known to forage for pollen on "hummingbird" flowers, and hummingbirds will forage from a wide range of species that are normally pollinated by bees. In many of these cases, there may be a poor fit to the flower, so the visitor will not consistently contact the reproductive structures. Even though mechanisms of floral isolation may be effective in many cases, it is important to note that even a small amount of inter-lineage pollen transfer may be adequate to generate hybrid offspring and to promote introgressive hybridization.

While the behaviors of animal pollinators foraging for food resources is often too inconsistent to provide strong reproductive isolation among plant lineages, there is much greater specificity for associations that are based on the reproductive success of the insect pollinator. Specific plant-pollinator relationships include sexual deception in orchids, with flowers mimicking female bees and wasps (pollination by *Pseudocopulation*; Schiestl and Johnson 2013), and tropical orchids offering floral fragrances that are collected by male euglossine bees for the attraction of mates (Hills et al. 1972). Sexually deceptive orchids have, in the genus *Ophrys* in the Mediterranean region and in the genus *Cryptostylis* in Australia, exhibited chemical and morphological characters that mimic female wasps. These relationships are species-specific, and flowering is timed to a period when males are already active, prior to female emergence in the spring. A similar level of specificity occurs in orchids of Central and South America that produce aromatic chemicals required by male euglossine bees for the attraction of females (Dressler 1968). Many of these species of euglossine orchids are visited by a single bee species, and for others, precise placement of pollinia enforces mechanical isolation.

Specific plant-pollinator relationships also exist in cases of brood-site dependence, with insects relying on flowers for rearing offspring, as with figs and fig wasps (Ramírez B. 1970) and the yucca and yucca moth (Pellmyr 2003). In most of these examples, the relationships between individual species of insects and plants tend to be specific. One exception is the case of fig wasps: detailed molecular analyses have revealed that the association between plants and pollinators are much more ambiguous than originally thought (Machado et al. 2005). For subspecies of Joshua tree (*Yucca brevifolia*), pollinator specificity is enforced by a close match between the morphology of flowers and the pollinating moth, so inappropriate brood site choice has severe negative consequences on the fitness of females (Smith et al. 2009). On the other hand, even these seemingly highly specialized relationships do not always provide complete reproductive isolation, as asymmetrical introgression is known to occur between these subspecies of Joshua tree (Starr et al. 2013).

From this discussion, we can conclude that floral isolation mechanisms are often imperfect, and most will allow some degree of interspecific pollen transfer. This is largely due to the behavioral characteristics of animal pollinators; while there may have been strong selection on plants to manipulate their pollinators to perform in specific manners, the aberrant behaviors of naïve individuals and the strong motivation for efficient foraging and successful reproduction may foil even the most complex floral isolation mechanisms. The exceptions to this generality may be associations where insects are responding to specific floral fragrances, such as the orchids, for which hybrids are rarely found in nature, even though they are often easily crossed under controlled conditions (Dressler 1990). Highly efficient floral isolation in *Ophrys* orchids combined with the learned avoidance behavior of male insects has been hypothesized to have been responsible for the rapid diversification of taxa in this group (Breitkopf et al. 2015), but this appears to be an exceptional example of diversification through floral isolation.

Evolution of Pollination Syndromes—The observation that many genera include species with different pollination syndromes has encouraged speculation that transitions to alternative pollinators are the primary mechanisms of diversification (Stebbins 1981; Grant 1994; Campbell 2008; van der Niet and Johnson 2012). On the other hand, it is clear that floral isolation is not necessary for divergence and reproductive isolation; there are many examples of closely related species that share the same pollinators and are isolated by other mechanisms (Kay and Sargent 2009; Schiestl and Schluter 2009). The most obvious example is the high level of diversity in the sunflower family (Asteraceae), where a wide range of generalist and specialist animals pollinate virtually all species. Similarly, the diversification in the grasses has been extensive even though nearly all of them are wind pollinated. Lineages sharing generalist pollen vectors are obviously not prone to divergence through floral isolation, but pollinator sharing is also known among closely related species that have specialized syndromes. For example, five species of sexually deceptive orchids in the genus *Cryptostylis* in Australia share the same wasp pollinator but are isolated by post-zygotic mechanisms (Schiestl et al. 2004). If floral isolation does not seem to be necessary or sufficient on its own to promote diversification, then how can we account for the numerous instances of diversification in pollination syndromes among closely related species?

Transitions to alternative pollination syndromes appear to be common for lineages that are associated with different habitats or geographic ranges (Baack et al. 2015). For example, a diversity of pollination syndromes has evolved in closely related taxa that are associated with different habitats in the Cape region of South Africa, the Andes of South America, and within the California Floristic Province (Kay and Sargent 2009). Note that most of these cases of transition to an alternative pollination syndrome include some degree of ecological isolation. With the possible exception of *Ophrys* and *Cryptostylis* orchids, no examples are available of plant species occupying the same habitat that are isolated solely by morphological or behavioral differences of flowers and pollinators. From these observations, we can infer that divergence is usually initiated by geographic or ecological isolation, and transitions to different pollination syndromes probably occur independently—either during or after divergence in different habitats.

Once a population becomes isolated from gene flow, the transition to an alternative pollination syndrome may be promoted if the primary pollinator becomes unavailable, and if an alternative pollinator is more reliable—an idea that Stebbins (1974) referred to as the *Most Effective Pollinator Principle (MEPP*; Kay and Sargent 2009). Thomson (2003) extended this principle by suggesting that poor pollinators are effectively parasites that reduce fitness gains through male function by wasting pollen (*Conspecific Pollen Loss*). Low pollinator availability will decrease fitness through both male and female function, but abundant pollinators can also have negative effects on plant fitness if they become unreliable, because of frequent interspecific pollen transfers (Morales and Traveset 2008).

The reliability of pollinators generally depends on the flowering community composition and the behavioral responses of pollinators. Plant species that are in the minority or are less rewarding may be subject to fitness loses due to *Heterospecific Pollen Deposition* that might interfere with fertilization of ovules, and conspecific pollen loss by deposition on the floral structures of other species. Competition for reliable pollinators among plants can be intense in some communities and is exasperated by the presence of highly rewarding species, which often seem to be introduced or invasive plants (Morales and Traveset 2008). If a more reliable pollinator were available in these circumstances, then we expect there would be selection for changes in floral morphology that favors the more effective pollinator while simultaneously discouraging a poor one.

Transitions to alternative pollination syndromes may be more likely if the geographic range of a plant species exceeds that of its pollinators. For example, *Ipomopsis tenuituba* suffers reduced seed set because of the low abundance of its hawkmoth pollinator in some parts of its range (Campbell 2008). This species is also visited by hummingbirds, which are generally more abundant. Campbell hypothesizes that selection could favor wider corolla tubes and other characters that would better accommodate hummingbirds. In this and other species suffering from poor pollinator service, the behavioral flexibility of animal pollinators—the same feature that makes ethological isolation barriers leaky—may generate opportunities for transitions to alternative pollination syndromes.

Pollination syndrome transitions may also be facilitated by the genetic architecture of floral and scent traits, which often seem to be governed primarily by genes of large effect (Thomson and Wilson 2008; Lowry et al. 2014; Chapter 7). Thomson and Wilson point out that pollination syndromes consist of multiple traits and occupy discrete locations in phenotypic space; except in cases of recent hybridization, there are generally no intermediates present. This suggests that rapid syndrome transitions may be facilitated by strong selection for effective pollination and simple genetic architectures, which has resulted in radiations in pollinator associations within a number of genera. Such radiations often appear to be enabled by particular morphologies, such as the fusion of petals into floral tubes or the development of nectar spurs (e.g., Hodges and Arnold 1994; Fig. 10.6; refer to color plate). While examples of pollination syndrome diversity within genera are plentiful, no examples of lineages that are in the process of transitioning to a different pollination syndrome are available. The lack of intermediate floral types is consistent with the scenario of pollination syndromes evolving during historical isolation—particularly during the Pleistocene. The future may hold more circumstances where selection favors transitions to alternative pollinators as many plants face novel challenges due to climate change, range shifts, and competition with introduced species.

10.6 POST-MATING ISOLATION

The pistil represents the one forum where plants have the greatest potential to influence the success of pollen donors and the paternity of their seeds. As described in Chapter 2, the separation between the location of pollen deposition (stigmas) and the ovules during the emergence of angiosperms from their seed-fern ancestors established the potential for plants to erect barriers to fertilization by pollen from unwanted sources. It should be clear from the preceding discussion, that low levels of interspecific pollen transfer are probably common in the large majority of flowering plants. The fitness consequences of heterospecific pollen deposition can be severe if foreign pollen interferes with conspecific pollen germination or growth or succeeds in fertilizing ovules, resulting in infertile or inviable offspring. The success or failure of pollen deposited on stigmas is determined by a complex set of interactions between microgametophytes and pistil tissues during germination and the growth of pollen tubes to the ovary (Chapter 9).

Barriers to extraneous fertilization that promote conspecific pollen precedence may develop passively as a side effect of divergence or may be a product of selection for the inhibition of foreign pollen. Successful pollen-tube growth requires intimate cooperation between the microgametophyte and the pistil, including the creation of appropriate germination conditions on the stigma, the provisioning and successful uptake of resources that are required for pollen-tube growth, and proper signaling to guide pollen tubes to the micropyle of ovules (Swanson et al. 2004). Failure of pollen may occur at any of these stages due to inhospitable conditions for germination or growth or lack of appropriate signaling that results in aberrant growth patterns of pollen tubes. Such an intricate sequence of conditions and interactions required for successful fertilization may be susceptible to small changes that result from the accumulation of fixed mutations in isolated populations. Particularly for species that are not routinely pollen limited, pistil traits responsible for pollen success may be free to evolve by genetic drift, while pollen traits would be under strong selection to track the changing pistil conditions (analogous to the evolution of reproductive isolation through sexual selection in animals; Fisher 1930; Lande 1981). This type of divergence would ultimately result in the failure of pollen from different lineages and species, which provides reproductive isolation resulting from conspecific pollen precedence that is due to *Incongruity* between the pollen and pistil phenotypes (Hogenboom 1975). Alternatively, heterospecific pollen inhibition may occur as a side effect of self-incompatibility (SI) mechanisms that result in the rejection of self-pollen. When one lineage is *Self-Compatible* (*SC*) and a second is self-incompatible, asymmetrical conspecific pollen precedence often results—a phenomenon referred to as *Unilateral Incompatibility* (Harrison and Darby 1955; Lewis and Crowe 1958; Harder et al. 1993; Rieseberg and Blackman 2010). In the following sections, we discuss the contributions of unilateral incompatibility and incongruity to *Heterospecific Incompatibility* (*HI*).

Unilateral Incompatibility—Self or interspecific pollen may be inhibited during germination or growth as a result of incompatibility reactions.

Self-incompatibility is due to pistil and pollen gene products (*S Genes* that segregate for multiple *S Alleles*) that interact to prevent the germination of self-pollen on the stigma or inhibit the growth of pollen tubes through the style (Brewbaker 1957; Takayama and Isogai 2005; Chapter 9). For interspecific crosses, inhibition of pollen tubes is often asymmetrical, with foreign pollen inhibited in the self-incompatible species but not in a closely related self-compatible species, thereby resulting in unilateral incompatibility.

Because the species that have the strongest heterospecific incompatibility are often also self-incompatible, it has been widely assumed that S-locus gene products are responsible for the rejection of foreign pollen (Swanson et al. 2004). Moreover, there are a number of examples where the site of pollen rejection is similar after self and interspecific crosses, and in one case, heterospecific incompatibility was established in a self-compatible species of *Nicotiana* (tobacco) by transforming it with the S gene (*S-RNase*) from a self-incompatible species (Swanson et al. 2004).

While there is some compelling evidence for the action of S alleles in heterospecific incompatibility, there are also difficulties with the hypothesis that the same mechanism responsible for self-incompatibility also acts to reject foreign pollen. In particular, the rejection of self-pollen requires matching between gene products in pollen and styles that are associated with the same S allele. Even though S alleles tend to be conserved across phylogenetic time scales (Igic and Kohn 2001), there are many different alleles, so it is extremely unlikely that any pair of individuals would have identical genotypes at the S locus. On the other hand, there are other gene products associated with SI such as those in the *HT* gene family, and some of these have been implicated in the rejection of interspecific pollen (Covey et al. 2010). While there is good evidence for an overlap between genetic mechanisms of self and heterospecific incompatibility in some species, this association is by no means universal (see the section on Incongruity). If these mechanisms are separate, they may respond to different selective pressures (i.e., selection for outcrossing may not be correlated with selection for isolation). The association between unilateral incompatibility and self-incompatibility makes it difficult to identify instances where heterospecific incompatibility might have been lost but instead has have been maintained by selection or where conspecific pollen precedence originated de novo because of selection for reinforcement of reproductive barriers.

Incongruity—Whether gene products associated with self-incompatibility are responsible for the rejection of heterospecific pollen seems to vary within and among genera and families. Hogenboom (1975) pointed out that the behavior of pollen tubes and the location of the region of inhibition in the style often differs for heterospecific and self-crosses, and proposed that interspecific pollen-tube failure was more frequently due to incongruity than the action of S alleles. This hypothesis is supported by several cases where treatments that normally disable self-incompatibility had no effect on heterospecific incompatibility. For example, strong heterospecific incompatibility was maintained in *Lilium longiflorum* (Easter lily) when styles were incubated at high temperature while the self-incompatibility response was eliminated (Ascher and Peloquin 1968). In other cases, treatments that successfully disabled the inhibition of self-pollen had an equal effect on pollen from a closely related species (Swanson et al. 2004), but note that such correlative evidence does not conclusively indicate that SI proteins are also responsible for heterospecific pollen inhibition. In general, there is a wide range of pollen growth responses after interlineage crosses, which suggests that there is no single genetic mechanisms responsible for the inhibition of heterospecific pollen.

A good example of the wide range of compatibility responses that are possible for interspecific crosses can be found in a group of self-compatible and -incompatible species in the genus *Solanum* (Pandey 1962; Fig. 10.8). For the majority of crosses there were clear responses of either full heterospecific compatibility or incompatibility expressed during pollen germination and growth in the style (17 and 27 crosses, respectively). However, a significant proportion of crosses displayed partial incompatibility (11 crosses) or full to partial post-zygotic failure of hybrid seed (2 and 12 crosses, respectively; Fig. 10.8). Conspecific pollen precedence differed among clonal lineages in some species and some self-compatible species exhibited the ability to reject interspecific pollen. This wide range of variation in heterospecific incompatibility responses suggests that a number of different mechanisms may be responsible for heterospecific incompatibility.

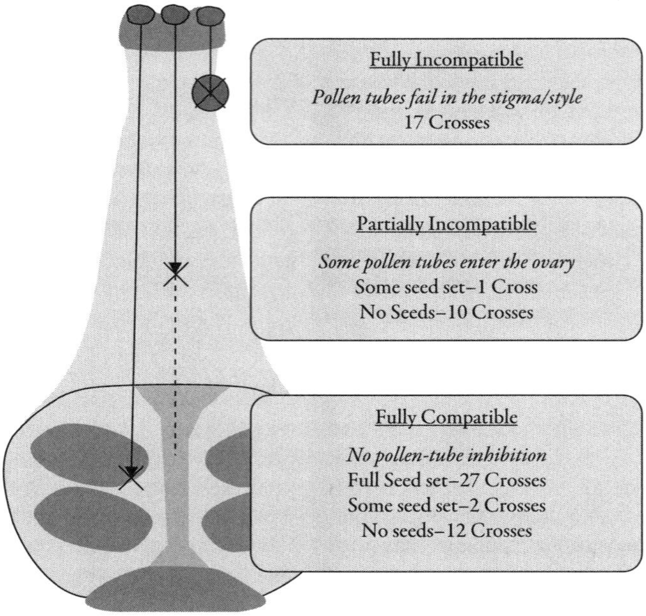

Fig. 10.8. Cross compatibility responses for 3 self-compatible and 11 self-incompatible species of *Solanum* from Mexico and South America. Summarized from Pandey (1962).

Evolution of Conspecific Pollen Precedence—Post-pollination discrimination among pollen from different donors should be favored under a number of circumstances, because it can allow mate selection without recipient fitness loss. Inhibition of pollen in the style may come at little or no cost, but once an ovule is fertilized by pollen from an undesirable mate, the fitness of the recipient may be compromised in a couple of ways. First, even if seed abortion occurs early, there is a resource cost for provisioning seeds that fail. Moreover, an aborted ovule may represent a fitness loss because it is likely to reduce the number of offspring produced (seed discounting—but see section 10.7). Consequently, we expect that mate discrimination mechanisms that occur prior to fertilization should be favored. Exclusion of interspecific pollen could occur during pollination, but as discussed earlier, floral isolation mechanisms are generally too imprecise to be reliable (Armbruster 2014). Hence, the interaction between pollen and styles is the next reproductive stage where stringent mechanisms of mate discrimination might evolve.

Given the pivotal importance of post-mating isolation mechanisms in plants, it is surprising how little we know about barriers that are based on pollen–style interactions. Comparisons of the growth and fertilization ability of conspecific and heterospecific pollen have been made for species in several genera, including *Lycopersicum, Nicotiana, Petunia, Datura, Solanum, Lilium*, and *Haplopappus* (see citations in Hogenboom 1975; Covey et al. 2010; Fig. 10.8), as well as for pairs of hybridizing species of *Iris* (Carney et al. 1994; Emms et al. 1996), *Silene* (Montgomery et al. 2010), *Piriqueta* (Wang and Cruzan 1998), *Helianthus* (Rieseberg et al. 1995), and *Erythranthe* (Diaz and Macnair 1999). Some general trends emerge from these studies. For example, some characteristic differences among flowers such as style length can sometimes predict cross compatibility (e.g., *Lilium* and *Datura*), but there are also exceptions to this trend (e.g., for Louisiana irises; Emms et al. 1996). For many species that are known to hybridize extensively in nature, conspecific pollen precedence tends to be weak, but strong barriers due to pollen–pistil interactions have been reported to contribute substantially to reproductive isolation in species of *Centaurium, Helianthus, Iris, Erythranthe, Orchis, Petunia, Quercus*, and *Trillium* (Lowry et al. 2008a; Baack et al. 2015).

Surprisingly, we currently have only a rudimentary understanding of broader phylogenetic patterns of conspecific pollen precedence and the processes affecting the evolution of heterospecific incompatibility. In particular, little information is available on the progress of the development of incongruity during divergence or on the potential for the evolution of reinforcement of reproductive barriers by heterospecific incompatibility. Comparisons of multiple species suggest that more distantly related taxa are generally less compatible (Hogenboom 1975), but none of these surveys have incorporated phylogenetic information with studies of conspecific pollen precedence. Similarly, there may be many cases where sympatry of closely related species may have favored the evolution of post-pollination isolating barriers because of selection for reinforcement, but documented cases of stronger heterospecific incompatibility for sympatric populations are currently unavailable.

Nonetheless, we can identify some trends in conspecific pollen precedence. For example, we generally find unilateral incompatibility for crosses between self-compatible and self-incompatible pairs of closely related species (unilateral incompatibility), but as discussed earlier, there are a number exceptions. Similarly, we might speculate that strong interspecific incompatibility in self-compatible plants provides evidence of selection for reinforcement; however, such cases are not conclusive since it is possible that conspecific pollen precedence may have been carried over from a self-incompatible ancestor and is not necessarily maintained by selection.

There are also circumstances where selection for conspecific pollen precedence may be weak; for example, in perennial plants that routinely overproduce flowers and ovules to increase display size for pollinator attraction (Stephenson 1981; Sutherland 1986; Chapter 9). In these cases, higher rates of ovule abortion due to interspecific pollination may not reduce total seed set, which is limited by resources rather than the availability of ovules. Similarly, Stebbins (1958) pointed out that the demographic cost of the loss of ovules to interspecific fertilization is high for annuals but much lower in perennial species, so selection for reproductive isolation should be much stronger in the former. If this were true, then we may expect heterospecific incompatibility to be favored primarily in annuals and other species that do not have high levels of seed and fruit abortion. Unfortunately, we do not have adequate information to make comparative tests of these ideas; such puzzles may best be solved through studies of pollen–style interactions by direct observation of pollen tubes and in controlled crosses that use genetically marked pollen (Chapter 9). These gaps in our knowledge represent challenges for future plant evolutionary biologists.

10.7 POST-ZYGOTIC ISOLATION

Once ovules are fertilized, the newly formed embryos are subject to several rounds of selection both before and after seed dispersal. Failure of interspecific fertilizations can occur at any stage of seed development or following dispersal during the germination, growth, and reproduction of hybrid offspring. Post-zygotic isolation differs from previous stages of reproduction because genomes from divergent lineages are combined into individuals where alleles may interact to produce *Genetic Incompatibilities* resulting in failure of the hybrid progeny or low hybrid fitness (hybrid breakdown). Incompatibilities produce reductions in fitness and are often delineated into *Intrinsic* and *Extrinsic* effects. Low fitness would be considered intrinsic if hybrids exhibited consistently poor growth, survival, or reproduction across a range of environments. Extrinsic sources of variation in fitness are ecological in nature and are evident as variation in performance across environments. In the example of mine-tailing ecotypes described earlier, mine soil genotypes and hybrids can survive and reproduce in the grassland habitat, so the low fitness of migrants and hybrids on contaminated soil is evidence of an extrinsic effect on fitness. On the other hand, abortion of seeds during development would generally be considered to be due to intrinsic factors. There is not always a clear distinction between intrinsic and extrinsic effects, as genetic incompatibilities that are normally considered intrinsic may be exaggerated under stressful conditions (e.g., seed abortion rates tend to be higher when the maternal plant is stressed; Wiens et al. 1987; Korbecka et al. 2002).

In this section, we discuss post-zygotic isolation in the context of seed abortion and the post-dispersal fitness of hybrid progeny. We first focus on models of *Genetic Incompatibility* as we examine the conditions that allow or inhibit the evolution of genetic changes contributing to intrinsic incompatibilities between lineages. We then expand our discussion to consider the factors that may promote the evolution of genetic incompatibilities and pre-zygotic reproductive isolation.

Single-Mutation Model—To understand the origin and evolution of genetic incompatibilities responsible for reproductive isolation, it is valuable to examine models of adaptive divergence. One of the most useful illustrations of divergence leading to speciation was proposed by Sewall Wright (1932), who utilized the metaphor of a *Fitness Landscape* (*Adaptive Landscape*; Chapter 11). In Wright's model, species occupy discrete fitness peaks that are separated by valleys of low fitness (Fig. 10.9). In this figure, the genetic landscape is depicted in a single dimension, but adaptive landscapes are generally visualized in two dimensions with each axis representing components of hyper-dimensional genetic space and elevation representing fitness. In the adaptive landscape model, divergence occurs in response to selection by local hill climbing and species are represented as clouds of points (populations) occurring close to the regions of highest fitness. The dispersion of population around peaks represents the opposing effects of genetic drift and selection. In Wright's model, reproductive isolation is accomplished, because any hybrids formed between divergent lineages will inevitably occur in regions of lower fitness (i.e., fitness valleys) due to underdominance effects at individual loci (*Underdominance Model* or *Single-Mutation Model*; Fig. 10.9).

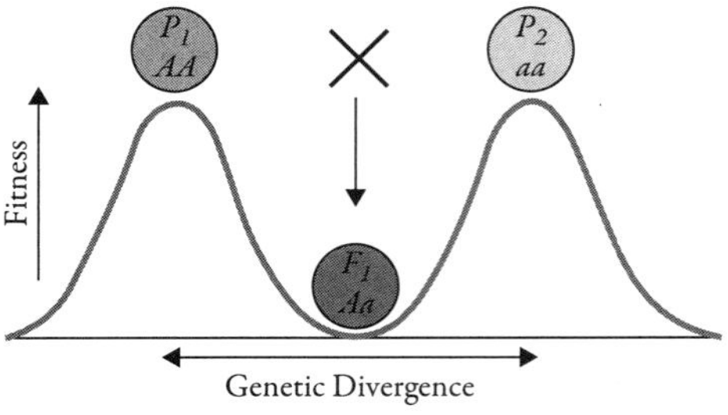

Fig. 10.9. The underdominance model of genetic incompatibility between divergent lineages. The two parental lineages P_1 and P_2 occupy fitness peaks and are separated by a valley of low fitness. First-generation hybrids inevitably have lower fitness than both parental lineages.

The primary difficulty with the underdominance model is that selection will oppose the spread of any mutations contributing to reproductive isolation. Consider the case where the ancestral lineage is fixed for the "a" allele (i.e., all individuals are homozygous aa). For reproductive isolation to evolve, one lineage needs to retain the aa genotype while a second becomes fixed for the AA genotype (Fig. 10.9). We expect that mutations would occur for only one of two alleles, so any individual carrying the new "A" allele would be heterozygous (Aa) and consequently would have low fitness. Such mutations conferring reproductive isolation would be opposed by strong selection, so they would be very unlikely to increase in frequency. On the other hand, we can circumvent this problem if we assume that reproductive isolation accumulates slowly and is due to many loci of small effect. In small populations, underdominant mutations that are subject to weak selection would have improved chances of becoming fixed through genetic drift. As more underdominant mutations of small effect accumulate, reproductive isolation would become stronger (i.e., the valley becomes deeper) until the lineages are completely isolated.

The most common examples of single-mutation underdominance contributing to reproductive isolation in plants are due to chromosomal rearrangements (Baack et al. 2015). Rearrangements include translocations that produce mosaics of chromosomes and inversions that generate chromosomes with genes in reverse order over a portion of the genome (Chapter 3). First-generation hybrids that are heterozygous for rearrangements are generally viable, but meiotic abnormalities will reduce their fertility to varying degrees depending on the size of the translocation or inversion. We may expect that smaller rearrangements could accumulate within lineages by drift as described here, but many of the documented cases indicate that translocations can have fairly substantial effects on the fertility of hybrids (Clausen 1962; Grant 1963). In these cases, it may be that the chromosomal rearrangement was associated with a fitness benefit in terms of the growth or survival of heterozygous individuals (Kirkpatrick 2010), so the mutation may have spread in spite of its negative effects on fertility (Baack et al. 2015). It also appears that chromosomal rearrangements are much more common among species of herbaceous perennials and are rare for sister species of long-lived woody perennials (Stebbins 1950). Short-lived species are more likely to be good colonizers, so rearrangements may become fixed in populations that are subject to population bottlenecks as a consequence of migration and selection in novel habitats. Ultimately, it is difficult to know whether these underdominant mutations were generated as an artifact of divergence or were actively selected to reinforce isolation barriers. In either case, the new mutation (rearrangement) would have to start out at low frequency and then spread through selection and genetic drift (Lowry 2012).

Parallel Mutations—An alternative version of the single-locus model of reproductive isolation posits that parallel mutations could occur in diverging lineages resulting in underdominance for combinations of derived alleles (Nei et al. 1983; Sweigart and Willis 2012). In this *Multi-Mutation Model* the ancestral population is fixed for a third allele (a for example). As the lineages diverge, independent mutations at the A locus (A' and A") are neutral in the presence of the ancestral allele (a) and may go to fixation by genetic drift. At the end of this process each lineage is characterized by different alleles (i.e., they are homozygous A'A' and A"A") that interact in the hybrid offspring (A'A" genotypes) to produce underdominance. For the multi-mutation model, we predict that the effects of single loci on the fitness of hybrids (reproductive isolation) can be quite large, even though the interacting alleles originally spread and fixed by genetic drift.

Evidence of parallel mutations leading to postzygotic reproductive isolation among lineages is available for several species, including the *OAK* gene in *Arabidopsis thaliana* and several genes that separate the rice varieties *Oryza sativa japonica* and *O. s. indica* (Sweigart and Willis 2012). These cases appear to be consistent with the multi-mutation model of neutral divergence for the evolution of underdominance at individual loci. For example, in *A. thaliana* the underdominant alleles are different by multiple amino acid substitutions, which suggests that independent neutral divergence pathways led to each of the derived alleles. All of these cases include loci of large effect, so the general assumption is that mutation and drift were responsible for the evolution of each of the alleles having underdominant effects in hybrids, and reproductive isolation is the outcome of neutral divergence and not selection.

Multilocus Complementation—It is generally assumed that selection will oppose the spread of mutations causing underdominance. The special circumstances leading to multi-mutational isolation at single loci appear to be uncommon, because it requires that multiple mutations occur in parallel in the same gene region in separate lineages. Recognition of the difficulties with the underdominance of genetic elements conferring low fitness in hybrids dates back to Darwin's *On the Origin of Species.* Although the title of his book would suggest otherwise, it is widely recognized and freely admitted by Darwin that he did little to resolve the "species problem" (Orr 1996). This gap in our understanding for the origin of species was the focus of contention in the late nineteenth century and through the first few decades of the twentieth century. The perception that geneticists had not solved the species problem led to the proposal that speciation occurred by "non-Mendelian" mechanisms, and was attributed to unknown factors in the cytoplasm by some and to chromosomal rearrangements by others (Orr 1996). In the latter case, these speculations were not too far off as chromosomal rearrangements continue to be considered important mechanisms of reproductive isolation in plants.

A resolution to the species problem based on allelic variation—the evolution of reproductive isolation—was proposed by William Bateson in a 1909 essay published in the compilation *Darwin and Modern Science* (Bateson 1909). Bateson's model assumes that two different complementary factors (i.e., at two different loci) arise independently in a pair of lineages and combine in hybrid offspring to produce sterility or inviability. This solution to the species problem remained hidden and unrecognized—even by Bateson himself who later proclaimed that "the origin and nature of *species* remains utterly mysterious" (Bateson 1922; reported as a statement by Bateson during the discussion following this lecture). It was not until 1936 that Dobzhansky "rediscovered" Bateson's solution to the species problem in a model that was more clearly expressed in genetic terms (Dobzhansky 1936) and that was later elaborated by Muller (1942)—Bateson's contribution remained lost until the end of the twentieth century (Orr 1996). This mechanism of reproductive isolation became widely known as the

Dobzhansky-Muller (*DM*) model, but is now more correctly referred to as the *Bateson-Dobzhansky-Muller* (*BDM*) model of incompatibility (*BDMI*).

The BDM model for the evolution of reproductive isolation assumes that interactions among alleles at different loci (epistasis) is the primary cause of genetic incompatibility expressed in hybrids. This mechanism differs from the multi-mutation model described earlier, because it requires at least two loci, and each lineage becomes fixed for a new mutation at a different locus (Fig. 10.10; refer to color plate). The new mutations are neutral in their native genetic backgrounds, so they become fixed by genetic drift, but they interact when combined in the F_1 and recombinant-generation hybrids; any individual carrying at least one copy of both the A and B alleles will have low fitness (Fig. 10.10; refer to color plate). If the effect of two or more BDMI loci is great enough, then hybrids will be inviable or sterile, and the lineages will be completely isolated.

In a more generalized version of the BDM model, we expect mutations to become fixed in neutral genetic backgrounds in each lineage and then hybrid fitness to be reduced due to epistasis among multiple lineage-specific mutations. As more mutations at loci contributing to BDM incompatibilities accumulate, the number of potential interactions among them increases exponentially (Orr 1995). Consequently, the increase in the strength of reproductive isolation due to BDM incompatibilities should be non-linear and accelerate with time since divergence—a phenomenon referred to as the "snowball effect."

Such non-linear scaling of genetic incompatibilities has been found in some genera. For example, the increase in the number of genomic regions contributing to decreased seed viability for crosses among species of *Solanum* appears to be exponential with increasing divergence (Moyle and Nakazato 2010). These results are consistent with the snowball effect and suggest that BDM incompatibilities are contributing to postzygotic isolation among these species. We assume BDMI mechanism of reproductive isolation is more common, because there is a much greater chance for the accumulation of mutations across loci than there would be for multiple mutations occurring at a single locus to result in underdominance (i.e., as for the single-locus and multi-mutational models)

Models and investigations on the evolution of BDM genetic incompatibilities have mostly focused on isolating barriers expressed in the F_1 hybrid generation. Genetic interactions reducing the fitness of first-generation hybrids are antagonistic such that the combination of independently derived alleles results in *Negative Epistasis* (Fig. 10.11). While Dobzhansky's 1936 paper is widely cited for its description of genetic incompatibility based on negative epistasis, the

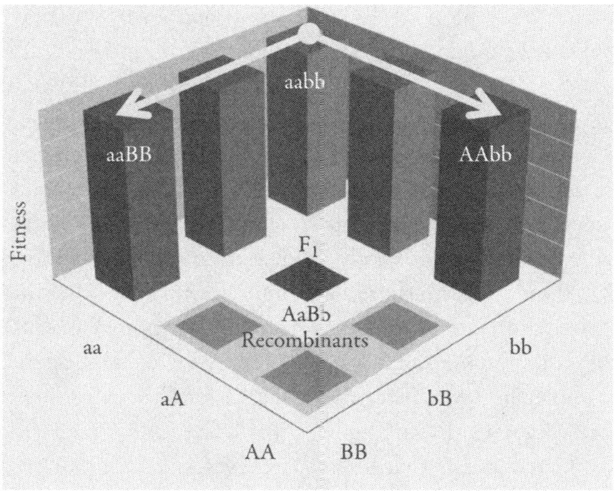

Fig. 10.10. Origin of Bateson-Dobzhanski-Muller incompatibilities. Two lineages diverge from a common ancestor (aabb) by genetic drift. The new alleles (A and B) characterizing each lineage have no fitness effects in their native genetic background, but epistasis between A and B in F_1 and other hybrids results in low fitness.

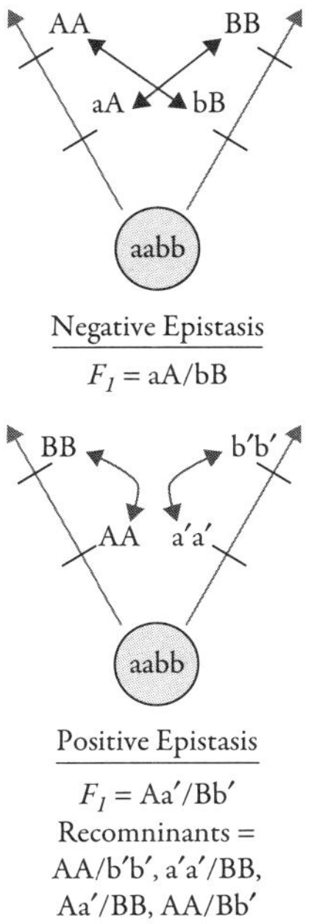

Fig. 10.11. Genetic incompatibilities based on negative and positive epistasis. The upper panel corresponds to the scenario depicted in Fig. 10.10. In the lower panel, positive epistasis develops due to fixed mutations at multiple loci in diverging lineages.

majority of this paper focuses on complementary genetic factors that collaborate to increase fitness in the backcross generations (Dobzhansky 1936). In particular, backcross hybrid males from crosses between races of *Drosophila pseudoobscura* display varying levels of fertility (testes size). In these crosses, male fertility increased with the number of conspecific genetic elements present in recombinant hybrids, indicating that there is *Positive Epistasis* among alleles derived from the same lineage (Fig. 10.11).

In the case of positive interactions among alleles across loci, we would not expect to see any fitness decline in the F_1 generation, because complementary alleles (e.g., A-B and a'-b') from each lineage are present across loci. In the next generation some recombinant hybrids will possess incongruous combinations of alleles (e.g., AA/b'b' and a'a'/BB), resulting in reduced fitness. Loss of hybrid fitness is due to the break-up of "co-adapted gene complexes" in recombinant hybrid generations. Note that recessive BDM incompatibilities would only be expressed when they were homozygous in a portion of recombinant hybrids, so it is not feasible to distinguish recessive antagonistic interactions from the disassociation of positively interacting alleles that are dominant. For our discussion, we will distinguish between interactions affecting F_1 and recombinant hybrid fitness, but we will not necessarily attribute the genetic mechanism of recombinant hybrid breakdown to epistasis that is complimentary/dominant or recessive.

Genetic modifications contributing to low F_2 hybrid fitness appear to accumulate earlier during divergence than those expressed in F_1 hybrids. This pattern is evident from the large number of cases where first-generation hybrids display high fitness and second-generation hybrids display lower fitness. For example, vigorous F_1 hybrids and low-fitness F_2 or backcross hybrids have been observed after crosses between subspecies of *Hemizonia angustifolia* and *Layia glandulosa*, between species of *Epilobium* (formerly *Zauschneria*), *Layia* (*L. gaillardioides* and *L. hieracioides*; Clausen 1962), and *Iris* (*I. fulva* and *I. brevicaulis*; Cruzan and Arnold 1994; Arnold 1997), and between morphotypes of *Piriqueta* (Cruzan 2005; Rhode and Cruzan 2005). This pattern was especially evident in the latter four examples where the F_1 vigor exceeded that of the parental species and most second-generation hybrids had slow growth and low rates of flowering. Second-generation hybrids in these cases displayed a wide range of fitness, with some being equally or more vigorous than the parentals, but in some cases (e.g., *Epilobium, Gossypium*), all of the recombinant hybrids displayed low growth rates and may have increased susceptibility to disease (Clausen 1962; Bomblies and Weigel 2007).

Polyploidy—Polyploidy is similar to the single-mutation isolating mechanisms described earlier, since a single change (whole genome duplication) is required to confer reproductive isolation. Genome duplication has traditionally been considered to be a very strong post-zygotic barrier (Stebbins 1950; Grant 1981), but recent studies suggest that ploidy differences do not always result in complete reproductive isolation. For example, in fireweed (*Chamerion angustifolium*), isolation due to differences in ploidy is estimated at 87% in sympatric populations (Husband and Sabara 2004). Isolation due to ploidy differences is even weaker among three species of Indian paintbrush (*Castilleja*) in the Colorado Rocky Mountains (Hersch-Green 2012).

In this case, fertile F_1 hybrids could be formed by crossing individuals of different ploidy within and among species. These three paintbrush species differ with respect to other characteristics, such as flower color and their ecological requirements, but hybrids are often present when two or more species are in close proximity.

While polyploidy resembles other types of single-mutation isolating mechanisms in some respects, the origin and evolutionary potential of newly formed polyploid individuals can differ in other important ways. For the single locus mutations and chromosome rearrangements discussed here, individuals carrying the new variant have lower vigor or fertility because of the intrinsic effects of their heterozygous condition. Whole-genome duplication, on the other hand, may produce vigorous and fully fertile individuals in a single mutational step. Newly formed polyploids may have lower fitness due to lower cross-compatibility with individuals having other ploidy levels, but this impediment can be overcome through self-fertilization and multiple whole-genome duplication events occurring in the same population. Since polyploid individuals are often self-compatible, they may have substantial capacity to increase in abundance and spread through the establishment of new populations (Ramsey and Schemske 2002). The fate of newly formed polyploid lineages may have more to do with their ability to succeed and remain isolated from gene flow while competing for resources in the same habitat occupied by their diploid progenitor.

Genetic Incompatibilities—Information on the mechanisms and architecture of genetic incompatibilities expressed in hybrids is available for a number of species (Bomblies and Weigel 2007; Lowry et al. 2008). While chromosomal rearrangements are easily observed by cytological investigation, the genetic mechanisms contributing to low fitness due to allelic variation are more difficult to identify. Nonetheless, information on the genetic architecture of genetic incompatibilities is available for several species, and some specific mechanisms of low viability and fertility of hybrids have been determined for a number of cases.

Many of the most thorough genetic analyses of post-zygotic reproductive barriers in plants have been conducted for species of agricultural importance (Bomblies and Weigel 2007). Low hybrid fitness in the F_1 and F_2 generations after crosses among crop plants has been referred to as "hybrid necrosis" and is evident as dwarfism, slow growth, chlorosis (loss of chlorophyll), and higher susceptibility to pathogens—many of the same symptoms that have been reported for genetic incompatibilities in wild species (Clausen 1962; Lowry et al. 2008).

Post-zygotic isolation in the F_1 generation is most often expressed during seed development or early stages of seedling growth and can be quite severe, resulting in very few or no viable seeds (Hollingshead 1930; Clausen 1962; Brys et al. 2014). In cases where viable F_1 hybrids are produced, they often displayed high rates of mortality or low fitness (Bomblies and Weigel 2007). The large majority of these cases of genetic incompatibilities are due to negative epistasis among a small number of loci (i.e., BDMI). Other than chromosome rearrangements, there are very few examples of underdominance at single loci. Relatively few studies have examined genetic incompatibilities affecting viability in the F_2 and later generations (e.g., Cruzan and Arnold 1994). Fitness declines in later generations are usually much less severe, and the fitness of some recombinant hybrids will often equal or exceed the fitness of the parental lineages (Scopece et al. 2010).

Some of the most extensive surveys of the viability and fertility of F_1 and F_2 hybrids are from crosses conducted within and among species by Clausen and colleagues in the mid-nineteenth century (reviewed in Clausen 1962). These authors often found that morphological, ecological, and taxonomic distinctions did not always predict the strength of post-zygotic barriers (Fig. 10.12; refer to color plate). In this example, plants from several populations representing three lineages that were then recognized as subspecies were crossed. In several cases, F_1 and F_2 hybrids from crosses within the putative subspecies were just as viable and fertile as crosses among subspecies. While there were no discernable patterns based on geographic location or taxonomic identity, there were some general patterns observed with respect to cross fertility. In particular, reduced growth performance often occurred in the F_2 hybrid generation, even if the F_1 hybrids were viable and fertile, but loss of fertility in F_1 hybrids was always accompanied by some degree of F_2 hybrid breakdown. In other words, evolution of

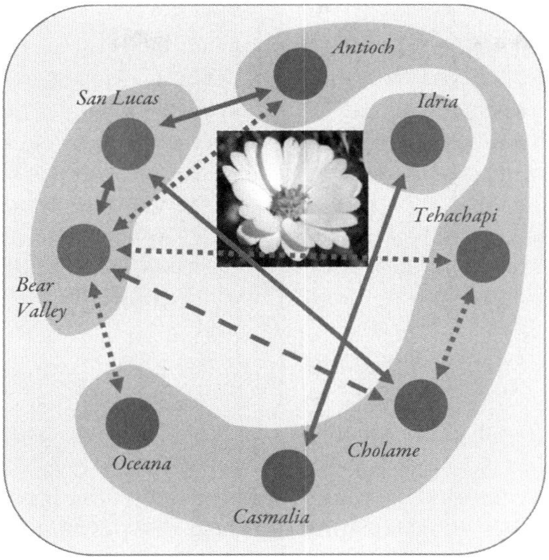

Fig. 10.12. Levels of post-zygotic isolation for pairs of populations within the *Layia glandulosa* complex. Solid arrows indicate that both the F_1 and F_2 hybrid generations are fully viable and fertile, dashed lines indicate lower viability for F_2 hybrids, and dotted lines indicate that F_1 hybrids and low fertility and F_2 hybrids had low viability. The orange, green, and blue colored regions correspond to the *lutea, discoidea*, and *typical* subspecies, which are no longer recognized. Modified from Clausen (1962). Photo by Stan Shebs.

post-zygotic barriers due to positive epistasis generally occur earlier during divergence, whereas negative epistasis tends to develop later and is expressed in the F_1 hybrid generation (Clausen 1962).

Cytoplasmic Male Sterility—While examination of hybrid viability and performance is rarely taken to the second generation, a number studies have examined the genetic basis of hybrid sterility in recombinant genotypes (Lowry et al. 2008; Rieseberg and Blackman 2010; Baack et al. 2015). In particular, *Cytoplasmic Male Sterility* (*CMS*) is phylogenetically widespread and contributes to post-zygotic isolation for a number of hybridizing species. Sterility is generally expressed as an absence of pollen or entire anthers and is due to rearrangements within the mitochondrial genome (chimeric ORFs) resulting in faulty gene products. Male sterility in these species is restored by a nuclear gene (*restorer of fertility—RF*), so plants carrying both chimeric ORFs and *RF* have normal male fertility. These two gene regions interact to produce a normal phenotype and recombinant hybrids inheriting chimeric ORFs and lacking *RF* will be male sterile, so this is a case of positive epistasis. Note that, like other cases of positive epistasis, some proportion of hybrids will have lower fitness, but normal phenotypes will also be present in recombinant populations.

10.8 EVOLUTION OF REPRODUCTIVE ISOLATION

The evolution of intrinsic reproductive isolating barriers must be preceded by the cessation of gene flow by other mechanisms. For homoploid lineages, initial divergence is usually promoted by geographic isolation that allows for the accumulation of genetic and phenotypic differences. Displacement of populations in different geographic locations is often accompanied by ecological divergence because plants dispersing to new locations are likely to encounter novel environmental conditions. As described in the last chapter, environments tend to change with distance across the landscape. Hence, during periods when species distributions are fragmented, selection favors local adaptation and consequently geographically isolated lineages became ecologically divergent. Mayr (1947, page 280) expressed this succinctly when he quipped, "All geographical races are also ecological races, and all ecological races are also geographical races."

Ecological differences that developed during periods of isolation are evident in the contemporary distributions of closely related species (Baack et al. 2015). Sister species often have broadly overlapping geographic distributions, but within these ranges, populations are rarely intermixed in the same habitat, as most species pairs display some degree of ecological divergence. Such ecological differentiation contributes to spatial isolation within overlapping ranges and to post-zygotic isolation due to extrinsic factors that reduce the fitness of hybrids in the parental habitats.

During periods of geographic isolation, genetic differences may accumulate by drift and contribute to intrinsic post-zygotic reproductive barriers. Some authors have expressed skepticism that genetic drift could be a primary driver for the development of isolating barriers because the time required would be too long (i.e., on the order of $4N_e$ generations for neutral divergence; Baack et al. 2015). With large population sizes, the rate of fixation of neutral alleles may be too slow to explain strong post-zygotic barriers between closely related species. Alternatively, genetic incompatibilities may develop as the byproducts of extrinsic selection due to linkage and hitchhiking or may be due to the pleiotropic effects of selected alleles. However, the association between ecological divergence and intrinsic genetic incompatibility seems to be weak (Baack et al. 2015), suggesting that the development of intrinsic post-zygotic isolation through hitchhiking during selection for adaptation is probably not common. On the other hand, the evolution of genetic incompatibilities among lineages by genetic drift may have been promoted by small population sizes. For example, range shifts and selection in novel environments would impose low effective population sizes because of bottlenecks during divergence. Under these conditions fixation of neutral mutations through genetic drift would be more rapid and could contribute to the evolution of post-zygotic reproductive isolation.

Reinforcement—Following periods of isolation the geographic distribution of divergent lineages will change in response to long-term climate cycles such that sister taxa may be brought into close proximity. When closely related species come into contact, the outcomes of their interactions will depend on the costs of interspecific pollen exchange and fertilization. If hybrid offspring have high fitness relative to the parental lineages, then introgression and complete admixture may proceed unabated. On the other hand, in cases where there is a high cost to interbreeding, we expect selection to favor stronger pre-zygotic reproductive isolation (Hopkins 2013). For sympatric populations, the development of isolation may be accelerated by selection for reinforcement, if there are high rates of interspecific pollen transfer or wastage of ovules due to strong post-zygotic failure of hybrid offspring (seed discounting). While it has been suggested that reinforcement may be unlikely because selection in areas of sympatry may be swamped by gene flow from allopatric regions, theoretical analyses predict that selection may promote reinforcement when populations are isolated from gene flow (Cain et al. 1999). A substantial number of cases of reinforcement have been documented (Howard 1993; Hopkins 2013), but most of these are between species of animals—conditions promoting selection for stronger reproductive barriers in plants may only occur under rather specific circumstances.

There are two potential causes of lower reproductive success for sympatric plant species. One that is often mentioned in discussions of reinforcement in plants is *Stigma Clogging* by interspecific pollen, which compromises the fertilization success of conspecific pollen. If stigma clogging were common, then we would expect that selection would favor mechanisms of floral isolation that would reduce interspecific pollen transfer. While the potential for stigma clogging is often identified as a source of selection promoting reinforcement, documented cases of interspecific pollen interference are rare (e.g., Thomson et al. 1981; Brown and Mitchell 2001). Currently, there are no clear-cut cases of selection for floral isolation due to stigma clogging, and as discussed earlier, evolution of different pollination syndromes is probably more often promoted by pollinator limitation in isolated populations.

Selection for reinforcement due to interspecific fertilization of ovules may be more common. Strong post-zygotic isolating barriers will lead to low hybrid fitness and may generate selection for reinforcement of pre-zygotic barriers. For example, in boreal regions of North America, conspecific pollen precedence is strengthened in populations of tetraploid fireweed (*Chamerion angustifolium*) that are in close contact with diploid populations (Baldwin and Husband 2011). In this case, there is strong selection for pre-zygotic isolation, because interspecific fertilization results in sterile triploid offspring. The effects of selection for reinforcement are evident, because artificially generated tetraploids do not have the same capacity to discriminate against pollen from diploid donors. Note that in this species and most others subject to polyploidy, genome duplication is not associated with ecological differentiation, so mixed-ploidy populations are common. Consequently, we expect that polyploidy in fireweed and other plants may frequently create conditions where selection favors the reinforcement of pre-zygotic barriers. This case of tetraploidy in fireweed is one of a few available examples where evidence of reinforcement has been clearly demonstrated, but reinforcement of reproductive barriers in mixed-ploidy populations may be common (Petit et al. 1999).

For divergent plant lineages of the same ploidy, there are very few well-documented cases that are consistent with reinforcement of reproductive barriers in areas of sympatry (Hopkins 2013). One of the few clear-cut examples of reinforcement is for annual species of *Phlox* from southern regions of North America. *Phlox drummondii* and *P. cuspidata* display geographic patterns of flower color variation that are consistent with reinforcement (Hopkins 2013; Fig. 10.13; refer to color plate). Both plants produce light-blue flowers in allopatric regions, but when ranges overlap in eastern Texas populations of the former species bear red flowers. Hybrids from crosses between these species are mostly sterile, and pollinators display strong flower color preferences in regions of overlap, which significantly reduces the frequency of hybrid seed production (floral isolation). While the level of reproductive isolation conferred by differences in flower color is far from complete, it still represents a substantial advantage. This is particularly important for this annual species, which does not produce large numbers of seeds. Furthermore, this case is consistent with Stebbins's (1958) proposal that reinforcement will be more common in annuals, and it conforms to our prediction that selection for reinforcement will be stronger in species where population growth is more limited by numbers of ovules available for conspecific fertilization than by the availability of resources for the provisioning of developing seeds.

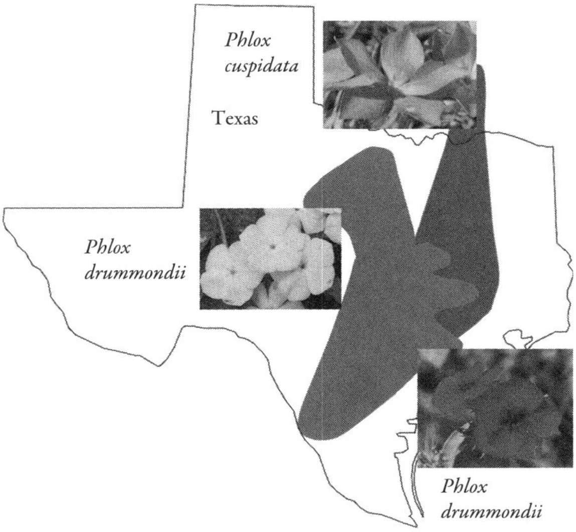

Fig. 10.13. Reinforcement in *Phlox drummondii* in Texas, USA. The range of *P. drummondii* is shown in blue, the sympatric range in purple, and the *P. cuspidata* range in red. Modified from Hopkins (2013). Photos by Хомелка, Varun Pabrai, and Joseph A. Marcus, Lady Bird Johnson Wildflower Center.

Selection for Reproductive Isolation—While examples of reinforcement of reproductive barriers in regions of sympatry are uncommon, the available examples suggest that it is possible for selection to strengthen isolation mechanisms. We predict that this will be particularly true for species that have strong post-zygotic isolation mechanisms and are more likely to suffer demographic costs from the loss of ovules due to interspecific fertilization (e.g., for annuals). The question remains, however, whether selection could strengthen reproductive isolation for entire species and not just within the regions of range overlap.

It may be difficult for isolating barriers that develop in regions of sympatry in response to selection for reinforcement to spread across the entire species range. Recall that the selection pressure that led to the establishment of isolating mechanisms would have been favored because of interspecific pollen transfer within the region of sympatry. It is possible that traits contributing to reproductive isolation could spread into allopatric regions if they conferred some selective advantage that was not directly associated with the prevention of interspecific mating, for example, if selection for reinforcement promoted a transition to a pollinator that was more effective across the species range. In this case, the traits would be favored outside regions of sympatry, because of the selective advantage of association with an alternative pollinator. While the origin of pre-zygotic isolation across a species because of reinforcement remains a possibility, it is less likely that the separate conditions required to promote the spread of a novel isolating mechanism (i.e., loss of fitness through hybridization and the availability of a more effective pollinator) would occur together than that a superior pollinator would be available.

One example of a plant that may be ripe for transition to an alternative pollinator due to selection for reinforcement is *Phlox pilosa* as described by Levin and Kerster (1967). This species displays a flower color polymorphism across its range with populations consisting primarily of pink-flowered plants along with a rare white-flowered morph. The white-flowered morph is prevalent in regions of sympatry, where it is easily differentiated from the pink-flowered plants of *P. glaberrima*, so the distribution of floral color morphs in this species is consistent with selection for reinforcement. While white-flowered plants receive less interspecific pollen than pink flowers in regions of sympatry, they also have lower seed set due to poor pollinator service. If a more effective pollinator for the white morph became available, it would elevate the plant's fitness and perhaps promote its spread across the species range. In this hypothetical example, the ultimate success of the white-flowered morph would depend on the appearance of a specialized pollinator, but at present, the white morph is likely to remain restricted to sympatric regions and may ultimately decline in abundance due to low seed set.

Evolution of Isolation Barriers—From the preceding discussion, it becomes clear that the origin and development of reproductive isolation in plants is unlikely to be a direct consequence of selection. Cases of reinforcement seem to be rare, and the spread of traits that strengthen barriers outside regions of sympatry is unlikely. Intrinsic barriers could develop as a side effect of adaptation to novel habitats, but the lack of a strong association between ecological divergence and genetic incompatibility suggests that the strengthening of intrinsic reproductive barriers through hitchhiking with genomic regions subject to selection is uncommon. Thus, reproductive isolation is much more likely to develop because of the accumulation of genetic differences that contribute to single- or multi-locus incompatibilities through genetic drift.

We can evaluate the effects of different genetic processes operating across levels of genetic distance between parents on offspring fitness by combining them into a single model that illustrates the evolution of reproductive isolation between divergent lineages (Fig. 10.14). In this figure, the combined effects of individual genetic influences (thin lines) are summarized as the overall fitness of offspring (thick line). As discussed in Chapter 4, populations of most plants are characterized by local inbreeding, so we expect inbreeding depression to be prevalent for close crosses (within-population crosses—dotted line in Fig. 10.14). Increasing the divergence between parents elevates the effects of heterosis (between-population crosses—solid line). As the time since a common ancestor for populations and lineages increases, we expect genetic incompatibilities to accumulate. Based on the information described here, it appears that multi-locus genetic incompatibilities expressed in recombinant hybrids (i.e., due to positive epistasis or recessive antagonistic interactions)

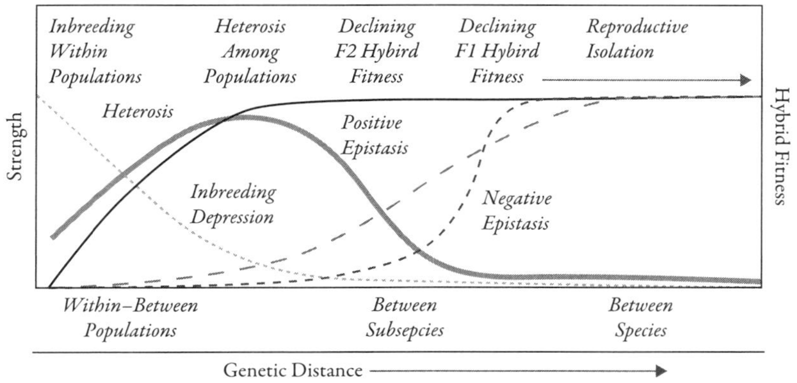

Fig. 10.14. A model of changes in the fitness of offspring (thick line) from parents separated by increasing genetic distance. Within species, the effects of inbreeding depression (dotted line) and heterosis (solid line) are prevalent. Between subspecies and species, the effects of genetic incompatibilities due to positive (long dashed line) and negative (short dashed line) epistasis have the largest influences on hybrid fitness.

accumulate faster than those expressed in the F_1 generation (Fig. 10.14 long dashed and short dashed lines, respectively). Consequently, at intermediate levels of divergence, F_1 hybrids may have high fitness while second-generation hybrids display a range of hybrid breakdown. Complete isolation between parents from different lineages is achieved once the fitness of first-generation hybrids approaches zero.

This model of the consequences of genetic divergence on offspring fitness accounts for multiple genetic effects that come into play for parents separated by a range of genetic distances. Within lineages, we may expect to see the combined influences of inbreeding depression and heterosis as the genetic distance between populations increases (Fig. 10.14). Between lineages, the effects of inbreeding are expected to be close to zero, but there can be strong influences of heterosis. For example, here we noted several cases where F_1 hybrids had more vigorous growth than their parental lineages (e.g., in *Piriqueta*; Rhode and Cruzan 2005). In the analyses of *Piriqueta* hybrids, the level of increase in F_1 hybrid fitness was equivalent to fitness increases seen for between-population crosses within lineages, which indicates that the enhanced vigor of plants is entirely due to heterosis (i.e., divergence is intermediate between populations and subspecies in Fig. 10.14). First-generation hybrids between *Piriqueta* morphotypes appear to have little or no hybrid breakdown, but backcross hybrids display a wide range of fitness (Cruzan 2005)—a pattern that appears to be common for a wide range of lineages that have diverged to intermediate levels and are not completely isolated.

Interbreeding between lineages that are not completely isolated can generate opportunities for genetic exchange through introgressive hybridization. In particular, crosses between recently diverged lineages may result in recombinant hybrids that display a wide range of fitness, but the effects of hybrid breakdown in the second and later generations are rarely severe enough to cause complete isolation. Consequently, a proportion of recombinant hybrids will often have high fitness and can facilitate genetic exchange between the parental lineages. Recurrent crossing within and between hybrids and parentals may result in stable admixed populations (hybrid zones) in geographic regions where closely related lineages come into contact. The processes governing the fate of hybrids and the dynamics of hybrid zones depends heavily on the influences of intrinsic and extrinsic mechanisms of post-zygotic reproductive isolation—a topic we will explore more thoroughly in the next section.

10.9 HYBRID ZONES

Throughout the history of life on Earth, changes in the geographic distributions of species have created opportunities for genetic exchange through hybridization. Gene flow among lineages during early stages of divergence appears to be common, and evidence of historic introgression has been unearthed through genomic analyses of a wide variety of organisms. The propensity for genetic exchange appears to be non-random across the genome, as only a minority of regions are typically affected by introgression (Abbott et al. 2013). By implication, we can surmise that these genetic exchanges have been important for diversification—perhaps vindicating the speculations of such scholars as Carl Linnaeus, Johannes Lotsy, Edgar Anderson, George Ledyard Stebbins, Verne Grant and others. Moreover, hybridization can promote evolution toward novel adaptive norms and has been important throughout evolutionary history (Lotsy 1916; Anderson and Stebbins 1954; Grant 1981; Arnold 1997).

Opportunities for introgression between any particular pair of lineages are necessarily ephemeral over evolutionary time scales. We expect these exchanges to eventually lead to complete fusion, as lineages meld, or alternatively, that reproductive isolation will eventually develop (Fig. 10.15). While individual instances of hybridization may be temporary, continued diversification will in turn generate novel genetic variation and new chances for opportunistic exchange. At any particular point in time, active cases of gene flow between divergent lineages are evidenced by regions of hybridization.

Tension Zones—Stable hybrid zones may develop when a balance is struck between selection and dispersal that results in regions of hybridization characterized by clines for diagnostic characteristics separating divergent lineages (Fig. 10.16). The dynamics of such dispersal-dependent hybrid zones (tension zones) are determined by the strength of selection against hybrids and the rates of dispersal from allopatric populations into the contact region (Barton and Hewitt 1985). Models of tension zones are built on the assumption of underdominance at individual loci resulting in low fitness for F_1 hybrids and provide predictions of hybrid zone characteristics under equilibrium conditions. The tension zone model was developed to explain the presence of narrow hybrid zones that do not seem to correspond to ecotones or environmental discontinuities (e.g., the fire-bellied toad tension zone in Europe; Szymura and Barton 1986), so the dynamics of tension zones are assumed to be controlled by intrinsic selection and to be independent of environmental conditions.

The tension zone model assumes that the region of hybridization is stabilized as selection limits the extent of introgression. Once two divergent lineages come into contact, the region of hybridization between them may begin to expand as clines for

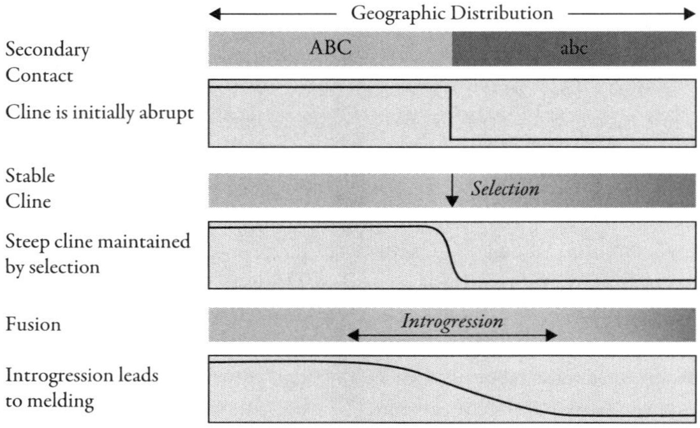

Fig. 10.15. Potential outcomes of secondary contact between closely related lineages. In the top panel reproductive isolation is complete so clines in diagnostic characters (ABC vs abc) remain sharp. In the middle panel, a stable hybrid zone results in steep clines that are maintained by selection. The bottom panel is a case where hybrids have high fitness across habitats so extensive introgression proceeds to promote the melding of two lineages.

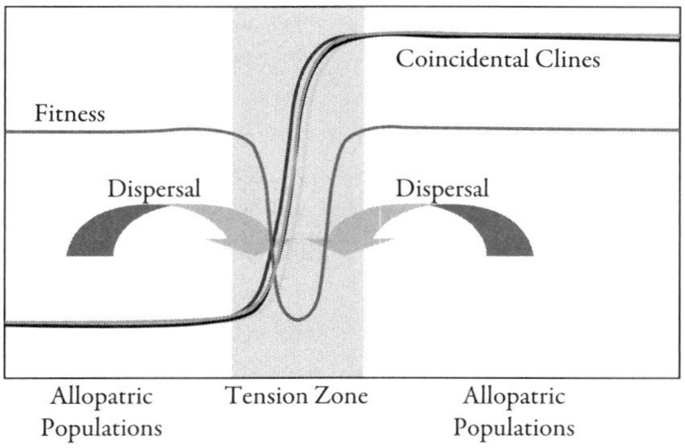

Fig. 10.16. Dispersal-dependent hybrid zone (tension zone). Selection against first-generation hybrids maintains sharp and coincidental clines for neutral genetic markers. The hybrid zone is maintained by a balance between dispersal from allopatric regions, and selection against hybrids.

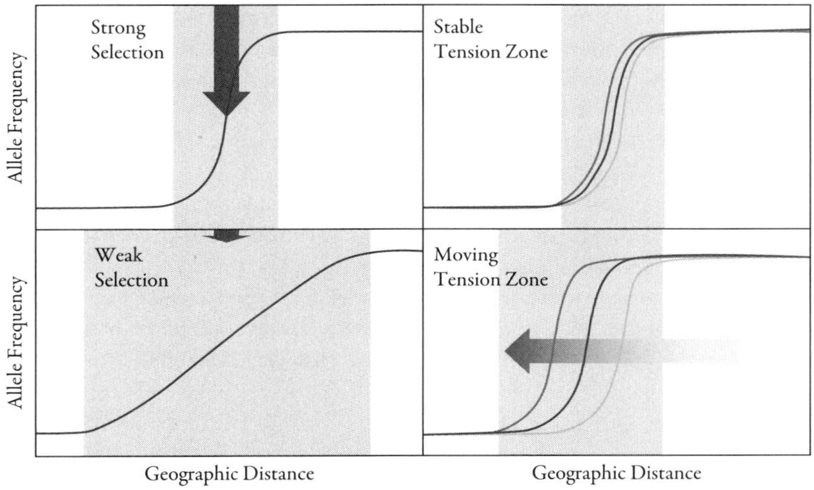

Fig. 10.17. Consequences of differing levels of selection and rates of dispersal on tension zones. Strong selection against first-generation hybrids maintains steep clines (upper left). With weaker selection cline slopes are more gradual and the region of introgression is wider (lower left). Stable tension zones have coincidental clines (upper right), while moving tension zones have staggered clines (lower right).

diagnostic characters form. The slope of clines will become more gradual until they approach equilibrium, which is assumed to be governed by the strength of selection against hybrids and the dispersal ability of the hybridizing lineages. Clines will remain steep if selection against hybrids is strong and dispersal distance is limited (Fig. 10.17 upper-left panel; refer to color plate). With weaker selection (s) or longer dispersal distances (expressed as the variance in dispersal distances from a single point—σ^2), cline slopes become more gradual and the tension zone is predicted to be wider (lower-left). The cline width (w) is predicted to depend only on rates of dispersal and the strength selection ($w = \sqrt{8}\ \sigma^2/s$; Barton and Hewitt 1985). This simple model predicts that the cline width will become greater as dispersal distance increases or selection weakens and will become narrower with limited dispersal or stronger selection.

A tension zone may shift position on a uniform landscape in response to asymmetrical rates of dispersal. As long as dispersal from the two allopatric regions is equal, the hybrid zone is predicted to be stationary over long periods of time. Under conditions of differential rates of dispersal due to lower density or population growth for one of the species, or due to intrinsically different dispersal abilities, the hybrid zone is expected to shift toward the species with the lower rate of dispersal or shorter average dispersal distance (Fig. 10.17 lower-right panel; refer to color plate). Such moving tension zones are characterized by staggered clines compared to coincidental clines for tension zones that are stationary (Fig. 10.17 upper-right panel; refer to color plate). The tension zone may continue to move until it is affected by extrinsic factors. For example, the moving clines may become caught in a trough of low density (e.g., due to an unfavorable environment). Alternatively, the tension zone may continue to move until one species is overcome and completely swamped by the other.

The tension zone model assumes that hybrid populations are maintained by gene flow from allopatric regions because hybrid individuals have higher rates of mortality. Parental offspring that disperse into the contact zone suffer low fitness because of high rates of interspecific mating resulting in offspring with low viability or fertility. If dispersal from allopatric regions were eliminated, then these hybrid populations would go extinct and the tension zone would cease to exist. In other words, hybrid zones are effectively demographic sinks where populations have negative growth rates. Consequently, tension zones can be identified as hybrid zones characterized by sharp clines and evidence of recent dispersal from allopatric regions (Barton and Hewitt 1985). Gene flow resulting in admixture between divergent lineages can be detected by elevated levels of gametic disequilibrium (Box 10.1; Fig. 10.19). Disequilibrium is generated in the hybrid zone as dispersal introduces excess haplotypes, characterized by pairs of alleles that are diagnostic or differ strongly in frequency between the hybridizing lineages. For unlinked loci, the disequilibrium declines by 50% each generation and will approach zero within a few generations. Consequently, elevated levels of gametic disequilibrium indicate recent gene flow from allopatric regions and combined with sharp clines, provide evidence for the presence of a tension zone.

Introgression Across Tension Zones—Gene flow is inhibited within tension zones, because hybrid populations subject to strong selection will have low density and limited rates of reproduction. Low fitness creates a "hybrid sink" with limited recruitment and few migrants, so the weight of gene flow is toward the tension zone, and there is limited dispersal from hybrid to allopatric populations. In the model discussed earlier, we assumed underdominance or negative epistasis (BMDI) would result in low F_1 hybrid fitness. Strong selection against F_1 hybrids would have an equal effect across genomes, which results in coincidental clines in stationary tension zones and severely limit levels of gene flow between lineages. Alternatively, a more general view of tension zones recognizes that clines may develop as a result of hybrid breakdown in recombinant generations (De La Torre et al. 2015). Unequal effects of selection across the genome in later hybrid generations would result in different rates of introgression and variation in the shapes of clines for different genomic regions. In such cases, F_1 hybrids may have high fitness, while most of the hybrid breakdown occurs in the recombinant generations. When genetic incompatibilities are expressed in recombinant hybrids, the strength of selection—and consequently the shape of clines and rates of introgression—varies for different genomic regions.

Gene flow past a region of steep clines is possible whenever viable and fertile backcross genotypes are generated (Gavrilets and Cruzan 1998). Introgression proceeds as recombination occurs across backcrossed generations and as recombinant individuals carrying neutral or beneficial alleles disperse outside the tension zone. Over longer periods of time, sharp clines may be maintained for genomic regions subject to strong selection, but continued backcrossing and recombination will facilitate introgression and generate more gradual clines for neutral and beneficial gene regions. The effect of more extensive introgression is sometimes depicted as "shoulders," with gradual slopes on either side of the sharp clines (stepped clines; Barton and Hewitt 1985). Introgression for specific genomic regions is facilitated by the phenotypic and genetic diversity of recombinant hybrids, which can be quite high—sometimes exceeding the range of variation observed between the original parental lineages—due to transgressive segreggation. Selection on recombinant hybrid generations results in *Transmission Ratio Distortion* such that alleles from one of the two parental lineages are favored (Tang

Box 10.1 Gametic Disequilibrium

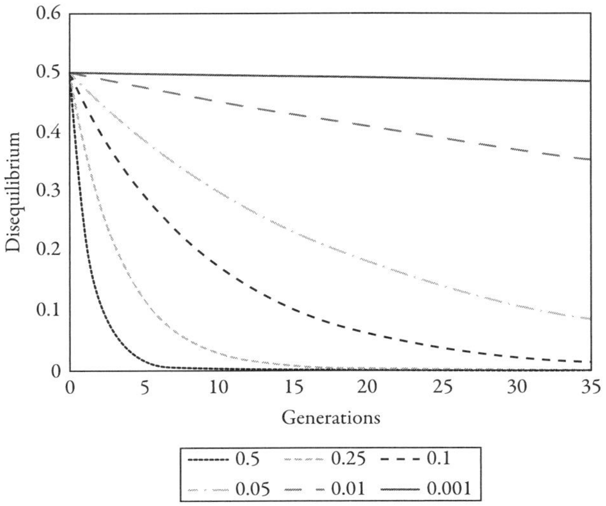

Fig. 10.18. The effects of recombination rate (r) on the rate of approach to zero for gametic disequilibrium. Recombination varies from 0.5 (light-blue line) to 0.001 (green line).

Gametic disequilibrium is defined as a non-random association among alleles at different loci (Hedrick 2005). Linkage disequilibrium is often used synonymously with gametic disequilibrium, but in this text, we reserve the former term for cases where recombination rates are lower than segregation rates of separate chromosomes (i.e. $r < 0.5$), indicating that the two loci are in close proximity on the same chromosome. Gametic disequilibrium is a more general term that reflects the statistical association among alleles across the genome. Disequilibrium between alleles at separate loci is analogous to departures from expected genotype frequencies at a single locus (i.e., the HW equilibrium). We use the product of allele frequencies to predict the frequency at which we expect to observe a combination of alleles under random association ($p_A p_B$ for alleles at the A and B loci). We can define the observed frequency of haplotypes or gametes carrying the both the A and B alleles as g_1; the sum of the expected value and D (disequilibrium), which is the departure from the expected frequency ($g_1 = p_A p_B + D$ for the A-B haplotype). For loci that are segregating for two alleles there are four possible haplotype frequencies:

$$g_1 = p_A p_B + D$$

$$g_2 = p_A p_b - D$$

$$g_3 = p_a p_B - D$$

$$g_4 = p_a p_b + D$$

By convention, the A-B and a-b allele combinations are coupling haplotypes, and A-b and a-B are repulsion haplotypes. This terminology defines associations among alleles in admixed populations between lineages fixed for A-B and a-b haplotypes. A population consisting entirely of F_1 hybrids would be in complete disequilibrium (only coupling haplotypes). As recombination occurs over subsequent generations, the frequency of repulsion haplotypes would increase as D asymptotically approaches zero. For example, under random mating with unlinked loci ($r = 0.5$), D is halved each generation:

$$D_t = (1-r)^t D_0$$

Where D_0 is the initial disequilibrium and D_t is the disequilibrium at generation t. For unlinked loci D is approximately zero within ten generations ($D \approx 0.00049$), but as the recombination rate declines the approach toward zero becomes much slower (Fig. 10.18).

Values of D can range from -0.25 to 0.25. For the F_1 population described here $D = 0.25$, since all alleles have frequencies of 0.5 (i.e. $g_1 = g_4 = 0.5$, and $g_2 = g_3 = 0$). Since complete disequilibrium is dependent on allele frequencies, it is often standardized for its maximum value ($D' = D / D_{max}$). D' ranges from zero to 1.0 and provides an equitable comparison among pairs of loci with alleles at different frequencies.

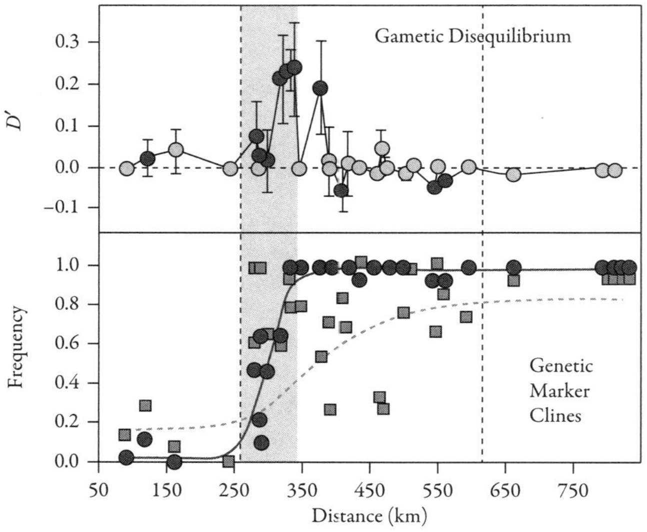

Fig. 10.19. Evidence for the presence of a tension zone in *Piriqueta caroliniana* ssp. *cistoides* in Florida. Populations with significantly positive gametic disequilibria indicating an excess of coupling haplotypes (red dots—upper panel) are concentrated in the region of sharp clines (shaded area). Examples of two clines (c28 in solid and c59 in dashed) for genetic markers that are nearly fixed in allopatric populations (lower panel). Geographic distance is from south to north. Four *viridis* markers had similar shapes to c59 (not shown). Modified from Cruzan (2005).

et al. 2010; Leppala et al. 2013), which promotes differential rates of introgression and varying shapes of clines among regions across the genome (De La Torre et al. 2015; de Lafontaine et al. 2015).

Variation in the fitness of recombinant hybrids contributes to the distortion of transmission among genomic regions, which translates into genomic variability for rates of introgression and cline shape. The effects of differences in the strength of selection across genomic regions can be seen for diagnostic genetic markers between morphotypes of *Piriqueta caroliniana* ssp. *cistoides* in Florida (Cruzan 2005; Fig. 10.19).

Patterns of marker variation in this broad hybrid zone are consistent with both a tension zone that is moving northward and with variation in the strength of selection across genomic regions that generates variable cline shapes among genetic markers. This conclusion comes from several observations. First, a tension zone is evident from the association between positive gametic disequilibria (*D'*) in populations and the presence of a sharp cline. Second, one of the two markers diagnostic for the *caroliniana* morphotype displays a sharp cline, while the second and all four clines for diagnostic *viridis* markers are much more gradual, and their centers are displaced northward (Martin and Cruzan 1999; Fig. 10.19). The high level of disequilibria within the tension zone indicates recent gene flow from allopatric regions, and the staggered clines, along with populations having high disequilibria, that are north of the tension zone are consistent with northward movement. Moreover, more extensive introgression is evident to the north of the tension zone as populations display varying frequencies for diagnostic markers extend across central Florida. While one marker (c28) appears to be associated with a genomic region under strong selection, clinal variation is much more gradual for the other five markers suggesting weaker selection, which allows more extensive northward introgression. Similar levels of variation in the shapes of clines and levels of introgression among SNP markers is also evident for *Picea* hybrid zones in eastern (de Lafontaine et al. 2015) and western (De La Torre et al. 2015) North America.

This *Piriqueta* hybrid zone is one of few available examples of a hybrid zone that displays diagnostic characteristics of a tension zone—that is, high levels of disequilibria that are associated with the location of a sharp cline. The variation in shape among clines indicates that the strength of selection differs among genomic regions. As mentioned earlier, F_1 hybrids between these *Piriqueta* morphotypes have high fitness, and recombinant hybrids vary widely in their ability to survive in the field. There are no obvious discontinuities associated with ecological conditions across this hybrid zone, so it is likely that genetic incompatibilities due to epistasis contribute to variation in the fitness of recombinant hybrids and are responsible for the maintenance of the sharp cline. Selection contributing to the transmission ratio distortion affecting cline shape may be caused by post-zygotic genetic incompatibilities, or it may be due to epistasis affecting pollen-tube growth in the style (e.g., Fishman et al. 2008). Ultimately, selection responsible for this tension zone appears to be affecting only a minority of genomic regions (one of six genetic markers assayed; Cruzan 2005), which apparently presents a relatively weak barrier to gene flow and allows for extensive northward introgression for a large proportion of the genome.

The tension zone model is built on the simplifying assumption of no extrinsic selection, but we also recognize that environmental conditions commonly affect the distribution of parental and hybrid genotypes. The tension zone model is clearly valuable, but we expect that it provides an incomplete picture of the dynamics of hybrid zones. For example, in the *Piriqueta* hybrid zone described earlier, some of the advanced-generation recombinant genotypes have higher fitness than parentals, so it is likely that selection has promoted introgression and the spread of hybrids across central Florida (Cruzan 2005). Tension zones may provide useful explanations for the distribution of variation for some portions of the genomic variation across hybrid zones, but the large majority of genomic regions are typically neutral or affected by extrinsic selection. Ecological conditions may have overriding effects on the distribution of hybrids, and the sedentary life form of plants results in an unbuffered exposure to variation in climates and soils, which may have overriding effects on patterns of introgression. In the next section, we explore the contributions of environmental conditions to hybrid zone dynamics.

10.10 ECOLOGICALLY-DETERMINED HYBRID ZONES

The tension zone model provides us with a set of expectations and tools for analyzing levels of introgression and the stability of hybrid zones, but the patterns of variation for alleles affected by intrinsic selection and dispersal can be further modified by variation in ecological conditions. The prevailing effects of environmental variation on outcomes of interspecific mating has been widely recognized since the mid-twentieth century with Anderson's (1948) proposal that hybrid zones could be maintained by "hybridization of the habitat." The key principle that Anderson promoted was that hybrid genotypes could not persist in nature unless they had access to adequate resources. In his view, these opportunities are afforded by environmental gradients that fall between the adaptive norms of species (ecotones) and by environmental perturbations (disturbances) that open up habitat for hybrid genotypes. Anderson's proposal quickly gained support from leading plant evolutionary biologists and continues to be influential (Stebbins 1959; Heiser 1973; Grant 1981; Arnold 1997). With improvements in sequencing technologies and bioinformatics, in particular, the prevalence of introgression is becoming clear (Abbott et al. 2013); however, many questions remain concerning the potential for ecological conditions to facilitate the establishment of hybrid populations and introgression among lineages. In this section, we discuss the potential for environmental conditions to affect the distribution and abundance of hybrid genotypes, including cases where hybrids have superior fitness under limited ecological conditions, or more broadly, across one or both parental habitats.

The Process of Introgression—Ecological conditions often determine the success of hybrids and consequently, the potential for introgression between lineages. For introgression to proceed, F_1 and recombinant generation hybrids must be afforded appropriate environments for their survival and reproduction. As discussed earlier, reproductive isolation based on extrinsic barriers can prevent the successful establishment of F_1 hybrids and genetic exchange between lineages. Typically, the largest extrinsic barrier to introgression often comes in the first generation for a couple of reasons. First, F_1 hybrids are more likely to have ecological requirements that are intermediate between the parental lineages, so the establishment of hybrids may be limited by the availability of transitional habitats. Second, F_1 hybrid seeds are necessarily produced by parental plants and may have poor performance in the local parental habitat. Even if hybrids can benefit from the effects of outbreeding and heterosis, the increase in vigor may not be sufficient to overcome performance deficits in habitats where the parental individuals are abundant. Unless the F_1 hybrids have sufficient fitness in habitats that are available within the range of seed dispersal from parental plants, the lineages will remain isolated and genetic exchange will be prevented.

Much of the research conducted on reproductive isolation and introgression focuses on F_1 hybrids, because this generation is the most critical for the establishment of hybrid populations and for the progress of genetic exchange between lineages. During secondary contact, F_1 hybrids will typically be formed after long-distance dispersal of pollen or seeds between populations of the two lineages (Fig. 10.2). Consequently, first-generation hybrids are likely to be produced in low frequency, and the large majority of recombinant hybrids will be generated by backcrossing. Backcross hybrids will express a wide range of phenotypes, so at least some proportion of them are likely to perform well in the parental habitat. Hence, even a very low frequency of interspecific crossing may initiate the establishment of recombinant hybrid populations. For example, in the Louisiana iris hybrid zone, F_1 hybrids are extremely rare, but populations of recombinant hybrids expressing diverse phenotypes are relatively common (Arnold 1997; Tang et al. 2010). Different iris species occasionally grow sympatrically without any evidence of interbreeding, but where hybridization has occurred, only recombinant individuals are present. Strong pre-zygotic isolation apparently prevents interbreeding among iris species (Arnold et al. 2012), but once F_1 hybrids become established, backcrossing ensues and the original parental populations are displaced. In this and other cases where closely related lineages are in close proximity, we expect that the infrequent production of viable and fertile F_1 hybrids within parental populations will trigger the generation of backcross hybrids, which will promote introgression between lineages.

Ecological Associations in Hybrid Zones—The prevalent influence of ecological conditions on levels of hybridization and introgression is evident from the distribution of parental and hybrid genotypes within and among habitats. In one location of the Louisiana iris hybrid zone just described, there was a non-random association between hybrid genotypes and parental habitats (Cruzan and Arnold 1993). Intermediate recombinant hybrids were absent, and recombinant hybrids had genetic marker profiles that were more similar to one parental species or the other. These hybrid genotypes were more abundant in the habitat of the parental species that they resembled most closely. Similar non-random associations between ecological conditions and hybrid genotypes have been documented for a number of cases, including: inland and coastal subspecies of *Gilia capitata* (Nagy 1997); a *Prunella* hybrid zone spanning a soil moisture gradient (Fritsche and Kaltz 2000); hybrid and parental populations of the fern genus *Polystichum* occupying understory habitats varying in soil moisture and light levels (Kentner and Mesler 2000); a great basin sagebrush hybrid zone distributed along an elevational gradient (Miglia et al. 2005); and a hybrid zone between riparian and non-riparian species of herbs in the genus *Ainsliaea* (Mitsui et al. 2011).

The persistence of stable ecological conditions over long periods combined with incomplete reproductive isolation may result in stable zones of introgression. Broad hybrid zones, apparently stabilized by ecological conditions, are known for a number of species pairs and appear to be particularly common for long-lived herbaceous and woody plants. For example, a broad region of hybridization occurs among three species of buckeye (*Aesculus*) in North America (Thomas et al. 2008). In this system, the lack of gametic disequilibria within the region of hybridization suggests that the persistence of hybrid populations is due to the availability of appropriate ecological conditions rather than ongoing gene flow, as would be expected under the tension zone model. In another example, the multiple regions of hybridization between Sitka and white spruce (*Picea*) are distributed along the pacific coast of Canada and Alaska (Hamilton and Aitken 2013). There, parallel hybrid zones in isolated glacial valleys displayed the same patterns of association with climatic variables and showed evidence of historical movement—probably in response to climate change. These observations highlight the importance of ecological conditions as the primary factors affecting the patterns and extent of hybridization; as long as the appropriate ecological conditions are available, hybrid populations will persist and will facilitate introgression between divergent lineages.

Hybrid Superiority—The superiority of hybrids and the potential for introgressive hybridization as a creative force has been recognized by plant biologists since the middle of the twentieth century (Fig. 10.20), but discussions of whether hybrids could display high fitness continued into the 1990s (Arnold and Hodges 1995). The prevailing view of hybrids as inferior mongrels that are unlikely to persist in nature has shifted dramatically as improvements in the resolution of genome-wide assays continued to reveal the prevalence of introgressive hybridization during early stages of divergence (Abbott et al. 2013; Harrison and Larson 2014). Our view of hybridization has become more focused on genic exchanges rather than as a genomic phenomenon, as we recognize that levels of introgression can be much more extensive than is evident from phenotypic variation. Having elucidated the abundant evidence of historical introgression in the genomes of individuals across the range of species, we are led to the conclusion that neutral and beneficial alleles must have migrated out of limited regions of secondary contact. In other words, at some point in the history of many species, recombinant genotypes must have spread widely to completely displace the original parental lineage. If this is the case, then hybrid superiority must be common.

The proposal that hybrids can have high fitness seems to contradict our observations of hybrid zones; in most examples, the F_1 and recombinant hybrid genotypes have relatively restricted distributions relative to the parental lineages and cases where hybrid genotypes are more broadly distributed appear to be rare. We expect that F_1 hybrids between closely related inbred lineages may have elevated fitness due to heterosis, and this increased heterozygosity may promote the establishment of early-generation hybrids in parental habitats. High fitness of first-generation hybrids may be common and would facilitate the generation of backcrossed hybrids; however, the effects of heterosis will dissipate in successive recombinant generations as heterozygosity declines. For extensive introgression to progress outside regions of secondary contact, some superior recombinant hybrids must be generated and succeed in displacing parentals under a broad range of conditions.

Analyses for a number of species has revealed that genomes are often permeable to introgression from closely related taxa and that these low levels of hybridization are not always accompanied by obvious phenotypic differences (Harrison and Larson 2014). For example, there was widespread evidence of introgression across genomes of a pair of hybridizing species of *Silene* without obvious phenotypic effects (Minder and Widmer 2008). Similarly, genomic analyses revealed evidence of introgression among several oak species across Europe (Gerber et al. 2014). Low levels of introgression appear to be possible even in cases where reproductive isolation between species is very

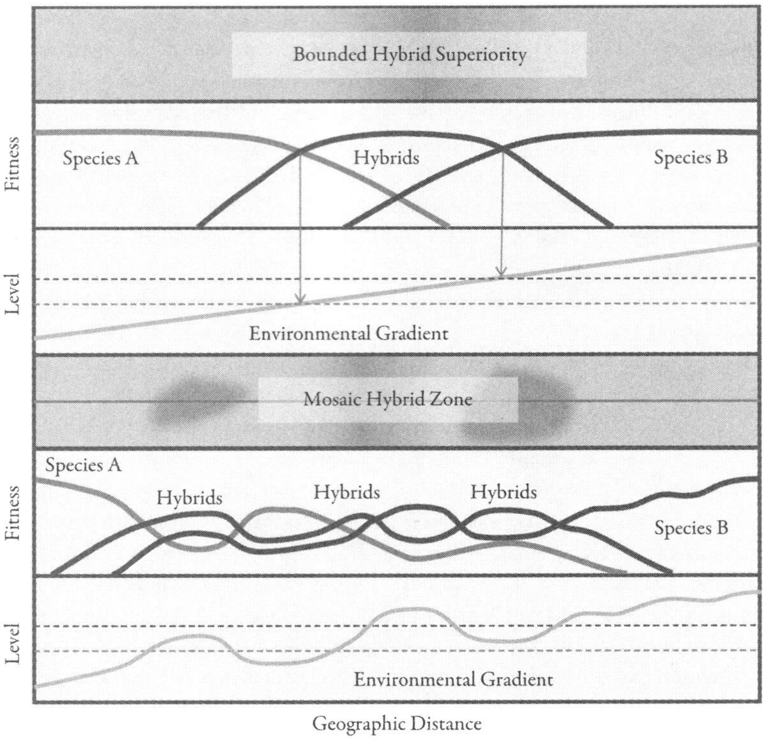

Fig. 10.20. A smooth environmental gradient limits the distribution of hybrids to ecotones in the Bounded Hybrid Superiority model (upper panel), and an undulating gradient results in a Mosaic Hybrid Zone (lower panel). The distribution of hybrids is limited to intermediate regions in both models (i.e. when environmental conditions fall between the horizontal dashed lines). The environmental gradient is based on a transect crossing the distribution map (horizontal line) in the mosaic hybrid zone.

strong. Sambatti et al. (2012) found reproductive isolation strong enough to reduce the production of F_1 hybrids to less than 0.1% of offspring produced each generation between a pair of annual sunflowers that display extensive evidence of introgression across the genome. The high level of interspecific gene flow in this case may be due to large effective populations sizes, but this case highlights the fact that introgression among species may be much more common than previously thought—even when reproductive isolation appears to be strong.

The apparent commonness of introgression contrasts with the rareness of hybrid superiority; extensive introgression requires that recombinant hybrids spread and displace the original parental populations. The only known cases of hybrids outcompeting their parental lineages over a large geographic range are hybridization between a pair of Galapagos finches (Grant and Grant 1996; Markert et al. 2004) and between the morphotypes of *Piriqueta* discussed earlier (Cruzan 2005). Of course, this apparent lack of evidence of hybrid superiority may be largely a perception problem; introgressed recombinant genotypes that are spreading within and among populations may be difficult to detect because they display little or no phenotypic differentiation. Such is the case for the invasion of slender false brome (*Brachypodium sylvaticum*) in western North America (Rosenthal et al. 2008). Invasion of this perennial bunchgrass across the region occurred after introductions from multiple source populations and is largely due to the spread of recombinant hybrids. Phenotypic differentiation between invasive and native populations of false brome is only evident for physiological and micro-morphological traits (Marchini et al. 2018). In this case, admixture among populations from different source regions was only detectable from genetic marker analyses, but hybridization associated with invasion success may be common, given the high frequency of invasive species that were initiated from multiple introductions from different native source regions (Chapter 8). We can expect that introgression will occur at some level whenever closely related lineages come into close contact due to either human-mediated or natural changes in their geographic distributions. In many of these cases, hybridization may not be obvious if hybrids occur at low frequencies or if phenotypic differentiation is not strong.

10.11 HISTORICAL INTROGRESSION

Opportunities for genetic exchange between lineages may occur during early stages of divergence and may continue until reproductive isolation is complete. As discussed earlier, the frequency of introgression over the history of diverging taxa may depend heavily on how often changes in geographic distributions and ecological conditions promote interbreeding and the production of successful hybrids. Given that glacial cycles throughout the Pleistocene generated cycles of fragmentation and range expansion across most continents, there may have been multiple opportunities for introgression within many clades of closely related lineages. The sum effects of these exchanges have influenced genomic composition and facilitated diversification in a many of groups of plants.

Detecting the effects of historical introgression requires evidence of the transfer of genomic regions or entire genomes (i.e., the non-recombining chloroplast or mitochondria genomes) that are characteristic of one species into a second species. For stable hybrid zones, we can detect ongoing introgression by identifying clines, but this task becomes more difficult in cases where introgression is complete and alleles have become fixed or nearly fixed across a region or the entire range of a species. Extensive introgression probably occurs more often for beneficial alleles that introgress and are spread by selection, but neutral alleles may also become fixed across a species' range either through genetic hitchhiking with alleles under selection or due to genetic drift and population bottlenecks during range expansion. Cases of complete introgression may be more difficult to identify with any degree of confidence, unless phylogenetic information is available to allow comparisons among taxa and with the inferred ancestral genomic composition. Alternatively, the availability of genome-wide scans has enabled several different methods that utilize statistical associations to detect historical introgression. In this section, we explore the evidence for historical introgression between species. We start with a discussion of changes in the associations among nuclear and cytoplasmic genomes between closely related species and then describe methods for the detection of introgressed regions across nuclear genomes.

Pollen Swamping—Prior to the availability of genome-wide sequence information, historical introgression could be inferred from observations of incongruous associations between cytoplasmic and nuclear genome regions. Exchange of nuclear and cytoplasmic genomes through hybridization is evident in cases where the chloroplast from a related species characterizes a portion of a range (*Chloroplast Capture*) and in cases of incongruities between phylogenies (reticulation) that are based on nuclear and cytoplasmic genomes (Rieseberg and Soltis 1991; Rieseberg and Wendel 1993; Fig. 10.21; refer to color plate). Differential introgression between nuclear and cytoplasmic genomes is thought to occur because chloroplasts and mitochondria are usually maternally inherited (i.e., from the seed parent) in flowering plants, while nuclear genomes are transferred to zygotes from both seed and pollen parents. Rates of movement of nuclear markers and chloroplast genomes can differ because of differences in the dispersal ability of seeds and pollen; for most plants, pollen dispersal distances exceed those of seeds (Chapter 8; Fig. 8.5; refer to color plate). Moreover, nuclear genomes are subject to movement via both pollen and seeds, so we expect that rates of gene flow for nuclear alleles will always be greater. These differential rates of gene flow suggest that chloroplast capture is probably more correctly referred to as *Pollen Swamping* as we expect nuclear genomes to overtake chloroplast genomes as a result of differential rates of gene flow (Petit et al. 2004).

Differential introgression of chloroplast and nuclear genomes was first identified in a portion of the geographic range of the common sunflower

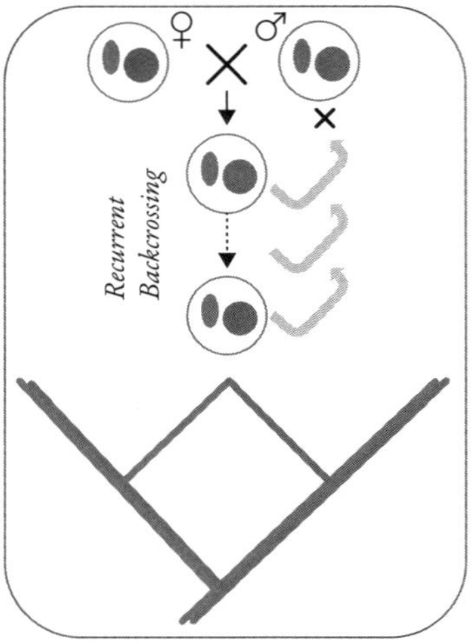

Fig. 10.21. Reticulate evolution. Hybridization followed by recurrent backcrossing into the pollen parent lineage results in a new lineage having a chloroplast genome more similar to the maternal lineage and a nuclear genome more similar to the paternal lineage. The resulting phylogenies based on chloroplast (green phylogeny) and nuclear (blue) genomes display alternative topologies.

(*Helianthus annuus* ssp. *texanus*). Populations in this region displayed morphological differences suggesting hybridization with *H. debilis* ssp. *cucumenifolius* (Rieseberg et al. 1990). Over most of its range *H. a. texanus* populations are characterized by *H. d. cucumenifolius* chloroplast DNA while there was no evidence of introgression of cpDNA markers in the reverse direction. Similarly, hybridization among North American (Whittemore and Schaal 1991) and European oaks (Petit et al. 2004) has led to the unequal introgression such that individual species may share chloroplast genomes over large portions of their ranges. Moreover, nuclear genomes of these species can remain largely intact; nuclear genomes are associated with taxonomic identity while several species may share the same chloroplast genomes. As expected from our understanding of pollen and seed dispersal, chloroplast genomes display geographic distributions suggesting lower rates of gene flow. In all of these cases, it appears that nuclear genomes have introgressed over the more stationary chloroplast haplotypes via pollen swamping.

Phylogenetic Reticulation—The historical effects of pollen swamping are evident from contrasting phylogenetic relationships estimated from chloroplast and nuclear gene regions (Rieseberg and Soltis 1991; Rieseberg and Wendel 1993; Fig. 10.21; refer to color plate). Reticulate evolution has been detected in a wide range of taxa: woody plants and trees such as oaks (*Quercus*), poplars (*Populus*), willows (*Salix*), and avocados (*Persea*); herbaceous annuals and perennials including sunflowers (*Helianthus*), cotton (*Gossypium*), alumroots (*Heuchera*), and mustards (*Brassica*); and grasses including wheat (*Triticum*) and corn (*Zea*). Most of these cases appear to have occurred in the last few million years, but more ancient pollen swamping events are evident in the beeches of the southern hemisphere (*Nothofagus*; Acosta and Premoli 2010). With the high rates of hybridization found across the seed plants, we expect that cases of historical pollen swamping may be quite common.

For pollen swamping to progress successfully, there needs to be little or no fitness loss from interlineage combinations of cytoplasmic and nuclear genomes. This is not always the case, as the effects of *Cytonuclear Epistasis* have been documented between a number of species pairs, causing lower rates of hybridization or asymmetrical introgression (e.g., Cruzan and Arnold 1998; Etterson et al. 2007). Differential patterns of introgression leading to *Cytonuclear Disequilibria* (Arnold 1993) can also be caused by asymmetrical reproductive isolation between species, such as unilateral incompatibility. Consequently, we expect that pollen swamping could be stalled due to asymmetrical mating success or low fitness of hybrids resulting from cytonuclear epistasis. Since cases of pollen swamping are common, it appears that the movement of nuclear genomes over the geographic ranges of chloroplast genomes often proceeds unabated and that negative cytonuclear epistasis is uncommon.

The origin of reticulate phylogenies requires that the chloroplast genome of one lineage becomes associated with the nuclear genome of another. Note that with long-distance pollen dispersal we would expect that nuclear genomes would have the potential to form hybrids in populations of a closely related lineage. If mating were random, then introgression would be minimal as the genome of the native lineage would be prevalent in the resulting hybrids. For pollen swamping to proceed and to eventually result in reticulate evolution, the invading nuclear genome would have to display some selective advantage in terms of mating or vigor in the local environment. A pattern of phylogenetic reticulation would develop only under conditions where the hybrid population remained isolated enough to eventually form a new lineage.

Genome-Wide Analyses of Historical Introgression—The advent of improved technologies and bioinformatic procedures for the collection and organization of genome-wide sequence data has led to a proliferation of analytical methods designed to detect the effects of historical introgression. Comparison of high densities of SNPs among multiple individuals has allowed dating of historical introgression between humans and archaic hominins (e.g., Sankararaman et al. 2012), and these methods are being applied to evaluate historical introgression in a wider variety of organisms including plants (e.g., Eaton and Ree 2013; Escudero et al. 2014). A common approach is to apply coalescence modeling combined with maximum likelihood or Bayesian inference to obtain the model that best explains SNP variation among divergent lineages (Sousa and Hey 2013). Introgression during divergence can be detected by comparing likelihoods over all possible genealogies generated across genomic regions (Chapter 11). Reticulate patterns among genomic regions would provide evidence of historical introgression as described earlier. These methods have the advantage of providing information on phylogenetic relationships while simultaneously testing for the effects of historical introgression, but they also require a large amount of sequence information across different genomic regions for multiple lineages.

The implementation of phylogenetic methods that analyze large numbers genomic regions to infer historical introgression is less common, because these methods can be computationally intensive and can be complicated by recombination within the regions being analyzed (Sousa and Hey 2013). Information on large numbers of SNPs can be more readily obtained, so most studies evaluating historical introgression have focused on statistical analysis of single nucleotide variation within and among lineages. One of these methods examines level of gametic disequilibria (linkage disequilibria; LD) among pairs of SNPs across the genome of individuals within a lineage to identify genomic regions that are likely to have been affected by historical introgression. Recall that recombination disassociates alleles, so levels of gametic disequilibrium among pairs of SNPs will tend to be low (Fig. 10.22). Most introgressing genomic regions will tend to be lost through recombination and genetic drift, unless they contain functional regions that confer fitness advantages. At the end of introgression, we expect the original genome of a species will be largely intact, but it may contain smaller regions of interspecific DNA. When compared to genomes that have not been subjected to introgression, levels of gametic disequilibria will be much stronger among SNPs within recently introgressed regions of the hybrid genomes, because they contain pairs of interspecific alleles that tend to segregate together. With high-densities of mapped SNPs, the size and location of introgressed genomic regions can be determined.

Linkage disequilibrium methods for the detection of historical introgression are not widely used, because they require access to complete genome sequences so the location of SNP markers can be mapped. An alternative statistical method, that does not require knowledge of the location of markers, tests for biased inheritance of SNP alleles from ancestral lineages due to historical introgression. The effects of historical introgression on SNP variation can be distinguished from *Lineage Sorting* using *Patterson's D* (D_p) statistic, which is also referred to as the *ABBA-BABA Test* (Sousa and Hey 2013; Fig. 10.23; refer to color plate). This test requires knowledge of the

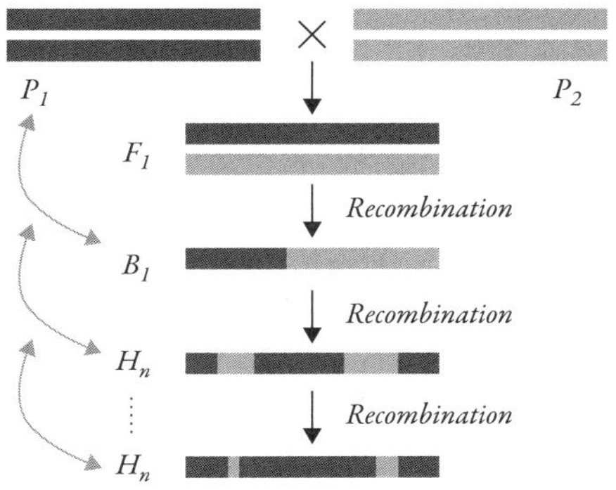

Fig. 10.22. The process of introgression over many generations is initiated by the production on backcrossed genotypes possessing recombined chromosomes. Over generations of backcrossing introgressed genomic regions become smaller as they stabilize.

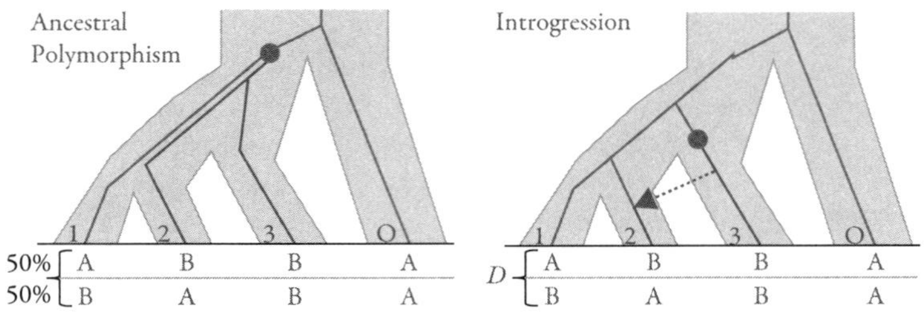

Fig. 10.23. Patterson's D statistic (the ABBA-BABA test) assumes that alleles not present in the outgroup (O—red, B alleles) have a 50% chance of appearing in lineages P_1 and P_2 (1 and 2; left panel); the ABBA and BABA patterns occur at equal frequencies. Historical introgression will result in asymmetrical inheritance (right panel) so in this example more alleles will follow the ABBA pattern than the BABA pattern of inheritance.

phylogenetic relationships among at least three taxa plus an outgroup (O) to root the tree. Alleles that are unique to P_3 relative to O may have been inherited by P_1 and P_2 when they originated (lineage sorting), or they may have been transferred via hybridization that occurred at various stages of divergence. The ABBA-BABA test assumes that, on the average, ancestral polymorphic alleles would have a 50% chance of appearing in P_1 and P_2 and that a significant departure from this 50/50 expectation would be evidence for historical introgression. Patterson's D statistic tests for asymmetry in the inheritance of SNP alleles between P_1 and P_2:

$$D_P = \frac{\sum C_{ABBA(i)} - C_{BABA(i)}}{\sum C_{ABBA(i)} + C_{BABA(i)}}$$

for $i = 1$ to n SNP alleles that occur in P_3 and not in O. Here, C_{ABBA} is the number of alleles that display the ABBA pattern, and C_{BABA} is the number of alleles that display the BABA pattern of inheritance (Fig. 10.23; refer to color plate). With only lineage sorting and complete isolation, we expect that $D_P \approx 0$, while historical introgression will result in $D_P > 0$ or $D_P < 0$ due to hybridization between P_3 with either P_2 or P_1, respectively (Fig. 10.23; refer to color plate). Note that this is not an absolute test for the occurrence of historical hybridization since equal levels of introgression could also result in symmetrical inheritance of alleles (i.e. $D_P \approx 0$).

Large amounts of SNP data can easily be generated for species lacking a reference genome, which has allowed application of the ABBA-BABA test to a wide range of species (e.g., Eaton and Ree 2013; Escudero et al. 2014). If we assume that introgression would only affect a minority of the genome then the same SNP data can be used to infer phylogenetic relationships. In clades with more than three taxa, different combinations of lineages can be compared to provide more robust inferences concerning the role of historical introgression during divergence. In one analysis among species of louseworts and elephant's heads (*Pedicularis*), introgression appears to have been common among most species within clades but not among species in different clades (Eaton and Ree 2013). In another study, low levels of introgression were detected among species in one clade of sedges (*Carex*; Escudero et al. 2014).

The ABBA-BABA test and the other methods described here continue to be refined, and new analytical procedures are constantly being developed. The power of these techniques lies in their ability to detect low levels of introgression that may not be evident from morphological variation or from the assessment of variation in low-density markers. From the studies that have been conducted thus far, it appears that historical introgression has been quite common—it may be the case that species escaping the effects of introgression during divergence are the exception rather than the rule.

10.12 THE PROCESS OF DIVERGENCE

Diversification requires mechanisms of isolation to limit the homogenizing effects of gene flow. Initial stages of isolation in plants are usually provided by geographic or ecological separation allowing *Allopatric* or parapatric divergence, and with the notable exception of polyploidy, isolating mechanisms are unlikely to develop during sympatry, because gene flow will generally overpower the effects of selection. We expect ecological divergence will typically accompany geographic departures as environmental conditions change with distance. Isolated populations are then free to diverge through genetic drift and responses to local selection, and given enough time, reproductive isolation may become complete. However, the development of strong reproductive barriers typically requires extended periods of time; with the possible exception of isolation after long-distance dispersal, diverging lineages are likely to encounter each other as their ranges adjust to long-term climatic changes, so opportunities for hybridization and introgression may be common.

From our discussions here, it appears that hybridization is quite common in plants and is an important mechanism for the promotion of adaptation and diversification. Even when isolating barriers are quite strong, low levels of gene flow among lineages may be possible and can infuse lineages with genetic variation that promotes adaptive transitions into novel habitats or improves plant fitness in the face of novel environmental challenges. Access to genomic information from a wide range of species has transformed our view of the genome and of the importance of introgressive hybridization, and it has become clear that genomes are subject to a mosaic of evolutionary processes. While some regions may be nearly neutral to selection, others may be affected by strong purifying selection or by positive selection on a beneficial allele that is increasing in frequency and spreading among populations. The origins of variation across the genome—whether by mutation or introgression from other lineages—and the general importance of genetic exchange among lineages are important areas of investigation where our knowledge remains incomplete.

At some point during divergence, isolation between lineages becomes strong enough that we begin to refer to the taxa as separate species. The identification of lineages as species can occur at a range of different levels of genetic divergence and may depend on a number of other considerations including the degree of ecological and morphological distinction. In the next chapter, we continue our discussion of evolutionary divergence as we take up the topics of speciation, species concepts, and phylogenetic diversification.

10.13 QUESTIONS FOR CONTEMPLATION AND DISCUSSION

1. Is there a useful distinction between gene flow and introgression? At what level of divergence do you think these two processes should be delineated?
2. What is introgressive hybridization? What is required for introgression to occur? Express your answer in a context that includes geographic distributions of lineages and dispersal.
3. What is transgressive variation? What are the probable causes, and how does it contribute to the evolution of novel phenotypes?
4. Make a list of premating isolation mechanisms. Which are usually the most effective? Under what conditions are mechanisms of floral isolation highly effective?
5. How do self-incompatibility, incongruity, and conspecific pollen precedence contribute to reproductive isolation?
6. What is the "species problem" that concerned Darwin and his contemporaries? What was the solution to this problem?
7. Why do many consider it unlikely that underdominance is important for the evolution of reproductive isolation? Under what circumstances could underdominance contribute to isolation? How is this concept related to chromosomal rearrangements?
8. Design an experiment that would determine whether post-zygotic isolation was due to positive or negative epistasis. How could you account for the effects of heterosis?
9. What is cytoplasmic male sterility and how does it contribute to reproductive isolation?
10. What is reinforcement? Under what conditions could selection for reinforcement precipitate species-wide divergence in floral traits?
11. Describe the tension zone model for stable hybrid zones. What conditions are required for the development of a tension zone? What information do you need to identify tension zones?

12. How do heterosis and extrinsic selection pressures facilitate or prevent hybridization and introgression?
13. What conditions are necessary for pollen swamping to occur? How does pollen swamping contribute to patterns of phylogenetic reticulation?
14. Describe an experiment that would detect low levels of introgression using genome-wide scans. What type of information do you need and what comparisons need to be made?
15. Describe how you could determine whether alleles shared between lineages originated from an ancestral polymorphism or from introgression during divergence.
16. How has the perception of the importance of hybridization for evolution changed over the last few hundred years? What type of evidence would you need to demonstrate the contribution of introgression to adaptive evolution?

11

SPECIATION AND DIVERSIFICATION

NO ONE DEFINITION has satisfied all naturalists; yet every naturalist knows vaguely what he means when he speaks of a species.

Charles Darwin (1859) On the Origin of Species, page 44.

Although we are limited in our ability to interpret the species of the present in terms of their past and their future, we can nevertheless utilize the study of the current stages in evolutionary differentiation successfully to reconstruct some of the gaps in the fragmentary record of the past.

Jens Clausen (1962) Stages in the Evolution of Plant Species, page 2.

11.1 THE SPECIES PROBLEM

The organization of genetic information into lineages that are independent and self-perpetuating is a necessary precondition for the origin and continued diversification of life. The process of evolution by natural selection requires entities that are isolated and capable of reliably transmitting genetic information across generations. Isolation is essential to ensure that advantages gained through mutation/selection processes are not contaminated by genetic exchange with competing lineages. At the extreme condition of universal exchange where individual lineages are not well defined (i.e., as in the early Proterozoic—Chapter 1), divergence and diversification would not be possible, because any incremental adaptation to local conditions would be swamped by gene flow. The establishment of isolated lineages allows the processes of adaptation and differentiation to generate increasing levels of specialization for novel modes of resource utilization. In heterogeneous environments, fragmentation of lineages is inevitable, as local conditions favor adaptive divergence and ultimately the formation of new lineages. At the core of this process are independent entities—what we refer to as *Species*—that are the product of evolutionary processes and constitute the fundamental units of the diversity of life. In this chapter, we are concerned with the utility of definitions of species, processes affecting the origin of species, their persistence, and phylogenetic diversification.

The criteria applied to define species has been a topic of discussion and contention for the last few centuries. On its surface, this question seems trivial—an oak is obviously different from a maple tree, grape vine, or a marigold—and it is widely recognized that there are different types of plants and animals. This is the essence of Darwin's point in the quote in this chapter's epigraph: biological entities that we can identify by their characteristics through observation and that are easily distinguished from others are generally referred to as species. We recognize the characteristics of species from our experience in the world—we can identify other individuals of our own species, and we can easily distinguish members of our own from other species. We also know from our collective human experience that individuals of the same species are capable of procreation with each other and do not usually mate with members of other species.

In general, we humans tend to categorize biological diversity using convenient labels that are based on our perception of the world, but these classifications may not always have the same degree of biological validity. For example, many people may consider different varieties of wine grapes to be distinct from each other and routinely refer to them using labels such as chardonnays, merlot, pinot noir, etc., but almost all wine grapes are varieties within a single biological species (*Vitis vinifera*). Similarly, we may not always recognize biological distinctions if they are not important to us; or as one student put it, "Grasses? You mean there's more than one type of grass?" In this text, we are concerned with defining species in a biological context—we endeavor to identify independent and genetically cohesive lineages that are characterized by their unique histories and their potential for adaptation and continued diversification. We start this discussion with a description of how plant species are classified and named.

Plant Systematics—The incredible diversity of plants that we see around the world today represents only a very small slice of the total number of plant species that have ever lived (Fig. 2.1; refer to color plate). Nearly all plant species have been named and classified based on their relatedness to other plants. Inferring historical relationships among species and classifying and naming individual species and groups are the goals of the field of *Plant Systematics*.

Plant systematists use phylogenetic methods—a topic we discuss later in this chapter—to uncover evolutionary relationships among living species (*Phylogenetic Systematics*). Traditionally, plant systematists used phenotypic traits that were assumed to be conserved and indicative of evolutionary relationships. This situation changed in the 1990s when DNA sequence data became available for a large number of species. Plant systematists changed the foundation of plant classification in 1998 when they adopted the *Angiosperm Phylogeny Group* (*APG*), which is based solely on phylogenetic relatedness determined by the analysis of genomic sequence data. This system continues to be refined—it has been replaced by APG II in 2003, APG III in 2009, and APG IV in 2016. In this text, we follow the most recent revision of the APG system. These revisions have resulted in the reclassification of many plant species at the family level (e.g., many species in the Scrophulariaceae were determined to be members of the Plantaginaceae and Phrymaceae). While individual species generally retain their original specific epithet, some genera have been reorganized to better reflect phylogenetic relatedness (e.g., many species originally in the genus *Mimulus* were reclassified as *Erythranthe*; Fig. 11.1; refer to color plate).

Individual species are named using the binomial nomenclature system that was originally developed by Carl Linnaeus in 1753 (see the Historical Perspectives section). This system identifies each species with two names (in the format, Genus species; e.g., *Erythranthe guttata*, the common monkeyflower). Modern systematists classify plant diversity using a hierarchical system with monophyletic clades of closely related species organized into genera (e.g., the genus *Erythranthe* includes *E. cardinalis*, scarlet monkeyflower; *E. alsinoides*, chickweed monkeyflower; and *E. lewisii*, great purple monkeyflower). Closely related genera are grouped into plant families (e.g., *Erythranthe* along with other genera are in the Phrymaceae, the lopseed family), and closely related families are grouped into orders (e.g., the order Lamiales encompasses twenty-one families including Lamiaceae, the mint family; Scrophulariaceae, the snapdragon family; and Phrymaceae; Fig. 11.1; refer to color plate). Plant orders are placed into larger clades (e.g., Asterids), which in turn are included in higher-level clades (e.g., Eudicots and then Angiosperms) until we reach a level that includes the entire diversity of streptophytes.

Historical Perspectives—Perceptions of the characteristics of species and the natural world have undergone considerable changes over the past few centuries. During the seventeenth century, the understanding of nature was muddled by fantastic stories of monsters and magical creatures that were considered by most people to be as real as the living plants and animals they encountered on a daily basis. Members of the "kingdoms" of animals, vegetables, and minerals were assumed to have the same elemental composition, so transformation of one type into another was fully within the realm of possibilities. This view was the foundation for alchemists' pursuit of the philosopher's stone—a substance that was thought to empower the *Transmutation* of one base metal to

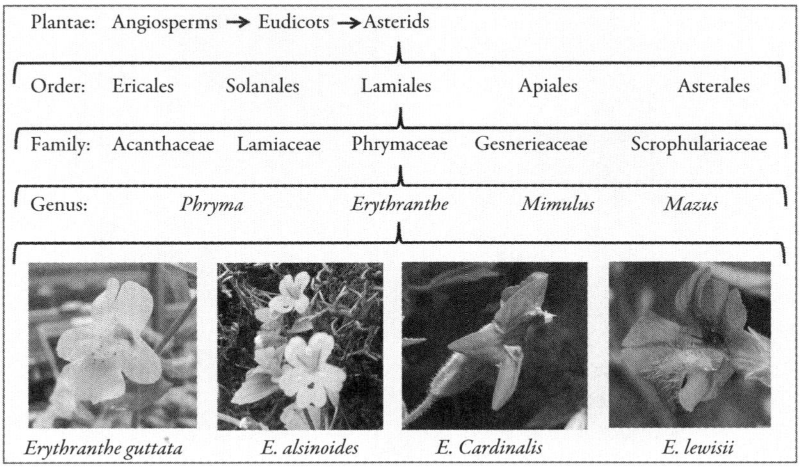

Fig. 11.1. The classification of the common monkeyflower, *Erythranthe guttata* and four other species in the genus. The family that includes *Erythranthe* is Phrymaceae, which is in the order Lamiales, and is included in the clade known as the Asterids. The Asterids are one of several clades within the Eudicots, which is one of several clades within the Angiosperms. Photos by J. Schowch, W. Siegmund, and M. Cruzan.

another (e.g., mercury to gold). Similarly, types of animals and plants were not necessarily viewed as fixed in their nature, as one kind might transform into another.

The idea that transmutation was real was not just restricted to the public—discussions of the possibility of transmutation of species were also popular among naturalists and scholars. These proposals were seemingly substantiated in the biological world by the observation of natural mutations of large effect. For example, spectacular examples of the spontaneous appearance of novel morphologies of several plant species have been reported, including a laciniate (finely-divided leaves) variety of *Mercurialis annua* (annual mercury), which normally bears leaves that are entire, and transformation of the normally *Zygomorphic* flowers of *Linaria vulgaris* (common toadflax) to a peloric variety (*Actinomorphic* flowers; Fig. 11.2; refer to color plate). We now know that the peloric form of *Linari*a is due to changes in methylation patterns that result in a transgenerational epimutation (Cubas et al. 1999; Grant-Downton and Dickinson 2006; Chapter 3), but at the time, such dramatic changes in morphology lent credence to claims of transmutation.

One widely known claim of transmutation was made by two respected botanists in Oxford (Jacob Bobart the Elder and his son Jacob the Younger) and was reported by Robert Sharrock in his book *History of the Propagation and Improvement of Vegetables by the Concurrence of Art and Nature* in 1660 (reported in Bateson 1913). These two gardeners described transmutation between crocus and gladiolus in cases where one type of bulb appeared in a garden that had been planted solely with the other. When the bulbs were dug up, Sharrock observed both types growing closely together, "but no bulb half of the one kind, and the other half of the other: But the changetime being past it was reason we should believe the report of good artists in matters of their own faculty" (Sharrock 1660, page 30). In other words, Sharrock interpreted his observations to mean that the transmutation was complete so bulbs had become separated, and he was inclined to trust the opinions of the Bobarts as they were professional botanists and considered experts in their craft.

It was during this time, when interpretations of the natural world were often a blend of science, art, and belief, that John Ray made the first attempt to delineate the nature of species in his book *Historia Plantarum* in 1686 (Bateson 1913). Ray was well aware of the claims of transmutation by the Bobarts and others, but he also saw consistency of form from one generation to the next in almost all the plants and animals with which he was familiar. Ray's definition recognized species as entities that were capable of perpetuating themselves

Fig. 11.2. The peloric (actinomorphic) epimutant of the normally zygomorphic common toadflax (*Linaria vulgaris*). From Grant-Downton and Dickinson (2006).

within limits of their own variation through the propagation of seed. He was fascinated by the variation that he observed within species, but his view of the natural world was colored by a belief in divine creation. Even in cases where novel varieties were formed and bred true (i.e., produced offspring of the same variety), he did not consider them to be separate species or to have any potential to become new species. Ray also did not completely discount the possibility of transmutation—like Sharrock, he acquiesced to the accounts of the Bobarts and others. Ray's writings reflected the prevailing views of the day; he accepted the fixity of the number of species that were created as described in Genesis. However, Ray's writings also reflected the widely held view that transmutations from one of the originally created species to another were within the realm of possibilities. In contrast, he considered variations within species and hybrids between them to be deviants from the norm—they were aberrations that were unlikely to persist in nature.

The eighteenth century was an exciting time for naturalists and scholars, as many newly discovered species from around the world were arriving in the museums of Europe. Much of the effort during this time was focused on describing and comparing the new species in the context of the diversity that had already been recorded from well-known regions. It was during this period that Carl Linnaeus, a young Swedish botanist, physician, and zoologist, began cataloging botanical diversity with the publication of his *Systema Naturae* in 1735 (Bateson 1913). Like other naturalists of his day, Linnaeus used a *Polynomial* system to provide accurate descriptions of each species. For example, the catnip plant was referred to as *Nepeta floribus interrupte spicatis pedunculatis*, which translates as "catmint with flowers in a stalked, interrupted spike." Each polynomial included enough Latin terms to distinguish a species from others in the same group. Polynomials worked quite well in cases where there were just two or three species in one genus, but these descriptions could become quite long and cumbersome for genera that contained many species—a problem that became even more evident as new specimens from around the world were described.

Linnaeus considered his classification of plants in *Systema Naturae* to be unnatural, as it was based solely on the number of stamens and was created primarily for the convenience of identification. It was not until 1753, with the publication of *Species Plantarium,* that he applied the now-familiar *Binomial Nomenclature,* and organized plant species into a system of classification that was more natural, as it was based on a number of floral and vegetative traits. In this new work, the organization of diversity was hierarchical, with species arranged into genera, which were grouped into families, themselves grouped into classes. Linnaeus developed his binomials from the shorthand version of plant names that he had previously developed for convenience in field surveys. In this abbreviated system of binomial nomenclature, names were simplified to include only the genus (a major group of similar types of plants) plus one specific name. For example, under the binomial system the specific name for catnip is now shortened to *Nepeta cataria*. Publication of *Species Plantarium* represents the first effort to consistently apply binomial nomenclature to a large group of organisms and established the field of plant systematics. Linnaeus's early work focused on plants; it was not until the tenth edition of this monograph that he started to include animals.

Linnaeus's views of biological diversity matured over the years. Early in his career, he declared that his mission in life was to continue the work of Adam from the book of Genesis—to complete the naming of all the plants and animals that God had created. He viewed species as immutable and fixed entities, and he assumed the challenge of divining the natural variation that had been established at the time of creation. In his later writings and lectures, Linnaeus seems to become more open to the possibility that varieties and hybrids have importance for their potential to generate novel species. We see hints of this change of view in the description of cultivated beets (*Beta vulgaris*) in *Species Plantarium*, with several types described (Briggs and Walters 1969), which Linnaeus suggests were derived from the wild beet (*B. maritima*). In another case, a student brought Linnaeus a peloric form of toadflax (*Linaria vulgaris*; Fig. 11.2; refer to color plate), which had appeared in a garden planted with the normal zygomorphic form. Linnaeus referred to this new form of toadflax as *Peloria* and considered it a new species. Here and in other writings, Linnaeus suggests the possibility that new varieties, and maybe even species, have originated through propagation by agriculturalists or are generated by their specific ecological conditions.

Linnaeus's hierarchical system of classification was a significant departure from the accepted order of nature and even included the classification of humans together with other primates, rather than being placed above the animals and just below the angels, as would be dictated according to the widely accepted *Scala Naturae* (The Chain of Being). By considering a nested arrangement of species within genera and genera within families, etc., Linnaeus established a framework for understanding natural variation that reflected relatedness, which represents the underpinnings of an evolutionary view of biological diversity.

Linnaeus's proclivity toward a more evolutionary view of nature can be seen in a series of lectures on "Natural Orders" that recognized the importance of varieties and the potential for diversification (Bateson 1913). In the preface, Linnaeus establishes principles of order based on what he considered the three sources of biological novelty: *Creator, Natura,* and *Casus* (creation, nature, and chance, respectively). Bateson interprets notes from these lectures to suggest that Linnaeus was promoting the idea that species were generated from intermixing within genera (*Natura*), and that representatives of each genus were the objects of creation. Variations within species and those created by hybridization were assigned to chance and accidents (*Casus*). In his lectures and writings, Linnaeus seems to interpret evolution-like processes from the patterns of relatedness that he observes within and among genera, but he considered the generation of species to be completed and not an ongoing process. Even though there are a good number of instances where Linnaeus makes statements that seem to anticipate the ideas of Lamarck, Darwin, and Wallace, he is usually recognized as a staunch advocate of the immutability of species.

The nineteenth century and the first half of the twentieth century saw little improvement in our understanding of the nature of species. Prior to Darwin, the discussion over classifications focused on separating variations that were assumed to represent accidents that caused deviations from the characteristics that define species (Linnaeus's *Casus*). Most naturalists viewed species as static entities—a *Typological* view of species—and their goal was to uncover and describe the original products of creation. Darwin brought a completely new perspective to these discussions in his *Origin of Species* and other writings, where he promoted the idea that speciation occurred as an outcome of gradual divergence in response to natural selection (Darwin 1859). In the late nineteenth century, some interpreted Darwin's theory to predict that species were endpoints of the selection process (a typological view), while others proposed that evolution was incompatible with the concept of species as static entities whether they be the product of evolution or creation (Bailey 1896).

Bailey, Darwin, and others recognized that divergence among lineages usually falls along a continuum and considered species naturally ambiguous. These authors suggested that species names were just labels applied for the sake of convenience—a view that is not useful for the delimitation of biological diversity into species, but one that is probably more realistic.

Darwin's theory provided a basis for understanding the evolution of species, but as described in Chapter 10, it was not widely acknowledged that his work had solved the species problem (Orr 1996). During the early twentieth century, there was controversy over whether the same mechanisms that promote adaptation within species would result in speciation, and these discussions continue into the present. These disagreements reflect different perceptions of the nature of species within the context of Darwin's theory; species are considered either as semi-permanent entities that are consistent over substantial periods of time or as dynamic lineages that are constantly undergoing change through neutral and adaptive processes.

Darwin naively thought that an evolutionary perspective would bring some resolution to arguments over classification schemes and criteria for the identification of species (Darwin 1859; Bateson 1913), but this has generally not been the case, and arguments remain just as intense since the widespread acceptance of his theory. Similarly, the advent of molecular phylogenetics and burgeoning availability of sequence data since the 1990s has contributed additional information to consider when defining species but has not presented a universal resolution to the species problem.

Much of the discussion and contention over species definitions stems from our limited view of ongoing evolutionary processes. In other words, we are seeing just a "snapshot in time" of the process of divergence, and the trajectory of future differentiation among lineages is generally not clear. As pointed out by Dobzhansky (1935), there is an incongruity between Darwin's view of species as dynamic entities and the static taxonomic descriptions that we apply in an attempt to categorize natural variation. More often than not, we find morphological discontinuities between lineages that provide reassurance that the species we are defining are real. On the other hand, it is not difficult to find examples of the influence of hybridization and species that are characterized by substantial levels of variation, both of which reflect gradual divergence—a situation that is especially common in plants.

In the next section, we discuss the salient issues surrounding definitions of species. We do not pretend to succeed in resolving the "species problem," but by delving into the various facets of the contentions among points of view, we hope to gain deeper insights into the nature of species and the boundaries that separate them.

11.2 MODERN SPECIES DEFINITIONS

The ultimate goal of species definitions—one that is probably not attainable—is to establish a universal set of criteria that can be applied across a wide range of organisms. There is general recognition that species are the fundamental units of biological diversification, but there is little agreement over how to define them. We recognize that lineage divergence is most often a gradual process (Fig. 10.1), so it may seem artificial to apply the status of "species" once a particular level of differentiation among lineages has been attained. On the other hand, we understand that the organization of genomic information into independent lineages is necessary for the development of biodiversity. We can conclude from this observation that the concept of species has biological validity and is not just a manifestation of human perception. Moreover, establishing criteria to define species facilitates our understanding of the nature of biological diversity, contributes to studies of diversity and the process of diversification, and enables effective communication within and among fields of science as well as with policy makers and the public.

While we recognize the importance of species as units of biological diversity and for their potential to contribute to diversification, we also acknowledge that there are several difficulties with the conceptual frameworks underpinning species definitions and their utility for the delimitation of species. To develop a biologically meaningful definition of species we need to establish of a set of criteria that reflect the characteristics of the independent lineages that the term "species" is meant to describe. The primary criteria that we might consider for the delimitation of species could include: 1) The presence of genetic or phenotypic characteristics that uniquely define lineages;

2) adequate levels of genetic exchange among individuals within the lineage for the maintenance of cohesion and homogeneity; and 3) mechanisms of reproductive isolation that are sufficient to maintain lineage independence and integrity. While these three criteria seem to be straightforward, note that they are based on subjective assessments rather than clear standards. Because species delimitation is a matter of degrees rather than discrete categories, we will find ourselves returning to these basic questions: how different is different enough (Criterion 1), and how much gene flow is too little (Criterion 2) or too much (Criterion 3)? These difficulties will be particularly apparent when we attempt to apply these criteria to patterns of natural diversity for groups that contain highly variable species that lack sharp genetic or phenotypic boundaries among divergent lineages.

We can begin to see the difficulties that might be encountered for species delimitation if we attempt to apply just one of the criteria outlined above. One approach toward a broadly useful species definition proposes to identify a level of genetic difference that confers adequate reproductive isolation (Criterion 3). Reproductive isolation was first proposed as a standard for defining species by Lotsy in the early twentieth century (Dobzhansky 1935). It is the foundation of the *Biological Species Concept* (*BSC*) that was proposed by Mayr (1942) and is a criterion that has been adopted by some plant biologists (e.g., Clausen et al. 1945; Stebbins 1950). On the surface, reproductive isolation seems to be a reasonable method for delineating lineages into species, but recall from our discussion in Chapter 10 that isolating barriers are often incomplete, and the degree of isolation can vary across a wide range. Simply identifying reproductive isolation as the primary characteristic defining species begs the question: how strong do isolating barriers need to be to ensure lineage integrity? This approach is further confounded by the recognition that the level of genetic divergence is not universally associated with the strength of reproductive barriers. Since the degree of genetic divergence is not a reliable indicator of lineage independence, the strength of isolation would have to be evaluated for all possible pairs of putative species within each group—a time-consuming task that may not be feasible for all groups. For this reason and others, some plant biologists, including many systematists, have rejected the BSC in favor of species definitions that focus on morphological traits or genetic similarity (Briggs and Walters 1969; Kress et al. 2005; Criterion 1).

Plants present us with a number of additional challenges for establishing a universal species definition. For example, many plant species consist of uniparental lineages because they are autogamous or, because they only reproduce asexually. Maternal lines within asexual or autogamous species are effectively isolated, so should we judge each lineage to be a separate species? Similarly, geographic isolation across populations of broadly distributed species is difficult to reconcile with the common assumption that some level of gene flow is necessary to ensure homogeneity and cohesion within a species (Criterion 2). This is a particularly important issue with plants, as their sedentary life forms impose spatial limitations on the extent of gene flow. As a further complication, morphological divergence in plants may or may not be associated with reproductive isolation. For example, diploid and tetraploid individuals that are morphologically indistinguishable but reproductively isolated may occur together in the same population (Soltis et al. 2007). With the variety of reproductive modes, the potential for geographic and ecological isolation, and the high frequency of hybridization and polyploidy, the number of issues challenging our concept of species may be greater for plants than for any other group of organisms. In the following, we discuss the functionality of different species concepts in general and in particular, as they apply to plants.

Morphological Species Concept—All modern species definitions focus on evolutionary processes, but they differ with respect to the types of evidence they use to determine whether a lineage is divergent enough to qualify as a species. Species delimitation based on morphological distinctiveness of individuals (Criterion 1) is referred to as the *Morphological Species Concept* (*MSC*), or sometimes as the *Taxonomic Species Concept*. Morphological traits have been used to define species since before the time of John Ray, but the way this evidence is viewed has changed since Darwin's theory of evolution has become widely accepted. Under the MSC, species are assumed divergent enough to be characterized by one or more phenotypic traits, but individuals within species may display minor levels of variation. Many early scholars considered the types of characteristics separating species to be intrinsically different from those varying within species, and the identification of traits that are more useful for the delimitation of species continues to be a major endeavor of plant systematists. For example, species within a genus may differ in their gross floral morphology, while the size or color of flowers and vegetative characters such as leaf size, shape, and indumentum may vary considerably within species. Ideally the traits defining species can be easily observed in the field, but there is growing interest in the application of *DNA Barcoding* technologies for the identification of species and quantification of biodiversity (Kress et al. 2005).

The identification and interpretation of variation in traits that display differences corresponding to the circumscription of species is the primary goal of taxonomists and systematists. Descriptions of species are based primarily on morphological traits but may also be informed by experimental evidence concerning phylogenetic distinctiveness and the prevalence of hybridization. The taxonomic labels emerging from these efforts are critically important for biological research and for communication of research results among scientists. Species names also have legal status in some nations and are important for the identification and conservation of rare, threatened, and endangered species.

The ultimate goal of systematists is to identify biological species, but these efforts may be hampered by the lack of correspondence between morphological traits and evolutionarily independent lineages. Separate lineages within a single species may not be easily detected in cases where *Cryptic Species* are indistinguishable based on taxonomic traits that are informative for delineating other species in the same genus. For example, *Erysimum nervosum* from Morocco includes two phylogenetically distinct lineages that can only be distinguished in the field by relatively subtle differences in flower color and shape (Abdelaziz et al. 2011). Similarly, populations of the same species that have different ploidy levels are usually reproductively isolated from each other (Chapter 10) but are not always classified as separate species (Soltis et al. 2007). In such cases of cryptic variation, evolutionarily independent lineages may not be reliably separated based on morphological characters, so levels of biological diversity would be underestimated. Cryptic variation also creates difficulties for studies comparing characteristics within or among species and for conservation efforts to restore or maintain biological diversity.

Morphological traits may be unreliable for delimiting species that are not separated by strong isolating barriers. When closely related species are in close contact, they may display gradients of variation and lack strong morphological boundaries among taxa. Clusters of lineages that are morphologically distinct and lack strong isolation mechanisms have been described as *Coenospecies* (Turesson 1922; Clausen et al. 1945; Stebbins 1950; Clausen 1962) but are more commonly referred to as *Syngameons* (Grant 1981; Arnold 1997). The term syngameon was first coined by Lotsy to describe "a habitually interbreeding community" of individuals that is reproductively isolated from other syngameons (referenced by Dobzhansky 1935, page 349). While Lotsy used this term to define species, later authors used it more generally to describe groups of closely related lineages that interbreed to varying degrees.

Syngameons are quite common in plants and have been described within a number of genera, including herbaceous (e.g., *Aquilegia, Iris, Erythranthe* (*Mimulus*), *Gilia,* and *Geum*) and woody (*Quercus, Nothophagus, Salix, Betula, Juniperus,* and *Pinus*). Lineages within syngameons are sometimes referred to as "bad species" by systematists because they are not separated by abrupt morphological discontinuities. In contrast, good species would be those that display sharp distinctions based on reliable taxonomic characteristics. Taxa within syngameons may have separate species names, but plant evolutionary biologists have often considered such delineations to be artificial, in which case the entire syngameon should be treated as a single species (Stebbins 1950; Grant 1981).

Different authors use the terms coenospecies and syngameon differently. Stebbins viewed the former to represent any group of species with incomplete reproductive barriers whether hybrids existed in nature or not. For example, Stebbins (1950) suggested that all species of columbine (*Aquilegia*) could be considered as a single coenospecies, because they could be crossed under cultivation, even though only a few species pairs are known to form natural hybrids and some interfertile species are separated by large geographic distances (Hodges and Arnold 1994). Grant (1981), on the other hand, limited the definition of a syngameon to groups of taxa that display extensive evidence of natural hybridization and lack strong morphological discontinuities separating species. By focusing on intrinsic mechanisms of isolation and disregarding ecological and geographic barriers (Chapter 10), Stebbins effectively adopted a strict interpretation of the BSC based on reproductive isolation.

The challenge presented by syngameons for species delimitation is clearly illustrated by the Pacific Coast irises of western North America (*Iris* section Apogon; Grant 1981; Fig. 11.3; refer to color plate). All of these species are morphologically distinct for at least some portion of their ranges, but hybridization and morphological gradients between species are common. At the center of this group is *Iris douglasiana*, which hybridizes with five other species and is one of two progenitors (along with *I. innominata*) of the putative hybrid species *I. thompsonii*. Lower levels of natural hybridization are known for several other species pairs in this group, but these patterns of apparent isolation may have more to do with opportunity than with the strength of isolation barriers; all members of the *Iris* section Apogon have the same chromosome number (N = 20) and are interfertile under cultivation (Grant 1981). These species of iris occur in different habitats and geographic locations, so hybridization is generally limited to cases where pairs of species come into close contact. One other interfertile member of the *Iris* section Apogon (*I. munzii*) does not naturally hybridize and was considered by Grant to be a separate species and not a member of the syngameon. On the other hand, note that since *I. munzii* is interfertile with other members of the group, according to Stebbins's definition, it would be included within the same coenospecies along with other members of the section Apogon.

Application of the MSC continues to be the most common method of delineating species because phenotypic traits are convenient and obvious characteristics that work well to define species. While morphology traits are the primary characters that taxonomists and systematists use to describe and delimit species, such decisions are generally advised by

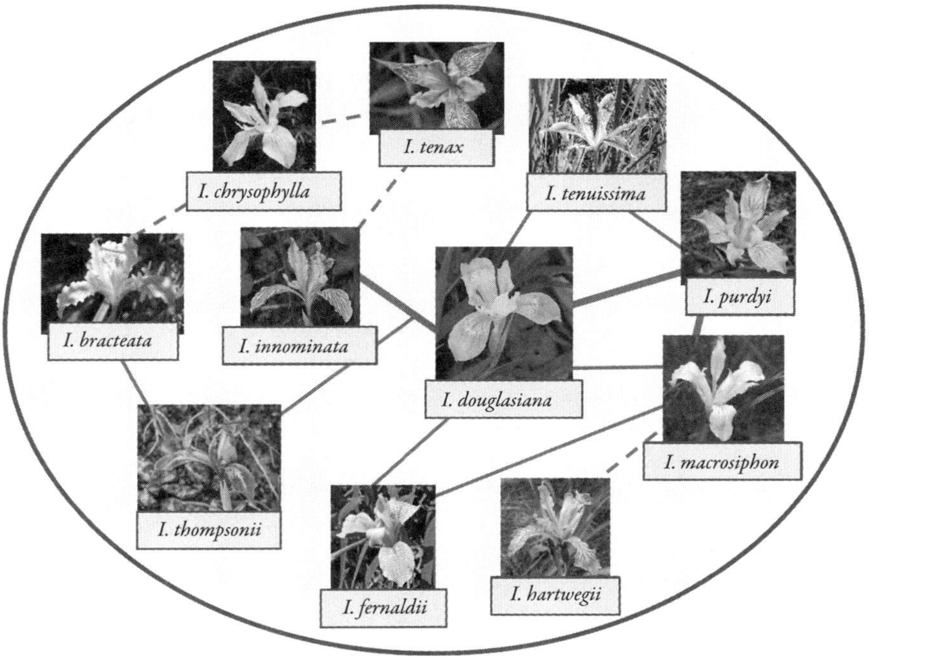

Fig. 11.3. The Pacific Coast iris syngameon (*Iris* section Apogon). The width of connecting lines indicates the level of natural hybridization between pairs of species, with thick lines indicating extensive hybridization and dashed lines indicating only occasional hybridization. *Iris thompsonii* is a homoploid hybrid species derived from *I. douglasiana* and *I. innominata*. Adapted from Grant (1981). Photos by Curtis Clark, Walter Siegmund, Tom Hilton, Eric in SF, Pbjamesphoto, Jared Zimmerman (WMF), Stickpen, Bill Bouton, Stan Shebs, and Tom Hilton.

other types of information that might be available, such as the extent of hybridization with other taxa, cytological differences in chromosome number and arrangement, and phylogenetic patterns inferred from genetic data. The MSC also has utility in fields such as paleobiology where only morphological information is available. This concept has been important for our understanding of delimitation among species known only from fossils and for characterizing the patterns of historical diversification of plants and animals.

The MSC focuses on phenotypic differences to define species (Criterion 1), but it only indirectly addresses cohesion processes that maintain morphological similarity (Criterion 2). This species concept does include some consideration of reproductive isolation (Criterion 3) in cases of extensive hybridization, but the presence of isolating barriers is not the primary focus of the MSC. Definitions of species based on morphological traits alone are convenient for a range of investigations and plant groups, but we recognize that these delineations do not necessarily reflect natural diversity and may not always correspond to evolutionarily independent lineages.

Biological Species Concept—One of the most widely recognized benchmarks used for distinguishing species is reproductive isolation as formalized in the BSC (biological species concept), which was introduced by Ernst Mayr in his book *Systematics and the Origin of Species* (Mayr 1942). The idea that interbreeding could be used to identify species boundaries was not new—discussions concerning shared reproduction or morphology as the most appropriate criterion for the identification of species had been discussed since the 1800s (Dobzhansky 1935). Dobzhansky and Mayr effectively changed the focus of this discussion by presenting the idea of species as a concept: species were presented as biological entities possessing sets of attributes that could be discerned, rather than as arbitrary taxonomic entities (Hey 2006). This view of species moved the discussion away from traits suitable for taxonomic descriptions and encouraged consideration of the fundamental properties that define species as independent lineages with evolutionary potential. Moreover, the species concept reinforced the idea that species represent fundamental units of biological organization much like the cell or the individual organism. Discussions of the theoretical aspects of species has led to a proliferation of species concepts, each of which attempts to provide evolutionary delineations of collections of organisms; the ultimate goal is to have a consistent method for identifying evolutionary units that reflect biological diversity and have potential for continued diversification.

The BSC was originally defined as "groups of actually or potentially interbreeding natural populations which are reproductively isolated from other such groups" (Mayr 1942, page xxi). This definition focuses on the process of divergence and recognizes that isolation from gene flow is critical to the maintenance of independent lineages (Criterion 3). Whether reproductive isolation develops abruptly (e.g., through polyploidy) or slowly as lineages accumulate genetic differences, Mayr and some plant biologists such as Stebbins (1950) considered isolation from gene flow to be a necessary condition for the definition of species.

In later versions of the BSC, Mayr dropped "actually or potentially" from the definition in response to a number of criticisms (Mayr 1969). This newer version of the BSC is more general, as it recognizes the validity of a broader range of reproductive barriers including ecological or geographic barriers. On the other hand, it generates difficulties because recognizing species as "groups of interbreeding natural populations" automatically defines geographically isolated populations of morphologically similar individuals as separate species. While both versions of the BSC satisfy Criterion 3 (reproductive isolation), these definitions promote the idea that genetic and phenotypic distinctions are not necessary for delineating species (Criterion 1), and they fail to adequately address mechanisms that maintain cohesion among populations for species with large geographic ranges (Criterion 2).

The two different versions of the BSC roughly correspond to the concepts of coenospecies and syngameon as discussed earlier; the coenospecies concept only recognizes intrinsic isolation mechanisms and is consistent with Mayr's original definition, while the syngameon concept includes extrinsic isolation and aligns best with the more modern version of the BSC. With the newer definition of the BSC, lineages may be considered separate species even though they may not have attained adequate levels of genetic differentiation for the development of strong intrinsic isolation barriers. The difficulty with this interpretation is that species may be subject to historical or future hybridization with other lineages as they change their geographic distributions. With this in mind, we may consider the original definition of the BSC as more appropriate, but this renders the BSC impractical in most situations, because each putative species would have to be crossed with all possible congeners to test the strength of intrinsic isolating barriers. Such large crossing experiments are rarely performed, particularly in long-lived woody species. The two definitions of the BSC are generally not distinguished and most authors continue to refer to Mayr's original definition.

While some plant biologists have viewed reproductive isolation as the ultimate arbiter of species status (e.g., Turesson 1922; Clausen et al. 1945; Stebbins 1950), many have rejected the BSC as unworkable (Briggs and Walters 1969). Most plant systematists have taken a more pragmatic approach to reproductive isolation; a lineage can be considered a "good species" if hybridization is limited in frequency or if hybrids are restricted to small regions relative to the ranges of species (e.g., hybrid zones along ecotones between habitats). A strict interpretation of the BSC is generally not applied, because it would dictate that any level of "actual or potential" hybridization would disqualify a lineage as a species. While this approach is consistent with Mayr's view that species occupy adaptive norms and hybridize infrequently (Chapter 10), it imposes some practical limitations on the utility of the BSC across different groups of organisms. During the second half of the twentieth century the BSC quickly gained wide acceptance among zoologists, but many plant biologists and systematists disputed the single criterion of reproductive isolation as adequate for the delimitation of species (Briggs and Walters 1969).

Phylogenetic Species Concept—The biological species concept focuses on processes contributing to divergence—reproductively isolated lineages would be expected to accumulate genetic differences over time—but it is not particularly meaningful for those who are more concerned with the outcomes of evolutionary processes. For example, a whole-genome duplication may result in an independent lineage that is reproductively isolated, but other than the duplicated genome, a recently formed tetraploid may be nearly indistinguishable from the diploids present in the same population. In other cases, recognized species may be genetically distinct but may maintain the ability to interbreed. The BSC has been criticized because the correspondence between reproductive isolation and the levels of genetic and morphological divergence are often inconsistent (Cracraft 1983; Donoghue 1985). Cracraft referred to species defined under the BSC as "polytypic," because they may contain a number of independent lineages that have potential for independent divergence. Donoghue considered the BSC impractical for species delimitation and criticized it for assuming that contemporary genetic exchange rather than shared history was responsible for cohesion of populations within a single lineage.

Several different versions of the *Phylogenetic Species Concept* (*PSC*) have been proposed (de Queiroz 1999; Mishler and Theriot 2000; Wheeler and Platnick 2000). All of these interpretations share a focus on the results (patterns) of evolution rather than characteristics that contribute to the process (e.g., reproductive isolation). One of the most widely cited definitions was proposed by Cracraft (1983, page 170) who suggested that "a species is the smallest diagnosable cluster of individual organisms within which there is a parental pattern of ancestry and decent." He considered species to be a natural level of biological organization in the same sense as cells and individual organisms and to be the primary units of evolutionary diversification.

A critical aspect of Cracraft's definition is the identification of objective and diagnostic phenotypic characteristics that defined species. He made a sharp distinction between the diagnostic traits that are useful for defining species because they display abrupt discontinuities and traits displaying clinal variation among subspecies. Cracraft's version of the PSC viewed subspecies designations as subjective, and he did not consider them entities having potential for evolutionary divergence. On the other hand, Cracraft assumed that application of the PSC would divide many species that are polytypic under the BSC into multiple species that could be delimited by diagnosable characters (Cracraft 1983). Under Cracraft's version of the PSC reproductive isolation was assumed necessary for phenotypic divergence, but it was not included as a diagnosable characteristic defining species. The implicit assumption in this version of the PSC is that common ancestry will result in groups of closely related populations that share diagnosable traits that can be used to distinguish species.

Donoghue's (1985) definition of the PSC differed from Cracraft's by focusing primarily on monophyly as a necessary condition for the delimitation of species (Fig. 11.4; refer to color plate). Around this time, an era of *Molecular Phylogenetics* was emerging, as new methods of sequence analysis were providing easier access to large numbers of traits. Interest in DNA-based phylogenies was growing, as the results of early analyses generally corresponded to those developed from phenotypic characters, and there was a seemingly endless number of traits available (e.g., Palmer and Zamir 1982; Hori et al. 1985; Hillis and Dixon 1991). By proposing that species could be defined as *the smallest discernable group of monophyletic organisms based on shared-derived traits* (paraphrased from Donoghue 1985), Donoghue established a species definition that effectively anticipated that phylogenies based on DNA sequence variation that would soon become widely available. His version of the PSC differed from Cracraft's by providing a definitive criterion for delimiting species; Donoghue's definition required that all individuals within a species had to be the only descendants of a single common ancestor (i.e., all intraspecific lineages would be monophyletic; Fig. 11.4; refer to color plate), and he assumed that such high-resolution phylogenies of populations within species would soon become available. In contrast, Cracraft's definition of the PSC was easier to apply, as it only required that a group be diagnosable by phenotypic traits and display an historical pattern of ancestry. His definition of the PSC also differed from Donoghue's because it potentially included cases of paraphyly and polyphyly (Fig. 11.4; refer to color plate). Unlike Cracraft, Donoghue's definition of species does not tolerate any subspecific classifications or any level of hybridization that would result in a phylogeny that was not monophyletic. The specific level of introgression that could be tolerated was not specified, but he clearly disqualified any product of hybridization (including polyploidy) that resulted in new lineages.

Other versions of the PSC recognize the overriding importance of the phylogenetic history of organisms (e.g., Mishler and Theriot 2000; Wheeler and Platnick 2000; and references within) but differ with respect to whether they require species to be

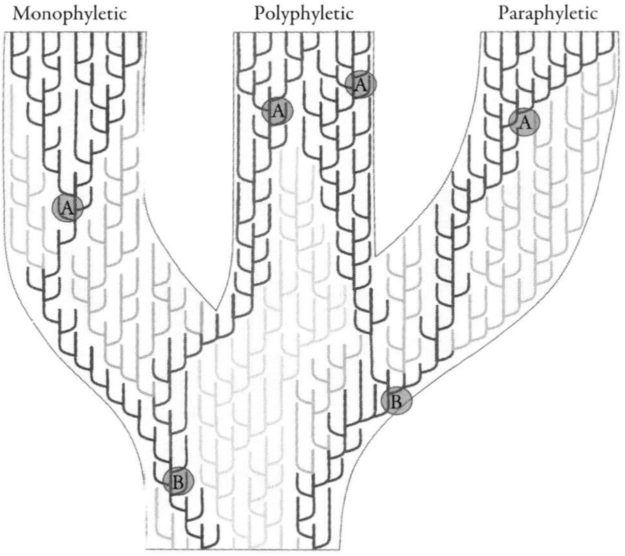

Fig. 11.4. Gene genealogies and lineage sorting within a phylogeny of three sister species delimited based on the morphological species concept. When all extant lineages within a species arise from a single ancestor the intraspecific phylogeny is monophyletic (left). A species consisting of two independently derived lineages is polyphyletic (center). A shared lineage between two species results in a paraphyletic pattern (right). Circled letters indicate points of coalescence of extant lineages for the most recent common ancestor of haplotypes within species (A), and the common ancestors of the major lineages (B).

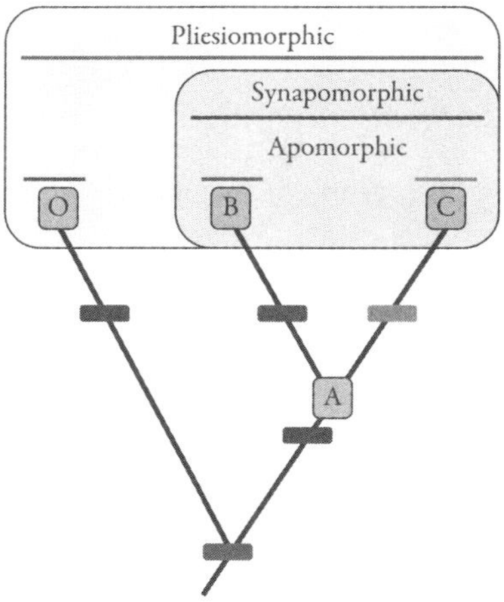

Fig. 11.5. Cladogram showing the distribution of pliesiomorphic traits (shared-ancestral), synapomorphic traits (shared-derived), and apomorphic traits (unique to one lineage). In this phylogeny the O taxon is the outgroup, A is a hypothetical ancestor (node), and B and C are sister taxa (members of the same clade).

characterized by traits that are unique to a phylogenetic lineage (*Apomorphic Traits*; Fig. 11.5; refer to color plate) and to display a monophyletic pattern of ancestry. One extreme view is represented by Mishler and Theriot (2000), who generally agreed with Donoghue (1985) by requiring all populations of the same species to be characterized by shared-derived traits (*Synapomorphic Traits*—i.e., these traits are apomorphic to a species). In contrast, the definition proposed by Wheeler and Platnick (2000, page 58) defines "species as the smallest aggregation of (sexual) populations or (asexual) lineages diagnosable by a unique combination of character states." Much like Cracraft (1983), they argue that species should be delimited based solely on "diagnosable" traits without the requirement of apomorphy.

The difference in the definitions of the PSC proposed by Mishler and Theriot (2000) and Wheeler and Platnick (2000) reflects contrasting views on the potential for *Phyletic Evolution*—the evolution of lineages along a single branch that results in differentiation over time. The definition proposed by Mishler and Theriot (2000) is consistent with a *Cladistic* point of view, which assumes that ancestral species (*Nodes* in the phylogenetic tree) cease to exist as they are replaced by two daughter lineages (*Dichotomous Branching*; Fig. 11.5; refer to color plate), each of which can be identified by apomorphic traits. Wheeler and Platnick (2000), on the other hand, suggest that some lineages may be characterized by stasis, so they retain the ancestral phenotype, while sister lineages become defined by novel traits. In both versions, species are characterized by diagnostic traits, so the main difference is whether they are assumed to have originated recently or whether traits could be carried over from an ancestor.

All definitions of the PSC recognize asexual species, but they differ with respect to their tolerance for hybridization. Asexual lineages within species can be viewed as another level within the hierarchy of classification (i.e., all asexual lineages within a species are expected to have a monophyletic origin). The inclusion of asexual reproduction for definitions of the PSC is consistent with the view that common ancestry, rather than gene flow is adequate to maintain cohesion within species (Criterion 2). On the other hand, the different versions of the PSC differ with respect to their recognition of the importance of reproductive isolation (Criterion 3). According to Wheeler and Platnick's (2000) and Donoghue's (1985) versions of the PSC, species are defined by discrete trait distributions, and any level of genetic exchange is not tolerated. In contrast, Mishler and Theriot (2000) recognized that varying levels of introgression commonly occur during divergence. From their point of view, a low level of hybridization does not present any difficulty as long as an adequate degree of monophyly is maintained such that species are characterized by apomorphic traits. With higher levels of introgression, species lose their independence and are grouped within the same syngameon.

The PSC is closely related to other species concepts that emphasize evolutionary history. For example, a species under the *Evolutionary Species Concept* (*ESC*) is defined as "an entity composed of organisms that maintains its identity from other such entities through time and over space and that has its own independent evolutionary fate and historical tendencies" (Wiley and Mayden 2000, page 73). Note that, like the PSC, this definition recognizes species as a natural unit of organization, but it differs because it requires reproductive isolation for the maintenance of lineage integrity.

In summary, all versions of the PSC emphasize the historical nature of species and rely on traceable patterns of diversification to provide guidelines for species delimitation. In practice, phenotypic traits that are unique to each lineage are used to delimit species, but versions of the PSC differ with respect to whether they assume diagnostic traits are always apomorphic. Versions of the PSC that take a strict cladistics view do not allow for phyletic speciation and always assume species can be identified by recently derived traits. These views of the PSC represent a significant departure from the BSC; in its purest form, the BSC does not consider phylogenetic history, so it effectively treats species as static entities. The BSC also assumes that the cohesion of lineages within species is maintained by genetic exchange, whereas the PSC generally considers the common history of populations to be adequate to explain homogeneity among individuals of the same species. Whether shared history or ongoing genetic exchange is more important for the maintenance of the cohesion of populations within species continues to be a question for ongoing speculation and research.

Cohesion Species Concept—An alternative approach to understanding the nature of species is to combine the evolutionary processes that promote the unification of populations into evolutionary lineages with evidence for phylogenetic independence. The *Cohesion Species Concept* (*CSC*) defines species as "the most inclusive group of individuals having genetic and/or demographic exchangeability" (Templeton 1989, page 25). Templeton elaborates on this definition to clarify that all populations within a species are expected to be members of a single lineage (i.e., they are characterized by a monophyletic gene genealogy; Fig. 11.4; refer to color plate). He argues that focusing on species cohesion (Criterion 2) reduces the need to evaluate reproductive barriers—a lineage that qualifies as a species under the CSC would necessarily have an independent history (i.e., strong reproductive isolation) and the potential for continued diversification. The CSC stands out, because it is one of the few species concepts that emphasizes ecological associations (demographic exchangeability) as a consideration for species delimitation. Templeton's CSC differs from various versions of the PSC because it focuses on intraspecific gene genealogies and emphasizes traits that reflect ecological variation over characteristics that are present primarily due to historical relationships.

Evaluation of species status under the CSC is based on the analysis of intraspecific gene genealogies and the geographic distribution of lineages. Templeton (2001) provides specific criteria for species delimitation by using assessments conducted within the context of nested clade analysis (Chapter 8). The first task is to identify monophyletic groups within the network phylogeny. These phylogenies are generally unrooted so monophyletic clusters can be identified by the presence of central haplotypes (lineages) associated with multiple tip haplotypes (e.g., clades 1-1 and 1-2 in Fig. 8.9). The monophyletic clusters are evaluated further to determine whether they exhibit largely non-overlapping geographic distributions, which would indicate historical fragmentation resulting in geographic isolation (Templeton 2001; Box 8.3). Once the presence of more than one geographically distinct monophyletic groups is confirmed, the level of genetic and ecological exchangeability within and among clades can be evaluated. This process amounts to comparing the similarity of phenotypic traits and habitat conditions to test for genetic and ecological exchangeability, respectively. Cohesion species are identified by examining the distribution of traits associated with morphological and ecological similarity among monophyletic groups; trait differences associated with genetic clusters are interpreted as evidence of genetic or ecological cohesion within each clade (Templeton et al. 2000). Templeton argues that these species delimitation analyses are conducted using statistical tests, so they lack the subjectivity that has plagued the application of other species concepts.

General Lineage Concept—The criteria that have been used to define species are also considered as the characteristics that are intrinsic to the definition of evolutionarily independent lineages. Historically, these species definitions have represented conceptual views of the entities that constitute biological diversity. Throughout the twentieth century, efforts to delimit species generally relied on meeting the criteria of a single species concept (Carstens et al. 2013). This situation changed with the introduction of the *General Lineage Concept* (*GLC*), which proposes that evolutionary history is a common aspect of all modern species concepts and attempts to unify all species concepts under a single set of characterizations (de Queiroz 1998, 2007). The GLC views criteria used in other species concepts as characteristics that develop during lineage divergence (see section 11.5). The GLC is not really a species concept; it is more of an analytical approach to species delimitation that considers different characteristics such as diagnostic phenotypic traits, reciprocal monophyly, and reproductive isolation as potentially sufficient for the delimitation of species. The GLC is broadly defined and differs from species concepts by incorporating a wide range of phylogenetic histories (not just monophyly), allowing for at least some degree of hybridization and including both sexual and asexual species. This approach to species delimitation has become increasingly popular and has been applied to a wide range of organisms (Hey 2006).

Since species could be delimited based on any single set of criteria under the GLC, this concept could potentially meet all three of the conditions described at the beginning of this section (i.e., 1) distinctive phenotypic traits, 2) genetic integrity within lineages, and 3) isolation among lineages). On the other hand, most delimitations conducted under the GLC use genetic and phylogenetic analyses to establish groups of individuals that display a history of cohesion and isolation. Once monophyletic groups are identified using molecular genetic data, additional analyses could be conducted to identify consistent phenotypic or ecological differences, or perhaps crosses could be used to assess the presence of intrinsic isolating barriers (Carstens et al. 2013). Ideally, several lines of evidence are brought to bear to delimit lineages and to assess their potential for long-term independence.

Summary of Species Concepts—There are at least a dozen more species concepts that have been proposed since the late 1990s that we have not discussed. With the exception of the GLC, all of these tend to focus on one or a few criteria, and many were developed with a particular set of organisms in mind. For example, the mate recognition concept works well for organisms with cognitive ability (i.e., animals) but is not so easily applied to the vast majority of eukaryotes. On the other hand, there is no single concept that works well for all organisms, which suggests that a single set of easily identified characteristics that define species across the diversity of life may be elusive.

Given our inability to identify a universally functional species definition one might ask why bother to discuss the species concepts that have been proposed. While this may seem like a futile exercise, there is much to be learned from the successes and failures of each definition; by exploring the characteristics of lineages that facilitate their integrity, independence, and potential for future diversification, we achieve a fuller understanding of what it the word species means in the context of evolution and biodiversity. In the next few sections, we discuss the process of speciation and methods to delimit species.

11.3 GEOGRAPHIC MODES OF SPECIATION

Isolation from gene flow is the single most critical factor responsible for the divergence of lineages. As discussed previously, populations that are secluded from the genetic influence of divergent lineages are better able to respond to selection for adaptation to the local environment and to accumulate, through mutation, genetic drift, and selection, alleles that may contribute to postzygotic isolation. The level of ecological divergence will depend on the strength of local selection, and in starkly contrasting environments, divergence in adaptive traits may become extensive enough for the establishment of strong ecological reproductive barriers. As discussed in Chapter 10, isolation also promotes divergence through drift, as mutations that are neutral in their native genetic backgrounds reduce the fitness of hybrids through negative epistasis and the breakup of co-adapted gene complexes. Reproductive barriers can develop through selection and drift, but it is generally thought that the development of isolation through mutation/drift will be much slower than through selection (Gavrilets 2003; Hendry et al. 2007). Whether the primary mechanism of divergence is selection or drift, some degree of isolation from gene flow is a necessary precondition to allow the emergence of novel lineages that ultimately may qualify as species.

The process of speciation is traditionally partitioned into categories that correspond to different levels of gene flow among lineages. At one extreme, *Allopatric Speciation* occurs for groups of populations that are geographically isolated, resulting in divergence with little or no gene flow. This is generally thought to be the most common mode of speciation (Grant 1981; Coyne and Orr 2004). The term "allopatric" (from Greek roots meaning other and fatherland) is derived from the field of biogeography, where it was used to explain the observation that different sets of species often occurred on either sides of geographic barriers. Such *Vicariance* events—the appearance of a mountain range or river or the separation of continental landmasses—are often described in generalized treatments as the primary source of allopatric distributions; however, this is probably an unrealistic view, since it is clear that the time scale for shifts in geographic distributions of species usually far outpaces major geological events (Chapter 8). For example, geographic ranges of temperate floras and faunas fluctuated in response to multiple glaciations over the last two million years, resulting in extended periods when individual species were subdivided into widely distributed refugia. Over longer time scales (between five and ten million years), we see the effects of major geological events such as the appearance of oceanic islands, the joining of major land masses (e.g., the joining of North and South America), and the uplift of major mountain ranges on species distributions (Cox and Moore 2010). These events promoted isolation and consequently, created ideal conditions for allopatric divergence and lineage diversification. While major geologic events have clearly contributed to large-scale patterns of diversification, processes occurring over shorter time scales, such as climatic shifts and ecological divergence, have probably had a much larger impact on species diversity and may be the source of the majority of speciation events contributing to contemporary biodiversity (Hendry et al. 2007; Schnitzler et al. 2012; Anacker and Strauss 2014).

As we consider scenarios for the geographic distribution of diverging populations, we can imagine a range of situations that result in different degrees of isolation. As long as there is some geographic separation between groups of populations then their distribution can be classified as allopatric. If the distributions of divergent lineages are closely adjacent to one another, then they are considered *Parapatric* (side-by-side), and we expect gene flow to be much higher. In the extreme case, the distributions of two diverging lineages may overlap to some degree, in which case we would describe them as being *Sympatric* (the same—individuals from the two lineages are within easy dispersal distance of each other). For sympatric and parapatric (and short-distance allopatric) distributions, levels of gene flow will be relatively high, and strong pre- or post-mating barriers would be required for the maintenance of lineage integrity and continued divergence. For example, distributions of divergent lineages in different soil types, such as the pasture and mine tailing ecotypes of *Agrostis* and *Anthoxanthum* (e.g., McNeilly and Antonovics 1968), would be considered to be parapatric, since their distributions do not overlap. On the other hand, these populations are within easy dispersal distance, so they could also be classified as sympatric. With the exception of instances of polyploidy, cases of sympatric divergence with substantial overlapping distributions are more difficult to find but for plants, may occur in fine-grained mosaics of starkly contrasting habitats that are defined by edaphic conditions.

Long-distance dispersal (extra-range dispersal) can also lead to the establishment of isolated populations that ultimately diverge into novel lineages. The resulting distribution is effectively allopatric, but this mechanism differs from the patterns of fragmentation of geographic ranges that are usually associated with allopatric speciation and is sometimes placed into its own category as *Peripatric Speciation* (Mayr 1942) or *Budding Speciation* (Gottlieb 2004; Crawford 2010). This idea was first proposed by Ernst Mayr (1942), who suggested that species were likely to be subjected to novel environments at the periphery of their ranges (parapatric peripheral isolates) or after long-distance dispersal. Grant (1981) expanded on this model to include ideas from Wright (1931; 1932) on the effects of genetic drift in small populations. Grant predicted that the combined effects of drift due to small population size and selection for local adaptation would result in "rapid and radical" divergence and termed this process *Quantum Speciation* (Grant 1981, pages 153–169). Modes of peripatric speciation, including budding and quantum speciation, may be quite common in plants (see section 11.6). This path to speciation is typified by divergence after long-distance dispersal to novel habitats including islands, and through peripheral isolates, which commonly occur in novel habitats (Gottlieb 2004; Grossenbacher et al. 2014).

Speciation with Gene Flow—With the emergence of genomic approaches the traditional categories of sympatric and parapatric speciation have been more generally described as cases of "speciation with gene flow" (Feder et al. 2012, page 342). This approach focuses on small genomic regions that display high levels of differentiation between lineages as evidence of selection (speciation islands; Chapter 5). Feder et al. (2012) suggest that speciation can proceed in response to selection on a small number of regions even as recurrent gene flow affects the majority of the genome. These authors predict that evidence for gene flow during speciation will be represented by a strong "L-shaped" distribution for the level of differentiation (F_{ST}—Chapter 4) between lineages. Under this model, a small minority of gene regions are expected to show large differences because of selection, while the majority the genome will be characterized by low values of F_{ST} because of the homogenizing effects of gene flow. While this interpretation seems reasonable, it may be difficult to make unambiguous conclusions because, as discussed in Chapter 5, a number of processes may cause genomic regions to display high levels of differentiation.

A putative example of speciation with gene flow has been described on Lord Howe Island, a small oceanic island east of Australia, for two species of endemic palms in the genus *Howea* (Savolainen et al. 2006; Babik et al. 2009). These authors suggest that these species must have diverged sympatrically because their current distributions overlap, and the pattern of genomic divergence is strongly L-shaped. Ancestors of these palms apparently colonized this island around 4.5 to 5.5 MYA. Early colonists presumably grew in soils of volcanic origin, which are common across the island. The split between these two lineages did not occur until much later during the Pleistocene, when *H. forsteriana* colonized calcareous soils that occur in limited regions (Savolainen et al. 2006). During the process of divergence, the calcareous species apparently lost protandry so presumably reproduces through self-fertilization more frequently than *H. belmoreana*. Soil types are highly interdigitated across the island, so these two species are often in close contact. In spite of their close proximity, the formation of hybrids is uncommon indicating strong reproductive isolation, but it is not clear whether the barriers are intrinsic or due to low rates of hybrid survival in the two soil types.

The interpretation of sympatric speciation for *Howea* palms has been questioned on a number of grounds (e.g., Stuessy 2006). Evidence for the effects of selection is inferred from the fixed allele differences for a few of the AFLP markers surveyed. While linkage to gene regions under selection is implied by higher divergence, the genomic positions of these genetic marker loci are unknown. Given that the low frequency of F_{ST} outliers and low divergence across these genomes falls short of providing sufficient evidence to conclude that speciation has occurred with ongoing gene flow, the argument for sympatric speciation hinges largely on the stability of geographic distributions over long periods of time. However, it appears that Lord Howe Island was much larger during the period when these two lineages diverged (Stuessy 2006), which makes it more possible that speciation occurred in geographically isolated populations. Moreover, the habitat associations for these two species of endemic palms are not absolute, suggesting that selection based on ecological conditions is not very strong—probably not strong enough to overcome ongoing gene flow among sympatric or parapatric populations. In other words, gene flow must have been minimal during speciation of the *Howea* palms. Sympatric speciation seems unlikely under these conditions. Moreover, these species differ for several traits, selection on each of which would have to have been strong enough to allow divergence in the face of gene flow. Additional information on the nature of the reproductive barriers and the strength of ecological associations may help resolve whether selection has been strong enough to overcome gene flow among sympatric populations, or whether speciation in *Howea* was more likely to have been allopatric.

11.4 HYBRID SPECIATION

The origin of novel lineages through hybridization has been a prevalent mode of speciation over the history of terrestrial plants. The idea that new species may have been derived via hybridization dates back at least to Linnaeus in his discussion of the origin of agricultural varieties covered earlier in this chapter (Grant 1981). Indeed, Linnaeus went so far as to suggest, "it is impossible to doubt that there are new species produced by hybrid generation" (translated in Grant 1981, page 245). The proposal that hybridization could be the source of novel species was echoed by a handful of authors during the nineteenth century, including Naudin (1863) and Kerner (1894). Not all of these authors appreciated the fact that recombination can result in wide variation and unexpected hybrid phenotypes (Grant 1981). For example, Linnaeus and

Kerner both assumed that hybrid variants would breed true to their morphology. Naudin, on the other hand, realized that some generations of breeding would be required for a consistent phenotype to emerge. During the early twentieth century, the possiblility of hybrid speciation was generally acknowledged, but it was thought to be uncommon because of the widely accepted view that nearly all available adaptive nodes (niches) were already occupied by existing species. These early authors considered only the possibility of hybrid speciation without changes in ploidy (*Homoploid Hybrid Speciation*), and did not know about the possibility of whole genome duplication (*Polyploid Speciation*). In this section we first discuss evidence of recent instances of polyploid speciation and then will review cases of hybrid speciation without changes in ploidy.

Polyploid Speciation—Up through the early twentieth century, the importance of changes in ploidy for speciation and diversification was not recognized. Appreciation of the significance of polyploidy grew rapidly after Öjvind Winge (1917, 1924) discovered that pairs of closely related plant species often displayed twofold differences in their chromosomes. Based on these comparisons, he speculated that doubling of the chromosomes could be a mechanism of speciation. Since the pioneering work of Winge, it has become clear that polyploid speciation has been prevalent for several major groups of plants including the ferns and angiosperms (Stebbins 1950; Grant 1981; Otto and Whitton 2000; Chapter 2). The high frequency of paleopolyploidy in ancestral lineages (Chapter 3) suggests that hybrid speciation through whole genome duplication has been an important mechanism of diversification.

While polyploidy has clearly been prevalent and important for diversification over the history of flowering plants, we have only a handful of examples of recent polyploid speciation. Most of these cases are associated with human modifications of the habitats or geographic distributions of species. For example, two polyploid species of *Cardamine* in the Swiss Alps have restricted distributions and are associated with human disturbance (Zozomova-Lihova et al. 2014). The hexaploid *C. schulzii* was derived by *Autopolyploidy* of the triploid *C. x insueta*. This species has limited genetic variation and appears to only reproduce through vegetative propagation. The triploid species appears to be derived from multiple instances of interbreeding between two diploid species of *Cardamine*. Because of its odd chromosome complement, the triploid species' capacity for sexual reproduction is very limited.

The *Allopolyploid, Senecio cambrensis,* has appeared multiple times in Wales, Edinburgh, and other locations in the British Isles by interbreeding between *S. vulgaris* and *S. squalidus* (Ashton and Abbott 1992). *S. squalidus* is a homoploid hybrid species, itself derived from hybrids that were introduced from Italy (see the Homoploid Hybrid Speciation section). Another allopolyploid species in Great Britain (*Spartina anglica*) originated from interbreeding between an introduced (*S. alterniflora*) and a native (*S. maritima*) species (Ayres and Strong 2001). The tetraploid *S. anglica* has probably been derived multiple times by doubling of the chromosomes of the infertile F_1 hybrid (*S. x townsendii*) of the two diploid species.

One of the best documented cases of polyploid speciation occurs in the sunflower genus *Tragopogon* in the Palouse region that extends from eastern Washington State into adjacent areas of Idaho in western North America (Soltis et al. 2004; Tate et al. 2009; Fig. 11.6; refer to color plate). In this case, three introduced diploid species have interbred to produce two allopolyploid hybrid species. *Tragopogon mirus* has been derived multiple times from crosses between the diploid species *T. porrifolius* and *T. dubius,* with *T. porrifolius* always acting as the maternal parent. In another case, two different floral forms of *Tragopogon miscellus* originated from hybridization between *T. pratensis* and *T. dubius* (Fig. 11.6; refer to color plate), with the more common short-liguled form resulting from crosses with *T. pratensis* as the maternal parent. A fairly accurate date can be placed on the origin of these species, because all three diploids did not occur together in North America until 1928, and the first herbarium specimens of the hybrid species were collected in 1949 (Ownbey 1950). The diploid parental species also occur together in Europe, but surprisingly, hybrid speciation has not been observed in the native range. It is likely that there are particular environmental conditions present in the introduced region that favor the persistence of the allopolyploid lineages.

From the handful of examples that we know of, it appears that the overriding factor associated with the appearance of polyploid species is exposure to novel environments. All of the species just

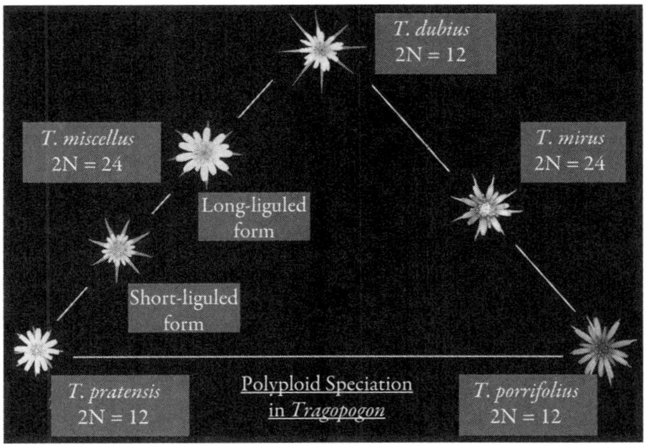

Fig. 11.6. Polyploid hybrid speciation in *Tragopogon* after introduction of three diploid species in Western North America. The short- and long-liguled forms of *T. miscellus* represent products of reciprocal crosses between *T. dubius* and *T. pratensis*. Modified from Tate et al. (2009).

described originated as a result of human activities, which makes their occurrence seem outside the range of natural variation. On the other hand, the circumstances surrounding the appearance of these species can provide insights into how natural processes may favor the origin of polyploid lineages. In particular, it appears that novel environmental conditions can promote polyploid speciation, and the importance of ecological opportunity for the establishment of novel lineages is a recurring theme in our discussions of speciation. Given that ecological transitions seem to be important for the persistence and divergence of polyploid lineages, it may not be surprising that we do not have more examples of recent polyploid speciation in undisturbed plant communities. This situation may be changing, as human-mediated climate change is likely to promote changes in geographic distributions and is likely to create more opportunities for the origin of novel polyploid species in the near future.

Whole genome duplication events have clearly been important for diversification, but there is not necessarily a direct correspondence between changes in ploidy and rates of speciation. It is generally true that some plant groups have a greater propensity toward polyploidy than others, and whole genome duplication is often associated with speciation (Otto and Whitton 2000; Soltis et al. 2009). On the other hand, higher ploidy levels are not associated with increased rates of diversification; changes in ploidy often lead to the establishment of new groups, but sister clades having lower ploidy levels tend to have similar speciation rates (Wood et al. 2009).

Given that whole genome duplication has been common throughout the history of flowering plants, it is perhaps surprising that contemporary instances of polyploid speciation are uncommon. However, it is important to recognize that even very low rates of polyploid speciation can translate into frequencies that are consistent with observed rates across phylogenetic timeframes. Comparison of ploidy within and among taxonomic groups has suggested that between 30% and 80% of angiosperm species have polyploid events in their ancestry (Soltis et al. 2009). On the other hand, patterns of ploidy observed across phylogenies do not accurately reflect historical rates of polyploid diversification (Wood et al. 2009). Most notably, as whole genome duplication lineages undergo diploidization resulting in a smaller number of chromosomes with bivalent pairing (Chapter 3), even taxonomic groups with low chromosome numbers are likely to have descended from polyploids. Furthermore, we may be missing much of the variation in ploidy within species, because novel polyploid lineages often lack distinguishing phenotypic characteristics. As discussed in Chapter 3, polyploid lineages may have some advantages over their diploid progenitors because they are often self-fertile and experience reduced effects of inbreeding depression (Chapter 5). As available technologies have permitted detailed analyses of genomic variation, the importance of polyploidy for speciation and diversification continues to become more evident.

Homoploid Hybrid Speciation—While instances of hybrid speciation through whole genome duplication are common, the importance of homoploid hybrid speciation is much less clear (Gross and Rieseberg 2005; Nolte and Tautz 2010; Yakimowski and Rieseberg 2014). We can distinguish homoploid hybrid speciation from allopatric divergence followed by introgression during secondary contact because: 1) hybrid species originate within the first few hybrid generations; and 2) the resulting species retain substantial genomic contributions from both of the parental lineages. For cases of homoploid hybrid speciation, we expect the genome of one species to consist of a mosaic of genomic regions derived from two progenitor species. Introgression during divergence, on the other hand, would result in strong genomic similarity between the derived genome and only one closely related species. In both cases, some minority of genomic regions may be divergent among all close relatives. In this section, we explore the mechanisms of homoploid hybrid speciation and discuss some well documented cases.

Grant (1981) identified two separate mechanisms of reproductive isolation during homoploid hybrid speciation. He referred to cases of reproductive barriers that were due to chromosomal rearrangements as cases of "recombinational speciation." In the second pathway to homoploid hybrid species, Grant assumed that reproductive isolation was a consequence of extrinsic barriers that developed due to ecological divergence or geographic separation. While homoploid hybrid speciation occurs infrequently in plants, with the exception of some well-documented examples such as in North American wolves (vonHoldt et al. 2011) and heliconius butterflies (Mallet 1993; Mallet 2007), it appears to be even more rare in animals. Interest in homoploid hybrid speciation waned for much of the twentieth century, but experienced a resurgence in the 1990s with the work of Rieseberg (1991a) and Arnold (1993) on homoploid hybrid species of *Helianthus* and *Iris*, respectively.

Much of the speculation about the hybrid origin of species has been motivated by the observation of taxa that display intermediate characteristics compared to other species. Kerner (1894), for example, focused on the example of *Rhododendron intermedium*, which is a putative hybrid species derived from *R. ferugineum* and *R. hirsutum*. Cases of intermediate morphology may be common—a search of the specific epithet *intermedium* in the US Department of Agriculture Plants Database (http://plants.usda.gov) yields hits for more than two hundred species in North America. This may be an underestimate since not all homoploid hybrids have intermediate phenotypes (e.g., *Iris thompsonii* in Fig. 11.3; refer to color plate). It is also true that inferring a hybrid origin from apparent intermediacy may not always be reliable. For example, Straw (1955) inferred the hybrid origin of two species of *Penstemon* on their intermediate morphology and close resemblances to F_1 hybrids (Fig. 11.7; refer to color plate). *Penstemon spectabilis* was hypothesized to have been derived from crosses between *P. centranthifolius* and *P. grinnellii*, and *P. clevelandii* appeared to have been derived from hybrids of *P. centranthifolius* and *P. spectabilis*; however, analyses with genetic markers only supports a homoploid hybrid speciation in the case *P. clevelandii* (Wolfe et al. 1998). There are many such cases of intermediate species that may have homoploid hybrid origins, but only a small number have been evaluated with genetic markers (Gross and Rieseberg 2005; Nolte and Tautz 2010).

Some of the best known cases of homoploid hybrid speciation occur in North American irises (*I. nelsonii* in Louisiana and *I. thompsonii* in northern California) and sunflowers (*Helianthus*) and in invasive species of *Senecio* in the British Isles (Abbott et al. 2010). The classic case of homoploid hybrid speciation in Louisiana irises was first proposed by Randolph (1966; Fig. 11.8; refer to color plate). From his observations of morphological and ecological characters, Randolph predicted that three species (*Iris hexagons, I. fulva,* and *I. brevicaulis*) had contributed to the origin of the hybrid species, *I. nelsonii*. Remarkably, this unique case of hybrid speciation from three parental taxa was later confirmed using genetic markers (Arnold 1993). This hybrid iris species occurs in an ecologically divergent habitat that is a unique combination of light availability and flooding conditions compared to the parental habitats (Taylor et al. 2011).

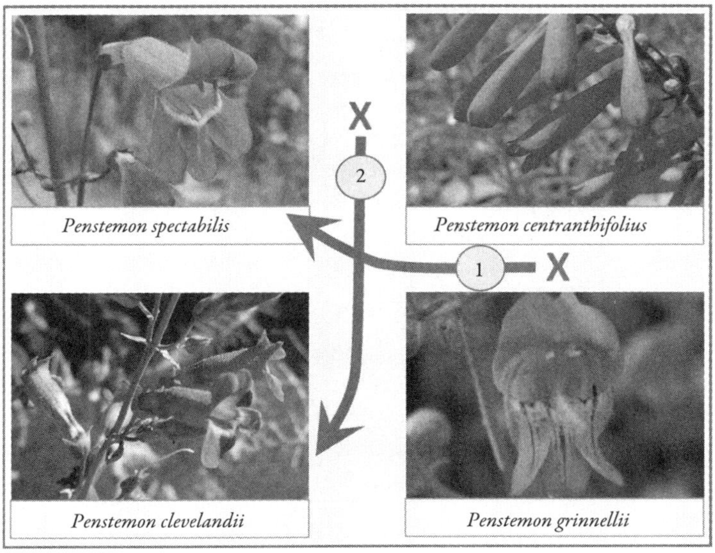

Fig. 11.7. The sequence of homoploid hybrid speciation in *Penstemon* proposed by Straw (1955). (1) *P. spectabilis* was hypothesized to have been derived from crossing between *P. centranthifolius* and *P. grinnellii*. (2) A second round of hybridization between *P. centranthifolius* and *P. spectabilis* results in the hybrid species *P. clevelandii*. Molecular genetic data support the hybrid origin of *P. clevelandii*, but not for *P. spectabilis* (Wolfe et al. 1998). Photos by Stickpen and Florian.

Fig. 11.8. The homoploid hybrid origin of *Iris nelsonii* from interbreeding of *I. brevicaulis*, *I. fulva*, and *I. hexagona*. Photos by Jax Strong, Mike Arnold, and Amy Bouck.

There is a surprisingly high propensity toward homoploid hybrid speciation for the combination of two widespread sunflower species, *Helianthus annuus* and *H. petiolaris* (Rieseberg 1997; Gross and Rieseberg 2005; Renaut et al. 2014; Fig. 11.9; refer to color plate). These taxa have generated three different hybrid species that occur in extreme habitats and have restricted geographic distributions

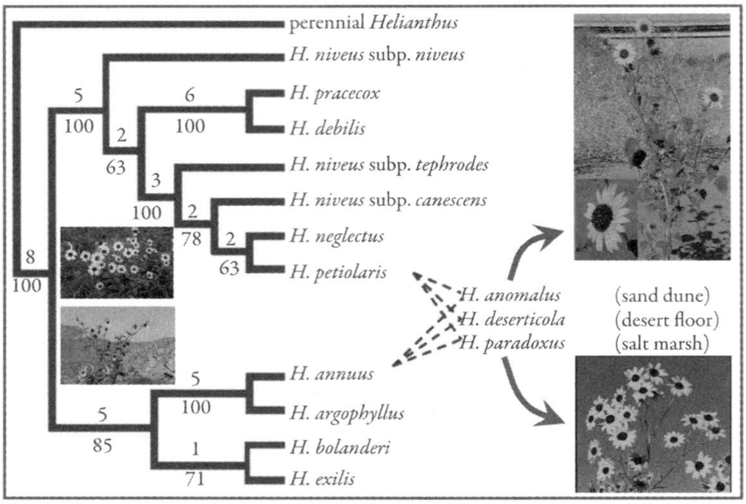

Fig. 11.9. Phylogenetic relationships and the origin of three homoploid hybrid species in the clade of North American annual *Helianthus*. The parental species *H. petiolaris* and *H. annuus* interbred to generate three hybrid species (dashed lines) that each occurs in a unique habitat. Modified from (Gross and Rieseberg (2005). Photos by Stan Shebs, Clarence A. Rechenthin @ USDA-NRCS PLANTS Database, and Mike.

relative to the parental lineages: *H. anomalus* occurs at the base of large sand dunes; *H. deserticola* on the desert floor; and *H. paradoxus* is found in salt marshes. These hybrid species are thought to have originated approximately sixty thousand to two hundred thousand years ago and are reproductively isolated by ecological barriers and chromosomal rearrangements. At least two of these hybrid species appear have originated from multiple hybrid lineages (Renaut et al. 2014).

The case of the homoploid hybrid *Senecio squalidus* (Oxford ragwort) is unique relative to the cases described above because it has originated very recently, after humans imposed geographic isolation. This species is the product of crosses between two species of *Senecio* (*S. aethnensis* and *S. chrysanthemifolius*) that naturally hybridize on Mount Etna in Sicily (Abbott et al. 2010). The new species is derived from hybrid plants that were introduced into Great Britain. This homoploid hybrid has become an aggressive invader that is spreading across England and Ireland. The establishment and spread of this new hybrid species may have been facilitated by environmental conditions, which are apparently intermediate between the habitats of the two parental species.

The origin of homoploid hybrid species is generally difficult unless the new lineage becomes reproductively isolated early during divergence. Unlike polyploid speciation, homoploid hybrids remain interfertile with parental taxa, so divergence could easily be prevented by recurrent gene flow. For example, in the case of *S. squalidus,* divergence of the hybrid derivative was apparently not possible until geographic isolation was established. It is less clear how reproductive isolation developed in the other cases discussed here. With the exception of *Helianthus,* these homoploid hybrid species are reproductively isolated primarily by extrinsic (ecological and geographic) rather than intrinsic barriers such as chromosomal rearrangements (Abbott et al. 2010). The three *Helianthus* hybrid species also display strong ecological divergence, so it is possible that initial isolation of these lineages was extrinsic and that chromosomal rearrangements accumulated during later stages of divergence; this scenario may be more likely since it is consistent with most cases of allopatric speciation where chromosomal rearrangements accumulate after divergence under other isolation mechanisms. On the other hand, population bottlenecks during the establishment of novel hybrid lineages may have facilitated the fixation of chromosomal rearrangements.

Early establishment of ecological barriers during the divergence of homoploid hybrid lineages may have been facilitated by the generation of *Transgressive* hybrid phenotypes (Gross and Rieseberg 2005). As discussed in Chapter 10, early-generation recombinant hybrids often display trait expression that falls outside the range found in the parental lineages. Expression of divergent traits could facilitate the establishment of hybrid genotypes in habitats that are outside the normal range of the parental species. The fact that hybrid species of *Helianthus* and *Iris nelsonii* occupy habitats that are novel seems to support the importance of divergent trait expression. Similarly, a homoploid hybrid species of pine (*Pinus densata*) survives under severe drought in habitats that its parental progenitors could not tolerate (Wang et al. 2001). These cases suggest that ecological divergence rather than chromosomal rearrangements is the key factor establishing reproductive isolation and facilitating homoploid hybrid speciation. If this were generally true, then we can speculate that anytime inter-mating between lineages generates a wide range of hybrid phenotypes, the potential exists for the establishment of novel hybrid lineages. Indeed, homoploid hybrid speciation may be an outcome of serendipity; the right combinations of divergent habitats and transgressive phenotypes may occur together only rarely, but when they do, the origination of novel hybrid lineages seems to be almost inevitable.

11.5 SPECIES DELIMITATION

Understanding the types of criteria used in different species concepts and modes of speciation improves our appreciation of the characteristics that we might use to identify evolutionarily independent lineages. To choose criteria to identify lineages that qualify as species, it is useful to evaluate how characteristics used for delimitation accumulate during divergence. In this section, we discuss how sets of criteria can be applied to delineate independent lineages. We first examine the process of divergence during speciation to understand how characteristics used in species delimitation accumulate over time. We then discuss applications of these criteria, as we explore some of the challenges we face when attempting species delimitation.

The Process of Divergence—From the discussion above it becomes apparent that there is a wide range of opinions on the appropriate set of criteria that should be used for delimiting species. The lack of consensus on a single species concept is partly due to the intrinsic nature of species as dynamic lineages that display increasing levels of independence and distinctiveness during divergence (Lowry 2012). Our goal is to establish a set of defining characteristics that reflect a level of divergence that indicates lineage independence. The difficulty with this approach is that criteria that reflect historical processes of divergence do not necessarily guarantee continued independence. This problem is amplified, because it is impossible to accurately predict the future; for all we know, a seemingly unremarkable population that displays only mild divergence may someday diversify into a completely novel clade of species.

The standards that we might choose to identify lineages that have unique evolutionary potential fall along a continuum, with populations at early stages of divergence—differing only in allele frequencies—at one extreme and lineages that are almost completely isolated by intrinsic reproductive barriers at the other. Whether we focus on the earliest discernable signs of "species" status or on later stages when our confidence in lineage independence is much higher will depend on our goals. Evolutionary biologists who focus on the role of species as the fundamental units of evolutionary diversification may want to establish criteria that identify lineages with unique evolutionary potential at an early stage of divergence. Many systematists, on the other hand, may value reliable sets of characters that are indicative of independence and provide landmarks for confident species delimitation. Choosing characteristics that are based only on early stages of divergence will result in assigning species status to some lineages that will ultimately interbreed and meld with others, while more conservative criteria will tend to underestimate diversity. We do not intend to attempt to establish a specific level of divergence for the delimitation of species, but rather, we endeavor to understand the pattern of emergence of different characteristics that occur during divergence and that can serve as tools for interpreting different degrees of lineage independence.

When considering criteria that could be used for species delimitation, it is instructive to evaluate the order in which distinctive characteristics are expected to accumulate during divergence (Fig. 11.10; refer to color plate). As discussed earlier, we assume that an initial mechanism of isolation such as ecological or geographic separation or an increase in ploidy, needs to be established to limit gene flow sufficiently to allow divergence. Other than changes in ploidy, ecological and geographic isolation are probably the most common mechanisms of initial isolation, but note that these are often not distinct since habitat divergence generally implies spatial separation. The primary difference is that ecological separation can occur over a relatively short spatial scale as populations diverge along environmental gradients (Hendry et al. 2007). It is also notable that ecological isolation is probably much more common in sedentary organisms such as plants; spatially restricted dispersal can result in local gene flow and will facilitate ecological divergence over small geographic scales.

The three scenarios illustrated in Fig. 11.10 (refer to color plate) differ with respect to the rate of emergence of different characteristics that might be used for species delimitation. For example, we expect that after isolation, genetic drift will result in the gradual divergence in allele frequencies for neutral markers and will allow for lineage sorting that eventually results in monophyly of gene genealogies (Fig. 11.4; refer to color plate). If we examine large numbers of gene regions for a pair of lineages, then we would expect to see a greater proportion of their genealogies to exhibit reciprocal monophyly as the time since divergence from a common ancestor increases. In the absence of population bottlenecks, higher levels of genetic differentiation will generally coincide with the time since initial isolation among lineages.

While we expect the level of genetic divergence at neutral genomic regions to generally reflect the time since separation, the effects of genetic drift can be amplified by colonization after long-distance dispersal, because fewer alleles and haplotypes are retained after population bottlenecks. With fewer haplotypes present in a population, less time will be required for lineage sorting to result in monophyly. Consequently, we expect neutral divergence and lineage sorting to start at higher levels after a long-distance dispersal event that results in geographic isolation (Fig. 11.10; refer to color plate). Note, however, that selection resulting in ecological isolation and the origin of a polyploid lineage could also cause population bottlenecks, so the time required for neutral

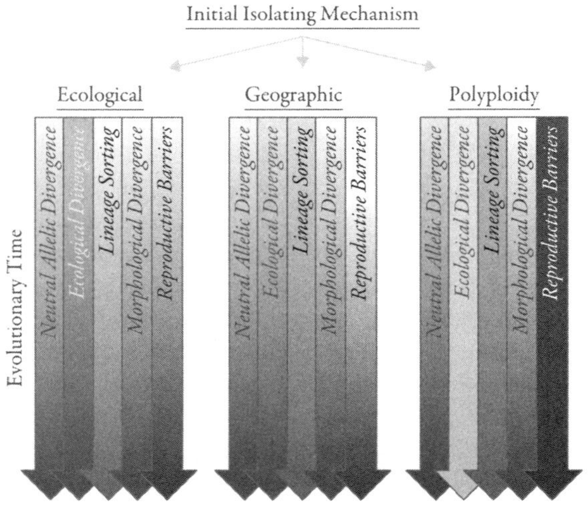

Fig. 11.10. Three scenarios for divergence after isolation due to ecological or geographic separation, or to an increase in ploidy. After isolation we expect divergence in neutral allele frequencies, habitat and ecologically-associated traits, and other morphological traits not associated with the local environment. During this time we also expect lineage sorting in gene genealogies that eventually results in monophyly for all members of the new lineage. Development of intrinsic reproductive barriers and floral isolation is expected to be slow after ecological and geographic isolation, and immediate in the case of a new polyploid lineage.

divergence and lineage sorting may be shorter under these scenarios. Continued gene flow among partially isolated lineages would result in greater genetic similarity and would delay or prevent the development of reciprocal monophyly of genealogies. Consequently, the level of allelic divergence and level of monophyly of genealogies should generally reflect the level of differentiation and independence among lineages.

Selection can drive divergence more rapidly than neutral processes in cases where populations are exposed to novel environmental conditions. Consequently, we expect larger differences among lineages for traits associated with local adaptation in cases of ecological isolation (Fig. 11.10; refer to color plate). In this scenario, transition into a novel habitat favors trait divergence and reduces gene flow with the parental lineage, because hybrids have low fitness or are restricted to ecotone regions. The strength of initial isolation due to habitat association will depend on the level of ecological divergence and the fitness of hybrid progeny across available habitats. For example, gene flow between ecologically divergent lineages will be lower when ecotones are abrupt and selection in each habitat is strong (Chapter 5). We also expect some ecological divergence under the geographic isolation scenario because environmental differences tend to be correlated with geographic distance.

Intrinsic reproductive barriers are generally expected to develop more slowly and may often be one of the last characteristics of speciation to emerge (Fig. 11.10; refer to color plate). The one exception to this generality is the case of polyploidy, where strong reproductive isolation is the consequence of chromosome doubling, so hybrid offspring are almost always infertile due to segregation abnormalities. In cases where initial isolation is afforded by ecological or geographic separation, reproductive isolation may take longer to develop; intrinsic barriers that are the consequence of accumulated genetic differences are expected to emerge more slowly because they are often dependent on the processes of mutation and genetic drift rather than selection (Hendry et al. 2007). One exception to this generality may be selection to accommodate a novel pollinator resulting in floral isolation. Reliance on an alternative pollinator will contribute to isolation when recently diverged lineages come into contact, but as discussed in Chapter 10, floral isolation mechanisms are generally more permeable than post-mating barriers. The other exception is the case of selection for the reinforcement of reproductive barriers, but as discussed in Chapter 10, characters that limit hybridization due to selection for reinforcement may be unlikely to spread outside regions of sympatry.

Assessing the characteristics that distinguish closely related lineages reminds us that the effectiveness of isolation barriers may be subject to change. In particular, while geographic isolation may be very effective at one point in time, ranges may shift in response to changing climatic conditions such that divergent lineages come into contact. Whether lineages can maintain their integrity when their geographic distributions overlap will depend on the type and strength of isolating barriers that have developed. For example, a lineage that has transitioned into a novel habitat may remain intact if the level of ecological divergence is great enough to limit gene flow. Similarly, lineages that have developed strong post-mating reproductive barriers may be more likely to persist. In general, lineages that have remained geographically separated for longer periods will have had more time to diverge ecologically and to develop genetic differences that contribute to post-mating incompatibilities that maintain lineage integrity during secondary contact.

Applying Divergence Criteria to Delimit Species—Comparison of sequence variation across multiple genomic regions has demonstrated the utility of coalescence- and clustering-based methodologies for the delimitation of species that generally fall under the GLC (Carstens et al. 2013). One of the more commonly used methods is the *General Mixed Yule Coalescent* model (GMYC; Pons et al. 2006; Reid and Carstens 2012). This approach attempts to identify transitions in the rate of coalescence, from higher frequencies of branching within lineages to lower rates of branching that indicate divergence of species across longer time scales. For example, in Fig. 11.4 (refer to color plate) the time for within-lineage coalescence is represented as the distance from the present (top endpoints of the branches) back to the most recent common ancestor of all extant haplotypes (i.e., ancestral lineages labelled A). Note that none of the lineages arising before recent common ancestors persist to the present time. The distance from these A ancestors back to the common ancestors among lineages (B) reflects levels of divergence among lineages during isolation. For the example in Fig. 11.4

(refer to color plate), GMYC analysis would identify cryptic species within the polyphyletic group, such that it would be divided into two independent species (four species total). Note that the within-lineage coalescence times would be increased further back in time if we allowed introgression among lineages; higher rates of gene flow among lineages would increase the chance that descendants of introgressed alleles would still be present. With high levels of gene flow among lineages, the within-lineage coalescence times would be extended back to the between-lineage coalescence times (B ancestors), and consequently, we would consider all four of these lineages to be included in a single species.

The GYMC method is often used to delimit species under the GLC. Many studies conducting species delimitation using GLC have reported discrepancies in conclusions made from criteria used under different species concepts. For example, there was a low correspondence between species identified under the MSC compared to those delimited using GMYC for marine green algae in the genus *Boodlea* (Leliaert et al. 2009). This group is highly variable, which apparently results from the evolution of parallel morphologies within divergent lineages. Consequently, a species that had been described under the MSC was found to consist of individuals from several different phylogenetic lineages. The taxonomic confusion in this case is probably due to a lack of easily observed morphological characters that can be used for species delimitation in this group. Hence, the poor correspondence between the morphological and phylogenetic circumscriptions of species may be a consequence of using traits that are subject to ecological selection for species delimitation. While this represents an extreme example, the discovery of both cryptic species and cases of parallel evolution appear to be common in a broad range of plant groups including: herbaceous plants in the genus *Silene* (Aydin et al. 2014) and *Erysimum* (Abdelaziz et al. 2011); epiphyllous liverworts in the *Cololejeunea lanciloba* complex (Yu et al. 2013); and tropical African trees in the genus *Carapa* (Duminil et al. 2012).

Other clustering algorithms, including the GMYC method, can provide detailed assessments of the presence of independent phylogenetic lineages. These methods differ with respect to the type of information used (e.g., allelic versus genealogical) and in their ability to resolve low levels of genetic differentiation; consequently, different delimitation methods do not always identify the same set of independent lineages (Carstens et al. 2013). It is also important for us to recognize that the methods for species delimitation under the GLC and most other species concepts cannot predict the fate of extant lineages. As mentioned earlier, lineages that have not diverged sufficiently to develop strong post-mating or ecological isolation mechanisms may be subject to genetic swamping if their distributions overlap other closely related lineages. In general, we expect that a large proportion of these apparently divergent lineages are ephemeral, because they will eventually succumb to extensive introgression and melding with other lineages, or because changes in environmental conditions will curtail population growth, ultimately leading to extinction. The consequences of the relatively frequent origin and low persistence of lineages is evident, as estimates of speciation rates at short time scales are often high compared to the much slower speciation rates over longer time scales (Rosenblum et al. 2012). From these observations, we can conclude that there will be a constant turnover in species over time through the processes of extinction and diversification and that only a small fraction of contemporary species will ultimately be responsible for future biological diversity.

11.6 MODES OF SPECIATION

From these discussions, it becomes clear that isolation from gene flow can facilitate continued divergence and the origin of independent lineages that can be considered species. The multiple pathways to species status illustrated in Fig. 11.10 (refer to color plate) suggest alternative modes for the initiation and promotion of speciation. If ecological divergence initiates isolation as in the case of budding or peripatric speciation (*Ecological Speciation*), then we expect that the development of characteristics leading to lineage independence may be different than when geographic isolation allows for divergence primarily by mutation/drift processes (*Genetic Speciation*). These two modes of speciation differ with respect to the predicted patterns of development of isolation barriers. In particular, we expect that reproductive isolation during ecological speciation may initially be due to associations with an alternative habitats or pollinators. If ecological barriers are strong enough to limit gene flow then other isolating mechanisms may develop. With continued isolation from gene flow, either mode will allow for the accumulation of genetic differences that contribute to intrinsic mechanisms of post-mating reproductive isolation (genetic incompatibility; Chapter 10). While the endpoint of these processes is similar, it is of interest to assess the prevalence of these two modes of speciation to more thoroughly understand the primary mechanisms that drive diversification.

While selection is often considered the primary driver of divergence, its contributions to the development of post-mating reproductive barriers are not well understood (Chapter 10). The process of speciation seems to require a combination of morphological, ecological, and karyotypic divergence (Stebbins 1950; Grant 1981), but no single one of these criteria is usually adequate to define species. Even though much of the discussion of speciation in plants has focused on ecological divergence, it is generally assumed that "good species" have strong post-mating barriers, which may be more likely to develop through neutral genetic processes. The view that intrinsic reproductive barriers are important criteria for the species status is reflected in the consideration of groups of potentially interbreeding lineages as coenospecies or syngameons as discussed earlier in this chapter.

Whether geographic isolation or ecological association with different habitats or pollinators is substantial enough to qualify lineages as species is a topic that continues to be discussed (Stebbins 1950; Grant 1981; Gavrilets 2003; Doebeli et al. 2005; Hendry et al. 2007; Lowry 2012). Over evolutionary timescales, geographic barriers may be viewed as ephemeral, as the ranges of divergent lineages may become sympatric at some time in the future. Similarly, ecological barriers that are based on association with different animals for pollen transfer may be considered permeable, and perhaps temporary, as specialized pollinators may not always be available (Kay and Sargent 2009), and reversals in pollination syndromes are possible (Lagomarsino et al. 2017). Association with different habitats can represent stronger isolating barriers, but the potential for continued gene flow may depend on whether the establishment of hybrids is promoted by the presence of gradual ecotones (Chapter 10). While isolating barriers developing during ecological divergence may often be temporary or imperfect, reproductive isolation based on genetic incompatibilities is generally considered more permanent, and for more diverged lineages, post-zygotic barriers can effectively prevent any degree of introgression (Stebbins 1950; Clausen 1962; Grant 1981; Baack et al. 2015).

Models of Speciation—Interest in the evolutionary processes responsible for the development of reproductive isolation dates back to Darwin's *On the Origin of Species* (1859), but as discussed in Chapter 10, Darwin never directly addressed the "species problem." The evolution of reproductive isolation seems unlikely because the desired outcome—low fitness of hybrids—is also the primary obstacle: selection would be expected to oppose the spread of any mutations contributing to the strengthening of reproductive barriers (i.e., underdominance at individual loci). One solution to this paradox assumes that novel mutations accumulating in each lineage are neutral or favored in their native genetic backgrounds but contribute to low fitness in hybrid offspring due to epistasis within and between the divergent genomes (the BDMI model; Chapter 10). Even though the BDMI model appeared to present a viable solution to the species problem, it remained widely ignored until late in the twentieth century (Coyne and Orr 2004).

An alternative approach to speciation, which received much more attention in the mid to late twentieth century, was Wright's *Shifting Balance Model* of evolution (Wright 1931, 1982; Coyne and Orr 2004). This model assumes that a combination of genetic drift and selection contribute to the process of speciation, and that large numbers of genes interact to generate isolated gene complexes of high fitness (*Adaptive Peaks*; Fig. 11.11; refer to color plate). Speciation occurs on a *Rugged Adaptive Landscape* as genetic drift displaces populations from adaptive peaks to promote transition to neighboring peaks through selection. The idea that fitness topographies across gene complexes could be illustrated as rugged adaptive landscapes reflected much of the thinking about the nature of species in the twentieth century. Namely, that species occupy regions of high fitness (corresponding to their niches), almost all the available niches (peaks) are occupied, and hybrids are expected to have low chances of success (i.e., because they fall into valleys of low fitness; Wright 1982). While the rugged adaptive landscape metaphor has remained popular with empiricists, theoreticians

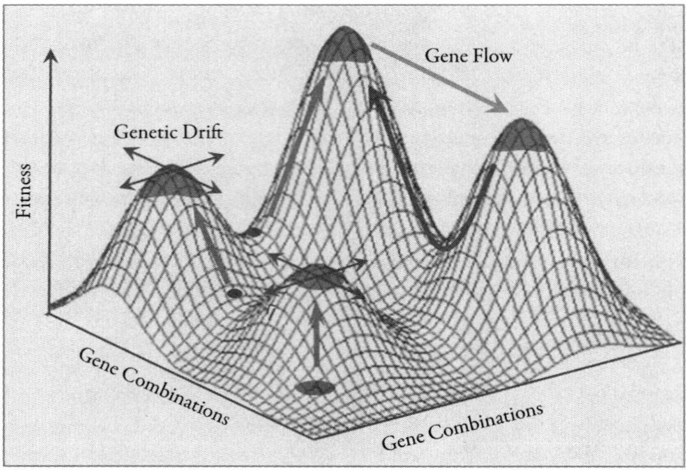

Fig. 11.11. A rugged adaptive landscape generated by interactions among large numbers of genes. The two horizontal axis represent different gene combinations and the vertical axis indicates fitness. Genic selection is responsible for moving a population (green circle) to the top of a local peak (pink arrow). Shifting balance evolution as described by Sewall Wright occurs as the green population shifts to higher peaks when it is dislodged from the local peak by genetic drift (red crossed arrows; Stage 1). Genic selection then moves the population up the slope of an alternative peak (Stage 2). Peak transitions continue by drift followed by genic selection until the highest peak is reached. The most fit gene complex then spreads to other populations by gene flow (blue arrow) until all populations occupy the highest adaptive peak (organismic selection—Stage 3).

have struggled to discover reasonable conditions that would allow peak transitions at high enough frequencies to explain observed rates of speciation (Gavrilets 2003).

In the late twentieth century, the adaptive landscape model was more thoroughly evaluated by Gavrilets and colleagues (Gavrilets 1997; Gavrilets et al. 1998). Like Wright, these authors assumed that large numbers of genes contribute to reproductive isolation, but the new approaches applied to this problem revealed that peak transitions could occur solely through genetic drift. Indeed, the landscape is not truly rugged because ridges interconnect peaks in networks of high fitness to produce a *Holey Adaptive Landscape*, which allows divergence and the evolution of reproductive isolation solely through mutation/drift processes. In the next few sections, we discuss models of speciation that rely on neutral genetic processes to initiate (shifting balance model) or to complete (holey adaptive landscape model) the speciation process. We will start by exploring adaptive landscapes and discuss how they have been applied to understand the evolution of reproductive isolation and speciation.

Adaptive Landscapes—The depiction of evolution as hill-climbing adaptive peaks has served as useful metaphor for the processes of adaptation and speciation since the early twentieth century. The idea of a "landscape" of adaptive values was first introduced by Sewall Wright (1932) to illustrate aspects of his shifting balance model of evolution. But since its introduction, the adaptive landscape has been interpreted in a variety of ways to demonstrate ideas concerning genetic and phenotypic aspects of adaptive evolution (Cruzan and Weinreich 2013), and adaptive landscapes continue to be useful tools for conveying conceptual aspects of adaptation and speciation.

Wright's views on adaptive evolution were shaped to a large degree by his own experiences of breeding experiments with small mammals and livestock (Wright 1982). In particular, Wright's observations led him to the conclusion that adaption was promoted by the expression of combinations of genes that worked in concert to produce the fittest phenotypes. This differed from what he referred to as a genic mode of adaptation—that adaption was due to loci that acted independently—a view that was favored by Fisher

(1930). Wright assumed that adaptive evolution involved large numbers of genes, and he struggled to find a way to accurately illustrate this process. He recognized that his depiction of a two-dimensional "gene combination landscape" was imperfect but was satisfied that it adequately illustrated salient features of adaption and speciation through shifting balance processes.

Wright's depiction of a rugged adaptive landscape is meant to reflect the effects of many genes interacting to produce multiple adaptive peaks separated by valleys and saddles of low fitness. On this landscape, populations or species are represented by clouds of points that are under the influence of a local peak. Hill climbing (adaptive evolution) occurs by allelic substitutions for genes of small effect. Wright referred to this as genic evolution: the process of fine-tuning the phenotype with respect to a local adaptive peak (Wright 1982). Populations are assumed to be constantly subject to genic evolution, as their position is frequently displaced away from adaptive peaks by genetic drift and due to shifts in the position of peaks that reflect fluctuations in the environment.

It is important to recognize that adaptive landscapes as described by Wright represent fitness responses in a single environment. If there were little or no epistasis such that adaptation was due to the independent effects of genes—as considered in the models of selection developed by Fisher (1930)—then only a single adaptive peak would be present. Wright assumed that fitness differences were primarily due to different gene combinations and that adaptive peaks on a rugged landscape were due to epistasis. In other words, each peak represents a different combination of alleles at multiple loci that interact to confer adaptation to the current environment: each peak represents a different solution to the same environmental challenges. Wright's depiction of the fitness topography as a rugged landscape remains a valuable metaphor for understanding the contributions of drift and selection to adaptation and is the centerpiece of Wright's shifting balance model of adaptation and speciation, which we will discuss in the next section.

The Shifting Balance Theory—Wright used his rugged genetic landscape to illustrate the process of adaptation as local hill climbing, by the transit of populations across saddles toward peaks of higher fitness. Wright, whose view of evolution departed from Fisher's, assumed that species were divided into many small population that were subject to genetic drift rather than consisting of a single large population that was primarily affected by selection. The importance of small population size for the shifting balance model of evolution is twofold. First, it provides a mechanism (genetic drift) for populations to escape a local peak to come under the influence of an alternative adaptive peak. Second, it allows evolution to conduct "many small experiments" (from Wade and Goodnight 1998 in reference to Wright's SBT) that may result in uniquely fit populations—some occupying higher adaptive peaks and some trapped on peaks of lower fitness. Increasing the number of populations elevates the chance that mutations allowing for peak transitions will arise in one population and spread across the species' range.

As mentioned earlier, Wright's view of adaptive evolution was influenced by his observations of outcomes of artificial selection to improve livestock. He observed a wide range of success among multiple lines that had been subjected to the same selection regime (Wright 1982) and assumed this outcome was due to the random fixation of different combinations of alleles within lines across a large number of interacting loci. Wright interpreted these results to suggest that adaptive evolution occurs at two levels: genic evolution being improvement within each line that improves fitness by small steps with respect to the local adaptive peak; and organismic selection being based on variation in fitness among lines (variation in adaptive peak elevation). These two processes represent the substance of shifting balance evolution; adaptation to the local environment can proceed incrementally through genic selection on a single adaptive peak and in larger jumps (peak transitions) through organismic selection as high-fitness gene complexes spread across populations.

According to the shifting balance theory (Wright 1931, 1932, 1982), adaption to a particular environment occurs in three stages (Fig. 11.11; refer to color plate). In the first stage, populations are dislodged from their local adaptive peak through genetic drift or possibly by fluctuations in the local environment. If random shifts in allele frequencies are large enough, then the population may come under the influence of a neighboring peak. The second stage of shifting balance comes into play as the population responds to the selection gradient presented by the new peak—selection moves the population upslope to the highest local region of fitness. If the new gene combination (peak) affords higher fitness then we expect more rapid population growth and the production of larger numbers of emigrants. The third stage of shifting balance ensues as the most fit gene complex (highest peak) spreads by gene flow until all the populations of a species come to occupy the same adaptive peak. Since many populations are subject to genetic drift and selection in the same environment, the entire landscape can be explored until all populations possess the most fit gene complex (i.e., the highest available peak).

The shifting balance theory of evolution has received a considerable amount of attention since it was first introduced in the early twentieth century (Coyne et al. 1997; Wade and Goodnight 1998; Whitlock and Phillips 2000). On the one hand, the division of species into many small populations is considered to better reflect natural conditions for many organisms; this is especially true for plants because dispersal of pollen and seeds is generally limited (Chapters 4 and 8). On the other hand, the chance of populations being displaced on the rugged adaptive landscape by drift to a large enough degree to come under the influence of an alternative adaptive peak seems remote. Theoretical analyses confirm this and suggest that the chance for peak transitions is vanishingly small except in the cases of very shallow saddles and small population sizes (Gavrilets 2003). Wright recognized this problem as well and suggested that the saddles defined by single pairs of interacting loci can be considered in isolation; with many such transitions, higher regions of the landscape can be reached, and eventually strong reproductive isolation can be attained (Wright 1982).

Another major obstacle to adaptive evolution through shifting balance processes is the high complexity of gene interactions that might be responsible for generating adaptive phenotypes (Coyne et al. 1997). When populations are distributed across multiple peaks on a rugged landscape, the propagules exported by a population occupying a high-fitness peak will recombine with genotypes from other populations. Recombinant offspring are likely to display a wide range of fitness, and continued interbreeding within and among populations could shift their positions to occupy low-fitness peaks. Capture of populations by intermediate peaks could be overcome with persistent gene flow from the high-fitness populations, but this could depend on the details of the landscape topography—it may be possible for populations to become trapped in regions where peaks are sparse and valleys are deep.

A third major criticism of Wright's model is that it is difficult to find convincing examples that meet all the requirements of shifting balance processes (Coyne et al. 1997). Possible exceptions are cases of chromosome rearrangements, which generally display underdominance and are commonly characteristic of different species (Grant 1981; Wright 1982). The fitness of individuals possessing rearrangements is frequency-dependent because there are few compatible mates available when modified plants first appear, but fitness is elevated as their frequency increases. While rearrangements and polyploidy share some aspects of shifting balance processes, they are effectively one-dimensional (i.e., based on a single locus, so no epistasis) with respect to the landscape, which is governed by gene interactions. Wright recognized the association of rearrangements with species, but he did not consider them to be an important component of adaptive evolution by shifting balance processes (Wright 1982). Moreover, these polymorphisms may sometimes have intrinsic advantages, in which case they can be driven to fixation by selection rather than through accidental capture and fixation by genetic drift (Kirkpatrick and Barton 2006).

While genetic drift may be unlikely to initiate transitions to alternative adaptive peaks, it may be possible for selective forces to drive the populations across saddles. For example, populations of *Piriqueta* apparently traverse through low-fitness regions of the genetic landscape during the origin of advanced-generation hybrids (Cruzan 2005). Two morphotypes of *Piriqueta* display contrasting leaf morphologies, with one (*P. caroliniana*) having broad and hairy leaves, and the second (*P. viridis*) having very narrow and glabrous leaves. First-generation hybrids from mating between these morphotypes display high fitness due to heterosis and are vigorous compared to inbred parental populations. Once F_1 hybrids are established in populations of the parental morphotypes, backcrossing results in recombinant backcross hybrids that display high variation in leaf morphology and generally have low fitness (Cruzan and Rhode 2004). The persistence of heterotic

F_1 hybrids apparently drives populations through saddles toward a single high-fitness phenotype that displays an intermediate leaf morphology. This process is apparently responsible for the spread of the advanced-generation hybrid phenotype that now occupies a large region in central Florida (Cruzan 2005). Note, however, that this example does not necessarily match expectations from the shifting balance model because the adaptive process is not initiated by genetic drift.

A possible example of evolution by shifting balance that is initiated by genetic drift is the transition to selfing in populations of *Eichhornia paniculata* (Brazilian water hyacinth; Barrett et al. 2009). This species is self-compatible, tristylous, and displays a range of outcrossing frequencies and morph compositions among populations in Brazil and Jamaica (Barrett et al. 1989; Cruzan and Barrett 2016). The loss of style morphs can occur during colonization, which leads to increased inbreeding that promotes the loss of genetic load and increases in selfing frequencies (Chapter 9). Jamaican populations have become fixed for a single style morph that is modified to promote selfing; one of the short anthers is extended to be adjacent to the mid-level stigma (Fig. 11.12; refer to color plate). Evolution of selfing populations in Jamaica matches our expectations for shifting balance because: 1) the process is initiated by genetic drift during colonization; 2) populations go through a period of low fitness during the purging or genetic load; and 3) movement toward a new adaptive peak (autogamous selfing) requires refinements accrued through (genic) selection. On the other hand, we do not know whether the selfing and outcrossing modes in this species are primarily due to gene interactions, which is a requirement for the manifestation of a rugged adaptive landscape (Wright 1982). In any case, evolution of selfing in *E. paniculata* may be one of the few documented examples of adaptive evolution through shifting balance processes.

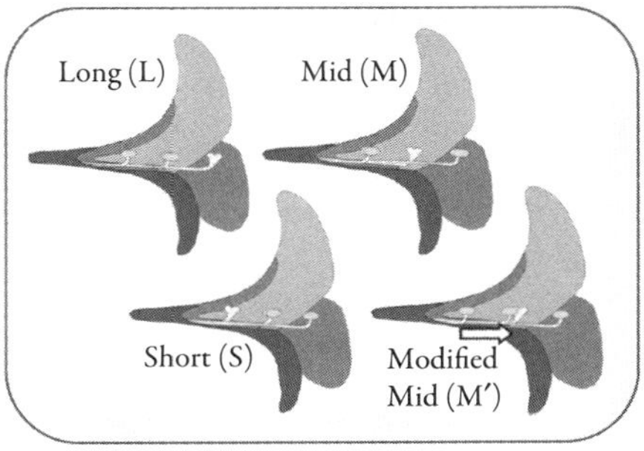

Fig. 11.12. Style morph variation in the tristylous *Eichhornia paniculata*. In the modified mid morph one or more short anthers are elongated so they are adjacent to the mid-level stigma. From Cruzan and Barrett (2016).

Rugged Landscapes as a Generalized Paradigm—In spite of continued disagreements over the importance of shifting balance in small populations versus selection in large populations (Coyne et al. 1997; Wade and Goodnight 1998; Whitlock and Phillips 2000), the rugged adaptive landscape continues to be a persuasive metaphor. The success of this paradigm may be due less to its generality and more to the fact that many authors have ignored Wright's original definition by interpreting adaptive landscapes for their own purposes (Cruzan and Weinreich 2013). In particular, peaks on the landscape are often assumed to represent adaptation to different environments, even though this interpretation represents a significant departure from the original model. The struggle to embrace Wright's adaptive landscape as originally intended may be a consequence of his particular view of species as homogeneous units occupying a single niche; the assumption that adaptive landscapes represent a single environment is difficult to reconcile with the range of habitat variation commonly seen within and among species. The lack of environmental heterogeneity in Wright's model has limited the utility of shifting balance for understanding adaptive processes that are commonly associated with ecotypic divergence and speciation. Indeed, under shifting balance, colonization of a novel habitat requires transitions from one adaptive landscape to another, and Wright's models do not provide any specific mechanisms to accommodate these transformations. In spite of the fact that the shifting balance model focuses on adaptation, it has not been particularly useful for understanding adaptive divergence and speciation.

Toward the end of the twentieth century, new interpretations of the effects of genetic interactions on adaptive landscape topographies emerged that may indicate the demise of the shifting balance model. Advanced analyses of multidimensional genetic systems reveal a drastically different genetic landscape—one that does not require perilous journeys through regions of low fitness for continued evolution and speciation. In the next section, we discuss a new paradigm—the holey adaptive landscape– that may be emerging as a more general model for the evolution of reproductive isolation and speciation.

Holey Adaptive Landscapes—The essential features of Wright's adaptive landscape include many genes contributing to fitness and fitness peaks that are generated by genetic interactions. The two-dimensional representation of adaptive genetic variation is clearly imperfect, but Wright, and most who have followed, assumed that a rugged landscape is an accurate depiction of the fitness topography. This situation changed when the characteristics of high-dimension fitness landscapes dominated by epistasis were examined by Gavrilets and colleagues using more advanced mathematical tools than those that had been available to Wright (Gavrilets 1997; Gavrilets et al. 1998). These authors discovered that as the number of dimensions (loci and alleles) increases, a threshold is reached where a "giant component" spreads (percolates) through the genetic hyperspace forming *Neutral Networks* (Gavrilets 2003). This "percolation threshold" can be achieved with a minimum of ten thousand loci contributing to fitness. It is obviously not possible to accurately illustrate this high-dimensional genetic landscape, but Gavrilets chose to depict this topography as a flat surface pockmarked by regions of low fitness (holes) to emphasize the presence of a neutral network of ridges connecting peaks (Fig. 11.13). This holey adaptive landscape facilitates the evolution of reproductive isolation through mutation/drift processes by divergence of lineages along the neutral network as lineages come to occupy the opposite sides of fitness holes.

It is important to note that the formation of neutral networks in the holey adaptive landscape is an emergent property of complex genetic interactions and does not require any special conditions. In essence, the holey adaptive landscape is an extension of the BDMI model for the evolution of reproductive isolation (Chapter 10; Gavrilets 2003). Dobzhansky originally depicted two interacting loci that formed a single ridge connecting two adaptive peaks (Fig. 10.10; refer to color plate; Dobzhansky 1936). The holey adaptive landscape is a consequence of many thousands such loci interacting to generate a fitness surface. As such, we predict that reproductive isolation is generated in the same fashion; mutations that are neutral in their native genetic backgrounds become fixed in diverging lineages and then interact within and among genomes to manifest low fitness in the F_1 and recombinant hybrid generations, which generates holes in the fitness landscape.

At first glance, Gavrilets' holey landscape does not seem to accommodate adaptive evolution, because the surface is flat (Fig. 11.13), but this topography is an artifact of the assumption of only two possible fitness levels that was used in the original model (Gavrilets 2003). Allowing variation in fitness generates regions of different elevation producing an undulating surface of nearly neutral networks; the network connectivity is expected to remain intact and allow for evolution

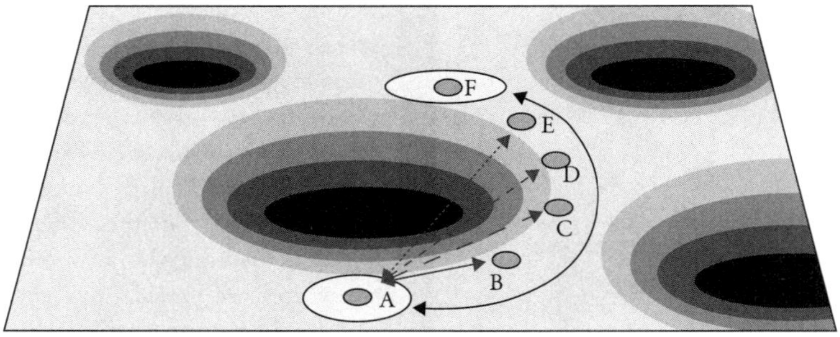

Fig. 11.13. A portion of a holey adaptive landscape showing the path of divergence between lineages that results in the development of complete reproductive isolation. Darker regions of the landscape indicate holes of low fitness. During initial stages of divergence, crosses between populations at locations A and B are vigorous, but viability of hybrids decline as one lineage accumulates mutations through genetic drift until lineages occupying locations A and F are completely isolated by post-mating barriers.

toward higher peaks with little or no loss in fitness. If we imagine that there is one optimal gene combination for a particular environment, then the landscape takes on a topography that is much more in line with Fisher's view of evolution toward a single adaptive peak (Fisher 1930). The main difference between Fisher's view and the holey adaptive landscape paradigm is that for the holey landscape, epistasis rather than independent gene effects determines the topography. Gavrilets does not make specific assumptions about demography, but genetic drift would be facilitated with limited effective population size, in which case a flock of small populations may approach the peak—each taking its own trajectory—rather than a single, large panmictic population. In any case, if the holey landscape represents an accurate view of adaptive evolution, then we no longer need to consider the difficulties associated with populations traversing saddles of low fitness; this new view of adaptive evolution may effectively supersede Wright's rugged landscape, and perhaps obviates the need to consider shifting balance processes (Gavrilets 1997, 2003).

Evolution on Adaptive Landscapes—The rugged and holey landscape models provide conceptual frameworks for understanding the process of adaptation and the evolution of reproductive isolation. These views of evolution both assume that the fitness topography is primarily due to epistasis, but they differ with respect to whether valleys of low fitness are expected to impede the evolution of populations toward regions of higher fitness. To understand which of these models is more representative of evolutionary divergence, it would be convenient to examine results from experimental analyses of fitness landscapes, but it is difficult to find direct tests of adaptive topographies in the literature. However, we can gain some insights into adaptive processes by examining the outcomes of *Experimental Evolution* trials (Elena and Lenski 2003). These experiments examine the evolution of populations under artificial conditions and make detailed analyses of genetic and phenotypic outcomes.

Experimental evolution trials have been typically conducted by exposing isolates from a single microbial population to the same challenging conditions (e.g., a novel food source or physical conditions) and tracking their responses over many thousands of generations. This treatment is analogous to a drastic change in the environment that places populations in a region of low fitness. In this scenario, we expect the adaptive responses of each isolated population to depend on the fitness topography; if neutral network connectivity is high then all populations should ultimately converge on the same solution (the highest adaptive peak), but if valleys are present then some populations may become trapped on peaks with lower fitness. Another possibility is that the adaptive landscape may present multiple solutions that are equivalent (separate peaks of equal elevation), in which case the specific mechanism of adaptation may differ, but the observed variance in fitness among populations at the end of experiment would be low.

Evidence from experimental evolution trials suggest that adaptive landscape topography may depend on the type of environment that organisms are exposed to. Studies using a range of microbes, including viruses, and prokaryotic and eukaryotic unicellular organisms, have generally found large initial responses when fitness is low, then the rate of evolution tapers off and reaches a plateau after several hundred to a few thousand generations (Elena and Lenski 2003). During early stages of evolution, mutations of large effect are favored, but incremental improvements in fitness can continue even after the population has plateaued. With multiple populations, the individual replicates often attain different fitness levels and display morphological differences, which suggests the presence of multiple peaks at different fitness elevations in a rugged adaptive landscape. On the other hand, this outcome seems to depend on the type of environment. Most notably, populations of microbes exposed to a substrate commonly encountered in native environments (e.g., sucrose) often converge on the same fitness level and exhibit parallel mutations compared to the ancestral population, which suggests that all replicates come to reside on a single adaptive peak (Elena and Lenski 2003; Melnyk and Kassen 2011). In contrast, exposure to a substrate that the organism may not normally be exposed to (e.g., maltose, xylose, etc.) is more likely to result in a range of fitness outcomes among replicates. This observation suggests that the prevalence of neutral networks depends on the environment; high connectivity may characterize landscapes in environments for which there is ancestral experience, but connections may become sparse—leading to a rugged topography—when populations are forced to respond to novel environments.

Without access to more experimental analyses of fitness topographies for a wider range of organisms, it is difficult to assess the relevance of the rugged and holey landscape paradigms for understanding adaptive evolution. From experimental evolution trials with microbes, the presence of neutral networks is evident from the facility of adaptation in some environments, but it is important to recognize that this process it not necessarily easy. The paths that connect adaptive peaks may be more analogous to narrow twisting paths than they are to broad highways. Another way to view this process is as a multi-dimensional puzzle: mutations may need to arise in a particular sequence to maintain high fitness—like the movements required to solve a Rubik's cube. This view of a limited number of narrow ridges leading to a fitness optimum by traversing the edges of fitness holes helps explain parallel mutations appearing in replicates exposed to the same environment. Unlike mistaken movements on the Rubik's cube, mutations accumulating along neutral networks may be guided by selection; mutations that fall off the path are culled, and progress is made when the appropriate mutations appear, leaving only the proper sequence of genetic changes that ultimately leads to the puzzle's solution (the highest peak). At the end of the process, all replicates may have obtained the same phenotype through convergent evolution such that they all possess the gene complex that confers the highest fitness in the present environment.

The high levels of connectivity that allow for convergence to the highest adaptive peak may depend on the degree of ancestral exposure to environments that are similar to the current one. Long periods of evolution within a range of similar environments may generate a larger battery of gene products that can be brought to bear for adaptation. Under this view, evolutionary potential can be considered as the number of interacting gene regions contributing to adaptation to novel conditions; pair-wise interactions may provide paths for continued evolution by drift and selection resulting in stepwise substitutions—similar to the pair-wise locus process of peak transitions that Wright discussed. In novel environments, the number of interacting genes may be fewer, which would restrict evolution by genetic drift. Under such conditions, shifting balance evolution may prevail if the nearby peaks offering improved solutions to the current environment lack connecting ridges and are separated by deep valleys.

Speciation on Adaptive Landscapes—The separation of populations on different adaptive peaks or on opposite sides of a fitness hole indicates some degree of reproductive isolation and the potential for further divergence leading to speciation. The difficulty for both the rugged and holey adaptive landscape views of evolution is that species generally occupy different habitats, and these models do not explicitly provide mechanisms for transitions into alternative environments. To accommodate ecological transitions, Wright suggested that shifting balance processes could facilitate adaptation when many small populations colonized a novel habitat (Wright 1982). We can infer that change in the environment would be expressed as shifts in the positions of adaptive peaks, and consequently, populations colonizing alternative environments may initially occupy regions of low fitness on the landscape. With many colonization events, the shifting balance process could begin anew as the newly divergent lineages transition into a novel habitat. We could imagine this process occurring, for example, in cases of budding speciation where propagules from the primary range of a species repeatedly colonize a novel habitat at the range periphery.

The holey adaptive landscape model presents a different view of speciation, as it is based on the evolution of reproductive isolation through the accumulation of BDM (Bateson-Dobzhansky-Muller) incompatibilities (Gavrilets 2003). While the shifting balance model focused on the process of adaptation to a particular environment, the fitness holes in Gavrilets' model are probably independent of environmental variation; in the large majority of cases BDM incompatibilities have been found to be intrinsic and not dependent on ecological conditions (Baack et al. 2015; Chapter 10). Consequently, the fitness topography is not expected to change substantially in response to variation in the environment, and the holey adaptive landscape is predicted to be stable across habitats.

While the holey adaptive landscape does not seem to directly address adaptive divergence (Doebeli et al. 2005), it may be more suitable as a generalized model for the evolution of intrinsic reproductive barriers. In cases where isolation is established by ecological or geographic mechanisms, BDM incompatibilities will be free to accumulate and contribute to the development of strong post-zygotic reproductive isolation.

Thus, the holey landscape along with the BDMI model can be viewed as the best available summary of our understanding of the development of post-zygotic reproductive isolation—the final stage of the speciation process—in lineages that are initially separated by ecological or geographical mechanisms (see the earlier section on The Process of Divergence).

11.7 ECOLOGICAL SPECIATION

Whether speciation more commonly occurs through mutation/drift processes in geographically isolated populations or is usually driven by selection as divergent populations adapt to alternative environments continues to be an active topic of discussion (Gavrilets 2003; Doebeli et al. 2005; Baack et al. 2015). We recognize that selection can drive reproductive isolation due to ecological divergence at a relatively fast pace—perhaps within thirty to one hundred generations (Hendry et al. 2007)—but we also acknowledge that ecological barriers are often less permanent and more permeable than post-zygotic incompatibilities. On the other hand, the development of reproductive isolation based on BDM incompatibilities is much slower because it depends on the appearance of mutations and fixation by genetic drift (Seehausen et al. 2014). Given that divergence among lineages can proceed much faster when it is driven by selection, we might expect that ecological speciation would be prevalent. This may be particularly true for plants because their sedentary life forms render them susceptible to the influences of local climatic and edaphic conditions. The combination of limited dispersal and close environmental association amplify the potential for plants to undergo ecological divergence, which may ultimately result in speciation. In this section, we examine the processes of ecological diversification and speciation.

Ecotypes and Stages of Speciation—If we accept that the emergence of species is often driven by selection then it seems reasonable to assume that ecotypic variation represents the initial stage of divergence. Darwin (1859) viewed the partitioning of species into varieties as an initial stage that might lead to distinct species (Fig. 8.1; refer to color plate). He envisioned that varieties "will continue to be exposed to the same conditions which made their parents variable . . ." and he considered that the basis for differences among varieties was heritable (Darwin 1859, page 109). The idea that intraspecific variation was the first stage of speciation was expanded on by Wallace soon after the publication of Darwin's *Origin of Species* (Wallace 1865). Wallace enumerated varieties within species as initial stages of divergence (Lowry 2012), but he did not describe the butterflies he observed in terms of their adaptation to specific habitats. In the late nineteenth century, the views of Darwin and Wallace—that differences among varieties within species were heritable and had the potential to initiate new species—were in the minority; most naturalists assumed that such small differences among forms within species were due to plasticity of the phenotype and were easily reversible by exposure to alternate environments (Chapter 8).

Through the early twentieth century, many biologists continued to hold a somewhat typological view of species. Variations within species were generally assumed to be environmentally induced and not of any particular evolutionary importance. This situation changed when first Turesson (1922) and later Clausen, Keck, and Heisey (1940) provided experimental evidence that minor morphological differences in vegetative characteristics among populations of the same species growing in different habitats (ecotypes) were genetically determined (Chapter 8). Clausen (1962) elaborated on Turesson's ideas of ecotypes as dynamic entities that may represent early stages of divergence and had the potential to ultimately become new species.

During the mid-twentieth century, there was much interest in ecotypic variation and the development of reproductive isolation—a time period described as a "second wave" of interest in ecological speciation (Lowry 2012). The early twenty-first century may represent a "third wave" of interest in the stages of speciation. New approaches have provided a more nuanced view of the process of ecological speciation: genes contributing to adaptation may respond to selection even as other genomic regions are affected primarily by gene flow (Nosil et al. 2009).

Within the ecological speciation paradigm, the ecotype appears as an early stage in the natural progression of divergence among lineages adapted to different habitats (Fig. 11.14). Much of the controversy over the ecotype concept during the last century centered on whether intraspecific variants could be categorized as discrete entities associated with specific habitats,

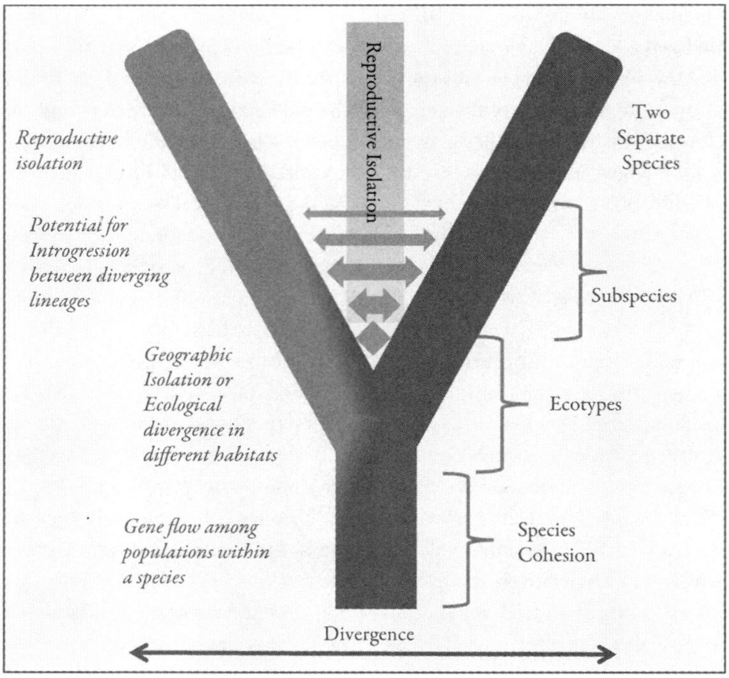

Fig. 11.14. The stages in the evolution of plant species from ecotypic variation.

or whether these classifications were artificial, as they represented the ends of a spectrum of divergent populations arrayed along an environmental gradient (Clausen 1962; Lowry 2012). For ecotypes to operate as early stages of speciation, they need to be isolated to a degree that continued divergence is possible. For plants, adaptation to local conditions is often facilitated by limited dispersal ability—even a relatively small physical separation may be adequate to allow populations to respond to local selection. Gene flow may be restricted further in cases where ecological divergence is substantial enough to reduce the viability of immigrants and hybrids (e.g., in the case of adaptation to contrasting soil conditions, elevation, or coastal/inland habitats; Rajakaruna 2004; Lowry et al. 2008; Givnish 2010; Halbritter et al. 2015). While reproductive isolation due to habitat associations can be quite strong, this stage of divergence has generally been considered to be reversible (Clausen 1962; Grant 1981; Lowry 2012; Seehausen et al. 2014). Even for cases of strong ecological isolation, many authors do not consider speciation complete until strong intrinsic post-zygotic isolation is established.

Ecotypic variation appears to be quite common, but the large majority of ecological variants probably never become established as independent lineages. Given the high level of variation that exists within many widespread species, it has generally been recognized that most of these ecotypes are ephemeral: they either become extinct through population decline or are assimilated through interbreeding as climatic conditions change over longer periods of time (Clausen 1962; Lowry 2012). This scenario is consistent with a more general pattern over phylogenetic history that includes large numbers of ephemeral species and high rates of species turnover (Rosenblum et al. 2012).

When ecological divergence occurs in habitats that are in close proximity, it has been suggested that a low rate of genetic exchange may occur even as divergence continues (Nosil et al. 2009). This model of "speciation with gene flow" has become synonymous with cases of sympatric, budding, or parapatric speciation, where some level of gene flow is likely to occur between populations in contrasting habitats. In this model, a small number of genomic regions diverge in response to selection for local adaptation, while the majority of the genome is subject to neutral genetic processes. On the other hand, genomic regions responsible for intrinsic post-zygotic isolation (such as BDM incompatibilities) are unlikely to coincide with those responsible for ecological adaptation. Consequently, even low levels of gene flow may be sufficient to overcome genetic drift for genomic regions that have the potential to contribute to genetic incompatibilities. In other words, "speciation with gene flow" may generally not be possible because the final step in the process—the establishment of strong intrinsic genetic incompatibilities—is opposed by gene flow among populations. The contributions of ecological divergence, gene flow, and the accumulation of genetic incompatibilities through mutation/drift processes to speciation in plants continue to be active areas of research (Clausen 1962; Grant 1981; Nosil et al. 2009; Lowry 2012; Seehausen et al. 2014)

A possible example of divergence with gene flow is provided by coastal and inland ecotypes of *Erythranthe guttata* (*Mimulus guttatus*) that occur in close proximity along the Oregon coast of North America (Lowry et al. 2008). These populations display substantial differences in ecological associations without the development of intrinsic reproductive isolation. Seeds and hybrids transplanted between habitats have low rates of survival, but intrinsic post-zygotic isolation has remained weak. Given the relatively close proximity of these ecotypes, adaptation to local conditions may have occurred in opposition to low rates of gene flow. Alternatively, their origin may have been recent enough that not enough time has passed to allow for the accumulation of post-zygotic incompatibilities. While such cases of parapatric and peripatric (budding) ecological divergence may be common in plants, divergent lineages such as the case of ecotypes of *Erythranthe* may have limited potential for continued diversification, unless barriers to gene flow are strong enough to allow the accumulation of genetic incompatibilities through mutation/drift processes.

To this point, we have discussed criteria for defining species and the processes that lead to the origin of new species. Over longer time frames, we expect speciation to be ongoing as novel habitats are exploited and as declining and extinct species are replaced by others. In the remainder of this chapter, we focus on processes and patterns of diversification, including factors affecting the potential for speciation, adaptive radiations, and patterns of evolution on islands. To start this discussion, we begin with an overview of methods to infer the evolutionary history of lineages—the field of phylogenetics.

11.8 QUANTIFYING DIVERSIFICATION

As we observe various types of organisms, it quickly becomes apparent that they can be grouped based on shared characteristics. At the coarsest level, we can easily make separations among major groups of eukaryotes such as plants, fungi, and different phyla of animals (e.g., mammals, birds, insects, etc.). But closer observation within these groups reveals more subtle similarities and differences that provide the basis for categories at finer scales. The notion that the order and classification of biological diversity can be inferred from such phenotypic similarities goes back to some of the earliest naturalists. In the eighteenth century, Linnaeus established a hierarchical system of classification that was based on shared traits and Buffon suggested that organisms had a reproductive history that defined their relationships to each other. The first visual depiction of a "tree" to represent relatedness among species may have been from Lamarck's *Philosophie Zoologique* (1809), but the idea of a "tree of life" dates to the medieval ages. Darwin first expressed a visual representation of the history and relatedness among species when he drew a branching "tree" as part of the musings in his notebook (Fig. 5.2). Apparently, Darwin felt that the visualization of evolutionary history as a tree diagram was important enough to be included as the only figure in *On the Origin of Species* (Darwin 1859; Fig. 8.1; refer to color plate).

While the depiction of relationships among species as trees persisted as a useful metaphor after Darwin, efforts to quantify degrees of similarity and relatedness that allowed the estimation of tree-like depictions of diversification did not begin until the middle of the twentieth century. *Phenetics* (*Numerical Taxonomy*), developed primarily by Robert Sokal and Peter Sneath (Sokal and Sneath 1963) in the late 1950s, brought the first effort to summarize similarities among species based on phenotypic traits as a *Dendrogram*. The idea that similarities among species could be quantified and then represented in a tree-like diagram apparently came about as a bet over a six-pack of beer among colleagues at the University of Kansas (R.R. Sokal, personal communication). The phenetic classification methods that these authors developed far exceeded the initial challenge they faced. In the late 1990s, phenetic methods were developed that often used clustering methods such as *Neighbor Joining* to assemble dendrograms that represent phenotypic similarity (Fig. 11.15). The process of neighbor joining starts by combining the two most similar taxa into a single node. Joined taxa are treated as a single unit in subsequent rounds as the most similar taxa are joined in each pass until all taxa have been included in the tree. Phenotypic similarity is represented in the resulting dendrogram as distance along branches.

Since phenetics utilizes overall phenotypic similarity without respect to the distribution of individual traits among species and groups, it does not always provide a representation of evolutionary relationships. For example, since all traits are treated equally, phenetics would not be able to identify traits that were due to convergent evolution. Moreover, phenetics does not distinguish between ancestral (*Plesiomorphic*), shared-derived (synapomorphic), and lineage-unique (apomorphic; Fig 11.5; refer to color plate) traits, so the branch lengths and the *Topology* of the resulting tree do not reflect evolutionary history. For example, a phenetic dendrogram of land plants would likely place all the bryophytes in monophyletic group with short branches, rather than long branches forming a grade of divergence as depicted in Fig. 2.1 (refer to color plate). Because of this limitation, many have considered phenetic approaches to have limited utility for understanding evolutionary relationships among organisms.

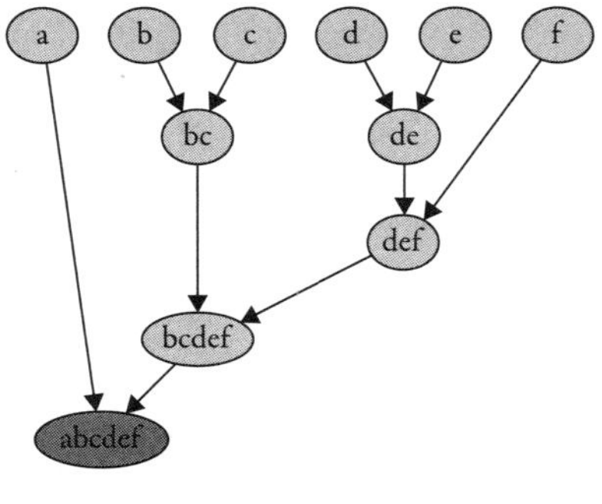

Fig. 11.15. An example of clustering using neighbor joining based on similarity. A distances matrix is calculated based on shared traits. In each round one pair of taxa or nodes (combined taxa) are joined until all taxa are included in the tree. Distances along branches represent phenotypic similarity.

The origin of modern phylogenetics and *Cladistics* is generally attributed to the German entomologist, Willi Hennig (1966). The methods proposed by Hennig differed from phenetics, because his aim was to use the distribution of traits among taxa to infer evolutionary history rather than to make taxonomic classifications. The cladistics method is applied by grouping species into clades based on synapomorphic traits. The resulting tree provides a depiction of the relative timing of divergence. Branch lengths are determined by numbers of trait differences among taxa and can be used to estimate phylogenetic distance among taxa. The cladistic methodology developed by Hennig and others in the late twentieth century provided a foundation for phylogenetic inference and the establishment of molecular phylogenetics.

When cladistics first become popular during the late twentieth century, only phenotypic traits were available for determining evolutionary relationships and phylogenies were inferred using *Parsimony*. The use of phenotypic variation for cladistics presents a number of difficulties, mostly because the availability of traits is limited and researchers were often required to choose traits that are "phylogenetically informative" over others to obtain well-resolved phylogenies. The resulting phylogenies may have been somewhat subjective, as they relied heavily on the taxonomic knowledge of researchers.

The reliability of cladistics methods and the robustness of the phylogenies produced improved dramatically with the availability of large amounts of DNA sequence data. While parsimony is still used, it has been largely replaced by *Maximum Likelihood* and *Bayesian* estimation methods (Box 11.1). These methods are particularly appropriate for sequence data, since different evolutionary models can be applied to account for a variety of circumstances, including differences between transition and transversion rates, heterogeneity of substitution rates

Box 11.1 Computational Phylogenetics

Cladistics as developed by Hennig (1966) and his followers conducts classification based on shared-derived traits (synapomorphies). This approach differs from phenetics in that the primary goal is to uncover evolutionary relationships rather than to group taxa based on phenotypic similarity. The algorithms used to estimate trees have changed as new types of data have become available.

Maximum Parsimony attempts to minimize the degree of homoplasy by finding the tree with the fewest number of trait changes occurring along branches. This method was originally used to estimate phylogenies based on phenotypic traits. The goal of this approach is to find the shortest tree that explains evolutionary relationships for a group of taxa. With small groups, an exhaustive search of all possible trees is conducted, but the number of trees that need to be evaluated increases exponentially, so algorithms have been developed to increase the efficiency of the search for the shortest tree when larger numbers of taxa are being evaluated. Any type of heritable phenotypic trait could be used for parsimony analysis, but all traits need to be classified into categories. This requirement can create some problems for many continuous traits that do not display sharp disjunctions among taxa.

Maximum Likelihood applies a model of evolution to find the probability distribution that best fits the observed data. This method is applied by finding the maximum of the likelihood function for the particular model being evaluated to identify the single best tree. Maximum likelihood became popular as an alternative to parsimony during the 1990s, because different evolutionary models could be applied to the analysis of DNA sequence variation. Tree estimation using maximum likelihood and parsimony is commonly conducted using the computer programs *PAUP* (Phylogenetic Analysis Using Parsimony, and other methods) and *MacClade* (Hall 2001).

Bayesian Inference implements a model of evolution based on a set of probabilities called *Priors* to find the set of trees that best explains the observed data based on their *Posterior Probabilities*. The primary advantage of Bayesian inference over maximum likelihood and parsimony is the ability to consider phylogenetic uncertainty—each tree generated has an associated probability for its fit to the observed data. Bayesian inference is implemented using *Monte Carlo Markov Chains (MCMC)*. Markov chains consist of single step advances with each step independent of previous steps taken. This method is implemented using a random function to identify the next possible step, calculating the probability of this step, and comparing the new state to other randomly generated states. If the probability if the proposed step is lower than the majority of randomly generated states, than it is accepted as part of the chain, and the process begins anew. Posterior probabilities are calculated for each possible tree based on the number of times they are visited during the meandering of the Markov chains.

Consensus Trees are estimated in maximum likelihood estimation by randomly subsampling the data (*Bootstrapping*) many thousands of times or by comparing the most probable trees generated by Bayesian inference. Under *Strict Consensus*, only the nodes appearing in all trees are retained, and all other nodes are collapsed into polytomies. More often, *Majority Rule Consensus* is applied by retaining nodes that appear in at least a certain fraction of the trees (usually 50%). Using maximum likelihood, support for each dichotomously branching node is provided by the proportion of bootstrap trees generated. Alternatively, the proportion of the most probable trees where the node appeared is determined from Bayesian inference. These are generally represented as a number between 50 and 100 (percent) appearing next to each node.

across sampled genomic regions, and synonymous and non-synonymous base changes within reading frames (Kelchner and Thomas 2007). Unlike phenotypic traits, all of the available base substitutions characterizing each taxon are generally used for inferring phylogenies. With large amounts of data, we expect a proportion to reflect convergent evolution resulting in homoplasy, but the majority of data usually provides a clear phylogenetic signal. Support for the resulting tree and individual branches are provided by bootstrap values (maximum likelihood) and posterior probabilities (Bayesian estimation) that reflect the level of consistency in the sequence data (Box 11.1).

Phylogenetic estimation methods will return a tree that best explains the data, but this representation may not correspond to the "true" evolutionary tree. Consequently, cladograms estimated from different gene regions represent hypotheses for the evolutionary relationships of the taxa included in the analysis. It is generally assumed that variation in lineage sorting will lead to alternative topologies, so phylogenies from several regions are often combined to develop more robust consensus trees (Wiley and Lieberman 2011; Yang and Rannala 2012).

Phylogenies for relationships among species differ from intraspecific network phylogenies (Chapter 8; Fig. 8.11) because the internal nodes of species phylogenies are hypothetical ancestors that have been lost through *Cladogenesis* and *Anagenesis* (phyletic evolution). Phylogenies may be estimated for different evolutionary scales, including relationships among species and subspecies within a genus, among genera in a family, among families in an order, and at higher levels of classification. In these phylogenies, tree topologies and branch lengths represent the degree of relatedness among taxa. To obtain estimates of divergence times in years, the phylogeny can be calibrated using fossils that are likely to correspond to nodes or using estimates of numbers of mutations per year (assuming a *Molecular Clock*) for the genomic region used to construct the phylogeny (Soltis et al. 2005). As discussed in Chapter 2, estimates of divergence times from molecular clocks are often much older than those determined from fossil evidence.

Phylogenies represent the foundation of our understanding of patterns of diversification and evolutionary relationships across the vast diversity of organisms inhabiting our planet. Phylogenetics and systematics represent major areas of investigation (Soltis et al. 2005; Wiley and Lieberman 2011; Yang and Rannala 2012), and the resulting trees are widely used by evolutionary biologists to conduct investigations within a *Phylogenetic Comparative Framework* (see exmples later in this chapter). These investigations allow us to ask questions concerning the patterns and rate of diversification, to examine the distribution and association of traits across phylogenies (Chapter 5), and to conduct studies to assess the effects of phenotypic traits or environments on patterns of diversification. In the remainder of this chapter, we apply the phylogenetic comparative framework to examine associations between life history and mating system traits with rates of diversification, patterns of diversification after long-distance dispersal, and patterns of evolution on oceanic islands.

11.9 PATTERNS OF DIVERSIFICATION

Excepting changes in ploidy, the process of diversification generally requires geographical separation, which may be coincidental with changes in environmental conditions and biotic interactions. Such cases of budding and allopatric speciation are considered quite common. In this section, we explore patterns of lineage divergence in the context of ecological opportunities and challenges afforded by the availability of habitats and resources. By examining contemporary geographic distributions of closely related species and patterns of vegetative and floral trait divergence, we may glean some insights into the roles of geographic isolation, habitat divergence, mating systems, and pollinator-mediated selection for lineage diversification and speciation.

Habitat Divergence—Given that plants are particularly sensitive to environmental variation, we might expect ecological divergence and speciation driven by selection in response to environmental variation to be common. One approach to assessing the prevalence of speciation through selection in alternative habitats is to examine levels of ecological diversification among closely related species. If selection resulting from changes in ecological associations is the primary driver of plant speciation, then we expect to see significant shifts in habitats or pollinator associations between pairs of *Sister Species* (two lineages formed from a single ancestral species; e.g., taxa B and C in Fig. 11.5; refer to color plate). It is important for applying this approach to have access to well-resolved phylogenies to ensure that the most recently derived pairs of species are compared.

If changes in ecological associations are primarily responsible for divergence, then we expect most sister pairs of species to differ in habitat traits; however, comparison of species across phylogenies often reveals evidence of *Phylogenetic Niche Conservatism*—the tendency for closely related species to display similar ecological associations (Wiens et al. 2010). If niche conservatism is strong, then phylogenetic inertia for morphological and physiological traits (i.e., due to evolutionary constraints; Chapter 6) may constrain lineages from making large ecological transitions, which would limit the potential for habitat shifts that might promote speciation. Consequently, we expect that the prevalence of ecological speciation will be the outcome of conflicts between two opposing forces: selection for ecological diversification on the one hand and constraints imposed by phylogenetic inertia on the other.

To make robust assessments of the prevalence of habitat transitions during speciation, it would be best to examine sister species pairs in diverse floras. Relatively comprehensive analyses of the prevalence of ecological speciation using phylogenetic comparisons are available for the Cape Floristic Region (FR) of South Africa (van der Niet and Johnson 2009) and for the California Floristic Province (Region) in North America (Anacker and Strauss 2014). These regions represent two of the most diverse floras in the world. Both have high levels of endemism and are characterized by edaphic and topographic complexity as well as high frequencies of fire. Ecological shifts between sister species are common in both floristic regions; a large majority of sister species pairs display some degree of ecological contrast. Ecological differences between sister species are lower in magnitude than comparisons among other species within the same genus, suggesting that niche conservatism may have limited the magnitude of recent ecological transitions. Some plant characteristics are apparently more constrained than others, as transitions in life history and growth form, flowering time, and altitude were less common in both regions. Changes in pollinator associations were more common in the Cape FR, while edaphic shifts were prevalent in the California FR, which may reflect a higher level of soil type heterogeneity in the latter. Some of the differences between these two studies may reflect differences in focus, as van der Niet and Johnson (2009) were very interested in differences in pollination syndromes and may not have divided soil types as finely as Anacker and Strauss (2014). In the latter study, pollinator associations were not explicitly considered, but changes in flower size occurred in some genera (e.g., *Erythranthe*), which usually reflect a transition toward smaller flowers and a selfing mating system (Chapter 9).

The geographic distributions and ecological associations of sister pairs in the Cape and California FRs provide insights into speciation modes. In both regions, the levels of range overlap between pairs of sister species is high but is somewhat less pronounced in the Cape FR (van der Niet and Johnson 2009; Anacker and Strauss 2014). We might interpret the high levels of sympatry to suggest that parapatric or budding speciation are more common than allopatric speciation. Indeed, Anacker and Strauss (2014) point out that the degree of overlap and the level of range asymmetry (i.e., the range of one species is much larger than the other) for these sister pairs of plant species in the California FR is generally much greater than for several groups of animals. They further suggest that the sedentary growth forms and restricted dispersal of plants facilitate budding speciation through habitat transitions. While budding speciation may be common for plants, it is important to recognize that the interpretation of range overlap largely depends on the geographic resolution of the analysis; plant species may be completely isolated in different habitats even though they grow in relatively close proximity to each other. It is also likely that the ranges of these species have changed since their divergence. On the other hand, the level of range overlap and asymmetry both decline with time since the time of divergence for sister species pairs in the California FR (Anacker and Strauss 2014). Assuming that geographic distributions have been relatively stable, this pattern is consistent with budding speciation; under this model, newly derived species would have high asymmetry and range overlap, and distributions would become more symmetrical as species continue to diverge.

Mating Systems and Diversification—One of the most evident evolutionary patterns across plant families is the frequent transition between selfing and outcrossing modes of reproduction (e.g., Figs. 11.16 and 11.17; refer to color plate). Across all plant families, species that predominantly self are less common (10% to 15%) than those that outcross or have mixed mating (i.e., with outcrossing frequencies between 0.2 and 0.8; Goodwillie et al. 2005). Selfing is particularly common in annuals and herbaceous

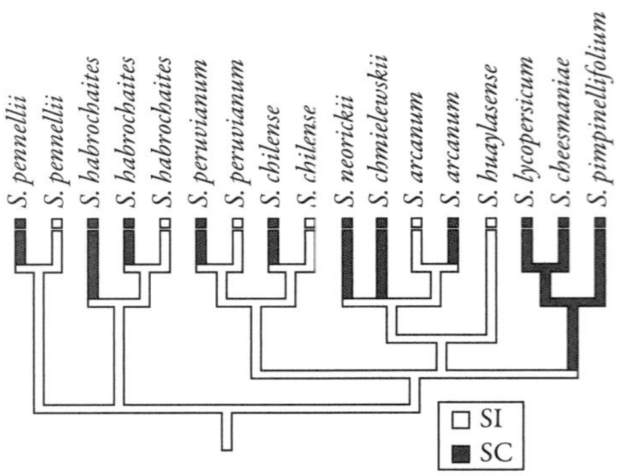

Fig. 11.16. Phylogenetic relationships among self-incompatible (SI—open bars) and self-compatible (SC—solid bars) taxa in *Solanum* sect. *Lycopersicon*. From Igic et al. (2008).

perennials that occupy ruderal and early successional communities and in species that colonize disturbed sites (Baker 1965; Cheptou 2012).

There are a number of advantages for self-fertilization (Chapter 9). First, plants capable of selfing have reproductive assurance when animal pollinators are unreliable, or after colonization of new sites. Moreover, self-fertilizing plants have a 50% transmission advantage for their gametes compared to outcrossers (Fisher 1941; Lloyd 1979; Lande and Schemske 1985). Under random mating we expect selfers to transmit two copies of their genome through their seeds plus one copy by fertilization of seeds on neighboring plants, while outcrossers would, on the average, transmit only two copies (one through their seeds and one through pollen; Chapter 9). Based on this transmission advantage (cost of outcrossing), we expect mutations elevating the propensity for selfing to increase in frequency in outcrossing populations. On the other hand, there can be severe disadvantages to selfing; most notably, outcrossing populations typically segregate for large numbers of recessive deleterious mutations that contribute to genetic load (Chapter 5). Offspring resulting from selfing are expected to have lower fitness (inbreeding depression; Chapters 4 and 5) compared to outbred individuals; increased homozygosity for inbred offspring results in larger numbers of loci that are homozygous for deleterious recessive alleles. It is generally thought that if genetic load is reduced by purging deleterious recessive alleles from populations then the fitness penalty for inbreeding will be alleviated, and increased selfing will be favored (Barrett 2017).

Evolutionary transitions from self-incompatibility (SI) to predominant selfing modes of reproduction are fairly common across herbaceous plants, but well-documented instances of the acquisition of physiological self-incompatibility appear to be rare or non-existent (for a possible exception see Chantha et al. 2013). Stebbins (1957) was the first to suggest that self-incompatibility was complex enough that it would be difficult to regain once it was lost. This proposal is supported by the observation of high frequencies of dioecious species that are descended from self-compatible colonizers of oceanic islands (see the section on Evolution on Islands later in this chapter). He further hypothesized that selfing lineages are most often short-lived evolutionary "dead ends" because they are effectively stuck with floral morphologies and physiologies that promote self-fertilization (Figs. 11.16 and 11.17; refer to color plate). Moreover, he predicted that selfing taxa lacked the genetic variation necessary to respond to environmental challenges, so they would be less likely to adapt to changing conditions. It is impressive that Stebbins made these inferences based on the existing theory of his day and without the advantages of more advanced surveys of genetic variation and evolutionary relationships among plant taxa based on DNA sequence data. As more information has been gathered through refined genetic methods, Stebbins' original predictions have been largely confirmed (Igic et al. 2008; Wright et al. 2013).

Even though Stebbins' intuition about the potential for lineages to regain self-incompatibility was generally correct, the reason that lineages might become "stuck" in the selfing mode was not fully appreciated until some thirty years later. In 1985, Lande and Schemske published a pair of papers that evaluated the theoretical conditions that promoted the evolution of selfing and supported the predictions of their model with a review of the literature on estimates of outcrossing frequencies (Lande and Schemske 1985; Schemske and Lande 1985). These authors suggested that the 50% transmission advantage for selfing creates a threshold for the effects of genetic load; if the fitness of selfed progeny is greater than half that of outcrossed progeny, then selfing will be more advantageous, and mutations that increased levels of self-fertilization would be expected to increase in

Fig. 11.17. Comparison of flower morphologies for sister species pairs where one reproduces primarily through outcrossing (left) and the other primarily through selfing. Photos by Dianne Fristrom, Walter Siegmund, Christophermluna, John Willis, Fritzflohrreynolds, and Alberto Salguero Quiles.

frequency (Chapter 9). They predicted that elevated levels of inbreeding would lead to further purging of genetic load and would result in evolution toward complete selfing. Based on the effects of genetic load and inbreeding, they predicted that either plants should be completely outcrossing or completely selfing, and any species with intermediate outcrossing frequencies are probably in transition. The results of their literature survey mostly supported this prediction, but more plants with mixed-mating frequencies were observed than expected (see Chapter 9 for a more thorough discussion).

While there is ongoing discussion about the evolution of intermediate outcrossing frequencies, the irreversibility of the loss of SI generally holds true; the reason that it is difficult to evolve toward higher levels of outcrossing is that ongoing inbreeding will continue to purge deleterious mutations as rapidly as they appear (Chapter 3). With high levels of selfing, we expect that it would be nearly impossible to accumulate enough genetic load to overcome the cost of outcrossing. Notable exceptions to this "rule" have occurred on oceanic islands where outcrossing lineages have been derived from selfing ancestors that were the original colonizers (see Evolution on Islands later in this chapter).

Stebbins (1957) may also have been mistaken concerning his expectation that selfing lineages would have low speciation potential. While it is generally true that selfing lineages have lower rates of diversification (Igic et al. 2008; Wright et al. 2013), this trend appears to be due to higher rates of extinction (μ_C), rather than lower rates of speciation (λ_C; Fig. 11.18; refer to color plate). In a survey of species of *Solanum*, Goldberg et al. (2010) found that rates of speciation were actually higher for selfers than for outcrossers, but this is more than offset by the loss of selfing lineages, so overall diversification remains greater for outcrossing species. The model used by Goldberg et al. included only unidirectional transitions from outcrossing-to-selfing lineages (Fig. 11.18C; refer to color plate), so they predicted that the number of outcrossing species would decline over time unless their loss was offset by higher rates of diversification (r_I). This appears to be generally true, but we should note outcrossing lineages would be lost only if evolution of selfing was primarily phyletic (along branches); if the evolution of selfing coincided with dichotomous branching, then the original outcrossing lineage would be preserved. Based on Goldberg et al. (2010), outcrossing lineages continue to diversify at a higher rate; while selfing lineages are more prolific, they are short-lived and occur at lower frequencies because of their higher rates of extinction.

The observation that selfing lineages have higher speciation rates than outcrossers may be explained by the implications of the transition to reproduction primarily through self-fertilization. An increases in selfing frequency has unique evolutionary consequences because it elevates the level of reproductive isolation (Chapter 10) and reduces the potential for gene flow to overcome the effects of selection and genetic drift in local populations (Goldberg et al. 2010; Wright et al. 2013). In general, we expect that the effective population size and levels of gene flow among populations will be negatively correlated with the frequency of self-fertilization—selfing effectively reduces levels of genetic exchange. Since selfing populations are effectively more isolated, there may be increased opportunities for divergence through genetic drift and local selection. Divergence of primarily selfing populations may be facilitated by low frequencies of outcrossing among selfing lineages, which allows for selection on new recombinant genotypes (Chapter 9). The contrast in speciation rates between selfing and outcrossing lineages of *Solanum* (Fig. 11.18; refer to color plate) suggests that isolation from gene flow may be the most critical population genetic process affecting rates of diversification.

Pollinator-Mediated Speciation—The diversity of pollination syndromes (Fig. 10.6; refer to color plate) is striking and suggests that changes in pollinator associations must be important for the promotion of diversification. This is a popular topic of discussion; based on the high diversity of pollination syndromes in some genera, some authors have suggested that pollinators are one of the primary selective forces driving ecological speciation (e.g., Sargent 2004; Givnish 2010; Van der Niet et al. 2014; Lagomarsino et al. 2017). While it is clear that transitions to alternative pollinators are common, it is unclear whether disruptive selection can be strong enough to overcome the effects of gene flow or if changes in pollination syndromes are more likely in geographically isolated populations. We can gain some insights into the prevalence of pollinator-mediated selection by examining the distribution of floral characters and habitat associations among recently derived species.

The studies of patterns of divergence in sister pairs of species discussed above provides some valuable information on the frequency of pollinator transitions (van der Niet and Johnson 2009; Lagomarsino et al. 2017). Floral diversity is high for the California FR, but it is particularly impressive in the Cape FR. If selection for association with alternative pollinators is a common mechanism for the evolution of reproductive isolation, then we expect floral characters to be more divergent for sympatric portions of the ranges of closely related species (*Reproductive Character Displacement*). In both regions, the degree of sympatry between recently diverged pairs of species is substantial, so we might expect divergence in floral traits in regions of sympatry to be common as selection favors intraspecific mating when hybrid fitness is low (i.e. selection for *Reinforcement* of reproductive barriers; Chapter 10). This pattern would be expected if selection for divergence in pollinator associations during secondary contact contributes to the strengthening of reproductive isolation. In contrast to this expectation, Anacker and Strauss (2014) found no evidence for reproductive character displacement in regions of sympatry between sister species in the California FR. There is some evidence of differences in flowering time between sympatric sister species in the Cape FR, but this pattern is limited to species in the

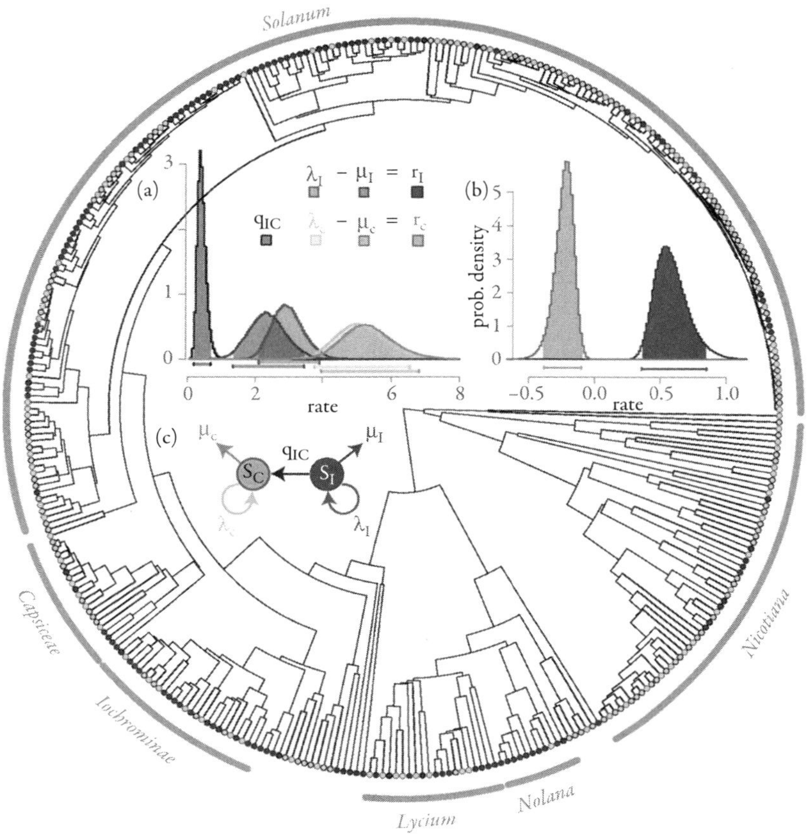

Fig. 11.18. Phylogenetic relationships among 356 species in the genus *Solanum*. Species are indicated as self-incompatible (SI; purple) or self-compatible (SC; turquoise) around the perimeter of the circle. A) Rates of speciation, extinction and transition from SI to SC species (q_{IC}; grey shading). SI rates of speciation (λ_I) and extinction (μ_I) are indicated by red and purple shading, respectively. SC rates of speciation (λ_C) and extinction (μ_C) are indicated by yellow and blue shading, respectively. B) Rates of diversification for SI (r_I) and SC (r_C) lineages are indicated by purple, and turquoise shading, respectively. C) The model of speciation, extinction, and transition used to estimate rates of diversification. Used by permission Goldberg et al. (2010).

Restionaceae, which are wind pollinated. For animal pollinated species in these two highly diverse regions, it appears that reinforcement does not contribute to the evolution of reproductive isolation between recently diverged species. It is possible that spatial separation due to habitat differences and post-zygotic isolation between sister species is typically strong enough in these regions to limit hybridization.

Patterns of diversification in the Cape FR provides some clarification of the role that pollinators are likely to have played in the process of speciation. For example, the large majority of sister species with differences in pollinator associations also display contrasting vegetative characteristics, which suggests that they are adapted to different habitats (van der Niet and Johnson 2009). This observation is consistent with a transition to alternative pollination syndromes after populations have become isolated in regions where their normal pollinators were unreliable or not present (most effective pollinator principle; Chapter 10). In contrast, cases of disruptive selection driving divergence in pollination syndromes in sympatry are probably rare, because selection is generally not strong enough to overcome genetic swamping by gene flow (e.g., such as in the case of floral tube length variation in a species of iris; Rymer et al. 2010). With the possible exception of a few highly specialized pollination syndromes (e.g., yucca moths or euglossid bees; Chapter 10), we expect reproductive isolation based on pollinator associations to be permeable and susceptible to reversal. Consequently, divergence in pollination syndromes should generally be considered as a symptom of plant diversification in conjunction with habitat or geographic isolation, since cases of pollinator-driven speciation are probably rare. The challenge for plant evolutionary biologists in the future is to develop a more thorough understanding of the conditions that favor divergence in pollinator associations and to elucidate how changes in pollination syndromes contribute to plant diversification.

11.10 RATES OF DIVERSIFICATION AND EXTINCTION

It has been recognized for some time that species are distributed unevenly among taxonomic groups with very few that display high diversity and the majority with one or a few taxa (Willis 1922). The number of species in a clade (e.g., genus, subfamily, or family) is a product of the counteracting effects of diversification and extinction; high rates of diversification combined with reduced extinction can lead to radiations, and conversely, slow diversification combined with higher extinction rates results in depauperate clades (*Depauperons*) or extinction (Donoghue and Sanderson 2015; Fig. 11.19). Most plant groups have an abundance of depauperons and very few highly diverse genera. As examples, the legume family (Fabaceae) contains around 29% monotypic genera (i.e., genera with a single species) and 10% diverse genera; the composites (Asteraceae) are 33% monotypic and 6% diverse; and the orchids (Orchidaceae) are 24% monotypic and 6% diverse (Scotland and Sanderson 2004). The low frequency of highly diverse clades is not an outcome of random processes, but probably reflects the acquisition of novel traits (key innovations), or the chance encounter of a range of available habitats with available resources (e.g., on oceanic islands). Both key innovations and underutilized resources can promote *Adaptive Radiations* that produce high-diversity clades.

Adaptive radiations are generally considered to result from the origin of multiple independent lineages over a short period of time (Schluter 2001), but the time frame of diversification may vary. Clades resulting from adaptive radiations stand out, because a large amount of diversification originated from a single common ancestor and the resulting species display specialized growth forms or ecological associations. The traditional view is that adaptive radiations can be promoted under circumstances where resources have been made available either 1) as a result of dispersal to oceanic islands or other regions characterized by depauperate floras (*Habitat Radiations*) or 2) by the appearance of key innovations that enable novel modes of resource utilization. In both cases diversification ensues as selection favors more efficient utilization of available resources through the changes in growth form (e.g., perennation and increased in stature on islands) or the colonization of underutilized habitats. An example of radiations after acquisition of a key innovation includes the development of defense traits or characteristics that strengthen associations with symbionts

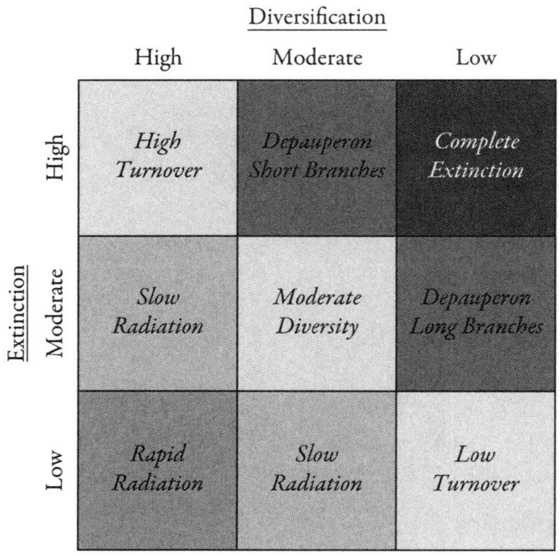

Fig. 11.19. Consequences of rates of diversification and extinction for clade dynamics and levels of diversity. Low diversity clades (depauperons) can result from a combination of moderate diversification and high extinction (short branch lengths) or low rates of diversification and moderate rates of extinction (long branch lengths). Moderate levels of diversity can result from low, high, or moderate rates of species turnover due to similar rates of speciation and extinction. Slow radiations (high diversity) can occur under moderate diversification if extinction rates are low, or high diversification if extinction rates are moderate. High rates of diversification combined with low extinction produces a high-diversity clade resulting from a rapid radiation.

(e.g., ants), which protect plants from herbivores (Weber and Agrawal 2014). Similarly, traits that promote associations with pollen and seed dispersers can facilitate reproductive isolation, which may lead to higher rates of diversification (Lengyel et al. 2009).

The level of species diversity within a clade can be viewed as an outcome of the combined rates of diversification and extinction (Fig. 11.19). High rates of diversification can overcome low to moderate extinction rates to generate high-diversity clades, but the radiation will be characteristically different depending on the relative rates of speciation and extinction. The traditional view is that a rapid radiation can occur when speciation rates are high and extinction rates are low resulting in a clade dominated by short branch lengths (Fig. 11.20; refer to color plate, upper-right phylogeny). This pattern may be most likely to occur when the diversification rate increases in response to the acquisition of one or more key innovations or to an increase in ecological opportunities after long-distance dispersal or climate change. Alternatively, radiations that occur more slowly with moderate diversification rates and low extinction rates are characterized by longer branch lengths (Fig. 11.20; refer to color plate, upper-left). Both of these radiation scenarios rely on low rates of extinction. Moderate rates of extinction combined with low rates of diversification will generally result in depauperate clades (lower panel). Under moderate extinction rates, low diversification results in a depauperon that could be maintained over a long period of time but may be in constant peril of complete extinction. Alternatively, high diversification rates may decline over time, leading to the gradual loss of lineages and a depauperon that has its origin from an older radiation (Fig. 11.20; refer to color plate, lower-right). This pattern fits many extant gymnosperm, lycophyte, pteridophyte, and bryophyte lineages that represent remnants left over from ancient radiations (e.g., *Welwitschia, Gnetum, Isoetes, Selaginella, Equisetum,* hornworts, and others; Chapter 2).

The circumstances responsible for the generation of a minority of species-rich genera within families are not well understood. Radiations caused

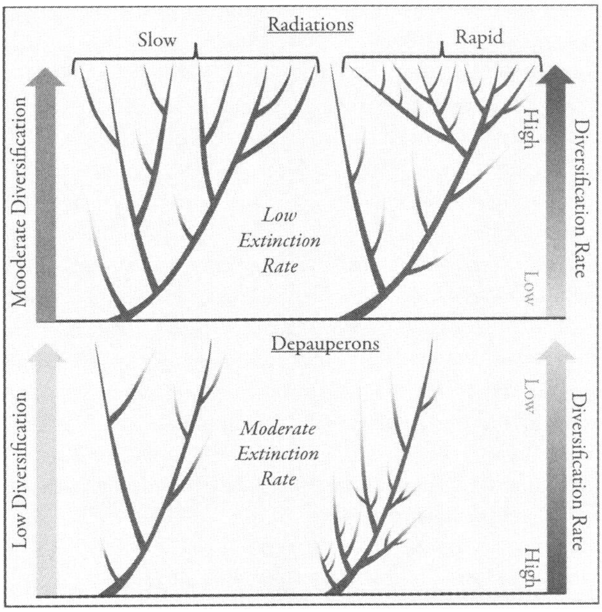

Fig. 11.20. The expected effects of different rates of diversification and extinction on clade diversity. When extinction rates are low, moderate to high speciation rates produce radiations (upper panel). Higher rates of extinction result in sparse phylogenies (depauperons; lower panel). Changes in rates of diversification produce a recent and rapid radiation (upper-right), or a depauperon left over from an ancient radiation (lower-right).

by single key innovations are probably rare, and more often high-diversity clades are the outcome of combinations of existing or novel plant traits along with ecological circumstances that favor diversification (Donoghue and Sanderson 2015). In some cases, the factors motivating radiations are clear. For example, adaptive radiations on oceanic islands with depauperate floras can be traced to long-distance dispersal events (Carlquist 1974; Givnish 2010) and others occurring in continental regions appear to be characterized by key innovations (Hodges and Arnold 1995; Donoghue and Sanderson 2015). On the other hand, many radiations are not associated with any obvious key innovations. For example, the genus *Astragalus* contains more than three thousand species across the northern hemisphere and does not possess unique traits compared to much less diverse genera in the legume family (Sanderson and Wojciechowski 1996). In fact, it may be the case that radiations lacking easily identified key innovations are more common. For example, genera containing over one thousand species include *Carex* (sedges; Cyperaceae), *Senecio* (ragworts and groundsels; Asteraceae), *Psychotria* (tropical understory trees; Rubiaceae), *Euphorbia* (spurges; Euphorbiaceae), *Begonia* (tropical understory herbs; Bigoniaceae), *Acacia* (mostly Australian shrubs and small trees; Fabaceae), *Solanum* (nightshades, annuals, shrubs, and small trees; Solanaceae), *Croton* (tropical herbs, shrubs, and small trees; Euphorbiaceae), *Eugenia* (woody evergreen trees and shrubs; Myrtaceae), *Piper* (tropical understory vines; Piperaceae), *Rhododendron* (temperate evergreen shrubs; Ericaceae), *Miconia* (tropical to temperate shrubs and small trees; Melastomataceae), *Peperomia* (tropical to subtropical perennial epiphytes; Piperaceae), and orchids in the genera *Dendrobium* and *Pleurothallis* (Orchidaceae); and there are at least thirty-five other genera with more than five hundred species (Frodin 2004). Surprisingly, species in these genera do not display diverse pollination syndromes and do not necessarily occur in a wide range of habits. The absence of unique traits that can be linked to the high diversity of these clades suggests that we have much to learn about the factors and circumstances responsible for the development of radiations.

The high diversity of some of the genera listed above (e.g., *Carex, Euphorbia, Croton,* and *Psychotria*) may be partly explained by their age and distribution across multiple continents, but other genera (e.g., *Acacia, Solanum,* and *Miconia*) have more restricted distributions. While these highly diverse genera may possess key innovations that have yet to be characterized, it is also important to recognize that at least a portion of the diversity differences may be due the absence of characters that allow for accurate delimitation. All of these genera consist of well-defined subclades, but reliable taxonomic characters that discriminate among these lower-level groups are lacking. In essence, there may be a discrepancy between our ability to perceive taxonomic differences and the evolutionary history of clades within these large genera. In any case, we need to exercise caution when comparing different taxonomic groups since they may differ in age, geographic distribution, and the refinement of their classification.

While unique traits or ecological circumstances responsible for promoting many of the large radiations described here have not been identified, there are a number of other cases where the influence of key innovations or ecological circumstances on clade diversity are more obvious. These radiations can generally be categorized by their association with key innovations or with events that facilitated diversification (e.g., adaptive radiations on oceanic islands). Clades with key innovations can be further divided into those innovations that affect reproductive isolation (e.g., pollination and seed dispersal) and those that provide improved herbivore resistance (the "escape-and-radiate" hypothesis; Ehrlich and Raven 1964; Thompson 1994), In the following subsections we discuss several types of traits and circumstances that have motivated clade diversification. We will hold off discussing insular radiations for the next section, which focuses on patterns and processes of evolution on islands.

Key Innovations—The origin of the embryophytes is marked by multiple periods of diversification that can be interpreted as adaptive radiations in response to key innovations. Some of these occurred quite early in the history of land plants; for example, apical growth and mutualisms with fungi were probably critical for colonization and diversification of early terrestrial plants (Brundrett 2002; Chapter 2). Much later, vertical growth along with improved vascularization, macrophyllous leaves, woody tissue, heterosporous reproductive modes, and bipolar growth for improved root development, all facilitated the

Devonian explosion of terrestrial plant life (Bateman and DiMichele 1994; Sperry 2003; Donoghue 2005). Multiple key traits responsible for the origin of flowers and the early diversification of angiosperms included closed carpels, displacement of the stigma away from the ovules, and the emergence of reduced gametophytic life stages. In the late Cretaceous, many flowering plant lineages underwent diversification along with animals that served as pollinators and seed dispersers (e.g., bees, birds, and perhaps early primates; Danforth et al. 2006; Chapter 2). The Miocene grassland expansion was apparently facilitated by the emergence of C_4 photosynthesis, which allowed diversification into warmer and drier regions (Spriggs et al. 2014). Considering these and many other traits that are unique to particular clades, it becomes clear that sequences of key innovations are primarily responsible for the diversification and dominance of flowering plants.

Given that key innovations have been so important for the evolution and early diversification of plants, it seems reasonable to expect that we should be able to identify unique traits that have facilitated the diversification of more recently derived clades. Examination of contemporary plant diversity reveals a number of genera and families that display characteristics of adaptive radiations promoted by key innovations. Much of the diversification during and since the Cretaceous occurred as plant lineages obtained traits that affected their interactions with animals. The most obvious of these are changes in floral morphology, such as bilateral symmetry and the fusion of petals to form tubular corollas, that have allowed increased pollinator specialization and diversification (Sargent 2004; Citerne et al. 2010). While innovations in floral morphology are often associated with radiations, it is not clear whether pollinator-mediated selection is driving diversification in these groups.

To understand the expression of some skepticism over the role of pollinator-mediated diversification of plants, we can examine one of the best-documented cases of a key innovation for floral morphology. The appearance of nectar bearing spurs in genera in the buttercup family (*Aquilegia*, the columbines; *Delphinium*, the larkspurs; and *Aconitum*, the monkshoods; Fig. 11.21; refer to color plate) is associated with adaptive radiation through pollinator specialization (Hodges and Arnold 1995). Most other genera in this family (e.g., *Ranunculus, Clematis, Anemone, Helleborus,* and others; Ranunculaceae) have bowl-shaped flowers that are radially symmetrical and do not display specialized pollination syndromes (Fig. 11.21; refer to color plate). All three genera with nectar spurs have distributions across the northern hemisphere and have relatively high diversity (around 70 species in *Aquilegia*, 300 in *Delphinium*, and 250 in *Aconitum*) resulting from parallel radiations on separate continents (Bastida et al. 2010; Fior et al. 2013). It is notable, however, that other genera in this family with unspecialized flowers are also very diverse (e.g., around 600 species in *Ranunculus*, 300 in *Clematis*, and 120 in *Anemone*). Moreover, phylogenetic analyses have revealed that diversification of *Aquilegia* in Europe has occurred primarily through habitat differentiation that is sometimes followed by transitions to alternative pollinators (Bastida et al. 2010). In North America a greater diversity of animal pollinators is available, and the range of pollination syndromes among columbines on this continent include hummingbirds as well as bees and moths. Geographic overlap of contemporary species tends to be greater in North America, but most species also display strong habitat differences. The pattern of diversification in *Aquilegia* seems to be similar to the comparison of speciation patterns in the California and Cape FRs discussed earlier; it is likely that divergence in pollination syndromes among species in the genus *Aquilegia* has been a symptom rather than a primary driver of diversification.

While it is tempting to attribute floral modifications as causal factors that drive divergence in plants such as the columbines and clades with bilateral floral symmetry (Sargent 2004), as discussed previously, it is probably more likely that pollinator transitions have occurred after geographic or habitat isolation. In other words, pollination syndrome shifts may have contributed more to decreasing rates of lineage extinction than to the process of divergence. In isolated populations, the loss of a specialized pollinator would be expected to have negative demographic consequences. In such cases, selection to accommodate a more abundant or effective pollen vector would be strong; any response in floral morphology that improved the effectiveness of pollen transfer could help rescue the population from extinction. Hence, the contribution of pollination syndrome transitions to clade diversity may be more about the resurrection of lineages that would

Fig. 11.21. Representative species of genera in the family Ranunculaceae. Upper photos are genera having mostly bowl-shaped flowers with radial symmetry. The lower three genera have nectar spurs. *Aquilegia* flowers are radially symmetrical while *Delphinium* and *Aconitum* flowers are bilateral. Photos by C. T. Johansson, Calibas, Rotational, NorCalHistory. Walter Siegmund, and Bernd Haynold.

otherwise have gone extinct; reducing rates of extinction will elevate clade diversity equivalently to increasing rates of diversification. While this scenario is speculative, such a pathway to pollinator specialization may have more credibility than the potential for pollinator-mediated selection to overcome the effects of gene flow among sympatric or parapatric populations. Given that higher rates of lineage retention also contribute to higher clade diversity, it is clear that floral specialization has promoted plant diversification; however, the roles of pollination syndromes and pollinator specialization have more to do with preventing extinctions of established lineages than the generation of new ones.

Reproductive Isolation—As we discussed earlier in this chapter, the process of speciation requires both phenotypic divergence and reproductive isolation. The emergence of reproductive barriers and trait evolution can work in conjunction as reductions in gene flow provide opportunities for divergence through drift and selection. While much of the research on radiations have emphasized morphological and ecological diversification, we recognize that reproductive isolation is pivotally important for the facilitation of divergence. Hence, the emergence of characteristics that facilitate the rapid development of isolation barriers among lineages may be interpreted as key innovations that promote diversification. We expect that the emergence of characteristics that limit gene flow will contribute to speciation by increasing opportunities for lineages exposed to contrasting selection regimes to continue to diverge.

Gene flow among lineages could be limited either by the adoption of a dispersal vector that engaged primarily in short-distance movements, or by the acquisition of characteristics that promote a rapid development of intrinsic reproductive isolation. The highly-diverse *Begonia* clade mentioned earlier is a possible example of the latter (Twyford et al. 2015). This is a tropical genus of more than 1,900 understory herbaceous species that are distributed across several continents. These species do not have strong habitat or floral divergence, but they often have limited geographic distributions and display a high degree of endemism. Crossing experiments combined with genetic analyses have demonstrated that intrinsic isolation barriers develop rapidly during divergence and contribute to low levels of hybridization and introgression between closely-related lineages (Twyford et al. 2015). It appears that the high diversity in this genus is due to a combination of clade age (i.e., the genus is distributed across several continents) and the propensity for the rapid development of strong reproductive isolation.

Reproductive isolation may also be facilitated by reduced dispersal ability, which would limit gene flow and allow populations to respond to local selection. To evaluate the effects of reduced dispersal potential for the promotion of clade diversification Lengyel et al. (2009) compared sister groups of ant-dispersed (myrmecochorous) species to non-myrmecochorous species. These authors identified ant-dispersed species by the presence of a lipid-rich food body (elaiosome) attached to each seed—ants detach the elaiosomes and discard the seeds after they have carried them into their nests. Lengyel et al. found a consistent pattern of increased diversity associated with ant dispersal across 101 lineages that included 241 genera and were distributed across six continents.

The approach used by Lengyel et al. (2009) to evaluate the effects of myrmecochory has a number of advantages over the study on *Begonia* described earlier. While the inference that more rapid evolution of reproductive isolation contributes to diversification is reasonable, this conclusion is not very robust, because the association is reported for a single clade and it is difficult to disentangle the effects of isolation from the confounding effects of shared phylogenic history. In contrast, the Lengyel et al. study circumvents these problems because it is conducted within a phylogenetic comparative framework that allows for multiple independent comparisons across a broad range of plant groups (Felsenstein 1985). A similar analysis using phylogenetic comparisons was used to demonstrate that clades with bilateral floral symmetry have had higher rates of diversification (Sargent 2004). This approach takes into account relatedness to reduce sample sizes to the points of origin of novel traits (i.e., elaiosomes or bilateral symmetry) within a phylogeny. Moreover, clade age is controlled for, because each pair of clades with or without the new trait were descended from a single common ancestor. Based on these results we can conclude that the transition to ant-dispersal by the acquisition of elaiosomes and to increased pollinator specialization through the evolution of bilateral symmetry are key innovations that promote clade diversification by limiting gene flow.

Herbivory and Diversification—In 1964, Ehrlich and Raven suggested that novel defense mechanisms could release plants from herbivore and pathogen pressure and consequently, promote rapid diversification (the *Escape and Radiate* hypothesis; Thompson 1994). Alleviation from herbivore pressure could increase population growth, which may reduce extinction rates and facilitate the adaptation to marginal habitats. For example, population growth rates decline when selection pressure is strong during adaptation to novel conditions (Chapter 5), so those that are released from herbivory may have improved chances of persistence. This hypothesis is supported by the observation that many major plant groups are characterized by having novel complements of anti-herbivore secondary metabolic compounds (Ehrlich and Raven 1964). Secondary metabolites that are unique to families and are thought to be important for herbivore defense include: the glucosinolates and myrosinase in the mustard family (Brassicaceae); fouranocoumarins and latex in the milkweed family (Asclepiadoideae); quinolizidine alkaloids in the legumes (Fabaceae); tropane and steroidal alkaloids in the nightshade/tomato family (Solanaceae); and iridoids and essential oils in the mint family (Lamiaceae; Wink 2003; Weber and Agrawal 2014; Suchan and Alvarez 2015). These innovations in herbivore defense arose early in the origin of each of these clades and may have provided relief from herbivore pressure for a long enough period to allow for the radiation of plant lineages (Thompson 1994). Eventually the herbivores adapted to the new armories of chemical defenses, so now these plant groups are associated with specialized herbivores. Whether release from herbivore pressure is adequate to promote clade diversification continues to be a topic of discussion and contention (Ehrlich and Raven 1964; Weber and Agrawal 2014; Suchan and Alvarez 2015).

While the evidence for the escape and radiate hypothesis is not clear in all cases, several studies have built a strong case for the contribution of release from herbivore pressure to clade diversification. One such investigation found consistently higher diversity for clades that were characterized by the presence of resin or latex canals compared to their sister groups (Farrell et al. 1991). This is a feature that has evolved independently in a wide range of plant lineages and effectively protects plants from of herbivores and pathogens.

One criticism of studies on the anti-herbivore role of secondary compounds such as the resin and latex canal survey just described is that these features may provide some additional adaptive advantages that are not related to protection from herbivory. Secondary metabolites are thought to have adaptive value in challenging and harsh environments (Hartmann 2007) and as antifungal, antibacterial, and antiviral compounds (Reichling 2010), so it is unclear whether release from herbivore pressure has been the primary cause of radiations in groups possessing these compounds.

A more direct test of the escape and radiate hypothesis was conducted by comparing clades characterized by extrafloral nectaries to sister groups lacking these structures (Weber and Agrawal 2014). These nectar-producing structures are typically located on sepals or leaves and attract ants that provide protection from insect herbivores. Weber and Agrawal (2014) surveyed more than one hundred families within which species have evolved extrafloral nectaries and found that clades characterized by these structures had more than twice the diversity of equal-aged clades that lack them. Extrafloral nectaries may be unlikely to provide any indirect benefits that are not related to herbivore defense, so this study may represent a strong case for Ehrlich and Raven's (1964) escape and radiate hypothesis.

Processes of Diversification—The most common roles of key innovations for diversification may be contributions to increased reproductive isolation and reduced rates of extinction. This is a departure from the traditional view, which assumes that novel traits become the targets of selection and drive divergence. For floral traits in particular, a number of authors have suggested that selection could promote divergence during sympatric or parapatric speciation or through reinforcement of reproductive isolation during secondary contact of partially diverged lineages (Grant 1981; Hodges and Arnold 1995; Sargent 2004; Kay and Sargent 2009). As discussed in Chapter 10, pollinator associations are usually not precise enough to limit gene flow or to provide strong enough selection to promote divergence of sympatric populations. Additional obstacles exist for speciation by reinforcement because of the restrictive conditions necessary for novel traits to spread out of regions of sympatry into allopatric populations. Given these difficulties, it is more likely that the promotion of diversification by key innovations occurs primarily in populations that are isolated by geographic or ecological barriers. Similarly, key innovations associated with dispersal and the ease of evolution of isolation mechanisms in diverging populations contribute to diversification by facilitating divergence through selection and genetic drift. As discussed earlier, the appearance of novel mechanisms of herbivore resistance may have demographic consequences that reduce extinction rates and facilitate adaptation to novel environments. Rather than driving divergence, it appears more likely that the large majority of key innovations contribute to clade diversification by reducing rates of extinction for isolated populations.

11.11 EVOLUTION ON ISLANDS

The unique faunas and floras found on many remote islands have fascinated naturalists and scholars since the time of Linnaeus. It is probably no coincidence that the inspiration for evolution by natural selection came to both Darwin and Wallace through their experiences with island biotas (Darwin 1859; Wallace 1880). Most species endemic to islands differ from their closest continental relatives in terms of vegetative morphology and growth form, and some—the Hawaiian species in particular—differ with respect to their floral form (Givnish 2010). Common trends among island endemics include transitions from annual to perennial life-histories, the evolution of shrubs and trees (referred to as *Insular Woodiness* by Carlquist 1974) in groups where all continental species are herbaceous (e.g., island endemics such as *Coreopsis* on the California Channel Islands; silverswords, *Schiedea*, and *Euphorbia* [previously *Chamaesyce*] in Hawaii, *Echium* in Macaronesia, and *Scalesia* in the Galapagos), loss of long-distance dispersal ability, and the evolution of two separate sexes (dioecy). These features are generally assumed to have evolved in lineages after they became established on islands and began to diversify. For example, the wood structure in island endemics is characteristically different from that of continental woody species (Carlquist 1974).

Self-compatibility is thought to greatly improve the chances for colonization after long-distance dispersal (*Baker's Law*; Cheptou 2012). This idea seems almost self-evident, as we can easily predict that the arrival of a single seed in an isolated location is much more likely than the immigration of two or more seeds within a short enough time span to allow interbreeding (Baker 1955, 1967). Consequently, we can predict that relatively few self-compatible annuals or short-lived perennials with high dispersal potential must have initially dominated the floras of remote islands. Following establishment, we expect that selection pressure would favor exploitation of the range habitats and modes of resource acquisition that were available because of the limited diversity in ecological range and growth forms displayed by the early colonists. Hence, perennial growth and woodiness allowed the overtopping of other species and led to the development of shrub and forest communities (Carlquist 1974). Reduced dispersal ability would be favored as nearly all seeds leaving the islands would be lost in the ocean. The fact that similar trait transitions have been observed for plant species that are endemic to a number of islands and archipelagos indicates that these trends are robust and repeatable.

Habitat Radiations—Some of the most spectacular examples of adaptive radiations have occurred after dispersal to oceanic islands where a single lineage diversifies into multiple species that may occupy different habitats, display a variety of growth forms, and be associated with different pollinators. Well documented adaptive radiations include plants in the Hawaiian Islands (e.g., Hawaiian lobelioids, the silversword alliance, and the genera *Schiedea* and *Euphorbia*; Fig. 11.22; refer to color plate), Micronesia (the genera *Aeonium, Aconium, Echium, Sonchus,* and *Argyranthemum*), and New Zealand (the genus *Hebe*; Givnish 2010). These collections of species are putatively monophyletic, indicating that all of the species present were probably derived from a single colonizing lineage. Species within each clade are typically distributed among different habitats and islands within the same archipelago, and in some cases, species occupy similar habitats on separate islands. There has been a substantial number of studies on phylogenetic relationships within and among insular and continental plant groups but surprisingly little experimental work on ecological associations and the adaptive value of traits. In the case of silverswords, we know that the defining phenotypic traits are maintained in common gardens indicating that they are genetically determined (i.e., rather than being due to phenotypic plasticity; Chapter 8). The adaptive values of functional traits have been assessed for a few other groups (Givnish 2010).

The Hawaiian archipelago stands out for the size and number of adaptive radiations found there and for a propensity toward diversification of floral forms in the Hawaiian lobelioids and mints (Lindqvist and Albert 2002; Antonelli 2008; Fig. 11.22; refer to color plate). Plants in these groups possess a wide diversity of floral forms and include associations with insect and avian pollinators. This high diversity of pollinator associations may be a consequence of the age and remoteness of the Hawaiian Islands, as well as the chance occurrence of bilateral symmetry and tubular flowers in colonizing lineages. The majority of radiations on these islands have occurred in the last five million years, but the lobelia radiation was initiated around 13 MYA (Givnish 2010). This is the most diverse group of plants on the islands, consisting of 128 species in several genera, followed by the Hawaiian mints with 57 species, the silversword alliance with more than 50 species, and plants in the genus *Schiedea* with 30 species.

In the case of the lobelioids, floral diversification would not have been possible but for the presence of specialized animal pollinators that radiated in parallel with the plants (i.e., in this case avian pollinators, the Hawaiian honeycreepers). The lobelioids may also represent a significant departure from the general trend of evolution of woodiness from an herbaceous ancestor. Plants in this group have a worldwide distribution, and the Hawaiian species seem to be closely related to several different tree species that occur in South America, Africa, and India (Antonelli 2008). All of the Hawaiian species are members of a basal *Polytomy* that surprisingly includes some South American, African, and Indian species. Several Hawaiian clades within this polytomy may reflect the long history of diversification of species in this archipelago. Similarly, the Hawaiian mints form a monophyletic clade that is resolved to various degrees into subclades and shows a close relationship to North American species in the genus *Stachys* (Lindqvist and Albert 2002). These phylogenetic analyses indicate that divergence has occurred in stages at a slower pace than might be expected for adaptive radiations that occur over a short time frame. Sadly, invasive herbivores and plants threaten many of the lobelioids, mints, and other Hawaiian endemics, and most honeycreepers are in severe decline or have become extinct.

As indicated for both the Hawaiian lobelioids and mints, not all adaptive radiations on islands have been rapid. A comparison between continental and Hawaiian woody plant groups reveals similar rates of

Fig. 11.22. Examples of the diversity within and among adaptive radiations in the Hawaiian Islands. Photos by David Eickhoff, Phyzome (United States Botanic Garden), David Eickhoff, Karl Magnacca, Karl Magnacca, Karl Magnacca, and Forest & Kim Starr.

trait evolution, but the silverswords represent a clear exception of this trend (Ackerly 2009). Comparison with three woody North American genera (the subfamily Arbutoideae in the Ericaceae and the genera *Acer* and *Ceanothus*) indicated rates of phenotypic diversification that are similar to the Hawaiian lobelioids, but trait divergence has been between two and three hundred times faster in the silversword alliance. It is important to note that Ackerly's analysis refers to the rate of phenotypic trait divergence, which presumably occurred in response to selection for adaptation to novel environments. Phylogenetic diversification based on neutral genomic regions in the silverswords displays a pattern of multiple clades and subclades that is similar to groups showing more gradual rates of phenotypic divergence, such as the Hawaiian lobelioids and mints (Baldwin and Sanderson 1998). These patterns of diversification reflect our expectation that individual lineages split off gradually in response to isolation after long-distance dispersal and colonization of novel habitats within the archipelago. The silversword alliance stands out because of the fast pace of phenotypic evolution, which may have been facilitated by polyploidy and gene duplication (Barrier et al. 1999). Spectacular examples of adaptive radiations in the Hawaiian Islands and many other archipelagos continue to offer a treasure trove of opportunities to investigate processes and patterns of phenotypic and genetic evolution.

Evolution of Dioecy on Islands—The high frequencies of dioecious species on oceanic islands has been more difficult to understand because, according to Baker's Law, the original colonists were most likely to have been self-compatible. Moreover, as discussed in Chapter 9 and earlier in this chapter, the lack of genetic load promotes the maintenance of selfing, so reversal to outcrossing is unlikely. The transition to outcrossing is made more difficult because once lost, physiological self-incompatibility appears to be difficult to regain (Igic et al. 2008). Presumably, outbreeding was favored as insular lineages diversified into novel habitats (Pannell 2015). Since the reestablishment of physiological self-incompatibility is less likely, the only other mechanism that would promote regular outcrossing was separation of sexes in different flowers and ultimately on different plants. While this scenario represents the best explanation for the observed high frequency of dioecy on remote islands, it contradicts the contention that selfing lineages are short-lived and represent evolutionary dead-ends (Wright et al. 2013).

The high frequency of dioecy on islands has been explained as a consequence of differences in the resistance for alternative evolutionary trajectories; the evolution of dioecy through changes in floral development is apparently more easily attained than the reversal of mutations that enabled self-compatibility (Igic et al. 2008). Baker (1967) illustrated the resistance to the recovery of self-incompatibility by describing the putative dispersal history of *Armeria maritima* from Europe to North American via arctic regions. In Europe, this species has large, showy flowers and is fully self-incompatible. It presumably moved northward by long-distance dispersal. Colonization of remote arctic regions was facilitated by selfing, so populations in northern Europe and North America have small flowers and are self-compatible. On the west coast of North America, populations were apparently influenced by selection for increased outbreeding, and these populations are *Gynodioecious* (females plus hermaphrodites; Fig. 9.1; refer to color plate). Hermaphrodites in these populations remain self-compatible, but female plants strictly outcross. Gynodioecy is commonly considered to be an intermediate step toward the evolution of dioecy (Thomson and Brunet 1990; Dufay et al. 2014), so these California populations of *Armeria* may be on a trajectory toward the eventual emergence of separate sexes (see Chapter 9 for a discussion of alternative hypotheses for the evolution of dioecy).

Dioecious plants are relatively common on many, but not all, remote islands. Frequencies of dioecy are generally low for nearly all continental regions—usually between 2% and 7%—but often range much higher (10% to 17%) on some islands (Sakai and Weller 1999). Islands with high frequencies of dioecious species include the Juan Fernandez archipelago off the coast of Chile (11%), Mauritius in the Indian Ocean (11%), the Hawaiian Islands (15%), New Zealand (12–13%), and a number of Polynesian islands including Norfolk Island (12%), Tonga (16%), Guam (13%), and Samoa (17%). However, the occurrence of dioecy is not universally high on islands as many, such as the Galapagos at 3%, have frequencies that are much more in line with continental regions.

Even though there tends to be high frequencies of dioecious clades on some islands, the general trend worldwide is for dioecious lineages to display lower rates of diversification than self-incompatible lineages (Heilbuth 2000). Two of the characteristics that seem to unify islands with higher frequencies of dioecious species are 1) they tend to be older—at least five million years—and 2) they are relatively remote from continents. The greater age and remote locations of islands may have been necessary to provide adequate time, in the absence of higher rates of immigration, for the transition to separate sexes in lineages that were ancestrally self-compatible and hermaphroditic. Hence, the higher frequencies of dioecious plants in some insular lineages may have been generated at a slow pace and are probably a consequence of the greater age and isolation of these islands.

While island age seems to be important for the occurrence of dioecy, we can also hypothesize that selection would be more likely to favor increased outcrossing to facilitate adaptation if a wide range of habitats were readily available. This hypothesis is supported by the observation that the frequency of dioecy across islands can be largely explained by distance from the equator and island elevation (Baker and Cox 1984). These authors found that low elevation islands, which presumably present a narrow range of habitats, had low frequencies of dimorphic species. Islands that are located closer to the equator and have large mountains probably offer a much wider range of climatic conditions both in terms of elevational gradients, and the effects of rain shadows that create arid regions (e.g., as on the big island of Hawaii).

The higher frequencies of dioecy on islands with more habitat diversity suggests that outbreeding may have been favored to facilitate adaptation and diversification. A possible example of the potential for dioecy to have facilitated adaptation to novel habitats is evident in the case of *Schiedea* in Hawaii (Sakai and Weller 1999). In this genus, the evolution of dioecy may have occurred at least twice, and all species that have separate sexes also occur in dry environments. These cases represent opportunities for comparative studies of the interactions between plant breeding systems and the potential for diversification.

The association between woodiness and dioecy on islands could be a consequence of the increased negative consequences of selfing in long-lived species. Outcrossing may be particularly advantageous in long-lived woody species, because it would avoid autogamy that results in homozygosity of the large numbers of deleterious mutations that are likely to accumulate during extended vegetative growth (Chapter 3). On the other hand, at least some woody species on islands remain hermaphroditic and self-compatible (Carlquist 1974), so it is not clear how they avoid the effects of additional genetic load accrued through somatic mutation. In spite of the large amount of attention that island floras have received over the last few hundred years, we obviously have much to learn from these fascinating and dynamic microcosms of evolution.

11.12 THE FUTURE OF PLANT DIVERSITY

In this chapter, we have promoted the idea of species as lineages that have potential for future diversification as well as a history of independence and integrity. From this perspective, we can imagine that future plant communities will be derived from species that we observe today, but the lineages surviving and diversifying into the future may be a non-random subset of current diversity. We may be able to make some predictions about the composition of future plant communities based on our knowledge of the selective pressures they are likely to face. In particular, the global scale of human-induced modifications to environments are likely to have substantial impacts on evolutionary trajectories. While contemporary diversity is tremendous, we recognize that all life is subject to threats that are manifestations of human activity, including the direct threat of habitat destruction and the indirect effects of human-induced climate change. Even though many plant communities appear to be stable in the short term, we expect to see declines in diversity as extinction rates increase and continue to outpace rates of speciation.

The potential for lineage survival and continued diversification may depend heavily on many plant characteristics, including their life forms, dispersal ability, and tolerance of extreme conditions through plasticity responses (Etterson and Shaw 2001; Burton et al. 2010; Nicotra et al. 2010). It is generally thought that changes in climates during the first few centuries of the new millennium may be much more abrupt than the large majority of climate shifts that have occurred over the history of life on Earth. Many organisms may be resilient even in the face of dramatic climate shifts; for example, most animals and many short-lived plants may be able to adjust their ranges in response to changing conditions. On the other hand, long-lived plant species and those restricted to highly fragmented habitats may be more susceptible to extinction. In the coming decades, we will witness a grand unnatural experiment on a scale that has no precedence; many changes will be alarming, and some species may not persist without human intervention; but as with all major climatic events over the history of life, the demise of some species will open windows of opportunity for others.

Even as human activities have often been harmful to natural plant communities, we have also created numerous openings for diversification. Over the past fifty thousand years or more, human activities have created open habitat through vegetation destruction and increased frequencies of fire. Human activities may have had the greatest impact on the evolution of plants in Europe, where the number of species associated with human-mediated disturbance is particularly high (Baker 1974). As humans improved their mobility over the last few centuries, they have carried other species along with them—sometimes intentionally and often by accident. Consequently, many species have been introduced to new geographic regions and new environments. Higher frequencies of long-distance movements have increased opportunities for admixture among divergent lineages, which may have facilitated adaptation to novel conditions. Ironically, the same types of activities that have led to the decline and demise of some species may set the stage for a new round of adaptation and diversification for others.

Within the next few thousand years, it could be the case that the floras of many geographic regions bear little resemblance to those that are present today. While we expect that plants with shorter generation times and greater dispersal ability will generally be favored, we do not expect anything as desperate as "a world full of weeds" with few or no woody species. If there is one thing that we have learned, it is that plants are not only exceptionally tolerant of short-term challenges, they are also resilient across evolutionary time scales. Just as we have seen on islands, we expect that resource opportunities in plant communities of the future would promote the emergence of novel woody floras—even in those regions that will be the most severely decimated by the effects of climate change. The forests of the future in many regions may be taxonomically different, but they are likely to fulfill the same ecological roles as the ones that are familiar to us now.

In the near future, conservationists and resource managers may be exploring new roles. We may soon find that preservation of existing plant communities is not feasible or even desirable. The challenge we face is to develop strategies that promote the long-term persistence of lineages and that create opportunities for future diversification. This may require a shift away from preservation of existing taxa and communities to the conservation of ecological and evolutionary processes. A new type of "conservation biology" is needed; one that focuses on opportunities for adaptation by elevating genetic diversity through gene flow and hybridization. Much of these efforts may include programs of artificial dispersal and the recreation of existing communities in novel locations (Scoble and Lowe 2010). But many of these experiments may not be very successful in the long term, because even our best climate models are not perfect predictors of future conditions.

Plant evolutionary biologists entering the field may be inclined to view the climate changes we are facing with some measure of desperation. On the other hand, we are witnessing events that may be unique to the history of life on Earth—this is an exciting and new period of discovery and innovation. Openings for novel approaches that promote understanding of evolutionary processes abound. The plant diversity of the future will certainly be different from what we know today, but these changes also present manifold opportunities to uncover new facets of our understanding of evolution and to develop novel approaches and concepts. The future of plant evolutionary biology is promising and bright; we only need to learn to adapt our science, management, and conservation strategies to a changing and dynamic natural world.

11.13 QUESTIONS FOR CONTEMPLATION AND DISCUSSION

1. List each of the species definitions discussed in this chapter and describe whether they account for the three criteria of phenotypic differentiation, reproductive isolation, and lineage integrity.
2. Search the literature for species concepts not discussed in this chapter and discuss how they might apply to plants.
3. Is reciprocal monophyly necessary for the definition and delimitation of species? Develop an argument for and against this criterion.
4. Is reproductive isolation necessary for the definition of species? Under what circumstances should intrinsic versus extrinsic reproductive barriers be used?
5. How is it possible that species with broad geographic distributions and relatively limited dispersal maintain lineage integrity?
6. What is the general lineage concept, and how can it be applied for species delimitation.
7. What is the species problem? Has it been solved? If so, then how?
8. Describe Wright's adaptive landscape. How does it differ from Fisher's view of adaptive evolution?
9. What are the three stages of Wright's shifting balance model of evolution. Provide examples of each stage.
10. How did Wright view the roles of genic and organismic evolution for the evolution of reproductive isolation? Are these useful concepts for understanding adaptive evolution and speciation?
11. How do rugged and holey adaptive landscapes differ? Describe the speciation process on each.

Under what circumstances might each type of landscape be more applicable?

12. How does ecological speciation differ from models assuming the development of genetic incompatibilities by drift? Which mode is more common? Do ecotypes and geographically isolated populations represent the initial stages of speciation?
13. Does selfing promote diversification or is it an evolutionary dead-end?
14. How do different rates of extinction and diversification lead to the development of adaptive radiations and depauperons?
15. Are key innovations engines of diversification or do they mostly retard rates of extinction? Develop an argument from each point of view.
16. Describe patterns of plant evolution on islands. What are the general trends? What circumstances favored these evolutionary transitions?
17. Why did island species develop dioecy to enforce outbreeding rather than regaining self-incompatibility? What are the impediments to the evolution of outcrossing from selfing ancestors? What circumstances made it possible for selfing lineages to transition to dioecy?

LITERATURE CITED

Abbott, R., D. Albach, S. Ansell, J. W. Arntzen, S. J. E. Baird, N. Bierne, J. W. Boughman, A. Brelsford, C. A. Buerkle, R. Buggs, R. K. Butlin, U. Dieckmann, F. Eroukhmanoff, A. Grill, S. H. Cahan, J. S. Hermansen, G. Hewitt, A. G. Hudson, C. Jiggins, J. Jones, B. Keller, T. Marczewski, J. Mallet, P. Martinez-Rodriguez, M. Moest, S. Mullen, R. Nichols, A. W. Nolte, C. Parisod, K. Pfennig, A. M. Rice, M. G. Ritchie, B. Seifert, C. M. Smadja, R. Stelkens, J. M. Szymura, R. Vainola, J. B. W. Wolf, and D. Zinner. 2013. Hybridization and speciation. Journal of Evolutionary Biology 26:229–246.

Abbott, R. J., M. J. Hegarty, S. J. Hiscock, and A. C. Brennan. 2010. Homoploid hybrid speciation in action. Taxon 59:1375–1386.

Abdelaziz, M., J. Lorite, A. J. Muñoz-Pajares, M. B. Herrador, F. Perfectti, and J. M. Gómez. 2011. Using complementary techniques to distinguish cryptic species: A new Erysimum (Brassicaceae) species from North Africa. American Journal of Botany 98:1049–1060.

Ackerly, D. 2009. Conservatism and diversification of plant functional traits: Evolutionary rates versus phylogenetic signal. Proceedings of the National Academy of Sciences of the United States of America 106:19699–19706.

Ackerly, D. D., S. A. Dudley, S. E. Sultan, J. Schmitt, J. S. Coleman, C. R. Linder, D. R. Sandquist, M. A. Geber, A. S. Evans, T. E. Dawson, and M. J. Lachowicz. 2000. The evolution of plant ecophysiological traits: Recent advances and future directions. BioScience 50:979–995.

Acosta, M. C., and A. C. Premoli. 2010. Evidence of chloroplast capture in South American *Nothofagus* (subgenus *Nothofagus*, Nothofagaceae). Molecular Phylogenetics and Evolution 54:235–242.

Addicott, J. F. 1986. Variation in the costs and benefits of mutualism: The interaction between yuccas and yucca moths. Oecologia 70:486–494.

Ahuja, M. R. 2005. Polyploidy in gymnosperms: Revisited. Silvae Genetica 54:59–69.

Al-Hiyaly, S. A., T. McNeilly, and A. D. Bradshaw. 1988. The effects of zinc contamination from electricity pylons—evolution in a replicated situation. New Phytologist 110:571–580.

Allard, R. W., A. L. Kahler, and B. S. Weir. 1972. The effect of selection on esterase allozymes in a barley population. Genetics 73:489–503.

Allen, A. M., and S. J. Hiscock. 2008. Evolution and phylogeny of self-incompatibility systems in angiosperms. Pp. 73–101 *in* Self-Incompatibility in Flowering Plants: Evolution, Diversity, and Mechanisms. Springer Berlin Heidelberg, Berlin, Heidelberg.

Allen, G. A., J. A. Antos, A. C. Worley, T. A. Suttill, and R. J. Hebda. 1996. Morphological and genetic variation in disjunct populations of the avalanche lily *Erythronium montanum*. Canadian Journal of Botany—Revue Canadienne De Botanique 74:403–412.

Ally, D., K. Ritland, and S. P. Otto. 2010. Aging in a long-lived clonal tree. PLoS Biology 8.

Alpert, P., and E. L. Simms. 2002. The relative advantages of plasticity and fixity in different environments: When is it good for a plant to adjust? Evolutionary Ecology 16:285–297.

Anacker, B. L. and S. Y. Strauss. 2014. The geography and ecology of plant speciation: Range overlap and niche divergence in sister species. Proceedings of the Royal Society B: Biological Sciences 281:20132980.

Andalo, C., M. B. Cruzan, C. Cazettes, B. Pujol, M. Burrus, and C. Thebaud. 2011. Post-pollination barriers do not explain the persistence of two distinct *Antirrhinum* subspecies with parapatric distribution. Plant Systematics and Evolution 286:223–234.

Anderson, E. 1948. Hybridization of the habitat. Evolution 2:1–9.

Anderson, E. 1949. Introgressive Hybridization. John, Wiley & Sons, New York.

Anderson, E., and G. L. Stebbins, Jr. 1954. Hybridization as an evolutionary stimulus. Evolution 8:378–388.

Andreasen, K., and B. G. Baldwin. 2001. Unequal evolutionary rates between annual and perennial lineages of checker mallows (Sidalcea, Malvaceae): Evidence from 18S–26S rDNA internal and external transcribed spacers. Molecular Biology and Evolution 18:936–944.

Angers, B., E. Castonguay, and R. Massicotte. 2010. Environmentally induced phenotypes and DNA methylation: How to deal with unpredictable conditions until the next generation and after. Molecular Ecology 19:1283–1295.

Angert, A. L., H. D. Bradshaw, and D. W. Schemske. 2008. Using experimental evolution to investigate geographic range limits in monkeyflowers. Evolution 62:2660–2675.

Anton, K. A., J. M. Rhode, and M. B. Cruzan. 2013. Pollinator-mediated selection on floral morphology: Evidence for transgressive evolution in derived hybrid populations. Journal of Evolutionary Biology 26:660–673.

Antonelli, A. 2008. Higher level phylogeny and evolutionary trends in Campanulaceae subfam. Lobelioideae: Molecular signal overshadows morphology. Molecular Phylogenetics and Evolution 46:1–18.

Antonovics, J. 1968. Evolution in closely adjacent plant populations. V. Evolution of self-fertility. Heredity 23:219–238.

Antonovics, J. 2006. Evolution in closely adjacent plant populations. X. Long-term persistence of prereproductive isolation at a mine boundary. Heredity 97:33–37.

Antonovics, J., and A. D. Bradshaw. 1970. Evolution in closely adjacent plant populations. VIII. Clinal patterns at a mine boundary. Heredity 25:349–362.

Arbogast, B. S., S. V. Edwards, J. Wakeley, P. Beerli, and J. B. Slowinski. 2002. Estimating divergence times from molecular data on phylogenetic and population genetic timescales. Annual Review of Ecology and Systematics 33:707–740.

Armbruster, W. S. 2014. Floral specialization and angiosperm diversity: Phenotypic divergence, fitness trade-offs, and realized pollination accuracy. AoB Plants 6: plu003.

Armbruster, W. S., and D. G. Rogers. 2004. Does pollen competition reduce the cost of inbreeding? American Journal of Botany 91:1939–1943.

Arnold, J. 1993. Cytonuclear disequilibria in hybrid zones. Annual Review of Ecology and Systematics 24:521–554.

Arnold, M. L. 1993. *Iris nelsonii*: Origin and genetic composition of a homoploid hybrid species. American Journal of Botany 80:577–583.

Arnold, M. L. 1994. Natural hybridization in Louisiana irises. BioScience 44:141–147.

Arnold, M. L. 1997. Natural Hybridization and Evolution. Oxford University Press, New York.

Arnold, M. L., E. S. Ballerini, and A. N. Brothers. 2012. Hybrid fitness, adaptation and evolutionary diversification: Lessons learned from Louisiana Irises. Heredity 108:159–166.

Arnold, M. L., and S. A. Hodges. 1995. Are natural hybrids fit or unfit relative to their parents? Trends in Ecology and Evolution 10:67–71.

Arnold, S. J. 1992. Constraints on phenotypic evolution. American Naturalist 140:S85–S107.

Arnold, S. J., R. Bürger, P. A. Hohenlohe, B. C. Ajie, and A. G. Jones. 2008. Understanding the evolution and stability of the G-Matrix. Evolution 62:2451–2461.

Aroca, R. 2012. Plant Responses to Drought Stress. Springer Science & Business Media.

Arunkumar, R., E. B. Josephs, R. J. Williamson, and S. I. Wright. 2013. Pollen-specific, but not sperm-specific, genes show stronger purifying selection and higher rates of positive selection than sporophytic genes in *Capsella grandiflora*. Molecular Biology and Evolution 30:2475–2486.

Ascher, P. D., and S. J. Peloquin. 1968. Pollen tube growth and incompatibility following intra- and inter-specific pollinations in *Lilium longiflorum*. American Journal of Botany 55:1230-&.

Ashton, P. A., and R. J. Abbott. 1992. Multiple origins and genetic diversity in the newly arisen allopolyploid species *Senecio cambrensis* Rosser (Compositae). Heredity 68:25–32.

Aubin-Horth, N., and S. C. P. Renn. 2009. Genomic reaction norms: Using integrative biology to understand molecular mechanisms of phenotypic plasticity. Molecular Ecology 18:3763–3780.

Avise, J., J. Arnold, R. Ball, B. Eldredge, T. Lamb, J. Neigel, C. Reeb, and N. Saunders. 1987. Intraspecific phylogeography: The mitochondrial DNA bridge between population genetics and systematics. Annual Review of Ecology and Systematics 18:489–522.

Avise, J. C. 2000. Phylogeography. Harvard University Press, Cambridge, MA.

Avise, J. C. 2004. Molecular Markers, Natural History, and Evolution. Chapman & Hall, New York.

Aydin, Z., T. Marcussen, A. S. Ertekin, and B. Oxelman. 2014. Marginal likelihood estimate comparisons to obtain optimal species delimitations in *Silene* sect. Cryptoneurae (Caryophyllaceae). PLoS One 9.

Ayres, D. R., and D. R. Strong. 2001. Origin and genetic diversity of *Spartina anglica* (Poaceae) using nuclear DNA markers. American Journal of Botany 88:1863–1867.

Baack, E., M. C. Melo, L. H. Rieseberg, and D. Ortiz-Barrientos. 2015. The origins of reproductive isolation in plants. New Phytologist 207:968–984.

Babik, W., R. K. Butlin, W. J. Baker, A. S. T. Papadopulos, M. Boulesteix, M. C. Anstett, C. Lexer, I. Hutton, and V. Savolainen. 2009. How sympatric is speciation in the *Howea* palms of Lord Howe Island? Molecular Ecology 18:3629–3638.

Bacles, C. F. E., A. J. Lowe, and R. A. Ennos. 2006. Effective seed dispersal across a fragmented landscape. Science 311:628–628.

Bada, J. L., C. Bigham, and S. L. Miller. 1994. Impact melting of frozen oceans on the early Earth—implications for the origin of life. Proceedings of the National Academy of Sciences of the United States of America 91:1248–1250.

Baer, C. F., M. M. Miyamoto, and D. R. Denver. 2007. Mutation rate variation in multicellular eukaryotes: Causes and consequences. Nature Reviews Genetics 8:619–631.

Bailey, L. H. 1896. The philosophy of species-making. Botanical Gazette 22:454–462.

Baird, N. A., P. D. Etter, T. S. Atwood, M. C. Currey, A. L. Shiver, Z. A. Lewis, E. U. Selker, W. A. Cresko, and E. A. Johnson. 2008. Rapid SNP discovery and genetic mapping using sequenced RAD markers. PLoS One 3:e3376.

Baker, H. G. 1951. Hybridization and natural gene-flow between higher plants. Biological Reviews of the Cambridge Philosophical Society 26:302–337.

Baker, H. G. 1955. Self-compatibility and establishment after "long-distance" dispersal. Evolution 9:347–348.

Baker, H. G. 1965. Characteristics and modes of origin of weeds. Pp. 147–169 *in* H. G. Baker and G. L. Stebbins, eds. The Genetics of Colonizing Species. Academic Press, New York.

Baker, H. G. 1967. Support for Baker's law—as a rule. Evolution 21:853–856.

Baker, H. G. 1974. The evolution of weeds. Annual Review of Ecology and Systematics 5:1–24.

Baker, H. G., and P. A. Cox. 1984. Further thoughts on dioecism and islands. Annals of the Missouri Botanical Garden 71:244–253.

Baker, H. G., and G. L. Stebbins, eds. 1965. The Genetics of Colonizing Species. Academic Press, New York.

Baker, J., M. Keller, and C. Preston. 2008. Genetic variability of wheel cactus (*Opuntia robusta* Wendl) in southern Australia: Implications for weed spread and biological control. Pp. 532 *in* R. D. Van Klinken, V. A. Osten, F. D. Panetta, and J. C. Scanian, eds. 16th Australian Weeds Conference Proceedings: Weed Management 2008: Hot Topics in the Tropics. Queensland Weeds Society, North Queensland, Australia.

Baker, R. L., L. C. Hileman, and P. K. Diggle. 2012. Patterns of shoot architecture in locally adapted populations are linked to intraspecific differences in gene regulation. New Phytologist 196:271–281.

Bakker, R. T. 1978. Dinosaur feeding behavior and origin of flowering plants. Nature 274:661–663.

Baldwin, B. G., and M. J. Sanderson. 1998. Age and rate of diversification of the Hawaiian silversword alliance (Compositae). Proceedings of the National Academy of Sciences of the United States of America 95:9402–9406.

Baldwin, J. M. 1896. A new factor in evolution. American Naturalist 30:441–451.

Baldwin, S. J., and B. C. Husband. 2011. Genome duplication and the evolution of conspecific pollen precedence. Proceedings of the Royal Society B: Biological Sciences 278:2011–2017.

Balloux, F., and N. Lugon-Moulin. 2002. The estimation of population differentiation with microsatellite markers. Molecular Ecology 11:155–165.

Banta, J. A., J. Dole, M. B. Cruzan, and M. Pigliucci. 2007. Evidence of local adaptation to coarse-grained environmental variation in *Arabidopsis thaliana*. Evolution 61:2419–2432.

Bar, M., and N. Ori. 2014. Leaf development and morphogenesis. Development 141:4219–4230.

Barow, M. 2006. Endopolyploidy in seed plants. Bioessays 28:271–281.

Barrett, P. M., and E. J. Rayfield. 2006. Ecological and evolutionary implications of dinosaur feeding behaviour. Trends in Ecology & Evolution 21:217–224.

Barrett, S. C. H. 2002. The evolution of plant sexual diversity. Nature Reviews Genetics 3:274–284.

Barrett, S. C. H. 2010. Darwin's legacy: The forms, function, and sexual diversity of flowers. Philosophical Transactions of the Royal Society B: Biological Sciences 365:351–368.

Barrett, S. C. H. 2017. Evolution of mating systems: Outcrossing versus selfing. Pp. 356–362 *in* J. B. Losos, ed. The Princeton Guide to Evolution. Princeton University Press, Princeton, NJ.

Barrett, S. C. H., and M. B. Cruzan. 1994. Incompatibility in heterostylous plants. Pp. 189–219 *in* E. G. Williams, A. E. Clarke, and R. B. Knox, eds. Genetic control of Self-incompatibility and Reproductive Development in Flowering Plants. Springer Netherlands, Dordrecht.

Barrett, S. C. H., and C. G. Eckert. 1990. Variation and evolution of mating systems in seed plants. Pp. 229–254 *in* S. Kawano, ed. Biological Approaches and Evolutionary Trends in Plants. Academic Press, London.

Barrett, S. C. H., and J. R. Kohn. 1991. Genetic and evolutionary consequences of small population size in plants—implications for conservation. Pp. 3–30 *in* D. A. Falk and K. E. Holsinger, eds. Genetics and Conservation of Rare Plants. Oxford University Press, New York.

Barrett, S. C. H., M. T. Morgan, and B. C. Husband. 1989. The dissolution of a complex genetic polymorphism: The evolution of self-fertilization in tristylous *Eichhornia paniculata* (Pontederiaceae). Evolution 43:1398–1416.

Barrett, S. C. H., R. W. Ness, and M. Vallejo-Marín. 2009. Evolutionary pathways to self-fertilization in a tristylous plant species. New Phytologist 183:546–556.

Barrier, M., B. G. Baldwin, R. H. Robichaux, and M. D. Purugganan. 1999. Interspecific hybrid ancestry of a plant adaptive radiation: Allopolyploidy of the Hawaiian silversword alliance (Asteraceae) inferred from floral homeotic gene duplications. Molecular Biology and Evolution 16:1105–1113.

Barringer, B. C. 2007. Polyploidy and self-fertilization in flowering plants. American Journal of Botany 94:1527–1533.

Barringer, B. C., E. A. Kulka, and L. F. Galloway. 2012. Reduced inbreeding depression in peripheral relative to central populations of a monocarpic herb. Journal of Evolutionary Biology 25:1200–1208.

Barton, N. H. 2013. Does hybridization influence speciation? Journal of Evolutionary Biology 26:267–269.

Barton, N. H., and B. O. Bengtsson. 1986. The barrier to genetic exchange between hybridizing populations. Heredity 56:357–376.

Barton, N. H., and G. M. Hewitt. 1985. Analysis of hybrid zones. Annual Review of Ecology and Systematics 16:113–148.

Baskin, C. C., and J. M. Baskin. 2014. Seeds: Ecology, Biogeography, and Evolution of Dormancy and Germination. Academic Press, New York.

Bastida, J. M., J. M. Alcantara, P. J. Rey, P. Vargas, and C. M. Herrera. 2010. Extended phylogeny of *Aquilegia*: The biogeographical and ecological patterns of two simultaneous but contrasting radiations. Plant Systematics and Evolution 284:171–185.

Bataillon, T. 2003. Shaking the "deleterious mutations" dogma? Trends in Ecology & Evolution 18:315–317.

Bateman, A. J. 1956. Cryptic self-incompatibility in the wallflower: *Cheiranthus cheiri* L. Heredity 10:257–261.

Bateman, R. M., and W. A. Dimichele. 1994. Heterospory: The most iterative key innovation in the evolutionary history of the plant kingdom. Biological Reviews of the Cambridge Philosophical Society 69:345–417.

Bateson, W. 1909. Heredity and variation in modern lights. Pp. 85–101 *in* A. C. Seward, ed. Darwin and Modern Science. Cambridge University Press, Cambridge, MA.

Bateson, W. 1913. Problems of Genetics. Yale University Press, New Haven.

Bateson, W. 1922. Evolutionary faith and modern doubts. Science 55:55–61.

Bawa, K. S., and C. J. Webb. 1984. Flower, fruit, and seed abortion in tropical forest trees: Implications for the evolution of paternal and maternal reproductive patterns. American Journal of Botany 71:736–751.

Baythavong, B. S. 2011. Linking the spatial scale of environmental variation and the evolution of phenotypic plasticity: Selection favors adaptive plasticity in fine-grained environments. American Naturalist 178:75–87.

Becker, B., and B. Marin. 2009. Streptophyte algae and the origin of embryophytes. Annals of Botany 103:999–1004.

Becker, C., and D. Weigel. 2012. Epigenetic variation: Origin and transgenerational inheritance. Current Opinion in Plant Biology 15:562–567.

Bedada, G., A. Westerbergh, E. Nevo, A. Korol, and K. J. Schmid. 2014. DNA sequence variation of wild barley *Hordeum spontaneum* (L.) across environmental gradients in Israel. Heredity 112:646–655.

Beerling, D. J. 2005. Leaf evolution: Gases, genes, and geochemistry. Annals of Botany 96:345–352.

Beers, E. P. 1997. Programmed cell death during plant growth and development. Cell Death Differentiation 4:649–661.

Bell, D. L., and L. F. Galloway. 2008. Population differentiation for plasticity to light in an annual herb: Adaptation and cost. American Journal of Botany 95:59–65.

Bell, G. 1982. The Masterpiece of Nature: The Evolution and Genetics of Sexuality. University of California Press, Berkeley.

Bell, G. 1985. On the function of flowers. Proceedings of the Royal Society London. Series B, Biological Sciences 224:223–265.

Bengtson, S., V. Belivanova, B. Rasmussen, and M. Whitehouse. 2009. The controversial "Cambrian" fossils of the Vindhyan are real but more than a billion years older. Proceedings of the National Academy of Sciences of the United States of America 106:7729–7734.

Bennett, M. D., and I. J. Leitch. 2011. Nuclear DNA amounts in angiosperms: Targets, trends and tomorrow. Annals of Botany 107:467–590.

Bennington, C. C., N. Fetcher, M. C. Vavrek, G. R. Shaver, K. J. Cummings, and J. B. McGraw. 2012. Home site advantage in two long-lived arctic plant species: Results from two thirty-year reciprocal transplant studies. Journal of Ecology 100:841–851.

Bergelson, J., and F. Roux. 2010. Towards identifying genes underlying ecologically relevant traits in *Arabidopsis thaliana*. Nature Reviews Genetics 11:867–879.

Bierzechudek, P. 1990. The demographic consequences of sexuality and apomixis in *Antennaria*. Pp. 293–308 *in* S. Kawano, ed. Biological Approaches to Evolutionary Trends in Plants, San Diego, CA.

Birchler, J. A., H. Yao, S. Chudalayandi, D. Vaiman, and R. A. Veitia. 2010. Heterosis. Plant Cell Online 22:2105–2112.

Birky, C. W. J. 1976. The inheritance of genes in mitochondria and chloroplasts. BioScience 26:26–32.

Bobiwash, K., S. T. Schultz, and D. J. Schoen. 2013. Somatic deleterious mutation rate in a woody plant: Estimation from phenotypic data. Heredity 111:338–344.

Bock, D. G., C. Caseys, R. D. Cousens, M. A. Hahn, S. M. Heredia, S. Hubner, K. G. Turner, K. D. Whitney, and L. H. Rieseberg. 2015. What we still don't know about invasion genetics. Molecular Ecology 24:2277–2297.

Bolmgren, K., and O. Eriksson. 2010. Seed mass and the evolution of fleshy fruits in angiosperms. Oikos 119:707–718.

Bomblies, K., and D. Weigel. 2007. Hybrid necrosis: Autoimmunity as a potential gene-flow barrier in plant species. Nature Reviews Genetics 8:382–393.

Bonfante, P., and A. Genre. 2008. Plants and arbuscular mycorrhizal fungi: An evolutionary-developmental perspective. Trends in Plant Science 13:492–498.

Borg, M., L. Brownfield, and D. Twell. 2009. Male gametophyte development: A molecular perspective. Journal of Experimental Botany 60:1465–1478.

Bossdorf, O., H. Auge, L. Lafuma, W. E. Rogers, E. Siemann, and D. Prati. 2005. Phenotypic and genetic differentiation between native and introduced plant populations. Oecologia 144:1–11.

Bowman, J. L. 2013. Walkabout on the long branches of plant evolution. Current Opinion in Plant Biology 16:70–77.

Boyko, A., and I. Kovalchuk. 2011. Genome instability and epigenetic modification—heritable responses to environmental stress? Current Opinion in Plant Biology 14:260–266.

Brachi, B., G. P. Morris, and J. O. Borevitz. 2011. Genome-wide association studies in plants: The missing heritability is in the field. Genome Biology 12.

Bradshaw, A. D. 1965. Evolutionary significance of phenotypic plasticity in plants. Advances in Genetics 13:115–156.

Bradshaw, A. D., and R. W. Snaydon. 1959. Population differentiation within plant species in response to soil factors. Nature 183:129–130.

Brady, K. U., A. R. Kruckeberg, and H. D. Bradshaw. 2005. Evolutionary ecology of plant adaptation to serpentine soils. Annual Review of Ecology Evolution and Systematics 36:243–266.

Bratteler, M., C. Lexer, and A. Widmer. 2006. Genetic architecture of traits associated with serpentine adaptation of Silene vulgaris. Journal of Evolutionary Biology 19:1149–1156.

Breese, E. L., M. D. Hayward, and A. C. Thomas. 1965. Somatic selection in perennial ryegrass. Heredity 20:367-&.

Breitkopf, H., R. E. Onstein, D. Cafasso, P. M. Schlueter, and S. Cozzolino. 2015. Multiple shifts to different pollinators fuelled rapid diversification in sexually deceptive *Ophrys* orchids. New Phytologist 207:377–389.

Brewbaker, J. L. 1957. Pollen cytology and self-incompatibility systems in plants. Heredity 48:271–277.

Briggs, D., and S. M. Walters. 1969. Plant Variation and Evolution. Weidenfeld and Nicolson, London.

Brocks, J. J., G. A. Logan, R. Buick, and R. E. Summons. 1999. Archean molecular fossils and the early rise of eukaryotes. Science 285:1033–1036.

Brooks, D. J., J. R. Fresco, A. M. Lesk, and M. Singh. 2002. Evolution of amino acid frequencies in proteins over deep time: Inferred order of introduction of amino acids into the genetic code. Molecular Biology and Evolution 19:1645–1655.

Broquet, T., and E. J. Petit. 2009. Molecular Estimation of Dispersal for Ecology and Population Genetics. Annual Review of Ecology, Evolution, and Systematics 40:193–216.

Brosius, J., and S. J. Gould. 1992. On "genomenclature": A comprehensive (and respectful) taxonomy for pseudogenes and other "junk DNA". Proceedings of the National Academy of Sciences of the United States of America 89:10706–10710.

Brown, A. H. D., M. T. Clegg, A. L. Kahler, and B. S. Weir, eds. 1990. Plant Population Genetics, Breeding, and Genetic Resources. Sinauer Associates, Inc, Sunderland, MA.

Brown, B., and R. Mitchell. 2001. Competition for pollination: Effects of pollen of an invasive plant on seed set of a native congener. Oecologia 129:43–49.

Brown, C. L., and H. E. Sommer. 1992. Shoot growth and histogenesis of trees possessing diverse patterns of shoot development. American Journal of Botany 79:335–346.

Brundrett, M. C. 2002. Coevolution of roots and mycorrhizas of land plants. New Phytologist 154:275–304.

Brys, R., A. V. Broeck, J. Mergeay, and H. Jacquemyn. 2014. The contribution of mating system variation to reproductive isolation in two closely related *Centaurium* species (Gentianaceae) with a generalized flower morphology. Evolution 68:1281–1293.

Buckler, E. S., J. B. Holland, P. J. Bradbury, C. B. Acharya, P. J. Brown, C. Browne, E. Ersoz, S. Flint-Garcia, A. Garcia, J. C. Glaubitz, M. M. Goodman, C. Harjes, K. Guill, D. E. Kroon, S. Larsson, N. K. Lepak, H. Li, S. E. Mitchell, G. Pressoir, J. A. Peiffer, M. O. Rosas, T. R. Rocheford, M. C. Romay, S. Romero, S. Salvo, H. S. Villeda, H. Sofia da Silva, Q. Sun, F. Tian, N. Upadyayula, D. Ware, H. Yates, J. Yu, Z. Zhang, S. Kresovich, and M. D. McMullen. 2009. The genetic architecture of maize flowering time. Science 325:714–718.

Bulmer, M. G. and G. A. Parker. 2002. The evolution of anisogamy: A game-theoretic approach. Proceedings of the Royal Society London. Series B, Biologial Sciences 269:2381–2388.

Burke, T. 1989. DNA fingerprinting and other methods for the study of mating success. Trends in Ecology & Evolution 4:139–144.

Burrell, A. M., A. K. Hawkins, and A. E. Pepper. 2012. Genetic analyses of nickel tolerance in a North American serpentine endemic plant, *Caulanthus amplexicaulis* var. *barbarae* (Brassicaceae). American Journal of Botany 99:1875–1883.

Burt, A. 2000. Perspective: Sex, recombination, and the efficacy of selection—was Weismann right? Evolution 54:337–351.

Burton, O. J., B. L. Phillips, and J. M. J. Travis. 2010. Trade-offs and the evolution of life-histories during range expansion. Ecology Letters 13:1210–1220.

Busch, J. W. 2006. Heterosis in an isolated, effectively small, and self-fertilizing population of the flowering plant *Leavenworthia alabamica*. Evolution 60:184–191.

Buss, L. W. 1983. Evolution, development, and the units of selection. Proceedings of the National Academy of Sciences of the United States of America—Biological Sciences 80:1387–1391.

Butterfield, N. J. 2000. *Bangiomorpha pubescens* n. gen., n. sp.: Implications for the evolution of sex, multicellularity, and the Mesoproterozoic/Neoproterozoic radiation of eukaryotes. Paleobiology 26:386–404.

Byers, D. L., and D. M. Waller. 1999. Do plant populations purge their genetic load? Effects of population size and mating history on inbreeding depression. Annual Review of Ecology and Systematics 30:479–513.

Cabin, R. J., R. J. Mitchell, and D. L. Marshall. 1998. Do surface plant and soil seed bank populations differ genetically? A multipopulation study of the desert mustard *Lesquerella fendleri* (Brassicaceae). American Journal of Botany 85:1098–1109.

Cain, M. L., V. Andreasen, and D. J. Howard. 1999. Reinforcing selection is effective under a relatively broad set of conditions in a mosaic hybrid zone. Evolution 53:1343–1353.

Cain, M. L., H. Damman, and A. Muir. 1998. Seed dispersal and the holocene migration of woodland herbs. Ecological Monographs 68:325–347.

Cain, M. L., B. G. Milligan, and A. E. Strand. 2000. Long-distance seed dispersal in plant populations. American Journal of Botany 87:1217–1227.

Cairns-Smith, A. G. 1982. Genetic Takeover and the Mineral Origins of Life. Cambridge University Press, Cambridge, UK.

Callahan, H. S., H. Maughan, and U. K. Steiner. 2008. Phenotypic plasticity, costs of phenotypes, and costs of plasticity toward an integrative view. Pp. 44–66 *in* C. D. Schlichting, and T. A. Mousseau, eds. Year in Evolutionary Biology, Wiley-Blackwell, New York.

Callahan, H. S., and M. Pigliucci. 2002. Shade-induced plasticity and its ecological significance in wild populations of *Arabidopsis thaliana*. Ecology 83:1965–1980.

Campbell, D. R. 1996. Evolution of floral traits in a hermaphroditic plant: Field measurements of heritabilities and genetic correlations. Evolution 50:1442–1453.

Campbell, D. R. 2000. Experimental tests of sex-allocation theory in plants. Trends in Ecology & Evolution 15:227–232.

Campbell, D. R. 2008. Pollinator shifts and the origin and loss of plant species. Annals of the Missouri Botanical Garden 95:264–274.

Campbell, D. R., N. M. Waser, and M. V. Price. 1996. Mechanisms of hummingbird-mediated selection for flower width in *Ipomopsis aggregata*. Ecology 77:1463–1472.

Carlquist, S. 1974. Island Biology. Columbia University Press, New York.

Carmichael, J. S., and W. E. Friedman. 1996. Double fertilization in *Gnetum gnemon* (Gnetaceae): Its bearing on the evolution of sexual reproduction within the Gnetales and the anthophyte clade. American Journal of Botany 83:767–780.

Carney, S. E., M. B. Cruzan, and M. L. Arnold. 1994. Reproductive interactions between hybridizing irises: Analyses of pollen-tube growth and fertilization success. American Journal of Botany 81:1169–1175.

Carstens, B. C., T. A. Pelletier, N. M. Reid, and J. D. Satler. 2013. How to fail at species delimitation. Molecular Ecology 22:4369–4383.

Carvalheira, G. M. G. 2000. Plant polytene chromosomes. Genetics and Molecular Biology 23:1043–1050.

Chang, Y., and S. W. Graham. 2011. Inferring the higher-order phylogeny of mosses (Bryophyta) and relatives using a large, multigene plastid data set. American Journal of Botany 98:839–849.

Chantha, S.-C., A. C. Herman, A. E. Platts, X. Vekemans, and D. J. Schoen. 2013. Secondary evolution of a self-incompatibility locus in the Brassicaceae genus *Leavenworthia*. PLoS Biology 11:e1001560.

Charlesworth, B., M. T. Morgan, and D. Charlesworth. 1993. The effect of deleterious mutations on neutral molecular variation. Genetics 134:1289–1303.

Charlesworth, D. 2003. Effects of inbreeding on the genetic diversity of populations. Philosophical Transactions of the Royal Society of London Series B: Biological Sciences 358:1051–1070.

Charlesworth, D., and B. Charlesworth. 1981. Allocation of resources to male and female functions in hermaphrodites. Biological Journal of the Linnean Society 15:57–74.

Charlesworth, D., and B. Charlesworth. 1992. The effects of selection in the gametophytic stage on mutational load. Evolution 46:703–720.

Charlesworth, D., and J. H. Willis. 2009. The genetics of inbreeding depression. Nature Reviews Genetics 10:783–796.

Charnov, E. L. 1982. The Theory of Sex Allocation. Princeton University Press, Princeton.

Chen, J., J. Ding, Y. Ouyang, H. Du, J. Yang, K. Cheng, J. Zhao, S. Qiu, X. Zhang, J. Yao, K. Liu, L. Wang, C. Xu, X. Li, Y. Xue, M. Xia, Q. Ji, J. Lu, M. Xu, and Q. Zhang. 2008. A triallelic system of S5 is a major regulator of the reproductive barrier and compatibility of indica–japonica hybrids in rice. Proceedings of the National Academy of Sciences of the United States of America 105:11436–11441.

Cheptou, P. O. 2012. Clarifying Baker's Law. Annals of Botany 109:633–641.

Chiappe, L. M. 2007. Glorified Dinosaurs: The Origin and Early Evolution of Birds. University of South Wales Press, Sydney.

Chun, Y. J., V. Le Corre, and F. Bretagnolle. 2011. Adaptive divergence for a fitness-related trait among invasive *Ambrosia artemisiifolia* populations in France. Molecular Ecology 20:1378–1388.

Citerne, H., F. Jabbour, S. Nadot, and C. Damerval. 2010. The evolution of floral symmetry. Pp. 85–137 *in* J. C. Kader, and M. Delseny, eds. Advances in Botanical Research, Vol 54.

Clara Castellanos, M., J. M. Alcantara, P. J. Rey, and J. M. Bastida. 2011. Intra-population comparison of vegetative and floral trait heritabilities estimated from molecular markers in wild *Aquilegia* populations. Molecular Ecology 20:3513–3524.

Clark, J. S., C. Fastie, G. Hurtt, S. T. Jackson, C. Johnson, G. A. King, M. Lewis, J. Lynch, S. Pacala, C. Prentice, E. W. Schupp, T. W. III, and P. Wyckoff. 1998. Reid's paradox of rapid plant migration. BioScience 48:13–24.

Clark, S. 1999. Polarized starlight and the handedness of life. American Scientist 87:336–343.

Clarke, J. T., R. C. M. Warnock, and P. C. J. Donoghue. 2011. Establishing a time-scale for plant evolution. New Phytologist 192:266–301.

Clausen, J. 1962. Stages in the Evolution of Plant Species. Cornell University Press, Ithaca, NY.

Clausen, J., D. D. Keck, and W. M. Hiesey. 1940. Experimental studies on the nature of species. I. The effect of varied environments on western American plants. Carnegie Institute of Washington Publication 520.

Clausen, J., D. D. Keck, and W. M. Hiesey. 1945. Experimental studies on the nature of species. II. Plant evolution through amphiploidy and autopolyploidy, with examples from the Madiinae. Carnegie Institute of Washington Publication 564.

Clausen, J., D. D. Keck, and W. M. Hiesey. 1947. Heredity of geographically and ecologically isolated races. American Naturalist 81:114–133.

Clausen, J., D. D. Keck, and W. M. Hiesey. 1948. Experimental studies on the nature of species. III. Environmental responses of climatic races of *Achillea*. Carnegie Institute of Washington Publication 581.

Claverie, J. M. 2006. Viruses take center stage in cellular evolution. Genome Biology 7:110.

Clegg, M. T., and R. W. Allard. 1972. Patterns of genetic differentiation in the slender wild oat species *Avena barabta*. Proceedings of the National Academy of Sciences of the United States of America 69:1820–1824.

Clement, M., D. Posada, and K. Crandall. 2000. TCS: A computer program to estimate gene genealogies. Molecular Ecology 9:1657–1660.

Coates, D. J. 1992. Genetic consequences of a bottleneck and spatial genetic structure in the triggerplant *Stylidium coroniforme* (Stylidiaceae). Heredity 69:512–520.

Cockerham, C. C. 1954. An extension of the concept of partitioning hereditary variance for analysis of covariances among relatives when epistasis is present. Genetics 39:859–882.

Colautti, R. I., and J. A. Lau. 2015. Contemporary evolution during invasion: Evidence for differentiation, natural selection, and local adaptation. Molecular Ecology 24:1999–2017.

Collard, B. C. Y., and D. J. Mackill. 2008. Marker-assisted selection: An approach for precision plant breeding in the twenty-first century. Philosophical Transactions of the Royal Society B: Biological Sciences 363:557–572.

Collevatti, R. G., S. G. Rabelo, and R. F. Vieira. 2009. Phylogeography and disjunct distribution in *Lychnophora ericoides* (Asteraceae), an endangered cerrado shrub species. Annals of Botany 104:655–664.

Comes, H., and R. Abbott. 1998. The relative importance of historical events and gene flow on the population structure of a Mediterranean ragwort, *Senecio gallicus* (Asteraceae). Evolution 52:355–367.

Conant, G. C., and K. H. Wolfe. 2008. Turning a hobby into a job: How duplicated genes find new functions. Nature Reviews Genetics 9:938–950.

Conner, J. K., K. Karoly, C. Stewart, V. A. Koelling, H. F. Sahli, and F. H. Shaw. 2011. Rapid independent trait evolution despite a strong pleiotropic genetic correlation. American Naturalist 178:429–441.

Cornelius, J. 1994. Heritabilities and additive genetic coefficients of variation in forest trees. Canadian Journal of Forest Research—Revue Canadienne De Recherche Forestiere 24:372–379.

Costa, M. M. R., S. Yang, J. Critchley, X. Feng, Y. Wilson, N. Langlade, L. Copsey, and A. Hudson. 2012. The genetic basis for natural variation in heteroblasty in *Antirrhinum*. New Phytologist 196:1251–1259.

Covey, P. A., K. Kondo, L. Welch, E. Frank, S. Sianta, A. Kumar, R. Nunez, G. Lopez-Casado, E. van der Knaap, J. K. Rose, B. A. McClure, and P. A. Bedinger. 2010. Multiple features that distinguish unilateral incongruity and self-incompatibility in the tomato clade. The Plant Journal For Cell and Molecular Biology 64:367–378.

Cox, C. B. and P. D. Moore. 2010. Biogeography: An Ecological and Evolutionary Approach. John Wiley & Sons, Inc., New York.

Cox, C. J., B. Goffinet, N. J. Wickett, S. B. Boles, and A. J. Shaw. 2010. Moss diversity: A molecular phylogenetic analysis of genera. Phytotaxa 9:175–195.

Coyne, J. A., and H. A. Orr. 2004. Speciation. Sinauer Associates, Inc, Sunderland, MA.

Coyne, J. A., M. Turelli, and N. H. Barton. 1997. Perspective: A critique of Sewall Wright's shifting balance theory of evolution. Evolution 51:643–671.

Cracraft, J. 1983. Species concepts and speciation analysis. Pp. 159–187 *in* R. F. Johnston, ed. Current Ornithology. Plenum Press, New York.

Crandall, K., and A. Templeton. 1993. Empirical tests of some predictions from coalescent theory with applications to intraspecific phylogeny reconstruction. Genetics 134:959–969.

Crane, P. R., P. Herendeen, and E. M. Friis. 2004. Fossils and plant phylogeny. American Journal of Botany 91:1683–1699.

Crawford, D. J. 2010. Progenitor-derivative species pairs and plant speciation. Taxon 59:1413–1423.

Crawford, T. J. 1984. The estimation of neighbourhood parameters for plant populations. Heredity 52:273–283.

Creighton, H. B., and B. McClintock. 1931. A correlation of cytological and genetical crossing-over in *Zea mays*. Proceedings of the National Academy of Sciences of the United States of America 17:492–497.

Crepet, W. L. 2000. Progress in understanding angiosperm history, success, and relationships: Darwin's abominably "perplexing phenomenon." Proceedings of the National Academy of Sciences of the United States of America 97:12939–12941.

Crnokrak, P., and S. C. H. Barrett. 2002. Perspective: Purging the genetic load: A review of the experimental evidence. Evolution 56:2347–2358.

Cronberg, N., R. Natcheva, and K. Hedlund. 2006. Microarthropods mediate sperm transfer in mosses. Science 313:1255–1255.

Cronn, R., B. J. Knaus, A. Liston, P. J. Maughan, M. Parks, J. V. Syring, and J. Udall. 2012. Targeted enrichment strategies for next-generation plant biology. American Journal of Botany 99:291–311.

Cronn, R., A. Liston, M. Parks, D. S. Gernandt, R. Shen, and T. Mockler. 2008. Multiplex sequencing of plant chloroplast genomes using Solexa sequencing-by-synthesis technology. Nucleic Acids Research 36:e122.

Crosby, J. L. 1949. Selection of an unfavourable gene-complex. Evolution 3:212–230.

Crow, J. F. 1948. Alternative hypotheses of hybrid vigor. Genetics 33:477–487.

Cruickshank, T. E., and M. W. Hahn. 2014. Re-analysis suggests that genomic islands of speciation are due to reduced diversity, not reduced gene flow. Molecular Ecology 23:3133–3157.

Cruzan, M. B. 1998. Genetic markers in plant evolutionary ecology. Ecology 79:400–412.

Cruzan, M. B., and J. D. Thomson. 1997. Effects of pre-dispersal selection on offspring growth and survival in *Erythronium grandiflorum*. Journal of Evolutionary Biology 10:295–314.

Cruzan, M. B. 1989. Pollen tube attrition in *Erythronium grandiflorum*. American Journal of Botany 76:562–570.

Cruzan, M. B. 2001. Population size and fragmentation thresholds for the maintenance of genetic diversity in the endemic *Scutellaria montana* (Lamiaceae). Evolution 55:1569–1580.

Cruzan, M. B. 2005. Patterns of introgression across an expanding hybrid zone: Analyzing historical patterns of gene flow using non-equilibrium approaches. New Phytologist 167:267–278.

Cruzan, M. B., and M. L. Arnold. 1993. Ecological and genetic associations in an *Iris* hybrid zone. Evolution 47:1432–1445.

Cruzan, M. B., and M. L. Arnold. 1994. Assortative mating and natural selection in an *Iris* hybrid zone. Evolution 48:1946–1958.

Cruzan, M. B., and M. L. Arnold. 1998. Consequences of cytonuclear epistasis and assortative mating for the genetic structure of hybrid populations. Heredity 82:36–45.

Cruzan, M. B., and S. C. H. Barrett. 1996. Post-pollination mechanisms influencing mating patterns and fecundity: An example from *Eichhornia paniculata*. American Naturalist 147:576–598.

Cruzan, M. B., and S. C. H. Barrett. 2016. Post-pollination discrimination between self and outcross pollen covaries with the mating system of a self-compatible flowering plant. American Journal of Botany 103:568–576.

Cruzan, M. B., P. B. Neal, and M. F. Wilson. 1988. Floral display in *Phyla incisa*: Consequences for male and female reproductive success. Evolution 42:505–515.

Cruzan, M. B., and J. M. Rhode. 2004. Experimental analysis of adaptive landscape topographies. Pp. 61–69 *in* Q. C. B. Cronk, J. Whitton, R. H. Ree, and I. E. P. Taylor, eds. Plant Adaptation: Molecular Genetics and Ecology. National Research Council of Canada, Vancouver, BC.

Cruzan, M. B., and A. R. Templeton. 2000. Paleoecology and coalescence: Phylogeographic analysis of hypotheses from the fossil record. Trends in Ecology and Evolution 15:491–496.

Cruzan, M. B., and D. M. Weinreich. 2013. Adaptive landscapes. Pp. 12–15 *in* S. Maloy, and K. Hughes, eds. Encyclopedia of Genetics. Academic Press, San Diego.

Cubas, P., C. Vincent, and E. Coen. 1999. An epigenetic mutation responsible for natural variation in floral symmetry. Nature 401:157–161.

Cutter, A. D., and B. A. Payseur. 2013. Genomic signatures of selection at linked sites: Unifying the disparity among species. Nature Reviews Genetics 14:262–274.

Cwynar, L. C., and G. M. MacDonald. 1987. Geographical variation of lodgepole pine in relation to population history. American Naturalist 129:463–469.

D'Hondt, B., P. Breyne, W. Van Landuyt, and M. Hoffmann. 2012. Genetic analysis reveals human-mediated long-distance dispersal among war cemeteries in *Trifolium micranthum*. Plant Ecology 213:1241–1250.

da Costa-Nunes, J. A. and U. Grossniklaus. 2004. Unveiling the gene-expression profile of pollen. Genome Biology 5:205.

Dalziel, A. C., S. M. Rogers, and P. M. Schulte. 2009. Linking genotypes to phenotypes and fitness: How mechanistic biology can inform molecular ecology. Molecular Ecology 18:4997–5017.

Danforth, B. N., S. Sipes, J. Fang, and S. G. Brady. 2006. The history of early bee diversification based on five genes plus morphology. Proceedings of the National Academy of Sciences of the United States of America 103:15118–15123.

Darlington, C. D. 1937. Recent Advances in Cytology. P. Blakiston's Son and Co., Philadelphia.

Darlington, C. D., and K. Mather. 1949. The Elements of Genetics. Allen and Unwin, Ltd., London.

Dartnell, L. 2008. Life's a beach on planet Earth. New Scientist 197:8–8.

Darwin, C. 1859. On the Origin of Species by Means of Natural Selection, or the Preservation of Favoured Races in the Struggle for Life. John Murray, London.

Darwin, C. 1872. On the Origin of Species by Means of Natural Selection, or the Preservation of Favoured Races in the Struggle for Life. John Murray, London.

Darwin, C. 1876. The Effects of Cross and Self Fertilization in the Vegetable Kingdom. John Murray, London.

Darwin, C. 1877. The Different Forms of Flowers on Plants of the Same Species. John Murray, London.

Darwin, E. 1800. Phytologia, or, The Philosophy of Agriculture and Gardening. J. Johnson, London.

Davey, J. W., P. A. Hohenlohe, P. D. Etter, J. Q. Boone, J. M. Catchen, and M. L. Blaxter. 2011. Genome-wide genetic marker discovery and genotyping using next-generation sequencing. Nature Reviews Genetics 12:499–510.

Davidson, A. M., M. Jennions, and A. B. Nicotra. 2011. Do invasive species show higher phenotypic plasticity than native species, and if so, is it adaptive? A meta-analysis. Ecology Letters 14:419–431.

Davies, T. J., E. M. Wolkovich, N. J. B. Kraft, N. Salamin, J. M. Allen, T. R. Ault, J. L. Betancourt, K. Bolmgren, E. E. Cleland, B. I. Cook, T. M. Crimmins, S. J. Mazer, G. J. McCabe, S. Pau, J. Regetz, M. D. Schwartz, and S. E. Travers. 2013. Phylogenetic conservatism in plant phenology. Journal of Ecology 101:1520–1530.

de Heredia, U. L., J. S. Carrion, P. Jimenez, C. Collada, and L. Gil. 2007. Molecular and palaeoecological evidence for multiple glacial refugia for evergreen oaks on the Iberian Peninsula. Journal of Biogeography 34:1505–1517.

de Kroon, H., H. Huber, J. F. Stuefer, and J. M. van Groenendael. 2005. A modular concept of phenotypic plasticity in plants. New Phytologist 166:73–82.

de Kroon, H., E. J. W. Visser, H. Huber, L. Mommer, and M. J. Hutchings. 2009. A modular concept of plant foraging behaviour: The interplay between local responses and systemic control. Plant Cell and Environment 32:704–712.

De La Torre, A., P. K. Ingvarsson, and S. N. Aitken. 2015. Genetic architecture and genomic patterns of gene flow between hybridizing species of *Picea*. Heredity 115:153–164.

de Lafontaine, G., J. Prunier, S. Gérardi, and J. Bousquet. 2015. Tracking the progression of speciation: Variable patterns of introgression across the genome provide insights on the species delimitation between progenitor-derivative spruces (*Picea mariana* × *P. rubens*). Molecular Ecology 24:5229–5247.

De Mita, S., J. Ronfort, H. I. McKhann, C. Poncet, R. El Malki, and T. Bataillon. 2007. Investigation of the demographic and selective forces shaping the nucleotide diversity of genes involved in Nod factor signaling in *Medicago truncatula*. Genetics 177:2123–2133.

de Queiroz, K. 1998. The general lineage concept of species, species criteria, and the process of speciation. Pp. 57–78 *in* D. J. Howard, and S. H. Berlocher, eds. Endless Forms: Species and Speciation. Oxford University Press, New York.

de Queiroz, K. 1999. The general lineage concept of species and the defining properties of the species category. Pp. 49–89 *in* R. A. Wilson, ed. Species. MIT Press, Boston, MA.

de Queiroz, K. 2007. Species concepts and species delimitation. Systematic Biology 56:879–886.

De Wit, P., M. H. Pespeni, J. T. Ladner, D. J. Barshis, F. Seneca, H. Jaris, N. O. Therkildsen, M. Morikawa, and S. R. Palumbi. 2012. The simple fool's guide to population genomics via RNA-Seq: An introduction to high-throughput sequencing data analysis. Molecular Ecology Resources 12:1058–1067.

Delcourt, H. R., and P. A. Delcourt. 1991. Quaternary Ecology: A Paleoecological Perspective. Chapman & Hall, London.

Delph, L. F., and J. K. Kelly. 2014. On the importance of balancing selection in plants. New Phytologist 201:45–56.

Demesure, B., N. Sodzi, and R. Petit. 1995. A set of universal primers for amplification of polymorphic non-coding regions of mitochondrial and chloroplast DNA in plants. Molecular Ecology 4:129–131.

DeRose-Wilson, L. J., and B. S. Gaut. 2007. Transcription-related mutations and GC content drive variation in nucleotide substitution rates across the genomes of *Arabidopsis thaliana* and *Arabidopsis lyrata*. BMC Evolutionary Biology 7:66.

DeWalt, S. J., E. Siemann, and W. E. Rogers. 2011. Geographic distribution of genetic variation among native and introduced populations of Chinese tallow tree, *Triadica sebifera* (Euphorbiaceae). American Journal of Botany 98:1128–1138.

DeWitt, T. J., A. Sih, and D. Sloan-Wilson. 1998. Costs and limits of phenotypic plasticity. Trends in Ecology & Evolution 13:77–81.

Diaz, A. and M. R. Macnair. 1999. Pollen tube competition as a mechanism of prezygotic reproductive isolation between *Mimulus nasutus* and its presumed progenitor *M. guttatus*. New Phytologist 144:471–478.

Diaz-Perez, A. J., M. Sequeira, A. Santos-Guerra, and P. Catalan. 2012. Divergence and biogeography of the recently evolved Macaronesian red *Festuca* (Gramineae) species inferred from coalescence-based analyses. Molecular Ecology 21:1702–1726.

Dickinson, W. J. 2008. Synergistic fitness interactions and a high frequency of beneficial changes among mutations accumulated under relaxed selection in *Saccharomyces cerevisiae*. Genetics 178:1571–1578.

Diggle, P. K. 2014. Modularity and intra-floral integration in metameric organisms: Plants are more than the sum of their parts. Philosophical Transactions of the Royal Society B: Biological Sciences 369:20130253.

Dixon, C. J., P. Schonswetter, P. Vargas, S. Ertl, and G. M. Schneeweiss. 2009. Bayesian hypothesis testing supports long-distance Pleistocene migrations in a European high mountain plant (*Androsace vitaliana*, Primulaceae). Molecular Phylogenetics and Evolution 53:580–591.

Dlugosch, K. M. and I. M. Parker. 2008. Invading populations of an ornamental shrub show rapid life history evolution despite genetic bottlenecks. Ecology Letters 11:701–709.

Do, C., R. S. Waples, D. Peel, G. M. Macbeth, B. J. Tillett, and J. R. Ovenden. 2014. NeEstimator v2: Re-implementation of software for the estimation of contemporary effective population size (N_e) from genetic data. Molecular Ecology Resources 14:209–214.

Dobes, C. H., T. Mitchell-Olds, and M. A. Koch. 2004. Extensive chloroplast haplotype variation indicates Pleistocene hybridization and radiation of North American *Arabis drummondii*, *A.* x *divaricarpa*, and *A. holboellii* (Brassicaceae). Molecular Ecology 13:349–370.

Dobzhansky, T. 1935. A critique of the species concept in biology. Philosophy of Science 2:344–355.

Dobzhansky, T. 1936. Studies on hybrid sterility. II. Localization of sterility factors in *Drosophila pseudoobscura* hybrids. Genetics 21:113–135.

Dobzhansky, T. H. 1937. Genetics and the Origin of Species. Columbia University Press, New York.

Dobzhansky, T. 1955. A review of some of the fundamental concepts and problems of population genetics. Cold Spring Harbor Symposium in Quantitative Biology 20:1–15.

Dobzhansky, T. 1957. On methods of evolutionary biology and anthropology: Part I. Biology. American Scientist 45:366A–392.

Dobzhansky, T. 1970. Genetics of the Evolutionary Process. Columbia University Press, New York.

Doebeli, M., U. Dieckmann, J. A. J. Metz, and D. Tautz. 2005. What we have also learned: Adaptive speciation is theoretically plausible. Evolution 59:691–695.

Doebley, J. 2004. The genetics of maize evolution. Annual Review of Genetics 38:37–59.

Donoghue, M. J. 1985. A critique of the biological species concept and recommendations for a phylogenetic alternative. Bryologist 88:172–181.

Donoghue, M. J. 2005. Key innovations, convergence, and success: Macroevolutionary lessons from plant phylogeny. Paleobiology 31:77–93.

Donoghue, M. J., and M. J. Sanderson. 2015. Confluence, synnovation, and depauperons in plant diversification. New Phytologist 207:260–274.

Donohue, K. 2004. Density-dependent multilevel selection in the Great Lakes sea rocket. Ecology 85:180–191.

Donohue, K. 2009. Completing the cycle: Maternal effects as the missing link in plant life histories. Philosophical Transactions of the Royal Society B: Biological Sciences 364:1059–1074.

Donohue, K., E. H. Pyle, D. Messiqua, M. S. Heschel, and J. Schmitt. 2001. Adaptive divergence in plasticity in natural populations of *Impatiens capensis* and its consequences for performance in novel habitats. Evolution 55:692–702.

Dorn, L. A., and T. Mitchell-Olds. 1991. Genetics of *Brassica campestris*.1. Genetic constraints on evolution of life-history characters. Evolution 45:371–379.

Dowling, T. E., and C. L. Secor. 1997. The role of hybridization and introgression in the diversification of animals. Annual Review of Ecology and Systematics 28:593–619.

Dressler, R. 1990. The Orchids: Natural History and Classification. Harvard University Press, Cambridge, MA.

Dressler, R. L. 1968. Pollination by euglossine bees. Evolution 22:202–210.

Duarte, J. M., L. Y. Cui, P. K. Wall, Q. Zhang, X. H. Zhang, J. Leebens-Mack, H. Ma, N. Altman, and C. W. dePamphilis. 2006. Expression pattern shifts following duplication indicative of subfunctionalization and neofunctionalization in regulatory genes of *Arabidopsis*. Molecular Biology and Evolution 23:469–478.

Dudley, S. A. 1996. Differing selection on plant physiological traits in response to environmental water availability: A test of adaptive hypotheses. Evolution 50:92–102.

Dufay, M., P. Champelovier, J. Kafer, J. P. Henry, S. Mousset, and G. A. B. Marais. 2014. An angiosperm-wide analysis of the gynodioecy-dioecy pathway. Annals of Botany 114:539–548.

Dminil, J., D. Kenfack, V. Viscosi, L. Grumiau, and O. J. Hardy. 2012. Testing species delimitation in sympatric species complexes: The case of an African tropical tree, *Carapa* spp. (Meliaceae). Molecular Phylogenetics and Evolution 62:275–285.

Dumolin-Lapègue, S., B. Demesure, S. Fineschi, V. Le-Corre, and R. J. Petit. 1997a. Phylogeographic structure of white oaks throughout the European continent. Genetics. 146:1475–1487.

Dumolin-Lapègue, S., M. H. Pemonge, and R. J. Petit. 1997b. An enlarged set of consensus primers for the study of organelle DNA in plants. Molecular Ecology 6:393–397.

Durand, E., M. I. Tenaillon, C. Ridel, D. Coubriche, P. Jamin, S. Jouanne, A. Ressayre, A. Charcosset, and C. Dillmann. 2010. Standing variation and new mutations both contribute to a fast response to selection for flowering time in maize inbreds. BMC Evolutionary Biology 10:2.

Durka, W., C. Nossol, E. Welk, E. Ruprecht, V. Wagner, K. Wesche, and I. Hensen. 2013. Extreme genetic depauperation and differentiation of both populations and species in Eurasian feather grasses (*Stipa*). Plant Systematics and Evolution 299:259–269.

Dyer, R. J., D. M. Chan, V. A. Gardiakos, and C. A. Meadows. 2012. Pollination graphs: Quantifying pollen pool covariance networks and the influence of intervening landscape on genetic connectivity in the North American understory tree, *Cornus florida* L. Landscape Ecology 27:239–251.

East, E. M. 1908. Inbreeding in corn. Reports of the Connecticut Agricultural Experiments Station 1907:419–428.

East, E. M. 1915. Studies on size inheritance in *Nicotiana*. Genetics 1:164–176.

East, E. M. 1936. Heterosis. Genetics 21:375–397.

Eaton, D. A. R., and R. H. Ree. 2013. Inferring phylogeny and introgression using RADseq data: An example from flowering plants (*Pedicularis*: Orobanchaceae). Systematic Biology 62:689–706.

Eckert, A. J., J. van Heerwaarden, J. L. Wegrzyn, C. D. Nelson, J. Ross-Ibarra, S. C. Gonzalez-Martinez, and D. B. Neale. 2010. Patterns of population structure and environmental associations to aridity across the range of loblolly pine (*Pinus taeda* L., Pinaceae). Genetics 185:969–982.

Eckert, C. G. 2001. The loss of sex in clonal plants. Evolutionary Ecology 15:501–520.

Eckert, C. G., K. E. Samis, and S. C. Lougheed. 2008. Genetic variation across species' geographical ranges: The central-marginal hypothesis and beyond. Molecular Ecology 17:1170–1188.

Edwards, C. E. 2015. Looking to the future of conservation genetics: The case for using quantitative genetic experiments to estimate the ability of rare plants to withstand climate change. American Journal of Botany 102:1011–1013.

Ehrenreich, I. M., and M. D. Purugganan. 2006. The molecular genetic basis of plant adaptation. American Journal of Botany 93:953–962.

Ehrlich, P. R., and P. H. Raven. 1964. Butterflies and plants: A study in coevolution. Evolution 18:586–608.

Eigen, M., and P. Schuster. 1978. Hypercycle—principle of natural self-organization. C. Realistic hypercycle. Naturwissenschaften 65:341–369.

Elena, S. F., and R. E. Lenski. 2003. Evolution experiments with microorganisms: The dynamics and genetic bases of adaptation. Nature Reviews Genetics 4:457–469.

Ellstrand, N. C., and D. R. Elam. 1993. Population genetic consequences of small population size: Implications for plant conservation. Annual Review of Ecology and Systematics 24:217–242.

Elshire, R. J., J. C. Glaubitz, Q. Sun, J. A. Poland, K. Kawamoto, E. S. Buckler, and S. E. Mitchell. 2011. A robust, simple genotyping-by-sequencing (GBS) approach for high diversity species. PLoS 6: e19379.

Emms, S. K., S. A. Hodges, and M. L. Arnold. 1996. Pollen-tube competition, siring success, and consistent asymmetric hybridization in Louisiana irises. Evolution 50:2201–2206.

Endress, P. K. 2011. Angiosperm ovules: Diversity, development, evolution. Annals of Botany 107:1465–1489.

Ennos, R. A. 1994. Estimating the relative rates of pollen and seed migration among plant populations. Heredity 72:250–259.

Erickson, D. L., C. B. Fenster, H. K. Stenoien, and D. Price. 2004. Quantitative trait locus analyses and the study of evolutionary process. Molecular Ecology 13:2505–2522.

Eriksson, O. 2008. Evolution of seed size and biotic seed dispersal in angiosperms: Paleoecological and neoecological evidence. International Journal of Plant Sciences 169:863–870.

Eriksson, O., E. M. Friis, and P. Lofgren. 2000. Seed size, fruit size, and dispersal systems in angiosperms from the early cretaceous to the late tertiary. American Naturalist 156:47–58.

Escudero, M., D. A. R. Eaton, M. Hahn, and A. L. Hipp. 2014. Genotyping-by-sequencing as a tool to infer phylogeny and ancestral hybridization: A case study in *Carex* (Cyperaceae). Molecular Phylogenetics and Evolution 79:359–367.

Essenberg, C. J. 2012. Explaining variation in the effect of floral density on pollinator visitation. American Naturalist 180:153–166.

Estoup, A., and T. Guillemaud. 2010. Reconstructing routes of invasion using genetic data: Why, how and so what? Molecular Ecology 19:4113–4130.

Etterson, J. R., S. R. Keller, and L. F. Galloway. 2007. Epistatic and cytonuclear interactions govern outbreeding depression in the autotetraploid *Campanulastrum americanum*. Evolution 61:2671–2683.

Etterson, J. R., and R. G. Shaw. 2001. Constraint to adaptive evolution in response to global warming. Science 294:151–154.

Ewens, W. J. 2004. Mathmatical Population Genetics, Volume I: Theoretical Foundations. Springer-Verlag, Berlin.

Excoffier, L., M. Foll, and R. J. Petit. 2009. Genetic consequences of range expansions. Annual Review of Ecology Evolution and Systematics 40:481–501.

Excoffier, L., and G. Heckel. 2006. Computer programs for population genetics data analysis: A survival guide. Nature Reviews Genetics 7:745–758.

Eyre-Walker, A., and P. D. Keightley. 2007. The distribution of fitness effects of new mutations. Nature Reviews Genetics 8:610–618.

Faegri, K., and L. van der Pijl. 1979. The Principles of Pollination Ecology. Pergamon, Oxford.

Fagerstrom, T., D. A. Briscoe, and P. Sunnucks. 1998. Evolution of mitotic cell-lineages in multicellular organisms. Trends in Ecology & Evolution 13:117–120.

Falconer, D., and T. MacKay. 1996. Introduction to Quantitative Genetics. Chapman & Hall, New York.

Farrell, B. D., D. E. Dussourd, and C. Mitter. 1991. Escalation of plant defense: Do latex and resin canals spur plant diversification? American Naturalist 138:881–900.

Feder, J. L., S. P. Egan, and P. Nosil. 2012. The genomics of speciation-with-gene-flow. Trends in Genetics 28:342–350.

Feder, J. L., and P. Nosil. 2010. The efficacy of divergence hitchhiking in generating genomic islands during ecological speciation. Evolution 64:1729–1747.

Fehlberg, S. D., and T. A. Ranker. 2009. Evolutionary history and phylogeography of *Encelia farinosa* (Asteraceae) from the Sonoran, Mojave, and Peninsular deserts. Molecular Phylogenetics and Evolution 50:326–335.

Felsenstein, J. 1985. Phylogenies and the comparitve method. American Naturalist 125:1–15.

Fenster, C. B., W. S. Armbruster, P. Wilson, M. R. Dudash, and J. D. Thomson. 2004. Pollination syndromes and floral specialization. Annual Review of Ecology Evolution and Systematics 35:375–403.

Finet, C., R. E. Timme, C. F. Delwiche, and F. Marletaz. 2010. Multigene phylogeny of the green lineage reveals the origin and diversification of land plants. Current Biology 20:2217–2222.

Finet, C., R. E. Timme, C. F. Delwiche, and F. Marletaz. 2012. Multigene phylogeny of the green lineage reveals the origin and diversification of land plants (vol 20, pg 2217, 2010). Current Biology 22:1456–1457.

Fior, S., M. G. Li, B. Oxelman, R. Viola, S. A. Hodges, L. Ometto, and C. Varotto. 2013. Spatiotemporal reconstruction of the Aquilegia rapid radiation through next-generation sequencing of rapidly evolving cpDNA regions. New Phytologist 198:579–592.

Fischer, M., and J. Stocklin. 1997. Local extinctions of plants in remnants of extensively used calcareous grasslands 1950–1985. Conservation Biology 11:727–737.

Fisher, R. A. 1918. The correlation between relatives on the supposition of Mendelian inheritance. Transactions of the Royal Society of Edinburgh 52:399–433.

Fisher, R. A. 1930. The Genetical Theory of Natural Selection. Clarendon Press, Oxford.

Fisher, R. A. 1941. Average excess and average effect of a gene substitution. Annals of Eugenics 11:53–63.

Fishman, L., J. Aagaard, and J. C. Tuthill. 2008. Toward the evolutionary genomics of gametophytic divergence: Patterns of transmission ratio distortion in monkeyflower (*Mimulus*) hybrids reveal a complex genetic basis for conspecific pollen precedence. Evolution 62:2958–2970.

Fishman, L., and R. Wyatt. 1999. Pollinator-mediated competition, reproductive character displacement, and the evolution of selfing in *Arenaria uniflora* (Caryophyllaceae). Evolution 53:1723–1733.

Fiz-Palacios, O., H. Schneider, J. Heinrichs, and V. Savolainen. 2011. Diversification of land plants: Insights from a family-level phylogenetic analysis. BMC Evolutionary Biology 11:341.

Flagel, L., J. Udall, D. Nettleton, and J. Wendel. 2008. Duplicate gene expression in allopolyploid Gossypium reveals two temporally distinct phases of expression evolution. BMC Biology 6:16.

Flagel, L. E., and J. F. Wendel. 2009. Gene duplication and evolutionary novelty in plants. New Phytologist 183:557–564.

Flatt, T. 2005. The evolutionary genetics of canalization. Quarterly Review of Biology 80:287–316.

Ford, E. B. 1971. Ecological Genetics. Methuen, London.

Forterre, P. 2006. Three RNA cells for ribosomal lineages and three DNA viruses to replicate their genomes: A hypothesis for the origin of cellular domain. Proceedings of the National Academy of Sciences of the United States of America 103:3669–3674.

Fox, S. W., J. R. Jungck, and T. Nakashima. 1974. From proteinoid microsphere to contemporary cell: Formation of internucleotide and peptide bonds by proteinoid particles. Origins of Life and Evolution of the Biosphere 5:227–237.

Franklin-Tong, V. E., and F. C. H. Franklin. 2003. The different mechanisms of gametophytic self-incompatibility. Philosophical Transactions of the Royal Society B: Biological Sciences 358:1025–1032.

Freeling, M., E. Lyons, B. Pedersen, M. Alam, R. Ming, and D. Lisch. 2008. Many or most genes in *Arabidopsis* transposed after the origin of the order Brassicales. Genome Research 18:1924–1937.

Freeling, M., and B. C. Thomas. 2006. Gene-balanced duplications, like tetraploidy, provide predictable drive to increase morphological complexity. Genome Research 16:805–814.

Frichot, E., S. D. Schoville, G. Bouchard, and O. François. 2013. Testing for associations between loci and environmental gradients using latent factor mixed models. Molecular Biology and Evolution 30:1687–1699.

Friedman, J., and S. C. H. Barrett. 2009. Wind of change: New insights on the ecology and evolution of pollination and mating in wind-pollinated plants. Annals of Botany 103:1515–1527.

Friedman, W. E. 2009. The meaning of Darwin's "abominable mystery." American Journal of Botany 96:5–21.

Friedman, W. E., E. N. Madrid, and J. H. Williams. 2008. Origin of the fittest and survival of the fittest: Relating female gametophyte development to endosperm genetics. International Journal of Plant Sciences 169:79–92.

Friedman, W. E., and J. H. Williams. 2004. Developmental evolution of the sexual process in ancient flowering plant lineages. Plant Cell Online 16:S119–S132.

Friis, E. M., K. R. Pedersen, and P. R. Crane. 2006. Cretaceous angiosperm flowers: Innovation and evolution in plant reproduction. Palaeogeography, Palaeoclimatology, Palaeoecology 232:251–293.

Fritsche, F., and O. Kaltz. 2000. Is the *Prunella* (Lamiaceae) hybrid zone structured by an environmental gradient? Evidence from a reciprocal transplant experiment. American Journal of Botany 87:995–1003.

Frodin, D. G. 2004. History and concepts of big plant genera. Taxon 53:753–776.

Furches, M. S., R. L. Small, and A. Furches. 2013. Genetic diversity in three endangered pitcher plant species (*Sarracenia*; Sarraceniaceae) is lower than widespread congeners. American Journal of Botany 100:2092–2101.

Gailing, O., M. R. Macnair, and K. Bachmann. 2004. QTL mapping for a trade-off between leaf and bud production in a recombinant inbred population of *Microseris douglasii* and *M. bigelovii* (Asteraceae, Lactuceae): A potential preadaptation for the colonization of serpentine soils. Plant Biology 6:440–446.

Galloway, L. F., and J. R. Etterson. 2007. Transgenerational plasticity is adaptive in the wild. Science 318:1134–1136.

Galloway, L. F., and C. B. Fenster. 1999. The effect of nuclear and cytoplasmic genes on fitness and local adaptation in an annual legume, *Chamaecrista fasciculata*. Evolution 53:1734–1743.

Garrick, R. C., A. Caccone, and P. Sunnucks. 2010. Inference of population history by coupling exploratory and model-driven phylogeographic analyses. International Journal of Molecular Sciences 11:1190–1227.

Garrick, R. C., J. D. Nason, C. A. Meadows, and R. J. Dyer. 2009. Not just vicariance: Phylogeography of a Sonoran Desert euphorb indicates a major role of range expansion along the Baja Peninsula. Molecular Ecology 18:1916–1931.

Gaudeul, M., T. Giraud, L. Kiss, and J. A. Shykoff. 2011. Nuclear and chloroplast microsatellites show multiple introductions in the worldwide invasion history of common ragweed, Ambrosia artemisiifolia. PLoS One 6:e17658.

Gaut, B., L. Yang, S. Takuno, and L. E. Eguiarte. 2011. The patterns and causes of variation in plant nucleotide substitution rates. Pp. 245–266 *in* D. J. Futuyma, H. B. Shaffer, and D. Simberloff, eds. Annual Review of Ecology, Evolution, and Systematics, Vol 42.

Gavrilets, S. 1997. Evolution and speciation on holey adaptive landscapes. Trends in Ecology and Evolution 12:307–312.

Gavrilets, S. 2003. Perspective: Models of speciation: What have we learned in forty years? Evolution 57:2197–2215.

Gavrilets, S., and M. B. Cruzan. 1998. Neutral gene flow across single locus clines. Evolution 52:1277–1284.

Gavrilets, S., H. Li, and M. D. Vose. 1998. Rapid speciation on holey adaptive landscapes. Proceedings of the Royal Society B: Biological Sciences 266:817–824.

Geber, M. A., and L. R. Griffen. 2003. Inheritance and natural selection on functional traits. International Journal of Plant Sciences 164:S21–S42.

Gerber, S., J. Chadœuf, F. Gugerli, M. Lascoux, J. Buiteveld, J. Cottrell, A. Dounavi, S. Fineschi, L. L. Forrest, J. Fogelqvist, P. G. Goicoechea, J. S. Jensen, D. Salvini, G. G. Vendramin, and A. Kremer. 2014. High rates of gene flow by pollen and seed in oak populations across Europe. PLoS One 9:e85130.

Gerrienne, P., and P. Gonez. 2011. Early evolution of life cycles in embryophytes: A focus on the fossil evidence of gametophyte/sporophyte size and morphological complexity. Journal of Systematics and Evolution 49:1–16.

Gerrienne, P., B. Meyer-Berthaud, M. Fairon-Demaret, M. Streel, and P. Steemans. 2004. Runcaria, a middle Devonian seed plant precursor. Science 306:856–858.

Geuten, K., and H. Coenen. 2013. Heterochronic genes in plant evolution and development. Frontiers in Plant Science 4:381.

Ghiselin, M. T. 1969. The evolution of hermaphroditism among animals. Quarterly Review of Biology l 44:189–208.

Ghiselin, M. T. 1974. The Economy of Nature and the Evolution of Sex. University of California Press, Berkeley.

Gibbs, P. E. 2014. Late-acting self-incompatibility: The pariah breeding system in flowering plants. New Phytologist 203:717–734.

Gitzendanner, M. A., and P. S. Soltis. 2000. Patterns of genetic variation in rare and widespread plant congeners. American Journal of Botany 87:783–792.

Givnish, T. J. 2010. Ecology of plant speciation. Taxon 59:1326–1366.

Glenn, T. C. and N. A. Schable. 2005. Isolating microsatellite DNA loci. Pp. 202–222 *in* E. A. Zimmer and E. H. Roalson, eds. Molecular Evolution: Producing the Biochemical Data, Part B. Elsevier Academic Press, San Diego, CA.

Glover, B. 2007. Understanding flowers and flowering: An integrated approach. Oxford University Press, New York.

Godoy, O., A. Saldana, N. Fuentes, F. Valladares, and E. Gianoli. 2011. Forests are not immune to plant invasions: Phenotypic plasticity and local adaptation allow *Prunella vulgaris* to colonize a temperate evergreen rainforest. Biological Invasions 13:1615–1625.

Godwin, I. D., E. A. B. Aitken, and L. W. Smith. 1997. Application of inter simple sequence repeat (ISSR) markers to plant genetics. Electrophoresis 18:1524–1528.

Goldberg, E. E., J. R. Kohn, R. Lande, K. A. Robertson, S. A. Smith, and B. Igić. 2010. Species selection maintains self-incompatibility. Science 330:493–495.

Goldsworthy, A. 1987. Why did nature select green plants? Nature 328:207–208.

Gomez, J. M., and R. Zamora. 1996. Wind pollination in high-mountain populations of *Hormathophylla spinosa* (Cruciferae). American Journal of Botany 83:580–585.

Gonzalez-Martinez, S. C., K. V. Krutovsky, and D. B. Neale. 2006. Forest-tree population genomics and adaptive evolution. New Phytologist 170:227–238.

Good-Avila, S. V., V. Souza, B. S. Gaut, and L. E. Eguiarte. 2006. Timing and rate of speciation in *Agave* (Agavaceae). Proceedings of the National Academy of Sciences of the United States of America 103:9124–9129.

Goodnight, C. 2013. On multilevel selection and kin selection: Contextual analysis meets direct fitness. Evolution 67:1539–1548.

Goodnight, C. J. 1988. Epistasis and the effect of founder events on the additive genetic variance. Evolution 42:441–454.

Goodwillie, C., S. Kalisz, and C. G. Eckert. 2005. The evolutionary enigma of mixed mating systems in plants: Occurrence, theoretical explanations, and empirical evidence. Annual Review of Ecology Evolution and Systematics 36:47–79.

Gossmann, T. I., D. Saleh, M. W. Schmid, M. A. Spence, and K. J. Schmid. 2016. Transcriptomes of plant gametophytes have a higher proportion of rapidly evolving and young genes than sporophytes. Molecular Biology and Evolution 33:1669–1678.

Gottlieb, L. D. 2004. Rethinking classic examples of recent speciation in plants. New Phytologist 161:71–82.

Gould, S. J. 1966. Allometry and size in ontogeny and phylogeny. Biological Reviews 41:587–680.

Gould, S. J. 1977. Ontogeny and Phylogeny. Harvard University Press, Cambridge, MA.

Gould, S. J., and N. Eldredge. 1977. Punctuated equilibrium: The tempo and mode of evolution reconsidered. Paleobiology 3:115–151.

Gould, S. J., and R. C. Lewontin. 1979. The spandrels of San Marco and the Panglossian paradigm: A critique of the adaptionist programme. Proceedings of the Royal Society of London B: Biological Sciences 205:581–598.

Gould, S. J., and E. S. Vrba. 1982. Exaptation—a missing term in the science of form. Paleobiology 8:4–15.

Graham, L. E., P. Arancibia-Avila, W. A. Taylor, P. K. Strother, and M. E. Cook. 2012. Aeroterrestrial *Coleochaete* (Streptophyta, Coleochaetales) models early plant adaptation to land. American Journal of Botany 99:130–144.

Grant, B. R., and P. R. Grant. 1996. High survival of Darwin's finch hybrids: Effects of beak morphology and diets. Ecology 77:500–509.

Grant, K. A., and V. Grant. 1964. Mechanical isolation of *Salvia apian* and *Salvia mellifera* (Labiatae). Evolution 18:196–212.

Grant, V. 1949. Pollination systems as isolating mechanisms in Angiosperms. Evolution 3:82–97.

Grant, V. 1953. The role of hybridization in the evolution of the leafy-stemmed gilias. Evolution 7:51–64.

Grant, V. 1963. The Origin of Adaptations. Columbia University Press, New York.

Grant, V. 1971. Plant Speciation. Columbia University Press, New York.

Grant, V. 1978. Genetics of Flowering Plants. Columbia University Press, New York.

Grant, V. 1981. Plant Speciation. Columbia University Press, New York.

Grant, V. 1993. Origin of floral isolation between ornithophilous and sphingophilous plant species. Proceedings of the National Academy of Sciences of the United States of America 90:7729–7733.

Grant, V. 1994. Modes and origins of mechanical and ethological isolation in angiosperms. Proceedings of the National Academy of Sciences of the United States of America 91:3–10.

Grant, V., and R. H. Flake. 1974. Solutions to the cost-of-selection dilemma. Proceedings of the National Academy of Sciences of the United States of America 71:3863–3865.

Grant-Downton, R. T., and H. G. Dickinson. 2006. Epigenetics and its implications for plant biology. 2. The "epigenetic epiphany": Epigenetics, evolution and beyond. Annals of Botany 97:11–27.

Grativol, C., A. S. Hemerly, and P. C. Gomes Ferreira. 2012. Genetic and epigenetic regulation of stress responses in natural plant populations. Biochimica Et Biophysica Acta-Gene Regulatory Mechanisms 1819:176–185.

Greb, S. F., and W. A. DiMichele. 2006. Wetlands Through Time. Geological Society of America, Boulder, CO.

Gribaldo, S., A. M. Poole, V. Daubin, P. Forterre, and C. Brochier-Armanet. 2010. The origin of eukaryotes and their relationship with the Archaea: Are we at a phylogenomic impasse? Nature Reviews Microbiology 8:743–752.

Griffin, S. R., and S. C. H. Barrett. 2004. Post-glacial history of *Trillium grandiflorum* (Melanthiaceae) in eastern North America: Inferences from phylogeography. American Journal of Botany 91:465–473.

Gross, B. L., and L. H. Rieseberg. 2005. The ecological genetics of homoploid hybrid speciation. Journal of Heredity 96:241–252.

Gross, C. L., P. A. Nelson, A. Haddadchi, and M. Fatemi. 2012. Somatic mutations contribute to genotypic diversity in sterile and fertile populations of the threatened shrub *Grevillea rhizomatosa* (Proteaceae). Annals of Botany 109:331–342.

Grossenbacher, D. L., S. D. Veloz, and J. P. Sexton. 2014. Niche and range size patterns suggest that speciation begins in small, ecologically diverged populations in North American monkeyflowers (*Mimulus* spp.). Evolution 68:1270–1280.

Grover, C. E., A. Salmon, and J. F. Wendel. 2012. Targeted sequence capture as a powerful tool for evolutionary analysis. American Journal of Botany 99:312–319.

Guerrant, E. O. 1982. Neotenic evolution of *Delphinium nudicaule* (Ranunculaceae): A hummingbird-pollinated larkspur. Evolution 36:699–712.

Gugger, P. F., A. Gonzalez-Rodriguez, H. Rodriguez-Correa, S. Sugita, and J. Cavender-Bares. 2011. Southward Pleistocene migration of Douglas-fir into Mexico: Phylogeography, ecological niche modeling, and conservation of "rear edge" populations. New Phytologist 189:1185–1199.

Gurevitch, J., S. M. Scheiner, and G. A. Fox. 2006. The Ecology of Plants. Sinauer Associates, Inc, Sunderland, MA.

Haag-Liautard, C., M. Dorris, X. Maside, S. Macaskill, D. L. Halligan, B. Charlesworth, and P. D. Keightley. 2007. Direct estimation of per nucleotide and genomic deleterious mutation rates in *Drosophila*. Nature 445:82–85.

Halbritter, A. H., R. Billeter, P. J. Edwards, and J. M. Alexander. 2015. Local adaptation at range edges: Comparing elevation and latitudinal gradients. Journal of Evolutionary Biology 28:1849–1860.

Haldane, J. B. S. 1932. The Causes of Evolution. Princeton University Press, Princeton.

Haldane, J. B. S. 1957. The cost of natural selection. Journal of Genetics 55:511–524.

Haldane, J. B. S. 1964. A defense of beanbag genetics. Perspectives in Biology and Medicine 7:343–359.

Hall, B. G. 2001. Phylogenetic Trees Made Easy. Sinauer Associates, Inc., Sunderland, MA.

Hall, D. W., and S. B. Joseph. 2010. A high frequency of beneficial mutations across multiple fitness components in *Saccharomyces cerevisiae*. Genetics 185:1397–1409.

Hall, M. C., D. B. Lowry, and J. H. Willis. 2010. Is local adaptation in *Mimulus guttatus* caused by trade-offs at individual loci? Molecular Ecology 19:2739–2753.

Halligan, D. L., and P. D. Keightley. 2009. Spontaneous Mutation Accumulation Studies in Evolutionary Genetics. Annual Review of Ecology Evolution and Systematics 40:151–172.

Hamilton, J. A., and S. N. Aitken. 2013. Genetic and morphological structure of a spruce hybrid (*Picea sitchensis* × *P. glauca*) zone along a climatic gradient. American Journal of Botany 100:1651–1662.

Hamilton, W. D. 1966. The moulding of senescence by natural selection. Journal of Theoretical Biology 12:12–45.

Hamrick, J. L., and R. W. Allard. 1972. Microgeographic variation in allozyme frequencies in *Avena barbata*. Proceedings of the National Academy of Sciences of the United States of America 69:2100–2104.

Hamrick, J. L., and M. J. W. Godt. 1990. Allozyme diversity in plant species. Pp. 43–63 *in* A. H. D. Brown, M. T. Clegg, A. L. Kahler, and B. S. Weir, eds. Plant Population Genetics, Breeding, and Genetic Resources. Sinauer Associates, Inc., Sunderland, MA.

Hamrick, J. L., and M. J. W. Godt. 1996. Effects of life history traits on genetic diversity in plant species. Philosophical Transactions: Biological Sciences 351:1291–1298.

Hamrick, J. L. and M. J. W. Godt. 1997. Effects of life history traits on genetic diversity in plant species. Pp. 102–118 *in* J. Silverton, M. Franco, and J. L. Harper, eds. Plant Life Histories: Ecology, Phylogeny, and Evolution. Cambridge University Press, Cambridge, UK.

Hamrick, J. L., and L. R. Holden. 1979. Influence of microhabitat heterogeneity on gene frequency distribution and gamete phase disequilibrium in *Avena barbata*. Evolution 33:521–533.

Hansen, T. F., C. Pelabon, and D. Houle. 2011. Heritability is not evolvability. Evolutionary Biology 38:258–277.

Harder, L. D., M. B. Cruzan, and J. D. Thompson. 1993. Unilateral incompatibility and the effects of interspecific pollination for *Erythronium americanum* and *Erythronium albidum* (Liliaceae). Canadian Journal of Botany 71:353–358.

Harder, L. D., and S. D. Johnson. 2009. Darwin's beautiful contrivances: Evolutionary and functional evidence for floral adaptation. New Phytologist 183:530–545.

Harder, L. D., S. A. Richards, and M. B. Routley. 2008. Effects of reproductive compensation, gamete discounting, and reproductive assurance on mating system diversity in hermaphrodites. Evolution 62:157–172.

Harder, L. D., and M. B. Routley. 2006. Pollen and ovule fates and reproductive performance by flowering plants. Pp. 61–80 *in* L. D. Harder and S. C. H. Barrett, eds. Ecology and Evolution of Flowers. Oxford University Press, New York.

Harder, L. D., and J. D. Thomson. 1989. Evolutionary options for maximizing pollen dispersal of animal-pollinated plants. American Naturalist 133:323–344.

Harder, L. D., and W. G. Wilson. 1998. A clarification of pollen discounting and its joint effects with inbreeding depression on mating system evolution. American Naturalist 152:684–695.

Hardy, G. H. 1908. Mendelian proportions in a mixed population. Science 28:49–50.

Harper, J. L. 1977. Population Biology of Plants. Academic Press, London.

Harrison, B. J., and L. A. Darby. 1955. Unilateral hybridization. Nature 176:982–982.

Harrison, R. G. 1990. Hybrid zones: Windows on evolutionary process. Oxford Surveys in Evolutionary Biology 7:69–128.

Harrison, R. G., and E. L. Larson. 2014. Hybridization, introgression, and the nature of species boundaries. Journal of Heredity 105:795–809.

Harrison, S., and N. Rajakaruna. 2011. Serpentine: The evolution and ecology of a model system. University of California Press, Berkeley.

Harte, C. 1994. *Oenothera*—Contributions of a plant to biology. Springer-Verlag, Gerlin.

Hartfield, M., and S. P. Otto. 2011. Recombinaiton and hitchhiking of deleterious alleles. Evolution 65:2421–2434.

Hartl, D. L., and A. G. Clark. 2006. Principles of Population Genetics. Sinauer Associates, Inc., Sunderland, MA.

Hartman, H., and A. Fedorov. 2002. The origin of the eukaryotic cell: A genomic investigation. Proceedings of the National Academy of Sciences of the United States of America 99:1420–1425.

Hartmann, T. 2007. From waste products to ecochemicals: Fifty years research of plant secondary metabolism. Phytochemistry 68:2831–2846.

Hastings, A., K. Cuddington, K. F. Davies, C. J. Dugaw, S. Elmendorf, A. Freestone, S. Harrison, M. Holland, J. Lambrinos, U. Malvadkar, B. A. Melbourne, K. Moore, C. Taylor, and D. Thomson. 2005. The spatial spread of invasions: New developments in theory and evidence. Ecology Letters 8:91–101.

Hastings, A., and S. Harrison. 1994. Metapopulation dynamics and genetics. Annual Review of Ecology and Systematics 25:167–188.

Heckman, D. S., D. M. Geiser, B. R. Eidell, R. L. Stauffer, N. L. Kardos, and S. B. Hedges. 2001. Molecular evidence for the early colonization of land by fungi and plants. Science 293:1129–1133.

Hedges, S. B. 2002. The origin and evolution of model organisms. Nature Reviews Genetics 3:838–849.

Hedrick, P. W. 1981. The establishment of chromosomal variants. Evolution 35:322–332.

Hedrick, P. W. 2005a. Genetics of Populations. Jones and Bartlett Publishers, Sudbury, MA.

Hedrick, P. W. 2005b. A standardized genetic differentiation measure. Evolution 59:1633–1638.

Hedrick, P. W. 2009. Genetics of Populations. Van Nostrand Reinhold, Portola Valley.

Heilbuth, J. C. 2000. Lower species richness in dioecious clades. American Naturalist 156:221–241.

Heinrich, B. 1979. "Majoring" and "minoring" by foraging bumblebees, *Bombus Vagans*: An experimental analysis. Ecology 60:246–255.

Heiser, C. B. 1949. Natural hybridization with particular reference to introgression. Botanical Review 15:645–687.

Heiser, C. B. 1973. Introgression reexamined. Botanical Review 39:347–366.

Hendry, A. P., P. Nosil, and L. H. Rieseberg. 2007. The speed of ecological speciation. Functional Ecology 21:455–464.

Hennig, W. 1966. Phylogenetic Systematics. University of Illinois Press, Urbana.

Herman, J. J., H. G. Spencer, K. Donohue, and S. E. Sultan. 2014. How stable "should" epigenetic modifications be? Insights from adaptive plasticity and bet hedging. Evolution 68:632–643.

Herrera, C. M., and P. Bazaga. 2011. Untangling individual variation in natural populations: Ecological, genetic, and epigenetic correlates of long-term inequality in herbivory. Molecular Ecology 20:1675–1688.

Herrera, C. M., and R. Jovani. 2010. Lognormal distribution of individual lifetime fecundity: Insights from a twenty-three-year study. Ecology 91:422–430.

Herrera, C. M., M. Medrano, and P. Bazaga. 2013. Epigenetic differentiation persists after male gametogenesis in natural populations of the perennial herb *Helleborus foetidus* (Ranunculaceae). PLoS One 8:e70730.

Hersch-Green, E. I. 2012. Polyploidy in Indian paintbrush (*Castilleja*; Orobanchaceae) species shapes but does not prevent gene flow across species boundaries. American Journal of Botany 99:1680–1690.

Hewitt, G. 2000. The genetic legacy of the Quaternary ice ages. Nature 405:907–913.

Hewitt, G. M. 1999. Post-glacial re-colonization of European biota. Biological Journal of the Linnean Society 68:1–2.

Hey, J. 2006. On the failure of modern species concepts. Trends in Ecology & Evolution 21:447–450.

Hey, J., and C. A. Machado. 2003. The study of structured populations: New hope for a difficult and divided science. Nature Reviews Genetics 4:535–543.

Heyman, J., T. Cools, F. Vandenbussche, K. S. Heyndrickx, J. Van Leene, I. Vercauteren, S. Vanderauwera, K. Vandepoele, G. De Jaeger, D. Van Der Straeten, and L. De Veylder. 2013. ERF115 controls root quiescent center cell division and stem cell replenishment. Science 342:860–863.

Hilbish, T. J., and R. K. Koehn. 1987. The adaptive importance of genetic variation. American Scientist 75:134–141.

Hill, W. G. 2010. Understanding and using quantitative genetic variation. Philosophical Transactions of the Royal Society B: Biological Sciences 365:73–85.

Hill, W. G., and A. Caballero. 1992. Artificial selection experiments. Annual Review of Ecology and Systematics 23:287–310.

Hill, W. G., and A. Robertson. 1966. The effect of linkage on limits to artificial selection. Genetics Research 8:269–294.

Hillis, D. M., and M. T. Dixon. 1991. Riosomal DNA—molecular evolution and phylogenetic inference. Quarterly Review of Biology 66:411–453.

Hills, H. G., N. H. Williams, and C. H. Dodson. 1972. Floral fragrances and isolating mechanisms in the genus *Catasetum* (Orchidaceae). Biotropica 4:61–76.

Hirao, A. S., and G. Kudo. 2004. Landscape genetics of alpine-snowbed plants: Comparisons along geographic and snowmelt gradients. Heredity 93:290–298.

Hiscock, S. J., and D. A. Tabah. 2003. The different mechanisms of sporophytic self-incompatibility. Philosophical Transactions of the Royal Society of London B: Biological Sciences 358:1037–1045.

Hodges, S. A., and M. L. Arnold. 1994. Columbines: A geographically widespread species flock. Proceedings of the National Academy of Sciences of the United States of America 91:5129–5132.

Hodges, S. A., and M. L. Arnold. 1995. Spurring plant diversification: Are floral nectar spurs a key innovation? Proceedings of the Royal Society of London B: Biological Sciences 262:343–348.

Hodgins, K. A., and L. Rieseberg. 2011. Genetic differentiation in life-history traits of introduced and native common ragweed (*Ambrosia artemisiifolia*) populations. Journal of Evolutionary Biology 24:2731–2749.

Hogenboom, N. G. 1975. Incompatibility and incongruity: Two different mechanisms for the non-functioning of intimate partner relationships. Proceedings of the Royal Society London B: Biological Sciences 188:361–375.

Hohenlohe, P. A., P. C. Phillips, and W. A. Cresko. 2010. Using population genomics to detect selection in natural populations: Key concepts and methodological considerations. International Journal of Plant Sciences 171:1059–1071.

Hojsgaard, D., S. Klatt, R. Baier, J. G. Carman, E. and Horandl. 2014. Taxonomy and biogeography of apomixis in angiosperms and associated biodiversity characteristics. Critical Reviews in Plant Sciences 33:414–427.

Hojsgaard, D., M. Pellino, T. Sharbel, and E. Horandl. 2015. Resolving genome evolution patterns in asexual plants. Pp. 119–153 *in* E. Horandl and M. S. Appelhans, eds. Next Generation Sequencing in Plant Systematics. Koeltz Scientific Books, Königstein, Germany.

Holderegger, R., D. Buehler, F. Gugerli, and S. Manel. 2010. Landscape genetics of plants. Trends in Plant Science 15:675–683.

Holeski, L. M., G. Jander, and A. A. Agrawal. 2012. Transgenerational defense induction and epigenetic inheritance in plants. Trends in Ecology & Evolution 27:618–626.

Hollingshead, L. 1930. A lethal factor in *Crepis* effective only in an interspecific hybrid. Genetics 15:114–140.

Holmquist, K. G., R. J. Mitchell, and J. D. Karron. 2012. Influence of pollinator grooming on pollen-mediated gene dispersal in *Mimulus ringens* (Phrymaceae). Plant Species Biology 27:77–85.

Holsinger, K. E. 2000. Reproductive systems and evolution in vascular plants. Proceedings of the National Academy of Sciences of the United States of America 97:7037–7042.

Holt, R. D., and T. H. Keitt. 2000. Alternative causes for range limits: A metapopulation perspective. Ecology Letters 3:41–47.

Honnay, O., B. Bossuyt, H. Jacquemyn, A. Shimono, and K. Uchiyama. 2008. Can a seed bank maintain the genetic variation in the above-ground plant population? Oikos 117:1–5.

Hopkins, R. 2013. Reinforcement in plants. New Phytologist 197:1095–1103.

Hörandl, E. 2006. The complex causality of geographical parthenogenesis. New Phytologist 171:525–538.

Hörandl, E., A.-C. Cosendai, and E. M. Temsch. 2008. Understanding the geographic distributions of apomictic plants: A case for a pluralistic approach. Plant Ecology & Diversity 1:309–320.

Hori, H., B. L. Lim, and S. Osawa. 1985. Evolution of green plants as deduced from 5S ribosomal DNA sequences. Proceedings of the National Academy of Sciences of the United States of America 82:820–823.

Horodyski, R. J., and L. P. Knauth. 1994. Life on land in the Precambrian. Science 263:494–498.

Hough, J., R. J. Williamson, and S. I. Wright. 2013. Patterns of selection in plant genomes. Annual Review of Ecology, Evolution, and Systematics, Vol 44:31–49.

Houle, D. 1989. Allozyme-associated heterosis in *Drosophila melanogaster*. Genetics 123:789–801.

Howard, D. J. 1993. Reinforcement: Origin, dynamics, and fate of an evolutionary hypothesis. Pp. 46–69 *in* R. G. Harrison, ed. Hybrid Zones and the Evolutionary Process. Oxford University Press, New York.

Howard, D. J. 1999. Conspecific sperm and pollen precedence and speciation. Annual Review of Ecology and Systematics 30:109–132.

Howe, H. F., and M. N. Miriti. 2004. When seed dispersal matters. BioScience 54:651–660.

Howe, H. F., and J. Smallwood. 1982. Ecology of seed dispersal. Annual Review of Ecology and Systematics 13:201–228.

Hudson, R. R. 1990. Gene genealogies and the coalescent process. Oxford Surveys in Evolutionary Biology 7:1–44.

Hunter, R. L., and C. L. Markert. 1957. Histochemical demonstration of enzymes separated by zone electrophoresis in starch gels. Science 125:1294–1295.

Husband, B. C., and H. A. Sabara. 2004. Reproductive isolation between autotetraploids and their diploid progenitors in fireweed, *Chamerion angustifolium* (Onagraceae). New Phytologist 161:703–713.

Husband, B. C., and D. W. Schemske. 1996. Evolution of the magnitude and timing of inbreeding depression in plants. Evolution 50:54–70.

Hutchison, D. W., and A. R. Templeton. 1999. Correlation of pairwise genetic and geographic distance measures: Inferring the relative influences of gene flow and drift on the distribution of genetic variability. Evolution 53:1898–1914.

Igic, B., and J. R. Kohn. 2001. Evolutionary relationships among self-incompatibility RNases. Proceedings of the National Academy of Sciences of the United States of America 98:13167–13171.

Igic, B., R. Lande, and J. R. Kohn. 2008. Loss of self-incompatibility and its evolutionary consequences. International Journal of Plant Sciences 169:93–104.

Imbert, E. 2002. Ecological consequences and ontogeny of seed heteromorphism. Perspectives in Plant Ecology Evolution and Systematics 5:13–36.

Ingrouille, M., and B. Eddie. 2006. Plants: Diversity and Evolution. Cambridge University Press, New York.

Ingvarsson, P. K., and M. C. Whitlock. 2000. Heterosis increases the effective migration rate. Proceedings of the Royal Society London B: Biological Sciences. Sci. 267:1321–1326.

Jablonski, D. 2008. Species Selection: Theory and Data. Annual Review of Ecology Evolution and Systematics 39:501–524.

Jackiw, R. N., G. Mandil, and H. A. Hager. 2015. A framework to guide the conservation of species hybrids based on ethical and ecological considerations. Conservation Biology 29:1040–1051.

Jakob, S. S., E. Martinez-Meyer, and F. R. Blattner. 2009. Phylogeographic analyses and paleodistribution modeling indicate pleistocene in situ survival of *Hordeum* species (Poaceae) in Southern Patagonia without genetic or spatial restriction. Molecular Biology and Evolution 26:907–923.

Jaquiery, J., T. Broquet, A. H. Hirzel, J. Yearsley, and N. Perrin. 2011. Inferring landscape effects on dispersal from genetic distances: How far can we go? Molecular Ecology 20:692–705.

Jarne, P. and J. R. Auld. 2006. Animals mix it up too: The distribution of self-fertilization among hermaphroditic animals. Evolution 60:1816–1824.

Jeffreys, A. J., V. Wilson, and S. L. Thein. 1985. Hypervariable minisatellite regions in human DNA. Nature 314:67–73.

Jiao, Y., J. Leebens-Mack, S. Ayyampalayam, J. E. Bowers, M. R. McKain, J. McNeal, M. Rolf, D. R. Ruzicka, E. Wafula, N. J. Wickett, X. Wu, Y. Zhang, J. Wang, Y. Zhang, E. J. Carpenter, M. K. Deyholos, T. M. Kutchan, A. S. Chanderbali, P. S. Soltis, D. W. Stevenson, R. McCombie, J. C. Pires, G. K.-S. Wong, D. E. Soltis, and C. W. dePamphilis. 2012. A genome triplication associated with early diversification of the core eudicots. Genome Biology 13:R3.

Jiao, Y., N. J. Wickett, S. Ayyampalayam, A. S. Chanderbali, L. Landherr, P. E. Ralph, L. P. Tomsho, Y. Hu, H. Liang, P. S. Soltis, D. E. Soltis, S. W. Clifton, S. E. Schlarbaum, S. C. Schuster, H. Ma, J. Leebens-Mack, and C. W. dePamphilis. 2011. Ancestral polyploidy in seed plants and angiosperms. Nature 473:97–U113.

Johnson, M. T. J., S. D. Smith, and M. D. Rausher. 2010. Effects of plant sex on range distributions and allocation to reproduction. New Phytologist 186:769–779.

Johnston, W. K., P. J. Unrau, M. S. Lawrence, M. E. Glasner, and D. P. Bartel. 2001. RNA-catalyzed RNA polymerization: Accurate and general RNA-templated primer extension. Science 292:1319–1325.

Jones, D. F. 1928. Selective Fertilization. University of Chicago Press, Chicago.

Jones, J. D. G., and J. L. Dangl. 2006. The plant immune system. Nature 444:323–329.

Jones, M. R., and J. M. Good. 2015. Targeted capture in evolutionary and ecological genomics. Molecular Ecology 25:185–202.

Jones, R., H. Ougham, H. Thomas, and S. Waaland. 2012. The Molecular Life of Plants. Wiley-Blackwell, Hoboken, NJ.

Jones-Rhoades, M. W., D. P. Bartel, and B. Bartel. 2006. MicroRNAs and their regulatory roles in plants. Annual Review of Plant Biology 57:19–53.

Joseph, S. B., and D. W. Hall. 2004. Spontaneous mutations in diploid *Saccharomyces cerevisiae*: More beneficial than expected. Genetics 168:1817–1825.

Jost, L. O. U. 2008. G_{ST} and its relatives do not measure differentiation. Molecular Ecology 17:4015–4026.

Käfer, J., G. A. B. Marais, and J. R. Pannell. 2017. On the rarity of dioecy in flowering plants. Molecular Ecology 26:1225–1241.

Kalisz, S., and M. D. Purugganan. 2004. Epialleles via DNA methylation: Consequences for plant evolution. Trends in Ecology & Evolution 19:309–314.

Karban, R., A. A. Agrawal, J. S. Thaler, and L. S. Adler. 1999. Induced plant responses and information content about risk of herbivory. Trends in Ecology & Evolution 14:443–447.

Karron, J. D., and R. J. Mitchell. 2012. Effects of floral display size on male and female reproductive success in *Mimulus ringens*. Annals of Botany 109:563–570.

Karron, J. D., R. Tucker, N. N. Thumser, and J. A. Reinartz. 1995. Comparison of pollinator flight movements and gene dispersal patterns in *Mimulus ringens*. Heredity 75:612–617.

Karst, A. L., and M. J. Lechowicz. 2007. Are correlations among foliar traits in ferns consistent with those in the seed plants? New Phytologist 173:306–312.

Kay, K. M. 2006. Reproductive isolation between two closely related hummingbird-pollinated neotropical gingers. Evolution 60:538–552.

Kay, K. M., and R. D. Sargent. 2009. The role of animal pollination in plant speciation: Integrating ecology, geography, and genetics. Annual Review of Ecology Evolution and Systematics 40:637–656.

Keeling, P. J. 2010. The endosymbiotic origin, diversification and fate of plastids. Philosophical Transactions of the Royal Society B: Biological Sciences 365:729–748.

Keeling, P. J., G. Burger, D. G. Durnford, B. F. Lang, R. W. Lee, R. E. Pearlman, A. J. Roger, and M. W. Gray. 2005. The tree of eukaryotes. Trends in Ecology & Evolution 20:670–676.

Kelchner, S. A., and M. A. Thomas. 2007. Model use in phylogenetics: Nine key questions. Trends in Ecology & Evolution 22:87–94.

Keller, B., J. D. Thomson, and E. Conti. 2014. Heterostyly promotes disassortative pollination and reduces sexual interference in Darwin's primroses: Evidence from experimental studies. Functional Ecology 28:1413–1425.

Keller, S. R., N. Levsen, M. S. Olson, and P. Tiffin. 2012. Local adaptation in the flowering-time gene network of balsam poplar, *Populus balsamifera* L. Molecular Biology and Evolution 29:3143–3152.

Keller, S. R., and D. R. Taylor. 2008. History, chance and adaptation during biological invasion: Separating stochastic phenotypic evolution from response to selection. Ecology Letters 11:852–866.

Kelley, S. E. 1989. Experimental studies of the evolutionary significance of sexual reproduction. 5. A field test of the sib-competition lottery hypothesis. Evolution 43:1054–1065.

Kempthorne, O. 1954. The correlation between relatives in a random mating population. Proceedings of the Royal Society of London B: Biological Sciences 143:102–113.

Kennedy, B. F., and E. Elle. 2008. The reproductive assurance benefit of selfing: Importance of flower size and population size. Oecologia 155:469–477.

Kentner, E. K., and M. R. Mesler. 2000. Evidence for natural selection in a fern hybrid zone. American Journal of Botany 87:1168–1174.

Kephart, S., and K. Theiss. 2004. Pollinator-mediated isolation in sympatric milkweeds (*Asclepias*): Do floral morphology and insect behavior influence species boundaries? New Phytologist 161:265–277.

Kerner, A. 1894. The Natural History of Plants, Volume 1 (English Translation). Blackie and Son, London.

Kerster, H. W., and D. A. Levin. 1968. Neighborhood size in *Lithospermum caroliniense*. Genetics 60:577–587.

Khanna-Chopra, R. 2012. Leaf senescence and abiotic stresses share reactive oxygen-species-mediated chloroplast degradation. Protoplasma 249:469–481.

Kimura, M. 1968. Evolutionary rate at the molecular level. Nature 217:624–626.

Kimura, M. 1983. The Neutral Theory of Molecular Evolution. Cambridge University Press, Cambridge, UK.

Kimura, M., and J. F. Crow. 1964. The number of alleles that can be maintained in a finite population. Genetics 49:725–738.

King, J. L., and T. H. Jukes. 1969. Non-Darwinian evolution. Science 164:788–798.

Kingman, J. 1982. The coalescent. Stochastic Processes and their Applications 13:235–248.

Kirkpatrick, M. 2010. How and why chromosome inversions evolve. PLoS Biology 8:e1000501.

Kirkpatrick, M., and N. Barton. 2006. Chromosome inversions, local adaptation, and speciation. Genetics 173:419–434.

Kirschbaum, M. U. F. 2011. Does enhanced photosynthesis enhance growth? Lessons learned from CO_2 enrichment studies. Plant Physiology 155:117–124.

Kitajima, K., S. S. Mulkey, M. Samaniego, and S. Joseph Wright. 2002. Decline of photosynthetic capacity with leaf age and position in two tropical pioneer tree species. American Journal of Botany 89:1925–1932.

Klekowski, E. J. 1998. Mutation rates in mangroves and other plants. Genetica 102.3:325–331.

Klekowski, E. J. 2003. Plant clonality, mutation, diplontic selection, and mutational meltdown. Biological Journal of the Linnean Society 79:61–67.

Kliber, A., and C. G. Eckert. 2004. Sequential decline in allocation among flowers within inflorescences: Proximate mechanisms and adaptive significance. Ecology 85:1675–1687.

Klinkhamer, P. G. L., T. J. deJong, and H. Metz. 1997. Sex and size in cosexual plants. Trends in Ecology & Evolution 12:260–265.

Klopfstein, S., M. Currat, and L. Excoffier. 2006. The fate of mutations surfing on the wave of a range expansion. Molecular Biology and Evolution 23:482–490.

Kluth, C., and H. Bruelheide. 2005. Central and peripheral *Hornungia petraea* populations: Patterns and dynamics. Journal of Ecology 93:584–595.

Knauth, L. P. 2005. Temperature and salinity history of the Precambrian ocean: Implications for the course of microbial evolution. Palaeogeography, Palaeoclimatology, Palaeoecology 219:53–69.

Knight, C. A., N. A. Molinari, and D. A. Petrov. 2005. The large genome constraint hypothesis: Evolution, ecology and phenotype. Annals of Botany 95:177–190.

Knight, T. A. 1799. An account of some experiments on the fecundation of vegetables. Philosophical Transactions of the Royal Society 89:195–204.

Knoll, A. H. 2003. Life on a Young Planet: The First Three Billion Years. Princeton University Press, Princeton.

Knoll, A. H., E. J. Javaux, D. Hewitt, and P. Cohen. 2006. Eukaryotic organisms in Proterozoic oceans. Philosophical Transactions of the Royal Society B: Biological Sciences 361:1023–1038.

Knowles, L. L. 2009. Statistical Phylogeography. Annual Review of Ecology Evolution and Systematics 40:593–612.

Knowles, L. L. and W. P. Maddison. 2002. Statistical phylogeography. Molecular Ecology 11:2623–2635.

Kohn, J. R. 2008. What genealogies of S-alleles tell us. Pp. 103–121 *in* V. E. Franklin-Tong, ed. Self-Incompatibility in Flowering Plants: Evolution, Diversity, and Mechanisms. Springer-Verlag, Berlin, Germany.

Kohn, J. R., and S. C. H. Barrett. 1992. Experimental studies on the functional significance of heterostyly. Evolution 46:43–55.

Kohrn, B. F., J. M. Persinger, and M. B. Cruzan. 2017. An efficient pipeline to generate data for studies in plastid population genomics and phylogeography. Applications in Plant Sciences 5:1700053.

Konieczny, A. and F. M. Ausubel. 1993. A procedure for mapping *Arabidopsis* mutations using co-dominant ecotype-specific PCR-based markers. The Plant Journal 4:403–410.

Korbecka, G., P. G. L. Klinkhamer, and K. Vrieling. 2002. Selective embryo abortion hypothesis revisited: A molecular approach. Plant Biology 4:298–310.

Korpelainen, H. 1998. Labile sex expression in plants. Biological Reviews 73:157–180.

Korte, A., and A. Farlow. 2013. The advantages and limitations of trait analysis with GWAS: A review. Plant Methods 9:29.

Krebs, S. L., and J. F. Hancock. 1990. Early-acting inbreeding depression and reproductive success in the highbush blueberry, *Vaccinium corymbosum* L. Theoretical and Applied Genetics 79:825–832.

Kress, W. J., K. J. Wurdack, E. A. Zimmer, L. A. Weigt, and D. H. Janzen. 2005. Use of DNA barcodes to identify flowering plants. Proceedings of the National Academy of Sciences of the United States of America 102:8369–8374.

Kutschera, U., and W. R. Briggs. 2009. From Charles Darwin's botanical country-house studies to modern plant biology. Plant Biology 11:785–795.

Labandeira, C. C. 1998. Early history of arthropod and vascular plant associations. Annual Review of Earth and Planetary Sciences 26:329–377.

Lagomarsino, L. P., E. J. Forrestel, N. Muchhala, and C. C. Davis. 2017. Repeated evolution of vertebrate pollination syndromes in a recently diverged Andean plant clade. Evolution 71:1970–1985.

Lamy, J. B., C. Plomion, A. Kremer, and S. Delzon. 2012. $Q(_{ST}) < F\text{-}_{ST}$ As a signature of canalization. Molecular Ecology 21:5646–5655.

Lande, R. 1979. Quantitative genetic analysis of multivariate evolution, applied to brain-body size allometry. Evolution 33:402–416.

Lande, R. 1981. Models of speciation by sexual selection on polygenic traits. Proceedings of the National Academy of Sciences of the United States of America 78:3721–3725.

Lande, R., and S. Arnold. 1983. The measurement of selection on correlated characters. Evolution 37:1210–1226.

Lande, R., and D. Schemske. 1985. The evolution of self-fertilization and inbreeding depression in plants. I. Genetic models. Evolution 39:24–40.

Lande, R., D. W. Schemske, and S. T. Schultz. 1994. High inbreeding depression, selective interference among loci, and the threshold selfing rate for purging recessive lethal mutations. Evolution 48:965–978.

Lanfear, R., H. Kokko, and A. Eyre-Walker. 2014. Population size and the rate of evolution. Trends in Ecology & Evolution 29:33–41.

Lankinen, Å., and M. C. Larsson. 2009. Conflicting selection pressures on reproductive functions and speciation in plants. Evolutionary Ecology 23:147–157.

Lankinen, Å., J. Maad, and W. S. Armbruster. 2009. Pollen-tube growth rates in *Collinsia heterophylla* (Plantaginaceae): One-donor crosses reveal heritability but no effect on sporophytic-offspring fitness. Annals of Botany 103:941–950.

Laufs, P., O. Grandjean, C. Jonak, K. Kieu, and J. Traas. 1998. Cellular parameters of the shoot apical meristem in *Arabidopsis*. Plant Cell 10:1375–1389.

Laurie, C. C., S. D. Chasalow, J. R. LeDeaux, R. McCarroll, D. Bush, B. Hauge, C. Lai, D. Clark, T. R. Rocheford, and J. W. Dudley. 2004. The genetic architecture of response to long-term artificial selection for oil concentration in the maize kernel. Genetics 168:2141–2155.

Lavergne, S., and J. Molofsky. 2007. Increased genetic variation and evolutionary potential drive the success of an invasive grass. Proceedings of the National Academy of Sciences of the United States of America 104:3883–3888.

Law, J. A., and S. E. Jacobsen. 2010. Establishing, maintaining, and modifying DNA methylation patterns in plants and animals. Nature Reviews Genetics 11:204–220.

Lawton, J. H. 1993. Range, population abundance, and conservation. Trends in Ecology & Evolution 8:409–413.

Lawton-Rauh, A. L., E. R. Alvarez-Buylla, and M. D. Purugganan. 2000. Molecular evolution of flower development. Trends in Ecology & Evolution 15:144–149.

Le Corre, V., and A. Kremer. 2012. The genetic differentiation at quantitative trait loci under local adaptation. Molecular Ecology 21:1548–1566.

Lee, C.-R., and T. Mitchell-Olds. 2011. Quantifying effects of environmental and geographical factors on patterns of genetic differentiation. Molecular Ecology 20:4631–4642.

Lee, E. K., A. Cibrian-Jaramillo, S. O. Kolokotronis, M. S. Katari, A. Stamatakis, M. Ott, J. C. Chiu, D. P. Little, D. W. Stevenson, W. R. McCombie, R. A. Martienssen, G. Coruzzi, and R. DeSalle. 2011. A Functional Phylogenomic View of the Seed Plants. PLoS Genetics 7:e1002411.

Lee, R. E. 2008. Phycology. Cambridge University Press, New York.

Lee, S. C., M. Ni, W. Li, C. Shertz, and J. Heitman. 2010. The evolution of sex: A perspective from the fungal kingdom. Microbiology and Molecular Biology Reviews 74:298–340.

Lee, T. 1988. Patterns of fruit and seed production. Pp. 179–202 *in* J. Lovett Doust and L. Lovett Doust, eds. Plant Reproductive Ecology—Patterns and Strategies. Oxford University Press, New York.

Leinonen, T., R. J. S. McCairns, R. B. O'Hara, and J. Merila. 2013. QST-FST comparisons: Evolutionary and ecological insights from genomic heterogeneity. Nature Reviews Genetics 14:179–190.

Leliaert, F., D. R. Smith, H. Moreau, M. D. Herron, H. Verbruggen, C. F. Delwiche, and O. De Clerck. 2012. Phylogeny and molecular evolution of the green algae. Critical Reviews in Plant Sciences 31:1–46.

Leliaert, F., H. Verbruggen, B. Wysor, and O. De Clerck. 2009. DNA taxonomy in morphologically plastic taxa: Algorithmic species delimitation in the *Boodlea* complex (Chlorophyta: Cladophorales). Molecular Phylogenetics and Evolution 53:122–133.

Lengyel, S., A. D. Gove, A. M. Latimer, J. D. Majer, and R. R. Dunn. 2009. Ants sow the seeds of global diversification in flowering plants. PLoS One 4:e5480.

Lenormand, T. 2002. Gene flow and the limits to natural selection. Trends in Ecology & Evolution 17:183–189.

Lepot, K., K. Benzerara, G. E. Brown, and P. Philippot. 2008. Microbially influenced formation of 2,724-million-year-old stromatolites. Nature Geoscience 1:118–121.

Leppala, J., F. Bokma, and O. Savolainen. 2013. Investigating Incipient Speciation in *Arabidopsis lyrata* from patterns of transmission ratio distortion. Genetics 194:697–708.

Lester, L., A. Meade, and M. Pagel. 2006. The slow road to the eukaryotic genome. Bioessays 28:57–64.

Levin, D. A. 2002. The role of chromosomal change in plant evolution. Oxford University Press, New York.

Levin, D. A., and H. W. Kerster. 1967. An analysis of interspecific pollen exchange in *Phlox*. American Naturalist 101:387–399.

Lewis, D., and L. K. Crowe. 1958. Unilateral interspecific incompatibility in flowering plants. Heredity 12:233–256.

Lewontin, R. C. 1970. The units of selection. Annual Review of Ecology and Sytematics 1:1–18.

Lewontin, R. C., and J. L. Hubby. 1966. A molecular approach to the study of heterozygosity in natural populations. II. Amount of variation and degree of heterozygosity in natural populations of *Drosophila pseudoobscura*. Genetics 54:595–609.

Lexer, C., R. A. Randell, and L. H. Rieseberg. 2003a. Experimental hybridization as a tool for studying selection in the wild. Ecology 84:1688–1699.

Lexer, C., M. E. Welch, O. Raymond, and L. H. Rieseberg. 2003b. The origin of ecological divergence in *Helianthus paradoxus* (Asteraceae): Selection on transgressive characters in a novel hybrid habitat. Evolution 57:1989–2000.

Lexer, C., R. O. Wüest, S. Mangili, M. Heuertz, K. N. Stölting, P. B. Pearman, F. Forest, N. Salamin, N. E. Zimmermann, and E. Bossolini. 2014. Genomics of the divergence continuum in an African plant biodiversity hotspot, I: Drivers of population divergence in *Restio capensis* (Restionaceae). Molecular Ecology 23:4373–4386.

Li, L., K. Lu, Z. Chen, T. Mu, Z. Hu, and X. Li. 2008. Dominance, overdominance and epistasis condition the heterosis in two heterotic rice hybrids. Genetics 180:1725–1742.

Li, P., and M. O. Johnston. 2000. Heterochrony in plant evolutionary studies through the twentieth century. Botanical Review 66:57–88.

Liepelt, S., R. Bialozyt, and B. Ziegenhagen. 2002. Wind-dispersed pollen mediates postglacial gene flow among refugia. Proceedings of the National Academy of Sciences of the United States of America 99:14590–14594.

Ligrone, R., J. G. Duckett, and K. S. Renzaglia. 2012. Major transitions in the evolution of early land plants: A bryological perspective. Annals of Botany 109:851–871.

Lim, P. O., H. J. Kim, and H. G. Nam. 2007. Leaf senescence. Annual Review of Plant Biology 58:115–136.

Lin, D., M. Hubbes, and L. Zsuffa. 1994. Differentiation of poplar and willow clones using RAPD fingerprints. Tree Physiology 14:1097–1105.

Lindqvist, C., and V. A. Albert. 2002. Origin of the Hawaiian endemic mints within North American *Stachys* (Lamiaceae). American Journal of Botany 89:1709–1724.

Linhart, Y. B., and M. C. Grant. 1996. Evolutionary significance of local genetic differentiation in plants. Annual Review of Ecology and Systematics 27:237–277.

Linkies, A., K. Graeber, C. Knight, and G. Leubner-Metzger. 2010. The evolution of seeds. New Phytologist 186:817–831.

Lipow, S. R., and R. Wyatt. 2000. Single gene control of postzygotic self-incompatibility in poke milkweed, *Asclepias exaltata* L. Genetics 154:893–907.

Liu, B., and J. F. Wendel. 2003. Epigenetic phenomena and the evolution of plant allopolyploids. Molecular Phylogenetics and Evolution 29:365–379.

Liu, J. Q., Y. S. Sun, X. J. Ge, L. M. Gao, and Y. X. Qiu. 2012. Phylogeographic studies of plants in China: Advances in the past and directions in the future. Journal of Systematics and Evolution 50:267–275.

Llaurens, V., L. Gonthier, and S. Billiard. 2009. The sheltered genetic load linked to the S locus in plants: New insights from theoretical and empirical approaches in sporophytic self-incompatibility. Genetics 183:1105–1118.

Lloyd, D. G. 1979. Some reproductive factors affecting the selection of self-fertilization in plants. American Naturalist 113:67–79.

Lloyd, D. G., and K. S. Bawa. 1984. Modification of the gender of seed plants in varying conditions. Evolutionary Biology 17:255–338.

Lloyd, D. G., and C. J. Webb. 1986. The avoidance of interference between the presentation of pollen and stigmas in angiosperms. I. Dichogamy. New Zealand Journal of Botany 24:135–162.

Loiselle, B. A., V. L. Sork, J. Nason, and C. Graham. 1995. Spatial genetic structure of a tropical understory shrub, *Psychotria officinalis* (Rubiaceae). American Journal of Botany 82:1420–1425.

Long, T. A., S. M. Brady, and P. N. Benfey. 2008. Systems approaches to identifying gene regulatory networks in plants. Annual Review of Cell and Developmental Biology 24:81–103.

Lopez-Gallego, C. and P. O'Neil. 2014. Genetic variation and the potential response to selection on leaf traits after habitat degradation in a long-lived cycad. Evolutionary Ecology 28:775–791.

Lopez-Garcia, P., and D. Moreira. 2006. Selective forces for the origin of the eukaryotic nucleus. Bioessays 28:525–533.

Lotsy, J. P. 1916. Evolution by Means of Hybridization. Martinus Nijhoff, The Hague.

Loveless, M. D., and J. L. Hamrick. 1984. Ecological determinants of genetic struction in plant populations. Annual Review of Ecology and Systematics 15:65–95.

Lowry, D. B. 2012. Ecotypes and the controversy over stages in the formation of new species. Biological Journal of the Linnean Society 106:241–257.

Lowry, D. B., M. C. Hall, D. E. Salt, and J. H. Willis. 2009. Genetic and physiological basis of adaptive salt tolerance divergence between coastal and inland *Mimulus guttatus*. New Phytologist 183:776–788.

Lowry, D. B., K. Hernandez, S. H. Taylor, E. Meyer, T. L. Logan, K. W. Barry, J. A. Chapman, D. S. Rokhsar,

J. Schmutz, and T. E. Juenger. 2014. The genetics of divergence and reproductive isolation between ecotypes of *Panicum hallii*. New Phytologist 205:402–414.

Lowry, D. B., J. L. Modliszewski, K. M. Wright, C. A. Wu, and J. H. Willis. 2008a. The strength and genetic basis of reproductive isolating barriers in flowering plants. Philosophical Transactions of the Royal Society B: Biological Sciences 363:3009–3021.

Lowry, D. B., R. C. Rockwood, and J. H. Willis. 2008b. Ecological reproductive isolation of coast and inland races of *Mimulus guttatus*. Evolution 62:2196–2214.

Loxdale, H. D., and G. Lushai. 2003. Rapid changes in clonal lines: The death of a "sacred cow." Biological Journal of the Linnean Society 79:3–16.

Lu, J. J., D. Y. Tan, J. M. Baskin, and C. C. Baskin. 2013. Trade-offs between seed dispersal and dormancy in an amphi-basicarpic cold desert annual. Annals of Botany 112:1815–1827.

Luikart, G., N. Ryman, D. Tallmon, M. Schwartz, and F. Allendorf. 2010. Estimation of census and effective population sizes: The increasing usefulness of DNA-based approaches. Conservation Genetics 11:355–373.

Luo, Y., and A. Widmer. 2013. Herkogamy and its effects on mating patterns in *Arabidopsis thaliana*. PLoS One 8:e57902.

Lush, J. L. 1937. Animal Breeding Plans. Iowa State Press, Ames.

Lynch, M. 1997. Mutation accumulation in nuclear, organelle, and prokaryotic transfer RNA genes. Molecular Biology and Evolution 14:914–925.

Lynch, M. 2006. The origins of eukaryotic gene structure. Molecular Biology and Evolution 23:450–468.

Lynch, M. 2010. Evolution of the mutation rate. Trends in Genetics 26:345–352.

Lynch, M., J. Blanchard, D. Houle, T. Kibota, S. Schultz, L. Vassilieva, and J. Willis. 1999. Perspective: Spontaneous deleterious mutation. Evolution 53:645–663.

Lynch, M., J. Conery, and R. Burger. 1995. Mutation accumulation and the extinction of small populations. American Naturalist 146:489–518.

Lynch, M., and J. S. Conery. 2003. The origins of genome complexity. Science 302:1401–1404.

Lynch, M., and W. Gabriel. 1983. Phenotypic evolution and parthenogenesis. American Naturalist 122:745–764.

Lynch, M., and B. Walsh. 1998. Genetics and Analysis of Quantitative Traits. Sinauer Associates, Inc., Sunderland, MA.

Mable, B. K., and S. P. Otto. 1998. The evolution of life cycles with haploid and diploid phases. Bioessays 20:453–462.

Macaya-Sanz, D., M. Heuertz, D. Lindtke, G. G. Vendramin, C. Lexer, and S. C. González-Martínez. 2016. Causes and consequences of large clonal assemblies in a poplar hybrid zone. Molecular Ecology 25:5330–5344.

Machado, C. A., N. Robbins, M. T. P. Gilbert, and E. A. Herre. 2005. Critical review of host specificity and its coevolutionary implications in the fig/fig-wasp mutualism. Proceedings of the National Academy of Sciences of the United States of America 102:6558–6565.

Mackay, T. F. C., E. A. Stone, and J. F. Ayroles. 2009. The genetics of quantitative traits: Challenges and prospects. Nature Reviews Genetics 10:565–577.

Magallón, S., and A. Castillo. 2009. Angiosperm diversification through time. American Journal of Botany 96:349–365.

Magallón, S., and K. W. Hilu. 2009. Land plants (Embryophyta). Pp. 133–137 *in* S. B. Hedges and S. Kumar, eds. The Timetree of Life. Oxford University Press, New York.

Majesky, L., R. J. Vasut, M. Kitner, and B. Travnicek. 2012. The pattern of genetic variability in apomictic clones of *Taraxacum officinale* indicates the alternation of asexual and sexual histories of apomicts. PLoS One 7:e41868.

Mallet, J. 1993. Speciation, raciation, and color pattern evolution in *Heliconius* butterflies: Evidence from hybrid zones. Pp. 226–260 *in* R. G. Harrison, ed. Hybrid Zones and the Evolutionary Process. Oxford University Press, NY.

Mallet, J. 2007. Hybrid speciation. Nature 446:279–283.

Mallet, J. 2008. Hybridization, ecological races, and the nature of species: Empirical evidence for the ease of speciation. Philosophical Transactions of the Royal Society B: Biological Sciences 363:2971–2986.

Mallet, J., and N. H. Barton. 1989. Strong natural selection in a warning-color hybrid zone. Evolution 43:421–431.

Marchini, G. L., N. Cole Sherlock, A. P. Ramakrishnan, D. M. Rosenthal, and M. B. Cruzan. 2016. Rapid purging of genetic load in a metapopulation and consequences for range expansion in an invasive plant. Biological Invasions 18:183–196.

Marchini, G. L., T. M. Arredondo, and M. B. Cruzan. 2018. Adaptive differentiation during establishment in a newly invasive species. (In review).

Markert, J. A., P. R. Grant, B. R. Grant, L. F. Keller, J. L. Coombs, and K. Petren. 2004. Neutral locus heterozygosity, inbreeding, and survival in Darwin's ground finches (*Geospiza fortis* and *G. scandens*). Heredity 92:306–315.

Martin, L. J., and M. B. Cruzan. 1999. Patterns of hybridization in the *Piriqueta caroliniana* complex in central Florida: Evidence for an expanding hybrid zone. Evolution 53:1037–1049.

Martin, N. H., and J. H. Willis. 2007. Ecological divergence associated with mating system causes nearly complete reproductive isolation between sympatric Mimulus species. Evolution 61:68–82.

Martin, R. D., C. Soligo, and S. Tavare. 2007. Primate origins: Implications of a cretaceous ancestry. Folia Primatologica 78:277–296.

Martin, W., and M. J. Russell. 2003. On the origins of cells: A hypothesis for the evolutionary transitions from abiotic geochemistry to chemoautotrophic prokaryotes, and from prokaryotes to nucleated cells. Philosophical Transactions of the Royal Society B: Biological Sciences 358:59–85.

Maskas, S. D., and M. B. Cruzan. 2000. Patterns of intraspecific diversification in the *Piriqueta caroliniana* complex in eastern North America and the Bahamas. Evolution 54:815–827.

Masterson, J. 1994. Stomatal size in fossil plants: Evidence for polyploidy in majority of angiosperms. Science 264:421–424.

Mather, K. 1949. Biometrical Genetics. Methuen, London.

Matsuki, S., and T. Koike. 2006. Comparison of leaf life span, photosynthesis, and defensive traits across seven species of deciduous broad-leaf tree seedlings. Annals of Botany 97:813–817.

Mauricio, R. 2001. Mapping quantitative trait loci in plants: Uses and caveats for evolutionary biology. Nature Reviews Genetics 2:370–381.

Maynard-Smith, J. 1978. The Evolution of Sex. Cambridge University Press, Cambridge, UK.

Maynard-Smith, J. 1982. Evolution and the Theory of Games. Cambridge University Press, Cambridge, UK.

Mayr, E. 1942. Systematics and the Origin of Species. Columbia University Press, New York.

Mayr E. 1947. Ecological factors in speciation. Evolution 1:263–288.

Mayr, E. 1963. Animal Species and Evolution. Harvard University Press, Cambridge.

Mayr, E. 1969. Principles of Systematic Zoology. McGraw-Hill, New York.

Mazer, S. J. and G. LeBuhn. 1999. Genmetic variation in life-history traits: Heritability estimates within and genetic differentiation among populations. Pp. 85–172 *in* T. O. Vuorisalo and P. K. Mutikainen, eds. Life History Evolution in Plants. Kluwer Academic Publishers, Dordrecht, The Netherlands.

McCauley, D. E. 1994. Contrasting the distribution of chloroplast DNA and allozyme polymorphism among local populations of *Silene alba*: Implications for studies of gene flow in plants. Proceedings of the National Academy of Sciences of the United States of America 91:8127–8131.

McCauley, D. E. 1997. The relative contributions of seed and pollen movement to the local genetic structure of *Silene alba*. Journal of Heredity 88:257–263.

McCue, K. A., and T. P. Holtsford. 1998. Seed bank influences on genetic diversity in the rare annual *Clarkia springvillensis* (Onagraceae). American Journal of Botany 85:30–36.

McDonald, M. J., D. P. Rice, and M. M. Desai. 2016. Sex speeds adaptation by altering the dynamics of molecular evolution. Nature 531:233–236.

McElwain, J. C., and S. W. Punyasena. 2007. Mass extinction events and the plant fossil record. Trends in Ecology & Evolution 22:548–557.

McGinley, M. A., D. H. Temme, and M. A. Geber. 1987. Parental investment in offspring in variable environments: Theoretical and empirical considerations. American Naturalist 130:370–398.

McKay, J. K., and R. G. Latta. 2002. Adaptive population divergence: Markers, QTL, and traits. Trends in Ecology & Evolution 17:285–291.

McKnight, T. D., K. Riha, and D. E. Shippen. 2002. Telomeres, telomerase, and stability of the plant genome. Plant Molecular Biology 48:331–337.

McKnight, T. D., and D. E. Shippen. 2004. Plant telomere biology. Plant Cell Online 16:794–803.

McNamara, K. J. 1996. Dating the origin of animals. Science 274:1995–1996.

McNeilly, T., and J. Antonovics. 1968. Evolution in closely adjacent plant populations. Heredity 23:205–218.

McRae, B. H. 2006. Isolation by resistance. Evolution 60:1551–1561.

McRae, B. H., B. G. Dickson, T. H. Keitt, and V. B. Shah. 2008. Using circuit theory to model connectivity in ecology, evolution, and conservation. Ecology 89:2712–2724.

Meagher, T. R. 1986. Analysis of paternity within a natural population of *Chamaelirrium luteum* I. Identification of most-likely parents. American Naturalist 128:199–215.

Medawar, P. 1946. Old age and natural death. Modern Quarterly 1:30–56.

Melnyk, A. H., and R. Kassen. 2011. Adaptive landscapes in evolving populations of *Pseudomonas fluorescens*. Evolution 65:3048–3059.

Melo, M. C., A. Grealy, B. Brittain, G. M. Walter, and D. Ortiz-Barrientos. 2014. Strong extrinsic reproductive isolation between parapatric populations of an Australian groundsel. New Phytologist 203:323–334.

Melser, C., and P. G. L. Klinkhamer. 2001. Selective seed abortion increases offspring survival in *Cynoglossum officinale* (Boraginaceae). American Journal of Botany 88:1033–1040.

Melzer, S., F. Lens, J. Gennen, S. Vanneste, A. Rohde, and T. Beeckman. 2008. Flowering-time genes modulate meristem determinacy and growth form in *Arabidopsis thaliana*. Nature Genetics 40:1489–1492.

Mena-Alí, J. I. and O. J. Rocha. 2005. Selective seed abortion affects the performance of the offspring in *Bauhinia ungulata*. Annals of Botany 95:1017–1023.

Merlo, L. M. F., J. W. Pepper, B. J. Reid, and C. C. Maley. 2006. Cancer as an evolutionary and ecological process. Nature Reviews Cancer 6:924–935.

Merrill, K. R., S. E. Meyer, and C. E. Coleman. 2012. Population genetic analysis of *Bromus tectorum* (Poaceae) indicates recent range expansion may be facilitated by specialist genotypes. American Journal of Botany 99:529–537.

Metlen, K. L., E. T. Aschehoug, and R. M. Callaway. 2009. Plant behavioural ecology: Dynamic plasticity in secondary metabolites. Plant Cell and Environment 32:641–653.

Meudt, H. M., and A. C. Clarke. 2007. Almost forgotten or latest practice? AFLP applications, analyses, and advances. Trends in Plant Science 12:106–117.

Miglia, K. J., E. D. McArthur, W. S. Moore, H. Wang, J. H. Graham, and D. C. Freeman. 2005. Nine-year reciprocal transplant experiment in the gardens of the basin and mountain big sagebrush (*Artemisia tridentata*: Asteraceae) hybrid zone of Salt Creek Canyon: The importance of multiple-year tracking of fitness. Biological Journal of the Linnean Society 86:213–225.

Miller, S. L. 1953. A production of amino acids under possible primitive Earth conditions. Science 117:528–529.

Minder, A. M., and A. Widmer 2008. A population genomic analysis of species boundaries: Neutral processes, adaptive divergence, and introgression between two hybridizing plant species. Molecular Ecology 17:1552–1563.

Mishler, B. D., and E. C. Theriot. 2000. The phylogenetic species concept (*sensu* Mishler and Theriot): Monophyly, apomorphy, and phylogenetic species concepts. Pp. 44–54 *in* Q. D. Wheeler and R. Meier, eds. Species Concepts and Phylogenetic Theory—A Debate. Columbia University Press, New York.

Mitchell, R. J., R. E. Irwin, R. J. Flanagan, and J. D. Karron. 2009. Ecology and evolution of plant-pollinator interactions. Annals of Botany 103:1355–1363.

Mitsui, Y., N. Nomura, Y. Isagi, H. Tobe, and H. Setoguchi. 2011. Ecological barriers to gene flow between riparian and forest species of *Ainsliaea* (Asteraceae). Evolution 65:335–349.

Mittell, E. A., S. Nakagawa, and J. D. Hadfield. 2015. Are molecular markers useful predictors of adaptive potential? Ecology Letters 18:772–778.

Mitton, J. B. 1989. Physiological and demographic variation associated with allozyme variation. Pp. 127–145 *in* D. E. Soltis and P. S. Soltis, eds. Isozymes in Plant Biology. Dioscorides Press, Portland, OR.

Mitton, J. B. 1993. Theory and data pertinent to the relationship between heterozygosity and fitness. Pp. 17–41 *in* N. Thornhill, ed. The Natural History of Inbreeding and Outbreeding. University of Chicago Press, Chicago.

Moeller, D. A., M. A. Geber, and P. Tiffin. 2011. Population genetics and the evolution of geographic range limits in an annual plant. American Naturalist 178:S44–S61.

Mogie, M. 1992. The Evolution of Asexual Reproduction in Plants. Chapman & Hall, London.

Mojica, J. P., and J. K. Kelly. 2010. Viability selection prior to trait expression is an essential component of natural selection. Proceedings of the Royal Society B: Biological Sciences 277:2945–2950.

Moles, A. T., D. D. Ackerly, C. O. Webb, J. C. Tweddle, J. B. Dickie, A. J. Pitman, and M. Westoby. 2005. Factors that shape seed mass evolution. Proceedings of the National Academy of Sciences of the United States of America 102:10540–10544.

Molina-Montenegro, M. A., C. Salgado-Luarte, R. Oses, and C. Torres-Díaz. 2013. Is physiological performance a good predictor for fitness? Insights from an invasive plant species. PLoS One 8:e76432.

Monro, K., and A. G. B. Poore. 2009. The potential for evolutionary responses to cell-lineage selection on growth form and its plasticity in a red seaweed. American Naturalist 173:151–163.

Montgomery, B. R., D. M. Soper, and L. F. Delph. 2010. Asymmetrical conspecific seed-siring advantage between *Silene latifolia* and *S. dioica*. Annals of Botany 105:595–605.

Morales, C. L., and A. Traveset. 2008. Interspecific pollen transfer: Magnitude, prevalence, and consequences for plant fitness. Critical Reviews in Plant Sciences 27:221–238.

Morgante, M., and A. M. Olivieri. 1993. PCR-amplified microsatellites as markers in plant genetics. Plant Journal 3:175–182.

Morin, P. A., G. Luikart, R. K. Wayne, and S. N. P. W. Grp. 2004. SNPs in ecology, evolution and conservation. Trends in Ecology & Evolution 19:208–216.

Morjan, C. L., and L. H. Rieseberg. 2004. How species evolve collectively: Implications of gene flow and selection for the spread of advantageous alleles. Molecular Ecology 13:1341–1356.

Morris, A. B., R. Baucom, and M. B. Cruzan. 2002. Stratified analysis of the soil seed bank in the cedar glade endemic *Astragalus bibullatus*: Evidence for historical changes in genetic structure. American Journal of Botany 89:29–36.

Moyle, L. C., M. Levine, M. L. Stanton, and J. W. Wright. 2012. Hybrid sterility over tens of meters between ecotypes adapted to serpentine and non-serpentine soils. Evolutionary Biology 39:207–218.

Moyle, L. C., and T. Nakazato. 2010. Hybrid incompatibility "snowballs" between *Solanum* species. Science 329:1521–1523.

Moyle, L. C., M. S. Olson, and P. Tiffin. 2004. Patterns of reproductive isolation in three angiosperm genera. Evolution 58:1195–1208.

Mueller, U. G., and L. L. Wolfenbarger. 1999. AFLP genotyping and fingerprinting. Trends in Ecological Evolution 14:389–394.

Mulcahy, D. L. 1979. Rise of the angiosperms: A genecological factor. Science 206:20–23.

Mulcahy, D. L., and G. B. Mulcahy. 1987. The effects of pollen competition. American Scientist 75:44–50.

Mulcahy, D. L., M. SariGorla, and G. B. Mulcahy. 1996. Pollen selection—past, present, and future. Sexual Plant Reproduction 9:353–356.

Müller, H. 1883. The Fertilization of Flowers. Macmillan, London.

Muller, H. J. 1932. Some genetic aspects of sex. American Naturalist 66:118–138.

Muller, H. J. 1942. Isolating mechanisms, evolution, and temperature. Biology Symposium 6:71–125.

Nagy, E. S. 1997. Frequency-dependent seed production and hybridization rates: Implications for gene flow between locally adapted plant populations. Evolution 51:703–714.

Nason, J. D., and J. L. Hamrick. 1997. Reproductive and genetic consequences of forest fragmentation: Two case studies of neotropical canopy trees. Journal of Heredity 88:264–276.

Nathan, R., G. Perry, J. T. Cronin, A. E. Strand, and M. L. Cain. 2003. Methods for estimating long-distance dispersal. Oikos 103:261–273.

Nathan, R., F. M. Schurr, O. Spiegel, O. Steinitz, A. Trakhtenbrot, and A. Tsoar. 2008. Mechanisms of long-distance seed dispersal. Trends in Ecology & Evolution 23:638–647.

Naudin, C. 1863. Nouvelles recherches sur l'hybridité dans les végétaux. Annales des sciences naturelles, série botanique 19:180–203.

Ne'eman, G., A. Jurgens, L. Newstrom-Lloyd, S. G. Potts, and A. Dafni. 2010. A framework for comparing pollinator performance: Effectiveness and efficiency. Biological Reviews 85:435–451.

Nei, M. 1973. Analysis of gene diversity in subdivided populations. Proceedings of the National Academy of Sciences of the United States of America 70:3321–3323.

Nei, M., T. Maruyama, and C. Wu. 1983. Models of evolution of reproductive isolation. Genetics 103:557–579.

Neigel, J. E. 2002. Is F_{ST} obsolete? Conservation Genetics 3:167–173.

Nelson, K. E., M. Levy, and S. L. Miller. 2000. Peptide nucleic acids rather than RNA may have been the first genetic molecule. Proceedings of the National Academy of Sciences of the United States of America 97:3868–3871.

Ni, M., M. Feretzaki, S. Sun, X. Wang, and J. Heitman. 2011. Sex in fungi. Annual Review of Genetics 45:405–430.

Nicotra, A. B., O. K. Atkin, S. P. Bonser, A. M. Davidson, E. J. Finnegan, U. Mathesius, P. Poot, M. D. Purugganan, C. L. Richards, F. Valladares, and M. van Kleunen. 2010. Plant phenotypic plasticity in a changing climate. Trends in Plant Science 15:684–692.

Nicotra, A. B., and A. Davidson. 2010. Adaptive phenotypic plasticity and plant water use. Functional Plant Biology 37:117–127.

Nielsen, L. R., H. R. Siegismund, and M. Philipp. 2003. Partial self-incompatibility in the polyploid endemic species *Scalesia affinis* (Asteraceae) from the Galápagos: Remnants of a self-incompatibility system? Botanical Journal of the Linnean Society 142:93–101.

Nielsen, R., and M. A. Beaumont. 2009. Statistical inferences in phylogeography. Molecular Ecology 18:1034–1047.

Nieto Feliner, G. 2014. Patterns and processes in plant phylogeography in the Mediterranean Basin: A review. Perspectives in Plant Ecology, Evolution, and Systematics 16:265–278.

Niklas, K. J. 1997. The Evolutionary Biology of Plants. University of Chicago Press, Chicago.

Nobel, P. 2005. Physiochemical and Environmental Plant Physiology. Academic Press, New York.

Nolte, A. W., and D. Tautz. 2010. Understanding the onset of hybrid speciation. Trends in Genetics 26:54–58.

Nosil, P., D. J. Funk, and D. Ortiz-Barrientos. 2009. Divergent selection and heterogeneous genomic divergence. Molecular Ecology 18:375–402.

Nunez-Farfan, J., and C. D. Schlichting. 2001. Evolution in changing environments: The "synthetic" work of Clausen, Keck, and Hiesey. Quarterly Review of Biology 76:433–457.

Nybom, H. 2004. Comparison of different nuclear DNA markers for estimating intraspecific genetic diversity in plants. Molecular Ecology 13:1143–1155.

Oakley, C. G. 2013. Small effective size limits performance in a novel environment. Evolutionary Applications 6:823–831.

Oakley, C. G. 2015. The influence of natural variation in population size on ecological and quantitative genetics of the endangered endemic plant *Hypericum cumulicola*. International Journal of Plant Sciences 176:11–19.

Oakley, C. G., K. S. Moriuchi, and A. A. Winn. 2007. The maintenance of outcrossing in predominantly selfing species: Ideas and evidence from cleistogamous species. Annual Review of Ecology Evolution and Systematics 38:437–457.

Oberle, B., and B. A. Schaal. 2011. Responses to historical climate change identify contemporary threats to diversity in Dodecatheon. Proceedings of the National Academy of Sciences of the United States of America 108:5655–5660.

Oetjen, K., and T. B. H. Reusch. 2007. Genome scans detect consistent divergent selection among subtidal versus intertidal populations of the marine angiosperm *Zostera marina*. Molecular Ecology 16:5156–5167.

Ohashi, K., and J. D. Thomson. 2009. Trapline foraging by pollinators: Its ontogeny, economics and possible consequences for plants. Annals of Botany 103:1365–1378.

Ohta, T. 1971. Associative overdominance caused by linked detrimental mutations. Genetical Research 18:277–286.

Ohta, T., and C. C. Cockerham. 1974. Detrimental genes with partial selfing and effects on a neutral locus. Genetics Research 23:191–200.

Oleszczuk, S., and A. J. Lukaszewski. 2014. The origin of unusual chromosome constitutions among newly formed allopolyploids. American Journal of Botany 101:318–326.

Olivieri, I. 2009. Alternative mechanisms of range expansion are associated with different changes of evolutionary potential. Trends in Ecology & Evolution 24:289–292.

Ollerton, J., R. Alarcón, N. M. Waser, M. V. Price, S. Watts, L. Cranmer, A. Hingston, C. I. Peter, and J. Rotenberry. 2009. A global test of the pollination syndrome hypothesis. Annals of Botany 103:1471–1480.

Ordano, M., J. Fornoni, K. Boege, and C. A. Dominguez. 2008. The adaptive value of phenotypic floral integration. New Phytologist 179:1183–1192.

Oro, J., T. Mills, and A. Lazcano. 1992. Comets and the formation of biochemical compounds on the primitive Earth: A review. Origins of Life and Evolution of the Biosphere 21:267–277.

Orr, H. A. 1995. The population genetics of speciation: The evolution of hybrid incompatibilities. Genetics 139:1805–1813.

Orr, H. A. 1996. Dobzhansky, Bateson, and the genetics of speciation. Genetics 144:1331–1335.

Orsini, L., J. Vanoverbeke, I. Swillen, J. Mergeay, and L. De Meester. 2013. Drivers of population genetic differentiation in the wild: Isolation by dispersal limitation, isolation by adaptation, and isolation by colonization. Molecular Ecology 22:5983–5999.

Ortegon-Campos, I., L. Abdala-Roberts, V. Parra-Tabla, J. C. Cervera, D. Marrufo-Zapata, and C. M. Herrera. 2012. Influence of multiple factors on plant local adaptation: Soil type and folivore effects in *Ruellia nudiflora* (Acanthaceae). Evolutionary Ecology 26:545–558.

Osborne, C. P., and D. J. Beerling. 2003. The penalty of a long, hot summer. Photosynthetic acclimation to high CO_2 and continuous light in "living fossil" conifers. Plant Physiology 133:803–812.

Ossowski, S., K. Schneeberger, J. I. Lucas-Lledo, N. Warthmann, R. M. Clark, R. G. Shaw, D. Weigel, and M. Lynch. 2010. The rate and molecular spectrum of spontaneous mutations in *Arabidopsis thaliana*. Science 327:92–94.

Ottaviano, E., M. S. Gorla, and E. Pe. 1982. Male gametophytic selection in maize. Theoretical and Applied Genetics 63:249–254.

Ottaviano, E., M. Sari-Gorla, and D. L. Mulcahy. 1980. Pollen tube growth rates in *Zea mays*: Implications for genetic improvement of crops. Science 210:437–438.

Otto, S. P. 2007. The evolutionary consequences of polyploidy. Cell 131:452–462.

Otto, S. P., and I. M. Hastings. 1998. Mutation and selection within the individual. Genetica 102-3:507–524.

Otto, S. P., and T. Lenormand. 2002. Resolving the paradox of sex and recombination. Nature Reviews Genetics 3:252–261.

Otto, S. P., and M. E. Orive. 1995. Evolutionary consequences of mutation and selection within an individual. Genetics 141:1173–1187.

Otto, S. P., M. F. Scott, and S. Immler. 2015. Evolution of haploid selection in predominantly diploid organisms. Proceedings of the National Academy of Sciences of the United States of America 112:15952–15957.

Otto, S. P., and J. Whitton. 2000. Polyploid incidence and evolution. Annual Review of Genetics 34:401–437.

Ouborg, N. J., Y. Piquot, and J. M. Van Groenendael. 1999. Population genetics, molecular markers, and the study of dispersal in plants. Journal of Ecology 87:551–568.

Ownbey, M. 1950. Natural hybridization and amphiploidy in the genus *Tragopogon*. American Journal of Botany 37:487–499.

Paetkau, D., R. Slade, M. Burden, and A. Estoup. 2004. Genetic assignment methods for the direct, real-time estimation of migration rate: A simulation-based exploration of accuracy and power. Molecular Ecology 13:55–65.

Paige, K. N., and T. G. Whitham. 1987. Overcompensation in response to mammalian herbivory: The advantage of being eaten. American Naturalist 129:407–416.

Pajoro, A., S. Biewers, E. Dougali, F. L. Valentim, M. A. Mendes, A. Porri, G. Coupland, Y. Van de Peer, A. D. J. van Dijk, L. Colombo, B. Davies, and G. C. Angenent. 2014. The (r)evolution of gene regulatory networks controlling *Arabidopsis* plant reproduction: A two-decade history. Journal of Experimental Botany 65:4731–4745.

Pal, C., B. Papp, and M. J. Lercher. 2006. An integrated view of protein evolution. Nature Reviews Genetics 7:337–348.

Palmer, J. D. and D. Zamir. 1982. Chloroplast DNA evolution and phylogenetic relationships in *Lycopersicon*. Proceedings of the National Academy of Sciences of the United States of America: Biological Sciences 79:5006–5010.

Pandey, K. K. 1962. Interspecific incompatibility in *Solanum* species. American Journal of Botany 49:874–882.

Pannell, J. R. 2015. Evolution of the mating system in colonizing plants. Molecular Ecology 24:2018–2037.

Pannell, J. R. 2017. Plant sex determination. Current Biology 27:R191–R197.

Pannell, J. R., and B. Charlesworth. 1999. Neutral genetic diversity in a metapopulation with recurrent local extinction and recolonization. Evolution 53:664–676.

Pannell, J. R., and B. Charlesworth. 2000. Effects of metapopulation processes on measures of genetic diversity. Philosophical Transactions of the Royal Society B: Biological Sciences 355:1851–1864.

Pannell, J. R., and P. D. Fields. 2014. Evolution in subdivided plant populations: Concepts, recent advances, and future directions. New Phytologist 201:417–432.

Pappert, R. A., J. L. Hamrick, and L. A. Donovan. 2000. Genetic variation in *Pueraria lobata* (Fabaceae), an introduced, clonal, invasive plant of the southeastern United States. American Journal of Botany 87:1240–1245.

Parfrey, L. W., D. J. G. Lahr, and L. A. Katz. 2008. The dynamic nature of eukaryotic genomes. Molecular Biology and Evolution 25:787–794.

Parker, E. T., H. J. Cleaves, J. P. Dworkin, D. P. Glavin, M. Callahan, A. Aubrey, A. Lazcano, and J. L. Bada. 2011. Primordial synthesis of amines and amino acids in a 1958 Miller H2S-rich spark discharge experiment. Proceedings of the National Academy of Sciences of the United States of America 108:5526–5531.

Pasonen, H. L., P. Pulkkinen, and M. Kapyla. 2001. Do pollen donors with fastest-growing pollen tubes sire the best offspring in an anemophilous tree, *Betula pendula* (Betulaceae)? American Journal of Botany 88:854–860.

Pauchard, A., and K. Shea. 2006. Integrating the study of non-native plant invasions across spatial scales. Biological Invasions 8:399–413.

Peakall, R., D. Ebert, L. J. Scott, P. F. Meagher, and C. A. Offord. 2003. Comparative genetic study confirms exceptionally low genetic variation in the ancient and endangered relictual conifer *Wollemia nobilis* (Araucariaceae). Molecular Ecology 12:2331–2343.

Pelabon, C., E. Albertsen, M. Falahati-Anbaran, J. Wright, and W. S. Armbruster. 2015. Does multiple paternity affect seed mass in angiosperms? An experimental test in *Dalechampia scandens*. Journal of Evolutionary Biology 28:1719–1733.

Pellino, M., D. Hojsgaard, T. Schmutzer, U. Scholz, E. Hörandl, H. Vogel, and T. F. Sharbel. 2013. Asexual genome evolution in the apomictic *Ranunculus auricomus* complex: Examining the effects of hybridization and mutation accumulation. Molecular Ecology 22:5908–5921.

Pellmyr, O. 2003. Yuccas, yucca moths, and coevolution: A review. Annals of the Missouri Botanical Garden 90:35–55.

Petit, C., F. Bretagnolle, and F. Felber. 1999. Evolutionary consequences of diploid-polyploid hybrid zones in wild species. Trends in Ecology & Evolution 14:306–311.

Petit, R. J., I. Aguinagalde, J. L. de Beaulieu, C. Bittkau, S. Brewer, R. Cheddadi, R. Ennos, S. Fineschi, D. Grivet, M. Lascoux, A. Mohanty, G. M. Muller-Starck, B. Demesure-Musch, A. Palme, J. P. Martin, S. Rendell, and G. G. Vendramin. 2003. Glacial refugia: Hotspots but not melting pots of genetic diversity. Science 300:1563–1565.

Petit, R. J., C. Bodenes, A. Ducousso, G. Roussel, and A. Kremer. 2004. Hybridization as a mechanism of invasion in oaks. New Phytologist 161:151–164.

Petit, R. J., and A. Hampe. 2006. Some evolutionary consequences of being a tree. Annual Review of Ecology Evolution and Systematics 37:187–214.
Petit, R. J., E. Pineau, B. Demesure, R. Bacilieri, A. Ducousso, and A. Kremer. 1997. Chloroplast DNA footprints of postglacial recolonization by oaks. Proceedings of the National Academy of Sciences of the United States of America 94:9996–10001.
Pfahler, P. L. 1967. Fertilization ability of maize pollen grains. II. Pollen genotype, female sporophyte, and pollen storage interactions. Genetics 37:513–521.
Pfeiffer, T., A. Klahr, A. Peterson, I. G. Levichev, and M. Schnittler. 2012. No sex at all? Extremely low genetic diversity in *Gagea spathacea* (Liliaceae) across Europe. Flora 207:372–378.
Picotte, J. J., D. M. Rosenthal, J. M. Rhode, and M. B. Cruzan. 2007. Plastic responses to temporal variation in moisture availability: Consequences for water use efficiency and plant performance. Oecologia 153:821–832.
Pigliucci, M. 2001. Phenotypic Plasticity: Beyond Nature and Nurture. Johns Hopkins University Press, Baltimore, MD.
Poethig, R. S. 1987. Clonal analysis of cell lineage patterns in plant development. American Journal of Botany 74:581–594.
Pons, J., T. G. Barraclough, J. Gomez-Zurita, A. Cardoso, D. P. Duran, S. Hazell, S. Kamoun, W. D. Sumlin, and A. P. Vogler. 2006. Sequence-based species delimitation for the DNA taxonomy of undescribed insects. Systematic Biology 55:595–609.
Powell, J. A. 2009. Coleoptera *in* V. H. Resh and T. C. Ting, eds. Encyclopedia of Insects. Academic Press, New York.
Powner, M. W., B. Gerland, and J. D. Sutherland. 2009. Synthesis of activated pyrimidine ribonucleotides in prebiotically plausible conditions. Nature 459:239–242.
Prasad, V., C. A. E. Strömberg, H. Alimohammadian, and A. Sahni. 2005. Dinosaur coprolites and the early evolution of grasses and grazers. Science 310:1177–1180.
Prentis, P. J., J. R. U. Wilson, E. E. Dormontt, D. M. Richardson, and A. J. Lowe. 2008. Adaptive evolution in invasive species. Trends in Plant Science 13:288–294.
Preston, J. C., and L. C. Hileman. 2009. Developmental genetics of floral symmetry evolution. Trends in Plant Science 14:147–154.
Proctor, M. C. F., and Z. Tuba. 2002. Poikilohydry and homoihydry: Antithesis or spectrum of possibilities? New Phytologist 156:327–349.
Provine, W. B. 1971. The Origins of Theoretical Population Genetics. University of Chicago Press, Chicago.
Provine, W. B. 2001. The Origins of Theoretical Population Genetics. University of Chicago Press, Chicago.
Pujol, B., S.-R. Zhou, J. S. Vilas, and J. R. Pannell. 2009. Reduced inbreeding depression after species range expansion. Proceedings of the National Academy of Sciences of the United States of America 106:15379–15383.
Purcell, S. M., N. R. Wray, J. L. Stone, P. M. Visscher, M. C. O'Donovan, P. F. Sullivan, and P. Sklar. 2009. Common polygenic variation contributes to risk of schizophrenia and bipolar disorder. Nature 460:748–752.
Qiu, Y.-L., A. B. Taylor, and H. A. McManus. 2012. Evolution of the life cycle in land plants. Journal of Systematics and Evolution 50:171–194.
Qiu, Y. L., L. B. Li, B. Wang, Z. D. Chen, V. Knoop, M. Groth-Malonek, O. Dombrovska, J. Lee, L. Kent, J. Rest, G. F. Estabrook, T. A. Hendry, D. W. Taylor, C. M. Testa, M. Ambros, B. Crandall-Stotler, R. J. Duff, M. Stech, W. Frey, D. Quandt, and C. C. Davis. 2006. The deepest divergences in land plants inferred from phylogenomic evidence. Proceedings of the National Academy of Sciences of the United States of America 103:15511–15516.
Qiu, Y. X., C. X. Fu, and H. P. Comes. 2011. Plant molecular phylogeography in China and adjacent regions: Tracing the genetic imprints of Quaternary climate and environmental change in the world's most diverse temperate flora. Molecular Phylogenetics and Evolution 59:225–244.
Rai, H. S., P. A. Reeves, R. Peakall, R. G. Olmstead, and S. W. Graham. 2008. Inference of higher-order conifer relationships from a multi-locus plastid data set. Botany-Botanique 86:658–669.
Rajakaruna, N. 2004. The edaphic factor in the origin of plant species. International Geology Review 46:471–478.
Ramakrishnan, A. P., S. E. Meyer, D. J. Fairbanks, and C. E. Coleman. 2006. Ecological significance of microsatellite variation in western North American populations of *Bromus tectorum*. Plant Species Biology 21:61–73.
Ramakrishnan, A. P., T. Musial, and M. B. Cruzan. 2010. Shifting dispersal modes at an expanding species' range margin. Molecular Ecology 19:1134–1146.
Ramanna, M. S., and E. Jacobsen. 2003. Relevance of sexual polyploidization for crop improvement: A review. Euphytica 133:3–18.
Ramírez B., W. 1970. Host Specificity of Fig Wasps (Agaonidae). Evolution 24:680–691.
Ramsey, J., H. D. Bradshaw, and D. W. Schemske. 2003. Components of reproductive isolation between the

monkeyflowers *Mimulus lewisii* and *M. cardinalis* (Phrymaceae). Evolution 57:1520–1534.

Ramsey, J., A. Robertson, and B. Husband. 2008. Rapid adaptive divergence in new world *Achillea*, an autopolyploid complex of ecological races. Evolution 62:639–653.

Ramsey, J., and D. W. Schemske. 1998. Pathways, mechanisms, and rates of polyploid formation in flowering plants. Annual Review of Ecology and Systematics 29:467–501.

Ramsey, J., and D. W. Schemske. 2002. Neopolyploidy in flowering plants. Annual Review of Ecology and Systematics 33:589–639.

Randolph, L. F. 1966. *Iris nelsonii*, a new species of Louisiana iris of hybrid origin. Baileya 14:143–169.

Rausch, J. H., and M. T. Morgan. 2005. The effect of self-fertilization, inbreeding depression, and population size on autopolyploid establishment. Evolution 59:1867–1875.

Rausher, M. D., and L. F. Delph. 2015. Commentary: When does understanding phenotypic evolution require identification of the underlying genes? Evolution 69:1655–1664.

Rauwolf, U., H. Golczyk, J. Meurer, R. G. Herrmann, and S. Greiner. 2008. Molecular Marker Systems for Oenothera Genetics. Genetics 180:1289–1306.

Redecker, D., R. Kodner, and L. E. Graham. 2000. Glomalean fungi from the Ordovician. Science 289:1920–1921.

Reed, D. H., and R. Frankham. 2001. How closely correlated are molecular and quantitative measures of genetic variation? A meta-analysis. Evolution 55:1095–1103.

Reeve, H. K., D. F. Westneat, and D. C. Quellers. 1992. Estimating average within-group relatedness from DNA fingerprints. Molecular Ecology 1:223–232.

Reichling, J. 2010. Plant-microbe interactions and secondary metabolites with antibacterial, antifungal and antiviral properties. Pp. 214–347 in M. Wink, ed. Annual Plant Reviews Volume 39: Functions and Biotechnology of Plant Secondary Metabolites. Wiley-Blackwell, Oxford, UK.

Reid, N. M., and B. C. Carstens. 2012. Phylogenetic estimation error can decrease the accuracy of species delimitation: A Bayesian implementation of the general mixed Yule-coalescent model. BMC Evolutionary Biology 12:196.

Rellstab, C., F. Gugerli, A. J. Eckert, A. M. Hancock, and R. Holderegger. 2015. A practical guide to environmental association analysis in landscape genomics. Molecular Ecology 24:4348–4370.

Remington, C. L. 1968. Suture zones of hybrid interaction between recently joined biotas. Evolutionary Biology 2:321–328.

Remy, W., T. N. Taylor, H. Hass, and H. Kerp. 1994. Four-hundred-million-year-old vesicular-arbuscular mycorrhizae. Proceedings of the National Academy of Sciences of the United States of America 91:11841–11843.

Renaut, S., C. J. Grassa, S. Yeaman, B. T. Moyers, Z. Lai, N. C. Kane, J. E. Bowers, J. M. Burke, and L. H. Rieseberg. 2013. Genomic islands of divergence are not affected by geography of speciation in sunflowers. Nature Communications 4:1827.

Renaut, S., H. C. Rowe, M. C. Ungerer, and L. H. Rieseberg. 2014. Genomics of homoploid hybrid speciation: Diversity and transcriptional activity of long terminal repeat retrotransposons in hybrid sunflowers. Philosophical Transactions of the Royal Society B: Biological Sciences 369:20130345.

Renner, S. S. 2014. The relative and absolute frequencies of angiosperm sexual systems: Dioecy, monoecy, gynodioecy, and an updated online database. American Journal of Botany 101:1588–1596.

Renzaglia, K. S., S. Schuette, R. J. Duff, R. Ligrone, A. J. Shaw, B. D. Mishler, and J. G. Duckett. 2007. Bryophyte phylogeny: Advancing the molecular and morphological frontiers. Bryologist 110:179–213.

Reynolds, M., M. J. Foulkes, G. A. Slafer, P. Berry, M. A. J. Parry, J. W. Snape, and W. J. Angus. 2009. Raising yield potential in wheat. Journal of Experimental Botany 60:1899–1918.

Rhode, J. M., and M. B. Cruzan. 2005. Contributions of heterosis and epistasis to hybrid fitness. American Naturalist 166:E124–E139.

Ribeiro, K. T., and G. W. Fernandes. 2000. Patterns of abundance of a narrow endemic species in a tropical and infertile montane habitat. Plant Ecology 147:205–218.

Ricca, M., and A. J. Shaw. 2010. Allopolyploidy and homoploid hybridization in the *Sphagnum subsecundum* complex (Sphagnaceae: Bryophyta). Biological Journal of the Linnean Society 99:135–151.

Richards, C. L., A. W. Schrey, and M. Pigliucci. 2012. Invasion of diverse habitats by few Japanese knotweed genotypes is correlated with epigenetic differentiation. Ecology Letters 15:1016–1025.

Richau, K. H., and H. Schat. 2009. Intraspecific variation of nickel and zinc accumulation and tolerance in the hyperaccumulator *Thlaspi caerulescens*. Plant and Soil 314:253–262.

Rieseberg, L. H. 1991a. Homoploid reticulate evolution in *Helianthus* (Asteraceae): Evidence from ribosomal genes. American Journal of Botany 78:1218–1237.

Rieseberg, L. H. 1991b. Hybridization in rare plants: Insights from case studies in *Cercocarpus* and *Helianthus*. Pp. 171–181 *in* D. A. Falk and K. E. Holsinger, eds. Genetics and Conservation of Rare Plants. Oxford University Press, New York.

Rieseberg, L. H. 1997. Hybrid origins of plant species. Annual Review of Ecology and Systematics 28:359–389.

Rieseberg, L. H., M. A. Archer, and R. K. Wayne. 1999. Transgressive segregation, adaptation, and speciation. Heredity 83:363–372.

Rieseberg, L. H., S. Beckstromsternberg, and K. Doan. 1990. *Helianthus annuus* ssp. *texanus* has chloroplast DNA and nuclear ribosomal RNA genes of *Helianthus debilis* ssp. *cucumerifolius*. Proceedings of the National Academy of Sciences of the United States of America 87:593–597.

Rieseberg, L. H., and B. K. Blackman. 2010. Speciation genes in plants. Annals of Botany 106:439–455.

Rieseberg, L. H., A. M. Desrochers, and S. J. Youn. 1995. Interspecific pollen competition as a reproductive barrier between sympatric species of *Helianthus* (Asteraceae). American Journal of Botany 82:515–519.

Rieseberg, L. H., and D. E. Soltis. 1991. Phylogenetic consequences of cytoplasmic gene flow in plants. Evolutionary Trends in Plants 5:65–84.

Rieseberg, L. H., D. E. Soltis, and J. D. Palmer. 1988. A molecular reexamination of introgression between *Helianthus annuus* and *H. bolanderi* (Compositae). Evolution 42:227–238.

Rieseberg, L. H., and J. F. Wendel. 1993. Introgression and its consequences in plants. Pp. 70–109 *in* R. G. Harrison, ed. Hybrid Zones and the Evolutionary Process. Oxford University Press, Oxford.

Rigney, L. P., J. D. Thomson, M. B. Cruzan, and J. Brunet. 1993. Differential success of pollen donors in a self-compatible lily. Evolution 47:915–924.

Ritland, K. 1996. A marker-based method for inferences about quantitative inheritance in natural populations. Evolution 50:1062–1073.

Ritland, K. 2000. Marker-inferred relatedness as a tool for detecting heritability in nature. Molecular Ecology 9:1195–1204.

Ritland, K. 2002. Extensions of models for the estimation of mating systems using n independent loci. Heredity 88:221–228.

Rivera, M. C. 2007. Genomic analyses and the origin of the eukaryotes. Chemistry & Biodiversity 4:2631–2638.

Rivera, M. C., and J. A. Lake. 2004. The ring of life provides evidence for a genome fusion origin of eukaryotes. Nature 431:152–155.

Roach, D. A., and R. D. Wulff. 1987. Maternal effects in plants. Annual Review of Ecology and Systematics 18:209–235.

Robakowski, P., Y. Li, and P. B. Reich. 2012. Local ecotypic and species range-related adaptation influence photosynthetic temperature optima in deciduous broadleaved trees. Plant Ecology 213:113–125.

Rockman, M. V. 2012. The QTN program and the alleles that matter for evolution: All that's gold does not glitter. Evolution 66:1–17.

Ronsheim, M. L. 1996. Evidence against a frequency-dependent advantage for sexual reproduction in *Allium vineale*. American Naturalist 147:718–734.

Rosas, F., and C. A. Domínguez. 2009. Male sterility, fitness gain curves, and the evolution of gender specialization from distyly in *Erythroxylum havanense*. Journal of Evolutionary Biology 22:50–59.

Rosenblum, E. B., B. A. Sarver, J. W. Brown, S. Des Roches, K. M. Hardwick, T. D. Hether, J. M. Eastman, M. W. Pennell, and L. J. Harmon. 2012. Goldilocks meets Santa Rosalia: An ephemeral speciation model explains patterns of diversification across time scales. Evolutionary Biology 39:255–261.

Rosenstiel, T. N., E. E. Shortlidge, A. N. Melnychenko, J. F. Pankow, and S. M. Eppley. 2012. Sex-specific volatile compounds influence microarthropod-mediated fertilization of moss. Nature 489:431–433.

Rosenthal, D. M., A. P. Ramakrishnan, and M. B. Cruzan. 2008. Evidence for multiple sources and intraspecific hybridization at early stages of the invasion of *Brachypodium sylvaticum* (Hudson) Beauv. in North America. Molecular Ecology 17:4657–4669.

Routley, M. B., R. I. Bertin, and B. C. Husband. 2004. Correlated evolution of dichogamy and Self-incompatibility: A phylogenetic perspective. International Journal of Plant Sciences 165:983–993.

Rowe, H. C., S. Renaut, and A. Guggisberg. 2011. RAD in the realm of next-generation sequencing technologies. Molecular Ecology 20:3499–3502.

Rozendaal, D. M. A., V. H. Hurtado, and L. Poorter. 2006. Plasticity in leaf traits of thirty-eight tropical tree species in response to light: Relationships with light demand and adult stature. Functional Ecology 20:207–216.

Rubinstein, C. V., P. Gerrienne, G. S. de la Puente, R. A. Astini, and P. Steemans. 2010. Early Middle Ordovician evidence for land plants in Argentina (eastern Gondwana). New Phytologist 188:365–369.

Runions, C. J., and M. A. Geber. 2000. Evolution of the self-pollinating flower in *Clarkia xantiana* (Onagraceae). I. Size and development of floral organs. American Journal of Botany 87:1439–1451.

Rutter, M. T., A. Roles, J. K. Conner, R. G. Shaw, F. H. Shaw, K. Schneeberger, S. Ossowski, D. Weigel, and C. B. Fenster. 2012. Fitness of *Arabidopsis thaliana* mutation accumulation lines whose spontaneous mutations are known. Evolution 66:2335–2339.

Rutter, M. T., F. H. Shaw, and C. B. Fenster. 2010. Spontaneous mutation parameters for *Arabidopsis thaliana* measured in the wild. Evolution 64:1825–1835.

Ruvkun, G. 2008. Tiny RNA: Where do we come from? What are we? Where are we going? Trends in Plant Science 13:313–316.

Rymer, P. D., S. D. Johnson, and V. Savolainen. 2010. Pollinator behaviour and plant speciation: Can assortative mating and disruptive selection maintain distinct floral morphs in sympatry? New Phytologist 188:426–436.

Sabour, D., and H. R. Scholer. 2012. Reprogramming and the mammalian germline: The Weismann barrier revisited. Current Opinion in Cell Biology 24:716–723.

Sadras, V. O., and R. F. Denison. 2009. Do plant parts compete for resources? An evolutionary viewpoint. New Phytologist 183:565–574.

Sagarin, R. D., and S. D. Gaines. 2002. The "abundant centre" distribution: To what extent is it a biogeographical rule? Ecology Letters 5:137–147.

Sahu, P. P., G. Pandey, N. Sharma, S. Puranik, M. Muthamilarasan, and M. Prasad. 2013. Epigenetic mechanisms of plant stress responses and adaptation. Plant Cell Reports 32:1151–1159.

Sakai, A. K., and S. G. Weller. 1999. Gender and sexual dimorphism in flowering plants: A review of terminology, biogeographic patterns, ecological correlates, and phylogenetic approaches. Pp. 1–25 *in* M. A. Geber, T. E. Dawson, and L. F. Delph, eds. Gender and Sexual Dimorphism in Flowering Plants. Springer-Verlag, New York.

Sambatti, J. B. M., J. L. Strasburg, D. Ortiz-Barrientos, E. J. Baack, and L. H. Rieseberg. 2012. Reconciling extremely strong barriers with high levels of gene exchange in annual sunflowers. Evolution 66:1459–1473.

Samis, K. E., and C. R. G. Eckert. 2007. Testing the abundant center model using range-wide demographic surveys of two coastal dune plants. Ecology 88:1747–1758.

Sanderson, M. J., and M. F. Wojciechowski. 1996. Diversification rates in a temperate legume clade: Are there "so many species" of *Astragalus* (Fabaceae). American Journal of Botany 83:1488–1502.

Sankararaman, S., N. Patterson, H. Li, S. Pääbo, and D. Reich. 2012. The date of interbreeding between Neandertals and modern humans. PLoS Genetics 8:e1002947.

Sargent, R. D. 2004. Floral symmetry affects speciation rates in angiosperms. Proceedings of the Royal Society B: Biological Sciences 271:603–608.

Savolainen, O., S. T. Kujala, C. Sokol, T. Pyhäjärvi, K. Avia, T. Knürr, K. Kärkkäinen, and S. Hicks. 2011. Adaptive potential of northernmost tree populations to climate change, with emphasis on scots pine (*Pinus sylvestris* L.). Journal of Heredity 102:526–536.

Savolainen, O., T. Pyhajarvi, and T. Knurr. 2007. Gene flow and local adaptation in trees. Annual Review of Ecology Evolution and Systematics 38:595–619.

Savolainen, V., M. C. Anstett, C. Lexer, I. Hutton, J. J. Clarkson, M. V. Norup, M. P. Powell, D. Springate, N. Salamin, and W. J. Baker. 2006. Sympatric speciation in palms on an oceanic island. Nature 441:210–213.

Sawers, R. J. H., M. J. Sheehan, and T. P. Brutnell. 2005. Cereal phytochromes: Targets of selection, targets for manipulation? Trends in Plant Science 10:138–143.

Schaal, B. A. 1980. Measurement of gene flow in *Lupinus texensis*. Nature 284:450–451.

Schaal, B. A., D. A. Hayworth, K. M. Olsen, J. T. Rauscher, and W. A. Smith. 1998. Phylogeographic studies in plants: Problems and prospects. Molecular Ecology 7:465–474.

Schaffer, W. M., and M. V. Schaffer. 1979. Adaptive significance of variations in reproductive habit in the Agavaceae. 2. Pollinator foraging behavior and selection for increased reproductive expenditure. Ecology 60:1051–1069.

Scheiner, S. M. 1993. Genetics and evolution of phenotypic plasticity. Annual Review of Ecology and Systematics 24:35–68.

Scheiner, S. M., and R. D. Holt. 2012. The genetics of phenotypic plasticity. X. Variation versus uncertainty. Ecology and Evolution 2:751–767.

Schemske, D. W. 2010. Adaptation and the origin of species. American Naturalist 176:S4–S25.

Schemske, D. W., and R. Lande. 1985. The evolution of self-fertilization and inbreeding depression in plants. II. Empirical observations. Evolution 39:41–52.

Schiestl, F. P., and S. D. Johnson. 2013. Pollinator-mediated evolution of floral signals. Trends in Ecology & Evolution 28:307–315.

Schiestl, F. P., R. Peakall, and J. Mant. 2004. Chemical communication in the sexually deceptive orchid genus *Cryptostylis*. Botanical Journal of the Linnean Society 144:199–205.

Schiestl, F. P., and P. M. Schluter. 2009. Floral isolation, specialized pollination, and pollinator behavior in orchids. Annual Review of Entomology 54:425–446.

Schlichting, C. D., and M. A. Wund. 2014. Phenotypic plasticity and epigenetic marking: An assessment of evidence for genetic accommodation. Evolution 68:656–672.

Schluter, D. 1988. Esimating the form of natural selection on a quantitative trait. Evolution 42:849–861.
Schluter, D. 2001. Ecology and the origin of species. Trends in Ecology & Evolution 16:372–380.
Schmitt, J., J. R. Stinchcombe, M. S. Heschel, and H. Huber. 2003. The adaptive evolution of plasticity: Phytochrome-mediated shade avoidance responses. Integrative and Comparative Biology 43:459–469.
Schmitt, J., and R. D. Wulff. 1993. Light spectral quality, phytochrome and plant competition. Trends in Ecology & Evolution 8:47–51.
Schmitz, R. J., and J. R. Ecker. 2012. Epigenetic and epigenomic variation in *Arabidopsis thaliana*. Trends in Plant Science 17:149–154.
Schnitzler, J., C. H. Graham, C. F. Dormann, K. Schiffers, and H. Peter Linder. 2012. Climatic niche evolution and species diversification in the Cape flora, South Africa. Journal of Biogeography 39:2201–2211.
Schoen, D. J. 2005. Deleterious mutation in related species of the plant genus *Amsinckia* with contrasting mating systems. Evolution 59:2370–2377.
Schoener, T. W. 2009. Ecological niche. Pp. 3–13 *in* S. A. Levin, S. R. Carpenter, H. C. J. Godfray, A. P. Kinzig, M. Loreau, J. B. Losos, B. Walker, and D. S. Wilcove, eds. The Princeton Guide to Ecology. Princeton University Press, Princeton.
Scholes, D. R., and K. N. Paige. 2014. Plasticity in ploidy underlies plant fitness compensation to herbivore damage. Molecular Ecology 23:4862–4870.
Schultz, S. T., and D. G. Scofield. 2009. Mutation accumulation in real branches: Fitness assays for genomic deleterious mutation rate and effect in large-statured plants. American Naturalist 174:163–175.
Schwartzman, D. W. 1999. Life, Temperature, and the Earth. Columbia University Press, New York.
Scoble, J., and A. J. Lowe. 2010. A case for incorporating phylogeography and landscape genetics into species distribution modelling approaches to improve climate adaptation and conservation planning. Diversity and Distributions 16:343–353.
Scofield, D. G. 2006. Medial pith cells per meter in twigs as a proxy for mitotic growth rate (Φ/m) in the apical meristem. American Journal of Botany 93:1740–1747.
Scopece, G., C. Lexer, A. Widmer, and S. Cozzolino. 2010. Polymorphism of postmating reproductive isolation within plant species. Taxon 59:1367–1374.
Scotland, R. W., and M. J. Sanderson. 2004. The significance of few versus many in the tree of life. Science 303:643–643.
Seavey, S. R., and K. S. Bawa. 1986. Late-acting self-incompatibility in angiosperms. Botanical Review 52:195–219.
Sede, S. M., M. V. Nicola, R. Pozner, and L. A. Johnson. 2012. Phylogeography and palaeodistribution modelling in the Patagonian steppe: The case of *Mulinum spinosum* (Apiaceae). Journal of Biogeography 39:1041–1057.
Seehausen, O. 2004. Hybridization and adaptive radiation. Trends in Ecology & Evolution 19:198–207.
Seehausen, O., R. K. Butlin, I. Keller, C. E. Wagner, J. W. Boughman, P. A. Hohenlohe, C. L. Peichel, G. P. Saetre, C. Bank, A. Brannstrom, A. Brelsford, C. S. Clarkson, F. Eroukhmanoff, J. L. Feder, M. C. Fischer, A. D. Foote, P. Franchini, C. D. Jiggins, F. C. Jones, A. K. Lindholm, K. Lucek, M. E. Maan, D. A. Marques, S. H. Martin, B. Matthews, J. I. Meier, M. Most, M. W. Nachman, E. Nonaka, D. J. Rennison, J. Schwarzer, E. T. Watson, A. M. Westram, and A. Widmer. 2014. Genomics and the origin of species. Nature Reviews Genetics 15:176–192.
Selkoe, K. A., and R. J. Toonen. 2006. Microsatellites for ecologists: A practical guide to using and evaluating microsatellite markers. Ecology Letters 9:615–629.
Seoighe, C., C. Gehring, and L. D. Hurst. 2005. Gametophytic selection in *Arabidopsis thaliana* supports the selective model of intron length reduction. PLoS Genetics 1:154–158.
Sexton, J. P., S. B. Hangartner, and A. A. Hoffmann. 2014. Genetic isolation by environment or distance: Which pattern of gene flow is most common? Evolution 68:1–15.
Sexton, J. P., P. J. McIntyre, A. L. Angert, and K. J. Rice. 2009. Evolution and ecology of species range limits. Annual Review of Ecology Evolution and Systematics 40:415–436.
Shafer, A. B. A., and J. B. W. Wolf. 2013. Widespread evidence for incipient ecological speciation: A meta-analysis of isolation-by-ecology. Ecology Letters 16:940–950.
Sharbel, T. F., B. Haubold, and T. Mitchell-Olds. 2000. Genetic isolation by distance in *Arabidopsis thaliana*: Biogeography and postglacial colonization of Europe. Molecular Ecology 9:2109–2118.
Sharma, V. K., and J. C. Fletcher. 2002. Maintenance of shoot and floral meristem cell proliferation and fate. Plant Physiology 129:31–39.
Sharp, N. P., and S. P. Otto. 2016. Evolution of sex: Using experimental genomics to select among competing theories. Bioessays 38:751–757.
Sharrock, R. 1660. *History of the Propagation and Improvement of Vegetables by the Concurrence of Art*

and Nature. Printed by A. Lichfield for T. Robinson, Oxford.

Shaw, A. J., P. Szovenyi, and B. Shaw. 2011. Bryophyte diversity and evolution: Windows into the early evolution of land plants. American Journal of Botany 98:352–369.

Shaw, F. H., C. J. Geyer, and R. G. Shaw. 2002. A comprehensive model of mutations affecting fitness and inferences for *Arabidopsis thaliana*. Evolution 56:453–463.

Shaw, J., and K. Renzaglia. 2004. Phylogeny and diversification of bryophytes. American Journal of Botany 91:1557–1581.

Shinozaki, K., K. Yamaguchi-Shinozaki, and M. Seki. 2003. Regulatory network of gene expression in the drought and cold stress responses. Current Opinion in Plant Biology 6:410–417.

Shull, G. H. 1908. The composition of a field of maize. American Breeders Association Reports 4:296–301.

Shull, G. H. 1948. What is heterosis? Genetics 33:439–446.

Sicard, A., and M. Lenhard. 2011. The selfing syndrome: A model for studying the genetic and evolutionary basis of morphological adaptation in plants. Annals of Botany 107:1433–1443.

Sillanpaa, M. J. 2011. On statistical methods for estimating heritability in wild populations. Molecular Ecology 20:1324–1332.

Simpson, G. G. 1944. Tempo and Mode in Evolution. Columbia University Press, New York.

Sims, H. J. 2012. The evolutionary diversification of seed size: Using the past to understand the present. Evolution 66:1636–1649.

Siol, M., S. I. Wright, and S. C. H. Barrett. 2010. The population genomics of plant adaptation. New Phytologist 188:313–332.

Slatkin, M. 1985. Rare alleles as indicators of gene flow. Evolution 39:53–65.

Slatkin, M. 1995. A measure of population subdivision based on microsatellite allele frequencies. Genetics 139:1463–1463.

Slatkin, M., and N. H. Barton. 1989. A comparison of three indirect methods for estimating average levels of gene flow. Evolution 43:1349–1368.

Sletvold, N., and J. Ågren. 2014. There is more to pollinator-mediated selection than pollen limitation. Evolution 68:1907–1918.

Sloan, D. B., A. J. Alverson, J. P. Chuckalovcak, M. Wu, D. E. McCauley, J. D. Palmer, and D. R. Taylor. 2012. Rapid evolution of enormous, multichromosomal genomes in flowering plant mitochondria with exceptionally high mutation rates. PLoS Biology 10:e1001241.

Smith, C. C., and Fretwell. 1974. The optimal balance between size and number of offspring. American Naturalist 108:499–506.

Smith, C. I., C. S. Drummond, W. Godsoe, J. B. Yoder, and O. Pellmyr. 2009. Host specificity and reproductive success of yucca moths (*Tegeticula* spp. Lepidoptera: Prodoxidae) mirror patterns of gene flow between host plant varieties of the Joshua tree (*Yucca brevifolia*: Agavaceae). Molecular Ecology 18:5218–5229.

Smith, E., and H. J. Morowitz. 2004. Universality in intermediary metabolism. Proceedings of the National Academy of Sciences of the United States of America 101:13168–13173.

Smith, L. M., K. Bomblies, and D. Weigel. 2011. Complex evolutionary events at a tandem cluster of *Arabidopsis thaliana* genes resulting in a single-locus genetic incompatibility. PLoS Genetics 7:e1002164.

Smith, S. A., and M. J. Donoghue. 2008. Rates of molecular evolution are linked to life history in flowering plants. Science 322:86–89.

Snaydon, R. W., and A. D. Bradshaw. 1961. Differential response to calcium within the species *Festuca ovina* L. New Phytologist 60:219–234.

Sokal, R. R., and P. H. A. Sneath. 1963. Principles of Numerical Taxonomy. W.H. Freeman, San Francisco.

Solana, J. 2013. Closing the circle of germline and stem cells: The Primordial Stem Cell hypothesis. EvoDevo 4:2.

Soltis, D. E., V. A. Albert, J. Leebens-Mack, C. D. Bell, A. H. Paterson, C. Zheng, D. Sankoff, C. W. dePamphilis, P. K. Wall, and P. S. Soltis. 2009. Polyploidy and angiosperm diversification. American Journal of Botany 96:336–348.

Soltis, D. E., R. J. A. Buggs, J. J. Doyle, and P. S. Soltis. 2010. What we still don't know about polyploidy. Taxon 59:1387–1403.

Soltis, D. E., H. Ma, M. W. Frohlich, P. S. Soltis, V. A. Albert, D. G. Oppenheimer, N. S. Altman, C. dePamphilis, and J. Leebens-Mack. 2007a. The floral genome: An evolutionary history of gene duplication and shifting patterns of gene expression. Trends in Plant Science 12:358–367.

Soltis, D. E., M. S. Mayer, P. S. Soltis, and M. Edgerton. 1991. Chloroplast-DNA variation in *Tellima grandiflora* (Saxigragaceae). American Journal of Botany 78:1379–1390.

Soltis, D. E., and P. S. Soltis, eds. 1989. Isozymes in Plant Biology. Dioscorides Press, Portland, OR.

Soltis, D. E., P. S. Soltis, P. K. Endress, and M. W. Chase. 2005. Phylogeny and Evolution of Angiosperms. Sinauer Associates Inc., Sunderland, MA.

Soltis, D. E., P. S. Soltis, J. C. Pires, A. Kovarik, J. A. Tate, and E. Mavrodiev. 2004. Recent and recurrent polyploidy in *Tragopogon* (Asteraceae): Cytogenetic, genomic, and genetic comparisons. Biological Journal of the Linnean Society 82:485–501.

Soltis, D. E., P. S. Soltis, D. W. Schemske, J. F. Hancock, J. N. Thompson, B. C. Husband, and W. S. Judd. 2007b. Autopolyploidy in angiosperms: Have we grossly underestimated the number of species? Taxon 56:13–30.

Soltis, D. F., M. A. Gitzendanner, D. D. Strenge, and P. S. Soltis. 1997. Chloroplast DNA intraspecific phylogeography of plants from the Pacific Northwest of North America. Plant Systematics and Evolution 206:353–373.

Soltis, P. S., D. E. Soltis, V. Savolainen, P. R. Crane, and T. G. Barraclough. 2002. Rate heterogeneity among lineages of tracheophytes: Integration of molecular and fossil data and evidence for molecular living fossils. Proceedings of the National Academy of Sciences of the United States of America 99:4430–4435.

Sorhannus, U., and M. Fox. 1999. Synonymous and nonsynonymous substitution rates in diatoms: A comparison between chloroplast and nuclear genes. Journal of Molecular Evolution 48:209–212.

Sork, V. L., S. N. Aitken, R. J. Dyer, A. J. Eckert, P. Legendre, and D. B. Neale. 2013. Putting the landscape into the genomics of trees: Approaches for understanding local adaptation and population responses to changing climate. Tree Genetics & Genomes 9:901–911.

Sork, V. L., and P. E. Smouse. 2006. Genetic analysis of landscape connectivity in tree populations. Landscape Ecology 21:821–836.

Sosa, V., E. Ruiz-Sanchez, and F. C. Rodriguez-Gomez. 2009. Hidden phylogeographic complexity in the Sierra Madre Oriental: The case of the Mexican tulip poppy *Hunnemannia fumariifolia* (Papaveraceae). Journal of Biogeography 36:18–27.

Sousa, V., and J. Hey. 2013. Understanding the origin of species with genome-scale data: Modelling gene flow. Nature Reviews Genetics 14:404–414.

Speakman, J. R. 2005. Body size, energy metabolism, and lifespan. Journal of Experimental Biology 208:1717–1730.

Sperry, J. S. 2003. Evolution of water transport and xylem structure. International Journal of Plant Sciences 164:S115–S127.

Spicer, R., and A. Groover. 2010. Evolution of development of vascular cambia and secondary growth. New Phytologist 186:577–592.

Spofford, J. B. 1969. Heterosis and the evolution of duplications. American Naturalist 103:407–432.

Sprengel, C. K. 1793. Das entdeckte Geheimnis der Natur im Bau und in der Befruchtung der Blumen. Vieweg sen, Berlin.

Sprengel, C. K. 1996. Discovery of the secret of nature in the structure and fertilization of flowers. Pp. 44–64 *in* D. G. Lloyd and S. C. H. Barrett, eds. Floral Biology: Studies of Floral Evolution in Animal-Pollinated Plants. Chapman & Hall, New York.

Spriggs, E. L., P.-A. Christin, and E. J. Edwards. 2014. C4 photosynthesis promoted species diversification during the Miocene grassland expansion. PLoS One 9:e97722.

Stanton, M. L. 1984. Seed variation in wild radish: Effect of seed size on components of seedling and adult fitness. Ecology 65:1105–1112.

Starr, T. N., K. E. Gadek, J. B. Yoder, R. Flatz, and C. I. Smith. 2013. Asymmetric hybridization and gene flow between Joshua trees (Agavaceae: *Yucca*) reflect differences in pollinator host specificity. Molecular Ecology 22:437–449.

Stebbins, G. L. 1950. Variation and Evolution in Plants. Columbia University Press, New York.

Stebbins, G. L. 1957. Self-fertilization and population variability in the higher plants. American Naturalist 91:337–354.

Stebbins, G. L. 1958. The inviability, weakness, and sterility of interspecific hybrids. Advances in Genetics 9:147–215.

Stebbins, G. L. 1959. The role of hybridization in evolution. Proceedings of the American Philosophical Society 103:231–251.

Stebbins, G. L. 1971. Chromosomal Evolution in Higher Plants. Addison-Wesley Publishing Company, Reading, MA.

Stebbins, G. L. 1974. Flowering Plants—Evolution above the Species Llevel. Harvard University Press, Cambridge, MA.

Stebbins, G. L. 1981. Why are there so many species of flowering plants? BioScience 31:573–577.

Stenoien, H. K. 2008. Slow molecular evolution in 18S rDNA, rbcL, and nad5 genes of mosses compared with higher plants. Journal of Evolutionary Biology 21:566–571.

Stephenson, A. G. 1981a. Flower and fruit abortion: Proximate causes and ultimate functions. Annual Review of Ecology and Systematics 12:253–279.

Stevens, L., C. J. Goodnight, and S. Kalisz. 1995. Multilevel selection in natural populations of *Impatiens capensis*. American Naturalist 145:513–526.

Stewart, W. N. 1983. Paleobotany and the evolution of plants. Cambridge University Press, New York.
Stiles, F. G. 1975. Ecology, flowering phenology, and hummingbird pollination of some Costa Rican *Heliconia* species. Ecology 56:285–301.
Stinchcombe, J. R., and H. E. Hoekstra. 2008. Combining population genomics and quantitative genetics: Finding the genes underlying ecologically important traits. Heredity 100:158–170.
Stock, A. J., B. E. Campitelli, and J. R. Stinchcombe. 2014. Quantitative genetic variance and multivariate clines in the ivyleaf morning glory, *Ipomoea hederacea*. Philosophical Transactions of the Royal Society B: Biological Sciences 369:20130259.
Storfer, A., M. A. Murphy, S. F. Spear, R. Holderegger, and L. P. Waits. 2010. Landscape genetics: Where are we now? Molecular Ecology 19:3496–3514.
Stout, A. B. 1928. Dichogamy in flowering plants. Bulletin of the Torrey Botanical Club 55:141–153.
Straub, S. C. K., M. Parks, K. Weitemier, M. Fishbein, R. C. Cronn, and A. Liston. 2012. Navigating the tip of the genomic iceberg: Next-generation sequencing for plant systematics. American Journal of Botany 99:349–364.
Straw, R. M. 1955. Hybridization, homogamy, and sympatric speciation. Evolution 9:441–444.
Stringer, C. 2012. Evolution: What makes a modern human. Nature 485:33–35.
Stuessy, T. F. 2006. Evolutionary biology: Sympatric plant speciation in islands? Nature 443:E12.
Stuessy, T. F., ed. 2008. Evolution and Speciation of Island Plants. Cambridge University Press, Cambridge, UK.
Stull, G. W., M. J. Moore, V. S. Mandala, N. A. Douglas, H.-R. Kates, X. Qi, S. F. Brockington, P. S. Soltis, D. E. Soltis, and M. A. Gitzendanner. 2013. A targeted enrichment strategy for massively parallel sequencing of Angiosperm plastid genomes. Applications in Plant Sciences 1:1200497.
Suchan, T., and N. Alvarez. 2015. Fifty years after Ehrlich and Raven, is there support for plant–insect coevolution as a major driver of species diversification? Entomologia Experimentalis et Applicata 157:98–112.
Sultan, S. E. 2003. Phenotypic plasticity in plants: A case study in ecological development. Evolution & Development 5:25–33.
Sun, J. H., Z. C. Li, D. K. Jewett, K. O. Britton, W. H. Ye, and X. J. Ge. 2005. Genetic diversity of *Pueraria lobata* (kudzu) and closely related taxa as revealed by inter-simple sequence repeat analysis. Weed Research 45:255–260.
Sutherland, S. 1986. Patterns of fruit-set: What controls fruit-flower ratios in plants. Evolution 40:117–128.
Swanson, R., A. F. Edlund, and D. Preuss. 2004. Species specificity in pollen-pistil interactions. Annual Review of Genetics 38:793–818.
Sweigart, A. L., and J. H. Willis. 2012. Molecular evolution and genetics of postzygotic reproductive isolation in plants. F1000 Biology Reports 4:23–23.
Szymura, J. M., and N. H. Barton. 1986. Genetic analysis of a hybrid zone between the fire-bellied toads, *Bombina bombina* and *Bombina variegata*, near Cracow in Southern Poland. Evolution 40:1141–1159.
Taberlet, P., L. Gielly, G. Pautou, and J. Bouvet. 1991. Universal primers for amplification of three non-coding regions of chloroplast DNA. Plant Molecular Biology 17:1105–1109.
Tajima, F. 1983. Evolutionary relationship of DNA sequences in finite populations. Genetics 105:437–460.
Takayama, S., and A. Isogai. 2005. Self-incompatibility in plants. Annual Review of Plant Biology 56:467–489.
Tallmon, D. A., G. Luikart, and R. S. Waples. 2004. The alluring simplicity and complex reality of genetic rescue. Trends in Ecology & Evolution 19:489–496.
Tang, S. X., R. A. Okashah, S. J. Knapp, M. L. Arnold, and N. H. Martin. 2010. Transmission ratio distortion results in asymmetric introgression in Louisiana iris. BMC Plant Biology 10:48.
Tate, J. A., V. V. Symonds, A. N. Doust, R. J. A. Buggs, E. Mavrodiev, L. C. Majure, P. S. Soltis, and D. E. Soltis. 2009. Synthetic polyploids of *Tragopogon miscellus* and *T. mirus* (Asteraceae): Sixty years after Ownbey's discovery. American Journal of Botany 96:979–988.
Tavare, S., C. R. Marshall, O. Will, C. Soligo, and R. D. Martin. 2002. Using the fossil record to estimate the age of the last common ancestor of extant primates. Nature 416:726–729.
Taylor, J. S., and J. Raes. 2004. Duplication and divergence: The evolution of new genes and old ideas. Annual Review of Genetics 38:615–643.
Taylor, S. A., E. L. Larson, and R. G. Harrison. 2015. Hybrid zones: Windows on climate change. Trends in Ecology & Evolution 30:398–406.
Taylor, S. J., R. W. Willard, J. P. Shaw, M. C. Dobson, and N. H. Martin. 2011. Differential response of the homoploid hybrid species *Iris nelsonii* (Iridaceae) and its progenitors to abiotic habitat conditions. American Journal of Botany 98:1309–1316.
Temme, D. H. 1986. Seed size variability: A consequence of variable genetic quality among offspring? Evolution 40:414–417.

Templeton, A. R. 1989. The meaning of species and speciation: A genetic perspective. Pp. 3–27 *in* O. Otte and J. A. Endler, eds. Speciation and Its Consquences. Sinauer Associates, Inc., Sunderland, MA.

Templeton, A. R. 1998. Nested clade analyses of phylogeographic data: Testing hypotheses about gene flow and population history. Molecular Ecology 7:381–397.

Templeton, A. R. 2001. Using phylogeographic analyses of gene trees to test species status and processes. Molecular Ecology 10:779–791.

Templeton, A. R. 2004. Statistical phylogeography: Methods of evaluating and minimizing inference errors. Molecular Ecology 13:789–809.

Templeton, A. R. 2009a. Statistical hypothesis testing in intraspecific phylogeography: Nested clade phylogeographical analysis versus approximate Bayesian computation. Molecular Ecology 18:319–331.

Templeton, A. R. 2009b. Why does a method that fails continue to be used? The answer. Evolution 63:807–812.

Templeton, A. R., K. A. Crandall, and C. F. Sing. 1992. A cladistic analysis of phenotypic associations and haplotypes inferred from restriction endonuclease mapping and sequence data. III. Cladogram estimation. Genetics 132:619–633.

Templeton, A. R., and D. A. Levin. 1979. Evolutionary consequences of seed pools. American Naturalist 114:232–249.

Templeton, A. R., S. D. Maskas, and M. B. Cruzan. 2000. Gene trees: A powerful tool for exploring the evolutionary biology of species and speciation. Plant Species Biology 15:211–222.

Theoharides, K. A., and J. S. Dukes. 2007. Plant invasion across space and time: Factors affecting nonindigenous species success during four stages of invasion. New Phytologist 176:256–273.

Thibert-Plante, X., and A. P. Hendry. 2010. When can ecological speciation be detected with neutral loci? Molecular Ecology 19:2301–2314.

Thien, L. B., P. Bernhardt, M. S. Devall, Z.-d. Chen, Y.-b. Luo, J.-H. Fan, L.-C. Yuan, and J. H. Williams. 2009. Pollination biology of basal angiosperms (ANITA grade). American Journal of Botany 96:166–182.

Thomas, D. T., A. R. Ahedor, C. F. Williams, C. Depamphilis, D. J. Crawford, and Q. Y. Xiang. 2008. Genetic analysis of a broad hybrid zone in *Aesculus* (Sapindaceae): Is there evidence of long-distance pollen dispersal? International Journal of Plant Sciences 169:647–657.

Thomas, H., L. Huang, M. Young, and H. Ougham. 2009. Evolution of plant senescence. BMC Evolutionary Biology 9:163.

Thomé, O. W. 1885. Flora von Deutschland Österreich und der Schweiz. Republished by Kurt Stuber, 2008, http://caliban.mpiz-koeln.mpg.de/thome/index.html.

Thompson, J. N. 1994. The Coevolutionary Process. University of Chicago Press, Chicago.

Thompson, J. N. 2003. When is it mutualism? (An American Society of Naturalists Presidential Address). American Naturalist 162:S1–S9.

Thomson, J. D. 1986. Pollen transport and deposition by bumble bees in *Erythronium*—influences of floral nectar and bee grooming. Journal of Ecology 74:329–341.

Thomson, J. D. 2006. Tactics for male reproductive success in plants: Contrasting insights of sex allocation theory and pollen presentation theory. Integrative and Comparative Biology 46:390–397.

Thomson, J. D., B. J. Andrews, and R. C. Plowright. 1981. The effect of a foreign pollen on ovule development in *Diervilla lonicera* (Caprifoliaceae). New Phytology 90:777–783.

Thomson, J. D., and J. Brunet. 1990. Hypotheses for the evolution of dioecy in seed plants. Trends in Ecology & Evolution 5:11–16.

Thomson, J. D., and B. A. Thomson. 1989. Dispersal of *Erythronium grandiflorum* pollen by bumblebees: Implications for gene flow and reproductive success. Evolution 43:657–661.

Thomson, J. D., and P. Wilson. 2008. Explaining evolutionary shifts between bee and hummingbird pollination: Convergence, divergence, and directionality. International Journal of Plant Sciences 169:23–38.

Tiffney, B. H. 2004. Vertebrate dispersal of seed plants through time. Annual Review of Ecology Evolution and Systematics 35:1–29.

Timme, R. E. and C. F. Delwiche. 2010. Uncovering the evolutionary origin of plant molecular processes: Comparison of Coleochaete (Coleochaetales) and Spirogyra (Zygnematales) transcriptomes. BMC Plant Biology 10:96.

Togashi, T., J. L. Bartelt, J. Yoshimura, K.-i. Tainaka, and P. A. Cox. 2012. Evolutionary trajectories explain the diversified evolution of isogamy and anisogamy in marine green algae. Proceedings of the National Academy of Sciences of the United States of America 109:13692–13697.

Tonsor, S. J., S. Kalisz, J. Fisher, and T. P. Holtsford. 1993. A life-history-based study of population genetic structure: Seed bank to adults in *Plantago lanceolata*. Evolution 47:833–843.

Travisano, M., and R. G. Shaw. 2013. Lost in the map. Evolution 67:305–314.

Treisman, M., and R. Dawkins. 1976. The "cost of meiosis": Is there any? Journal of Theoretical Biology 63:479–484.

Tremblay, N. O., and D. J. Schoen. 1999. Molecular phylogeography of *Dryas integrifolia*: Glacial refugia and postglacial recolonization. Molecular Ecology 8:1187–1198.

Trinks, H., W. Schroder, and C. K. Biebricher. 2005. Ice and the origin of life. Origins of Life and Evolution of the Biosphere 35:429–445.

Tsvuura, Z., M. E. Griffiths, R. M. Gunton, and M. J. Lawes. 2011. Predator satiation and recruitment in a mast-fruiting monocarpic forest herb. Annals of Botany 107:379–387.

Turesson, G. 1922. The genotypic response of the plant species to the habitat. Hereditas 3:211–350.

Turmel, M., J. F. Pombert, P. Charlebois, C. Otis, and C. Lemieux. 2007. The green algal ancestry of land plants as revealed by the chloroplast genome. International Journal of Plant Sciences 168:679–689.

Turnbull, L. A., L. Santamaria, T. Martorell, J. Rallo, and A. Hector. 2006. Seed size variability: From carob to carats. Biology Letters 2:397–400.

Twyford, A. D., C. A. Kidner, and R. A. Ennos. 2015. Maintenance of species boundaries in a neotropical radiation of Begonia. Molecular Ecology 24:4982–4993.

Udovic, D. 1981. Determinants of fruit set in *Yucca whipplei*: Reproductive expenditure versus pollinator availability. Oecologia 48:389–399.

Uyenoyama, M. 1986. Inbreeding and the cost of meiosis: The evolution of selfing in populations practicing biparental inbreeding. Evolution 40:388–404.

Uyenoyama, M. K., K. E. Holsinger, and D. M. Waller. 1993. Ecological and genetic factors directing the evolution of self-fertilization. Pp. 327–381 *in* D. J. Futuyma, and J. Antonovics, eds. Oxford Surveys in Evolutionary Biology, Oxford University Press, Oxford.

Valen, L. V. 1963. Haldane's dilemma, evolutionary rates, and heterosis. The American Naturalist 97:185–190.

van der Niet, T., and S. D. Johnson. 2009. Patterns of plant speciation in the Cape floristic region. Molecular Phylogenetics and Evolution 51:85–93.

van der Niet, T., and S. D. Johnson. 2012. Phylogenetic evidence for pollinator-driven diversification of angiosperms. Trends in Ecology & Evolution 27:353–361.

Van der Niet, T., R. Peakall, and S. D. Johnson. 2014. Pollinator-driven ecological speciation in plants: New evidence and future perspectives. Annals of Botany 113:199–212.

van Dijk, P. J. 2003. Ecological and evolutionary opportunities of apomixis: Insights from *Taraxacum* and *Chondrilla*. Philosophical Transactions of the Royal Society of London Series B: Biological Sciences 358:1113–1120.

van Kleunen, M., and M. Fischer. 2005. Constraints on the evolution of adaptive phenotypic plasticity in plants. New Phytologist 166:49–60.

Van Valen, L. M. 1973. A new evolutionary law. Evolutionary Theory 1:1–30.

Vandepitte, K., H. Jacquemyn, I. Roldan-Ruiz, and O. Honnay. 2007. Landscape genetics of the self-compatible forest herb *Geum urbanum*: Effects of habitat age, fragmentation, and local environment. Molecular Ecology 16:4171–4179.

Vasek, F. C. 1980. Creosote bush: Long-lived clones in the Mojave Desert. American Journal of Botany 67:246–255.

Vavilov, N. I., and D. T. Love. 1992. Origin and Geography of Cultivated Plants. Cambridge University Press, New York.

Vega-Frutis, R., R. Macias-Ordonez, R. Guevara, and L. Fromhage. 2014. Sex change in plants and animals: A unified perspective. Journal of Evolutionary Biology 27:667–675.

Vekemans, X., and O. J. Hardy. 2004. New insights from fine-scale spatial genetic structure analyses in plant populations. Molecular Ecology 13:921–935.

Vendramin, G. G., B. Degen, R. J. Petit, M. Anzidei, A. Madaghiele, and B. Ziegenhagen. 1999. High level of variation at *Abies alba* chloroplast microsatellite loci in Europe. Molecular Ecology 8:1117–1126.

Vesteg, M., and J. Krajcovic. 2011. The falsifiability of the models for the origin of eukaryotes. Current Genetics 57:367–390.

Vijverberg, K., and K. Bachmann. 1999. Molecular evolution of a tandemly repeated trnF(GAA) gene in the chloroplast genomes of *Microseris* (Asteraceae) and the use of structural mutations in phylogenetic analyses. Molecular Biology and Evolution 16:1329–1340.

Vik, U., M. H. Jorgensen, H. Kauserud, I. Nordal, and A. K. Brysting. 2010. Microsatellite markers show decreasing diversity but unchanged level of clonality in *Dryas octopetala* (Rosaceae) with increasing latitude. American Journal of Botany 97:988–997.

Villarreal, J. C., and S. S. Renner. 2012. Hornwort pyrenoids, carbon-concentrating structures, evolved and were lost at least five times during the last 100 million years. Proceedings of the National Academy of Sciences of the United States of America 109:18873–18878.

Visscher, P. M., W. G. Hill, and N. R. Wray. 2008. Heritability in the genomics era—concepts and misconceptions. Nature Reviews Genetics 9:255–266.

Vitalis, R., S. Glemin, and I. Olivieri. 2004. When genes go to sleep: The population genetic consequences of seed dormancy and monocarpic perenniality. American Naturalist 163:295–311.

Vogel, S. 1996. Christian Konrad Sprengel's theory of the flower: The cradle of floral ecology. Pp. 44–64 *in* D. G. Lloyd and S. C. H. Barrett, eds. Floral Biology: Studies of Floral Evolution in Animal-Pollinated Plants. Chapman & Hall, New York.

Von der Lippe, M., and I. Kowarik. 2007. Long-distance dispersal of plants by vehicles as a driver of plant invasions. Conservation Biology 21:986–996.

von Frisch, K. 1967. The Dance Language and Orientation of Bees. Harvard University Press, Cambridge, MA.

vonHoldt, B. M., J. P. Pollinger, D. A. Earl, J. C. Knowles, A. R. Boyko, H. Parker, E. Geffen, M. Pilot, W. Jedrzejewski, B. Jedrzejewska, V. Sidorovich, C. Greco, E. Randi, M. Musiani, R. Kays, C. D. Bustamante, E. A. Ostrander, J. Novembre, and R. K. Wayne. 2011. A genome-wide perspective on the evolutionary history of enigmatic wolf-like canids. Genome Research 21:1294–1305.

Vos, P., R. Hogers, M. Bleeker, M. Reijans, T. v. d. Lee, M. Hornes, A. Frijters, J. Pot, J. Peleman, M. Kulper, and M. Zabeau. 1995. AFLP: A new technique for DNA fingerprinting. Nucleic Acids Research 23:4407–4414.

Wächtershäuser, G. 2007. On the chemistry and evolution of the pioneer organism. Chemistry & Biodiversity 4:584–602.

Wade, M. J., and C. J. Goodnight. 1998. The theories of Fisher and Wright in the context of metapopulations: When nature does many small experiments. Evolution 52:1537–1553.

Wagner, G. P. 1996. Homologues, natural kinds, and the evolution of modularity. American Zoologist 36:36–43.

Walbot, V., and M. M. S. Evans. 2003. Unique features of the plant life cycle and their consequences. Nature Reviews Genetics 4:369–379.

Walker, D., and J. C. Avise. 1998. Principles of phylogeography as illustrated by freshwater and terrestrial turtles in the southeastern United States. Annual Review of Ecology and Systematics 29:23–58.

Wallace, A. R. 1865. On the phenomena of variation and geographic distribution as illustrated by the Papilonidae of Malayan region. Transactions of the Linnean Society of London 25:1–71.

Wallace, A. R. 1880. Island Life. Richard Clay and Sons, Ltd., London.

Wallace, A. R. 1889. Darwinism. MacMillan and Co., London.

Walsh, N. E., and D. Charlesworth. 1992. Evolutionary interpretations of differences in pollen tube growth rates. Quarterly Review of Biology 67:19–37.

Wang, B., L. H. Yeun, J. Y. Xue, Y. Liu, J. M. Ane, and Y. L. Qiu. 2010. Presence of three mycorrhizal genes in the common ancestor of land plants suggests a key role of mycorrhizas in the colonization of land by plants. New Phytologist 186:514–525.

Wang, J., and M. B. Cruzan. 1998. Interspecific mating in the *Piriqueta caroliniana* (Turneraceae) complex: Effects of pollen load size and composition. American Journal of Botany 85:1172–1179.

Wang, R.-L., A. Stec, J. Hey, L. Lukens, and J. Doebley. 1999. The limits of selection during maize domestication. Nature 398:236–239.

Wang, X. R., A. E. Szmidt, and O. Savolainen. 2001. Genetic composition and diploid hybrid speciation of a high mountain pine, *Pinus densata*, native to the Tibetan plateau. Genetics 159:337–346.

Wang, Z. W., C. Wang, L. Gao, S. Y. Mei, Y. Zhou, C. P. Xiang, and T. Wang. 2013. Heterozygous alleles restore male fertility to cytoplasmic male-sterile radish (*Raphanus sativus* L.): A case of overdominance. Journal of Experimental Botany 64:2041–2048.

Waples, R. S. 2006. Seed banks, salmon, and sleeping genes: Effective population size in semelparous, age-structured species with fluctuating abundance. American Naturalist 167:118–135.

Waser, N. M., L. Chittka, M. V. Price, N. M. Williams, and J. Ollerton. 1996. Generalization in pollination systems, and why it matters. Ecology 77:1043–1060.

Watanabe, Y., J. E. J. Martini, and H. Ohmoto. 2000. Geochemical evidence for terrestrial ecosystems 2.6 billion years ago. Nature 408:574–578.

Watson, M. A., and B. B. Casper. 1984. Morphogenetic constraints on patterns of carbon distribution in plants. Annual Review of Ecology and Systematics 15:233–258.

Webb, C. J., and D. G. Lloyd. 1986. The avoidance of interference between the presentation of pollen and stigmas in angiosperms. II. Herkogamy. New Zealand Journal of Botany 24:163–178.

Weber, M. G., and A. A. Agrawal. 2014. Defense mutualisms enhance plant diversification. Proceedings of the National Academy of Sciences of the United States of America 111:16442–16447.

Webster, M. T., and L. D. Hurst. 2012. Direct and indirect consequences of meiotic recombination: Implications for genome evolution. Trends in Genetics 28:101–109.

Weinberg, W. 1908. Über den nachweis der vererbung beim menschen. Jahreshefte des Vereins für vaterländische Naturkunde in Württemberg 64:368–382.

Weismann, A. 1889. Essays on Heredity and Kindred Biological Subjects. Oxford University Press, Oxford, UK.

Weldon, W. F. R. 1893. On certain correlated variations in *Carcinus mocenas*. Proceedings of the Royal Society 54:329.

Welsh, J., and M. McClelland. 1990. Fingerprinting genomes using PCR with arbitrary primers. Nucleic Acids Research 18:7213–7218.

Wenny, D. G. 2001. Advantages of seed dispersal: A re-evaluation of directed dispersal. Evolutionary Ecology Research 3:51–74.

West-Eberhard, M. J. 2005. Developmental plasticity and the origin of species differences. Proceedings of the National Academy of Sciences of the United States of America 102:6543–6549.

Westoby, M., and B. Rice. 1982. Evolution of the seed plants and inclusive fitness of plant tissues. Evolution 36:713–724.

Wheeler, Q. D., and N. I. Platnick. 2000. The phylogenetic species concept (sensu Wheeler and Platnick). Pp. 55–69 *in* Q. D. Wheeler and R. Meier, eds. Species Concepts and Phylogenetic Theory: A Debate. Columbia University Press, New York.

Whibley, A. C., N. B. Langlade, C. Andalo, A. I. Hanna, A. Bangham, C. Thébaud, and E. Coen. 2006. Evolutionary paths underlying flower color variation in *Antirrhinum*. Science 313:963–966.

White, J. 1979. The plant as a metapopulation. Annual Review of Ecology and Systematics 10:109–145.

White, T. L., W. T. Adams, and D. B. Neale. 2007. Forest Genetics. CABI Publishing, Cambridge, UK.

Whitehouse, H. L. K. 1950. Multiple-allelomorph incompatibility of pollen and style in the evolution of the angiosperms. Annals of Botany 14:199–216.

Whitham, T. G., and C. N. Slobodchikoff. 1981. Evolution by individuals, plant-herbivore interactions, and mosaics of genetic variability: The adaptive significance of somatic mutations in plants. Oecologia 49:287–292.

Whitlock, C. and P. J. Bartlein. 1997. Vegetation and climate change in northwest North America during the past 125 KYR. Nature 388:57–61.

Whitlock, M. C. 2002. Selection, load, and inbreeding depression in a large metapopullation. Genetics 160:1191–1202.

Whitlock, M. C. 2008. Evolutionary inference from Q_{ST}. Molecular Ecology 17:1885–1896.

Whitlock, M. C. 2011. G'_{ST} and D do not replace F_{ST}. Molecular Ecology 20:1083–1091.

Whitlock, M. C., and D. E. McCauley. 1999. Indirect measures of gene flow and migration: F_{ST} <> 1/($4Nm$ + 1). Heredity 82:117–125.

Whitlock, M. C. and P. C. Phillips. 2000. The exquisite corpse: A shifting view of the shifting balance. Trends in Ecological Evolution 15:347–348.

Whittemore, A. T., and B. A. Schaal. 1991. Interspecific gene flow in sympatric oaks. Proceedings of the National Academy of Sciences of the United States of America 88:2450–2544.

Wichmann, M. C., M. J. Alexander, M. B. Soons, S. Galsworthy, L. Dunne, R. Gould, C. Fairfax, M. Niggemann, R. S. Hails, and J. M. Bullock. 2009. Human-mediated dispersal of seeds over long distances. Proceedings of the Royal Society B: Biological Sciences 276:523–532.

Wiens, D., C. L. Calvin, C. A. Wilson, C. I. Davern, D. Frank, and S. R. Seavey. 1987. Reproductive success, spontaneous embryo abortion, and genetic load in flowering plants. Oecologia 71:501–509.

Wiens, J. J., D. D. Ackerly, A. P. Allen, B. L. Anacker, L. B. Buckley, H. V. Cornell, E. I. Damschen, T. Jonathan Davies, J.-A. Grytnes, S. P. Harrison, B. A. Hawkins, R. D. Holt, C. M. McCain, and P. R. Stephens. 2010. Niche conservatism as an emerging principle in ecology and conservation biology. Ecology Letters 13:1310–1324.

Wiley, E. O., and B. S. Lieberman. 2011. Phylogenetics: Theory and Practice of Phylogenetic Systematics. Wiley-Blackwell, New York.

Wiley, E. O., and R. L. Mayden. 2000. The evolutionary species concept. Pp. 70–92 *in* Q. D. Wheeler and R. Meier, eds. Species Concepts and Phylogenetic Theory: A Debate. Columbia University Press, New York.

Williams, G. C. 1957. Pleiotropy, natural selection, and the evolution of senescence. Evolution 11:398–411.

Williams, G. C. 1966. Adaptive and Natural Selection: A Critique of Some Current Evolutionary Thought. Princeton University Press, Princeton.

Williams, G. C. 1975. Sex and Evolution. Princeton University Press, Princeton.

Williams, J. G. K., A. R. Kubelik, K. L. Livak, J. A. Rafalski, and S. V. Tingey. 1990. DNA polymorphisms amplified by arbitrary primers are useful as genetic markers. Nucleic Acids Research 18:6531–6535.

Williams, J. H. 2009. *Amborella trichopoda* (Amborellaceae) and the evolutionary developmental origins of the angiosperm progamic phase. American Journal of Botany 96:144–165.

Williamson, M. H. 1996. Biological Invasions. Chapman Hall, New York.

Willing, E.-M., C. Dreyer, and C. van Oosterhout. 2012. Estimates of genetic differentiation measured by F(ST) do not necessarily require large sample sizes when using many SNP markers. PLoS One 7:e42649.

Willis, J. C. 1922. Age and Area. Cambridge University Press, London.

Willis, J. H. 1999. The role of genes of large effect on inbreeding depression in *Mimulus guttatus*. Evolution 53:1678–1691.

Willis, K. J., and J. C. McElwain. 2002. The Evolution of Plants. Oxford University Press, New York.

Wilson, J. R. U., E. E. Dormontt, P. J. Prentis, A. J. Lowe, and D. M. Richardson. 2009. Something in the way you move: Dispersal pathways affect invasion success. Trends in Ecology & Evolution 24:136–144.

Winge, Ö. 1917. The chromosomes: Their numbers and general importance. Comptes Rendus des Travaux du Laboratoire Carlsberg 13:131–275.

Winge, Ö. 1924. Contributions to the knowledge of chromosome numbers in plants. La Cellule 35:303–324.

Wink, M. 2003. Evolution of secondary metabolites from an ecological and molecular phylogenetic perspective. Phytochemistry 64:3–19.

Winn, A. A., E. Elle, S. Kalisz, P.-O. Cheptou, C. G. Eckert, C. Goodwillie, M. O. Johnston, D. A. Moeller, R. H. Ree, R. D. Sargent, and M. Vallejo-Marin. 2011. Analysis of inbreeding depression in mixed-mating plants provides evidence for selective interference and stable mixed mating. Evolution 65:3339–3359.

Winsor, J. A., S. Peretz, and A. G. Stephenson. 2000. Pollen competition in a natural population of *Cucurbita foetidissima* (Cucurbitaceae). American Journal of Botany 87:527–532.

Winther, J. L., and W. E. Friedman. 2008. Arbuscular mycorrhizal associations in Lycopodiaceae. New Phytologist 177:790–801.

Wodniok, S., H. Brinkmann, G. Glockner, A. J. Heidel, H. Philippe, M. Melkonian, and B. Becker. 2011. Origin of land plants: Do conjugating green algae hold the key? BMC Evolutionary Biology 11:104.

Wolfe, A. D., Q. Y. Xiang, and S. R. Kephart. 1998. Diploid hybrid speciation in *Penstemon* (Scrophulariaceae). Proceedings of the National Academy of Sciences of the United States of America 95:5112–5115.

Wolfe, K. H., W. H. Li, and P. M. Sharp. 1987. Rates of nucleotide substitution vary greatly among plant mitochondrial, chloroplast, and nuclear DNAs. Proceedings of the National Academy of Sciences of the United States of America 84:9054–9058.

Wood, T. E., N. Takebayashi, M. S. Barker, I. Mayrose, P. B. Greenspoon, and L. H. Rieseberg. 2009. The frequency of polyploid speciation in vascular plants. Proceedings of the National Academy of Sciences of the United States of America 106:13875–13879.

Wright, S. 1931. Evolution in Mendelian populations. Genetics 16:97–159.

Wright, S. 1932. The roles of mutation, inbreeding, crossbreeding, and selection in evolution. Proceedings of the Sixth International Congress of Genetics 1:356–366.

Wright, S. 1940. Breeding structure of populations in relation to speciation. American Naturalist 74:232–248.

Wright, S. 1941. On the probability of fixation of reciprocal translocations. American Naturalist 75:513–522.

Wright, S. 1943. Isolation by distance. Genetics 28:114–138.

Wright, S. 1946. Isolation by distance under diverse systems of mating. Genetics 31:39–59.

Wright, S. 1950. Genetical structure of populations. Nature 166:247–249.

Wright, S. 1965. The interpretation of population structure by F statistics with special regard to systems of mating. Evolution 19:395–420.

Wright, S. 1982. The shifting balance theory and macroevolution. Annual Review of Genetics 16:1–19.

Wright, S. I., and B. S. Gaut. 2005. Molecular population genetics and the search for adaptive evolution in plants. Molecular Biology and Evolution 22:506–519.

Wright, S. I., S. Kalisz, and T. Slotte. 2013. Evolutionary consequences of self-fertilization in plants. Proceedings of the Royal Society B: Biological Sciences 280:20130133.

Wright, S. I., C. B. K. Yau, M. Looseley, and B. C. Meyers. 2004. Effects of gene expression on molecular evolution in *Arabidopsis thaliana* and *A. lyrata*. Molecular Biology and Evolution 21:1719–1726.

Wu, C. S., S. M. Chaw, and Y. Y. Huang. 2013. Chloroplast phylogenomics indicates that *Ginkgo biloba* is sister to cycads. Genome Biology and Evolution 5:243–254.

Wu, L., A. D. Bradshaw, and D. A. Thurman. 1975. The potential for evolution of heavy metal tolerance in plants. III. The rapid evolution of copper tolerance in *Agrostis stolonifera*. Heredity 34:165–187.

Wyatt, R., I. J. Odrzykoski, A. Stoneburner, H. W. Bass, and G. A. Galau. 1988. Allopolyploidy in bryophytes—multiple origins of *Plagiomnium medium*. Proceedings of the National Academy of Sciences of the United States of America 85:5601–5604.

Xia, Z., Y. Tsubokura, M. Hoshi, M. Hanawa, C. Yano, K. Okamura, T. A. Ahmed, T. Anai, S. Watanabe, M. Hayashi, T. Kawai, K. G. Hossain, H. Masaki, K. Asai, N. Yamanaka, N. Kubo, K. Kadowaki, Y. Nagamura, M. Yano, T. Sasaki, and K. Harada.

2007. An integrated high-density linkage map of soybean with RFLP, SSR, STS, and AFLP markers using a single F2 population. DNA Research 14:257–269.

Xu, C. Y., M. H. Julien, M. Fatemi, C. Girod, R. D. Van Klinken, C. L. Gross, and S. J. Novak. 2010. Phenotypic divergence during the invasion of *Phyla canescens* in Australia and France: Evidence for selection-driven evolution. Ecology Letters 13:32–44.

Yakimowski, S. B., and C. G. Eckert. 2008. Populations do not become less genetically diverse or more differentiated towards the northern limit of the geographical range in clonal *Vaccinium stamineum* (Ericaceae). New Phytologist 180:534–544.

Yakimowski, S. B., and L. H. Rieseberg. 2014. The role of homoploid hybridization in evolution: A century of studies synthesizing genetics and ecology. American Journal of Botany 101:1247–1258.

Yamasaki, M., S. I. Wright, and M. D. McMullen. 2007. Genomic screening for artificial selection during domestication and improvement in maize. Annals of Botany 100:967–973.

Yang, S., J. G. Bishop, and M. S. Webster. 2008. Colonization genetics of an animal-dispersed plant (*Vaccinium membranaceum*) at Mount St Helens, Washington. Molecular Ecology 17:731–740.

Yang, Z., and B. Rannala. 2012. Molecular phylogenetics: Principles and practice. Nature Reviews Genetics 13:303–314.

Yang, Z. H., and J. P. Bielawski. 2000. Statistical methods for detecting molecular adaptation. Trends in Ecology & Evolution 15:496–503.

Yoo, M. J., A. S. Chanderbali, N. S. Altman, P. S. Soltis, and D. E. Soltis. 2010. Evolutionary trends in the floral transcriptome: Insights from one of the basalmost angiosperms, the water lily *Nuphar advena* (Nymphaeaceae). Plant Journal 64:687–698.

Yoon, H. S., J. D. Hackett, C. Ciniglia, G. Pinto, and D. Bhattacharya. 2004. A molecular timeline for the origin of photosynthetic eukaryotes. Molecular Biology and Evolution 21:809–818.

Young, A., T. Boyle, and T. Brown. 1996. The population genetic consequences of habitat fragmentation for plants. Trends in Ecology & Evolution 11:413–418.

Young, A. G., H. G. Merriam, and S. I. Warwick. 1993. The effects of forest fragmentation on genetic variation in *Acer saccharum* Marsh. (sugar maple) populations. Heredity 71:277–289.

Young, H. A., G. Sarath, and C. M. Tobias. 2012. Karyotype variation is indicative of subgenomic and ecotypic differentiation in switchgrass. BMC Plant Biology 12:117.

Young, T. P. and C. K. Augspurger. 1991. Ecology and evolution of long-lived semelparous plants. Trends in Ecology & Evolution 6:285–289.

Yu, Y., J. Heinrichs, R. L. Zhu, and H. Schneider. 2013. Empirical evidence supporting frequent cryptic speciation in epiphyllous liverworts: A case study of the *Cololejeunea lanciloba* complex. PLoS One 8:e84124.

Yuan, X. L., S. H. Xiao, and T. N. Taylor. 2005. Lichen-like symbiosis 600 million years ago. Science 308:1017–1020.

Yue, J.-X., J. Li, D. Wang, H. Araki, D. Tian, and S. Yang. 2010. Genome-wide investigation reveals high evolutionary rates in annual model plants. BMC Plant Biology 10:1–12.

Zalapa, J. E., H. Cuevas, H. Y. Zhu, S. Steffan, D. Senalik, E. Zeldin, B. McCown, R. Harbut, and P. Simon. 2012. Using next-generation sequencing approaches to isolate simple sequence repeat (SSR) loci in the plant sciences. American Journal of Botany 99:193–208.

Zavada, M. S., and T. N. Taylor. 1986. The role of self-incompatibility and sexual selection in the gymnosperm-angiosperm transition—a hypothesis. American Naturalist 128:538–550.

Zhai, W., R. Nielsen, and M. Slatkin. 2009. An investigation of the statistical power of neutrality tests based on comparative and population genetic data. Molecular Biology and Evolution 26:273–283.

Zhang, D. Y. 2006. Evolutionarily stable reproductive investment and sex allocation in plants. Pp. 41–60 *in* L. D. Harder and S. C. H. Barrett, eds. Ecology and Evolution of Flowers. Oxford University Press, New York.

Zhang, J. Z. 2003. Evolution by gene duplication: An update. Trends in Ecology & Evolution 18:292–298.

Zhang, Y.-Y., D.-Y. Zhang, and S. C. H. Barrett. 2010. Genetic uniformity characterizes the invasive spread of water hyacinth (*Eichhornia crassipes*), a clonal aquatic plant. Molecular Ecology 19:1774–1786.

Zhang, Z.-Q., X.-F. Zhu, H. Sun, Y.-P. Yang, and S. C. H. Barrett. 2014. Size-dependent gender modification in *Lilium apertum* (Liliaceae): Does this species exhibit gender diphasy? Annals of Botany 114:441–453.

Zhou, H., I. J. Kimsey, E. N. Nikolova, B. Sathyamoorthy, G. Grazioli, J. McSally, T. Bai, C. H. Wunderlich, C. Kreutz, I. Andricioaei, and H. M. Al-Hashimi. 2016. m1A and m1G disrupt A-RNA structure through the intrinsic instability of Hoogsteen base pairs. Nature Structural & Molecular Biology 23:803–810.

Zhu, J.-K. 2002. Salt and drought stress signal transduction in plants. Annual Review of Plant Biology 53:247–273.

Zotz, G., K. Wilhelm, and A. Becker. 2011. Heteroblasty: A Review. Botanical Review 77:109–151.

Zou, C., M. D. Lehti-Shiu, M. Thomashow, and S.-H. Shiu. 2009. Evolution of stress-regulated gene expression in duplicate genes of *Arabidopsis thaliana*. PLoS Genetics 5:e1000581.

Zozomova-Lihova, J., K. Krak, T. Mandakova, K. K. Shimizu, S. Spaniel, P. Vit, and M. A. Lysak. 2014. Multiple hybridization events in *Cardamine* (Brassicaceae) during the last 150 years: Revisiting a textbook example of neoallopolyploidy. Annals of Botany 113:817–830.

Zuckerkandl, E. and L. Pauling. 1965. Evolutionary divergence and convergence in proteins. Pp. 97–166 *in* V. Bryson and H. J. Vogel, eds. Evolving Genes and Proteins. Academic Press, New York.

INDEX